Introduction to
Solid Modeling Using SOLIDWORKS® 2019

William E. Howard
East Carolina University

Joseph C. Musto
Milwaukee School of Engineering

Mc Graw Hill

INTRODUCTION TO SOLID MODELING USING SOLIDWORKS 2019
Published by McGraw-Hill Education, 2 Penn Plaza, New York, NY 10121.

This book is printed on acid-free paper.

1 2 3 4 5 6 7 8 9 QVS 21 20 19

ISBN 978-1-260-11330-3
MHID 1-260-11330-2

Cover Image: *William E. Howard*

mheducation.com/highered

About the Authors

Ed Howard is an Associate Professor in the Department of Engineering at East Carolina University, where he teaches classes in solid modeling, engineering computations, solid mechanics, and composite materials. Prior to joining ECU, Ed taught at Milwaukee School of Engineering. He holds a B.S. in Civil Engineering and an M.S. in Engineering Mechanics from Virginia Tech, and a Ph.D. in Mechanical Engineering from Marquette University.

Ed worked in design, analysis, and project engineering for 14 years before beginning his academic career. He worked for Thiokol Corporation in Brigham City, UT; Spaulding Composites Company in Smyrna, TN, and Sta-Rite Industries in Delavan, WI. He is a registered Professional Engineer in Wisconsin.

Joe Musto is a Professor in the Mechanical Engineering Department at Milwaukee School of Engineering, where he teaches in the areas of machine design, solid modeling, and numerical methods. He holds a B.S. degree from Clarkson University, and both an M.Eng. and Ph.D. from Rensselaer Polytechnic Institute, all in mechanical engineering. He is a registered Professional Engineer in Wisconsin.

Prior to joining the faculty at Milwaukee School of Engineering, he held industrial positions with Brady Corporation (Milwaukee, WI) and Eastman Kodak Company (Rochester, NY). He has been using and teaching solid modeling using SOLIDWORKS since 1998.

Joe and Ed, together with Rick Williams of Auburn University, are the authors of Engineering Computations: An Introduction using MATLAB® and Excel®, part of the McGraw-Hill "Best" Series.

CONTENTS

SPECIAL FEATURES

DESIGN INTENT

FUTURE STUDY

PREFACE

As design engineers and engineering professors, the authors have witnessed incredible changes in the way that products are designed and manufactured. One of the biggest changes over the past 30 years has been the development and widespread usage of solid modeling software. When we first saw solid modeling, it was used only by large companies. The cost of the software and the powerful computer workstations required to run it, along with the complexity of using the software, limited its use. As the cost of computing hardware dropped, solid modeling software was developed for personal computers. In 1995, the SOLIDWORKS® Corporation released the initial version of SOLIDWORKS® software, the first solid modeling program written for the Microsoft Windows operating system. Since then, the use of solid modeling has become an indispensable tool for almost any company, large or small, that designs a product.

While 2-D drawings can be an effective tool to document and communicate design details, a solid model's usefulness extends throughout the design process. The solid model data can be saved in a format from which a physical model can be made with a 3-D printer. Structural, thermal, and fluid flow analysis can be performed with finite element analysis (FEA) programs. The images on the cover of this book show the solid model of a heat sink, a device used to remove heat from a device such as a computer processor. As a fan forces cool air over the fins of the heat sink, heat is transferred from the base of the part, which is in contact with the processor, to the air. The color plot produced with the SOLIDWORKS Simulation add-in program shows the distribution of temperature throughout the part, based on the thermal properties of the part material and the forced air. Both physical prototypes and simulation results allow designers to make changes early in the design process.

SOLIDWORKS is a registered trademark of Dassault Systémes SolidWorks Corporation.

Motivation for This Text

When we saw a demonstration of the SOLIDWORKS software in 1998, we were both instantly hooked. Not only was the utility of the software obvious, but the program was easy to learn and fun to use. Since then, we have shared our enthusiasm with the program with hundreds of students in classes at Milwaukee School of Engineering and East Carolina University, in summer programs with high school students, and in informal training sessions. Most of the material in this book began as tutorials that we developed for these purposes. We continue to be amazed at how quickly students at all levels can learn the basics of the program, and by the sophisticated projects that many students develop after only a short time using the software.

While anyone desiring to learn the SOLIDWORKS program can use this book, we have added specific elements for beginning engineering students. With these elements, we have attempted to introduce students to the design process and to relate solid modeling to subjects that most engineering students will study later. We hope that the combination of the tutorial style approach to teaching the functionality of the software together with the integration of the material into the overall study of engineering will motivate student interest not only in the SOLIDWORKS software but in the profession of engineering.

Philosophy of This Text

The development of powerful and integrated solid modeling software has continued the evolution of computer-aided design packages from drafting/graphical communication tools to full-fledged engineering design and analysis tools. A solid model is more than simply a drawing of an engineering component; it is a true virtual representation of the part, which can be manipulated, combined with other parts into complex assemblies, used directly for analysis, and used to drive the manufacturing equipment that will be used to produce the part.

This text was developed to exploit this emerging role of solid modeling as an integral part of the engineering design process; while proficiency in the software will be achieved through the exercises provided in the text, the traditional "training" exercises will be augmented with information on the integration of solid modeling into the engineering design process. These topics include:

- The exploitation of the parametric features of a solid model, to not only provide an accurate graphical representation of a part but also to effectively capture an engineer's design intent,
- The use of solid models as an analysis tool, useful for determining properties of components as well as for virtual prototyping of mechanisms and systems,
- The integration of solid modeling with component manufacturing, including the generation of molds, sheet metal patterns, and rapid prototyping files from component models.

Through the introduction of these topics, students will be shown not only the powerful modeling features of the SOLIDWORKS program, but also the role of the software as a full-fledged integrated engineering design tool.

The Use of This Text

This text primarily consists of chapter-long tutorials, which introduce both basic concepts in solid modeling (such as part modeling, drawing creation, and assembly modeling) and more advanced applications of solid modeling in engineering analysis and design (such as mechanism modeling, mold creation, sheet metal bending, and rapid prototyping). Each tutorial is organized as "keystroke-level" instructions, designed to teach the use of the software.

While these tutorials offer a level of detail appropriate for new professional users, this text was developed to be used as part of an introductory engineering course, taught around the use of solid modeling as an integrated engineering design and analysis tool. Since the intended audience is undergraduate students new to the field of engineering, the text contains features that help to integrate the concepts learned in solid modeling into the overall study of engineering. These features include:

- *Video Examples:* Short video tutorials accompany multiple chapters. These videos introduce students to the concepts of solid modeling and the SOLIDWORKS commands that they will use in the chapter following the step-by-step tutorials. These videos cover:

  — Getting started with modeling (Chapter 1);
  — Making 2-D drawings (Chapter 2);
  — Using symmetry when creating parts (Chapter 3);
  — Creating parts with lofts and sweeps (Chapter 4);
  — Making assemblies from part files (Chapter 6);
  — Making parts with 3-D printing (Chapter 13);
  — Setting up the SOLIDWORKS interface (Appendix A).

- *Design Intent Boxes:* These are intended to augment the "keystroke-level" tutorials to include the rationale behind the sequence of operations chosen to create a model.
- *Future Study Boxes:* These link the material contained in the chapters to topics that will be seen later in the academic and professional careers of new engineering students. They are intended to motivate interest in advanced study in engineering, and to place the material seen in the tutorials within the context of the profession.

While these features are intended to provide additional motivation and context for beginning engineering students, they are self-contained, and may be omitted by professionals who wish to use this text purely for the software tutorials.

New in This Edition

This new edition of the text has been fully updated for the SOLIDWORKS 2019 software package. All tutorials and figures have been modified for the new version of the software. Additionally, all videos have been updated to reflect the latest software.

The Organization of This Text

The organization of the chapters of the book reflects the authors' preferences in teaching the material, but allows for several different options. We have found that

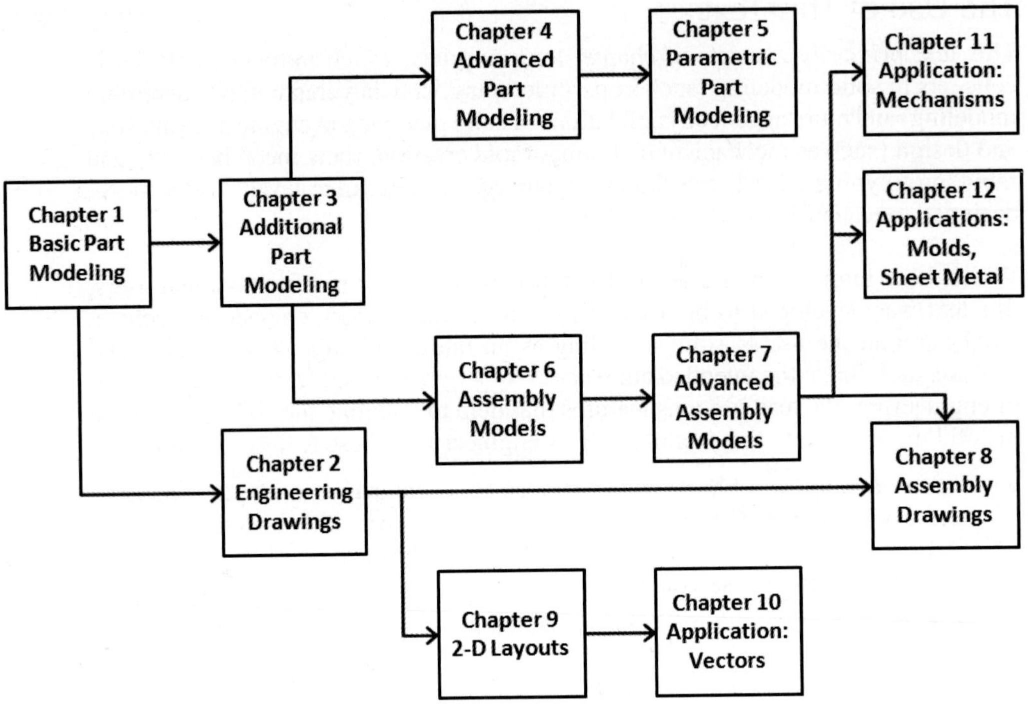

covering drawings early in the course is helpful in that we can have students turn in drawings rather than parts as homework assignments. The eDrawings feature, which is covered in Chapter 2, is especially useful in that eDrawings files are small (easy to e-mail), self-contained (not linked to the part file), and can be easily marked up with the editing tools contained in the eDrawings program.

The flowchart above illustrates the relations between chapters, and can be used to map alternative plans for coverage of the material. For example, if it is desired to cover assemblies as soon as possible (as might be desired in a course that includes a project) then the chapters can be covered in the order 1-3-4-6-7-2-8, with the remaining chapters covered in any order desired. An instructor who prefers to cover parts, assemblies, and drawings in that order may cover the chapters in the order 1-3-4-5-6-7-2-8 (skipping section 5.4 until after Chapter 2 is covered), again with the remaining chapters covered in any order.

Chapters 9 and 10 may be omitted in a standard solid modeling course; however, they can be valuable in an introductory engineering course. Engineering students will almost certainly find use at some point for the 2-D layout and vector mechanics applications introduced in these chapters. Chapter 13 is intended to wrap up the course with a discussion of how solid modeling is used as a tool in the product development cycle. Appendix A summarizes the recommended settings to the SOLIDWORKS program that are used throughout the book, while Appendix B shows options for customizing the SOLIDWORKS interface.

Resources for Instructors

Additional resources are available on the web at *www.mhhe.com/howard2019.* Included on the website are tutorials for three popular SOLIDWORKS Add-Ins: SOLIDWORKS Simulation®, SOLIDWORKS Motion™, and PhotoView 360™, the video examples, and the book figures in PowerPoint format. Instructors can also access PowerPoint files for each chapter and model files for all tutorials and end-of-chapter problems as well as a teaching guide (password-protected; contact your McGraw-Hill representative for access).

Acknowledgments

We are grateful to our friends at McGraw-Hill, especially Suzy Bainbridge and Heather Ervolino, for their support and encouragement during this project. In particular, we offer special thanks to Karen Fleckenstein of Fleck's Communications, Inc. who did the page layouts. Also, thanks to Tim Maruna, who encouraged us to initiate this project.

At SOLIDWORKS Corporation, Marie Planchard has provided continuous support for the project. The authors are also appreciative of the support of our SOLIDWORKS resellers, Computer Aided Technology, Inc. and TriMech Solutions.

We also want to thank the reviewers whose comments have undoubtedly made the book better.

Many of our students and colleagues used early versions of the manuscript and materials that eventually became this text. We thank them for their patience and helpful feedback along the way.

Ed Howard
Joe Musto

PART ONE

Learning SOLIDWORKS®

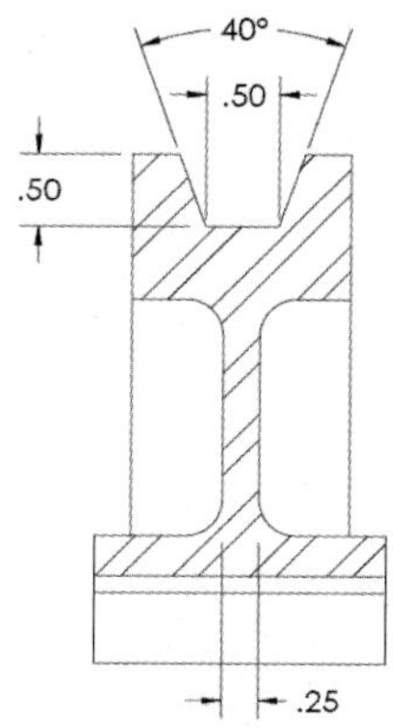

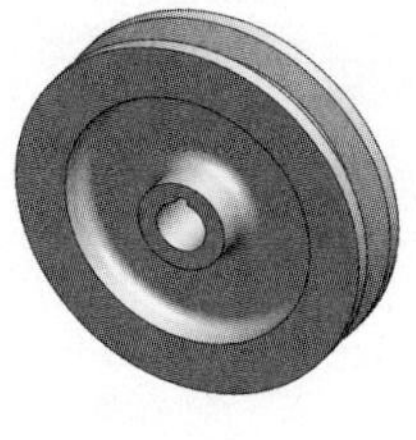

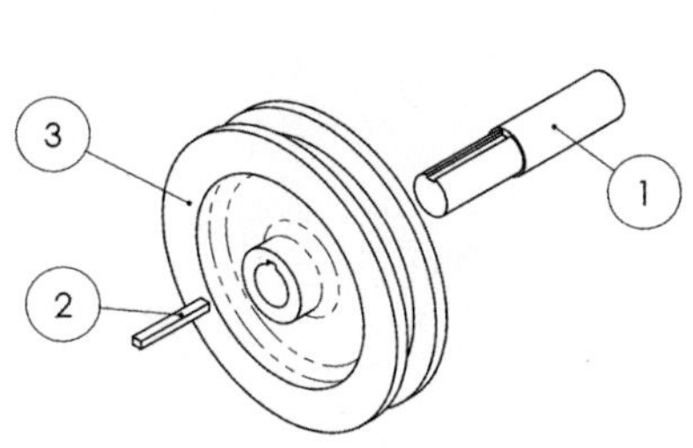

CHAPTER 1

Basic Part Modeling Techniques

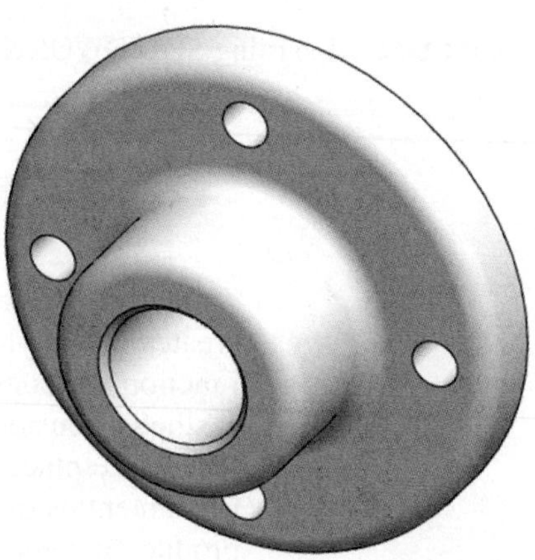

Introduction

Solid modeling has become an essential tool for most companies that design mechanical structures and machines. As recently as the 1990's, this would have been hard to imagine. While 3-D modeling software existed, it was very expensive and required high-end computer workstations to run. An investment of $50,000 or more was required for every workstation with software, not including training of the operator. As a result, only a few industries used solid modeling, and the trained operators tended to work exclusively with the software. The dramatic performance improvements and price drops of computer hardware, along with increased competition among software vendors, have significantly lowered the cost barrier for companies to enter the solid modeling age. The software has also become much easier to use, so that engineers who have many other job functions can use solid modeling when required without needing to become software specialists. The SOLIDWORKS® program was among the first solid modeling programs to be written exclusively for the Microsoft Windows environment. Since its initial release in 1995, it has been adopted by thousands of companies worldwide. This text is laid out as a series of tutorials that cover most of the basic features of the SOLIDWORKS program. Although these tutorials will be of use to anyone desiring to learn the software, they are written primarily for freshmen engineering students. Accordingly, topics in engineering design are introduced along the way. "Future Study" boxes give a preview of coursework that engineering students will encounter later, and relate that coursework to the solid modeling tutorials. In this first chapter, we will learn how to make two simple parts with SOLIDWORKS software.

Chapter Objectives

In this chapter, you will:

- be introduced to the role of solid modeling in engineering design,
- learn how to create 2-D sketches and create 3-D extruded and revolved geometry from these sketches,
- use dimensions and relations to define the geometry of 2-D sketches,
- add fillets, chamfers, and circular patterns of features to part models,
- learn how to modify part models, and
- define the material and find the mass properties of part models.

SOLIDWORKS is a registered trademark of Dassault Systémes SolidWorks Corporation.

1.1 Engineering Design and Solid Modeling

The term *design* is used to describe many endeavors. A clothing designer creates new styles of apparel. An industrial designer creates the overall look and function of consumer products. Many design functions concentrate mainly on aesthetic considerations—how the product looks, and how it will be accepted in the marketplace. The term *engineering design* is applied to a process in which fundamentals of math and science are applied to the creation or modification of a product to meet a set of objectives.

Engineering design is only one part of the creation of a new product. Consider a company making consumer products, for example bicycles. A marketing department determines the likely customer acceptance of a new bike model and outlines the requirements for the new design. Industrial designers work on the preliminary design of the bike to produce a design that combines functionality and styling that customers will like. Manufacturing engineers must consider how the components of the product are made and assembled. A purchasing department will determine if some components will be more economical to buy than to make. Stress analysts will predict whether the bike will survive the forces and environment that it will experience in service. A model shop may need to build a physical prototype for marketing use or to test functionality.

During the years immediately following World War II, most American companies performed the tasks described above more or less sequentially. That is, the design engineer did not get involved in the process until the specifications were completed, the manufacturing engineers started once the design was finalized, and so on. From the 1970s through the 1990s, the concept of *concurrent engineering* became widespread. Concurrent engineering refers to the process in which engineering tasks are performed simultaneously rather than sequentially. The primary benefits of concurrent engineering are shorter product development times and lower development costs. The challenges of implementing concurrent engineering are mostly in communications—engineering groups must be continuously informed of the actions of the other groups.

Solid modeling is an important tool in concurrent engineering in that the various engineering groups work from a common database: the solid model. In a 2-D CAD (Computer-Aided Design) environment, the design engineer produced sketches of the component, and a draftsman produced 2-D design drawings. These drawings were forwarded to the other engineering organizations, where much of the information was then duplicated. For example, a toolmaker created a tool design from scratch, using the drawings as the basis. A stress analyst created a finite element model, again starting from scratch. A model builder created a physical prototype by hand from the drawing parameters. With a solid model, the tool, finite element model, and rapid prototype model are all created directly from the solid model file. In addition to the time savings of avoiding the steps of recreating the design for the various functions, many errors are avoided by having everyone working from a common database. Although 2-D drawings are usually still required, since they are the best way to document dimensions and tolerances, they are linked directly to the solid model and are easy to update as the solid model is changed.

A mechanical engineering system (assembly) may be composed of thousands of components (parts). The detailed design of each component is important to the operation of the system. In this chapter, we will step through the creation of simple components. In future chapters, we will learn how to make 2-D drawings from a part file, and how to put components together in an assembly file.

1.2 Part Modeling Tutorial: Flange

This tutorial will lead you through the creation of a simple solid part. The part, a flange, is shown in **Figure 1.1** and is described by the 2-D drawing in **Figure 1.2**.

FIGURE 1.1

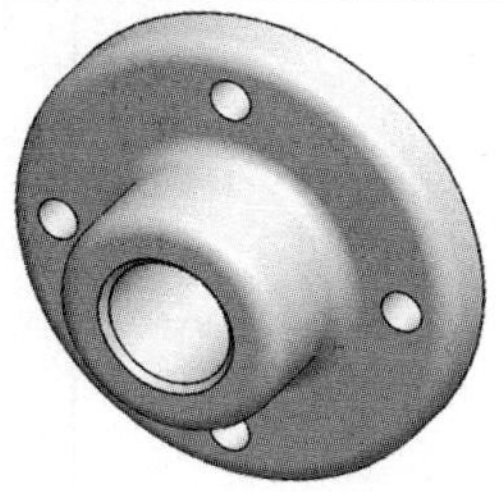

FIGURE 1.2

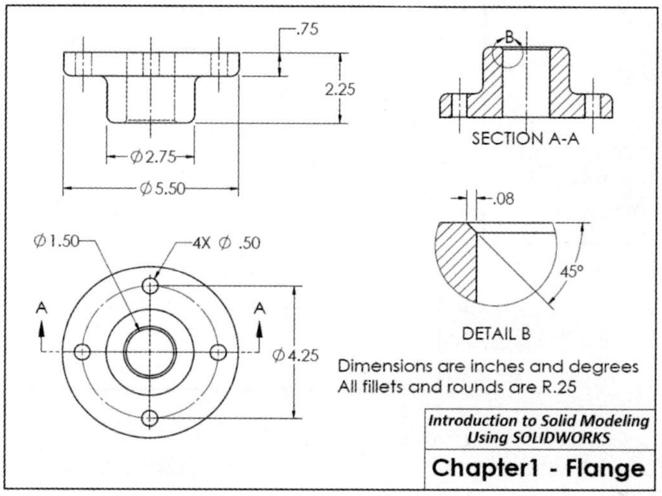

Begin by double-clicking the SOLIDWORKS icon on your desktop. The Welcome dialog box opens, as shown in Figure 1.3. From this box, we can begin a new document (part, assembly, or drawing) or select a recently-opened document. Click Part from the New group. If the Units and Dimension Standard box appears, as shown in Figure 1.4, select "IPS" as the units and "ANSI" as the standard. Click OK.

FIGURE 1.3

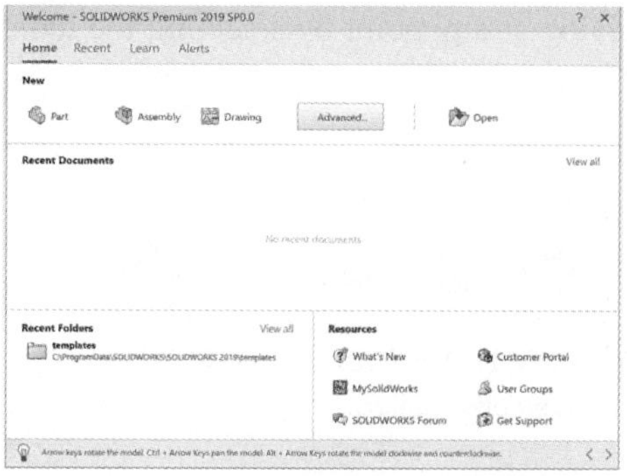

FIGURE 1.4

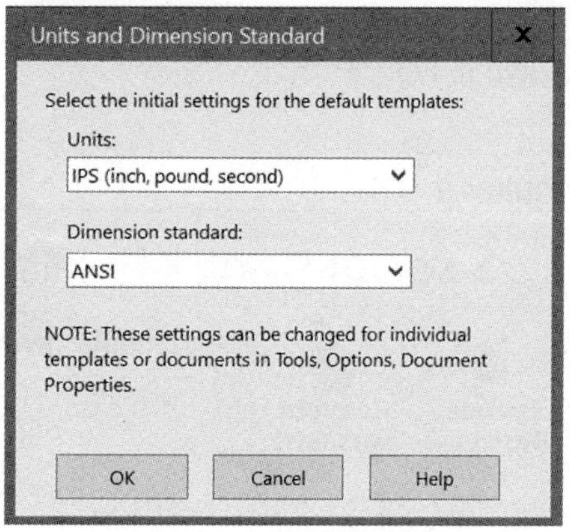

VIDEO EXAMPLE 1

In this chapter, we begin by making and dimensioning 2-D sketches and then creating 3-D features from extrusions of the sketches.

Creation of the simple part shown here, with the dimensions as shown in the drawing to the right, is demonstrated in a video at *www.mhhe.com/howard2019*. (We will learn to make drawings from 3-D parts in Chapter 2.)

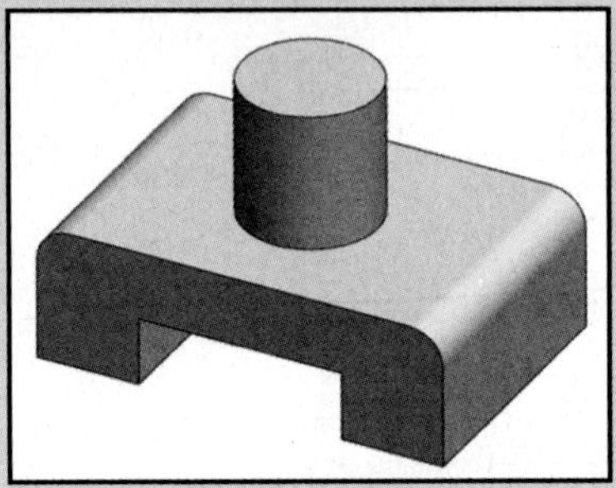

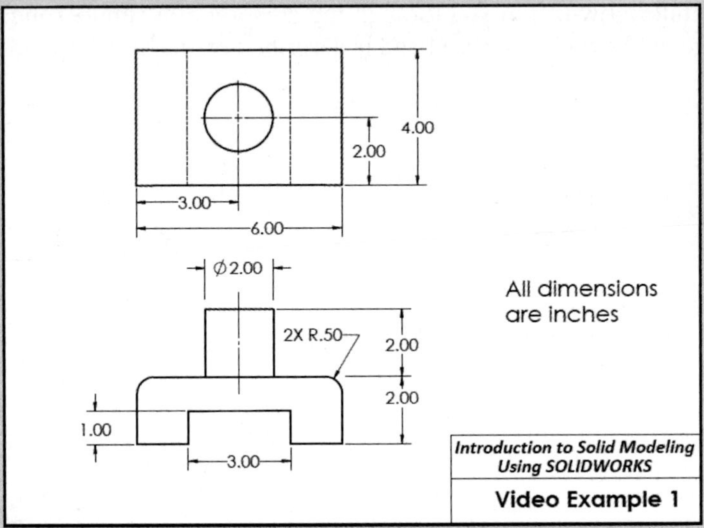

In this chapter, we will be making adjustments to the SOLIDWORKS interface. These adjustments are summarized in Appendix A and in Video Example 7, which is available at www.mhhe.com/howard2019.

The Units and Dimension Standard box only appears the first time SOLIDWORKS is opened. The selections become the default values for all new files. In this chapter, we will see how to set these values for individual files and to change the default values.

Note that you can return to the Welcome dialog at any point by selecting the icon shown in **Figure 1.5**.

FIGURE 1.5

Before we begin modeling the flange, we will establish a consistent setup of the SOLIDWORKS environment. The default screen layout is shown in **Figure 1.6**. The graphics area occupies most of the screen. The part, drawing, or assembly will be displayed in this area. At the top of the screen is the Menu Bar, which contains the Main Menu and a toolbar with several commonly-used tools such as Save, Print, and Redo. Note that if you pass the cursor over the SOLIDWORKS button in the Menu Bar, the Main Menu will "fly out," or be temporarily displayed, as shown in **Figure 1.7**. The fly-out feature is designed to save room on the screen. However, since we will be using the menu often, we will disable the fly-out so that the menu is always displayed.

Move the cursor over the SOLIDWORKS button to display the menu. Click on the pushpin icon at the right side of the menu, as shown in Figure 1.8, to lock the display of the menu.

FIGURE 1.6

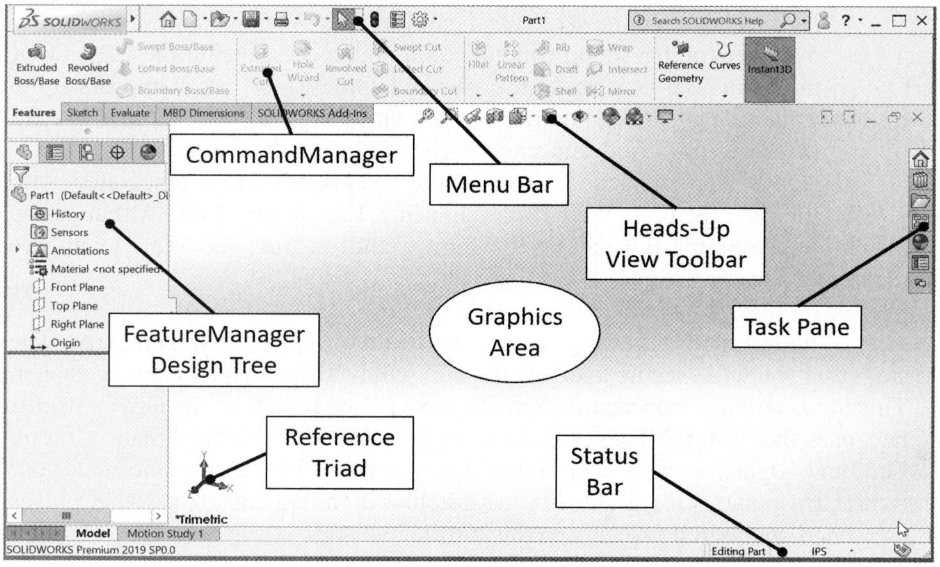

FIGURE 1.7

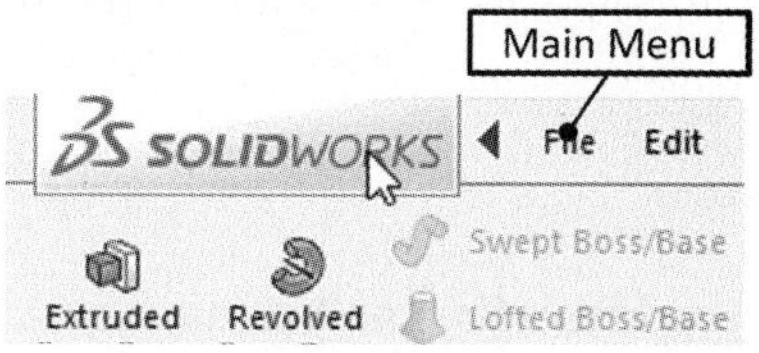

FIGURE 1.8

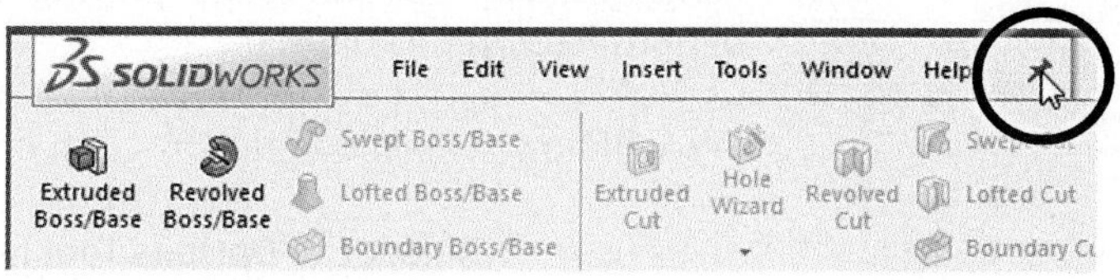

The CommandManager contains most of the tools that you will use to create parts. When working in the part mode, there are two categories of tools that we will use extensively: Sketch tools used in creating 2-D sketches, and Features tools used to create and modify 3-D features. Clicking on the Sketch and Features tabs at the bottom of the CommandManager, as shown in **Figure 1.9**, changes the tools on the CommandManager to those of the selected group. By default, there are several other groups available besides the Sketch and Features groups. To simplify the interface, we will hide these groups for now.

FIGURE 1.9

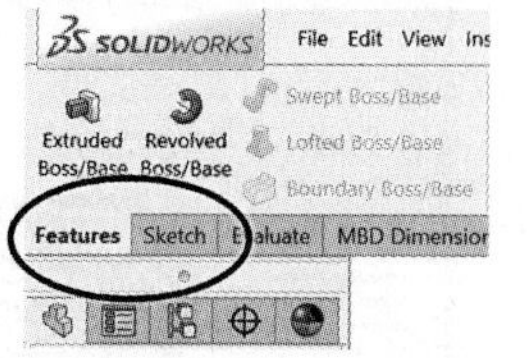

FIGURE 1.10

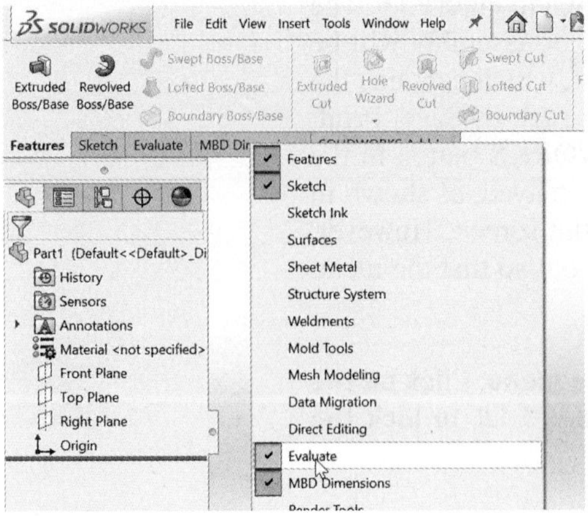

Right-click on one of the CommandManager tabs. A list of available groups is displayed, with a check mark shown beside each active group (Figure 1.10). Click on any of the active groups other than Features and Sketch. This will clear the check mark and turn off the display of that group. Repeat until only the Features and Sketch groups remain active.

At the right side of the screen is the Task Pane. The Task Pane is a fly-out interface for accessing files and online resources. We will not use the Task Pane that often, but since it takes up very little room in its normal collapsed state, we will leave it on. If you would like to turn it off completely, select View: User Interface from the Main Menu and click on Task Pane.

FIGURE 1.11

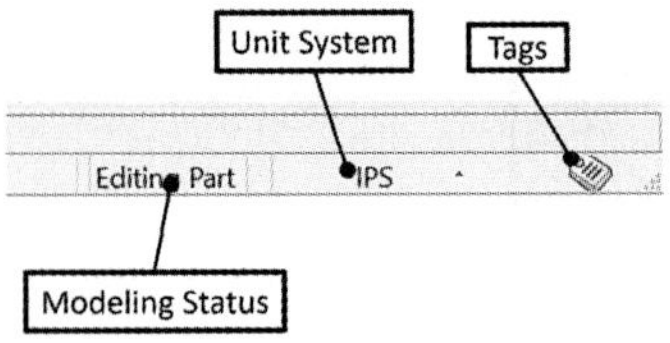

At the bottom of the screen is the Status Bar. When you move the cursor over any toolbar icon or menu command, a message on the left side of the Status Bar describes the command. Other information appears at the right side of the Status Bar, as shown in **Figure 1.11**. The unit system in use is displayed and can be changed directly from the Status Bar. Another feature, called Tags, allows keywords to be associated with files and features. We will not be using Tags in this book. Although the display of the Status Bar can be toggled off and on from the View menu, we recommend leaving it on.

FIGURE 1.12

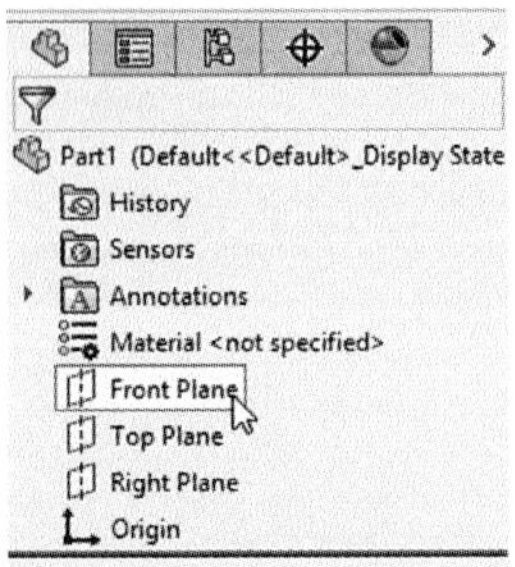

Just to the left of the drawing area is the FeatureManager® Design Tree. The steps that you will execute to create the part will be listed in the FeatureManager. This information is important when the part is to be modified. When you open a new part, the FeatureManager lists an origin and three predefined planes (Front, Top, and Right), as shown in **Figure 1.12**. As you select each plane with your mouse, the plane is highlighted in the graphics area. We can create other planes as needed, and will do so later in this tutorial.

FIGURE 1.13

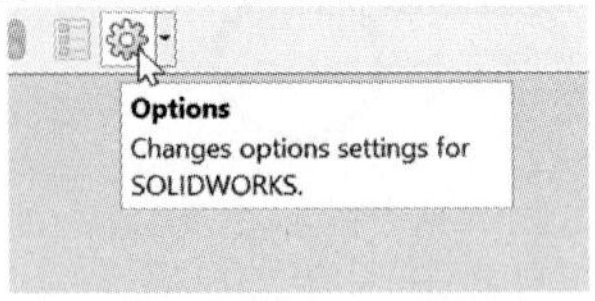

At the top of the graphics area is the Heads-Up View Toolbar. This toolbar contains many options for displaying your model. We will explore these options later in this tutorial.

We will now set some of the program options.

Select the Options Tool from the Menu Bar toolbar, as shown in Figure 1.13. (You can also access the options from the Main Menu, by selecting Tools: Options.)

FIGURE 1.14

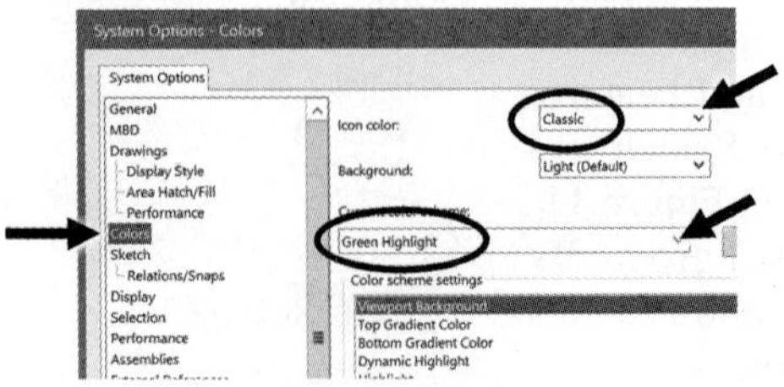

The dialog box contains settings for both the system and for the specific document that is open.

Under the System Options tab, choose Colors and change the icon color to "Classic" and the color scheme to "Green Highlight," as shown in Figure 1.14.

The Classic option for icon colors makes many of the icons display in colors other than the default blue and black, making them easier to recognize for new users. The Green Highlight scheme causes currently selected items to be highlighted in green, as the name implies. The default option is for selected items to be highlighted in light blue. Since another shade of blue is used for other purposes, green highlighting is used in this book to avoid confusion. Since these changes were made to the System Options, they will remain in effect for future SOLIDWORKS sessions. The changes below, which will be made to the Document Properties, will apply only to the current part model.

Select the Document Properties tab. In the list of options, Drafting Standard will be highlighted. Select ANSI from the pull-down menu, as shown in Figure 1.15.

FIGURE 1.15

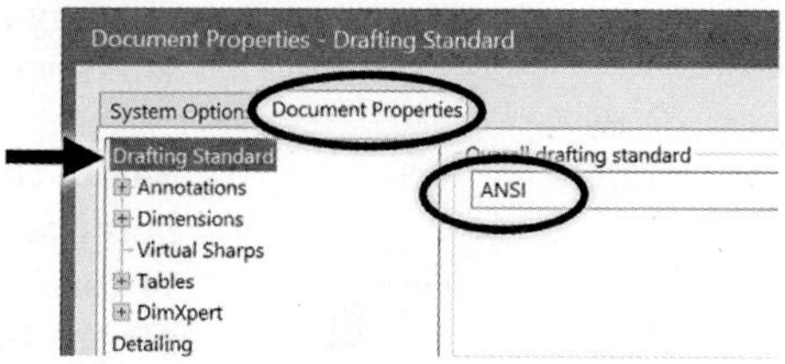

ANSI is the American National Standards Institute, an organization that formulates and publishes the standard drawing practices used by most companies in the United States. European companies are more likely to use the standards of ISO, the International Organization for Standardization.

Also under the Document Properties tab, select Dimensions. Use the pull-down menu by the Primary precision box to set the number of decimal places to 3 (.123), as shown in Figure 1.16. (Ignore the message that the drafting standard has been changed to "ANSI-MODIFIED.") Select Grid/Snap and check the box labeled "Display Grid," as shown in Figure 1.17. Also, select Units and set the unit system to IPS (inches, pounds, and seconds), the primary length precision to .123, and the precision for angles to None, as shown in Figure 1.18.

FIGURE 1.16

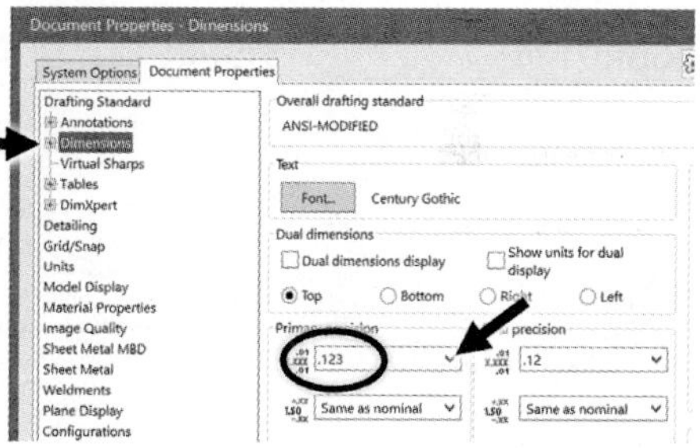

FIGURE 1.17

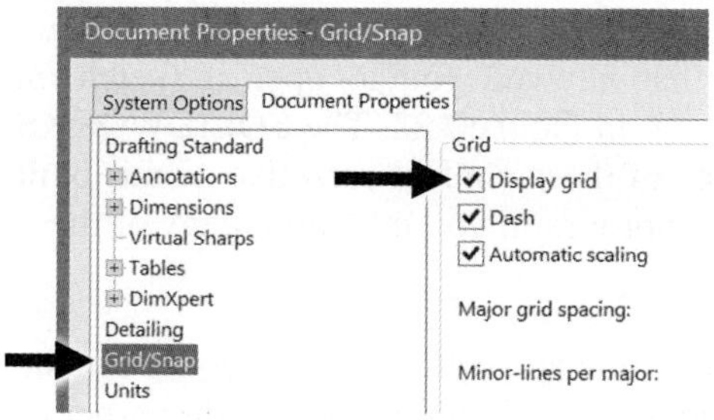

FIGURE 1.18

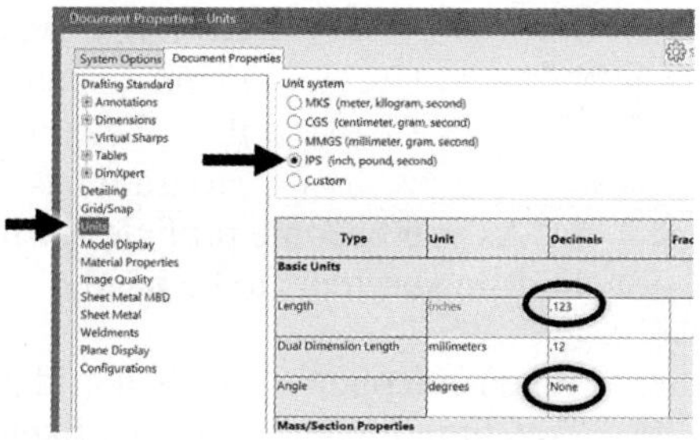

Note that there are "Dual Dimension" units that can be set in the Units options. For some drawings, you may want to show dimensions in both US units (inches) and SI units (typically millimeters). Since we will not use dual dimensions for this part, it is not necessary to change those settings. Also note that we have set the decimal display to .123 in two separate locations. The display of decimal places can be changed at either location. You may also change the font style and size of the dimension text by clicking the Font button. For clarity, the figures in this book were made with a dimension font larger than the default size.

Click OK to close the dialog box.

Any of the options just set can be changed at any time during the modeling process. Later in this chapter, we will learn how to create a *template* that allows us to begin a new part with our preferred settings in place.

We will make two more changes to the default settings before beginning our part. A feature called "Instant 3D" allows for changes to be made by clicking and dragging on model faces, without entering dimensions from the keyboard. While this feature can be handy for experienced users, it is recommended that new users avoid using Instant 3D in order to prevent unintended changes to the model. Similarly, a feature called "Instant 2D" allows for dimensions in sketches to be changed by clicking and dragging rather than entering a numerical value. This feature will also be turned off.

FIGURE 1.19

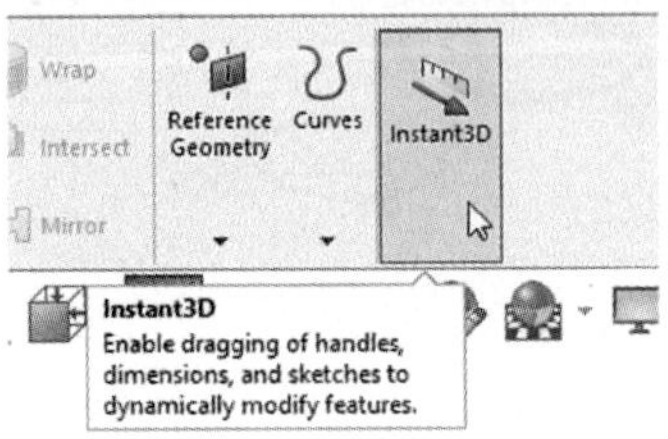

Select the Features tab of the CommandManager. If the Instant 3D Tool is turned on (the icon will be "depressed," as shown in Figure 1.19), click to turn it off. Select the Sketch tab of the CommandManager and turn off the Instant 2D Tool as well.

We start the construction of the flange by sketching a circle and extruding it into a 3-D disk.

FIGURE 1.20

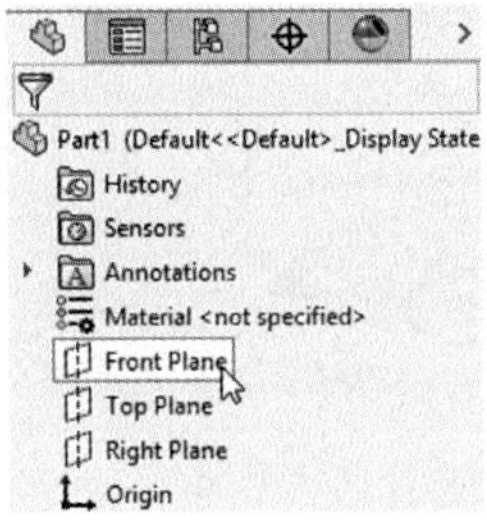

FIGURE 1.21

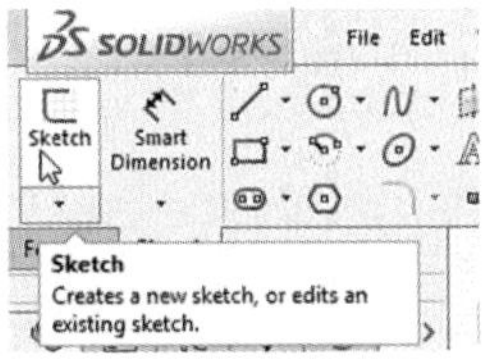

Select the Front Plane by clicking on it in the FeatureManager Design Tree, as shown in Figure 1.20.

The Front Plane will be highlighted in green. The color green indicates that an item is the currently selected entity (since we chose the "Green Highlight" color scheme).

FIGURE 1.22

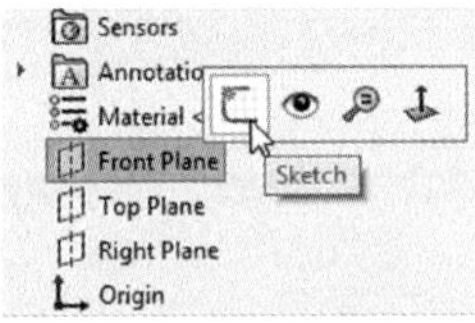

Begin a sketch by selecting the Sketch tab of the CommandManager, and then the Sketch Tool, as shown in Figure 1.21.

Note that when you selected the Front Plane, a pop-up menu appeared that allowed you to open a sketch on that plane, as shown in **Figure 1.22**. The SOLIDWORKS program has many of these context-sensitive menus built in. As you become proficient with the program, you may find many of these built-in shortcuts to be handy.

FIGURE 1.23

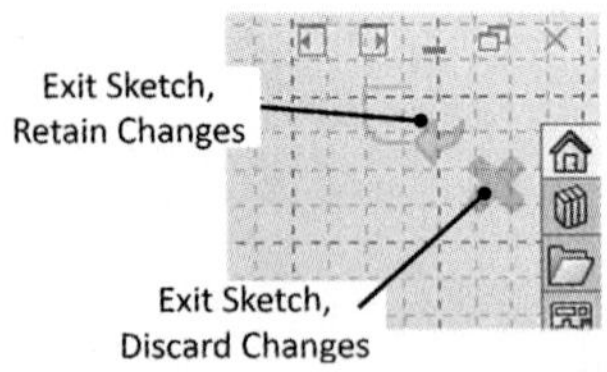

When you open a sketch, a grid pattern appears, signifying that you are in the sketching mode. Also, Exit Sketch icons appear in the upper-right corner of the screen, as shown in **Figure 1.23**.

FIGURE 1.24

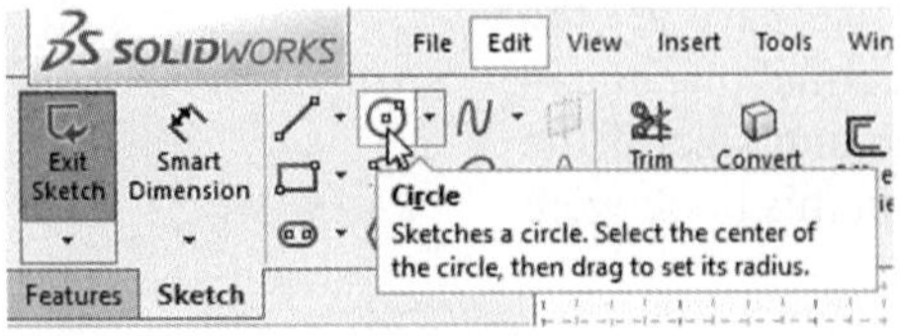

Select the Circle Tool from the Sketch group of the CommandManager, as shown in Figure 1.24.

When selecting any tool which has a pull-down menu (designated by the down arrow to the right of the icon), use caution to be sure that you are selecting the proper tool. In the case of the Circle Tool, there are two possible methods for defining the circle: by the center point and a point on the perimeter, or by three points on the perimeter. Clicking on the down arrow displays these options, so that the proper tool can be selected (**Figure 1.25**). By default, the option for defining the circle by locating the center point and a point on the perimeter is selected by clicking on the Circle Tool without accessing the pull-down menu. However, if the last option selected was to define the circle by three points on the perimeter, then that option becomes the default for the next selection. When that occurs, the icon shown for the Circle Tool will change, as shown in **Figure 1.26**. Because many of the icons are similar and are very small, you should use caution with tools that have pull-down menus.

FIGURE 1.25

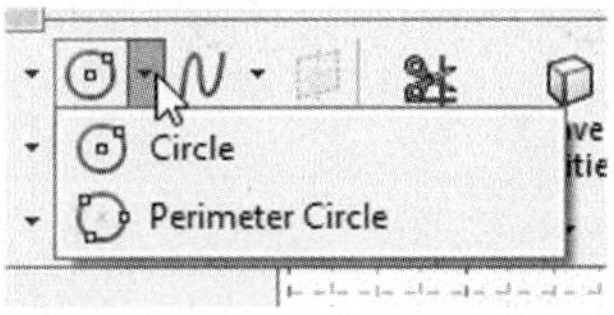

FIGURE 1.26

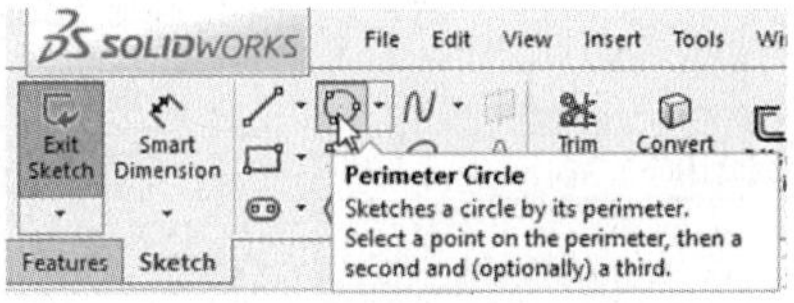

You can check to see that you have selected the proper tool by looking at the PropertyManager, which appears in the area where the FeatureManager is normally shown whenever a tool is activated or an object is selected. The PropertyManager now shows the two alternative methods for defining a circle (**Figure 1.27**). If we selected the wrong tool accidentally, then we can change the method for defining the circle in the PropertyManager.

FIGURE 1.27

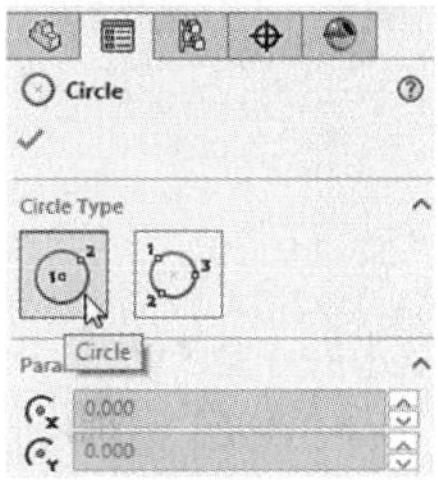

In the PropertyManager, make sure that the icon representing the first construction method is selected, as shown in Figure 1.27. If it is not, then click it to select it.

Notice as you move the cursor into the drawing area that it changes appearance into a pencil icon with a circle next to it, as shown in **Figure 1.28**. This lets you know that the Circle Tool is active.

FIGURE 1.28

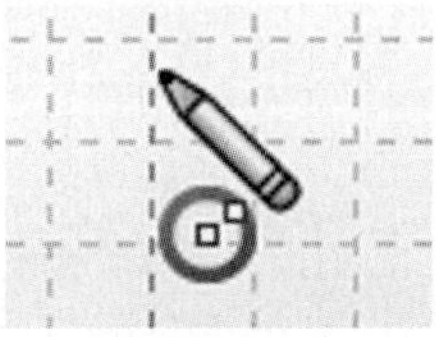

Move the tip of the pencil icon toward the origin until a red dot appears at the origin, as shown in Figure 1.29; this indicates that you will snap to an existing point (in this case the origin) when you click with the mouse. Also, note the small icon next to the origin that signifies a coincident relation: the origin and the center point of the circle will share the same location.

FIGURE 1.29

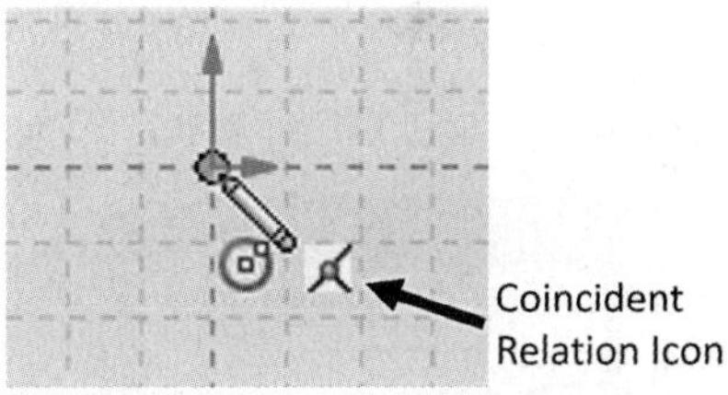

A snap adds a relation to the positions of two entities. In this example, when you snap to the origin, the circle will be centered at the exact coordinates of x = 0 and y = 0. The relation added when one entity is created by snapping to another can be edited later, if desired. The addition of a snap automatically is a nice feature of the SOLIDWORKS program: snaps are intuitive. It is not necessary to enter the numerical coordinates of the center of the circle.

FIGURE 1.30

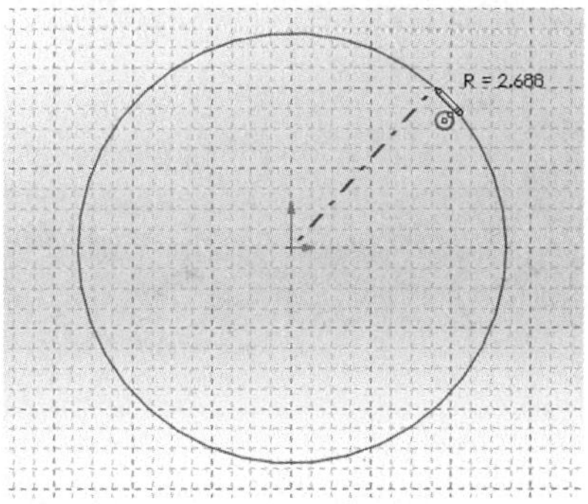

With the center point highlighted as in Figure 1.29, click the left mouse button to place the center of the circle at the origin. Drag the mouse outward to create a circle, as shown in Figure 1.30. Click the left mouse button again to define a point on the perimeter and create the circle. The size of the circle drawn is not important; we will add a dimension to define its diameter precisely.

The circle will appear in green, indicating that it is the currently selected item.

Press the Esc key twice to close the Circle Tool and deselect the circle just drawn.

The circle should now appear in blue. In the Status Bar at the bottom of the screen, notice that "Under Defined" appears. This is because we have not set the diameter of the circle yet. When a sketch does not contain enough dimensions and/or relations to define its size and position in space, it is said to be under defined, and is denoted by blue entities.

Other possible conditions of the sketch are "Fully Defined," when the sketch contains exactly enough dimensions and/or relations to define its size and position in space (denoted by black entities), and "Over Defined," where the sketch has at least one dimension or relation that contradicts or is redundant to the other dimensions and relations (denoted by red entities). Over defined sketches should be avoided.

FIGURE 1.31

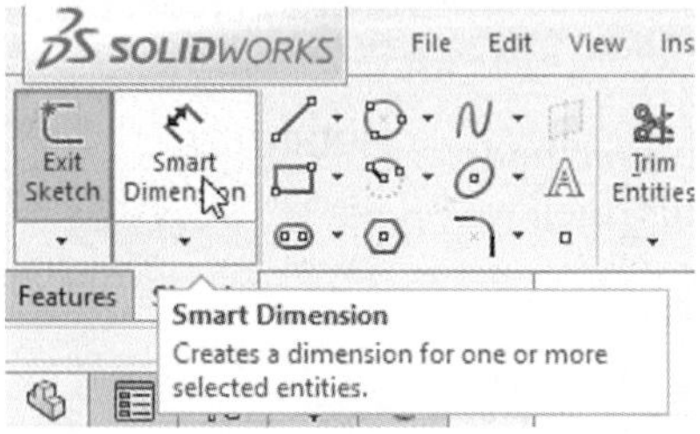

Also note that the area of the circle is shaded. By default, closed contours within sketches are shown shaded. The shading can be toggled on and off by clicking the Shaded Sketch Contours Tool from the Sketch group of the CommandManager or from Tools: Sketch Settings from the Main Menu. Most of the figures in this book will be shown with the contour shading turned off for clarity.

FIGURE 1.32

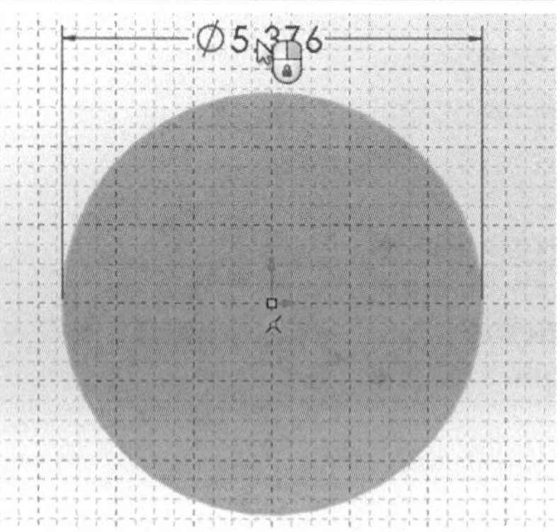

Select the Smart Dimension Tool from the CommandManager, as shown in Figure 1.31. Click with the left mouse button anywhere on the circle. A dimension will be added to the diameter of the circle. Drag the dimension to a convenient location, as shown in Figure 1.32. When the dimension is where you want to put it, left-click again to place the dimension.

A dialog box displaying the SOLIDWORKS name of the dimension and prompting for its value will be displayed, as shown in Figure 1.33. Enter "5.5" in the box. (You don't need to enter any units, since inches are the default units, set earlier.) Press the Enter key or click on the check mark to update the dimension. Press the Esc key to turn off the Smart Dimension Tool and deselect the dimension.

FIGURE 1.33

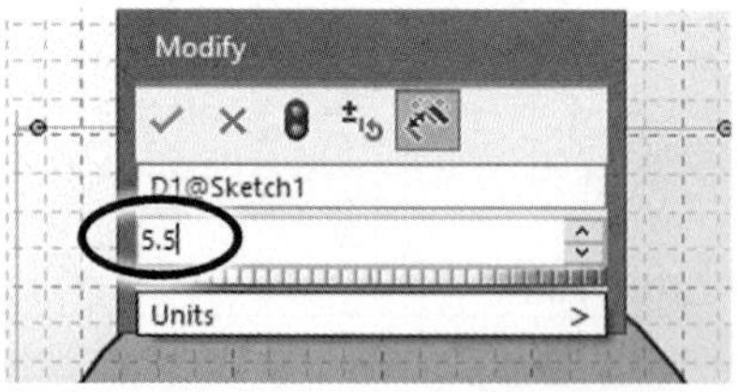

Notice that the circle is redrawn to the correct dimension, as shown in **Figure 1.34**. (Note that the figures in the book use a larger font size than the default value. We will see how to change the font size later in this chapter.) The dimension in inches is displayed, and the circle is black. Notice at the bottom of the screen in the Status Bar that the sketch is now Fully Defined. (Note: If we had not snapped to the origin for the circle's center, the sketch would still be under defined because the circle's location within the Front Plane would not be specified.)

FIGURE 1.34

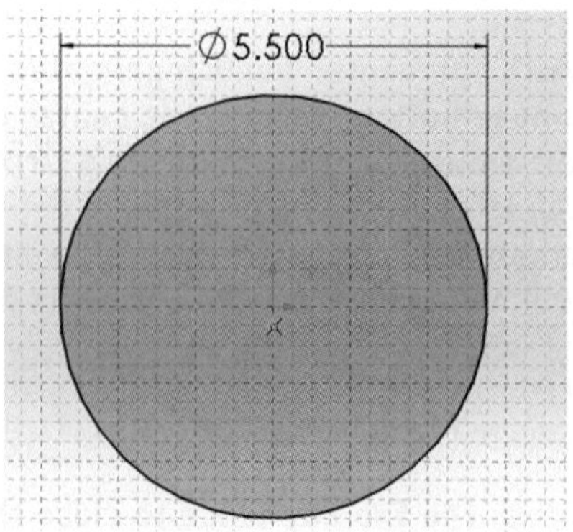

If you double-click on the dimension value, the dialog box reappears and you can change the dimension. Try this, and then use the Undo Tool from the main menu toolbar, shown in Figure 1.35, to return the dimension to 5.5 inches.

FIGURE 1.35

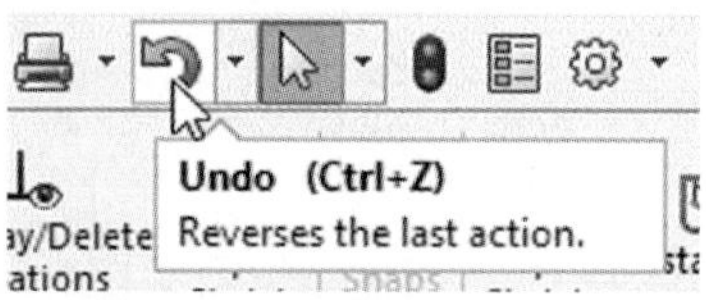

Next to the center of the circle, an icon shows that a relation is applied. By moving the cursor over the relation icon, details about the relation can be viewed, as shown in Figure 1.36. Relations can be deleted by clicking on the relation icon to select it, and then pressing the Delete key. The display of sketch relations can be toggled on and off by selecting View: Hide/Show from the Main Menu and clicking Sketch Relations. Some experienced users prefer to not show the relations because they can cause a sketch to appear cluttered, but new users are advised to keep the relation display turned on.

FIGURE 1.36

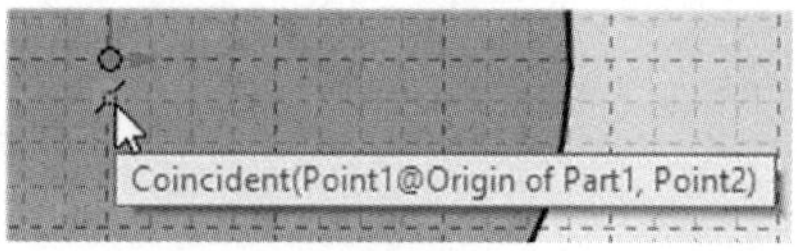

Now we are ready to turn this 2-D sketch into a 3-D part with the Extruded Boss/Base Tool.

Select the Features tab of the CommandManager, and the Extruded Boss/Base Tool, as shown in Figure 1.37.

FIGURE 1.37

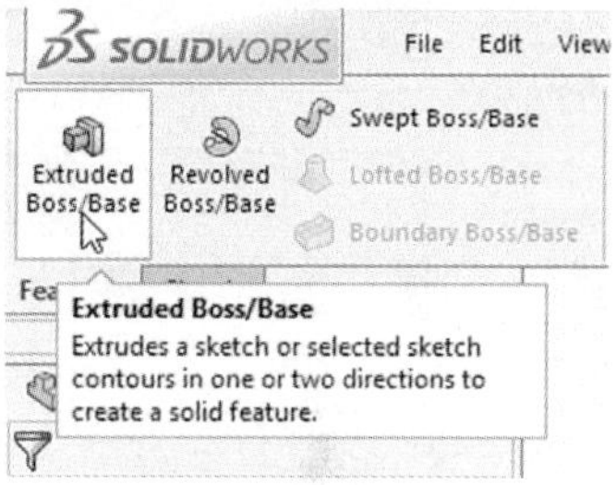

The base feature is the first solid feature created. Any subsequent solid features are called bosses. Note that the view of the part changes to display a 3-D preview of the extruded solid. On the left side of the screen, the PropertyManager is now active. The PropertyManager allows the properties of the selected entity to be viewed and edited. There are several options available for the extrusion, including adding draft (taper) to the part, but for now we only need to adjust the depth of the extrusion.

FIGURE 1.38

Set the depth of the extrusion to 0.75 inches, as shown in Figure 1.38. Press Enter, and the preview in the graphics area will be updated to reflect the new thickness, as shown in Figure 1.39. Click on the check mark (OK) in the PropertyManager, as shown in Figure 1.40, and the circle is extruded into a solid disk, as shown in Figure 1.41.

FIGURE 1.39

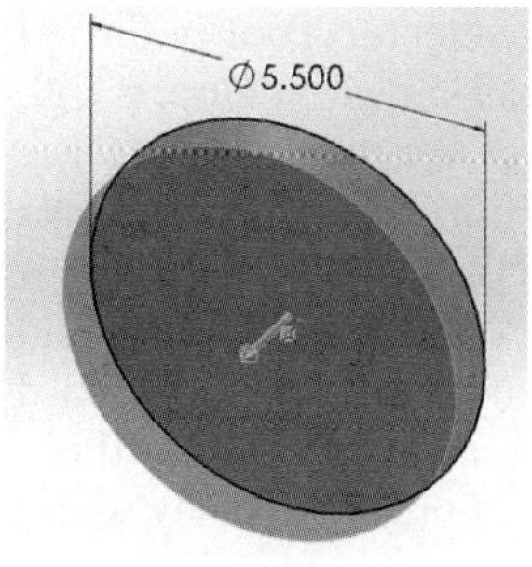

FIGURE 1.40

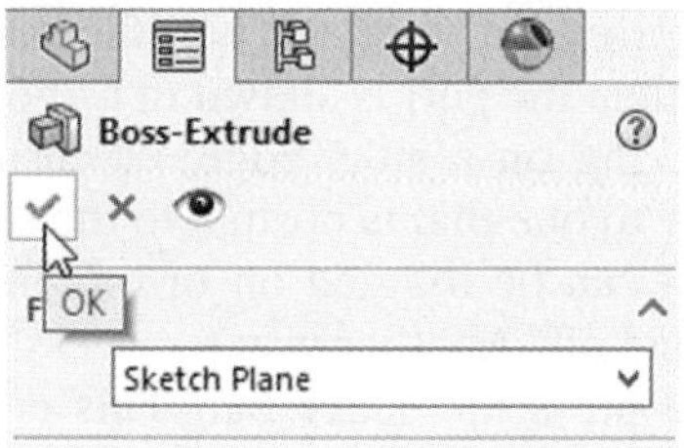

FIGURE 1.41

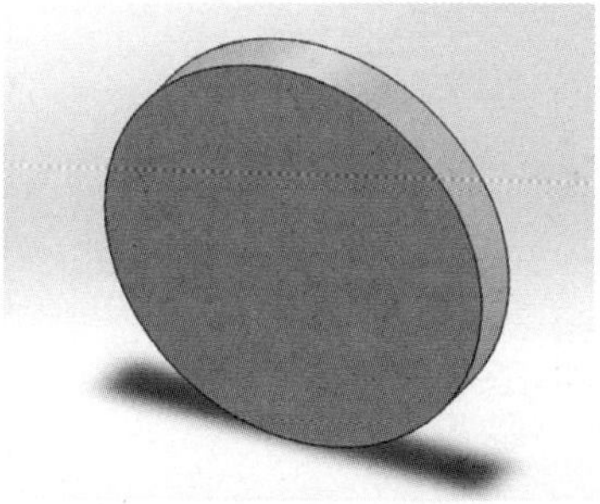

FIGURE 1.42

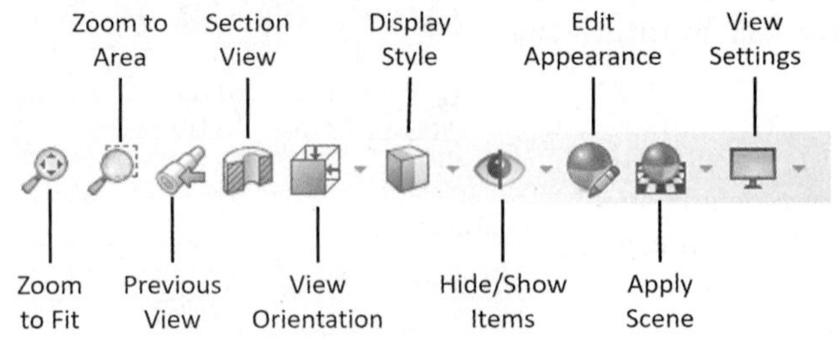

Now that we have a solid part, we can examine the functions of the viewing tools. The viewing tools are located on the Heads-Up View Toolbar at the top of the graphics area. The default configuration of the Heads-Up View Toolbar is shown in **Figure 1.42**. The Zoom to Fit Tool adjusts the zoom so that the entire model can be viewed. The Zoom to Area Tool allows a viewing window to be selected by dragging out an area on the screen. The Previous View Tool returns the view orientation and zoom level to the configuration prior to the most recent change of view. The Section View Tool displays a cross-section of the part. We will use this tool in later chapters.

FIGURE 1.43

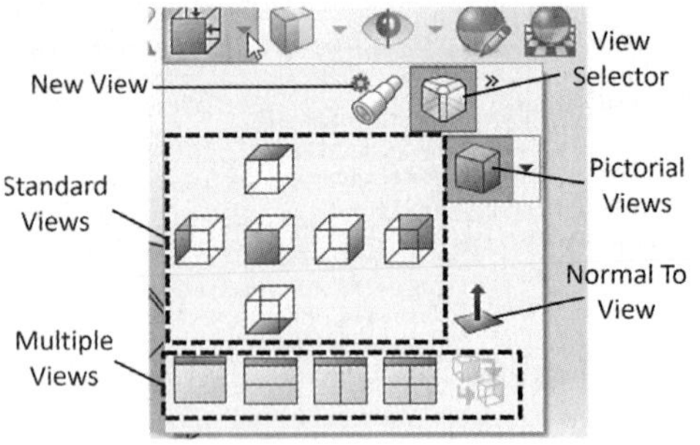

The View Orientation Tool opens a menu of standard view options, as shown in **Figure 1.43**. The six principal or standard views—Front, Back, Top, Bottom, Left, and Right—can be displayed by clicking on the appropriate icon. At the right side of the menu is a pull-down menu of three pictorial views: Isometric, Trimetric, and Dimetric, as shown in **Figure 1.44**. In an isometric view, the view orientation is such that the angles between the displayed edges of a cube are equal, as shown in **Figure 1.45**. In a dimetric view, two of the angles are equal, and in a trimetric view all angles are different. The Trimetric View in SOLIDWORKS emphasizes the display of the front of the part, and is the default pictorial view. (It is good practice to orient a model so that the front view is the view that is most descriptive of the part.) At the bottom of the menu are tools for displaying multiple views in separate windows on the screen. We will demonstrate their use later in this chapter. The Normal To Tool aligns the view to be perpendicular to a selected plane or surface. This tool is useful when sketching in a plane that is not perpendicular to any of the principal views. When the View Selector Tool is turned on, then the part is shown in a "box," as shown in **Figure 1.46**. Clicking on a side, edge, or corner of the box changes the view to one that is normal to the selection. The View Selector Tool can be toggled on or off; in this book we will leave it turned off. Another way to access the View Orientation Menu is to press the space bar. This is an example of a keyboard shortcut. Another useful keyboard shortcut is to press the z key to zoom out from a model view and press the z key while holding the Shift key down to zoom in.

FIGURE 1.44

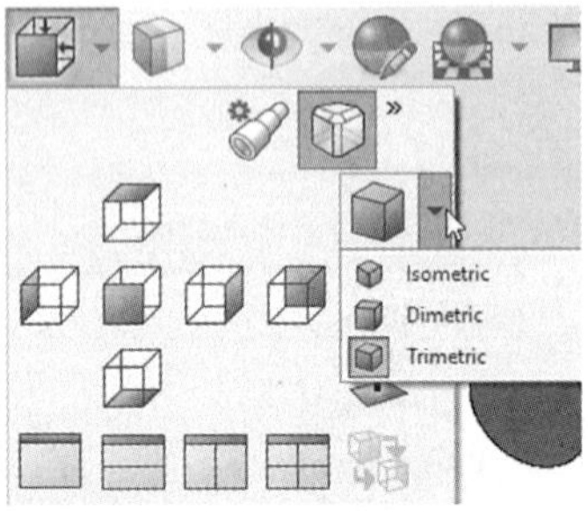

FIGURE 1.45

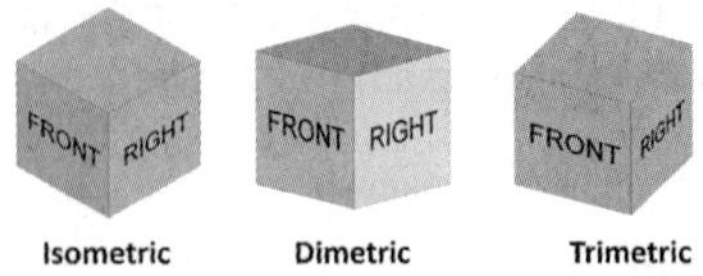

FIGURE 1.46

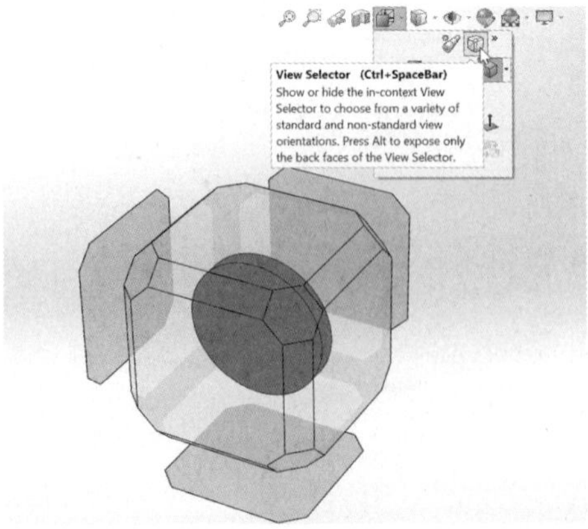

The Display Style Tool opens a pull-down menu of options for displaying the model, as shown in **Figure 1.47**. There are two shaded modes, with or without the edges shown by lines, and three wireframe modes, with hidden edges removed, shown as dashed lines, or shown as solid lines. Usually, we work with one of the shaded modes, but for some operations, displaying the model in wireframe mode is preferred.

FIGURE 1.47

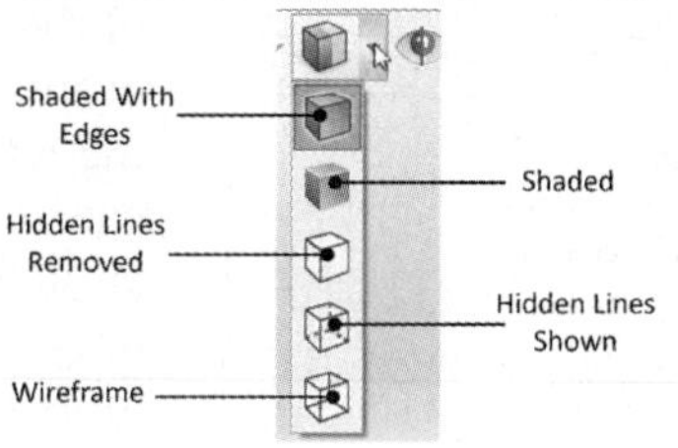

The Hide/Show Items Tool allows you to toggle on or off the display of several items, such as the origin, planes, axes, etc. For now, we will skip over this tool, and will explore its use later in this chapter.

The Edit Appearance Tool allows you to change the color and optical properties (such as transparency and reflectivity) of the entire model or selected model features. When this tool is selected, the PropertyManager displays the selected feature(s), as shown in **Figure 1.48**, and palettes from which a new color can be selected. If no features are selected prior to selecting the Edit Appearance Tool, then by default the change will apply to the entire model. It is recommended that light colors be used, as dark colors can make some features and selections difficult to see. The Advanced tab contains additional options for displaying the model, such as applying textures to selected surfaces or making the part transparent or translucent. We will explore some of these options later. As shown in **Figure 1.48**, the Task Pane at the right of the screen is also displayed when the Edit Appearance Tool is shown. Colors and textures can be applied by dragging and dropping them from the task pane to the part.

FIGURE 1.48

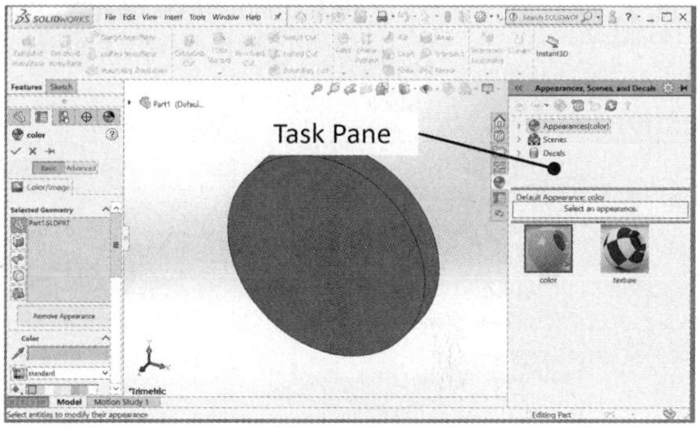

The Apply Scene Tool allows you to select backgrounds and lighting options from a pre-defined menu, as shown in **Figure 1.49**. In this book, we will use the Plain White scene. Some of the scenes contain background graphics rather than just colors. A SOLIDWORKS Add-In program, PhotoView 360, can also be used to produce photo-realistic display images of the model in various scenes.

FIGURE 1.49

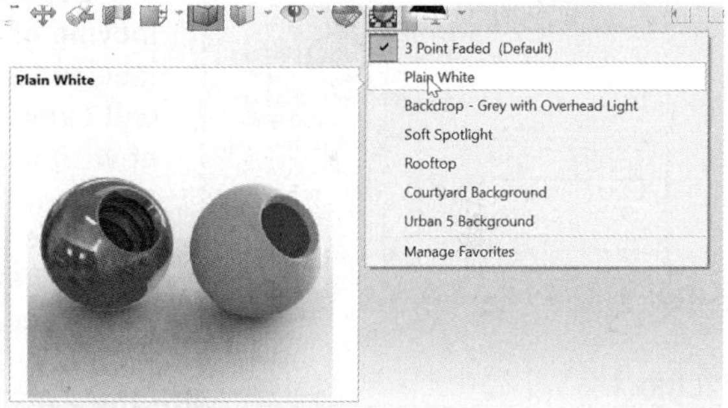

The View Settings Tool, shown in **Figure 1.50**, allows you to add shadows to either of the shaded modes. Also, the model can be shown in perspective mode. In perspective mode, sight lines converge at a single point (the vanishing point), producing a more realistic view. However, most engineering views are produced with parallel sight lines (these are called *orthographic* projections). Normally, we will leave the Perspective View option turned off. There is also a Cartoon View option, which is intended to make a model appear as though it were hand-drawn.

FIGURE 1.50

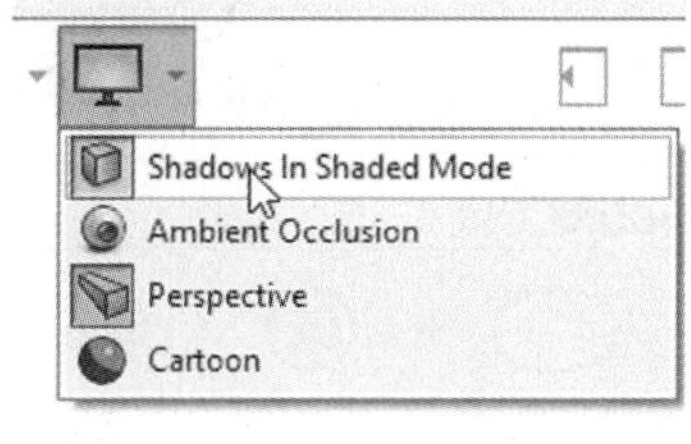

FIGURE 1.51

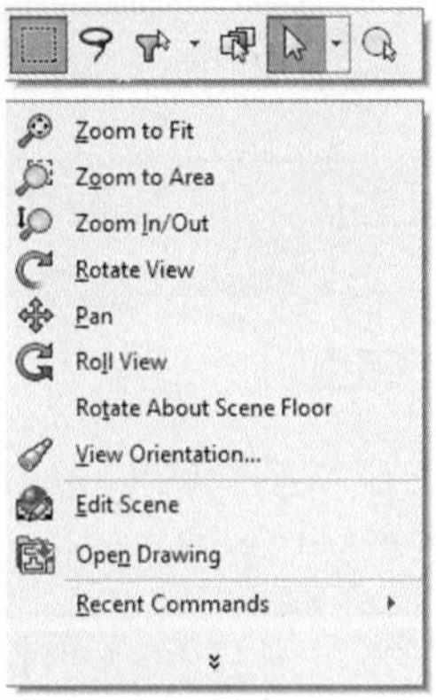

Several other viewing tools can be accessed by right-clicking in the white space of the graphics area. The menu shown in **Figure 1.51** is displayed. A particularly useful tool is the Rotate View Tool. After selecting this tool, you can hold down the left mouse button and move the mouse to rotate the model view so that you can see all sides of the model. The Pan Tool can be used to move the model around in the graphics area, again by clicking and holding the left button while dragging the mouse. Note that when using either of these tools (or several other tools selected from this menu), that the tool remains active until it is turned off by pressing the Esc key.

Because the Rotate View and Pan Tools are used often, we will make them available on the Heads-Up View Toolbar. Also, since the Trimetric View is the default pictorial view, we will add it to the toolbar so that we do not have to go through the View Orientation Tool pull-down menu to select it.

FIGURE 1.52

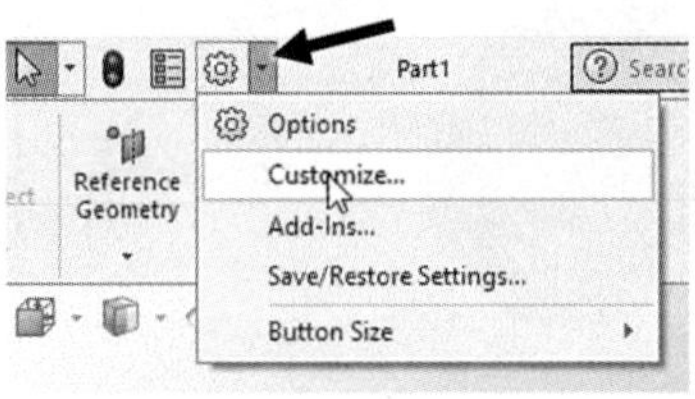

FIGURE 1.53

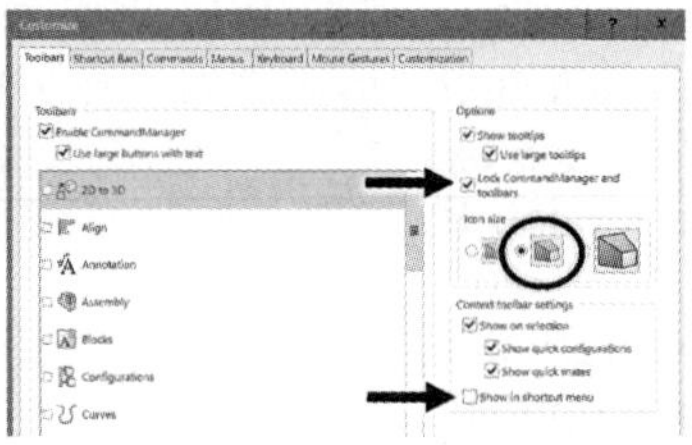

Click the arrow beside the Options Tool and select Customize from the menu, as shown in Figure 1.52. If desired, check the larger Icon size option, as shown in Figure 1.53, so that all of the icons in toolbars and the CommandManager are easier to see. You will see a message that some options will not take effect until SOLIDWORKS is restarted; click OK to close this message. Check the box labeled "Lock the CommandManager and toolbars" to prevent unintentional moving of these items around the screen. Also, uncheck the box labeled "Show in shortcut menu," as shown in Figure 1.53. This will cause several menu items to be displayed with text instead of with icons. This option is discussed further in Appendix B.

FIGURE 1.54

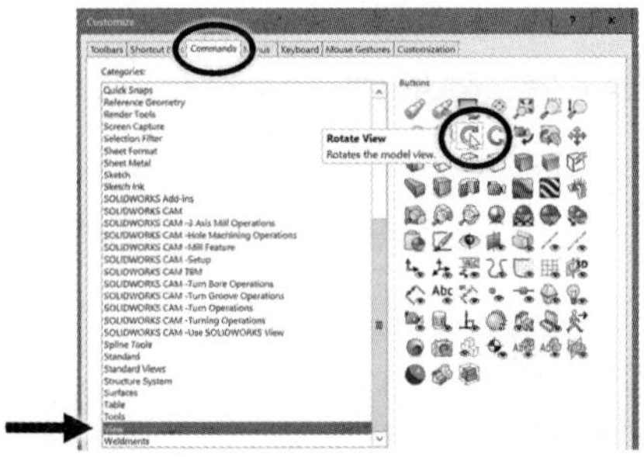

Click the Commands tab, and select the View group. Locate the Rotate View Tool, as shown in Figure 1.54. Click and drag the tool to the desired location on the Heads-Up View Toolbar, as shown in Figure 1.55. Release the mouse button to place the tool. Repeat with the Pan Tool from the View group and the Trimetric View Tool from the Standard Views group. The edited toolbar is shown in Figure 1.56. Click OK to close the Customize box.

FIGURE 1.55

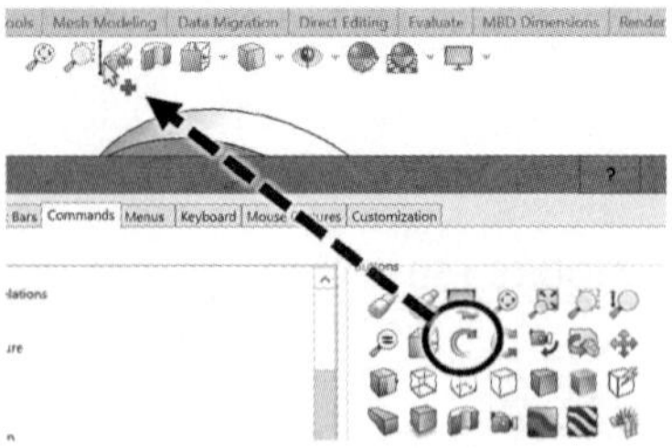

Experiment with the zoom and viewing options. When finished, select a shaded solid display (either with or without edges displayed) and the Trimetric View.

FIGURE 1.56

Now we are ready to add to our part. The next feature we will add is the 2.75-inch-diameter boss. We could sketch the circle to be extruded on the front or back face of the existing part or in the Front Plane, but instead we will create a new plane that is 2.25 inches away from the Front Plane. There are several reasons why we might want to define the part in this manner. One is that we may want to add draft, a slope to the sides of a feature that allows it to be extracted from a mold. If so, then we want our 2.75-inch dimension to apply at the top of the boss, allowing the diameter to get larger closer to its base.

Select the Reference Geometry Tool from the Features group of the CommandManager. From the menu that appears, select Plane (see Figure 1.57).

FIGURE 1.57

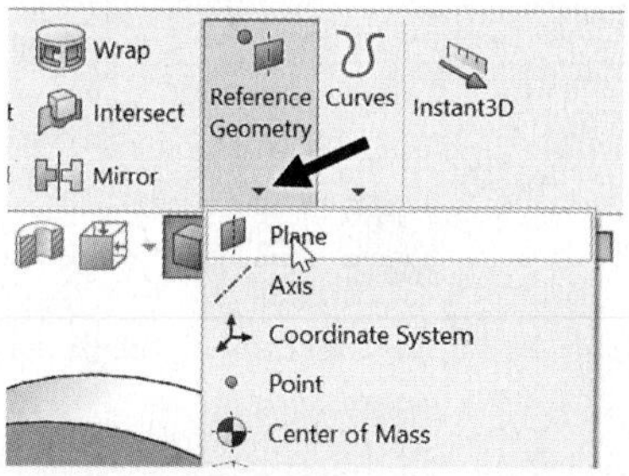

Note that the FeatureManager has been replaced in its usual position by the PropertyManager, where the parameters of the new plane will be defined. However, the FeatureManager is still visible as a "fly out" list to the right of the PropertyManager. By default, the FeatureManager is shown collapsed; that is, only the name of the part is shown. The full FeatureManager can be shown by clicking on the arrow next to the part name.

FIGURE 1.58

Click the arrow next to the part name (Part1) to expand the FeatureManager, as shown in Figure 1.58. Click on the Front Plane to select it. In the box defining the offset distance, enter 2.25, as shown in Figure 1.59. Note that the new plane is previewed in the graphics area. Click the check mark and the new plane, labeled Plane1, is created, as shown in Figure 1.60.

FIGURE 1.59

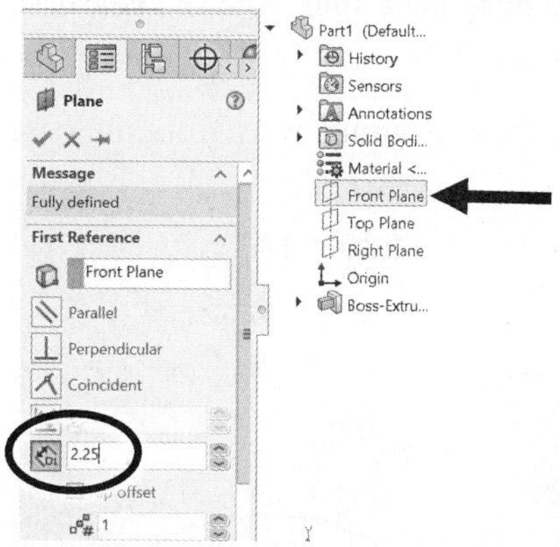

FIGURE 1.60

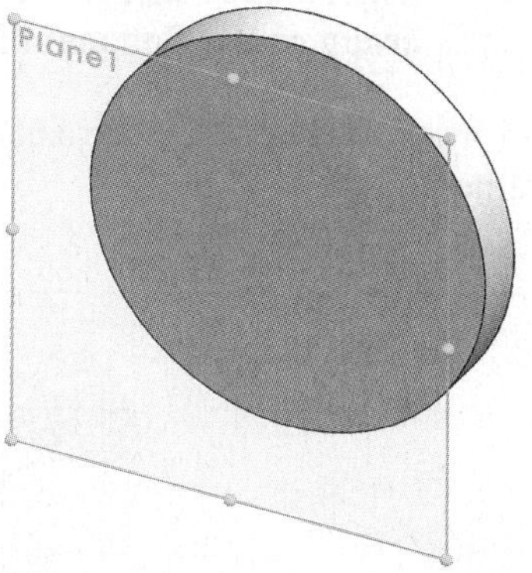

There are several options for creating a new plane. When we selected the Front Plane, the option for defining the new plane parallel to the selected plane was selected by default. Of course, the Flip box could have been checked if we wanted the new plane to be behind the Front Plane.

Now with Plane1 selected (highlighted in green), click the Sketch tab of the CommandManager and select the Circle Tool.

Remember that there are two ways to define a circle; either by defining its mid-point and a point on the perimeter or by defining three points on the perimeter. When you create a circle with either method, then that method becomes the default method of construction. If you hold the cursor over the icon momentarily, the current tool

FIGURE 1.61

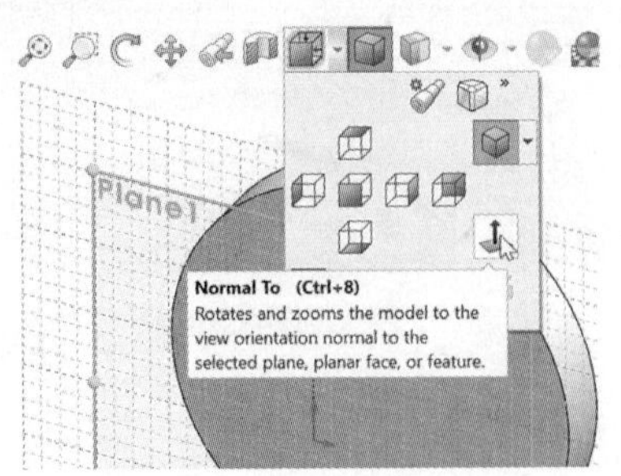

and its description are displayed in the Tooltips popup. If the correct tool is current, then you can select it by clicking on the icon, without using the pull-down menu.

Also note that a sketch was opened when you selected the Circle Tool. When a plane or face is selected, choosing a drawing tool from the Sketch group opens a new sketch and activates the tool.

From the View Orientation Tool, select the Normal To View, as shown in Figure 1.61.

FIGURE 1.62

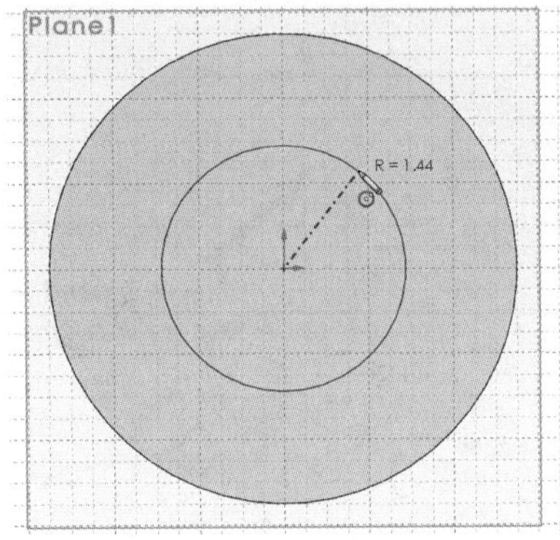

FIGURE 1.63

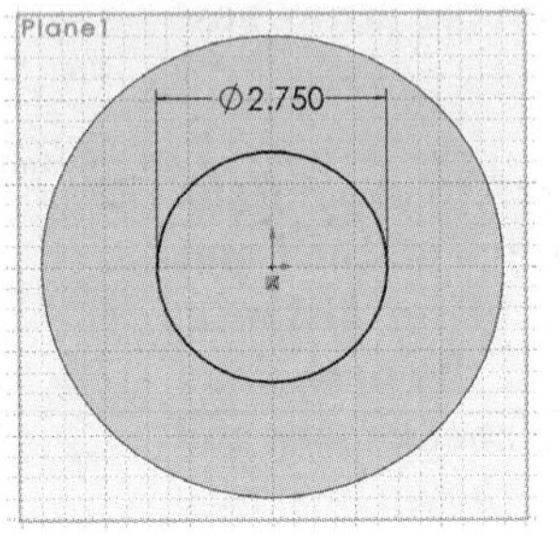

Notice that in this case, the Normal To View is the same as the Front View.

Move the cursor to the origin and click to place the center of the circle at the origin. Drag out a circle, as shown in Figure 1.62, and click to complete the circle.

Select the Smart Dimension Tool from the Sketch group of the CommandManager. Add a 2.75-inch dimension to the diameter, as shown in Figure 1.63. Switch to the Trimetric View (Figure 1.64). Select the Extruded Boss/Base Tool from the Features group of the CommandManager.

As shown by the preview (Figure 1.65), the extrusion is going away from the base feature rather than toward it.

FIGURE 1.64

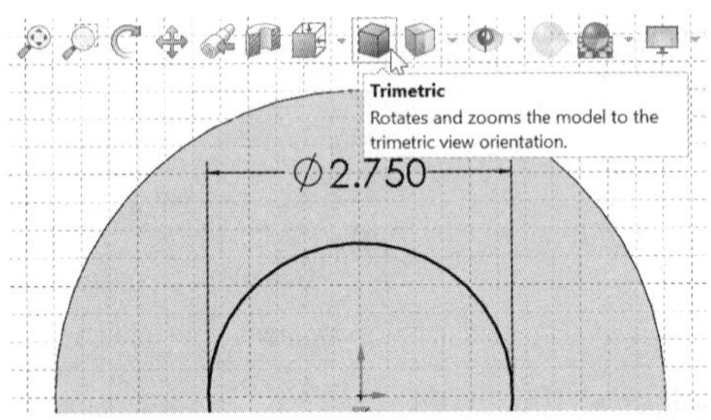

FIGURE 1.65

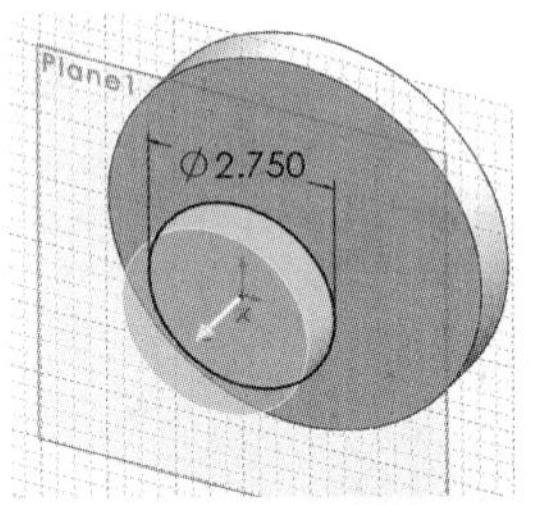

FIGURE 1.66

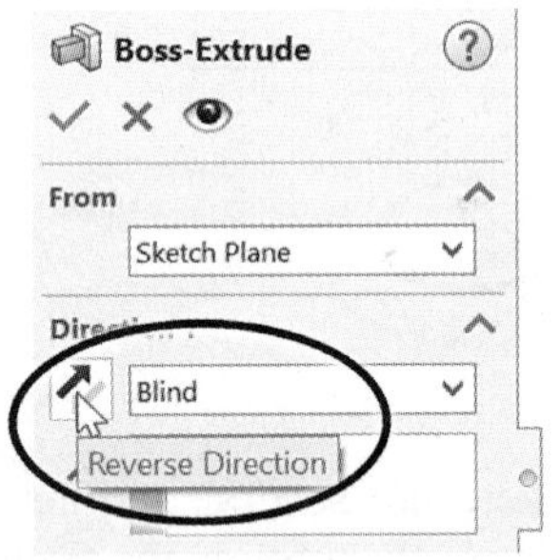

FIGURE 1.67

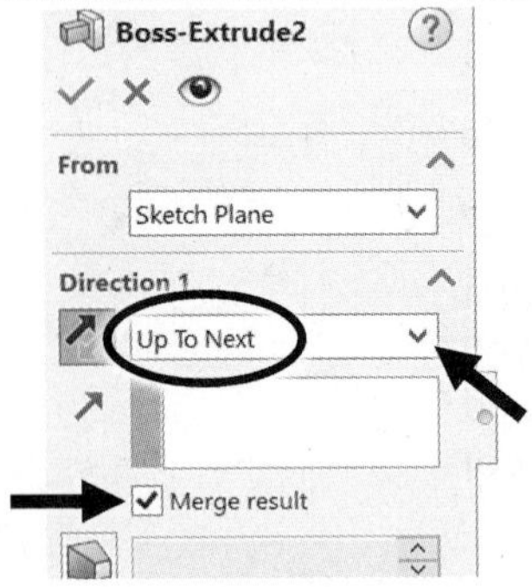

FIGURE 1.68

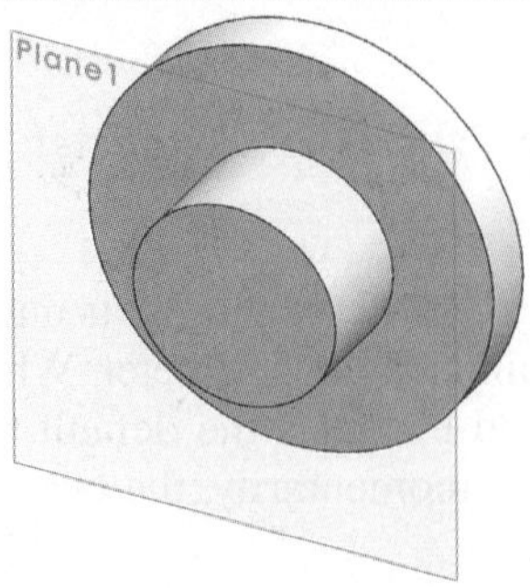

In the PropertyManager, click the Reverse Direction button so that the extrusion is directed toward the base (Figure 1.66). In the pull-down menu, select Up To Next as the type of extrusion, as shown in Figure 1.67. Click the check mark to complete the extrusion, which is shown in Figure 1.68.

DESIGN INTENT **Planning the Model**

As we build the 2.75-inch diameter boss of our flange, we can choose from two existing planes/surfaces or construct a new plane. The choice of constructing a new plane in order to allow us to add draft to our part is an example of design intent. There are many definitions of design intent. Ours is:

Design intent is the consideration of the end use of a part, and possible changes to the part, when creating a solid model.

Throughout this book, we will identify examples of considering design intent when modeling.

Choosing "Up To Next" as the type of extrusion instead of defining the distance allows for changes to be made easily. If we later decide to change the distance between Plane1 and the Front Plane, then the part will rebuild correctly, as the boss extrusion will still extend back to the base feature. Also notice that by default, the "Merge result" option is checked when adding a new feature. With this option checked, the two features that we have created combine to form a single solid body. If the option is unchecked, then the features represent separate bodies.

We can turn off the display of Plane1.

From the Hide/Show Items Tool, click on the View Planes icon to toggle off the display of planes, as shown in Figure 1.69.

FIGURE 1.69

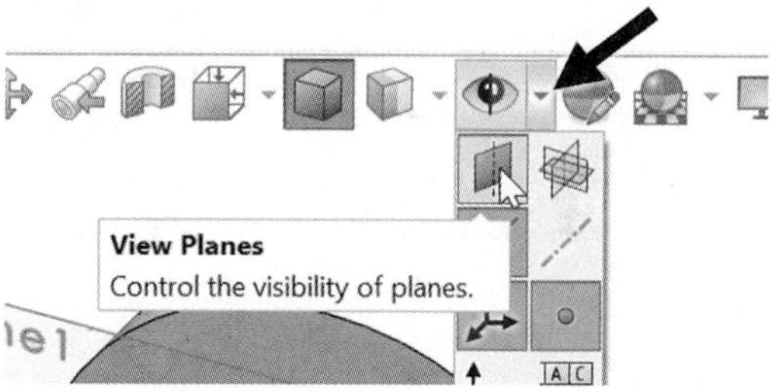

Plane1 still exists in the model, but turning off the display of planes results in a less cluttered model view. Note that Plane1 could also have been hidden by right-clicking its entry in the FeatureManager and selecting Hide.

It is a good idea to save your work periodically.

Choose File: Save from the Main Menu. Save the part with the name "Flange." The file type will be "sldprt."

Note that the new file name appears in the Menu Bar and at the top of the FeatureManager design tree.

Next we will add the center hole. This time we will select a face to define our sketch plane. As you move the cursor over the front surface, notice that a square icon appears. This indicates that a surface will be selected when you click with the left mouse button. Similarly, a line icon indicates that an edge will be selected.

FIGURE 1.70

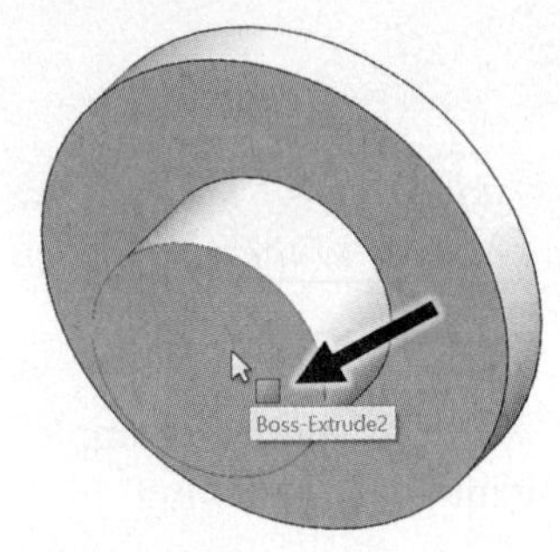

FIGURE 1.71

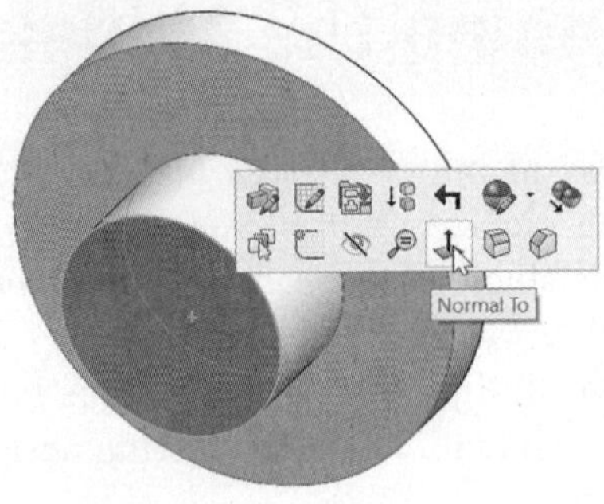

Move the cursor over the front face, so that the square icon appears, as shown in Figure 1.70. Click to select this face; it will be highlighted in green. A pop-up toolbar, called a Context Toolbar, will appear. Click the Normal To View, as shown in Figure 1.71.

(Note: If the Context Toolbar does not appear when you select the face, then select the Customize Tool and make sure that the box labeled "Show on selection" is checked, as shown in Figure 1.53.)

FIGURE 1.72

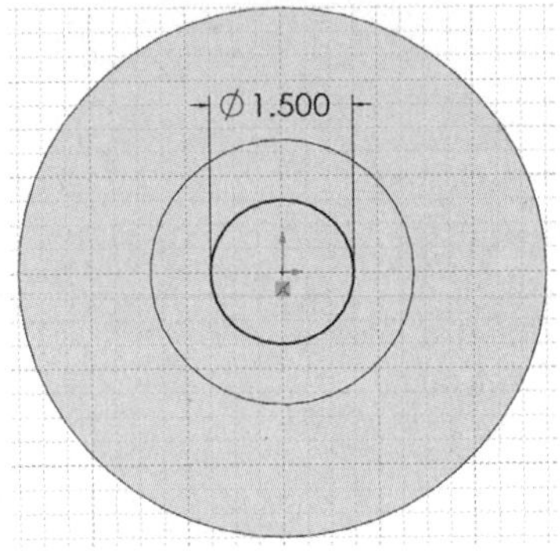

Click the Sketch tab of the CommandManager and select the Circle Tool. Drag out a circle centered at the origin. Select the Smart Dimension Tool and dimension the circle diameter as 1.5 inches, as shown in Figure 1.72. Click the Features tab of the CommandManager and select the Extruded Cut Tool, as shown in Figure 1.73. Select the type as Through All in the PropertyManager, as shown in Figure 1.74. Click the check mark to complete the cut. Switch to the Trimetric View. The result of this operation is shown in Figure 1.75.

FIGURE 1.73

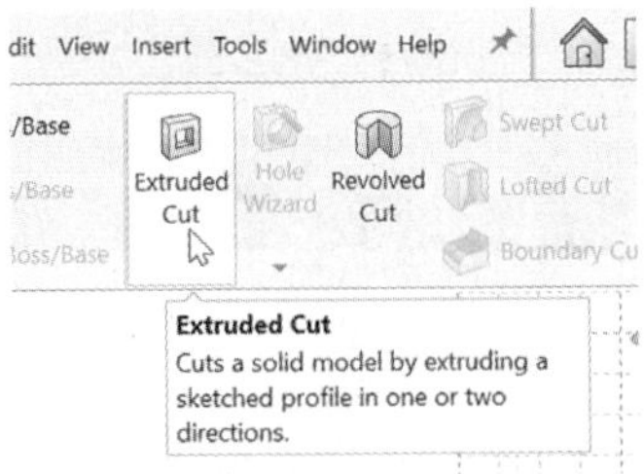

FIGURE 1.74

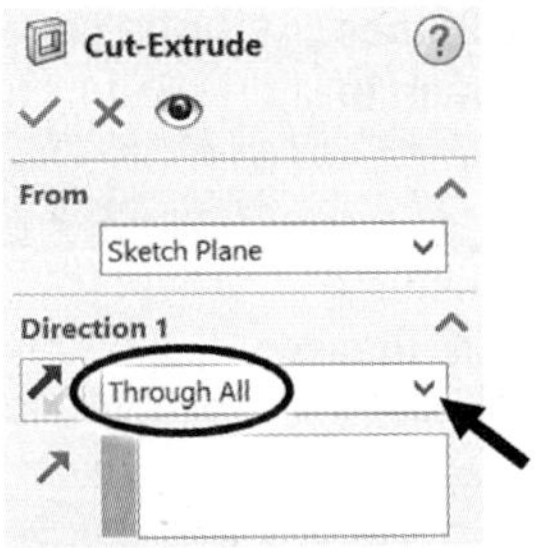

FIGURE 1.75

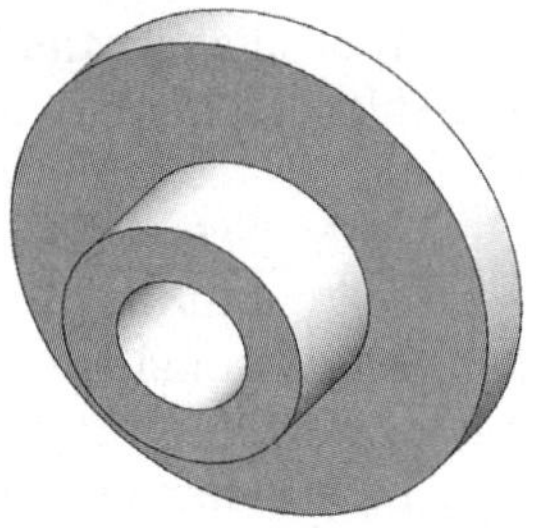

FIGURE 1.76

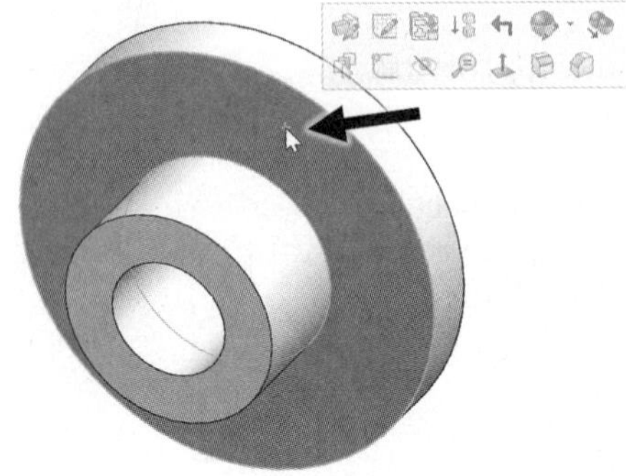

We will now add the four bolt holes.

Select the surface shown in Figure 1.76. Switch to the Normal To View. Click the Sketch tab of the CommandManager, and select the Circle Tool. Drag out a circle centered at the origin. In the PropertyManager, check the "For construction" box, as shown in Figure 1.77.

FIGURE 1.77

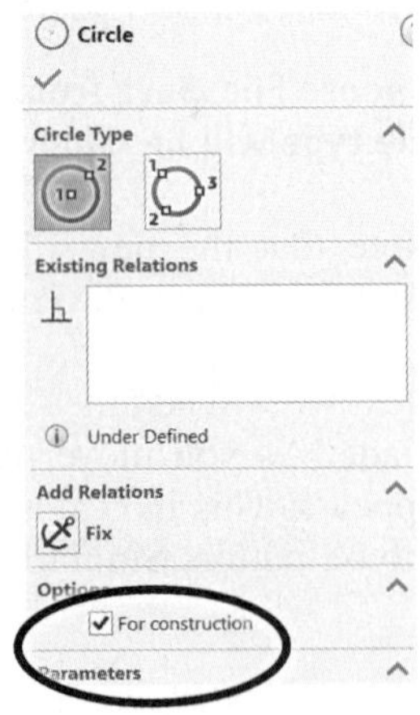

Construction entities help you locate and size sketch parameters, and are indicated by dashed-dotted lines. The circle just drawn represents the bolt circle.

FIGURE 1.78

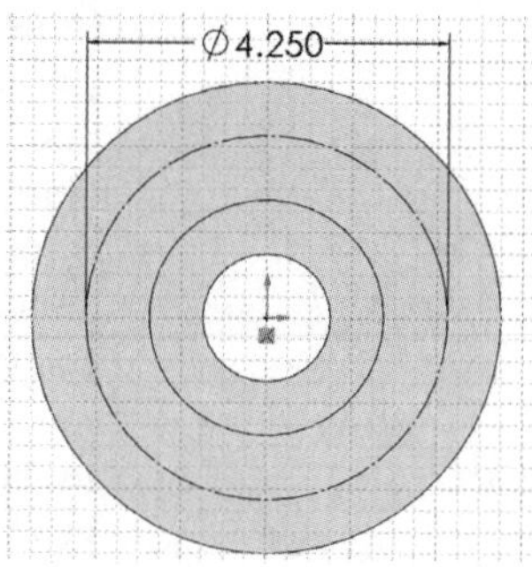

Select the Smart Dimension Tool. Add a 4.25-inch diameter dimension to the circle, as shown in Figure 1.78. Select the Circle Tool. Move the cursor to the top quadrant point on the construction circle, as shown in Figure 1.79. Note the red diamond that appears, along with the coincident and vertical relation icons. Drag out a circle, as shown in Figure 1.80. Select the Smart Dimension Tool and add a diameter dimension of 0.50 inches, as shown in Figure 1.81.

FIGURE 1.79

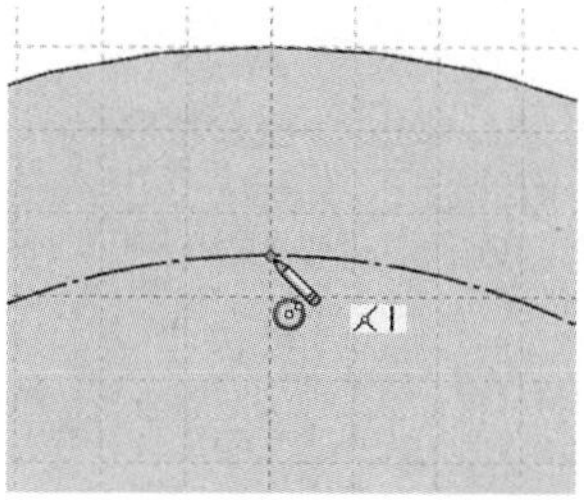

FIGURE 1.80

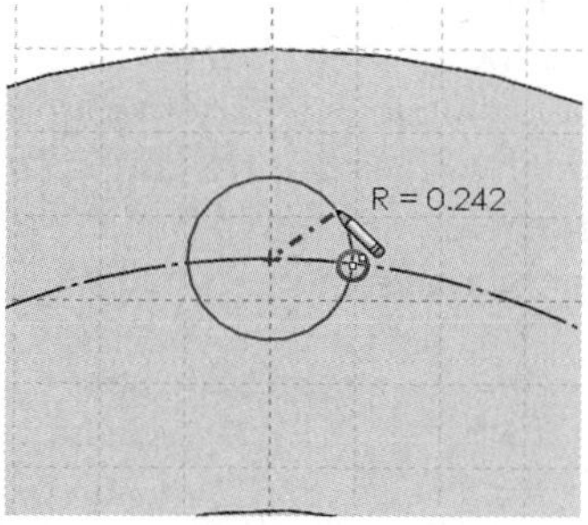

FIGURE 1.81

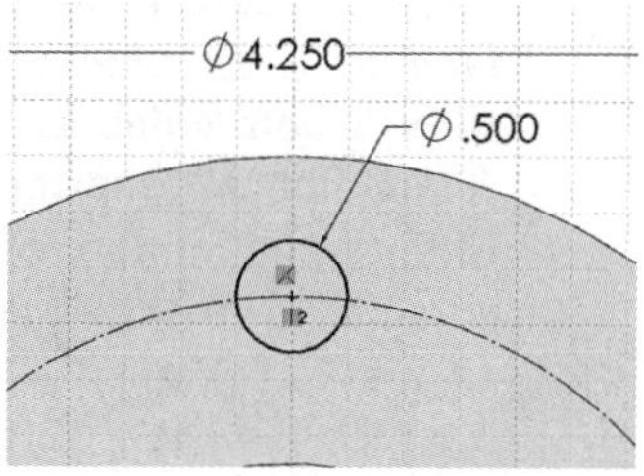

The sketch is fully defined, since the center of the circle just drawn has been located at the top quadrant point of the bolt circle.

Select the Extruded Cut Tool from the Features group of the CommandManager, and extrude a hole with a type of Through All. Click the check mark.

The first bolt hole is now in place, as shown in the trimetric view in Figure 1.82. Note that when we selected the Extruded Cut tool, only the small circle was used as the geometry of the cut. The bolt circle, because it was identified as construction geometry, was not included as a sketch entity to be extruded.

FIGURE 1.82

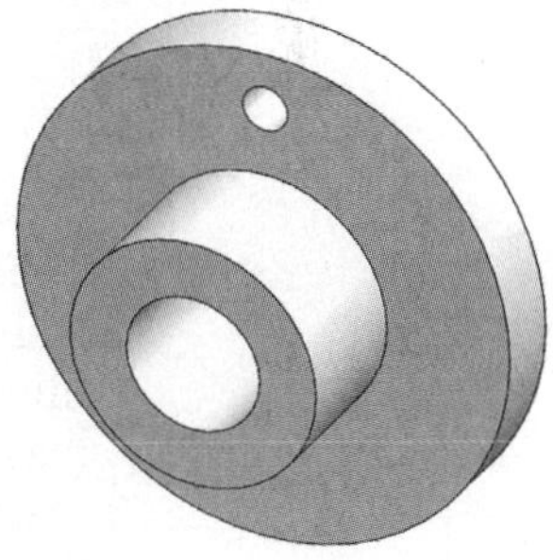

Notice that in the FeatureManager, all of our procedures are being recorded. The names of the features are not particularly descriptive; the four features that we have created so far were all created by extrusions, and so are named "Boss-Extrude1," "Cut-Extrude1," etc. To more easily identify features for later modifications, we can rename features.

Click once on "Cut-Extrude2" in the FeatureManager to select and highlight the name. Click again to allow editing of the name. (Use two separate mouse clicks, not a double-click.) Type "Bolt Hole" to rename the feature, as shown in Figure 1.83. Press the Enter key to accept the new name.

FIGURE 1.83

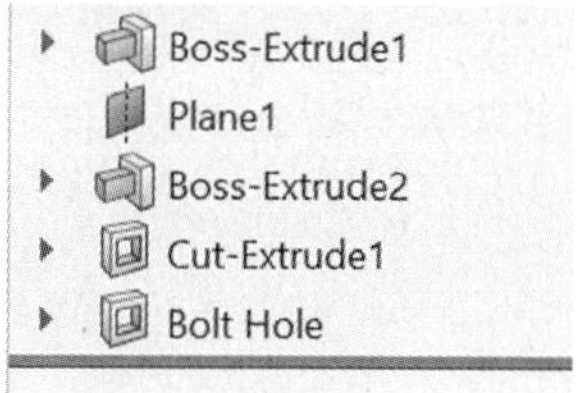

We could create the other three holes separately, but it is easier to copy the single hole into a circular pattern. Also, since our design intent is for the holes to exist in a circular pattern, it makes sense to construct them that way. If we later change the diameter of the holes, the diameter of the bolt circle, or the number of holes, it will be easy to do if we have created them in a pattern.

Make sure that the first bolt hole is selected. Click the Features tab of the CommandManager, and click the arrow under the Linear Pattern Tool to reveal a menu of pattern tools, as shown in Figure 1.84. Choose the Circular Pattern Tool.

FIGURE 1.84

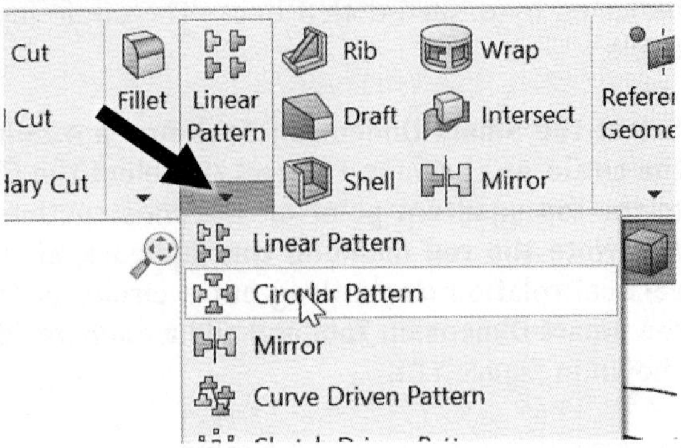

In the PropertyManager, click in the top box (Pattern Axis) to activate it, as shown in Figure 1.85. To define the axis of the pattern, select a cylindrical face or a circular edge (other than a bolt hole), as shown in Figure 1.86. In the PropertyManager, check the "Equal spacing" option which will cause the angle to be changed to 360 degrees. Change the number of holes to 4, as shown in Figure 1.87. Click the check mark to complete the pattern, which is shown in Figure 1.88.

FIGURE 1.85

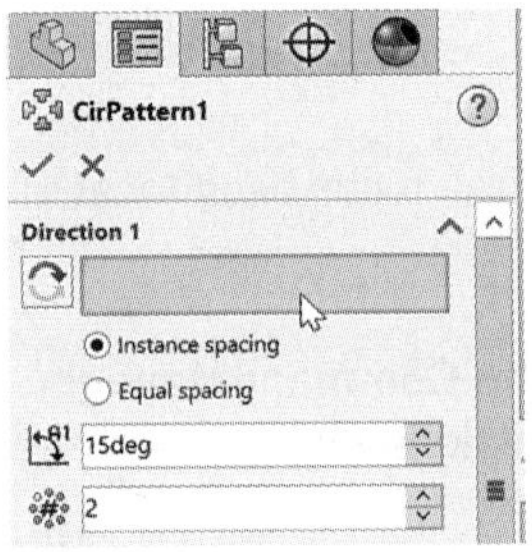

FIGURE 1.86

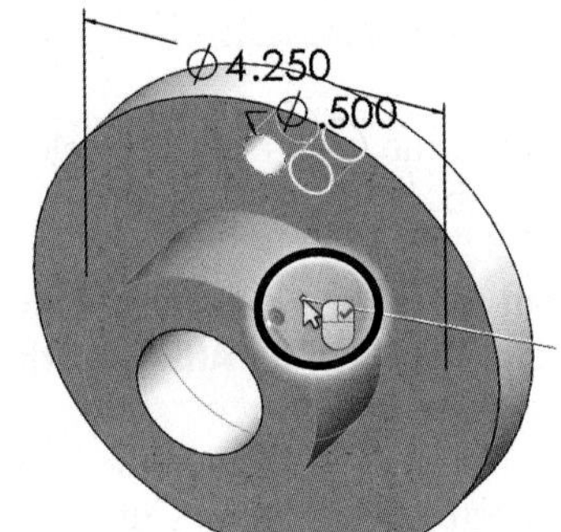

FIGURE 1.87

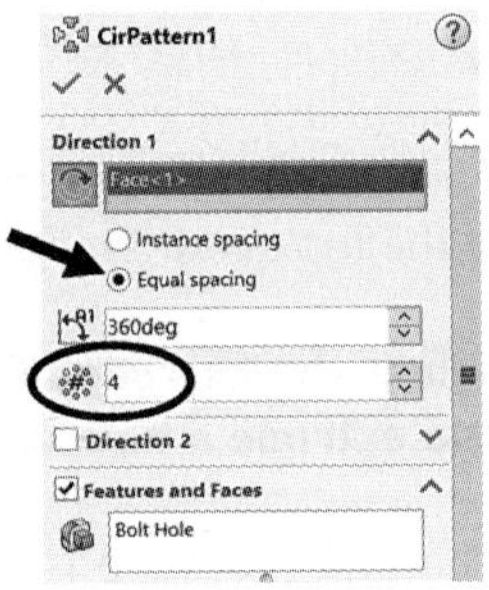

Now let's finish the flange by adding fillets to three of the sharp edges. A fillet is a feature that rounds off a sharp edge. Actually, a fillet is a rounded edge created by adding material, while a round is created by removing material. Fillets and rounds are created with the SOLIDWORKS software by the same command.

FIGURE 1.88

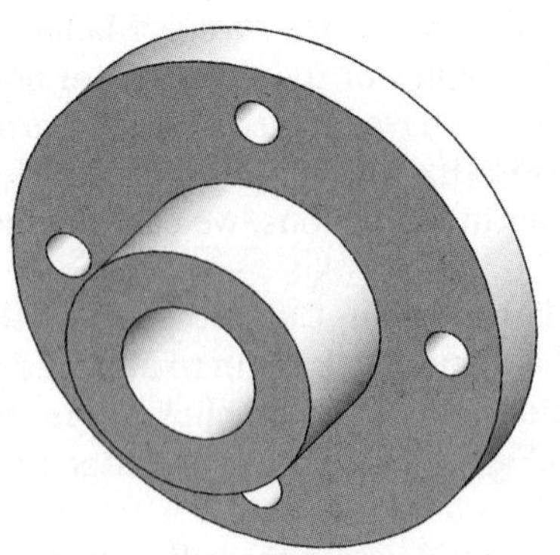

From the Features group of the CommandManager, select the Fillet Tool, as shown in Figure 1.89. Select the three edges indicated in Figure 1.90 to be filleted. (Be sure to see the line next to the cursor, as shown in Figure 1.91, to indicate that an edge and not a face is being selected. If a face is selected, then all of the edges of that face will be filleted.) In the PropertyManager, enter the radius as 0.25 inches, as shown in Figure 1.92. Check the Full preview option to see the fillets that will be created. Click the check mark to add the fillets, which are shown in Figure 1.93.

FIGURE 1.89

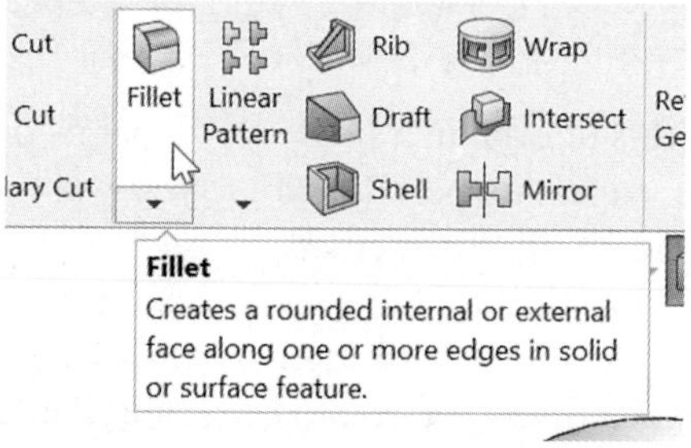

FIGURE 1.90

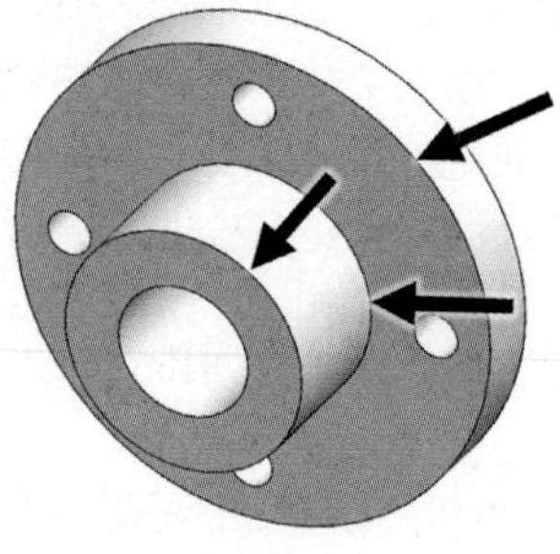

FIGURE 1.91

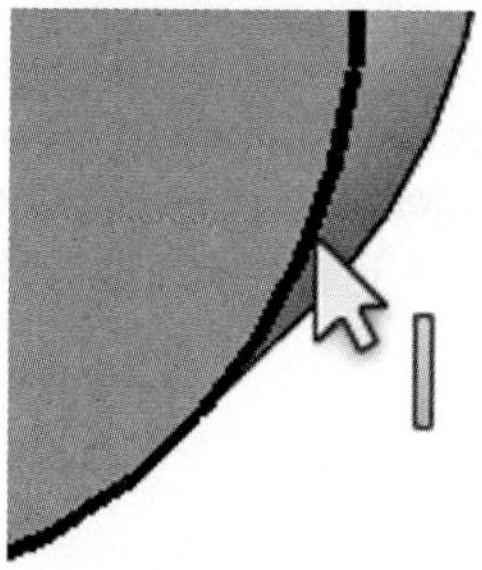

FIGURE 1.92

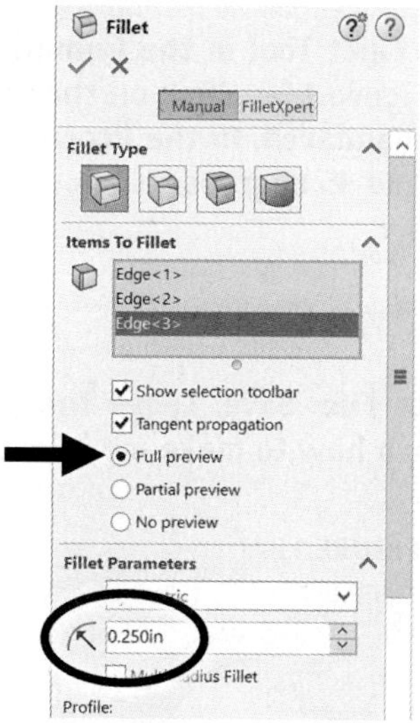

FIGURE 1.93

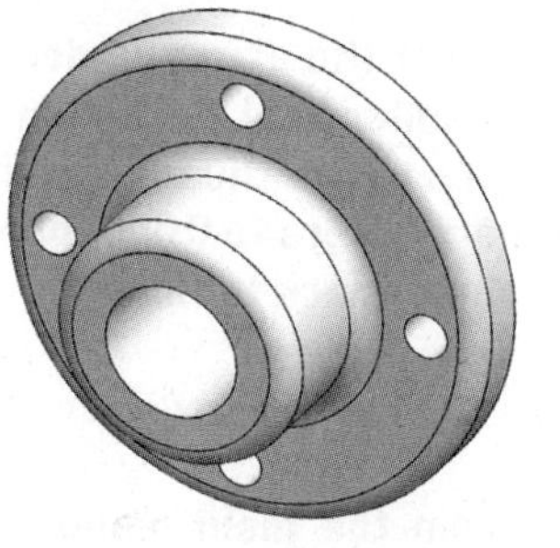

Notice that the intersections of the fillets and the cylindrical and flat surfaces are shown. These interfaces are called tangent edges. Display of tangent edges is often undesirable. Their display can be controlled from the Options menu.

FIGURE 1.94

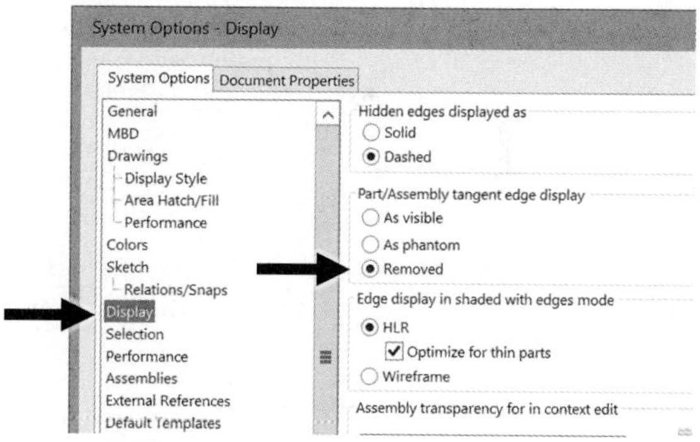

Select the Options Tool. Under the System Options tab, under Display, choose Removed as the Part/Assembly tangent edge display option, as shown in Figure 1.94. Click OK.

The part should appear as in **Figure 1.95**.

FIGURE 1.95

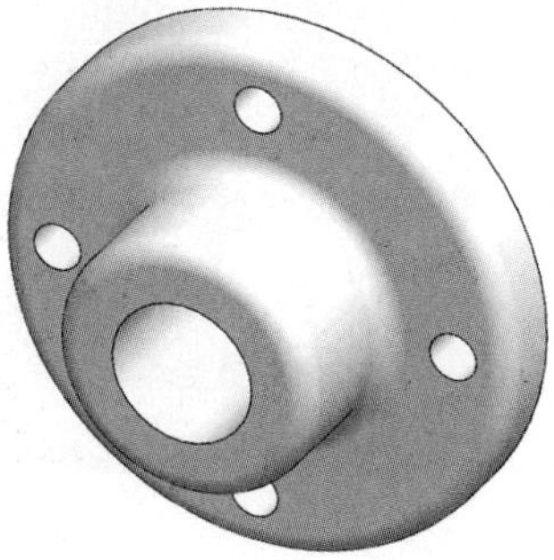

DESIGN INTENT Selecting a Modeling Technique

The three fillets are added in this tutorial in a single step by selecting the three edges to be filleted within a single fillet command. With this method, only the first fillet is dimensioned. Another way to add the fillets is to close the Fillet Tool after each fillet is created, so that the fillets are created in three separate steps. The preferred method depends on how you wish to edit the fillet radii. If you want all of the fillets to always have the same radius, then the first method allows one value to be changed for all three fillets. If you prefer to edit the fillets separately, then the second method provides an editable dimension for each fillet.

Now we can add the chamfer to the center hole. A chamfer is a conical feature formed by removing material from an edge.

Select the arrow under the Fillet Tool in the CommandManager, and select the Chamfer Tool, as shown in Figure 1.96. Click on the edge shown in Figure 1.97 to select it as the edge to be chamfered. In the PropertyManager, set the chamfer parameters to 0.080 inches and 45 degrees, as shown in Figure 1.98, and click the check mark to finish.

The finished part is shown in **Figure 1.99**.

From the main menu, select File: Save. Leave the part file open for the next section, in which we will learn how to make modifications to the part.

FIGURE 1.96

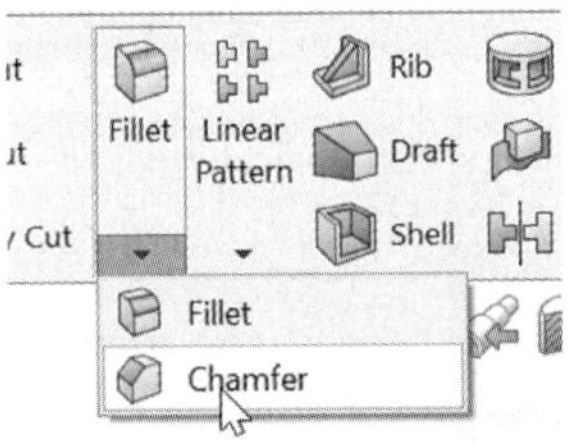

FIGURE 1.97

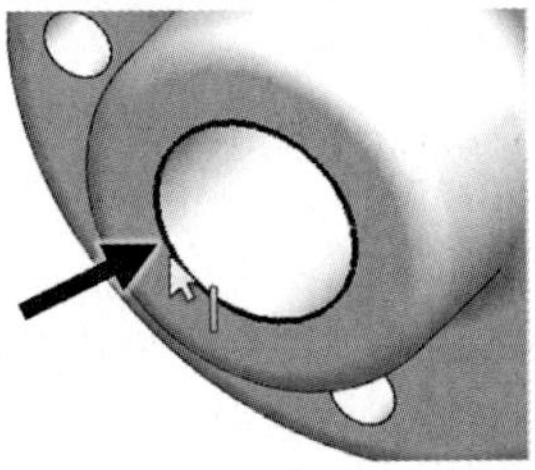

FIGURE 1.98

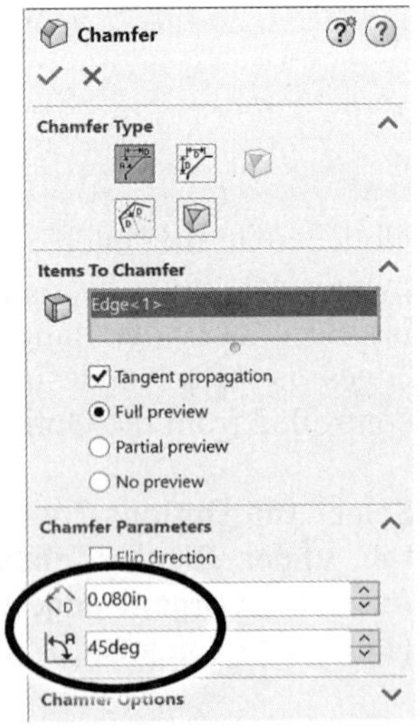

FIGURE 1.99

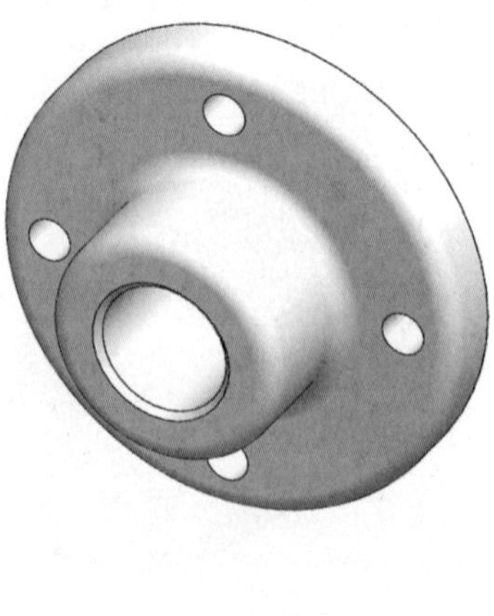

1.3 Modifying the Flange

FIGURE 1.100

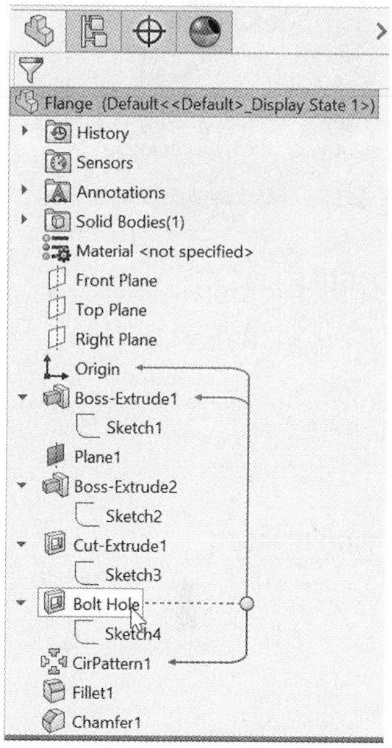

One of the main advantages of solid modeling is the ability to make changes easily. As we have observed, the FeatureManager has recorded all of the operations required to make the flange, as shown in **Figure 1.100**. If we click on the arrow next to each feature, we see that the sketch associated with each feature is stored as well. Relationships between features can be displayed by selecting View: User Interface from the Main Menu and turning on the two Visualization options shown in **Figure 1.101**. For example, the first bolt hole is the basis for the hole pattern. In **Figure 1.100**, this relationship is displayed with an arrow from the "parent" feature (Bolt Hole) to the "child" feature (CirPattern1). Similarly, since the sketch defining the position of the bolt hole is on a surface of the initial solid disk (Boss-Extrude1) and is dimensioned relative to the origin, the hole is shown as a "child" of Boss-Extrude1 and the origin.

Let's change the first item that we created by increasing the diameter of the base from 5.5 to 7 inches.

Right-click Sketch1 in the FeatureManager, and select Edit Sketch.

Note that if Edit Sketch does not appear in the menu, then an icon for editing the sketch appears in the Context toolbar at the top of the menu. Earlier in the chapter, we selected the Customize tool and cleared the check box labeled "Show in shortcut menus." This causes commands such as Edit Sketch, Edit Feature, Hide, etc., to appear as entries in the menu rather than as icons at the top of the menu. If you missed this step earlier, then it is recommended that you clear the check box now. (See Appendix B for more information about customizing the SOLIDWORKS interface.)

Double-click the 5.5-inch dimension, and change it to 7.0 inches, as shown in Figure 1.102. When you close the sketch by clicking on the Exit Sketch Tool in the Sketch group of the CommandManager, the part will be updated to the new dimension, as shown in Figure 1.103.

An even easier way to edit the sketch dimensions or the extrude depth is illustrated next.

FIGURE 1.101

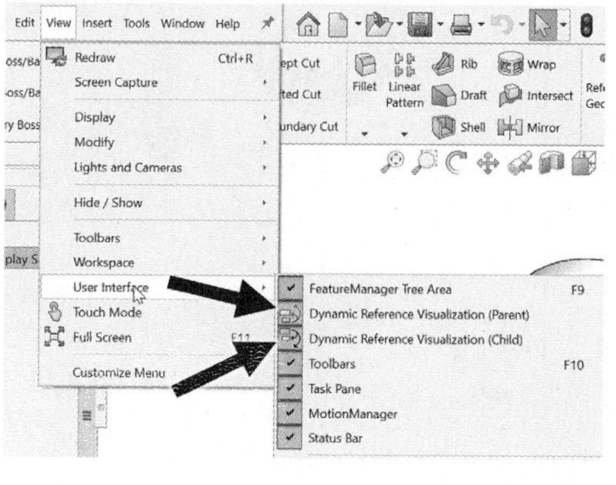

FIGURE 1.102

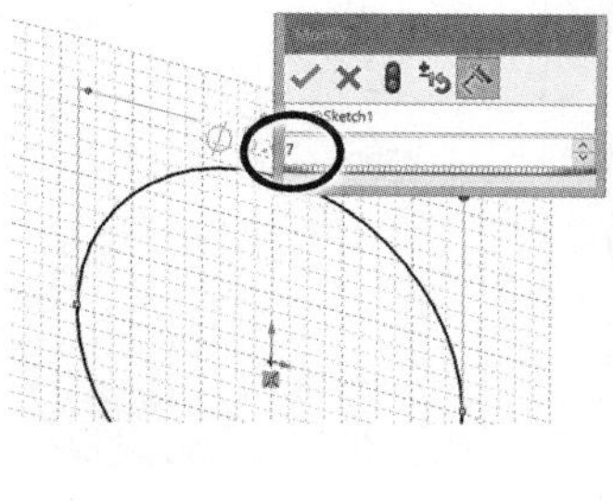

FIGURE 1.103

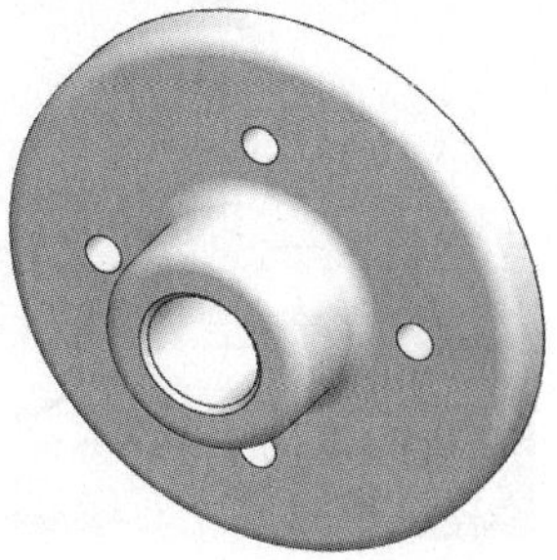

FIGURE 1.104

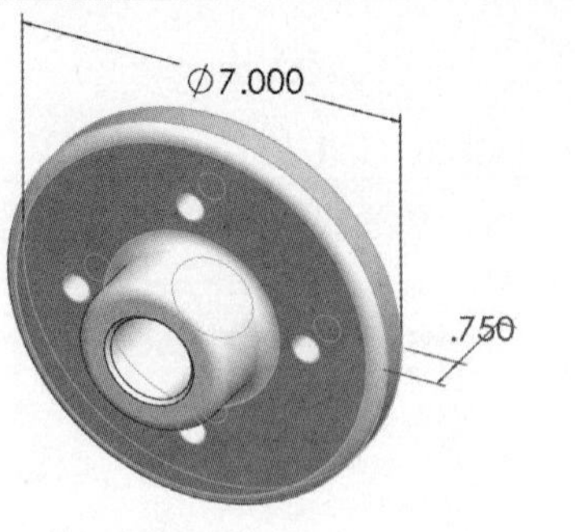

Double-click the icon next to Boss-Extrude1 in the FeatureManager. All of the dimensions used to create the feature are displayed, as shown in Figure 1.104. The sketch dimensions are shown in black, while the feature dimensions (in this case the extrude depth) are shown in blue. (Note that the dimensions in the figure are oriented so that they are aligned to be parallel with the bottom of the screen rather than with the dimension lines. To show the dimensions in this manner, select Options: System Options: Display and check the box labeled "Display dimensions flat to screen.") Double-click the diameter dimension and change it back to 5.5 inches. Click the Rebuild Tool, as shown in Figure 1.105.

FIGURE 1.105

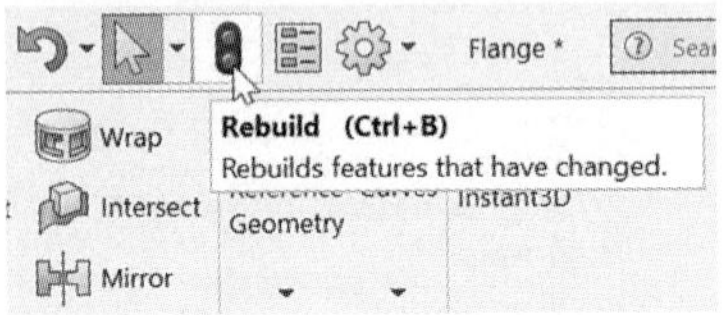

To add draft to the boss, select Boss-Extrude2 from the FeatureManager, right-click and select Edit Feature, as shown in Figure 1.106. In the PropertyManager, turn the draft on (see Figure 1.107) and set the angle to 3 degrees. Check the "Draft outward" box so that the boss increases in size as it is extruded. Click the check mark to finish.

FIGURE 1.106

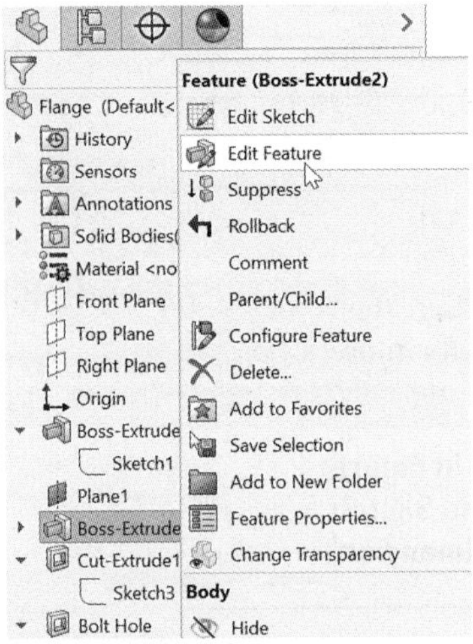

FIGURE 1.107

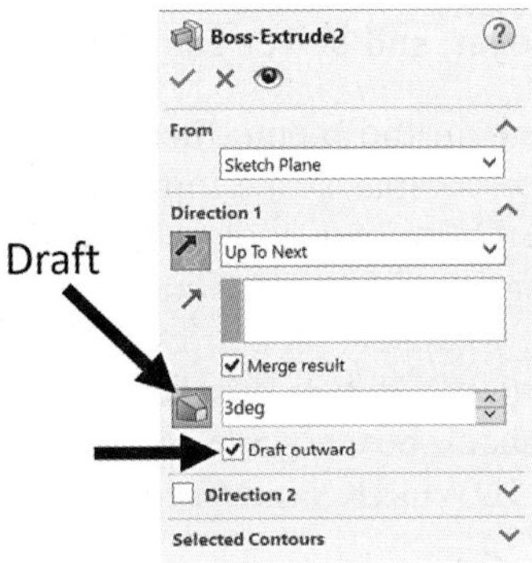

The draft will be easier to see from a top or side view. You can show the Front, Top, and Right Views along with the current (Trimetric) view with the Four-View option.

Select the Four-View window from the View Orientation Tool of Heads-Up View Toolbar, as shown in Figure 1.108.

The drafted feature can be seen clearly in the Top and Right Views, as shown in **Figure 1.109**. Note that **Figure 1.109** shows the Top View above the Front View, and the Right View to the right of the Front View. Views oriented in this manner are referred to as third-angle projections. If the views on your screen

FIGURE 1.108

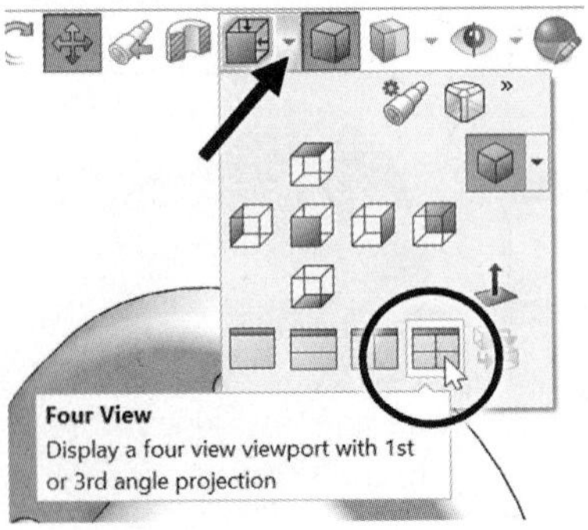

FIGURE 1.109

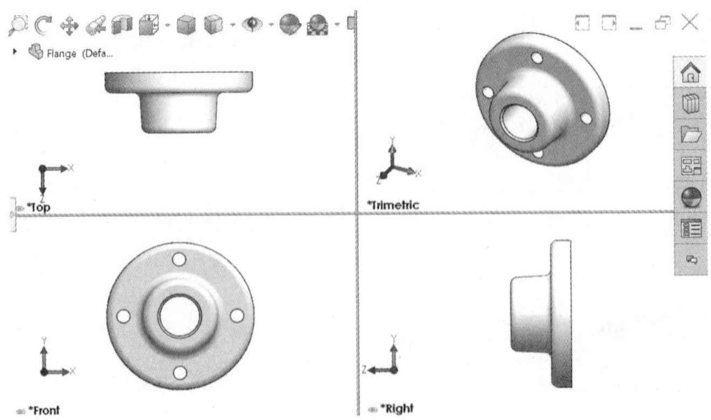

are oriented with the Top View below the Front View, then you are seeing first-angle projections, which are typical of European drawings. To switch from first-angle to third-angle projections, select Options: System Options: Display and choose Third Angle as the option for the Four-View viewport. After doing so, you will need to select the Four-View window again to refresh the views.

FIGURE 1.110

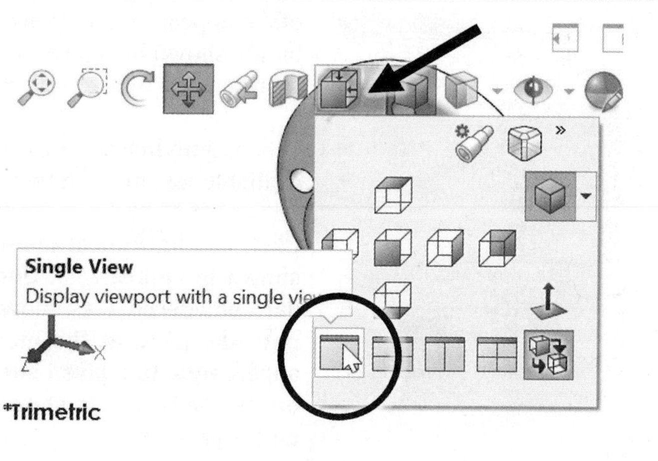

To revert to a single view, click in the window displaying the Trimetric View, and select Single View from the View Orientation Tool of the Heads-Up View Toolbar, as shown in Figure 1.110.

Finally, right-click on CirPattern1 in the FeatureManager and select Edit Feature. Change the number of holes from four to six, as shown in Figure 1.111.

FIGURE 1.111

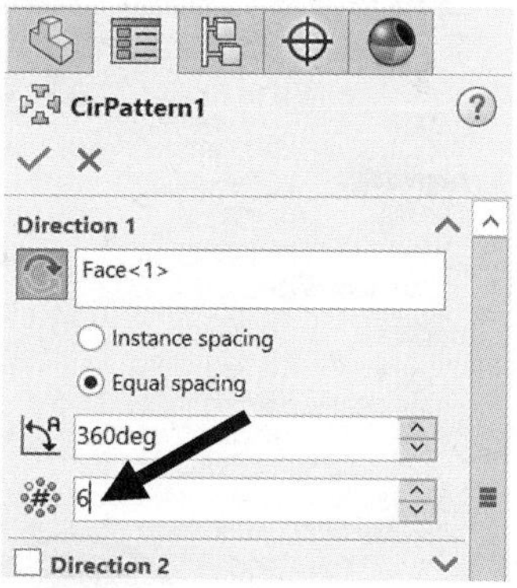

The modified part is shown in Figure 1.112.

FIGURE 1.112

These last two changes illustrate the importance of considering design intent when modeling. If the first boss had been extruded from the base feature, then adding draft would have required us to change the diameter of the boss, calculating the diameter that will result in a 2.75-inch diameter at the top of the boss when draft is included. By sketching in a plane at the top of the boss, the critical 2.75-inch dimension can be maintained easily. Also, by constructing the holes as a circular pattern instead of individually, the number of holes could be modified easily.

Click the Undo Tool to reverse the previous command.

We can change the appearance of a part or of individual features and faces of a part with the Edit Appearance Tool.

Press the Esc key to cancel any selections that may be active. Select the Edit Appearance Tool from the Heads-Up View Toolbar, as shown in Figure 1.113. In the PropertyManager, note that since no specific entities have been selected, the entire part will take on the selected appearance. Select a color from the color palette, as shown in Figure 1.114, and click the check mark.

FIGURE 1.113

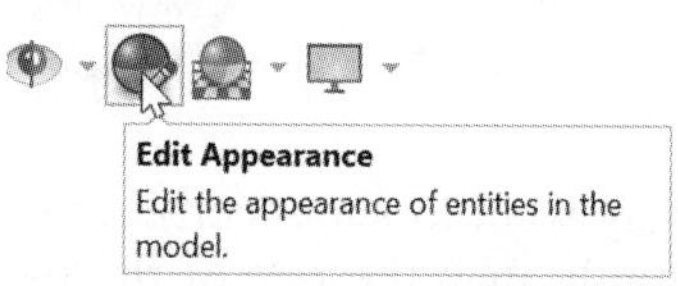

FIGURE 1.114

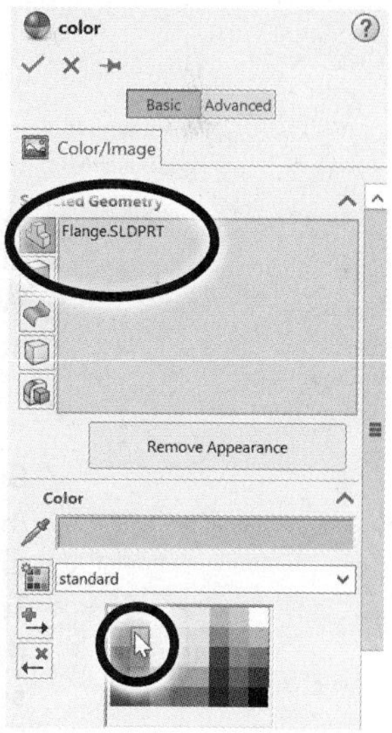

FIGURE 1.115

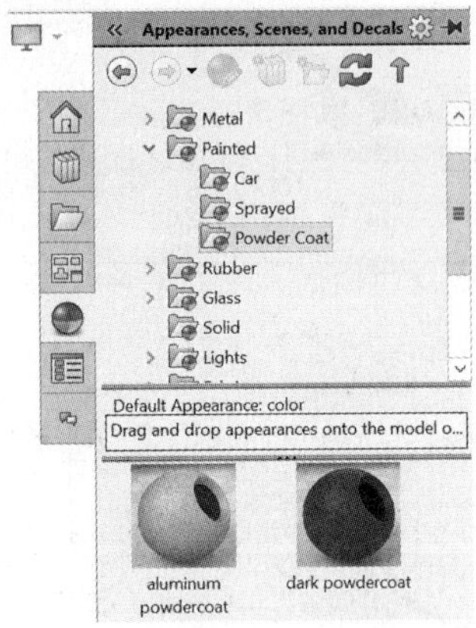

The entire flange will now be shown in the selected color. Note that many other appearance options can be selected by clicking on the Advanced button shown in **Figure 1.114**. These include modifying the reflectivity or the transparency of a component or applying a surface texture.

When applying textures, the Task Pane is useful in that previews of the available textures can be viewed.

Click on the Appearances, Scenes, and Decals tab of the Task Pane, as shown in Figure 1.115. Under Appearances: Painted: Powder Coat, select dark powdercoat, as shown in Figure 1.116. Click and drag the appearance onto the part. In the menu that appears, you can choose to apply the appearance to a given surface, a feature, a body, or the entire part. Click on the Part icon, as shown in Figure 1.117, to apply the appearance to the entire part.

FIGURE 1.116

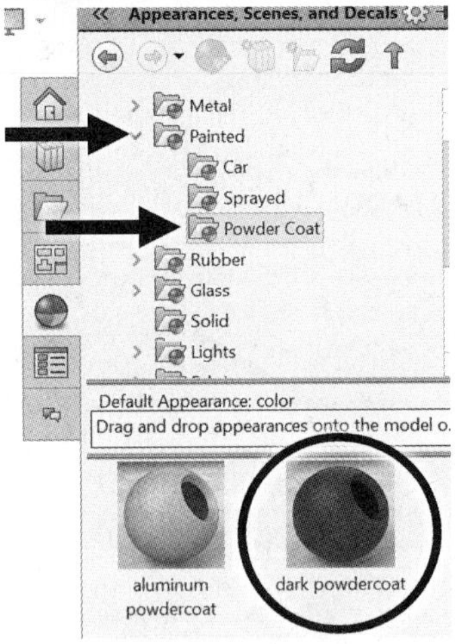

FIGURE 1.117

We may want to show certain faces differently than the rest of the part. We will show the surfaces of the holes and the chamfer as machined steel.

Select the surfaces of the four bolt holes, the center hole, and the chamfer, as shown in Figure 1.118. To select multiple entities, hold down the Ctrl key as you make your selections. Click on the Appearances, Scenes, and Decals tab of the Task Pane. Select Appearances: Metal: Steel, and double-click on Machined

FIGURE 1.118

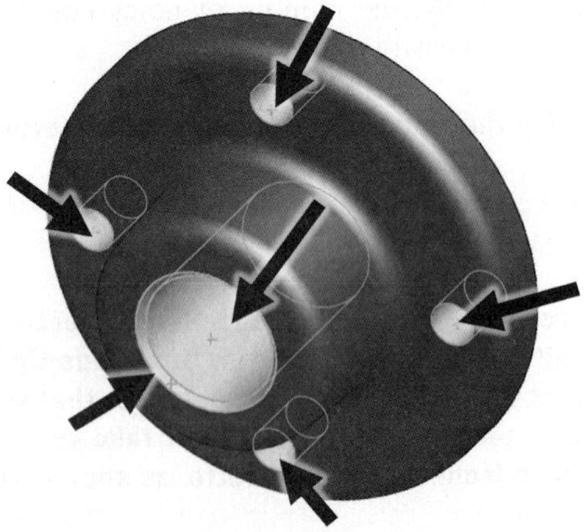

Steel, as shown in Figure 1.119. The selected surfaces are now shown with the selected texture, as shown in Figure 1.120.

FIGURE 1.119

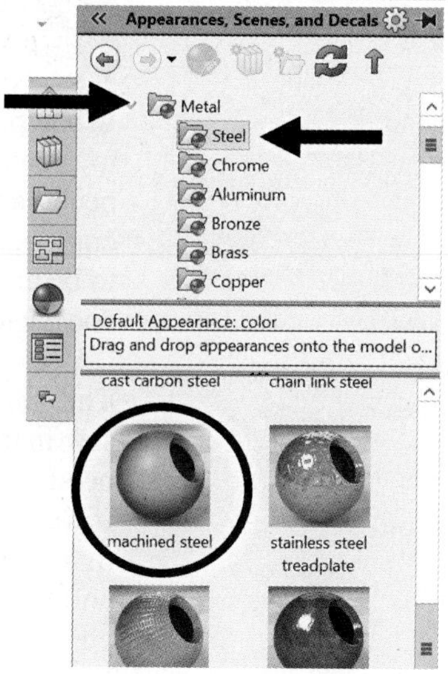

The colors applied to a model can be viewed and/or edited from the DisplayManager. The DisplayManager can be viewed by clicking on its icon above the FeatureManager. As shown in **Figure 1.121**, there are several icons that can be used to display tools and options in the space normally occupied by the FeatureManager. These include the PropertyManager, which as we have seen is automatically displayed when one or more model entities are selected, the ConfigurationManager, which is used to select a specific configuration of a model (as will be discussed in Chapter 3), the DimXpertManager, which is used to apply dimensions and geometric tolerances to a model (the DimXpertManager is not discussed in this text), and the DisplayManager. When the DisplayManager is selected, three options are available: Appearances, Decals, and Scene, Lights, Cameras. A fourth option, PhotoView 360, is grayed out unless the PhotoView 360 add-in is activated. This add-in program allows photo-realistic renderings of models to be made. It is not discussed in this text, but a tutorial is available at the book's website: www.mhhe.com/howard2019.

Select the DisplayManager and click on the Appearances icon. Change the Sort Order to Hierarchy, and expand the items as shown in Figure 1.122.

FIGURE 1.120

FIGURE 1.121

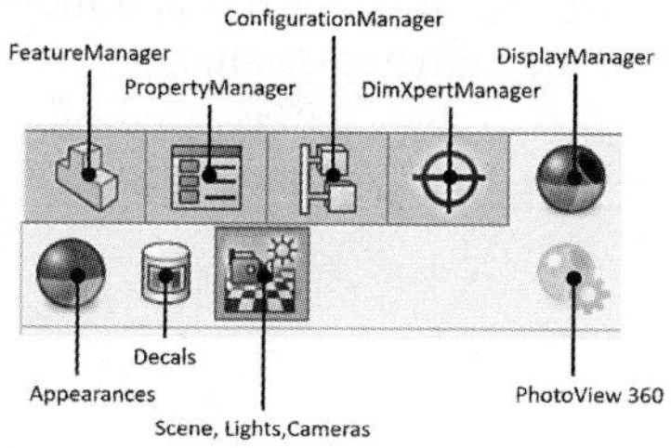

FIGURE 1.122

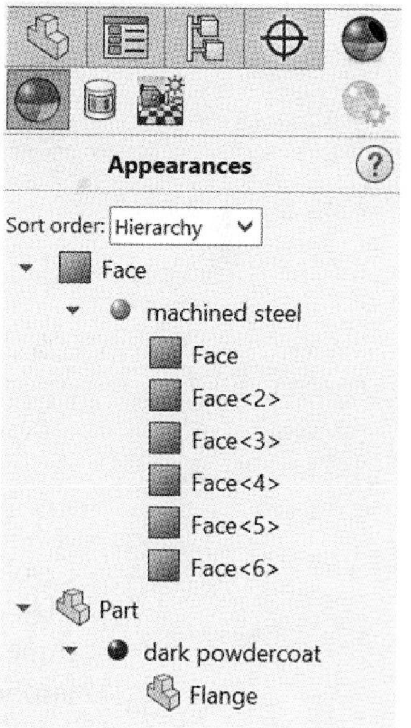

Note that the faces are shown first in the hierarchal order, even though the colors were applied to the faces after the color was applied to the entire model. In the hierarchy of appearances, appearances applied to faces take priority over those applied to features or the entire model, and appearances applied to features take priority over those applied to the entire model. In the DisplayManager, the appearances can be edited and/or deleted by right-clicking on the corresponding entry (machined steel or dark powdercoat in this example) and choosing the desired action from the menu.

Close the part window by clicking on the X in the upper-right corner of the part window. Do not save the changes to the file.

1.4 Using Dimensions and Sketch Relations

In the previous tutorial, we used a combination of dimensions and sketch relations to create fully defined sketches for our model features. While it is not absolutely necessary to use fully defined sketches, it is good design practice. After all, an engineering design of a component must include sufficient detail for the component to be analyzed and eventually built. Using fully defined sketches helps to ensure the complete definition of the geometry of the component.

The Smart Dimension Tool is used to add numerical dimensions to a sketch. As we saw in the previous tutorial, the tool is "smart" in that the type of dimension does not need to be specified. When we clicked on a circle, a diameter dimension was created. If we click on a line, then a linear dimension is created as shown in **Figure 1.123**. Recall that two mouse clicks are required—one to identify the entity to be dimensioned, and the second at the location where the dimension is to be placed. Note that if the cursor is dragged away from the line in a direction roughly perpendicular to the line, then the resulting dimension defines the length of the line. However, if the cursor is dragged away horizontally, then a dimension defining the vertical distance between the endpoints is created, as shown in **Figure 1.124**. Similarly, dragging the cursor vertically results in a dimension defining the horizontal distance between endpoints is created, as shown in **Figure 1.125**. Note the "lock" icon beside the cursor before the dimension is placed. If the dimension is in the desired alignment, then clicking the right mouse button causes this alignment to be maintained until the dimension is placed. This is not usually necessary. As long as the dimension is placed in its proper orientation, it can be moved to a more desirable location by simply clicking and dragging on the numerical value.

FIGURE 1.123

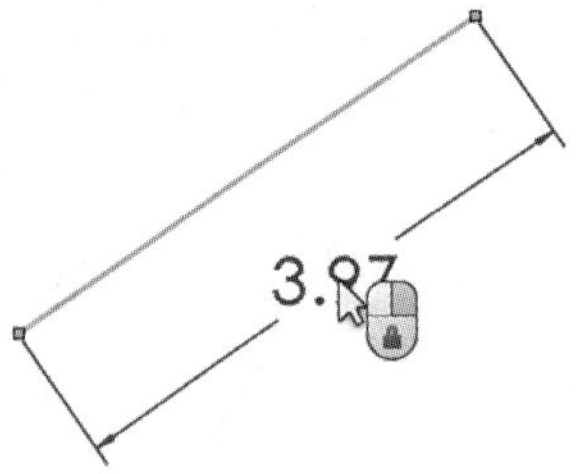

FIGURE 1.124

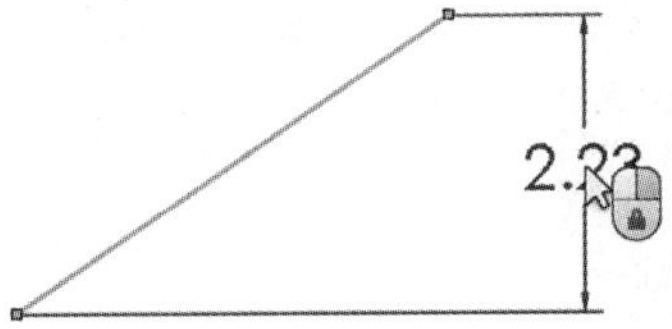

FIGURE 1.125

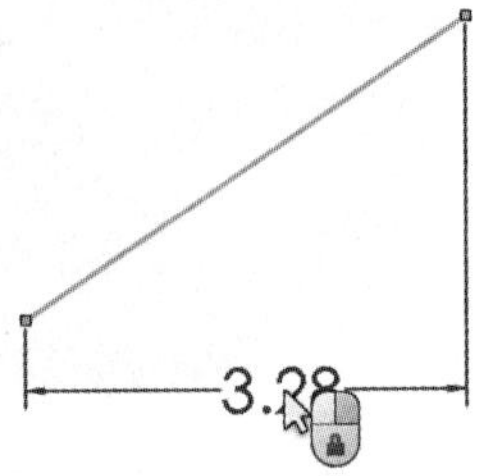

Circles, lines, and arcs can all be dimensioned by clicking once on the entity and then clicking away from the entity to place the dimension. (Arcs are automatically dimensioned with a radius rather than a diameter.) These are examples of dimensions applied to single entities. The Smart Dimension Tool also allows for dimensions relating two entities to one another to be created. For example, consider the two parallel lines shown in **Figure 1.126**. With the Smart Dimension Tool selected, the first mouse click selects one of the lines. If the second mouse click is in the graphics area away from any other entity, then a linear dimension for the length of the line is created, as discussed above. However, if the second mouse click is on another entity, then a dimension is created between the two entities, and a third mouse click is required to place the dimension. In this example, the dimension created is the distance between the two lines, as shown in **Figure 1.126**. If the two lines are not parallel, then the same mouse clicks create an angular dimension, as shown in **Figure 1.127**.

FIGURE 1.126

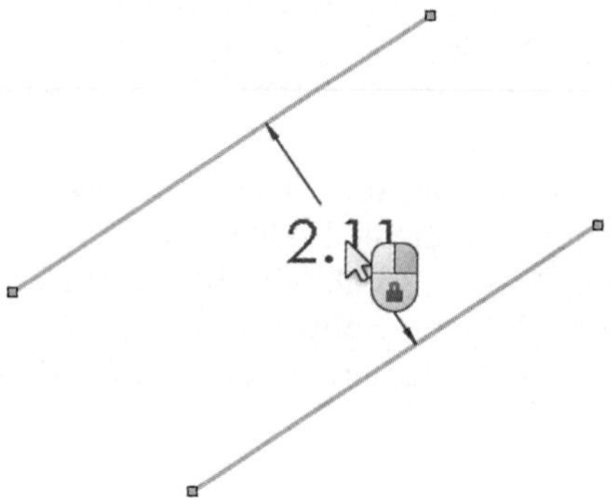

FIGURE 1.127

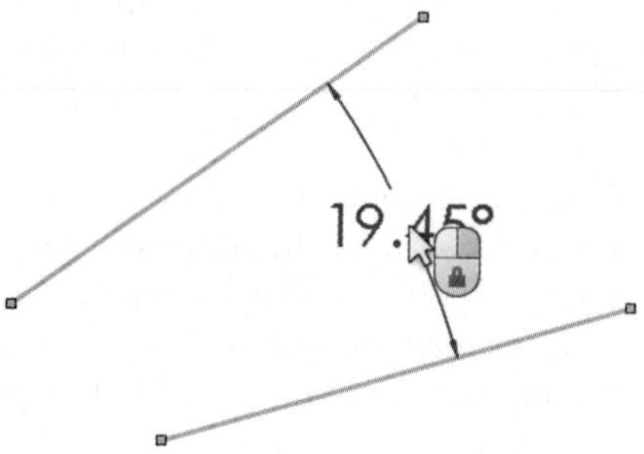

When a circle is one of the two entities selected, then the resulting dimension is always to the center of the circle, as shown in **Figure 1.128**. It is not necessary to select the center point of the circle; clicking on the perimeter of the circle and the line creates the dimension to the circle's center.

When a centerline is one of the entities selected, then the resulting dimension can define either the distance from the centerline to the second entity or the distance from the second entity to a mirror image of itself on the other side of the centerline. For example, consider the horizontal line and centerline shown in **Figure 1.129**. Clicking on the centerline and line creates a linear dimension, since they are parallel to each other. If the next mouse click is made between the two entities, then the dimension as shown in **Figure 1.129** is created. However, if the cursor is dragged to the other side of the centerline before clicking to place it, then the dimension as shown in **Figure 1.130** is created. This method of dimensioning is especially useful when working with revolved geometry, such as the pulley of the next section, in that it allows for the diameters of revolved features to be specified rather than their radii. Since diameters of physical parts can be directly measured, defining a component using diameters is good design practice.

FIGURE 1.128

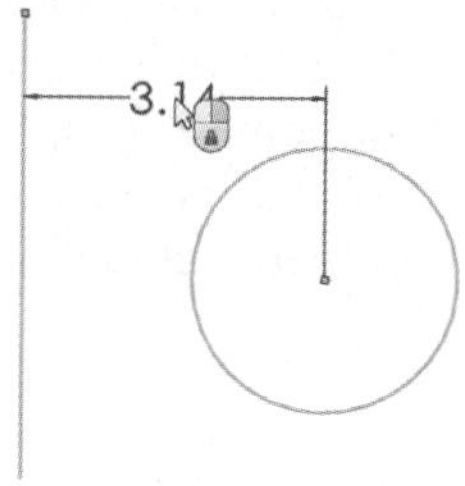

FIGURE 1.129

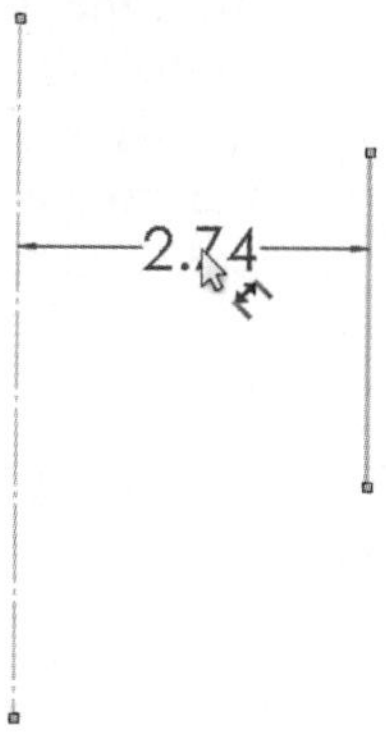

FIGURE 1.130

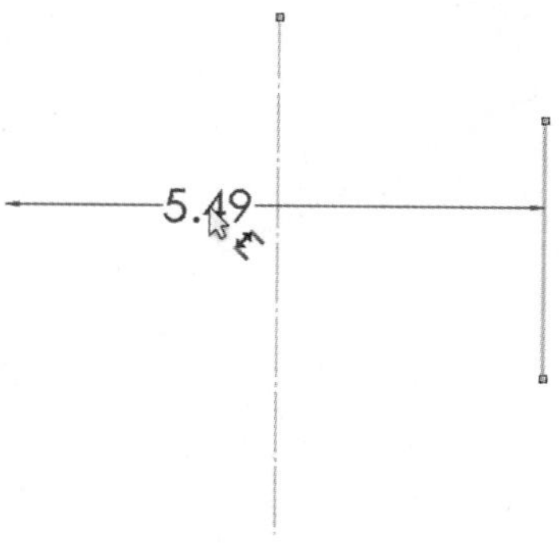

A fully defined sketch is usually not possible without sketch relations. In the case of the circles used in the flange, the location of the center points had to be specified in order for the sketches to be fully defined. In each sketch, the relation defining the center of the circle and the origin as being *coincident* was added automatically through a snap—the cursor was moved close to the origin before the first mouse click and the center of the circle "snapped" to the origin. In the case of the sketch defining the bolt hole, the snap was made to a quadrant point of the bolt circle. These are examples of *automatic relations*. By default, SOLIDWORKS creates these automatic relations. This feature can be turned off by selecting Tools: Sketch Settings: Automatic Relations from the Main Menu, but most users will not find a reason to do so. In addition, automatic relations are created when specifying an entity's geometry. For example, when drawing a line, a small icon beside the line indicates that the line will be horizontal or vertical, as shown in **Figure 1.131**. When the line is completed, it will have a horizontal or vertical relation associated with it, as indicated by the icon shown in **Figure 1.132**.

FIGURE 1.131

FIGURE 1.132

FIGURE 1.133

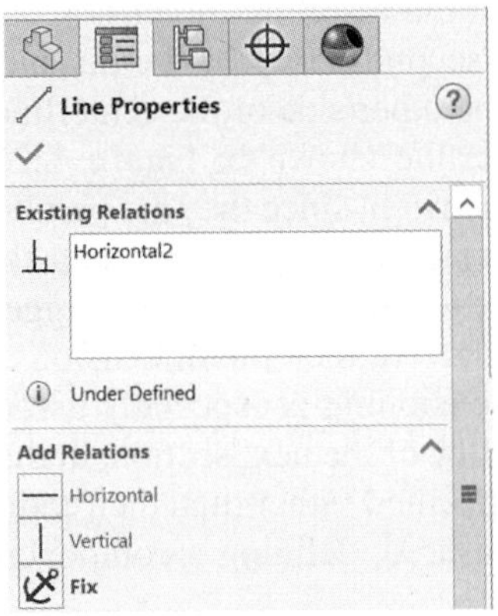

When an entity is selected, its associated relations are shown in the PropertyManager, as shown in **Figure 1.133**. In the PropertyManager, relations can be deleted by selecting them and pressing the Delete key or added by clicking the appropriate icon. Of course, relations must be compatible with each other and with any dimensions existing in the sketch. For example, clicking the Vertical icon in this case would result in an error, since a line cannot be both horizontal and vertical.

Horizontal and vertical relations, along with Fix, which simply fixes the location of an entity within the sketch, are relations that are applied to single entities. Most relations apply to multiple entities. For example, **Figure 1.134** illustrates the addition of a new line to an existing line. With the Line Tool selected, moving close to the midpoint of the first line causes the second line's first point to snap to the midpoint. As the line is dragged out, there are dashed guidelines

FIGURE 1.134

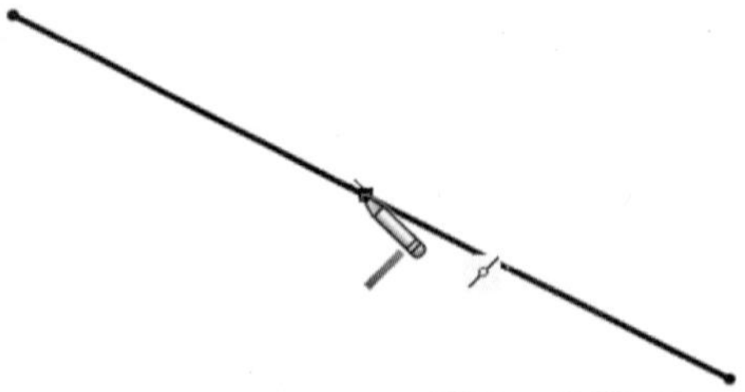

parallel and perpendicular to the first line displayed on the screen, as shown in **Figure 1.135**. If the second mouse click is made close to the perpendicular guideline, then a perpendicular relation between the two lines is created. As shown in **Figure 1.136**, the sketch relation icons indicate the midpoint relation between the first line and the endpoint of the second line, and the perpendicular relation between the two lines.

FIGURE 1.135

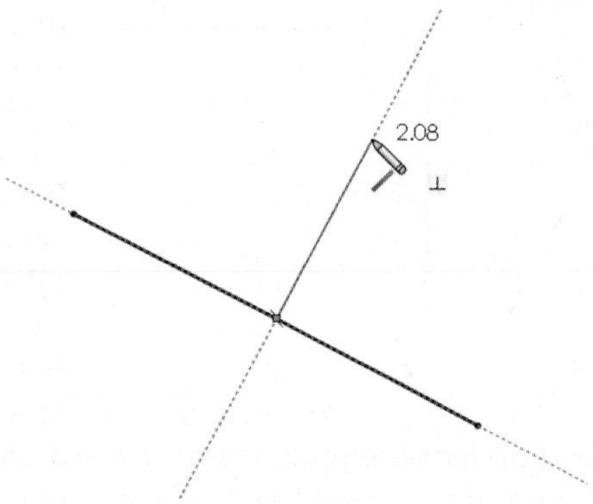

FIGURE 1.136

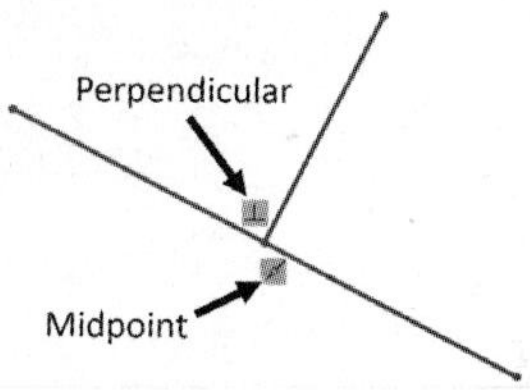

Relations can also be added manually. For example, consider the two lines in **Figure 1.137**. Clicking on the first line selects it and shows its properties in the PropertyManager. A vertical relation can be added by clicking the Vertical icon in the PropertyManager or in the context toolbar that pops up when the line is selected. If we want to merge endpoints of the two lines, then we click on the first endpoint to select it. Then, while holding down the Ctrl key, we select the other endpoint, as shown in **Figure 1.138**. As in most Windows programs, the Ctrl key allows multiple entities to be selected. The Merge relation can then be applied by clicking the Merge icon in the context toolbar or in the PropertyManager, as shown in **Figure 1.139**. (We could also merge these points by dragging the endpoint of the second line until it snaps to the endpoint of the first line, creating an automatic relation.)

FIGURE 1.137

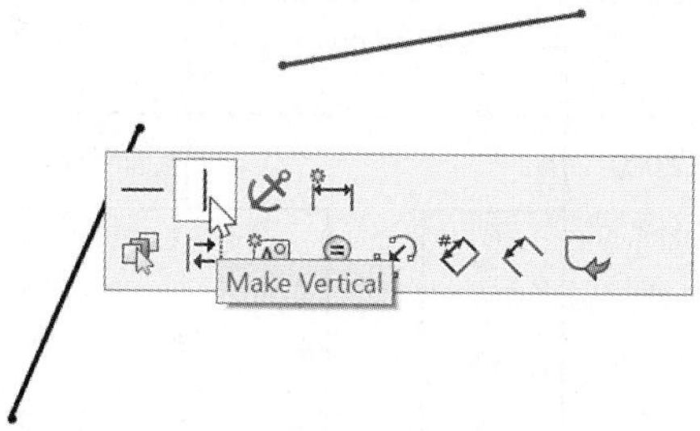

FIGURE 1.138

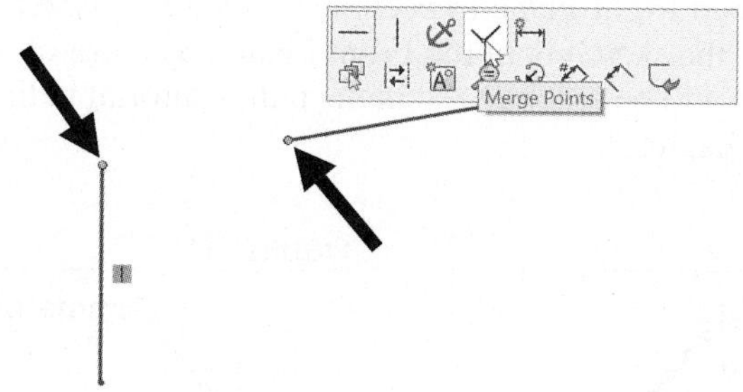

FIGURE 1.139

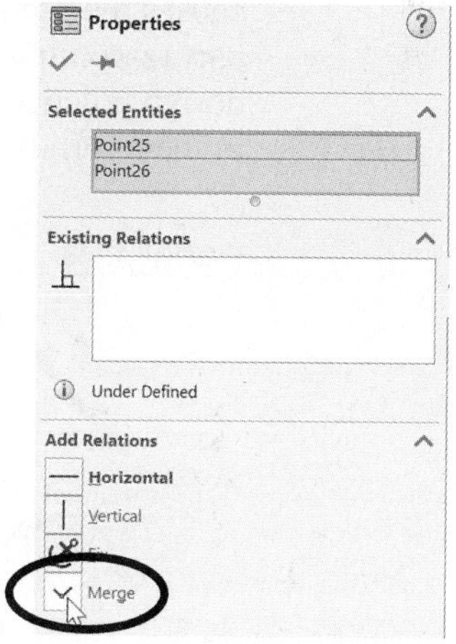

FIGURE 1.140

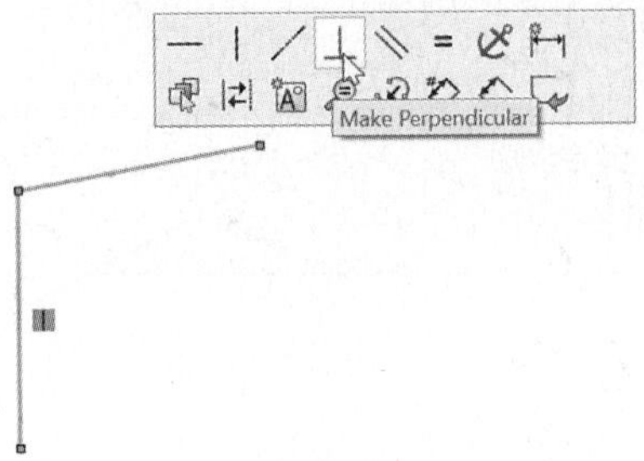

FIGURE 1.141

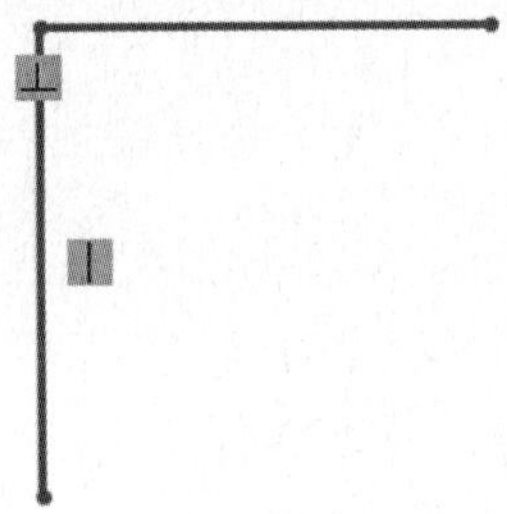

Now both lines can be selected, and a perpendicular relation added, as shown in **Figure 1.140**. The result is shown in **Figure 1.141**.

It should be noted that the display of the sketch relation icons can be toggled on and off by selecting View: Hide/Show: Sketch Relations from the Main Menu. There may be occasions where a sketch becomes so cluttered that turning off the display of the relation icons temporarily is desired, but in most cases displaying the icons is helpful when creating and editing sketches.

FIGURE 1.142

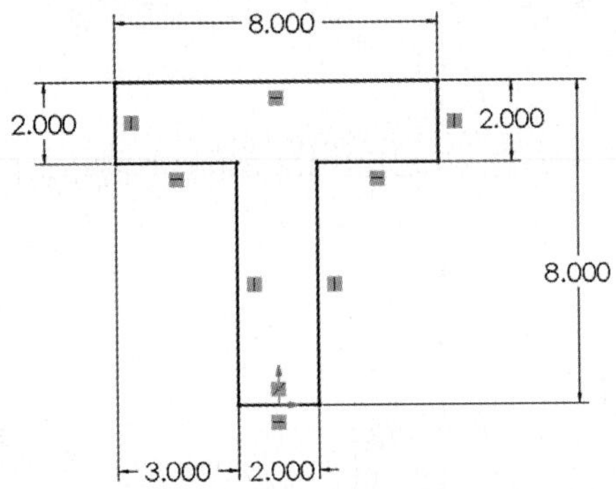

As we have noted, both dimensions and relations are used to fully define sketch geometry. As a general rule, we try to use as few dimensions as possible and rely on relations to complete the geometry definition. For example, consider the T-beam section of **Figure 1.142**. The section consists of horizontal and vertical lines, and the bottom line's midpoint is fixed to the origin. In order to fully define the sketch, six dimensions are required. However, consider the design intent of the part to be made from this sketch. We probably desire the two "legs" at the top of the section to be the same thickness and width. Therefore, we can delete two dimensions and replace them with the relations shown in **Figure 1.143**. The advantage of this approach is that if we make a change, say to the thickness of the legs, then we have only one dimension to change, and our design intent of equal thicknesses is maintained. Another solution is shown in **Figure 1.144**, where the relation of a single point to the origin has been replaced by a vertical centerline and a symmetric relation of two sides about the centerline. Either of these solutions causes the pre-defined Right Plane to become a *plane of symmetry* of the resulting part (assuming that the sketch is in the Front Plane). The use of symmetry is good design practice, and will be emphasized in the pulley tutorial in the next section and in the tutorials of Chapter 3.

FIGURE 1.143

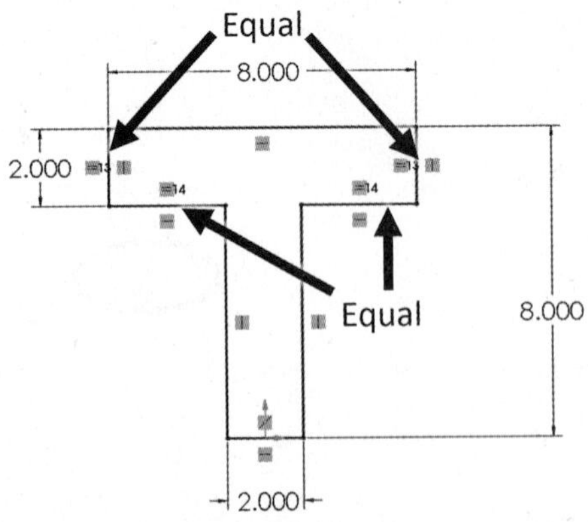

FIGURE 1.144

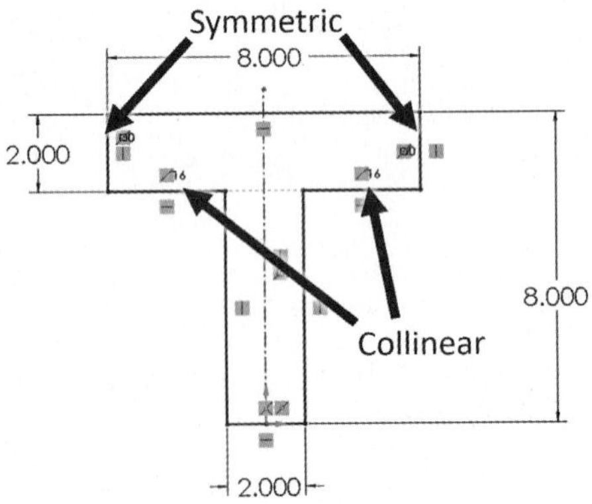

A list of the relation icons is shown in **Figure 1.145**. All are common in 2-D sketches except for the last two: the Pierce relation is used in multiple-sketch applications such as sweeps and lofts, and the Along X-Axis relation is used in 3-D sketches (there are similar Along Y-Axis and Along Z-Axis relations).

FIGURE 1.145

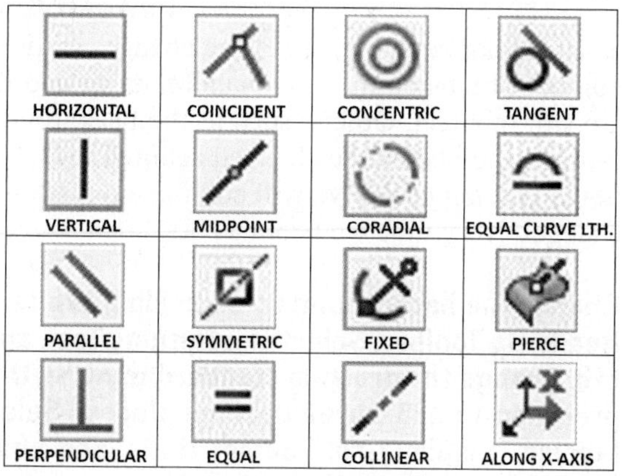

1.5 A Part Created with Revolved Geometry

The flange created earlier utilized extruded features. In this exercise, we will use *revolved* features to create the pulley shown in **Figure 1.146**. We will sketch features of the cross-section of the pulley, and then revolve those features around a centerline to create solids and cuts. The final feature will be a keyway, which will be made with an extruded cut. Dimensions of the pulley are detailed in the 2-D drawing of **Figure 1.147**.

FIGURE 1.146

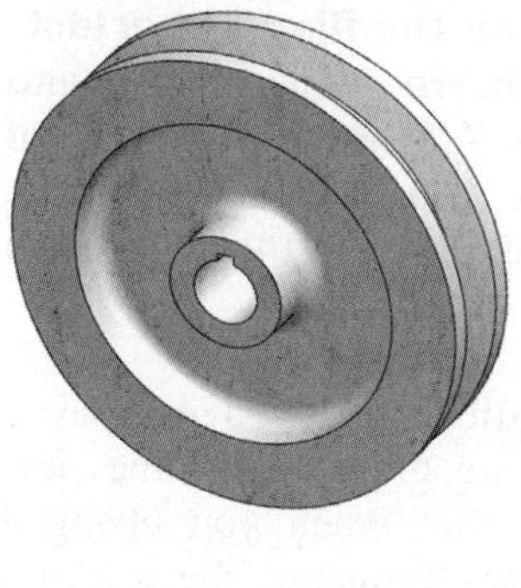

FIGURE 1.147

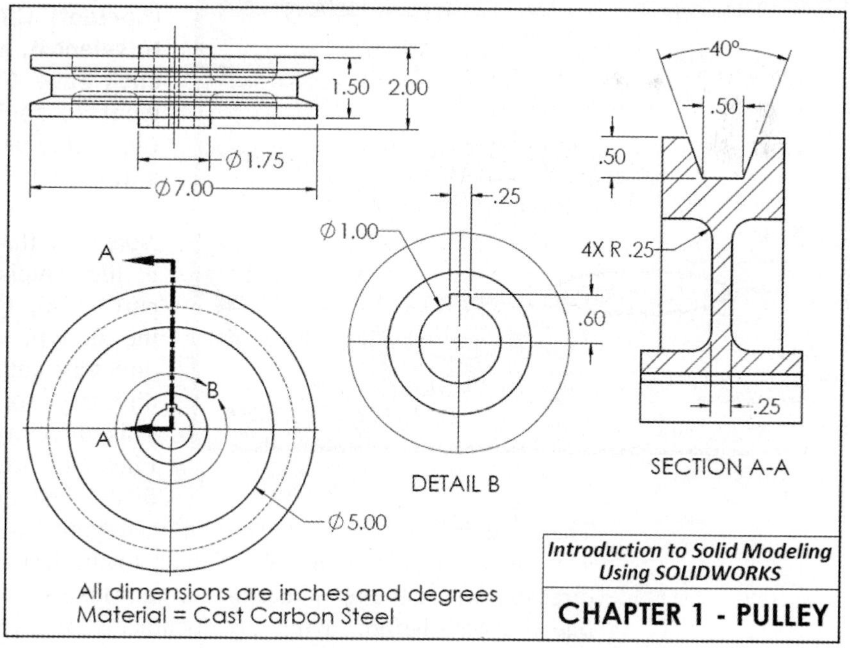

Open a new part.

You will notice that some of the changes made earlier to the interface, such as the number of tabs on the CommandManager and the tools on the Heads-Up Toolbar, are still in effect. Others, such as the background color, have reverted to the initial settings and need to be changed again. These document-specific settings are stored in the Part *template*. We will edit the template so that we do not have to make these changes every time we begin a part.

Change the background color to Plain White from the Apply Scene Tool of the Heads-Up Toolbar. Select the Options Tool, and click the Document Properties tab. Change the drafting standard to ANSI. Under Dimensions, set the Primary precision to .123 (three decimal places). Select Grid/Snap, and check the box labeled "Display Grid." Select Units, and change the unit system to IPS, the number of decimals for length dimensions to .123, and the number of decimals for angles to None.

Note that we have set the number of decimal places in two separate locations. The number of places can be changed in either location; by setting both to .123 we ensure that the template setting will be stored correctly.

Any of the other settings under the Document Properties tab can be stored in the template. For example, you may want to change the font used for dimensions or turn the display of the grid off (most of the figures in this book are made with a larger font and with the grid off for clarity).

FIGURE 1.148

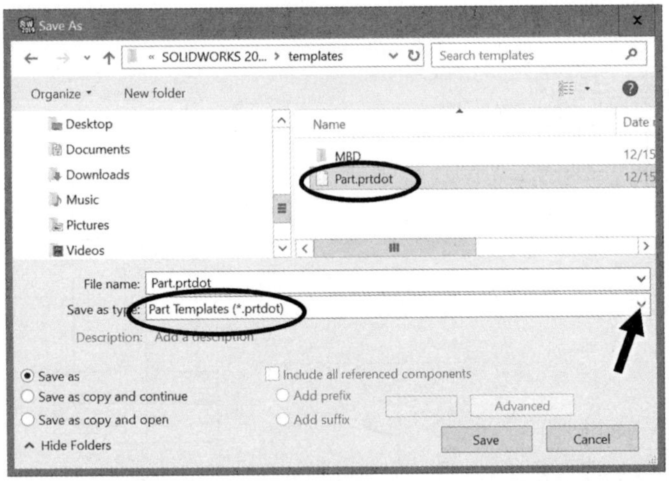

From the main menu, select File: Save As. Change the file type to Part Templates (*.prtdot). Click on the file "Part.prtdot" to select it, as shown in Figure 1.148, and click Save. Click Yes when asked if you want to replace the existing template. Close this part document and then open a new part.

Note that the settings that you just saved in the template are effective for the new part. Also, note that when you changed the file type to Part Template, the working directory automatically changed to default directory for SOLIDWORKS templates. This directory is located in the Program Files or Program Data directory on the drive containing your operating system. Therefore, it is not a good location to store your work. The next time that you save a document, the directory will default to the last one accessed; in this case the directory where the templates are saved. Make sure to change the directory to the one you want before saving any documents. We will note this when we save the pulley file later in this section.

Click on the Right Plane to select it, as shown in Figure 1.149. From the Sketch group of the CommandManager, click the arrow beside the Line Tool and select the Centerline Tool, as shown in Figure 1.150. Draw a vertical centerline up from the origin, as shown in Figure 1.151.

FIGURE 1.149

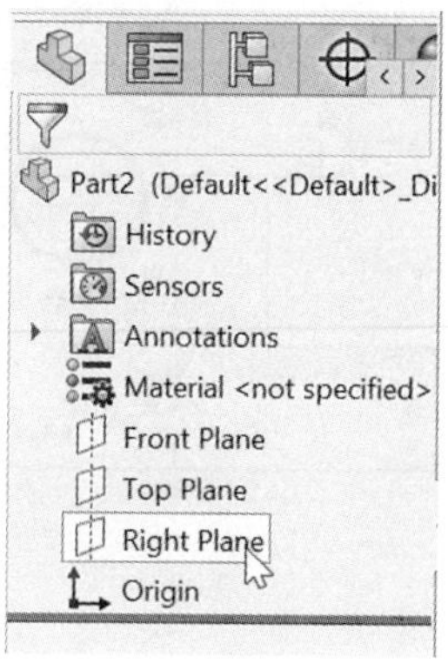

This centerline will allow us to take advantage of the symmetry of the cross-section.

Click the arrow beside the Rectangle Tool, and select the Center Rectangle Tool, as shown in Figure 1.152. Click above the origin on the centerline, as shown in Figure 1.153, to set the center of the rectangle. Then drag out a corner of the rectangle, as shown in Figure 1.154. The size is not important, but keep the entire rectangle above the origin. Repeat to create a second rectangle above the first, as shown in Figure 1.155.

FIGURE 1.150

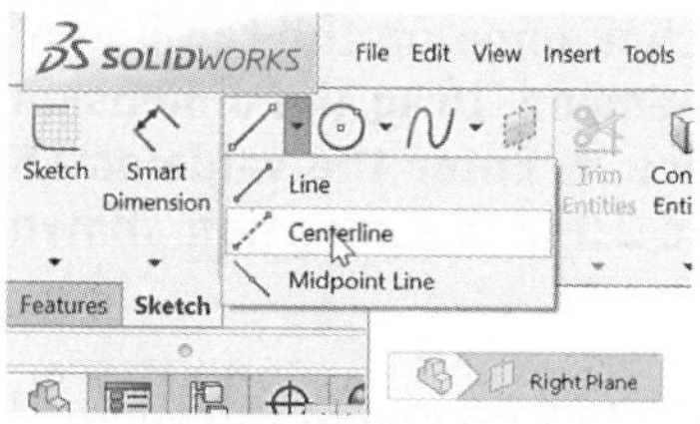

FIGURE 1.151

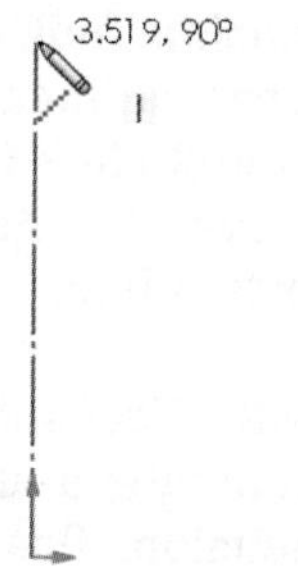

FIGURE 1.152

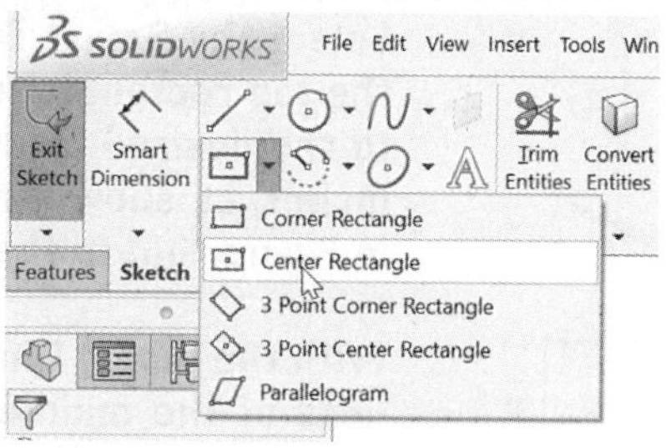

FIGURE 1.153

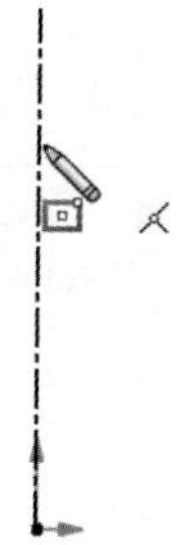

FIGURE 1.154

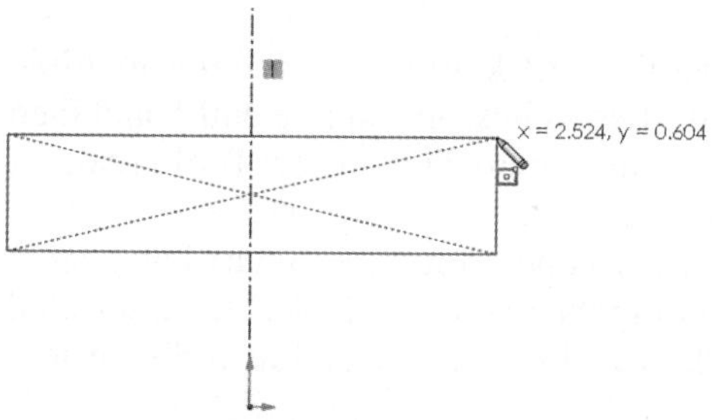

FIGURE 1.155

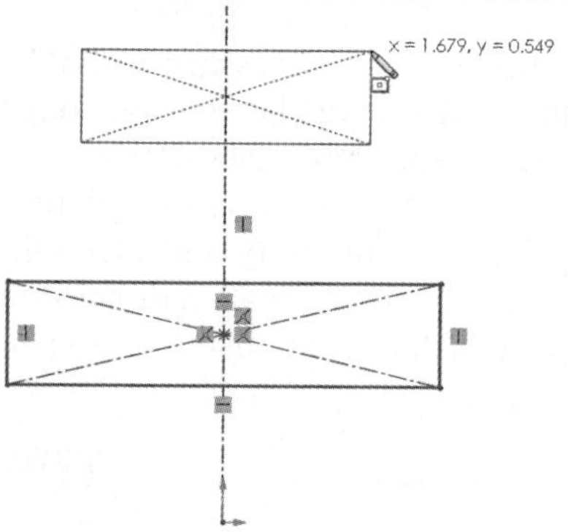

FIGURE 1.156

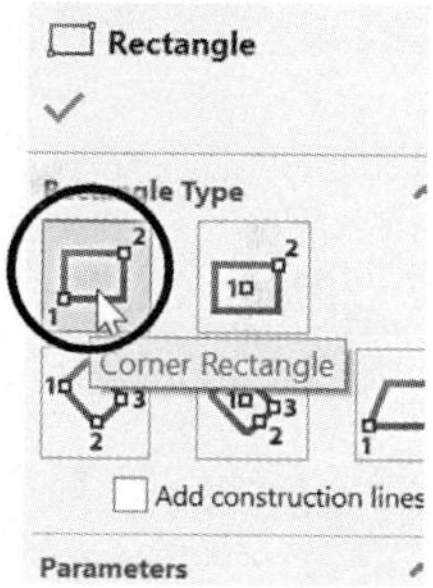

FIGURE 1.157

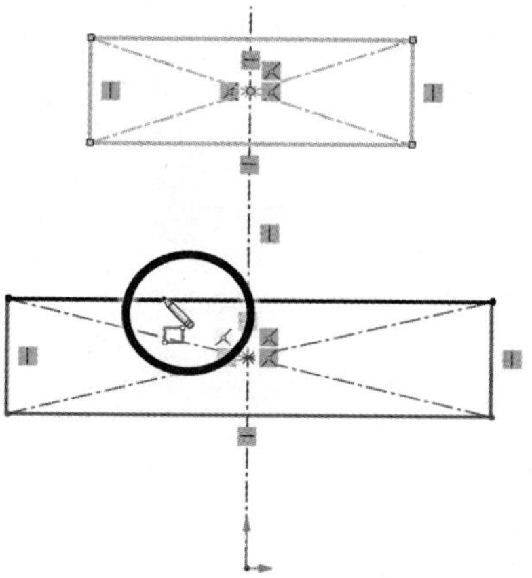

In the PropertyManager, change the type of rectangle from Center to Corner, as shown in Figure 1.156. Move the cursor to the top line of the bottom rectangle (but not to a corner or midpoint; when the entire bottom line turns red, then you are snapping to the line and not to a specific point), as shown in Figure 1.157, and click to place one corner of the rectangle on

FIGURE 1.158

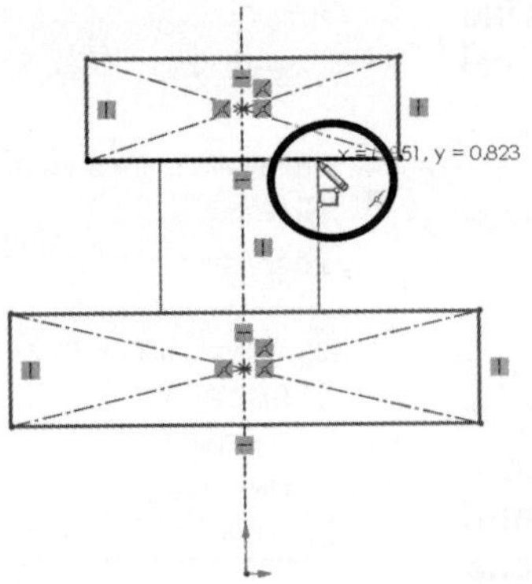

this line. Drag the rectangle up until the opposite corner is along the bottom line of the top rectangle, as shown in Figure 1.158. Click to complete the rectangle, and press Esc.

When we used the Center Rectangle Tool and placed the center points along the vertical centerline, those rectangles became symmetric about the centerline. However, to join the first two rectangles together, we used a Corner Rectangle, which is not centered on the centerline. Therefore, we need to add the symmetry of this rectangle manually.

FIGURE 1.159

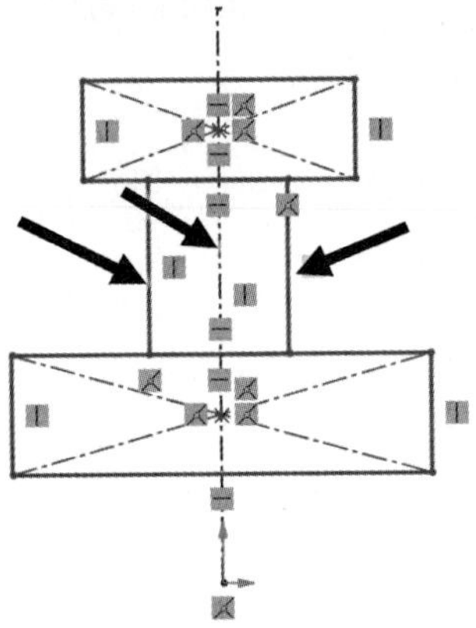

Click on the two lines and the centerline shown in Figure 1.159 to select them, remembering to hold down the Ctrl key when making multiple selections. Click the Make Symmetric icon from the context toolbar, as shown in Figure 1.160 (or the Symmetric icon in the PropertyManager).

Select the Smart Dimension Tool, and click once on the top line of the top rectangle to create a linear dimension. Drag the dimension to the desired location and click to place it. Enter the value as 1.5 inches, as shown in Figure 1.161. Add the 2.0-inch dimension shown in Figure 1.161 to the bottom line.

With the Smart Dimension Tool selected, click on one of the vertical lines of the middle rectangle, and then on the other vertical line, creating a linear dimension. Drag the dimension to the desired location, and then click to place it. Enter the value as 0.25 inches, as shown in Figure 1.162.

Note that it is not necessary to hold down the Ctrl key when selecting multiple entities for the Smart Dimension command. If you click on a single entity and then click away from any other entity to place it, the Smart Dimension Tool creates a dimension from that entity (length of a line, diameter of a circle, radius of an arc). If you click on an entity and then click on a second entity, the Smart Dimension Tool will attempt to create a dimension relating the two entities. When we selected two parallel lines, the distance between the two lines was added as a dimension.

FIGURE 1.160

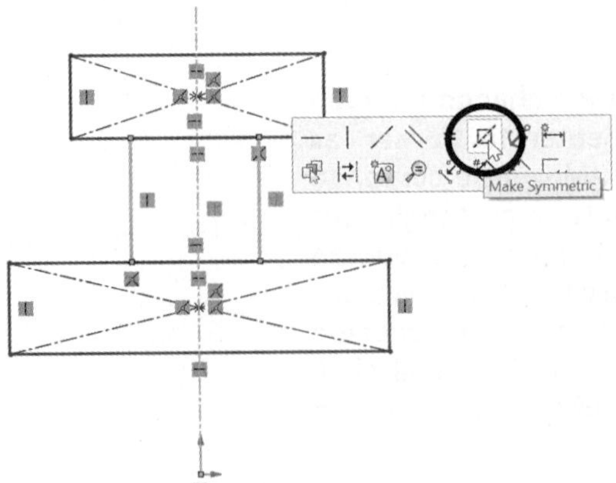

FIGURE 1.161

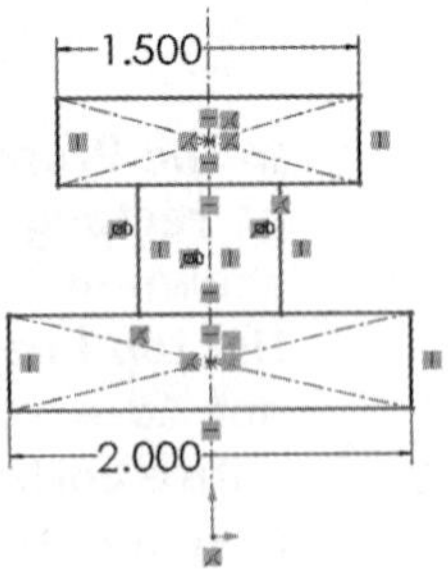

FIGURE 1.162

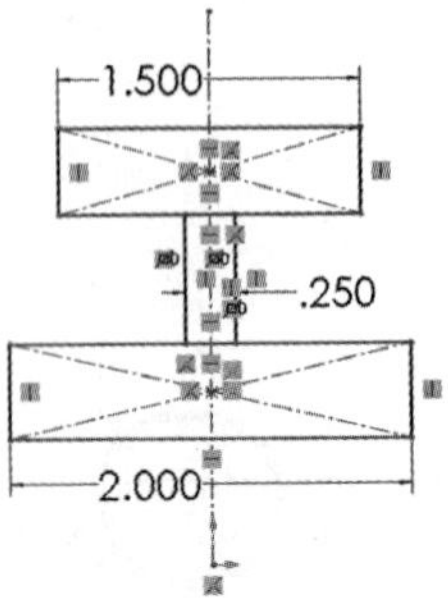

As noted in the previous section, if we select two non-parallel lines, an angular dimension will be created.

If you refer back to **Figure 1.147**, you will see that the other dimensions used to define this cross section are diameter dimensions. In order to place the diameter dimensions, we need to establish the centerline which will become the axis of revolution for the resulting solid.

Select the Centerline Tool, and drag a horizontal centerline from the origin, as shown in Figure 1.163. Select the Smart Dimension Tool. Click on the top line of the sketch and then on the horizontal centerline. Before clicking to place the dimension, recall from the previous section that if the dimension is placed above the centerline, a radius dimension is created, while if the dimension is placed below the centerline, a diameter dimension is created. Click below the centerline to place the dimension, as shown in Figure 1.164, and enter the value of the dimension as 7.0 inches. Repeat to add the 5.0-, 1.75-, and 1.0-inch diameter dimensions shown in Figure 1.165. Note that after creating the first diameter dimension, simply clicking on the next item to be dimensioned above the centerline will cause a diameter dimension to be created by default. If the lowest line moves below the centerline as you are adding the dimensions, then you can simply click and drag it above the centerline before adding its diameter dimension. Also note that after placing the dimensions below the centerline, you can click and drag the numerical value above the centerline if desired, and the dimension will remain as a diameter dimension.

FIGURE 1.163

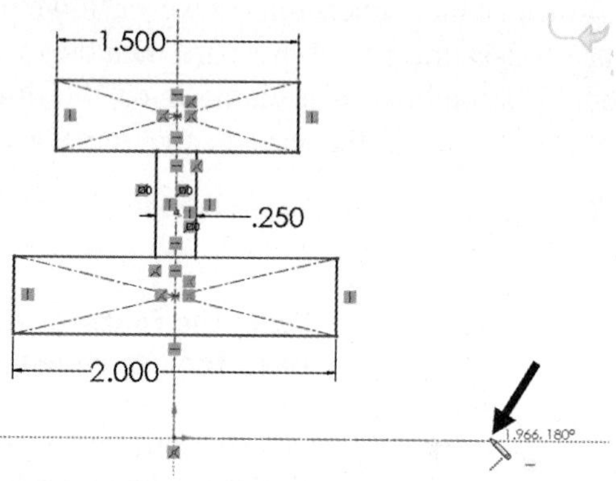

The sketch should be fully defined.

FIGURE 1.164

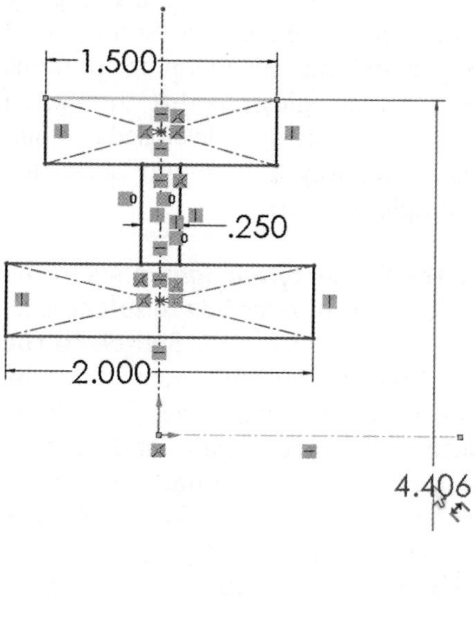

FIGURE 1.165

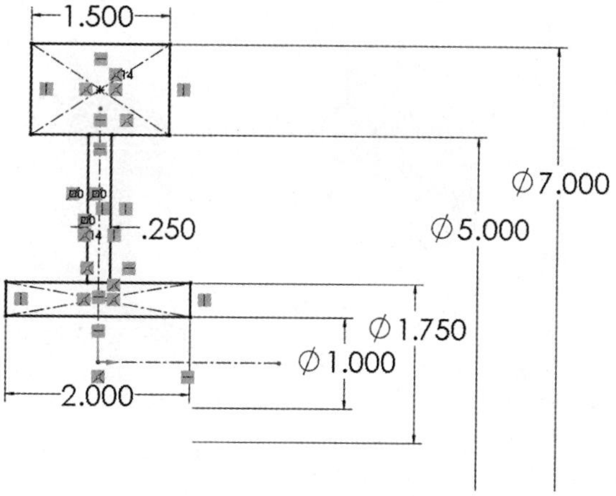

DESIGN INTENT Planning for Other Uses of the Model

When adding dimensions to the initial sketch of the pulley, we could specify some of the dimensions as either diameters or radii. For example, the 7-inch-diameter dimension that defines the overall size of the pulley could just as easily be entered as a 3.5-inch-radius dimension. However, if we plan to make a 2-D drawing of this part, then the diameter should be defined as a diameter on the drawing. By dimensioning the part in the same way that we will dimension the drawing, then dimensions can be imported directly from the part file. This prevents us from having to add dimensions manually or override a dimension's properties. You should consider all of the possible future uses of a model (e.g., drawing creation, stress analysis, 3-D printing, etc.) when choosing the best way to create and dimension a part.

From the Features group of the CommandManager, select the Revolved Boss/Base Tool, as shown in Figure 1.166.

FIGURE 1.166

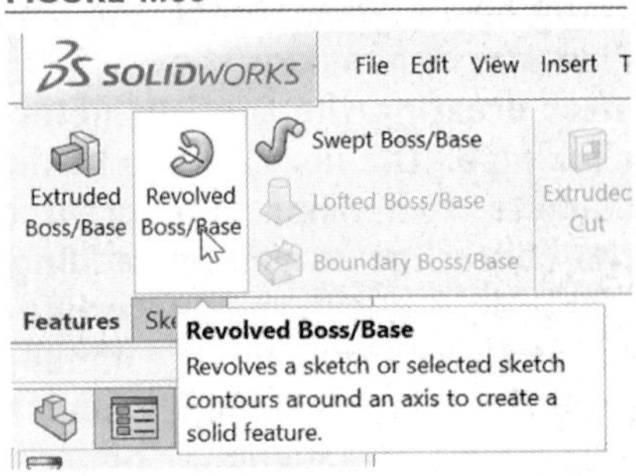

Note that no preview is displayed on the screen as it was when we selected the Extrude Boss/Base tool earlier in the chapter. The reason for this is that there are two ambiguities in our sketch that must be defined:

1. There is more than one enclosed region (contour) within our sketch, so we need to define which of these regions will be revolved.
2. There are two centerlines in the sketch, so we need to define which centerline is the axis of revolution.

FIGURE 1.167

FIGURE 1.168

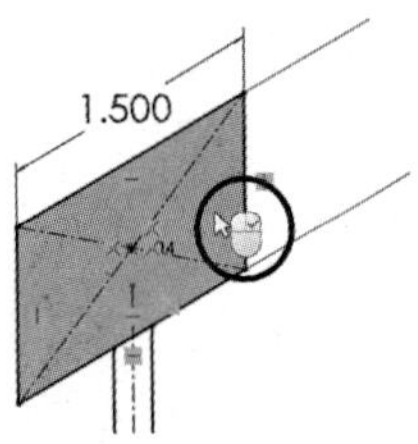

FIGURE 1.169

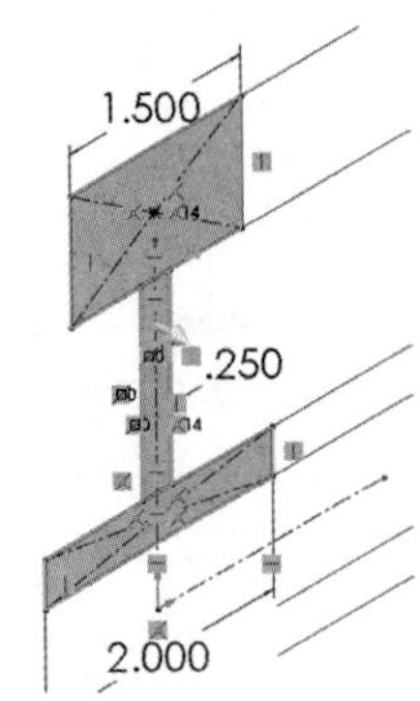

One way to approach the first ambiguity would be to use the Trim Entities Tool to remove the overlapping portions of the rectangles, so that the sketch consists of only a single closed contour. An easier way is to simply select the multiple contours.

In the PropertyManager, click in the Selected Contours box to select it, as shown in Figure 1.167. Switch to the Isometric View and zoom in to the three rectangles. Move the cursor into the top rectangle, and click to select the rectangular region, as shown in Figure 1.168. Repeat for the other two rectangular regions, as shown in Figure 1.169.

Click in the Axis of Revolution box in the PropertyManager to select it, as shown in Figure 1.170. Click on the horizontal centerline, as shown in Figure 1.171.

FIGURE 1.170

A preview will now be displayed, as shown in **Figure 1.172**. In the PropertyManager, we can change the number of degrees of the revolution if we want less than a fully revolved part. Since we want to revolve the section a full 360 degrees, we can accept the default value.

FIGURE 1.171

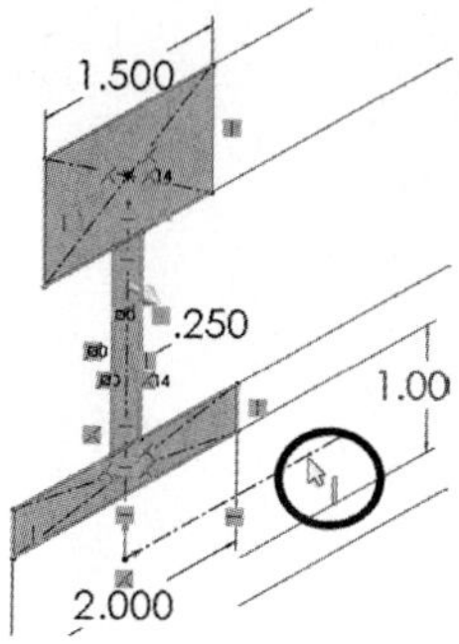

FIGURE 1.172

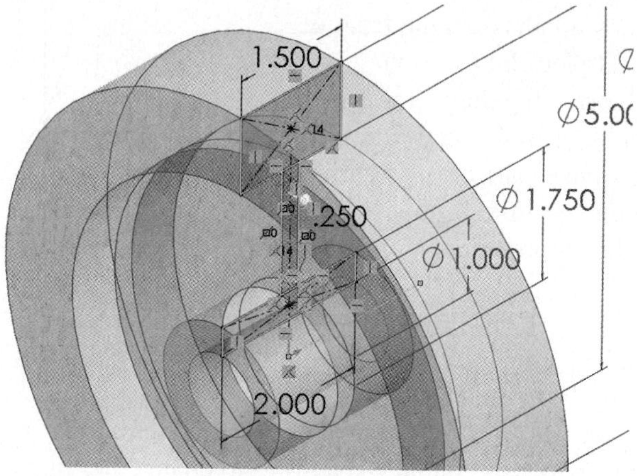

Click the check mark to complete the revolution.

The resulting solid part is shown in **Figure 1.173**.

FIGURE 1.173

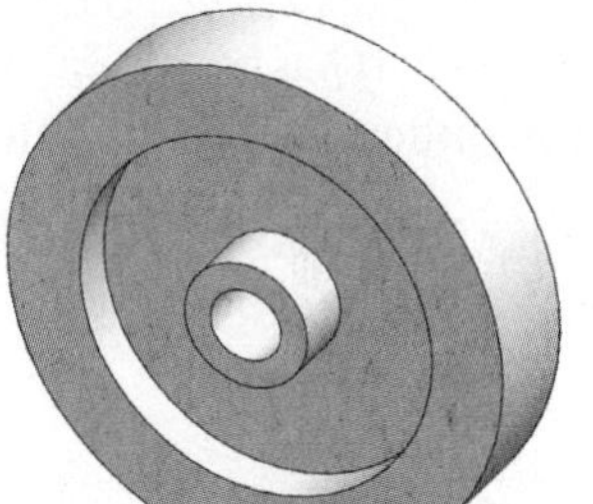

Select File: Save from the main menu. Make sure to change the file directory to the location where you want to save the file, since the default path will be to the directory containing the template files. If necessary, change the file type from Part Template to Part. Save the file with the name "Pulley."

Select the Right Plane, and select Normal To from the Context Toolbar, as shown in Figure 1.174. Select the Centerline Tool, and create vertical and horizontal centerlines from the origin, as shown in Figure 1.175.

When using either the Centerline Tool or Line Tool, note that there are two different methods for drawing line segments. If you click once to set the first endpoint, move the cursor and then click again to set the other endpoint, you can then move the cursor to create another line segment beginning at the previous endpoint. This technique allows you to create continuous line segments. Double-clicking after the last endpoint has been placed allows you to start a new group

FIGURE 1.174

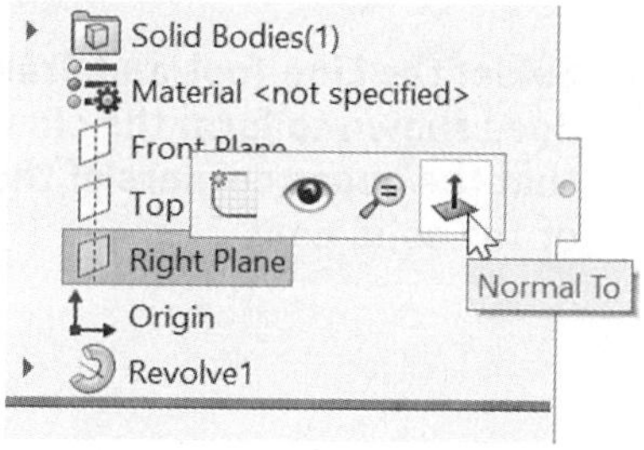

FIGURE 1.175

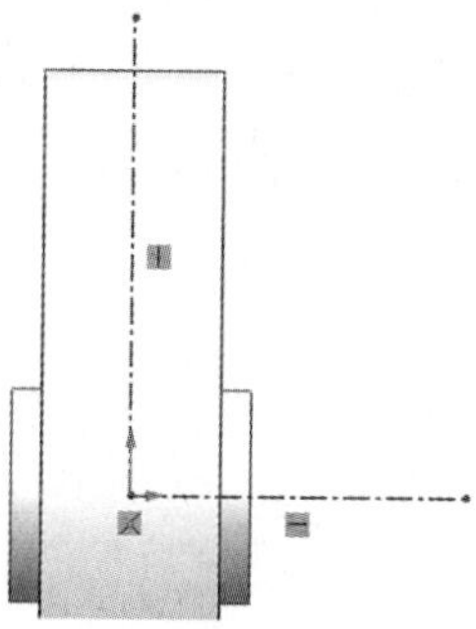

DESIGN INTENT Choosing the Initial Sketch Plane

The choice of the initial sketch plane for a part should be considered before beginning the first sketch. Standard practice is to orient the part so that the Front View provides the best visualization of the part of all of the principal views (Front, Back, Top, Bottom, Right, and Left). Therefore, the Trimetric View, which emphasizes the front of the part, provides the best pictorial (3-D) view. As we will see in Chapter 2, orienting the part properly will make the creation of a multi-view drawing easier as well. For the flange part created earlier in this chapter, sketching the initial circle in the Front Plane resulted in the proper orientation of the flange.

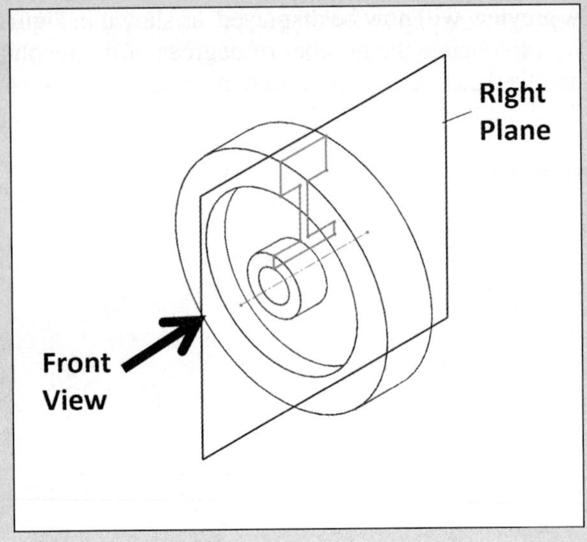

In the case of the pulley, which is created by revolved features, the circular profile of the part is the most descriptive view, so we chose the Right Plane as the sketch plane for our initial sketch. We could have also sketched in the Top Plane and achieved the same result.

Once a part has been created in a certain orientation, it is difficult to re-orient it. There is no command to rotate the part relative to the axes. It is possible to change the plane in which a sketch is contained by right-clicking the sketch in the FeatureManager and selecting Edit Sketch Plane. However, this may cause errors in subsequent features. Therefore, it is good practice to carefully consider the orientation of the part before selecting the initial sketch plane.

FIGURE 1.176

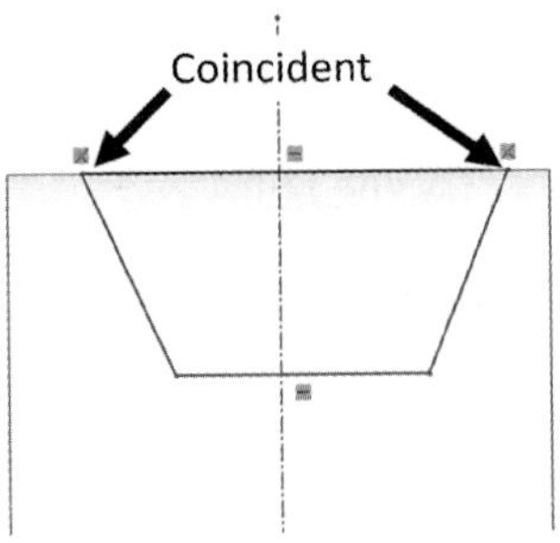

of segments, while pressing the Esc key turns off the Line or Centerline Tool. The second method is used to create discontinuous line segments. Click to place the first endpoint, and hold the mouse button while dragging to the other end. Releasing the mouse button places the second endpoint. The cursor can then be moved to the starting point of a new segment.

Select the Line Tool, and draw the two horizontal lines and two diagonal lines shown to form the closed shape shown in Figure 1.176, making sure that the upper corners of the shape are coincident with the upper edge of the solid part.

Select the two diagonal lines and the vertical centerline, using the Ctrl key to make multiple selections, as shown in Figure 1.177. Add a symmetric relation.

Select the Smart Dimension Tool. Click on both of the diagonal lines to create an angular dimension. Set its value to 40 degrees. Add the two linear dimensions shown in Figure 1.178.

FIGURE 1.177

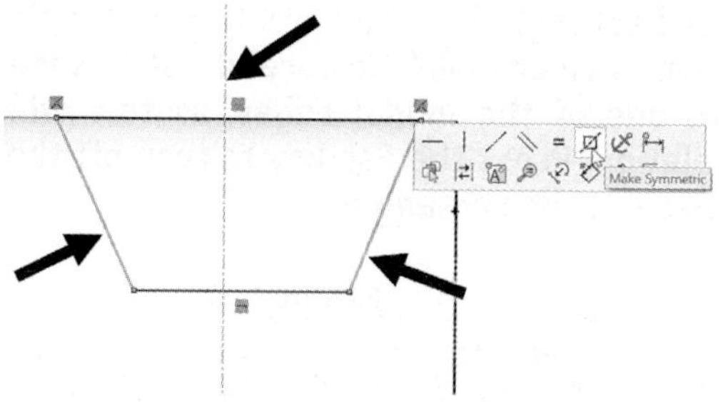

FIGURE 1.178

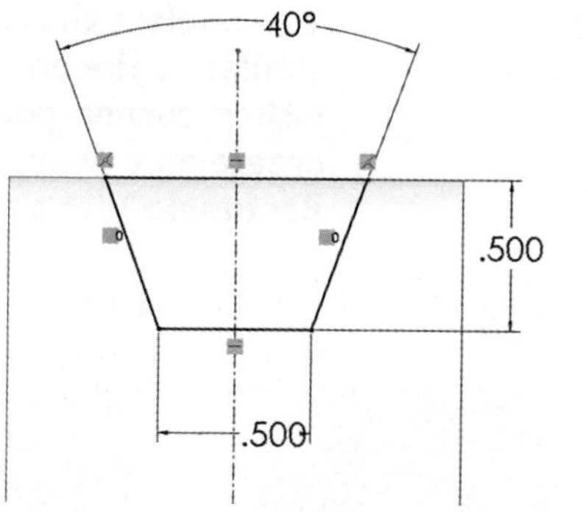

The sketch should be fully defined.

From the Features group of the CommandManager, select the Revolved Cut Tool, as shown in Figure 1.179. Click on the horizontal centerline to select it as the axis of revolution, as shown in Figure 1.180. Click the check mark to complete the cut, which is shown in Figure 1.181.

FIGURE 1.179

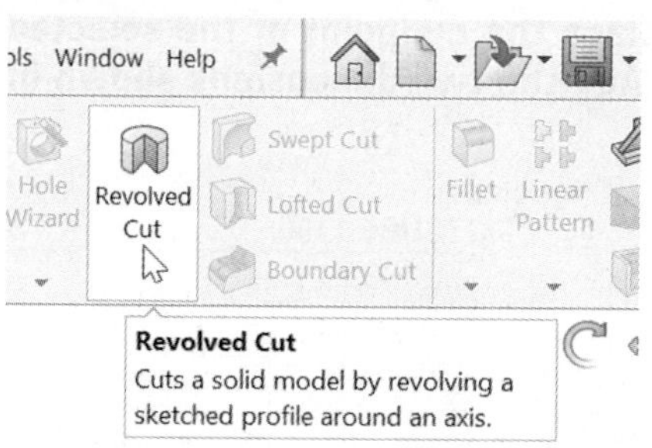

FIGURE 1.180

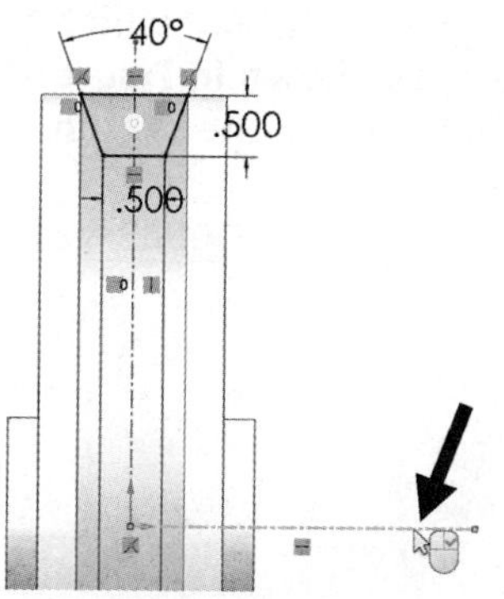

FIGURE 1.181

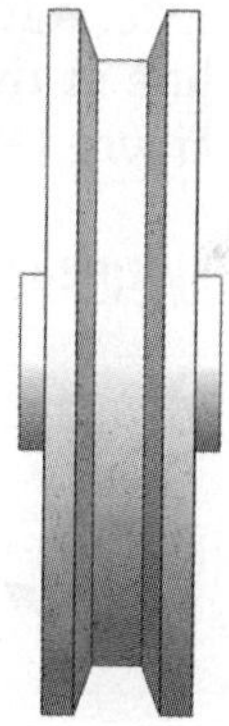

Select the Fillet Tool. In the PropertyManager, set the fillet radius to 0.25 inches. Switch to the Trimetric View, and select the two edges shown in Figure 1.182 (shown with the No preview option selected in the PropertyManager; it may be easier to select the edges with No preview selected). Select the Rotate View Tool, shown in Figure 1.183. Rotate the view until the back side of the pulley is visible, then press the Esc key to turn off

FIGURE 1.182

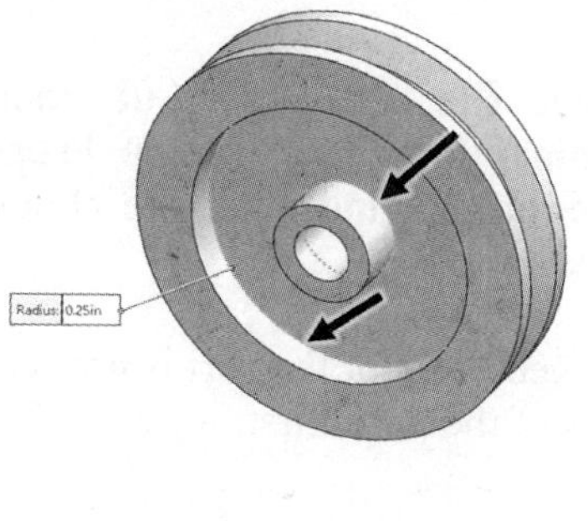

FIGURE 1.183

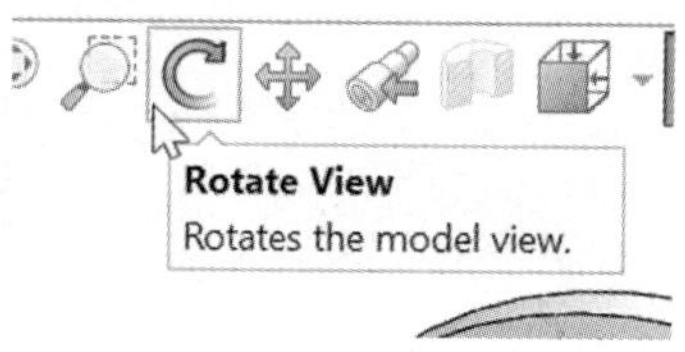

FIGURE 1.184

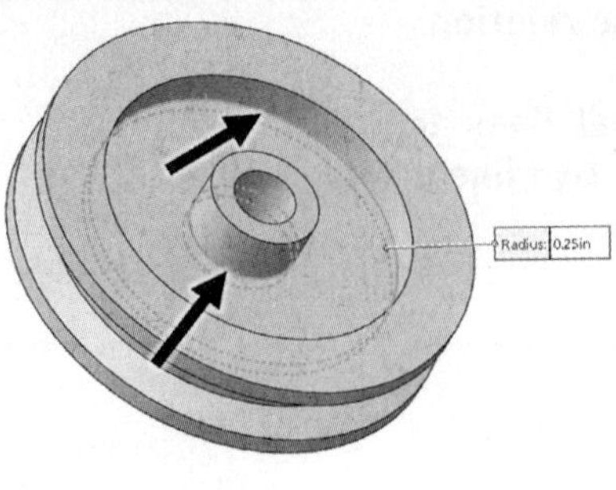

the Rotate View Tool. Click on the two edges on the back side to be filleted, as shown in Figure 1.184. Click the check mark to add the fillets.

The filleted part is shown in Figure 1.185. We will complete the part by adding the keyway.

Select the face shown in Figure 1.186, and select the Normal To View. Select the Corner Rectangle Tool, and create a rectangle similar to the one shown in Figure 1.187. Be careful not to snap either corner point to one of the model edges, as this will create an unwanted relation. Press the Esc key to turn off the Rectangle Tool and de-select the rectangle.

FIGURE 1.185

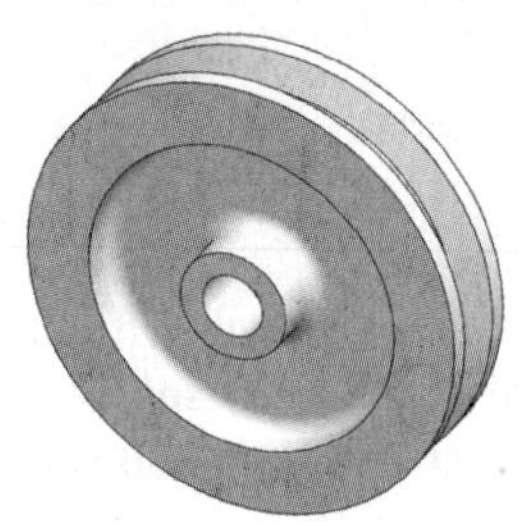

FIGURE 1.186

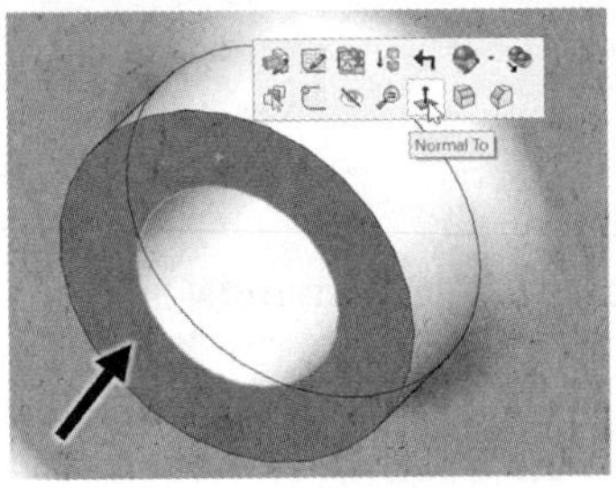

FIGURE 1.187

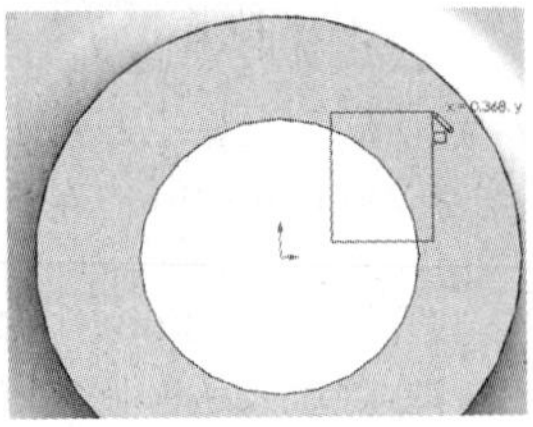

Select the bottom line of the rectangle and the origin, as shown in Figure 1.188. Add a midpoint relation, which will place the midpoint of the selected line at the origin, as shown in Figure 1.189. Add the two dimensions shown in Figure 1.190 to fully define the sketch.

FIGURE 1.188

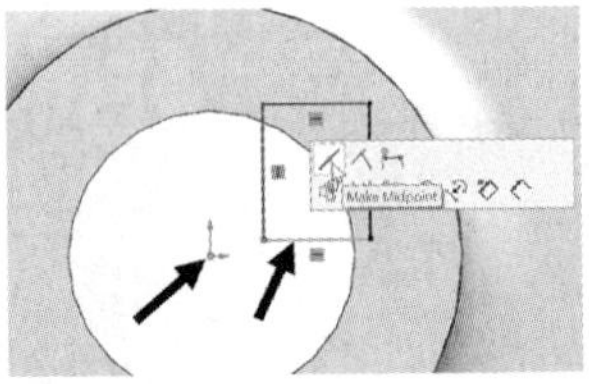

FIGURE 1.189

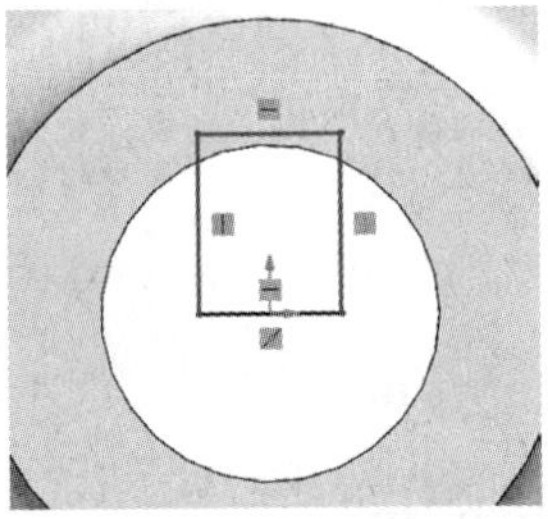

FIGURE 1.190

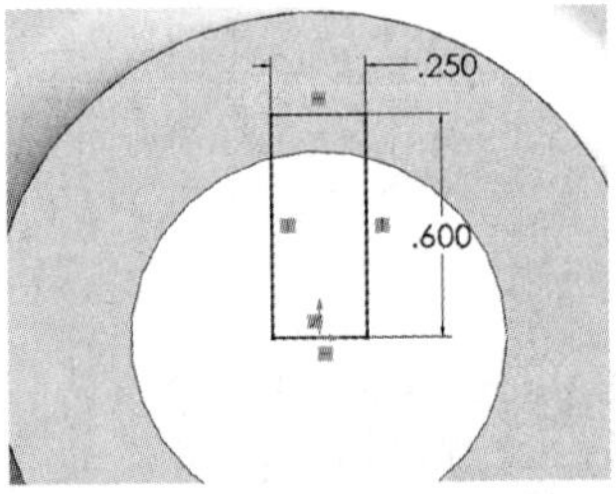

FIGURE 1.191

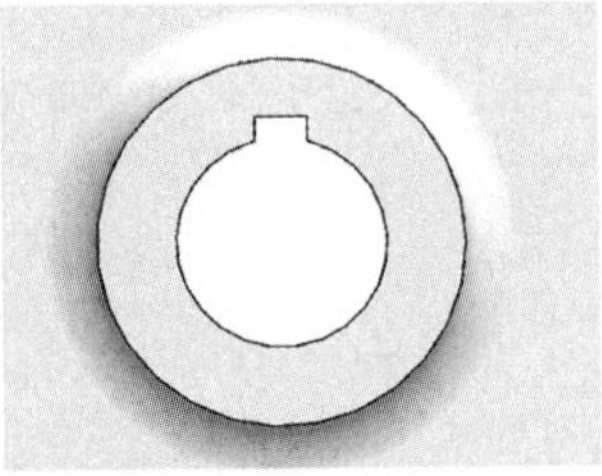

Select the Extruded Cut Tool from the Features group of the CommandManager. In the PropertyManager, set the type of cut to Through All, and click the check mark to complete the cut, which is shown in Figure 1.191.

The geometry of the part is now complete. Often, we will want to know the mass of the part. First, we must define the part's material.

In the FeatureManager, right-click on Material and select Edit Material, as shown in Figure 1.192. From the list of available materials, select Cast Carbon Steel from the Steel group, as shown in Figure 1.193. To view the properties in English units, select English (IPS) from the pull-down menu.

Note the properties for this material are displayed in **Figure 1.193**. For the calculation of mass, the density, approximately 0.28 pounds-mass per cubic inch, is used. The other properties listed are used in stress and thermal analyses.

FIGURE 1.192

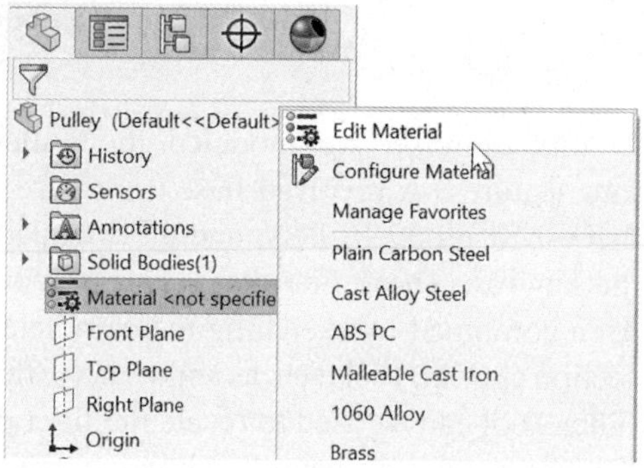

FIGURE 1.193

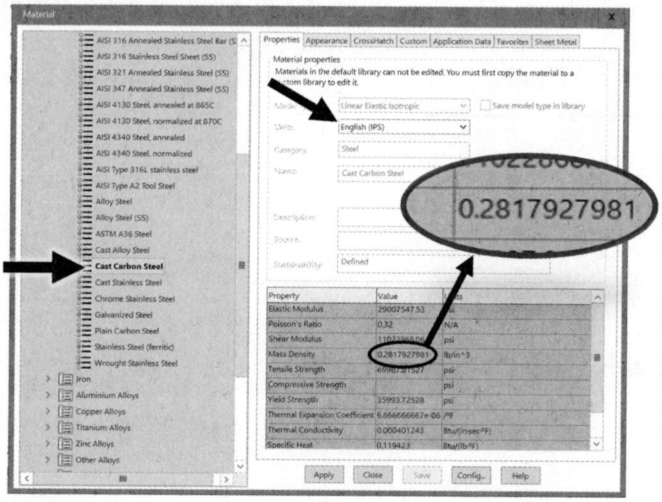

Click Apply to accept the material choice, and close the Material box. From the Main Menu, select Tools: Evaluate: Mass Properties, as shown in Figure 1.194.

FIGURE 1.194

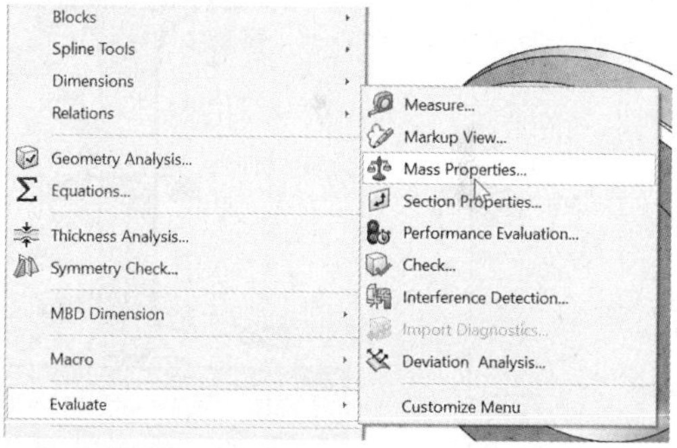

Mass properties are displayed, as shown in **Figure 1.195**. Note the mass of 8.264 pounds (more precisely, the mass is 8.264 pounds-mass and the weight is 8.264 pounds); the volume and surface area are also calculated. The part's center of mass is indicated on the part, and its coordinates are listed in the Mass Properties window. For this part, the center of mass is a small distance (0.001 inches) below the origin, as the keyway causes the part to be slightly asymmetric. If more decimal places are desired for any of the quantities displayed, then within the Mass Properties dialog box choose Options: Use custom settings and choose the desired number of decimal places. You can also change the default number of decimal places displayed by choosing Tools: Options: Document Properties: Units and changing the value beside Length under Mass/Section

FIGURE 1.195

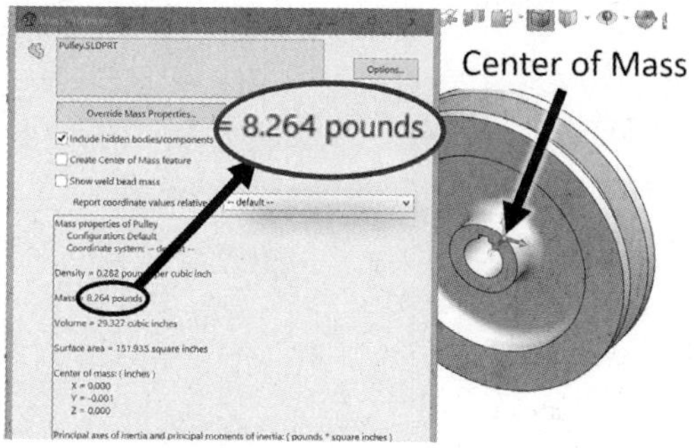

DESIGN INTENT **Keeping It Simple**

In this exercise, we modeled the pulley using four features: A revolved base, a revolved cut for the V-belt groove, fillets, and an extruded cut for the keyway. These first three features could have been combined by sketching the completed cross-section as a single sketch, as shown here (the Sketch Fillet Tool can be used to create the fillet profiles). Which method is better? Although the finished parts will be identical, breaking a part's creation into simple steps is usually the best approach. Many of the errors that are encountered when making parts are related to sketches—duplicate entities, open contours, endpoints of lines and arcs that do not meet, and under defined geometry are examples of the types of errors that are associated with sketches. (Using the Shaded Sketch Contours Option is helpful in finding errors in complex sketches, as only closed contours will be shaded.) The more complex the sketch, the more likely these errors are to occur, and the harder they are to troubleshoot. Also, design changes are easier to make with a part created from many simple steps.

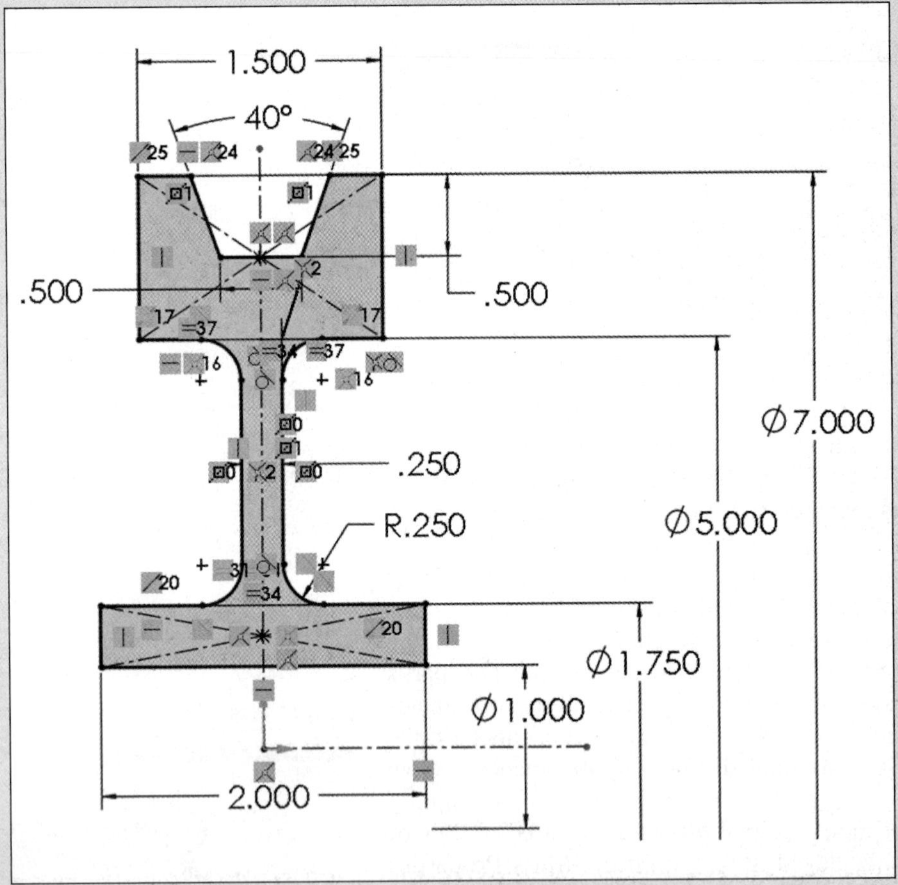

FUTURE STUDY

Dynamics (Kinetics)

In physics, you learned Newton's Second Law:

$$F = ma$$

or the forces (F) acting on a body equal the mass of the body (m) times its acceleration (a). For mechanical engineers designing machines with moving components, application of this law results in the forces required to move a body in a particular manner. In the field of engineering materials, the development of lightweight, strong materials allows moving components to be moved at extremely high accelerations, resulting in higher performance.

The form of Newton's Second Law written above applies to bodies moving in linear or translational motions. Most machine components also move in rotational motions. Newton's Second Law for rotational motions is:

$$T = I\alpha$$

or the torques (T) acting on a body equal the mass moment of inertia (I) times the angular acceleration (α) of the body. This mass moment of inertia is a function of the body's mass and the way in which the mass is distributed relative to the axis of rotation. For example, consider the two wheels shown here. The two wheels are the same diameter and have the same weight, but the wheel on the left has a mass moment of inertia almost twice that of the wheel on the right. If the wheels were mounted onto shafts, it would take twice as much torque to bring the wheel on the left up to speed as it would the wheel on the right (over the same period of time). While a low mass moment of inertia is usually desirable, a notable exception is a flywheel. The inertia of a flywheel in an engine rotating at a high speed makes it difficult to slow down. As a result, the rotational speed of the engine remains smooth despite small fluctuations in power.

The calculation of mass and mass moments of inertia for complex shapes can be tedious. Therefore, the use of solid modeling software to perform these calculations saves time for engineers performing these types of analysis.

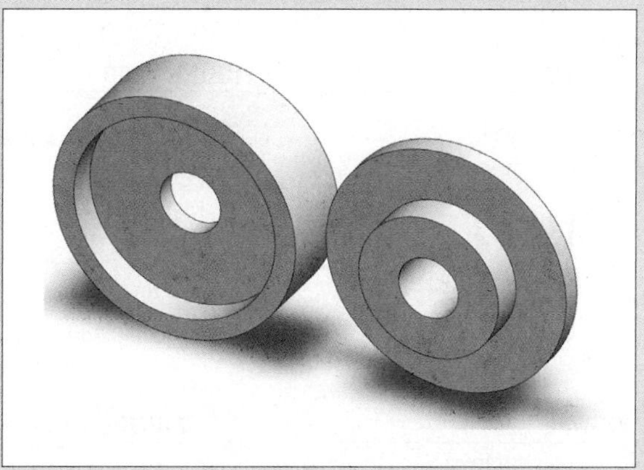

Properties. Also displayed are the part's moments of inertia, which are important in dynamic analysis and are often very cumbersome to calculate by hand (see the Future Study box).

Close the Mass Properties window, and save the part file.

Often, we will be interested in seeing how changes we make to a part's geometry will affect its mass. A *sensor* allows mass changes to be displayed continuously.

FIGURE 1.196

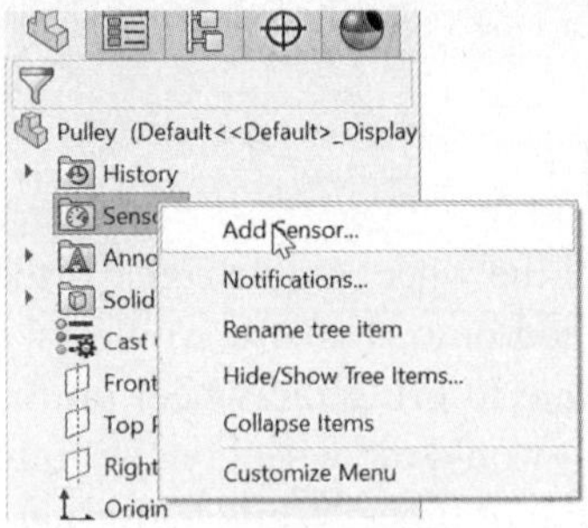

Right-click Sensors in the FeatureManager, and select Add Sensor, as shown in Figure 1.196.

Set the type of sensor to Mass Properties: Mass, as shown in Figure 1.197, and click the check mark.

Click the arrow beside the Sensors entry in the FeatureManager to display the new sensor. The mass of the part will now be displayed, as shown in **Figure 1.198**.

FIGURE 1.197

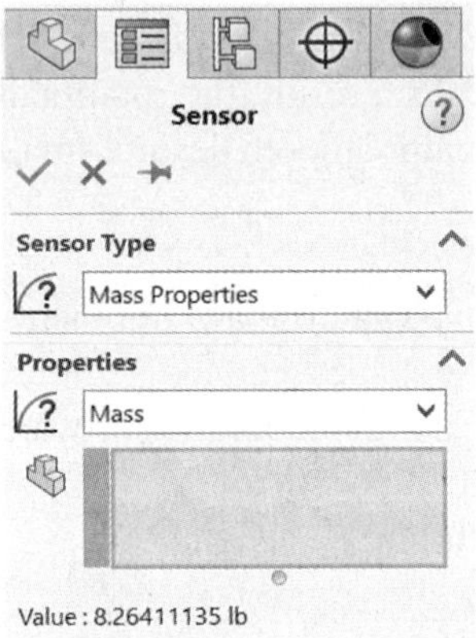

FIGURE 1.198

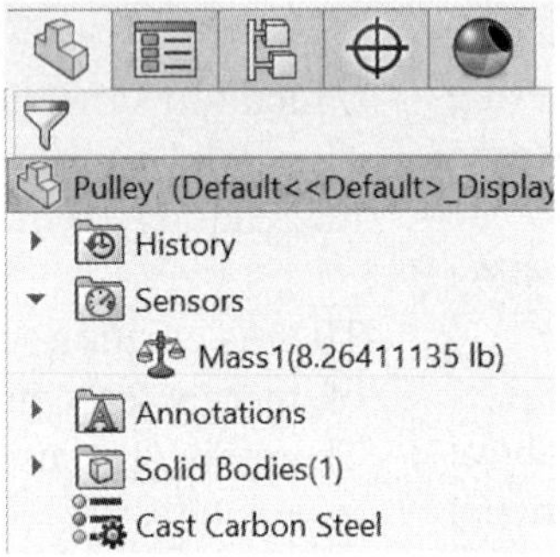

Note that sensors may be used to track other types of data, such as stress analysis and motion analysis results. Within the Mass Properties group, the volume, surface area, and coordinates of the center of mass can be tracked with sensors.

Double-click the V-belt Groove feature (Cut-Revolve1) in the FeatureManager. Double-click the 0.5-inch depth dimension, as shown in Figure 1.199 (shown here from the Right View), and change it to 0.4 inches. Rebuild the model.

FIGURE 1.199

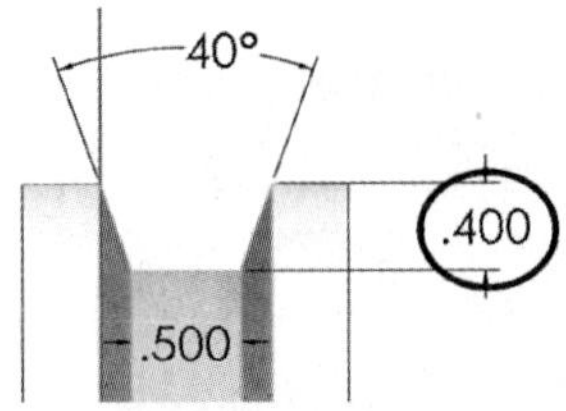

FIGURE 1.200

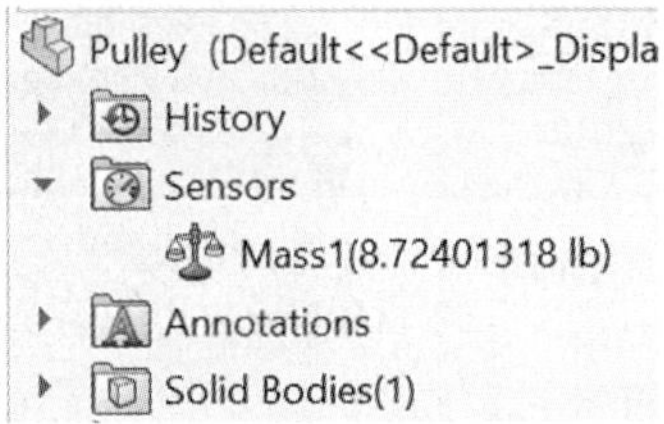

Note that the mass reported by the sensor has changed to reflect the change to the model, as shown in **Figure 1.200**.

Suppose we want to find the mass for a casting of the pulley without the V-belt groove and the keyway. An easy way to do this is to *suppress* those features from the calculations.

Right-click the V-belt Groove (Cut-Revolve1) and select Suppress, as shown in Figure 1.201. Do the same for the keyway (Cut-Extrude1).

FIGURE 1.201

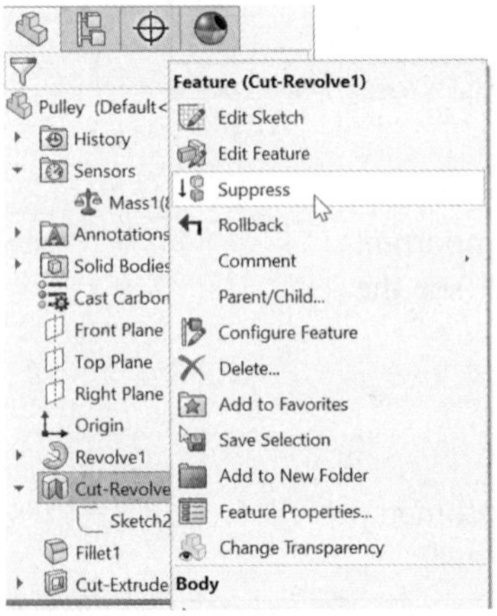

FIGURE 1.202

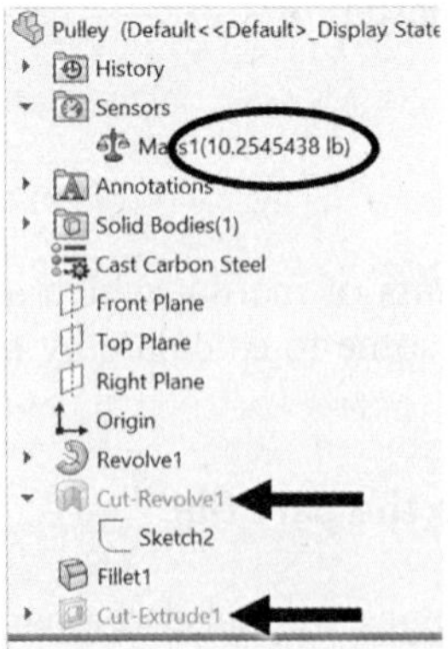

Note that the suppressed entities are shown in gray, as shown in **Figure 1.202**, and the weight of the casting is shown as 10.25 pounds. The suppressed entities can be restored to the model by right-clicking and selecting Unsuppress.

Close the part file without saving any of the changes made.

PROBLEMS

P1.1 Create a solid model of the stepped shaft, using Extruded Boss/Base and Fillet features.

FIGURE P1.1A

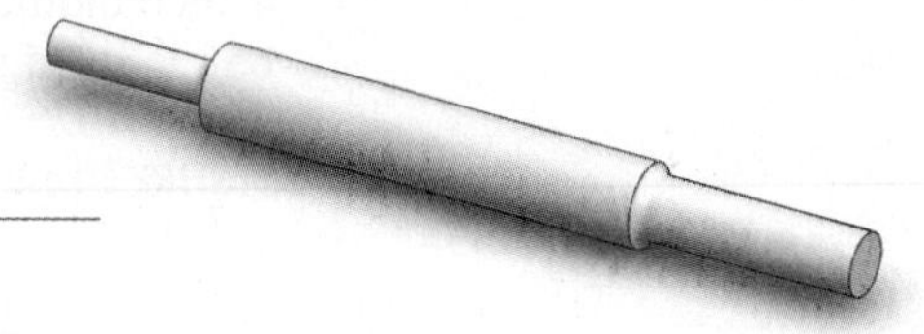

FIGURE P1.1B

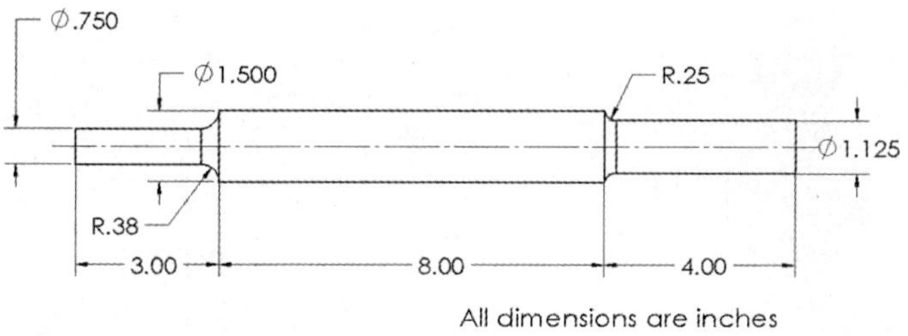

P1.2 Make a solid model of each of the parts shown in **Figure P1.2**. Let one grid space equal one inch. The cut on the bottom of part C and the hole in part D both go completely through the part. Use fully defined sketches.

FIGURE P1.2

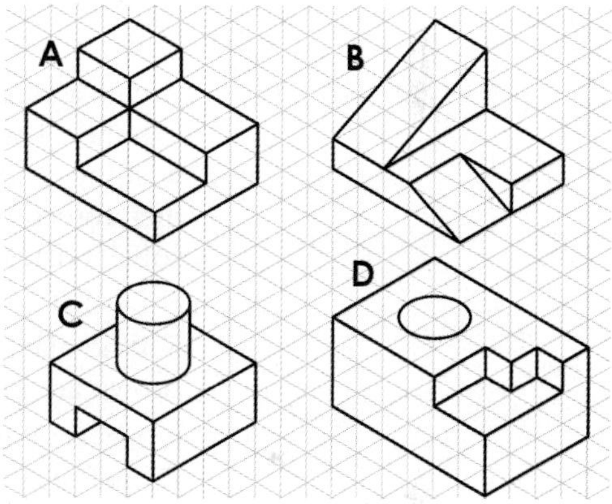

P1.3 Create a model of the torus shown in **Figure P1.3A**. Dimensions are shown in **Figure P1.3B**. Find the formula for the volume of a torus, and compare the value calculated from the formula with that obtained from the Mass Properties Tool.

FIGURE P1.3A

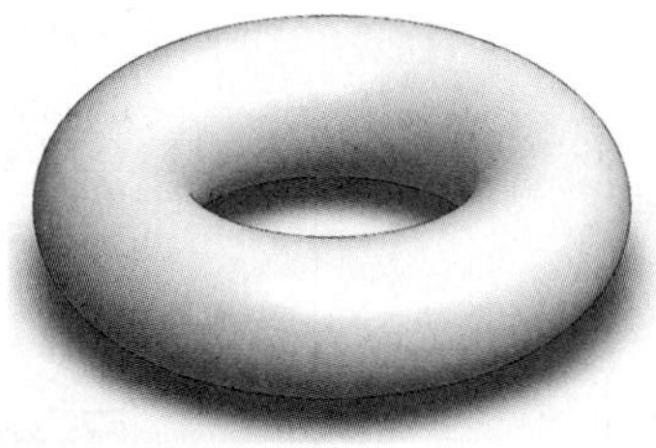

FIGURE P1.3B

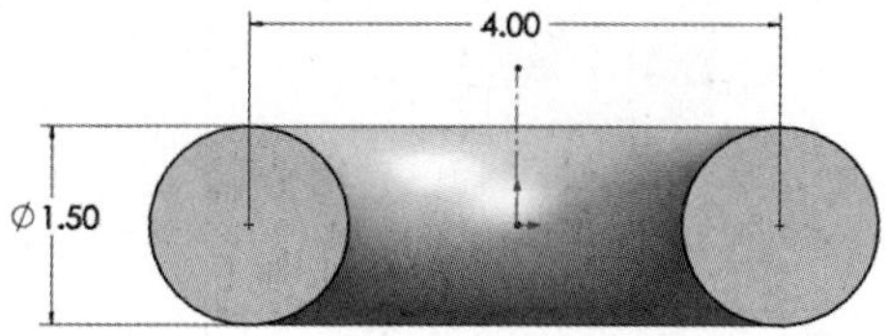

P1.4 Construct a solid model of the wooden lamp base shown in **Figure P1.4A**. It can be constructed by revolving the sketch shown in **Figure P1.4B** (dimensions in inches). Add 0.25″ fillets at exterior and interior edges, except at the top and bottom edges. If the base will be turned down from 4-inch diameter stock on a lathe, determine the volume of wood that must be removed to make the base.

FIGURE P1.4A

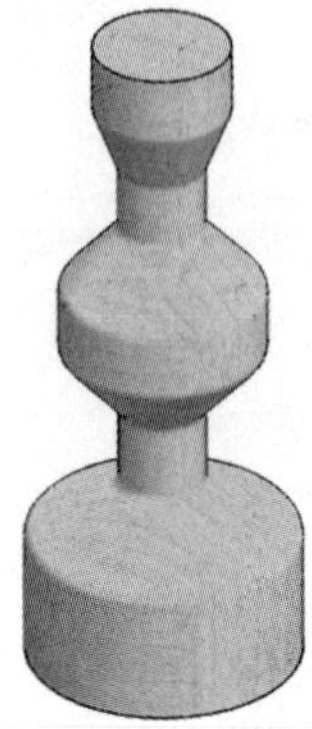

FIGURE P1.4B

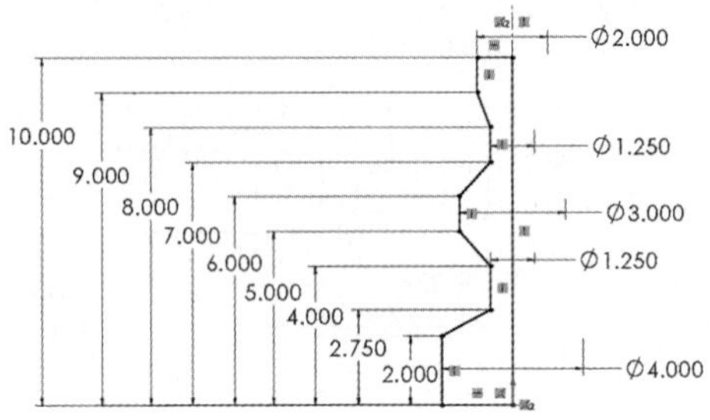

P1.5 Create a solid model of this plastic pipe tee. A tee is used to connect pipes together. The type of tee shown here is used to join pipes with solvent welding. A chemical is applied to the inside of the socket, and the pipe is then forced into the socket. The solvent softens the plastic, and when the solvent dries, a strong, permanent joint is created. The sockets are tapered slightly to allow for a tight fit with the pipe.

Set the material to PVC Rigid and find the weight of the tee. Dimensions shown are inches.

(Answer: Weight = 0.1044 pounds. Click the Options tab in the Mass Properties box and increase the number of decimal places, if necessary.)

FIGURE P1.5A

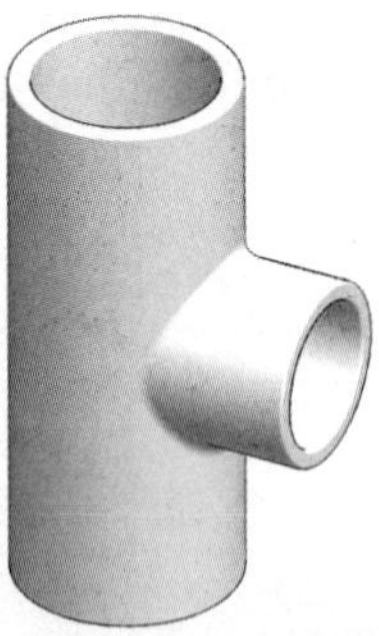

FIGURE P1.5B

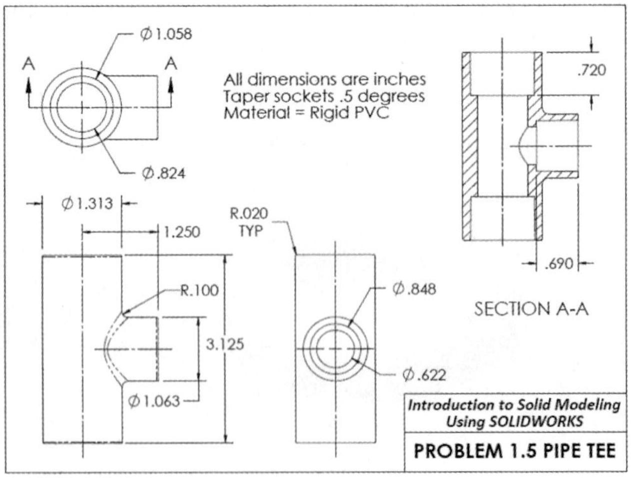

P1.6 Create a solid model of this 2-mm-thick steel bicycle disk brake rotor. Calculate the mass of the part, using AISI 304 steel as the material.

Note: Although this part can be modeled with a single extrusion from a complex sketch, you will find it easier to extrude a solid disk and then use a series of simple extruded cuts and circular patterns.

FIGURE P1.6A

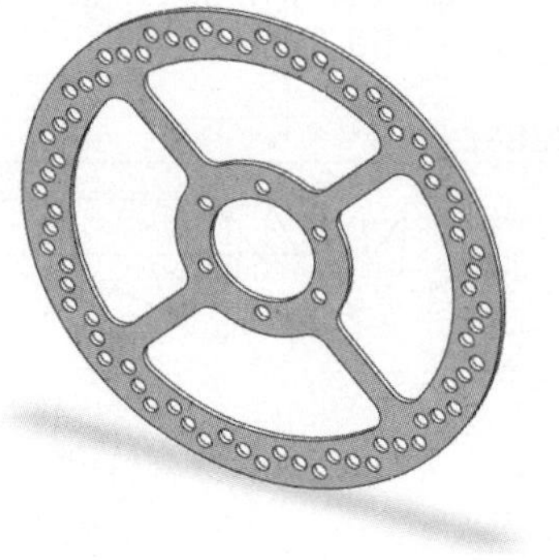

FIGURE P1.6B

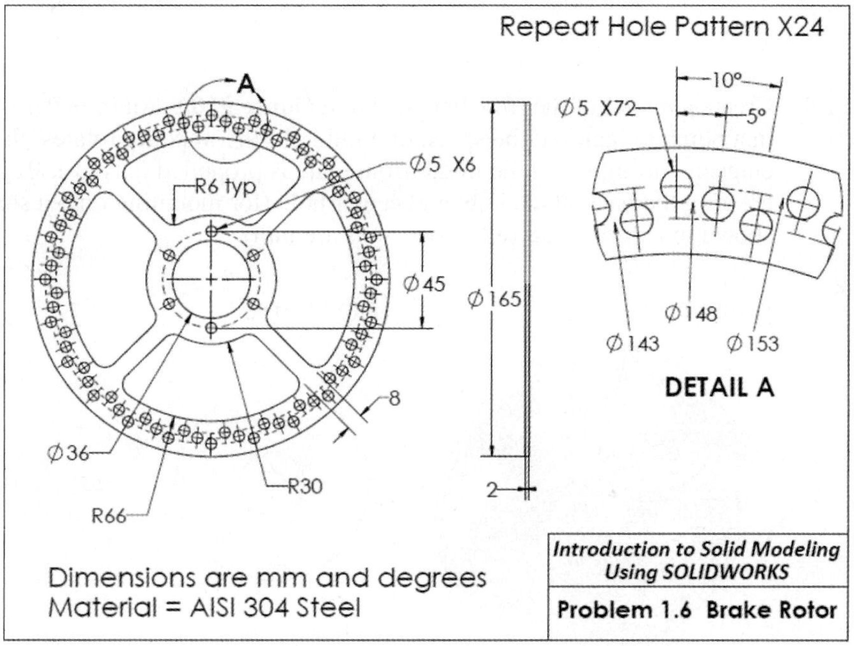

P1.7 Create a solid model of the pleated filter element. Dimensions shown are inches, and the diameter dimensions shown are nominal dimensions. There are 36 pleats in the part, and the part is 8 inches long.

To create this part, start a sketch with two construction circles representing the nominal inner and outer diameters. Add and dimension two lines representing one pleat, and use a circular pattern to copy these two lines into the other 35 positions. Select the Extruded Boss Tool. Since the sketch is an open contour, a thin-feature extrusion will be created. Set the thickness to 0.02 inches, and set the type as Mid-Plane.

FIGURE P1.7A

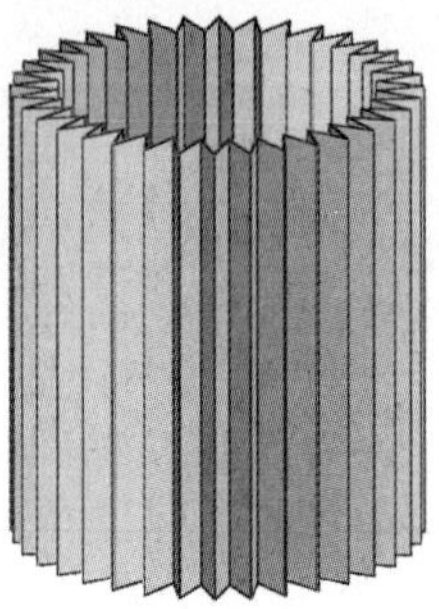

FIGURE P1.7B

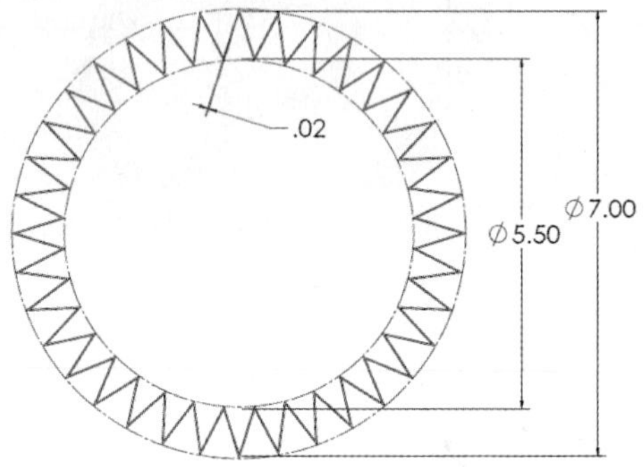

P1.8 Create a model of the impeller shown in **Figure P1.8A**. An impeller is used in a pump to increase the speed of a fluid. As the impeller rotates, fluid entering through the hole in the front plate is propelled outward. Begin by creating the back plate, hub, and center hole (for mounting onto a shaft), as shown in **Figure P1.8B** (all dimensions are mm).

FIGURE P1.8A

FIGURE P1.8B

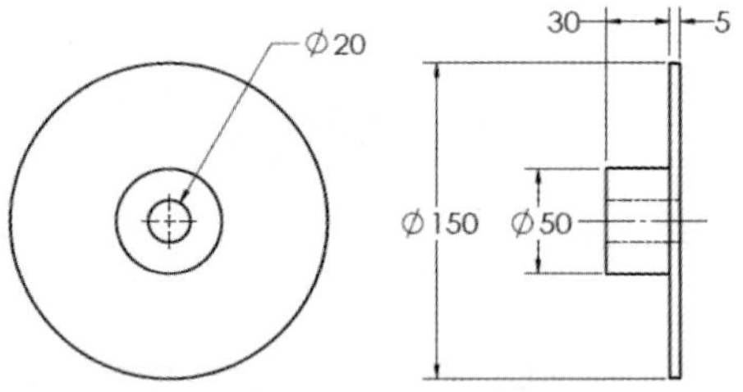

1. Open a sketch on the front face of the back plate. Select the Convert Entities Tool from the Sketch Group of the CommandManager, and click on the two circular edges indicated in **Figure P1.8C**. Click the check mark to create circular entities in the sketch. Draw two lines from the edge of the hub to the edge of the back plate. The top line should be tangent to the edge of the hub. The bottom line should be parallel to the top line. Add a 5-mm dimension between the two lines. Although the radial alignment of the lines is not critical, making one of the lines horizontal will make the sketch fully defined.

FIGURE P1.8C

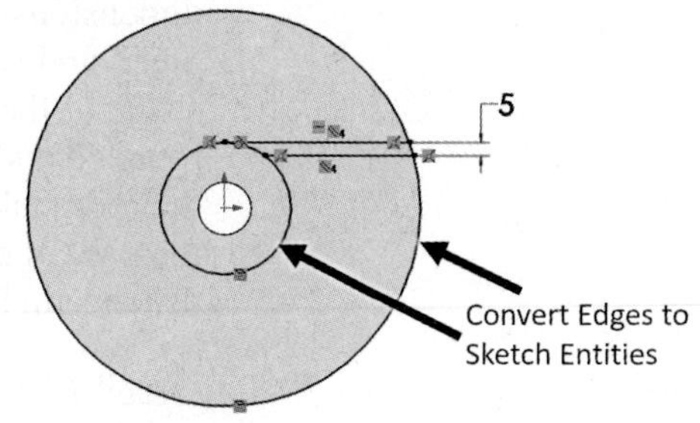

2. Extrude the sketch contour bounded by the lines and circles to the same 30-mm depth as the hub (try using "Up to Surface" as the type of extrusion) to create the first vane on the impeller. Add a 5-mm fillet at the base of the impeller, and use the Circular Pattern Tool to create a pattern of six vanes and radii, as shown in **Figure P1.8D**.

FIGURE P1.8D

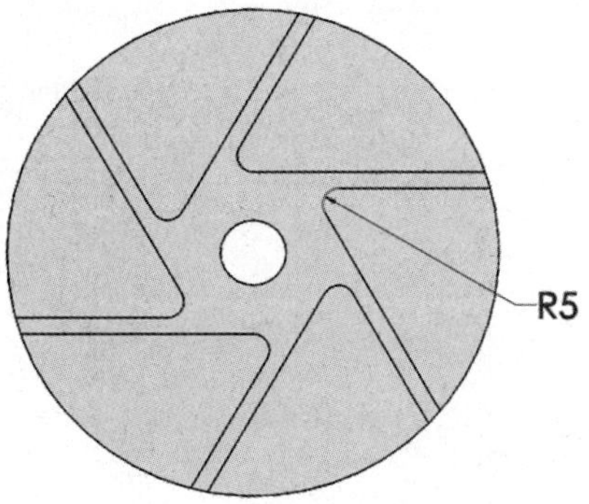

3. Add the 5-mm-thick front plate, which has a 75-mm-diameter "eye" to allow fluid to enter.

P1.9 Create the body of the flange shown in **Figure 1.1** and detailed in **Figure 1.2** by sketching its cross-section in the Right Plane and revolving it about the flange's axis. Include the fillets in the sketch, using the Sketch Fillet Tool.

P1.10 Create a solid model of the connecting rod shown in **Figure P1.10A**. Use the dimensions shown in **Figure P1.10B**. Assign the material to be Cast Carbon Steel, and add a sensor to determine the mass. Notes: When sketching the shape of the rod, use tangent relations between the diagonal edges and the rounded ends, as shown in **Figure P1.10C**. To maintain the symmetry of the rod, make your sketches in the Front Plane and choose Midplane as the type of extrusion. This will cause the sketch to be extruded an equal distance in both directions from the Front Plane.

FIGURE P1.10A

FIGURE P1.10B

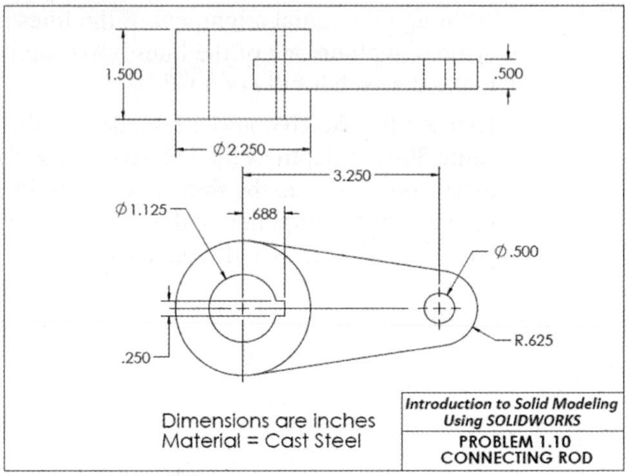

FIGURE P1.10C

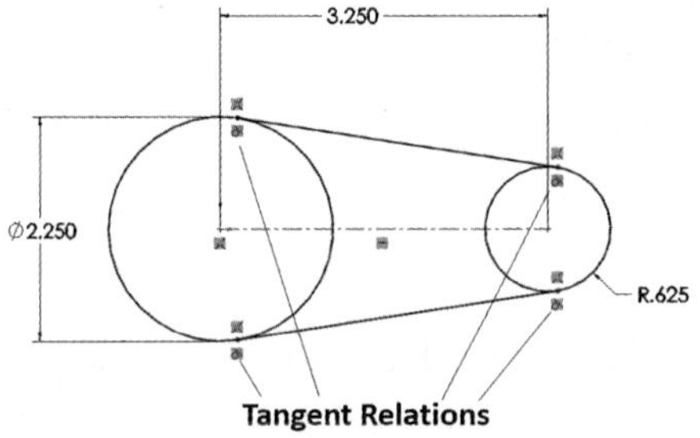

P1.11 Consider the connecting rod model created in **P1.10**. Due to weight restrictions on the design, the weight of the connecting rod must be reduced to 1.70 lb. Add a recessed pocket like the one shown in **Figure P1.11** to both sides of the connecting rod (to maintain symmetry), and select appropriate dimensions to reduce the weight of the part to an acceptable level. Notes: Experiment with the Offset Entities Tool when creating the recessed pocket. This tool allows for new entities to be created at a specified distance away from selected lines or arcs. Also, try the Mirror Tool to create the second pocket, with the Front Plane as the mirror plane. These tools will be explained in detail in later chapters.

FIGURE P1.11

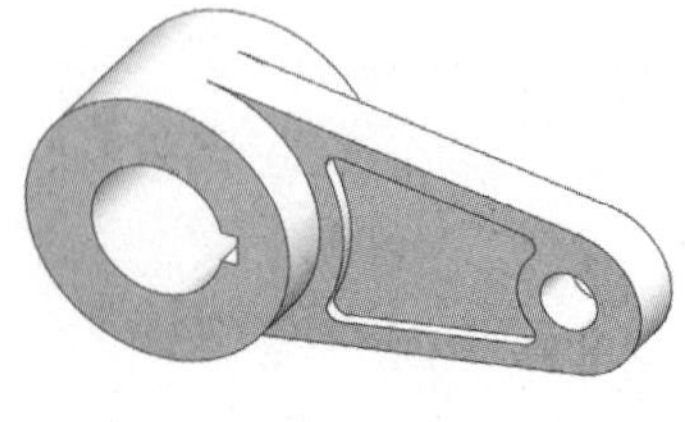

CHAPTER 2

Engineering Drawings

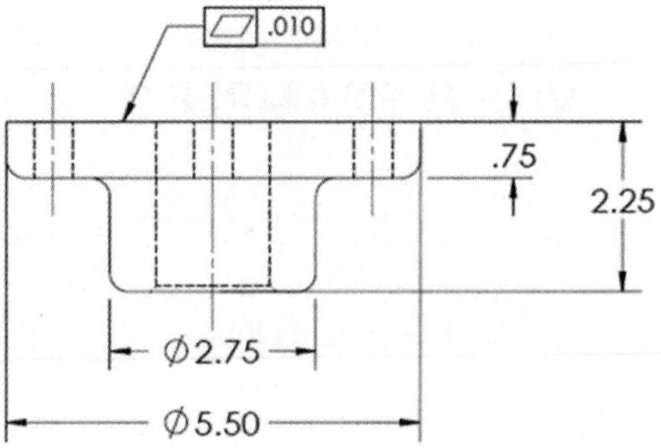

Introduction

Even companies that design parts exclusively with solid modeling software still need to produce 2-D drawings. A multiple-view 2-D drawing is the best way to document the design, showing all of the dimensions necessary to produce the part. Since no manufacturing process can produce "perfect" parts, the tolerances (variations from the stated dimension of a part) allowed for important dimensions are also shown on 2-D drawings.

The SOLIDWORKS® program allows for 2-D drawings to be quickly and easily produced from 3-D models. The drawing will be fully associative with the part file, so that changes to the part will be automatically reflected in the drawing, and vice versa.

Chapter Objectives

In this chapter, you will:

- make a 2-D drawing from a SOLIDWORKS part,
- create a custom drawing sheet format, and,
- use eDrawings® software to create a drawing file that allows for easy file sharing and collaborative editing.

2.1 Drawing Tutorial

In this tutorial, a 2-D drawing of the flange from Chapter 1 will be made.

Open the SOLIDWORKS program. From the Welcome dialog box, select Drawing, as shown in Figure 2.1. (You can also open a new drawing by selecting the New Document Tool, as shown in Figure 2.2, and then selecting Drawing as the type of document.)

FIGURE 2.1

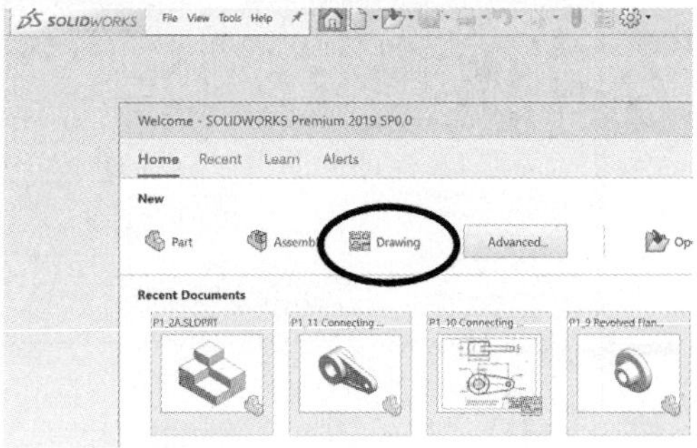

FIGURE 2.2

SOLIDWORKS is a registered trademark of Dassault Systémes SolidWorks Corporation.
eDrawings is a trademark of Dassault Systémes SolidWorks Corporation.

VIDEO EXAMPLE 2

In this chapter, a 2-D drawing is created from the flange part of Chapter 1. The video at *www.mhhe.com/howard2019* demonstrates the creation of a drawing of the 3-D part made in Video Example 1. Steps shown in this video are selecting drawing views, importing dimensions, moving dimensions between views, adding notes, and adding a sheet format (title block and border).

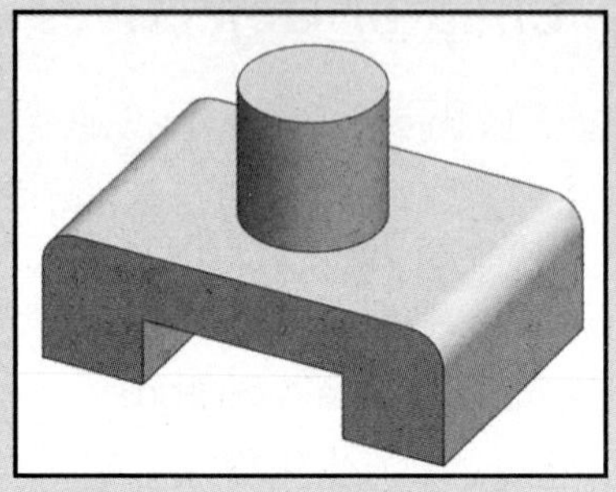

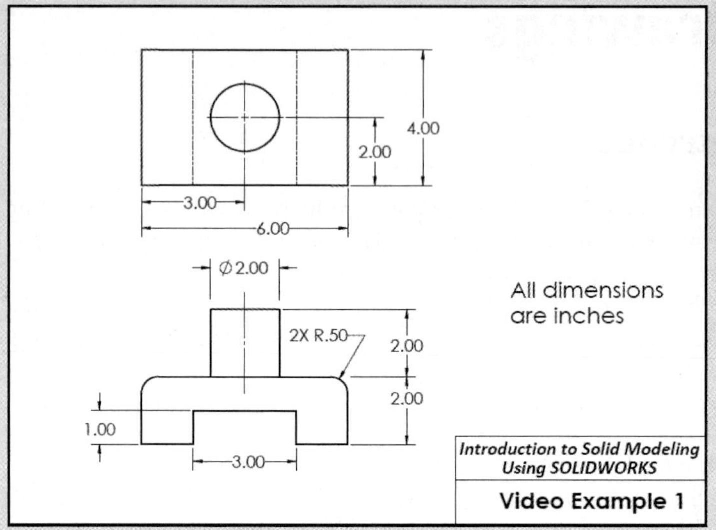

FIGURE 2.3

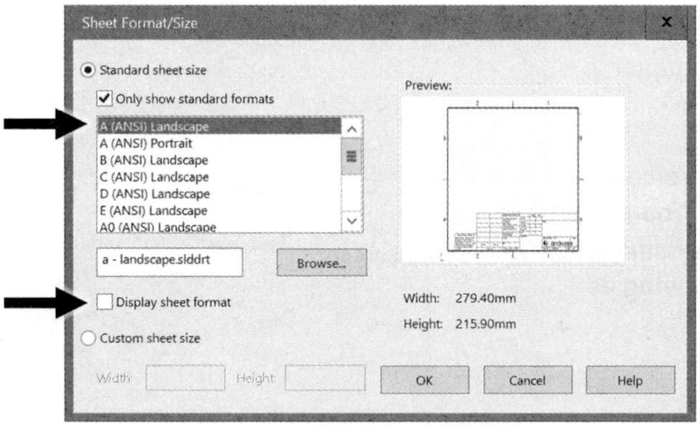

As if you were making a drawing by hand, your first step is to select the size and type of paper to be used. Later we will look at using sheet formats with a title block, but for now let's use a blank 8-1/2 × 11-inch sheet (A-Landscape).

Choose a paper size of A-Landscape, and click on the box labeled "Display sheet format" to clear the check mark, as shown in Figure 2.3.

Click OK to close the dialog box. Click on the X (Cancel) in the Model View PropertyManager, as shown in Figure 2.4.

FIGURE 2.4

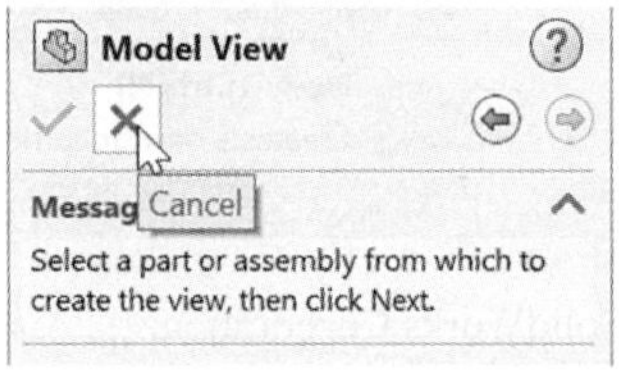

The Model View command starts automatically for a new drawing, but we will set a few options before bringing in the flange model data.

Select the Options Tool, as shown in Figure 2.5. Under the System Options tab, select Drawings: Display Style, and set the hidden line and tangent displays as shown in Figure 2.6.

FIGURE 2.5

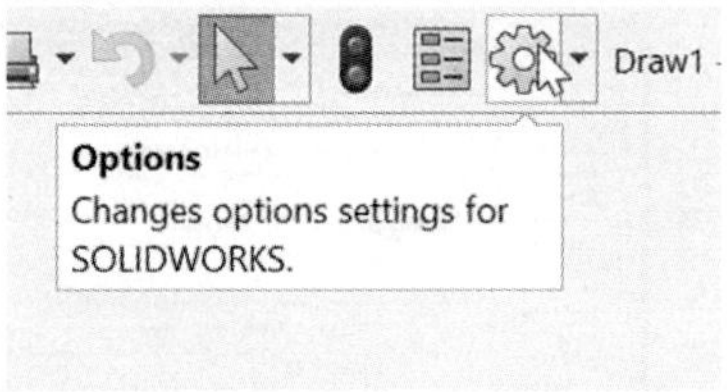

Hidden lines are usually displayed in standard drawing orthographic views, but not in section or detail views. Tangent edges are usually not shown in drawing views. Whether or not to show hidden and tangent edges is sometimes dependent on the complexity of the part. As we will see later, we can change these display options for each model view.

Select Colors: Drawings, Paper Color. Select Edit, as shown in Figure 2.7. Select white for the paper color and click OK.

FIGURE 2.6

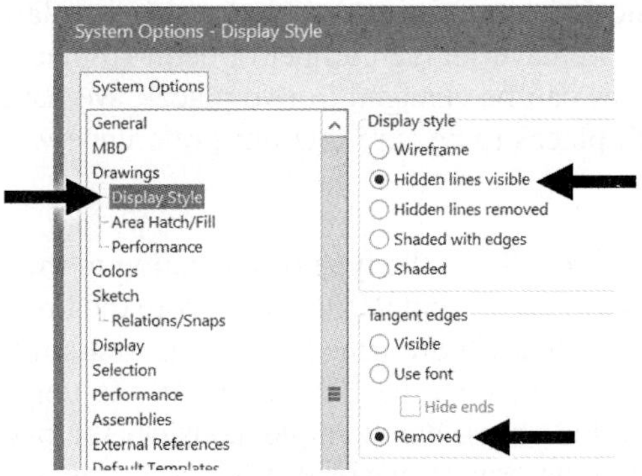

FIGURE 2.7

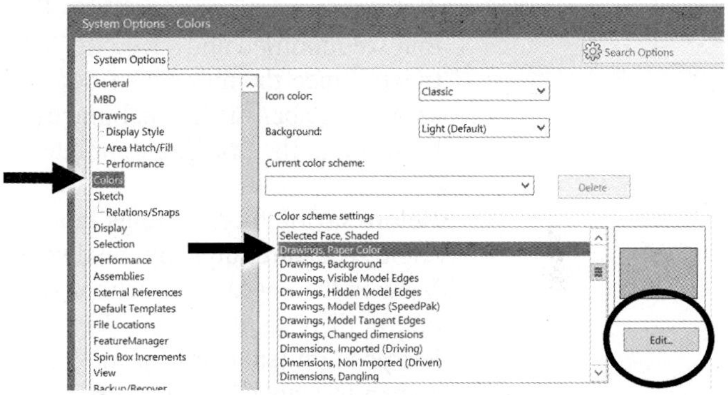

Under the Document Properties tab, select Drafting Standard and set the dimensioning standard to ANSI, as shown in Figure 2.8. Select Dimensions, and make sure that the number of decimal places is set to 2 (.12), as shown in Figure 2.9. Select Detailing, and turn on the automatic display of centerlines, as shown in Figure 2.10. Select Units, and set the Unit system to IPS (inch, pound, second), and the

FIGURE 2.8

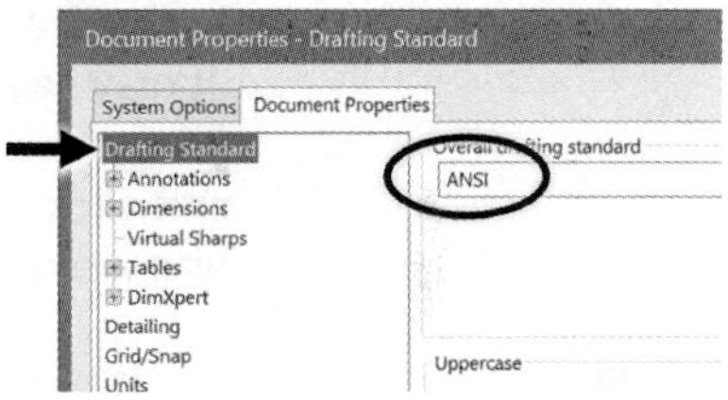

FIGURE 2.9

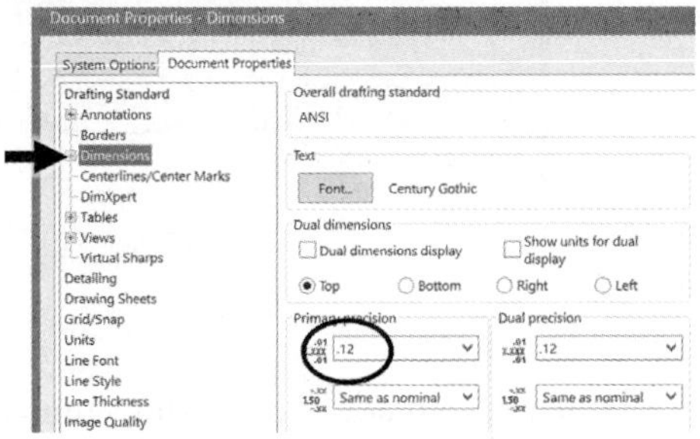

FIGURE 2.10

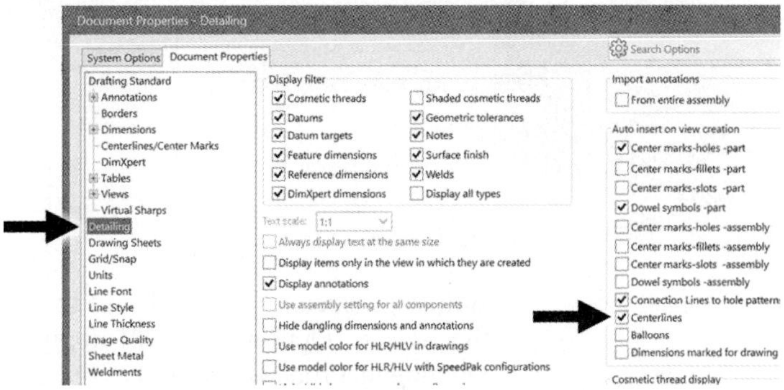

FIGURE 2.11

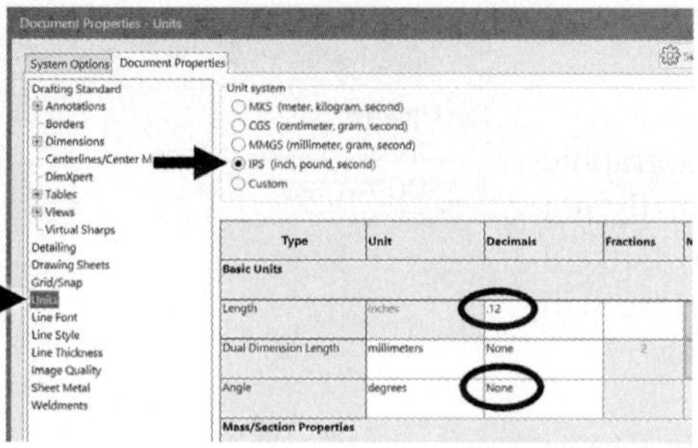

decimal places to 2, as shown in Figure 2.11. If you see a message that the drafting standard has been changed to ANSI-Modified, ignore it. Set the decimal places for angles to None. Click OK.

Although the part was modeled with the decimal places set to 3, the number of decimals does not affect the accuracy of the model, only the way dimensions are displayed. For a drawing, the number of decimal places should be related to the tolerance level of most of the dimensions. Note that there is an option for setting the units of dual dimension lengths. Often, drawings show one set of dimensions in the primary unit system and a second set of dimensions in brackets. In this way, both US and SI units can be displayed on a single drawing. Since we will not be using dual dimensions, we do not need to change the default setting. As with the part template that we modified and saved in Chapter 1, we have set the number of decimal places twice. Since the number of decimal places can be changed in two places, we have set the number of decimal places in both places to be sure that our preference will be saved in the drawing template.

Before saving these settings in a template, we will set the type of projection to third angle. The third-angle option displays the Front, Top, and Right Views, with the Front View in the lower left position (standard for most US drawings), while the first-angle option displays the Front, Top, and Left Views, with the Front View in the upper left position (standard for most European drawings). For example, consider the part shown in Figure 2.12. The first- and third-angle projections of this part are shown in Figure 2.13.

In the drawing space, right-click and select Properties from the menu that appears. (It may be necessary to click the double-arrow at the bottom of the menu to show the Properties option.) Set the Type of projection to Third angle, as shown in Figure 2.14, and click Apply Changes to close the Sheet Properties box (or Cancel if Third Angle was already selected).

FIGURE 2.12

FIGURE 2.13

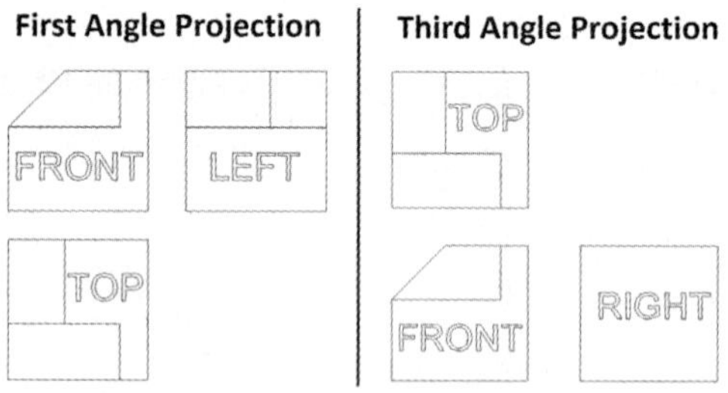

FIGURE 2.14

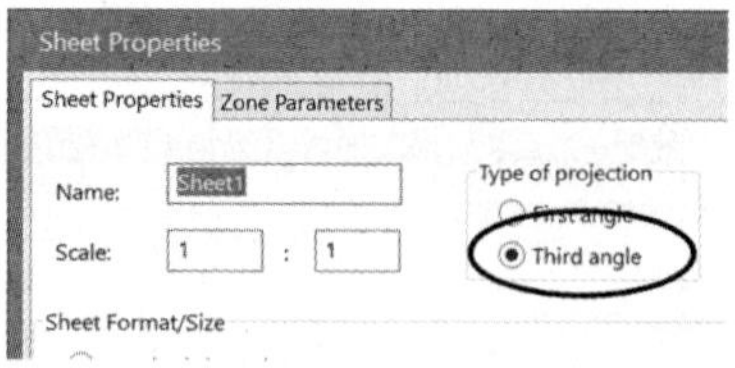

Before proceeding, we will save the settings that we have made. The settings under the System Options tab will apply to any future drawings, but those under the Document Properties tab, as well as the First/Third Angle Setting, will revert to their previous values unless saved. These settings are stored in a template file. When we start a new document and select Part, Drawing, or Assembly, we are opening the template associated with that type of file. In Chapter 1, we modified the Part template. We will now perform a similar operation to modify the Drawing template, saving our changes for future drawings.

Select File: Save As from the menu. From the Save as type pull-down menu, select Drawing Templates, as shown in Figure 2.15. The default template name, Drawing.drwdot, will be displayed in the file list. Click on the template name to select it, and click Save. When asked if you want to overwrite the existing file, click Yes. Exit the drawing, and open a new drawing document. As before, choose an A-Size Landscape sheet with the Display sheet format box unchecked.

FIGURE 2.15

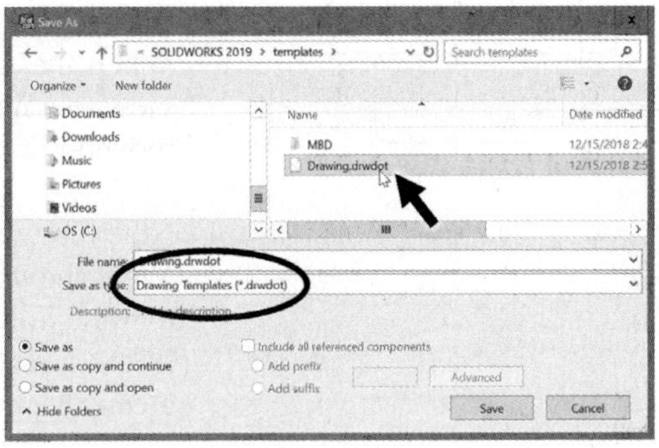

Note that the settings that you specified are applied when you open a new drawing. One note of caution: as with the Part template in Chapter 1, the next time you save a file, the default folder will be the one where the templates are stored. Be sure to change the folder to the one where you want to store the file.

When we modeled a part in Chapter 1, we found most of the commands we needed in the CommandManager. In the drawing mode, the CommandManager has three groups of commands that we will use regularly. To keep the interface as simple as possible, we will hide the other groups.

FIGURE 2.16

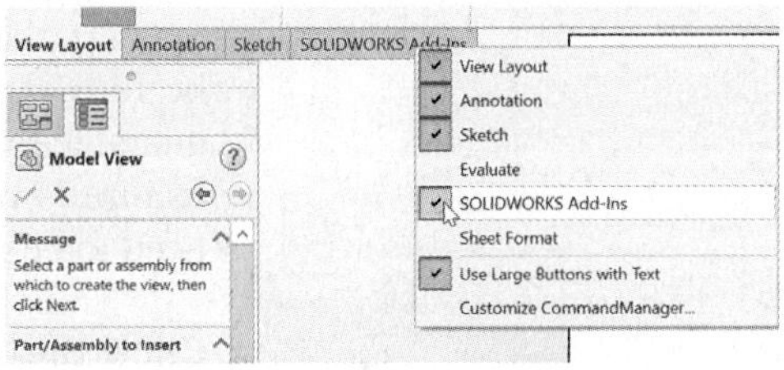

Right-click on any of the CommandManager tabs. Click on the name of each group other than View Layout, Annotation, and Sketch to clear the check mark and turn off display of that group, as shown in Figure 2.16.

We are now ready to import the geometry from the part file.

By default, the Model View Tool opens automatically when you begin a drawing. You can also access the Model View Tool from the View Layout group of the CommandManager, as shown in Figure 2.17.

FIGURE 2.17

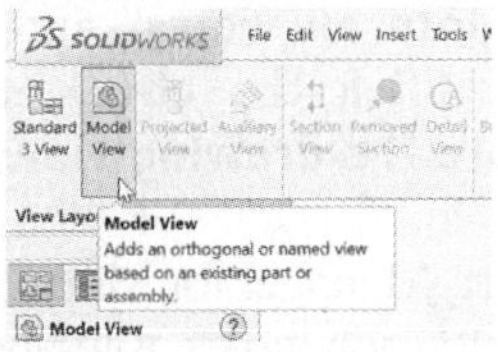

FIGURE 2.18

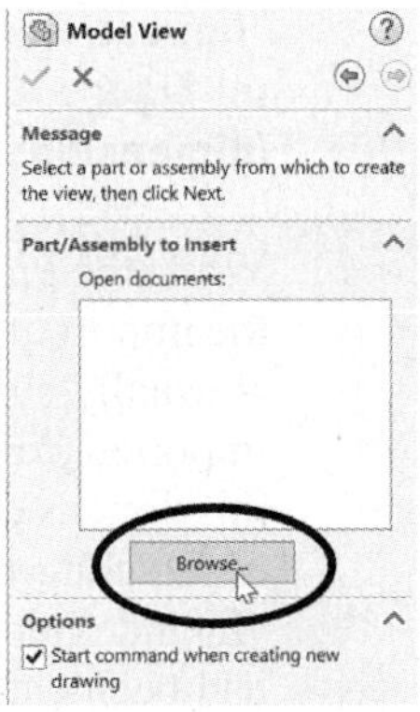

FIGURE 2.19

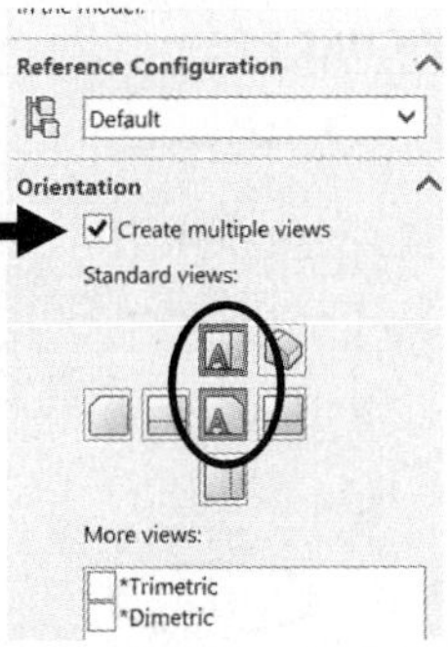

If you have the flange open in another window, then it will appear in the PropertyManager. If so, double-click the file name to select it. If not, then click the Browse . . . button, as shown in Figure 2.18, and find the flange part file where you stored it. In the PropertyManager, select Create multiple views and click on the Front and Top Views to select them, as shown in Figure 2.19. If any other views are selected (such as the Trimetric View), click the box next to the view name to clear the check mark and de-select the view. Click the check mark to place the views.

Note the "A" designation shown on the Front and Top Views in **Figure 2.19**. This indicates that model dimensions are associated with these views.

FIGURE 2.20

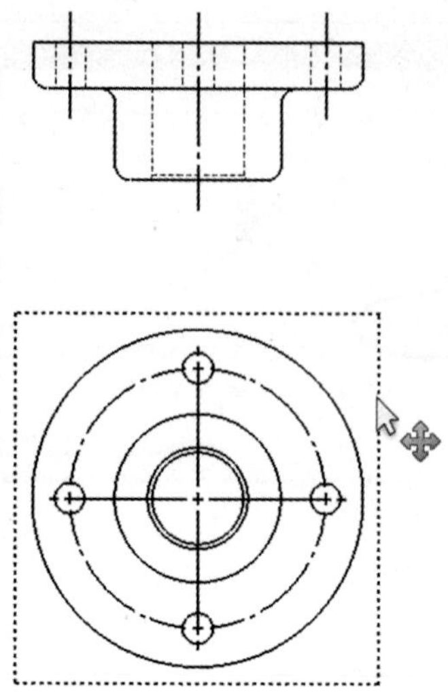

As you pass the cursor over each drawing view, notice that a rectangular box that defines the boundary of that view appears. Moving the cursor over the drawing view boundary displays the move arrows, as shown in **Figure 2.20**. When the move arrows are displayed, you can click and drag the drawing view to a new location. Note that the views remain in alignment as they are moved.

Note that we need only the Front and Top Views for this part, since the Right View shows no information that cannot be seen in the Top View. Also, note that the drawing scale was automatically selected so that the drawing views fit on the sheet. You can override the scale by right-clicking in the drawing area, selecting Properties, and defining a new scale.

FIGURE 2.21

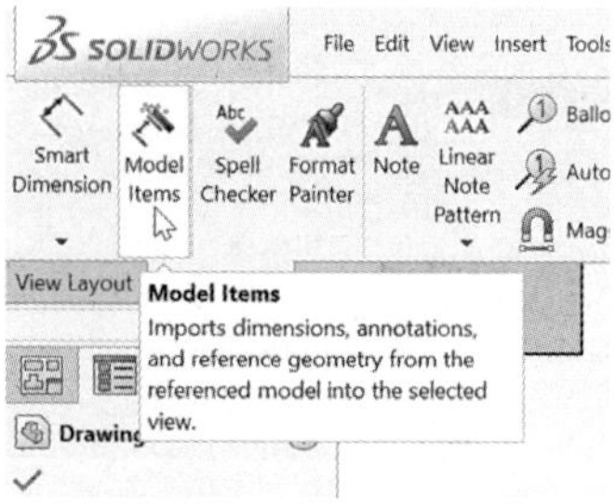

While we can add dimensions manually using the Smart Dimension Tool, we usually prefer to import the dimensions from the part model. Doing so has the advantage that imported dimensions can be changed and changes will be reflected in the model. Dimensions added to the drawing with the Smart Dimension Tool will be *driven* dimensions, meaning that their values will be updated when changes are made to the model, but the dimensions cannot be used to change the model from the drawing. Another advantage of importing dimensions is that if fully defined sketches are used to create the part, then importing the dimensions helps to assure that the drawing will be fully defined as well. (However, using fully defined sketches does not guarantee that the drawing will be fully defined. In particular, if a model feature is positioned with relations, then additional dimensions will often need to be added to the drawing to precisely locate that feature.)

FIGURE 2.22

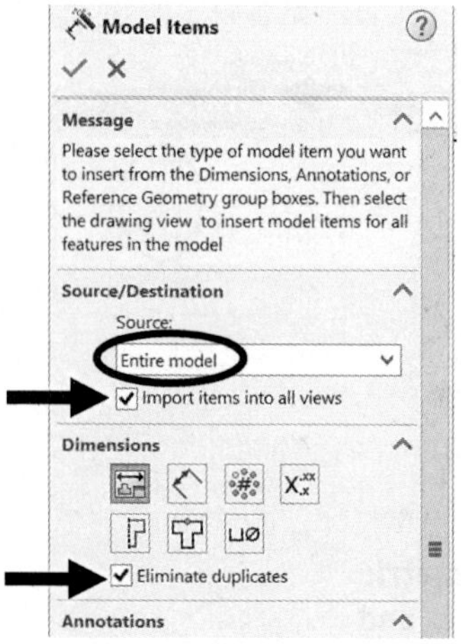

Click the Annotation tab of the CommandManager, and select the Model Items Tool as shown in Figure 2.21. In the PropertyManager, select Entire model as the source. Also check the boxes labeled "Import items into all views" and "Eliminate duplicates," as shown in Figure 2.22. Click the check mark to import the model items (dimensions) into the drawing views.

While the PropertyManager for Model Items contains options for controlling the import of dimension, notes, tolerances, etc., accepting the default options is usually sufficient. Note that all dimensions used to create the model are imported except for one: the 45-degree angle of the chamfer. The reason that this dimension is not imported is that this dimension would have to be applied to a feature that is hidden in the current drawing views. If the box labeled "Include items from hidden features" in the Model Items PropertyManager had been checked, then this dimension would have been imported. Instead, we will import the dimension into a detail view that we will add later. Also note that the R.25 dimension defining the fillet radii may be shown on a different fillet, depending on which edge you selected first when creating the fillets in the model.

You can click and drag on the value of any dimension to reposition it within its associated drawing view.

Click and drag the dimensions on the drawing sheet so that they can all be clearly seen, as shown in Figure 2.23. Note that a dimension will often automatically align with other entities when you drag it to move its position. Holding down the Alt key while dragging the dimensions turns off the auto-alignment and allows you to place the dimensions more precisely.

FIGURE 2.23

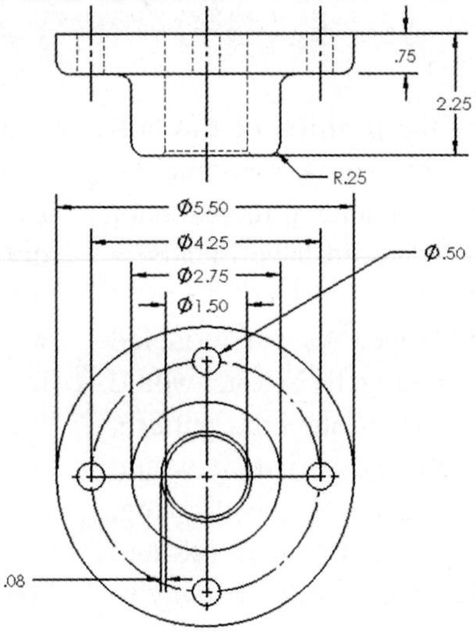

FIGURE 2.24

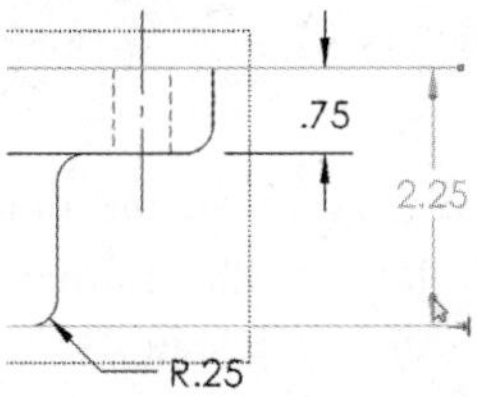

FIGURE 2.25

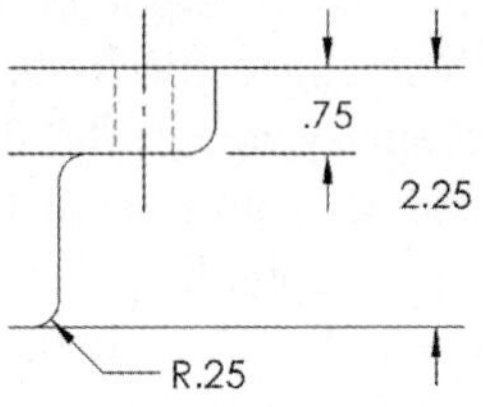

You can also change how the arrows are displayed on linear dimensions, such as for the 2.25-inch dimension shown in **Figure 2.24**. When you click once on the dimension value to select it, the dimension will appear in green and dots will appear on the arrowheads. When you click on one of these dots, the arrowheads will switch from outside to inside or vice versa, as shown in **Figure 2.25**.

Dimensions are shown in the view in which they were created in the part. For example, the 5.5-inch and 2.75-inch diameters are shown in the Front View, since they were created from circles sketched in the Front Plane (or sketch planes parallel to the Front Plane). However, standard drawing practice is for the diameter of a solid cylindrical feature to be shown in a view normal to the one in which it appears as a circle. Therefore, we want these dimensions in the Top View. Rather than deleting these dimensions from the Front View and adding them to the Top View, we will simply move them.

FIGURE 2.26

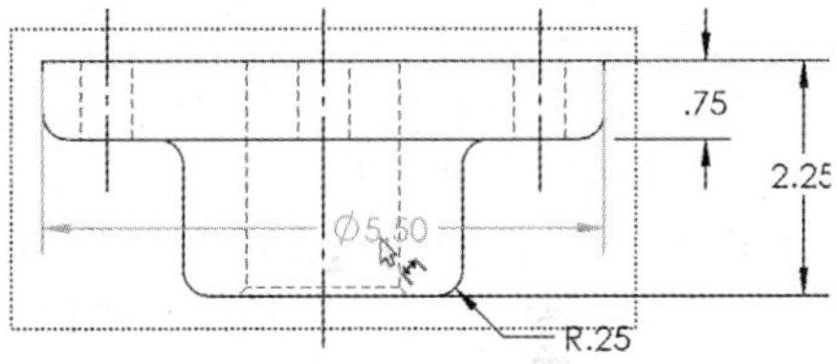

Press and hold down the Shift key. Click and drag the 5.5-inch dimension until the cursor is within the boundaries of the Top View. Release the mouse button, and the dimension will be placed in the Top View, as shown in Figure 2.26. (Make sure to release the mouse button before releasing the Shift key; otherwise, the move operation will not work.) Repeat for the 2.75-inch dimension.

FIGURE 2.27

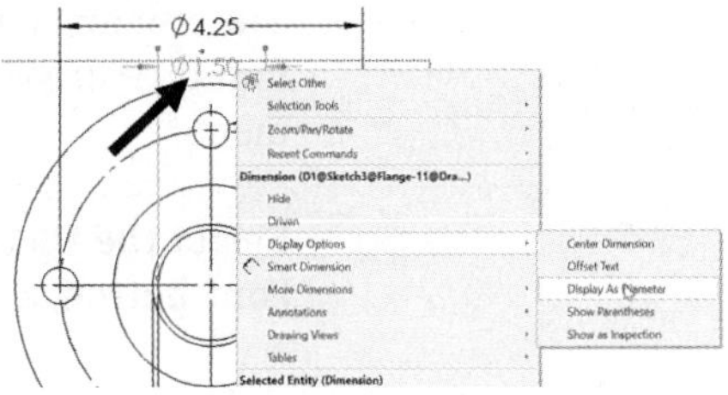

Hole diameters should be shown with leaders, rather than as linear dimensions. Depending on how you dimensioned the center hole in the part model, the 1.50-inch dimension may be imported as a linear dimension, as shown in **Figure 2.23**, or with a leader.

FIGURE 2.28

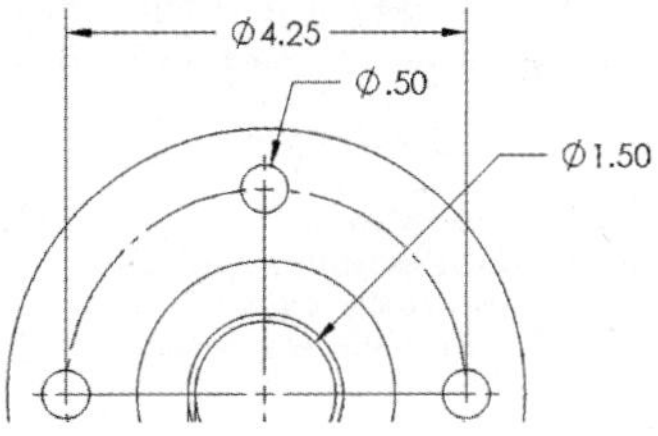

If the 1.5-inch diameter dimension appears as a linear dimension, right-click on the dimension value, and choose Display Options: Display as Diameter, as shown in Figure 2.27. Click and drag the dimension to a convenient location, as shown in Figure 2.28.

DESIGN INTENT **Exploiting Associativity**

In Chapter 1, we defined the position of the bolt holes in the flange by creating a construction circle and specifying the diameter dimension of the circle. Alternatively, we could have located a single hole by specifying its radial location from the center point of the flange. As we modeled the flange, either dimensioning technique would properly locate the first hole in our hole pattern, and it would seem that either alternative would be acceptable. When dimensioning the drawing, however, we see that drawing a bolt circle and dimensioning its diameter is the preferred way to show the radial position of the bolt holes. If the distance were defined by a radial dimension, then we could simply delete the radial dimension in the drawing, add a construction circle, and add a new dimension. However, the new dimension would not be associative with the part. That is, changing the added 4.25-inch dimension would not change the bolt hole positions, and changing the radial dimension in the part would not cause the drawing dimension to be updated. Therefore, maintaining full associativity between the part and the drawing would require editing of the part file.

FIGURE 2.29

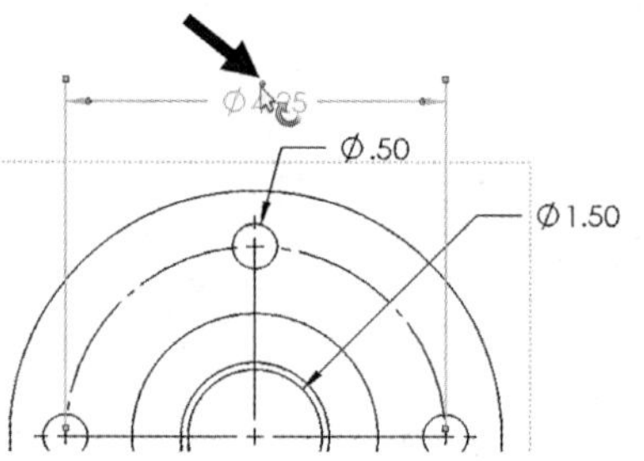

The orientation of diameter dimensions can be changed by clicking and dragging the green dot that appears directly above the dimension value when it is selected. For example, if the 4.25-inch diameter dimension is imported with the dimension line horizontal, it can be rotated 90 degrees by clicking and dragging the green dot, as shown in **Figure 2.29**. (This will prevent the leaders of the hole diameters from crossing the extension lines of 4.25-inch dimension.)

The options for changing the appearance of dimensions, dimension lines, and leaders are numerous, and we will cover only a few in this text. By right-clicking on any dimension and selecting Display Options or Properties, you will find that it is easy to change the appearance of the dimension.

The font for any dimension can be changed individually, but if you want to change the font type or size for all dimensions, you can do that from the Document Properties menu.

Select the Options Tool. Under Document Properties, select Dimensions, click the Font button, and select a font and font size. The font size may be input in inches or points. The figures in the remainder of this chapter were made with the default font (Century Gothic) at a size of 18 points. Click OK to apply the selected font.

FIGURE 2.30

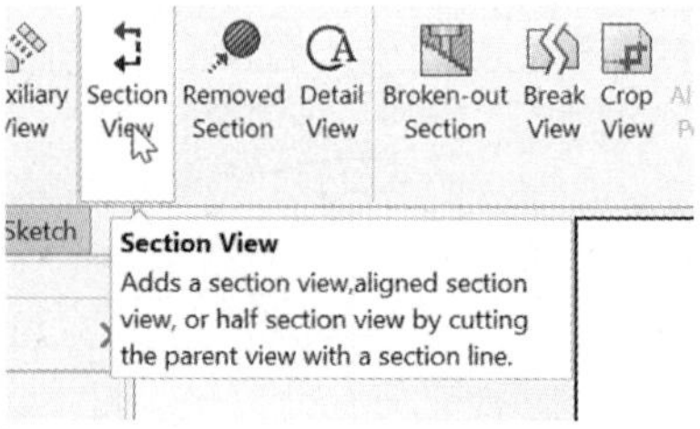

A section view will help show the center hole and the chamfer more clearly.

Click the View Layout tab on the CommandManager, and select the Section View Tool as shown in Figure 2.30.

In the PropertyManager, make sure that Section is selected (so that a full section view—one that passes completely through the part—will be created). Select the horizontal section option, as shown in Figure 2.31.

FIGURE 2.31

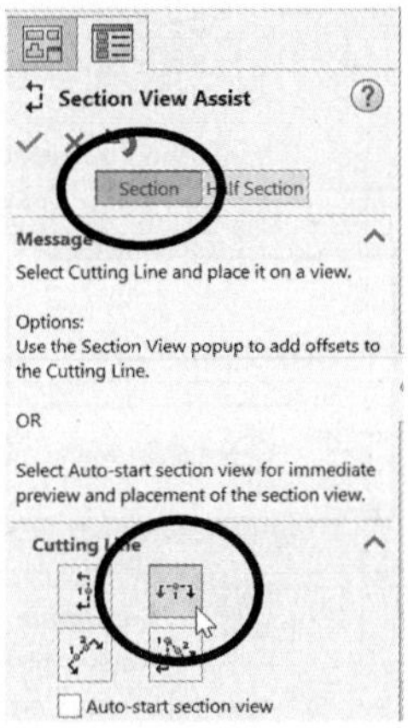

Move the cursor into the Front View. A preview of the section line will appear. Note the line will pass completely through the part. Move the cursor to the center mark of the view, which will cause the section to pass through the center of the part, as shown in Figure 2.32. Click to place the section line. A pop-up menu will appear, with options for creating an offset section line, as shown in Figure 2.33. Since we are creating a simple full section, just click the check mark.

FIGURE 2.32

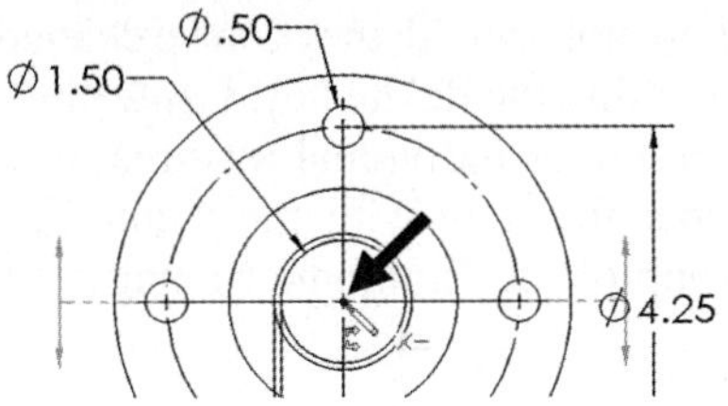

FIGURE 2.33

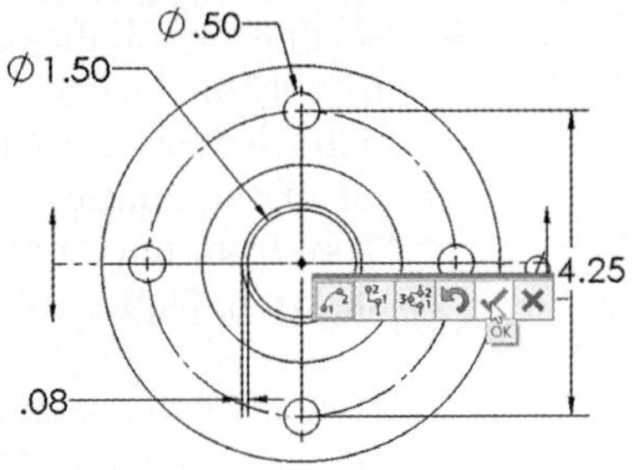

Move the cursor to place the section view. By default, the view will be kept in alignment with the Front View. Drag the section view away from the Front View, and click once to place the view, as shown in Figure 2.34. The section view arrows should point up, so that the section view is in the desired orientation. If the arrows point down, then click the Flip Direction button, as shown in Figure 2.34.

FIGURE 2.34

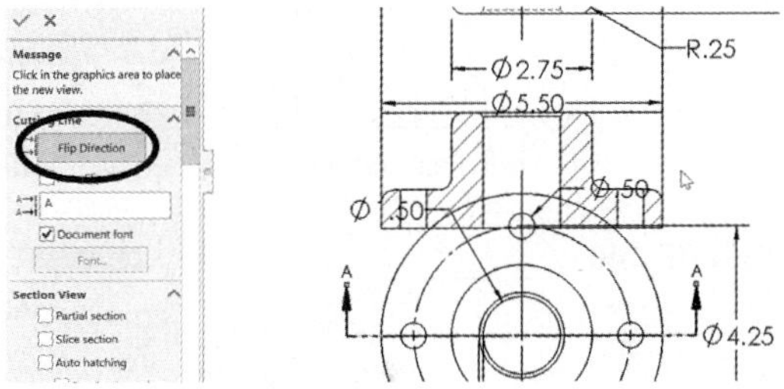

Right-click within the section view boundary box, and from the menu that appears, select Alignment: Break Alignment, as shown in Figure 2.35. Move the cursor over the edge of the section view boundary box so that the move arrows appear. Click and drag the view to a new location on the sheet, as shown in Figure 2.36.

FIGURE 2.35

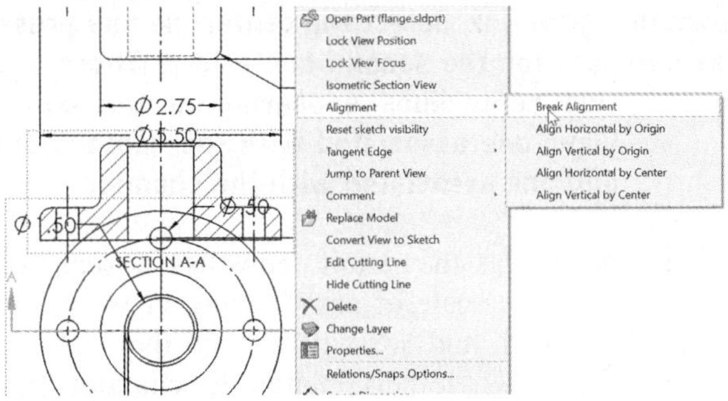

FIGURE 2.36

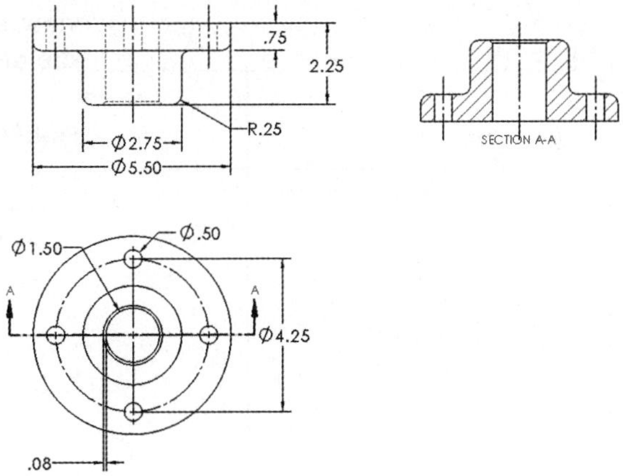

Note that the Drawing Heads-Up View Toolbar contains some of the same icons as the Part Heads-Up View Toolbar, and also includes a Zoom to Sheet Tool, as shown in **Figure 2.37**. When this tool is selected, the entire drawing sheet is displayed below the Heads-Up View Toolbar. You may add tools to the Heads-Up View Toolbar in the same manner as presented in Chapter 1. One tool that you may find useful to add is the Pan Tool, which allows you to view different regions of the drawing without changing the zoom level.

FIGURE 2.37

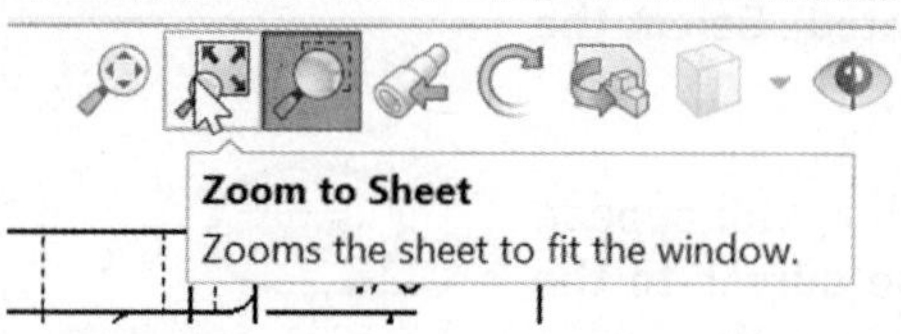

Since we specified centerlines to be added automatically when we set the options for the drawing template, they appear in the section view. If the centerlines are not wanted, they can be deleted by clicking on each one and pressing the delete key. Centerlines may also be added manually to either individual features or to all of the entities within a specific view with the Centerline Tool from the Annotation group of the CommandManager, as shown in **Figure 2.38**.

FIGURE 2.38

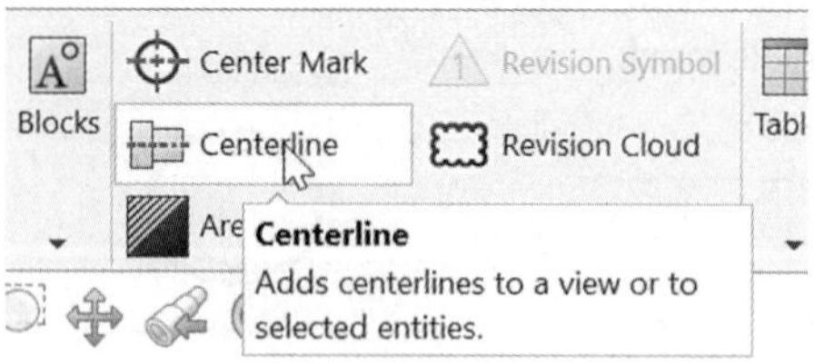

Note that the section view was added without hidden edges shown, even though displaying hidden edges was set as an option, and hidden edges are displayed in the Front and Top Views. Hidden edges are not typically shown in section views, but the display of hidden edges in any view can be changed by selecting the view and then choosing the display style from the Heads-Up View Toolbar, as shown in **Figure 2.39**. We will leave the hidden lines removed for our section view.

FIGURE 2.39

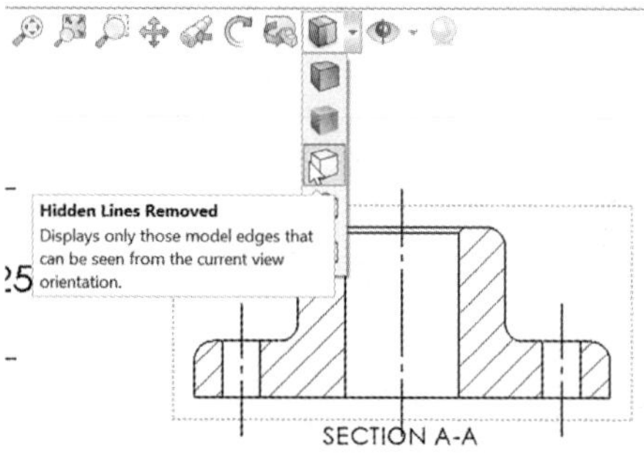

The chamfer is difficult to see at the default scale (1:2). A detail view that enlarges the chamfer region will be helpful.

FIGURE 2.40

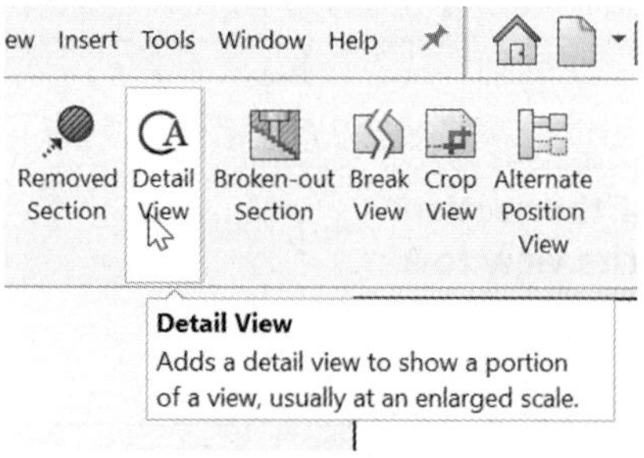

Select the Detail View Tool from the View Layout group of the CommandManager, as shown in Figure 2.40.

The Circle Tool will be activated automatically. Drag out a circle around the area to be included in the detail view, as shown in Figure 2.41. Move the cursor to the position where you want the detail view to appear, and click to place the view, as shown in Figure 2.42. Select the centerline and press the delete key. Repeat for the second centerline. (There are two centerlines appearing in the detail view: one associated with the center hole and one associated with the chamfer.)

FIGURE 2.41

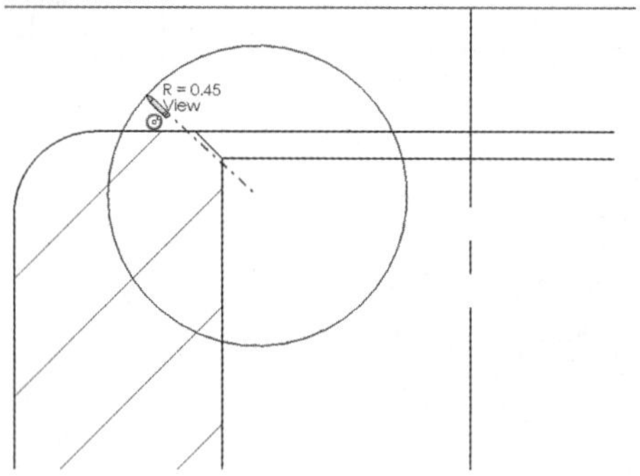

FIGURE 2.42

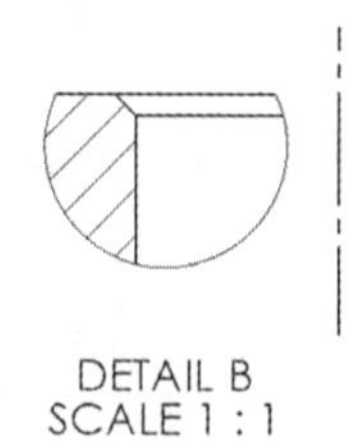

Note that the detail view has a scale of 1:1. The scale of any drawing view can be changed, and we will change the scale of the detail view to enlarge the chamfer area further.

With the detail view selected, scroll down in the PropertyManager to the Scale settings. Check the "Use custom scale" option and change the scale to 2:1, as shown in Figure 2.43. Click the check mark to apply the scale.

FIGURE 2.43

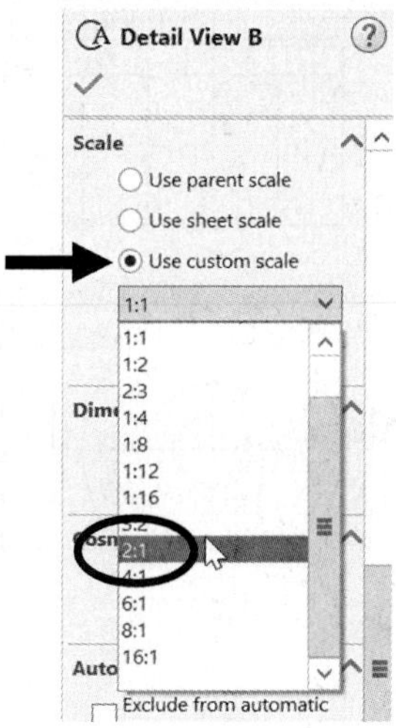

Dimensions can be imported into section and detail views. In this case, the 45-degree angle of the chamfer can be seen best in the detail view. Also, dimensions should not be applied to hidden features. Therefore, the 45-degree dimension was not imported along with the other dimensions to either the Front or Top Views.

Click in the white space of the detail view boundary box to select this view, and select the Model Items Tool from the Annotation group of the CommandManager, as shown in Figure 2.44. Click the check mark to import dimensions into the view. The 45-degree dimension will be added, as shown in Figure 2.45. Press Esc to de-select the detail view.

We would also like to show the 0.080-inch width of the chamfer on the detail view, but that dimension has already been imported into the Front View. We will move the dimension from the Front View to the detail view.

Press and hold the Shift key. Click and drag the 0.080-inch chamfer dimension from the Front View into the detail view boundary box. When you release the mouse button, the dimension will be moved into the detail view. Release the Shift key, and drag the dimension to its final position, as shown in Figure 2.46.

FIGURE 2.44

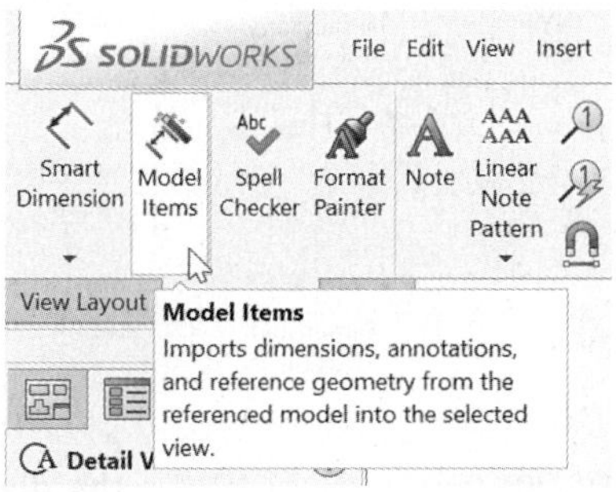

FIGURE 2.45

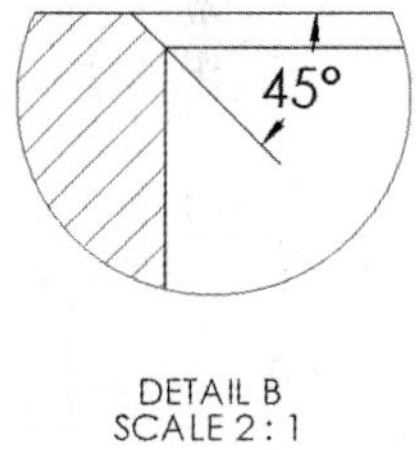

FIGURE 2.46

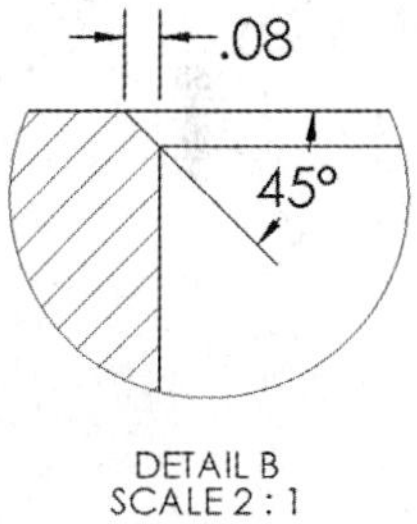

The labels (A's and B) on the section lines and the detail view circle, as well as the captions of the views, are shown with the default font size. These font sizes can be changed in the Document Options, or can be changed individually.

Click on the section line to select it, as shown in Figure 2.47. In the PropertyManager, clear the check box labeled "Document font," click the Font button, and increase the size of the font. When you click OK, you will be prompted to apply this change onto the Section View as well. Click Yes. Repeat for the detail view label and caption. Note that it may be necessary to rebuild the drawing for the increased font size to appear. Also, if you want to change the letter associated with the view, you can also do this from the PropertyManager. If you want to change the style of the section line, select Options: Document Properties: Views: Section and select a different line style and/or thickness.

FIGURE 2.47

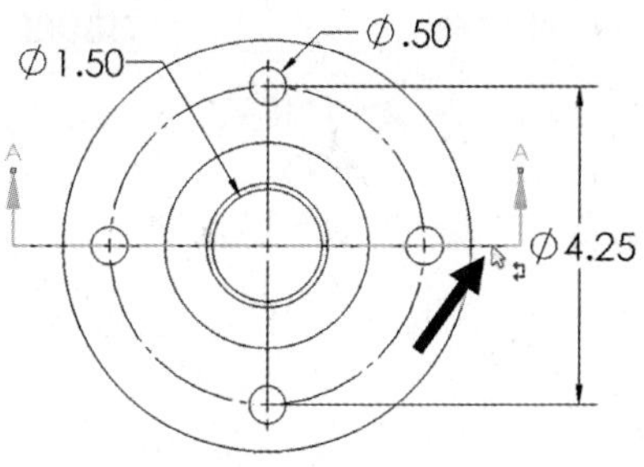

FIGURE 2.48

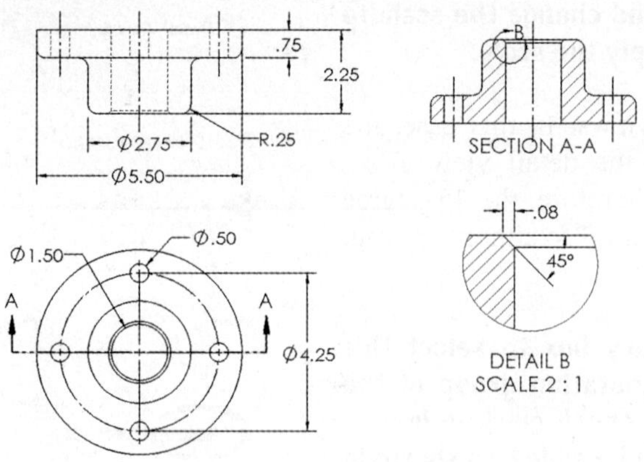

Your drawing should now appear as in **Figure 2.48**. This is a good time to save your work.

Select File: Save and save the drawing as "Flange Drawing," with a file type of .slddrw (SOLIDWORKS Drawing). Since the last Save operation stored the drawing template to the default directory, make sure to save the drawing file to your desired location. If prompted to save the part file as well, click OK.

As we noted earlier, the number of decimal places displayed in the drawing will relate to the tolerances applied to each dimension. Often, certain dimensions are more critical than others and will need to have tolerance values specified. For example, suppose that the center 1.5-inch diameter hole needs to have a tight tolerance to allow for a good fit with the part that will be inserted into the hole.

FIGURE 2.49

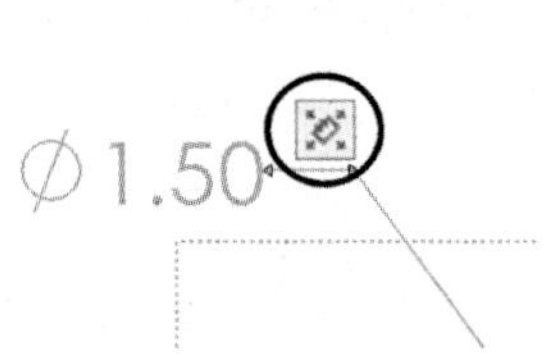

FIGURE 2.50

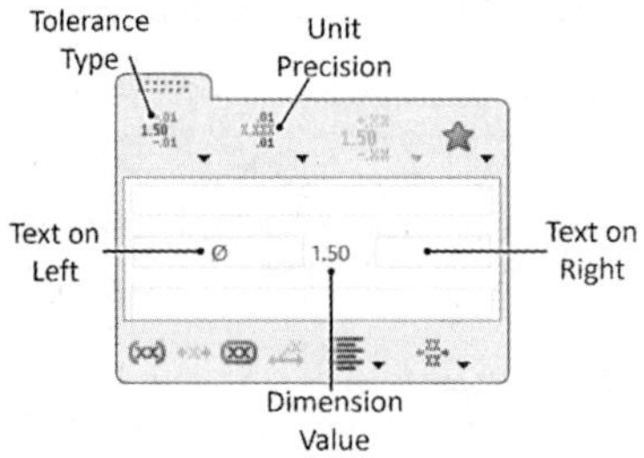

Click on the number portion of the 1.50-inch dimension. An icon representing the Dimension Palette will appear, as shown in Figure 2.49. Move the cursor over the icon, and the Dimension Palette will be expanded. The Dimension Palette contains many tools for modifying the appearance of the dimension, as shown in Figure 2.50. The dimension value can be modified by adding text to the left (such as the diameter symbols that appeared automatically with the diameter dimensions) or to the right (such as "Typ" to show the dimension is "typical" of the dimension of similar features). Also, the number of decimal places (unit precision) can be modified and tolerances can be added. In the Dimension Palette, use the pull-down Unit Precision menu to set the number of decimal places to three (.123), as shown in Figure 2.51. Set the Tolerance Type to Bilateral, as shown in Figure 2.52. Click on the numerical value of the plus tolerance, and enter a value of .003, as shown in Figure 2.53. Click outside of the Dimension Palette to close it.

FIGURE 2.51

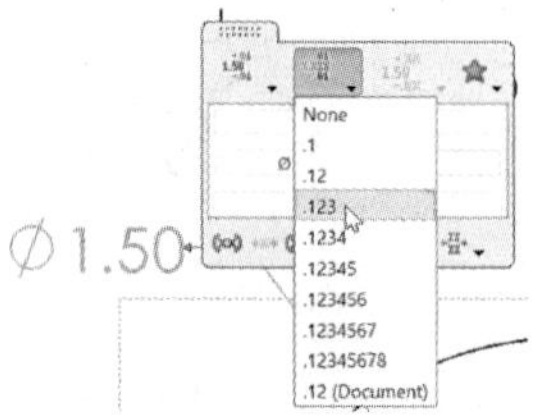

FIGURE 2.52

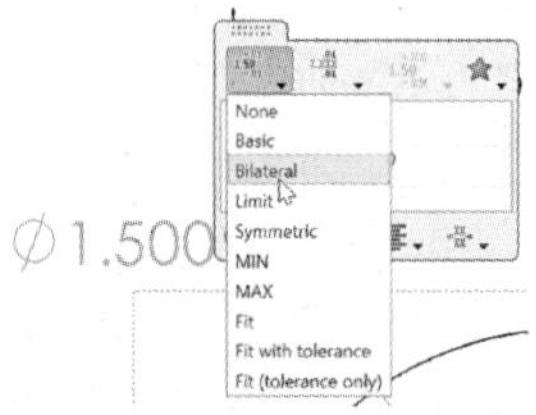

FIGURE 2.53

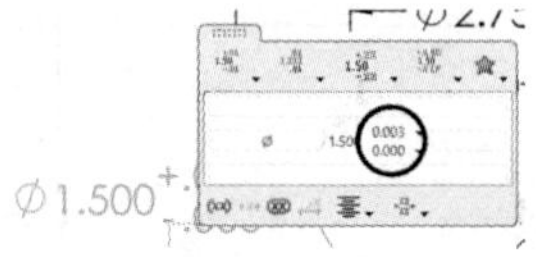

FIGURE 2.54

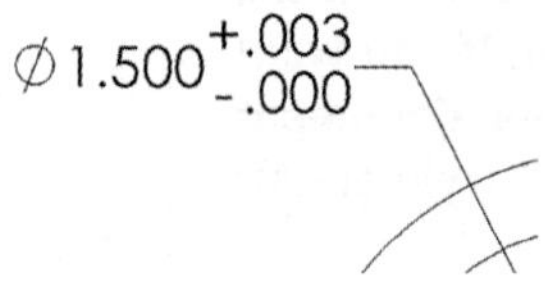

The dimension should now appear as in **Figure 2.54**, indicating that the diameter can be no smaller than the 1.500-inch nominal dimension, but can be up to 0.003 inches larger.

FUTURE STUDY

Manufacturing Processes, Geometric Dimensioning and Tolerancing, and Metrology

In this example, we placed a relatively tight tolerance on the diameter of the center hole so that a tight fit could be realized between the flange and the part that fits into the hole. How do you know what the tolerance should be? Perhaps a better question is: what tolerance is required for the part to function as designed in an assembly? The answer to this question will in many cases dictate the manufacturing process that can be used. Many engineering students study manufacturing processes, but it is impossible to learn the specific details of all of the manufacturing processes in use today. On the job, engineers should learn as much as possible about the manufacturing processes used by their company and its suppliers, in order to understand what tolerances are practical and economical.

Applying dimensions and tolerances to engineering drawings is a more complex topic than it first appears. For example, consider a simple part such as a pin that is required to fit into a hole. If we simply define the diameter of the pin and place tolerances on the diameter, are we sure that the pin will fit into the hole? What if the pin is bent slightly? Its diameter might be within the limits defined by the tolerance, but it may not work in its intended purpose. The process of Geometric Dimensioning and Tolerancing (GD&T) allows a designer to specify the acceptable condition of a part, considering its function. In the example we just discussed, a straightness tolerance might be required. In the drawing tutorial in this chapter, we added a flatness specification to a surface that was important for sealing. Proper application of GD&T standards can actually reduce the cost of making many parts, since they allow control of the important features of a part more efficiently than simply tightening the tolerance values of all dimensions. At many companies, the detailing of drawings is performed by a drafting department, and engineers document designs with less-formal drawings and sketches. At many smaller companies, however, engineers detail their own drawings. For these engineers, study of GD&T standards and practices is necessary.

A related area of study is metrology, the science of measurements. Many features are difficult to measure accurately with traditional methods, and computer-controlled Coordinate Measuring Machines (CMMs) are now widely used in industry. Using statistical methods with measured data, quality assurance engineers track variations and work to control the processes that are used to produce the components. This method of Statistical Process Control (SPC) allows problems to be detected and corrected before defective parts are produced.

In addition to dimensions with tolerances, *geometric tolerances* are often added to drawings in order to fully define the acceptable limits of a part's geometry. For example, suppose that the back surface of the flange is to mate to another part with a gasket between the two parts to create a seal. If the back of the flange is warped, it may not allow for a proper seal, even if all dimensional tolerances are met. To ensure a good surface for sealing, we might need to add a flatness specification.

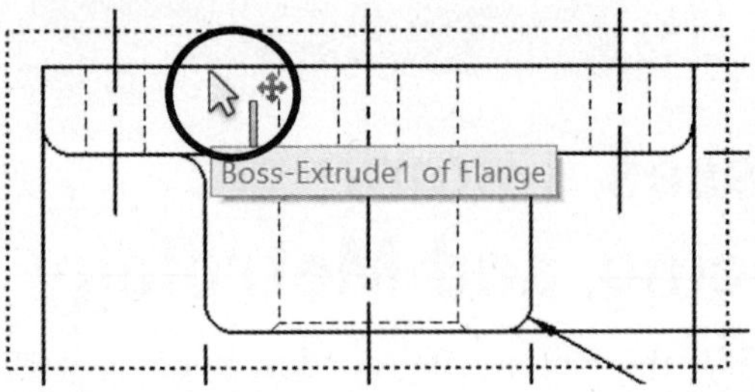

FIGURE 2.55

In the Top View, click on the line corresponding to the back surface of the flange, as shown in Figure 2.55. Select the Geometric Tolerance Tool from the Annotation group of the CommandManager, as shown in Figure 2.56.

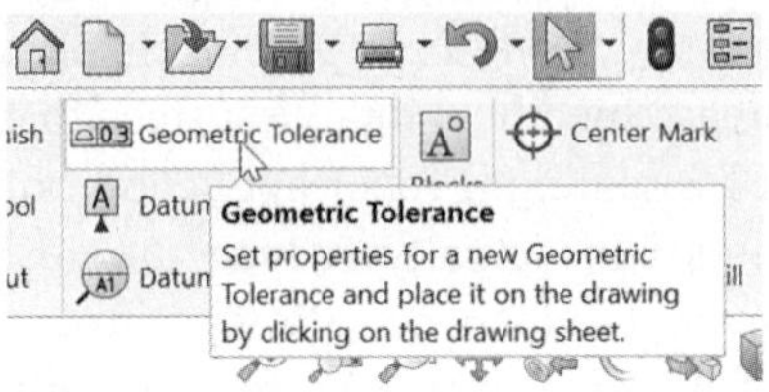

FIGURE 2.56

In the dialog box, select the Flatness symbol from the pull-down symbol menu, as shown in Figure 2.57. Enter 0.010 as the value for Tolerance 1, and a preview of the tolerance call-out will appear, as shown in Figure 2.58. Click to place the annotation in the desired location. Click OK to apply the tolerance, and drag the call-out box to the desired location, as shown in Figure 2.59. If desired, change the font size in the PropertyManager.

Since we are not using a title block with the unit system and default tolerances shown, we will specify them in a note.

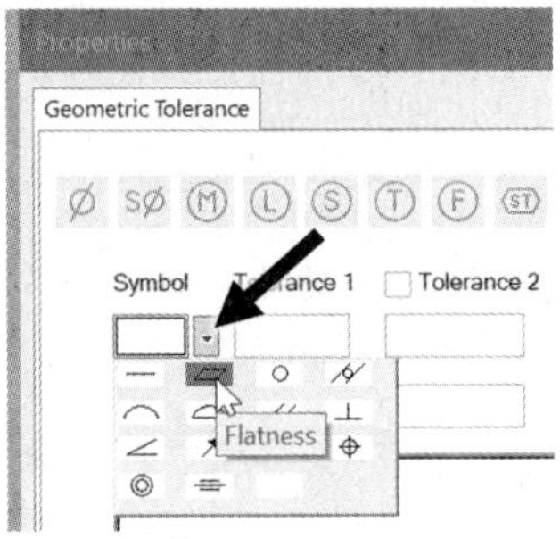

FIGURE 2.57

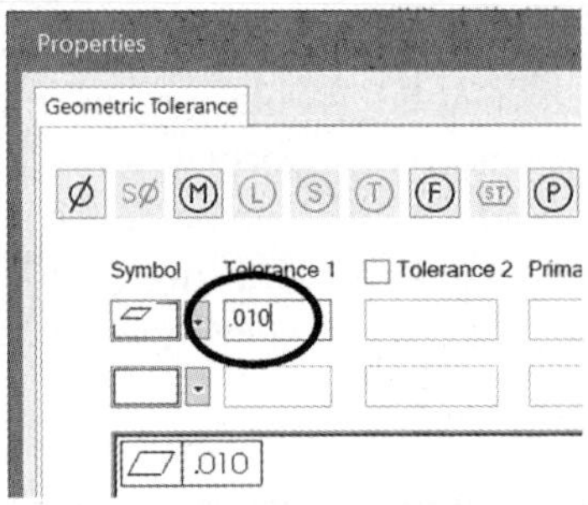

FIGURE 2.58

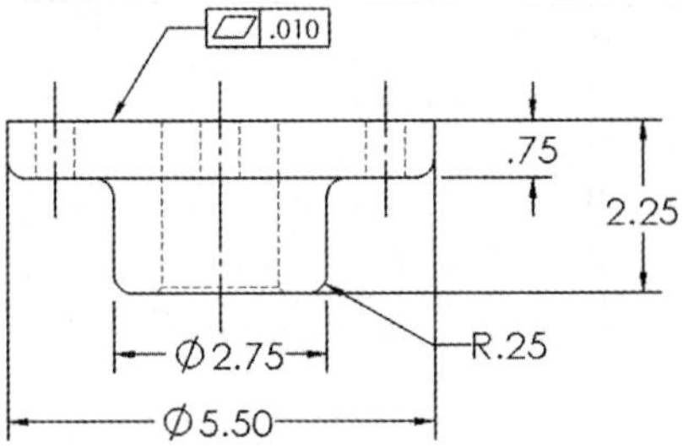

FIGURE 2.59

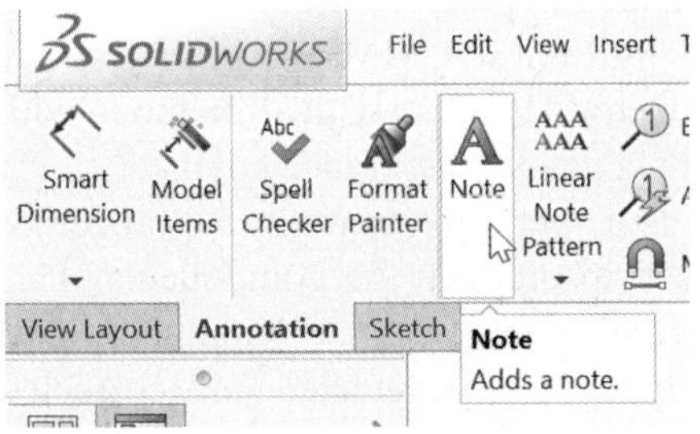

FIGURE 2.60

Select the Note Tool from the Annotation group of the CommandManager, as shown in Figure 2.60. Drag the cursor to the approximate location where the note will be placed (near the bottom of the drawing), and click. Do not click directly on an item in the drawing, or the note will appear with a leader to that item. Choose a font type and size from the toolbar that appears. In the text box that appears at the location where you clicked, begin typing the text of the notes shown in Figure 2.61.

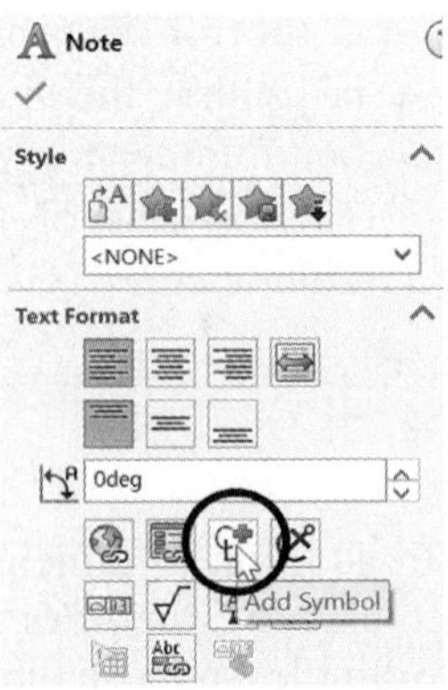

FIGURE 2.62

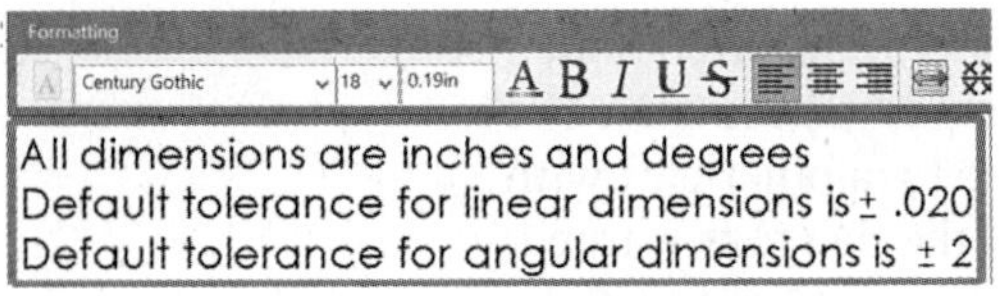

FIGURE 2.61

Use the Enter key to start a new line. The ± symbol is inserted by clicking on the Add Symbol Tool in the PropertyManager (Figure 2.62) and selecting Plus/Minus from the list of available

symbols, as shown in Figure 2.63. Click outside of the text box to place the note. Press the Esc key to turn off the Note Tool; otherwise, subsequent mouse clicks will place duplicate notes on the drawing.

If desired, you can change the font type and size by double-clicking the note and changing the font parameters on the toolbar that appears.

Since we defined all of the fillets with a single fillet command in the part, only one of the fillets is dimensioned in the drawing. We might prefer to call out the fillet radius in a note, yet maintain associativity with the part.

Double-click on the note. Move the cursor to the end of the last line and press the Enter key to move to a new line. Type in "All fillets and rounds are," and click on the .25-inch dimension. The dimension's value will appear in the text box. Click outside of the text box to end editing of the note. The completed note is shown in Figure 2.64.

If you change the value of the radius, then the value in the note will change accordingly. Now that the fillet radius is shown in a note, displaying it on the drawing is unnecessary. However, deleting the dimension will produce an error, since the note links to the value of the dimension. Therefore, we will hide the dimension.

Right-click on the R.25 dimension and select Hide from the menu, as shown in Figure 2.65.

If you want to show the dimension later, you can do so by selecting View: Hide/Show: Annotations from the main menu. Any hidden dimensions will show up light, as shown in **Figure 2.66**, and clicking on them will change them back to visible dimensions. Similarly, clicking on visible dimensions will hide them. This is a good method for hiding several dimensions at once. When finished, press the Esc key to return to the normal editing mode.

A pictorial view such as a trimetric view can often be helpful in interpreting a 2-D drawing.

Select the Model View Tool from the View Layout group of the CommandManager. The part can be selected by double-clicking the part name in the PropertyManager. In the PropertyManager, select the Trimetric View as the orientation of the new model view, and the Hidden Lines Removed (wireframe) mode as the display style, as shown in Figure 2.67. Move the cursor to the desired location of the new view, and click to place the view.

FIGURE 2.63

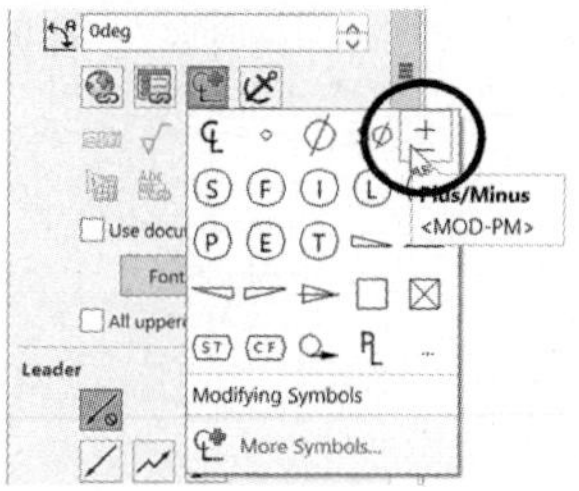

FIGURE 2.64

All dimensions are inches and degrees
Default tolerance for linear dimensions is ± .020
Default tolerance for angular dimensions is ± 2
All fillets and rounds are R.25

FIGURE 2.65

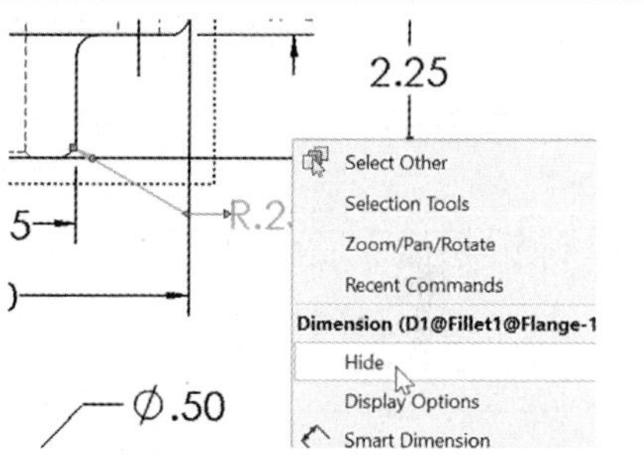

FIGURE 2.66

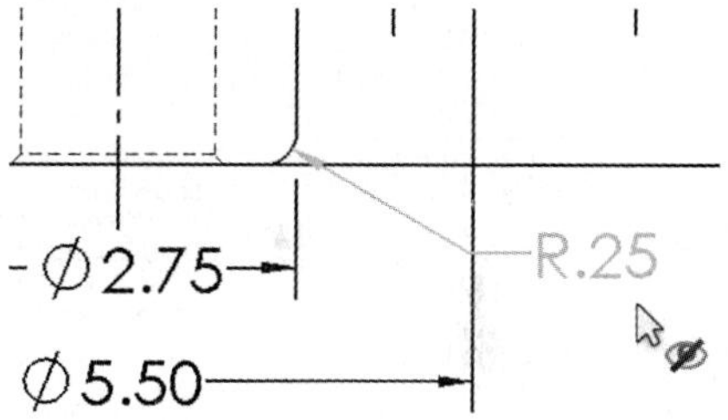

FIGURE 2.67

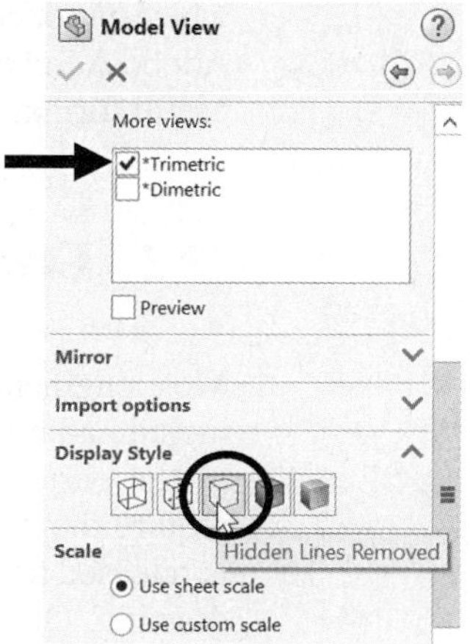

FIGURE 2.68

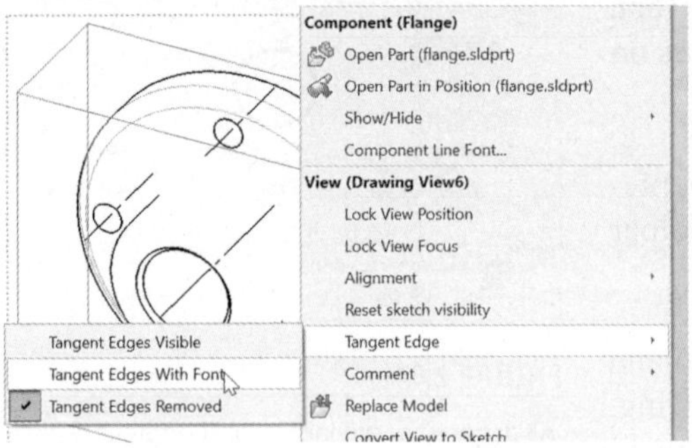

While not displaying the tangent edges for the other views results in cleaner and less cluttered views, the Trimetric View will look much better with the tangent edges displayed.

Right-click on the Trimetric View, and select Tangent Edge: Tangent Edges with Font, as shown in Figure 2.68.

Move the drawing views and notes around the sheet so that they can all be seen clearly. Add a note identifying the scale as 1:2. Select the .50-inch dimension of the bolt hole and in the Dimension text box of the PropertyManager, add "4X" before the diameter symbol, indicating that the dimension applies to four holes. Delete the centerlines from the Trimetric View. To see how the drawing will look when printed, select Print Preview from the File menu.

FIGURE 2.69

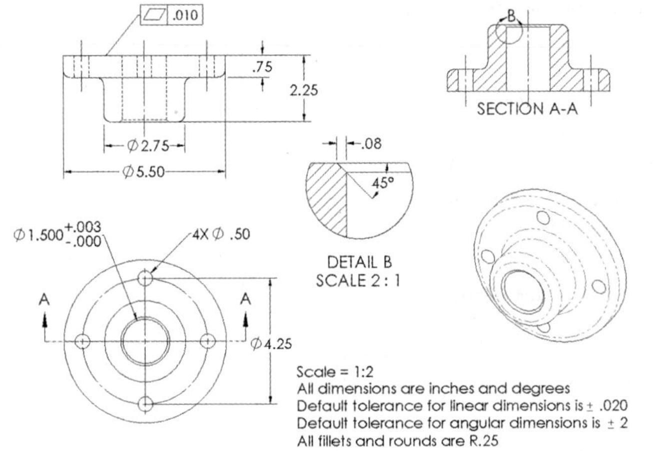

Your drawing should appear similar to Figure 2.69.

Save the drawing file.

Open the Flange part file and experiment with changing dimensions, and noticing the updating of the drawing. Similarly, change some drawing dimensions and observe the changes in the part. Close the part and drawing files without saving any of the changes.

A SOLIDWORKS drawing can be saved in several other formats in addition to the .SLDDRW file type. These include .dxf, a file standard developed by Autodesk for exchange with AutoCAD and other 2-D CAD programs, and .jpg and .tif image files. A file type that is especially convenient for sharing is the Adobe .pdf format, which produces a high-quality image that can be viewed by anyone with the free Adobe Acrobat Reader. Later in the chapter, we will create an eDrawing, which is a useful format for collaboration within groups working on a common project.

2.2 Creating a Drawing Sheet Format

Most engineering drawings include a title block and a border. The title block can include a large amount of information. In addition to basic data, such as the name of the part, the part number, the scale, the date created, etc., many companies require the names or initials of reviewers who must approve the drawing before it is released to users. Some require a revision history that tracks changes made to the drawing. Others require the listing of information about where the part described on the drawing is used in assemblies.

The SOLIDWORKS program includes several default title blocks and borders, called *sheet formats*. These sheet formats can be edited to fit the particular needs of a company. The sheet formats are different for every standard paper size. In this section, we will create a simple sheet format for an A-size drawing. Our format will include some basic information without taking up much room on the sheet.

Open a new drawing. Choose A-Landscape as the paper size and check the "Display sheet format" box. If the border and title block do not appear when you click OK, right-click on the drawing sheet, select Properties, check the Display Sheet Format box again, choose Reload, and then Apply Changes.

Check the X in the PropertyManager to close the Model View command.

Note that you can prevent the Model View command from starting automatically by clearing the check mark from the box labeled "Start command when creating new drawing" in the PropertyManager. However, since we will usually be creating a drawing from an existing part, we will allow the command to continue to start automatically.

Right-click in the drawing space. Select Edit Sheet Format, as shown in Figure 2.70.

FIGURE 2.70

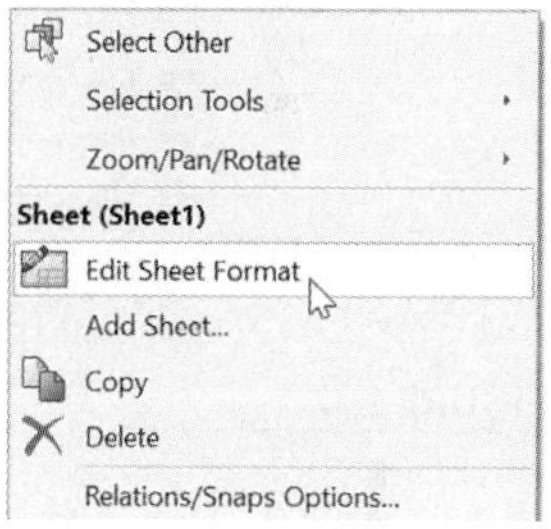

In a drawing, you can toggle between editing the drawing itself and editing the sheet format, which contains the title block and border.

From the Main Menu, choose Edit: Select All, and then press the Delete key.

Note that the border, which includes numbers and letters used to identify zones within a drawing, is still present, as shown in **Figure 2.71**. On larger drawings, the zone identifiers are helpful in locating items on the sheet, but for our small drawing sheet, they are not necessary.

In the PropertyManager, expand the Sheet1 and Sheet Format entries. Right-click the border (Border1) and select Delete Border, as shown in Figure 2.72. Click Yes to confirm that you want to delete the border.

We now have a blank sheet to create our custom border and title block.

FIGURE 2.71

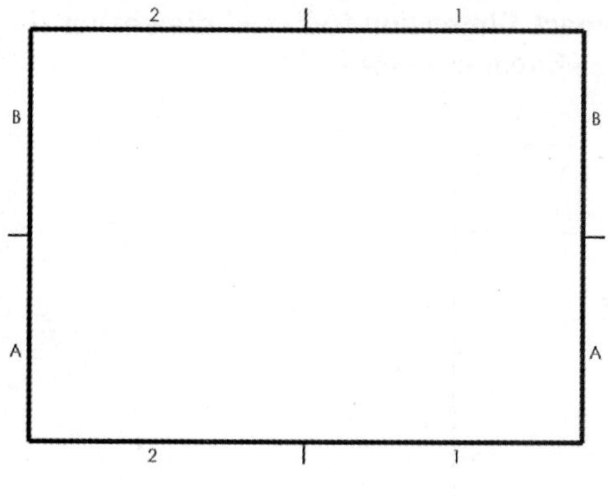

FIGURE 2.72

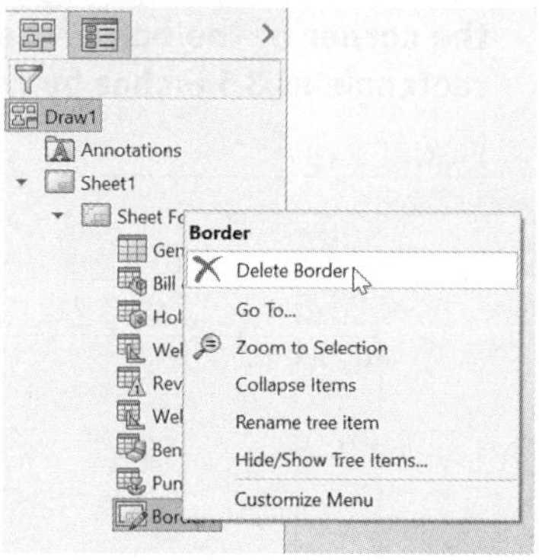

FIGURE 2.73

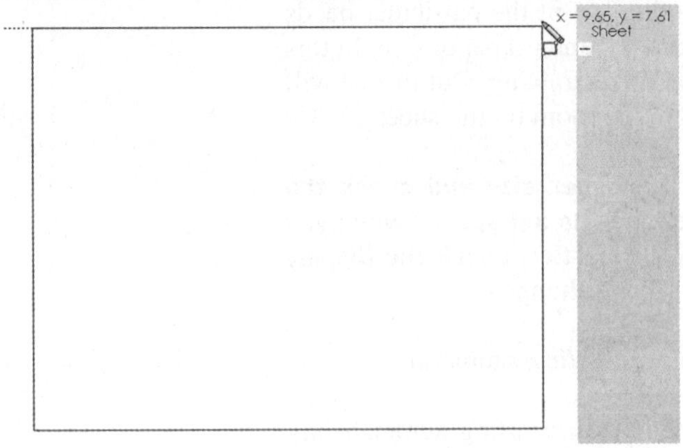

Select the Corner Rectangle Tool from the Sketch group of the CommandManager. Drag out a rectangle that fills most of the page, as shown in Figure 2.73.

Right-click on the menu bar or the CommandManager, and select Line Format from the list of available toolbars, as shown in Figure 2.74. Locate the Line Format Toolbar (by default, it will appear at the bottom left of the screen), and select Line Thickness, as shown in Figure 2.75. With the rectangle still selected, choose a heavier line weight, as shown in Figure 2.76. Close the Line Format Toolbar by right-clicking on it and selecting its name again.

FIGURE 2.74

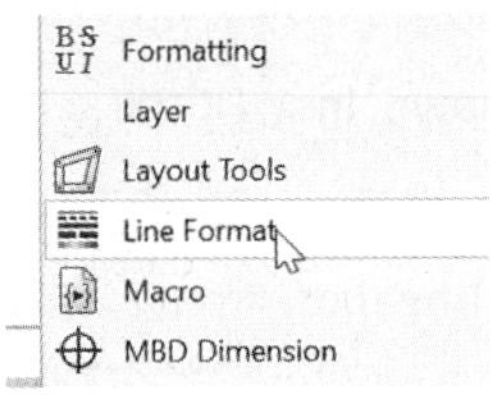

FIGURE 2.75

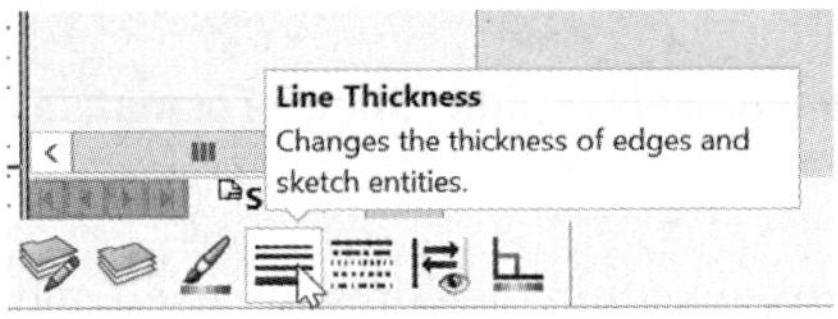

FIGURE 2.76

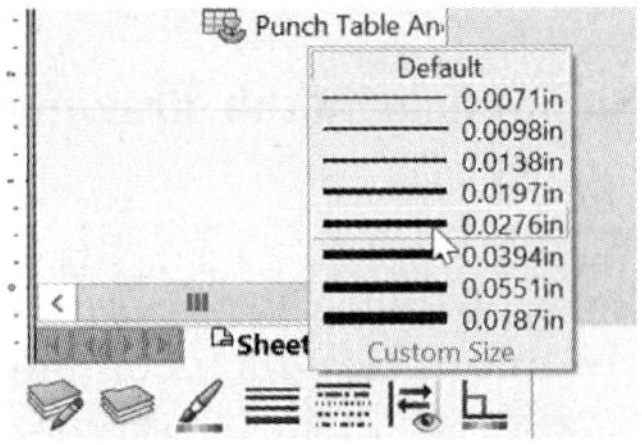

FIGURE 2.77

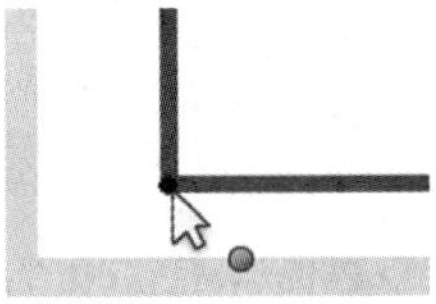

If you choose a new line format with entities selected, as we just did, the new format applies only to the selected items. If you select a line format with no entities selected, then the new line format becomes the default for the entities added subsequent to the selection.

Click on the lower left corner of the rectangle to select it, as shown in Figure 2.77. In the PropertyManager, input its x and y coordinates as 0.375 inches each, as shown in Figure 2.78. Click on the Fix icon.

FIGURE 2.78

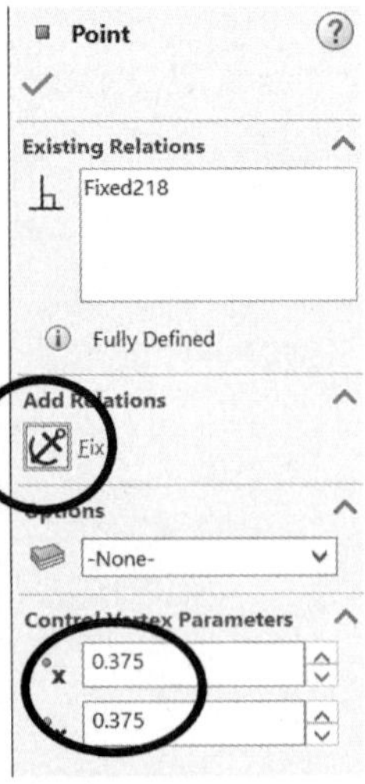

Select the upper right corner of the rectangle. Set its coordinates to x = 10.625 inches and y = 8.125 inches. Fix this point.

Zoom in to the lower right portion of the sheet. Drag out a rectangle from the corner of the border. Select the Smart Dimension tool, and dimension the rectangle as 3.5 inches by 1.4 inches, as shown in Figure 2.79.

FIGURE 2.79

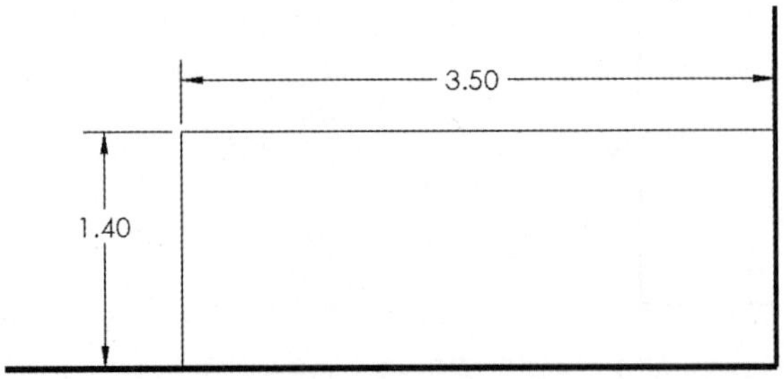

Select the Line Tool, and add the horizontal and vertical lines shown in Figure 2.80. Be sure that the endpoints snap to lines and not specific points (such as midpoints). Note that for clarity, sketch relations are not shown here. Make sure that all lines are horizontal or vertical by viewing each line's properties in the PropertyManager.

FIGURE 2.80

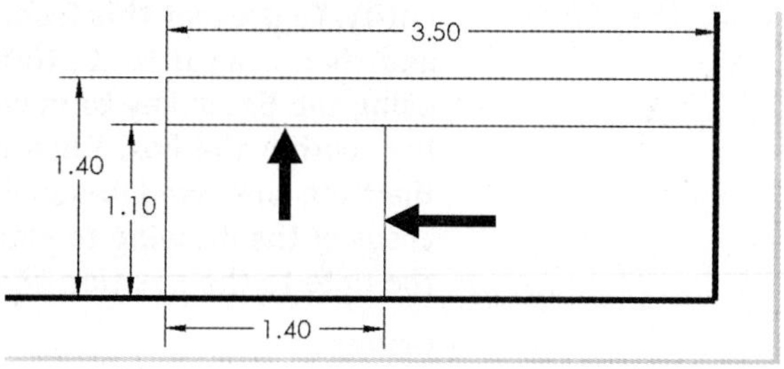

Select the Smart Dimension Tool, and add the 1.1-inch and 1.4-inch dimensions shown.

Select the Line Tool, and add the two horizontal lines shown in Figure 2.81. Use the Smart Dimensions Tool to add the .5-inch and .75-inch dimensions.

FIGURE 2.81

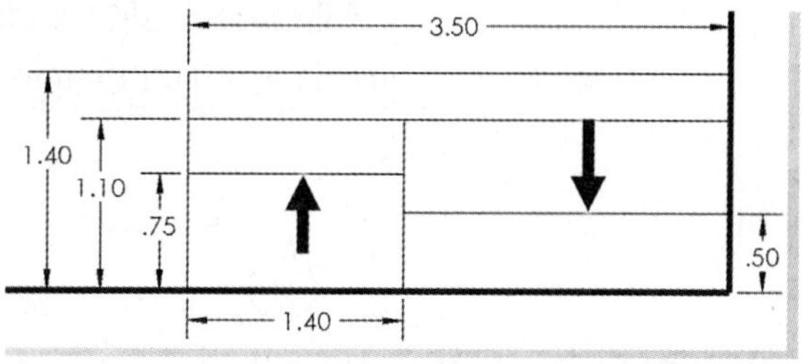

Add the other two lines shown in Figure 2.82, and add the .75-inch and .25-inch dimensions. The drawing should now be fully defined.

We will now hide the dimensions just added.

FIGURE 2.82

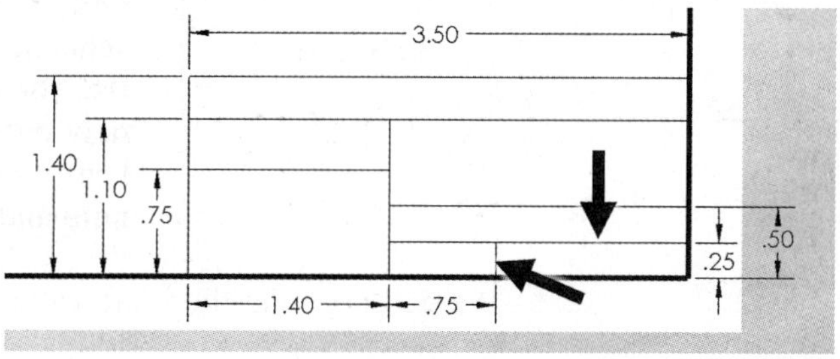

From the main menu, select View: Hide/Show: Annotations, as shown in Figure 2.83. Click on each dimension, and it will turn gray, as shown in Figure 2.84. When all dimensions are selected, press the Esc key to hide them.

Select the Note Tool from the Annotation group of the CommandManager. Click in any blank area to place the note, and set the font as desired and the size to 10 point. (Recall

FIGURE 2.83

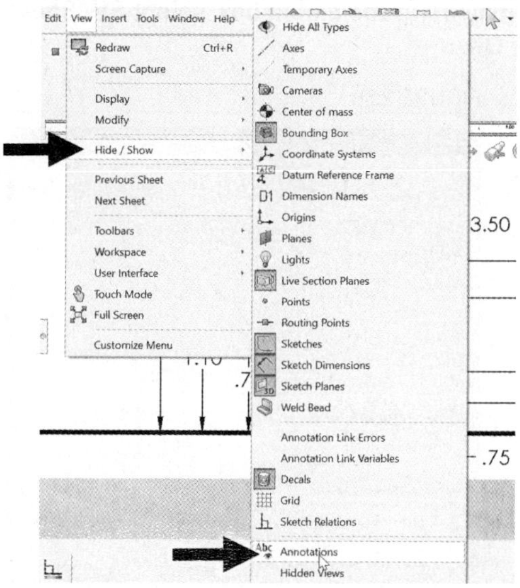

FIGURE 2.84

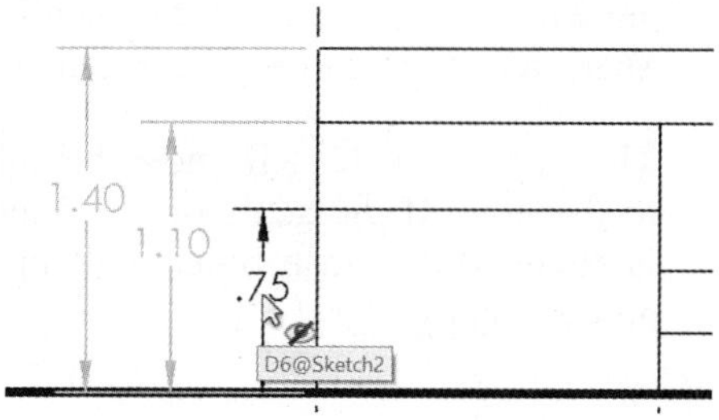

that if you click on a sketch entity, the note will have a leader pointing to that entity. To prevent this from happening, create the note well away from any entity and then drag it to its final position.) Type in the text as shown in Figure 2.85, using the Enter key to move to a new line. Choose the Center tool to center the text within the box. Click outside of the text box, and press the Esc key to end the Note command (if you do not press the Esc key, then you can click in different areas of the drawing to place the same note in multiple locations). Click and drag the note to the position shown in Figure 2.86.

FIGURE 2.85

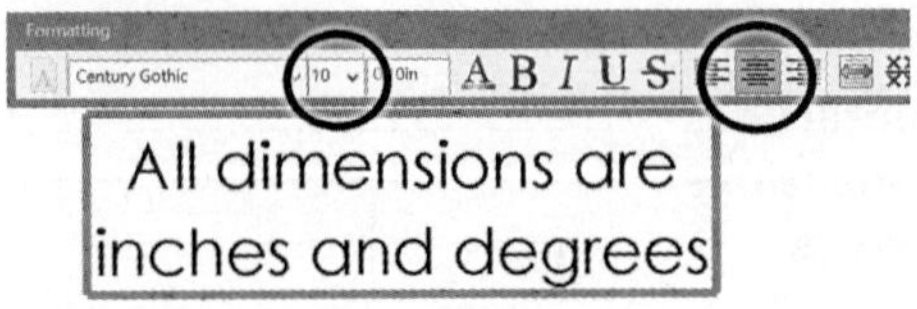

FIGURE 2.86

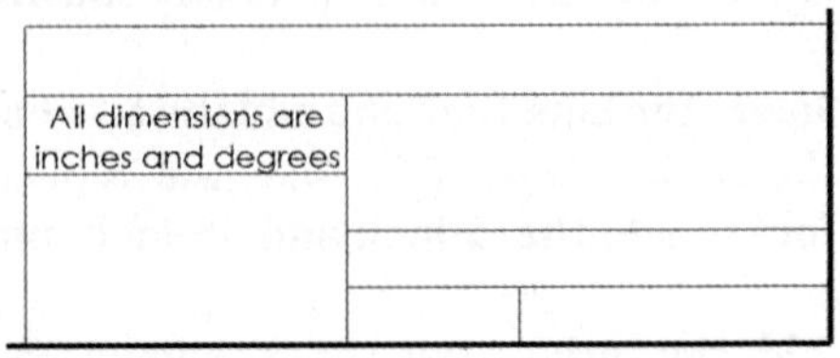

FIGURE 2.87

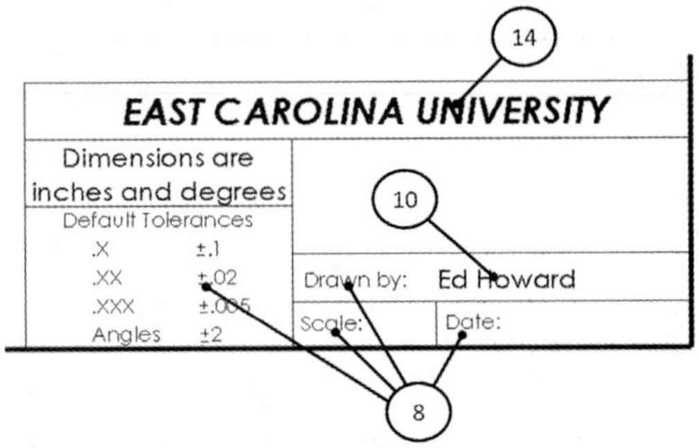

Note that it is sometimes difficult to move a text entity to the position desired because of auto-aligning features. To have more precise control when locating a note, press and hold the Alt key while moving the text. This temporarily suspends the auto-aligning.

Add the rest of the text shown in Figure 2.87, adding your school/company name and your name to the title block. Use the font sizes shown in Figure 2.87. When opening a new note, select the font size before typing in the text. Font sizes can be easily changed by double-clicking each note and selecting a new font size.

We will now add the scale value to the drawing, linking it to the drawing sheet's scale.

Open a new note. In the PropertyManager, select the Link to Property icon, as shown in Figure 2.88. From the pull-down menu in the dialog box, select SW-Sheet Scale, as shown in Figure 2.89, and click OK.

FIGURE 2.88

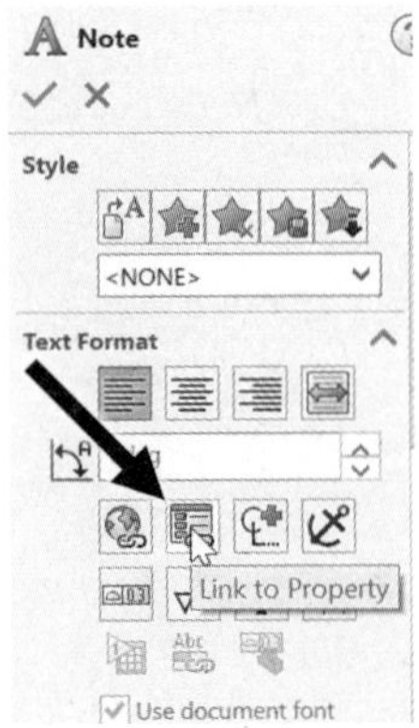

The sheet's scale will now be added to the note. If the scale of the drawing is changed, the title block will update automatically.

FIGURE 2.89

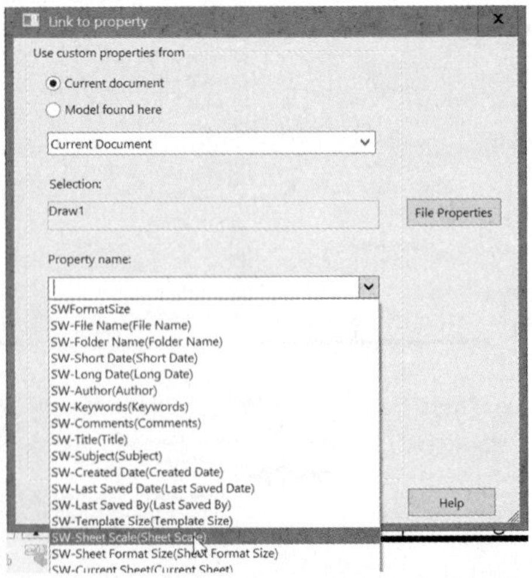

Click outside the text box, and press the Esc key to end the Note command. Move the scale note into the correct location in the title block, as shown in Figure 2.90. Right-click in the drawing area and select Edit Sheet, as shown in Figure 2.91.

FIGURE 2.90

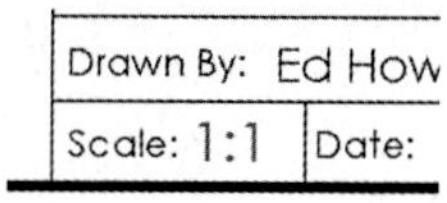

The title block and border are now in the background, and cannot be edited without toggling back to Edit Sheet Format.

From the main menu, select File: Save Sheet Format, as shown in Figure 2.92. Give the sheet a unique name and save it to the default directory, as shown in Figure 2.93. Close the drawing. Select Don't Save when asked if you want to save the drawing file.

FIGURE 2.91

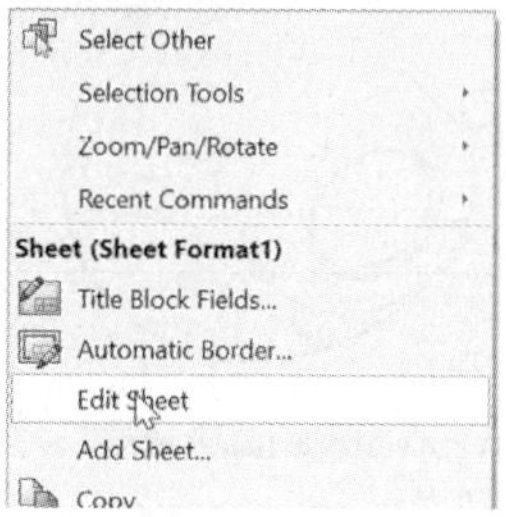

FIGURE 2.92

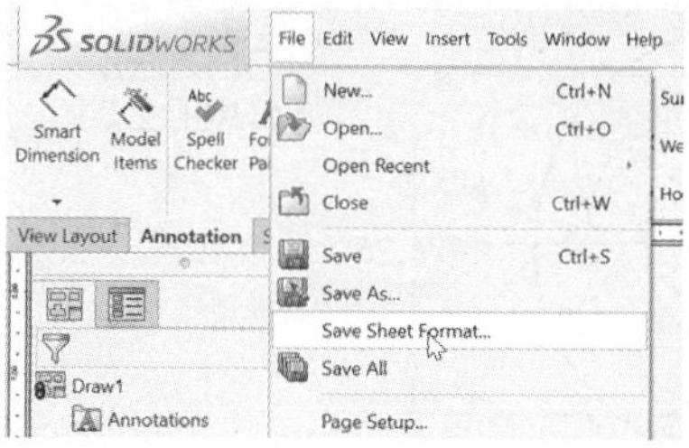

FIGURE 2.93

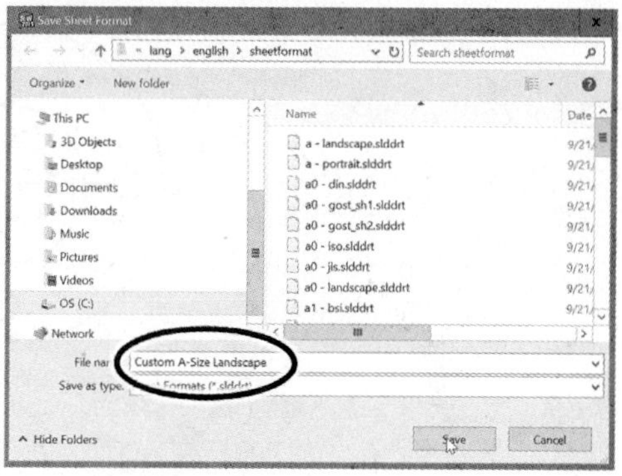

The drawing sheet format just created can be selected when beginning a new drawing, and can be applied to existing drawings. We will now apply the format to the flange drawing.

Open the flange drawing created earlier. Right-click in the drawing space (away from any drawing view) and select Properties. (You may need to click on the arrow at the bottom of the menu to have Properties appear. If so, you can make Properties always appear by selecting Customize Menu at the bottom of the menu and checking the box beside Properties.) In the Sheet Properties box, check Standard sheet size and uncheck the box labeled "Only show standard format." Select the sheet format just created from the pull-down list, as shown in Figure 2.94. Check the box labeled "Display sheet format" and click Apply Changes.

FIGURE 2.94

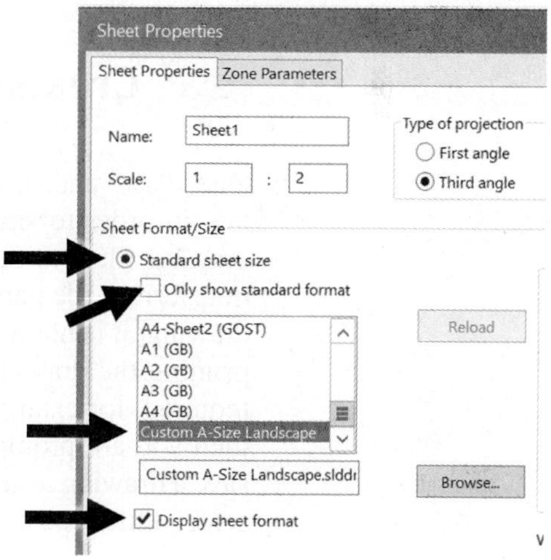

FIGURE 2.95

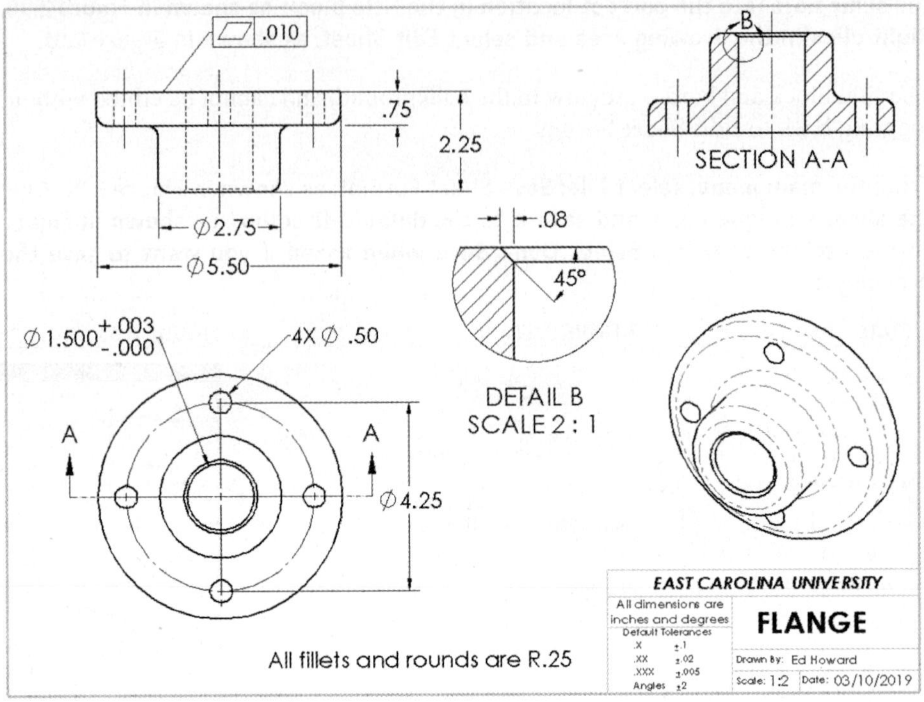

Move the drawing views as necessary. Delete the notes defining the scale, units, and tolerances, as these values are now covered in the title block. (Note that the scale in the title block has automatically changed to 1:2.) Add notes showing the drawing title and the date. The finished drawing is shown in Figure 2.95. Save the drawing for use in the eDrawings tutorial that follows.

2.3 Creating an eDrawing

One disadvantage of using drawings that are fully associative with part files is that in order to share the file with other users, they must have access to both the drawing file and the associated part file. One option for sharing drawings without transferring the part file is to create a *detached drawing*. When you open a detached drawing, it is not necessary to have the model file loaded. In addition to viewing and printing the drawing, annotations can be added to the drawing. When the model is required, for example if you attempt to edit a dimension imported from the model, then you are prompted to load the model from which the drawing was created. To save a drawing as a detached drawing, choose File: Save As from the main menu and select a file type as "Detached Drawing."

Another option for sharing drawings with others, including those who do not have access to SOLIDWORKS software, is to create an eDrawing. The eDrawings Viewer can be downloaded for free, and the eDrawings Publisher also allows the creation of executable files that can be read on any computer. Even within organizations using SOLIDWORKS software, eDrawings can be used to share information. Markup tools allow users to make comments and corrections, and the small file sizes of eDrawings make exchange of data via e-mail more practical.

To save a SOLIDWORKS drawing as an eDrawing, simply select File: Save As from the main menu and select eDrawing (.edrw) as the file type or publish the eDrawing directly from the SOLIDWORKS environment, as we will demonstrate in the following instructions. The eDrawing file format is very efficient, and file sizes produced are small.

The eDrawings file is smaller than most image files (.bmp, .tif, etc.) that could be used to electronically communicate a drawing, but as we will see, an eDrawing is more than a 2-D image. The viewer can rotate any of the drawing views in 3-D space.

Open the flange drawing file that you saved in the previous section. Select File: Publish to eDrawings from the main menu, as shown in Figure 2.96. A dialog box will appear, from which you can select options such as whether or not to allow users to measure directly from the eDrawing, include the shaded data (which allows for the 3-D animations that we will demonstrate), or set a password that is required to open the file. You can also select specific sheets to include from a multi-sheet drawing. All of these options give the originator control over how much data to share with other users. Click OK to accept the default options. An eDrawing is created and the eDrawings Publisher is opened, as shown in Figure 2.97.

FIGURE 2.96

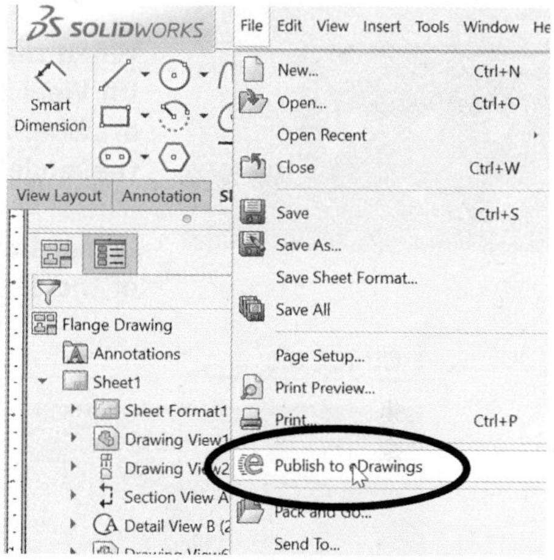

Note: While the eDrawings program can be accessed easily from the File menu, it is a stand-alone program that can also be opened from the Start menu of your computer or from a desktop icon.

If the shaded data was included in the options, then by default, the eDrawing views are displayed in the shaded mode. The display mode can easily be changed to show line drawings.

FIGURE 2.97

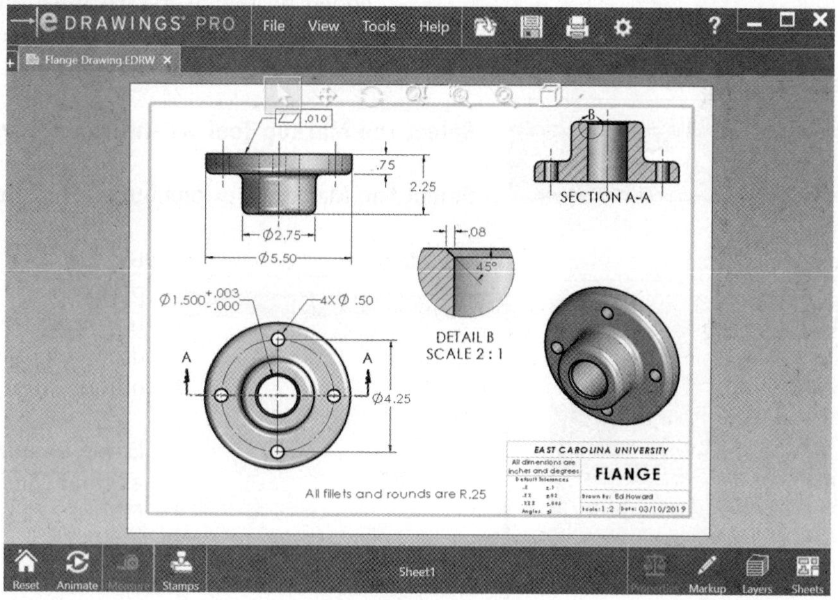

FIGURE 2.98

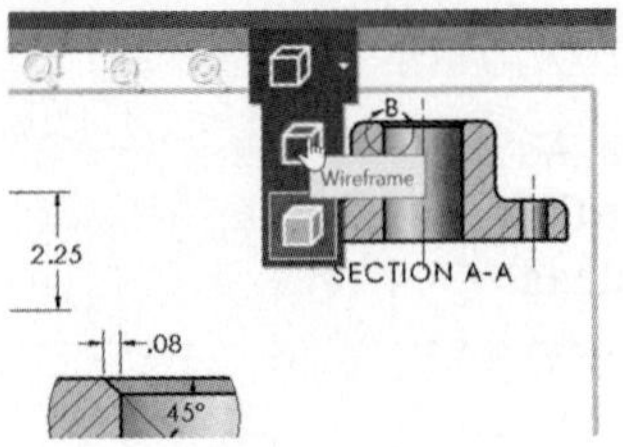

From the Heads-Up View Toolbar, select the Wireframe mode from the Display Style options, as shown in Figure 2.98.

The eDrawings Viewer allows drawings to be animated. The Play Tool automatically switches from one view to another. The Next Tool allows you to control the animation steps.

Choose the Animate icon and then click on the right arrow to change the view from one to another, including the section and detail views, as shown in Figure 2.99.

FIGURE 2.99

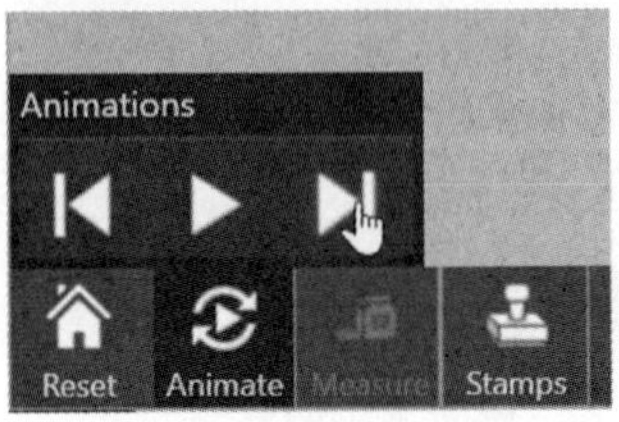

When the Section View is active, select the Rotate Tool from the Heads-Up View Toolbar, as shown in Figure 2.100. (You may need to click on the drawing view in order to activate the Rotate Tool.) Click and drag to rotate the model view, as shown in Figure 2.101.

Select the Reset Tool, as shown in Figure 2.102 to return to an overview of the drawing.

FIGURE 2.100

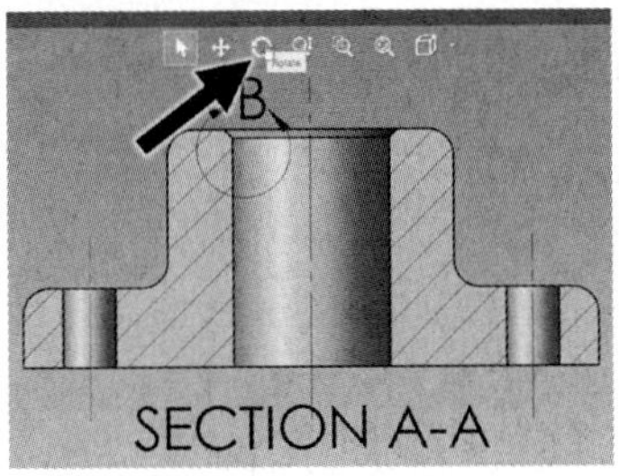

FIGURE 2.101

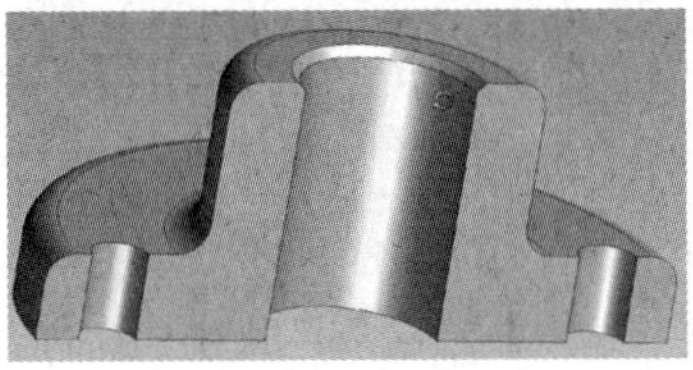

FIGURE 2.102

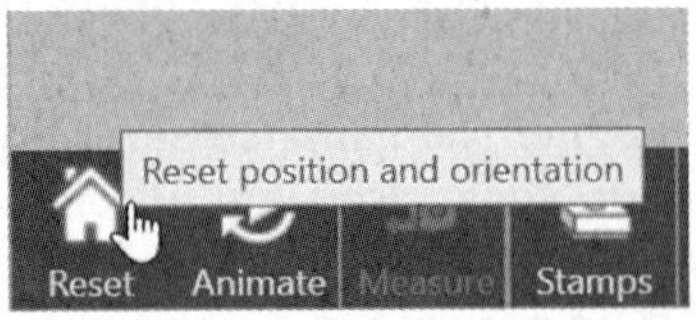

An important feature of eDrawings is that comments can be added by reviewers. To illustrate, we will add a comment for draft (taper to allow the part to be easily removed from a mold) to be added to the outer surface of the 2.75-inch-diameter boss.

FIGURE 2.103

Select the Markup Tool, as shown in Figure 2.103.

Select the Markup Options button, as shown in Figure 2.104.

FIGURE 2.104

FIGURE 2.105

In the box that appears, you can set preferences for your comments. Several reviewers can add comments to the same drawing, and each reviewer can have a unique color.

Enter a name for the reviewer, and select a color and font type/size, as shown in Figure 2.105. Click OK.

From the markup tools, select New Comment, as shown in Figure 2.106. Then select Labels and Text with leader, as shown in Figure 2.107. Click on the edge to be drafted in the Top View, as shown in Figure 2.108 (shown in wireframe mode for clarity). Click the location where the notation will be placed, and in the pop-up box that appears, type in the text of the comment, as shown in Figure 2.109. Click the check mark to close the box. The comment as it appears on the drawing is shown in Figure 2.110.

FIGURE 2.106

Comments are listed in the Markup box, as shown in **Figure 2.111**, allowing the person making the corrections to the drawing to see who made each comment. Clicking on each comment shows the date and time that the comment was made and allows you to add a description or to delete the comment. In this way, eDrawings allow efficient collaboration among different groups working on a project.

Files can be e-mailed directly from the eDrawings Publisher.

FIGURE 2.107

FIGURE 2.108

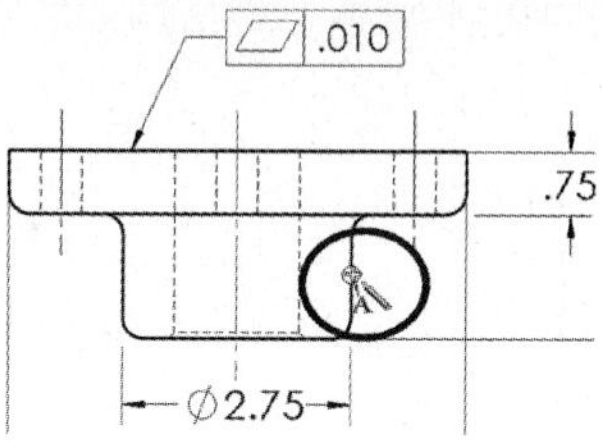

FIGURE 2.109

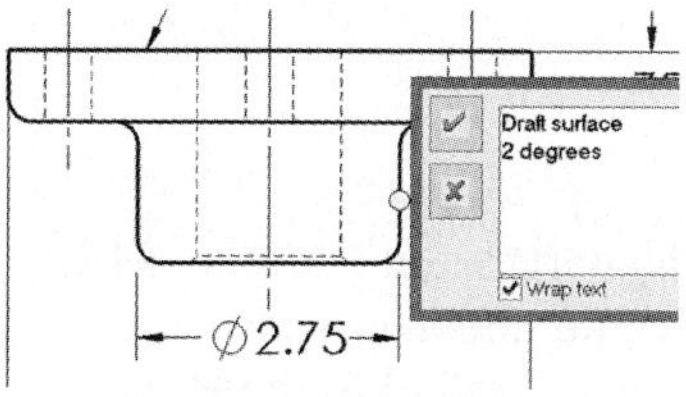

FIGURE 2.110

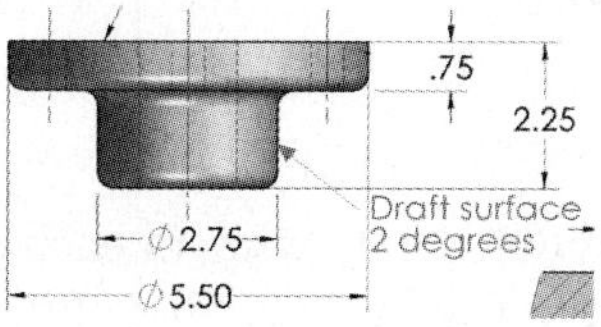

FIGURE 2.111

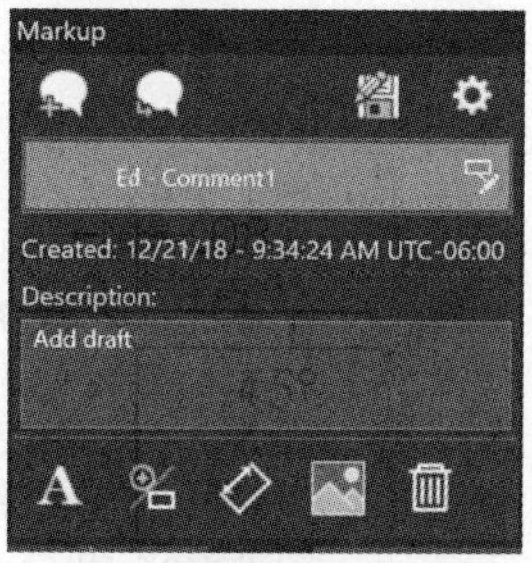

Select the Send Tool from the File menu, as shown in Figure 2.112.

FIGURE 2.112

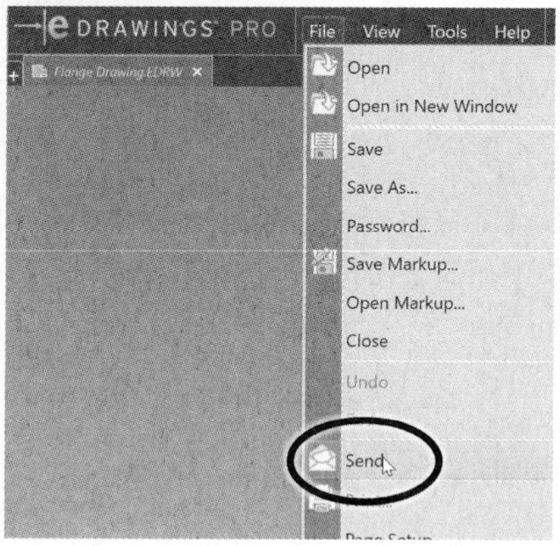

Two options are available, as shown in **Figure 2.113**. If the file is sent as an eDrawings file, then the recipient must have eDrawings or the free eDrawings viewer installed. If the drawing is sent as an HTML file, then the eDrawings viewer is installed automatically when the recipient opens the file.

Select Cancel to close the Send box. Select File: Save from the main menu, save the eDrawing as an .edrw file, and close the eDrawings Publisher.

FIGURE 2.113

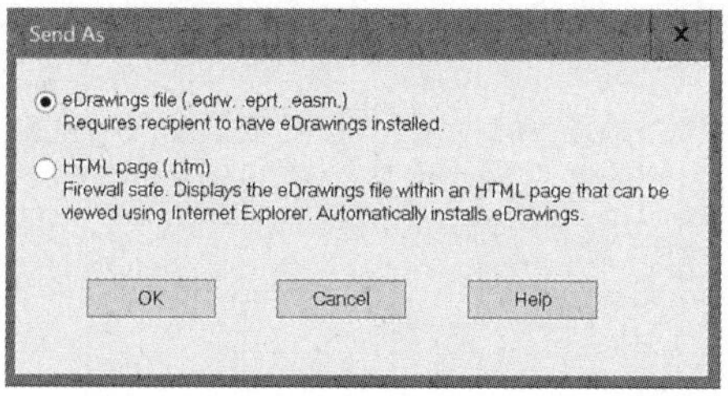

PROBLEMS

P2.1 Create a solid model of the part shown in **Figure P2.1**. The dimensions shown are inches. Create a detailed 2-D drawing, with Front, Top, and Right views.

FIGURE P2.1

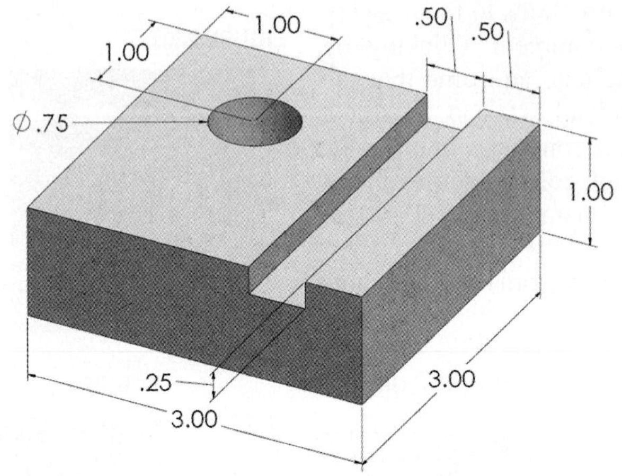

P2.2 Create a SOLIDWORKS drawing of each of the parts from **Problem P1.2**, which are shown in **Figure P2.2**. Show Front, Right, and Top Views on each drawing. For part D, also show a section view parallel to the Right View and through the center of the hole. Let one grid space equal one inch. The cut on the bottom of part C and the hole in part D both go completely through the part.

FIGURE P2.2

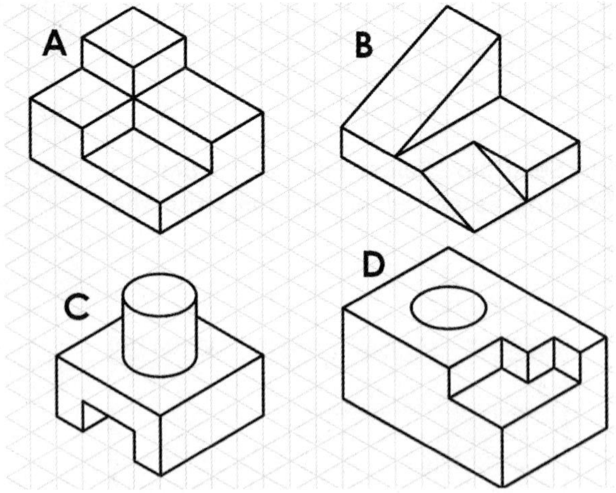

P2.3 Create a SOLIDWORKS drawing of the pulley created in Chapter 1. Include Front and Right Views in your drawing, as well as a section view. Show a detailed view of the keyway region.

FIGURE P2.3

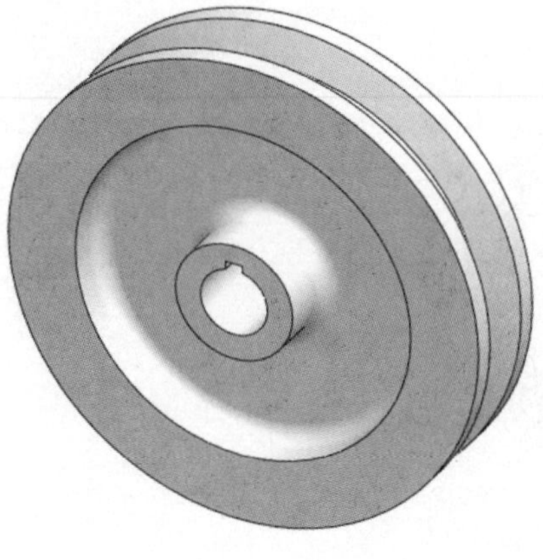

P2.4 Create an eDrawing of the pulley from **Problem P2.3**.

P2.5 Create a SOLIDWORKS drawing of the plastic pipe tee described in **Problem P1.5**. Include Front and Right Views in your drawing, as well as a section view.

FIGURE P2.5

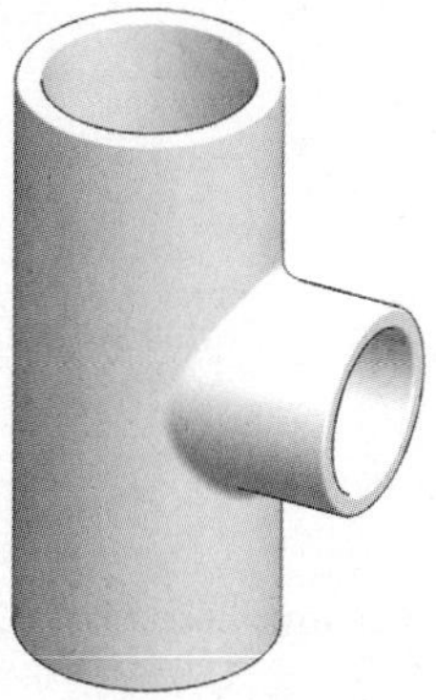

P2.6 Create an eDrawing of the pipe tee from **Problem P2.5**.

P2.7 Create a SOLIDWORKS drawing of the brake rotor described in **Problem P1.6**. Include Front and Right Views in your drawing.

FIGURE P2.7

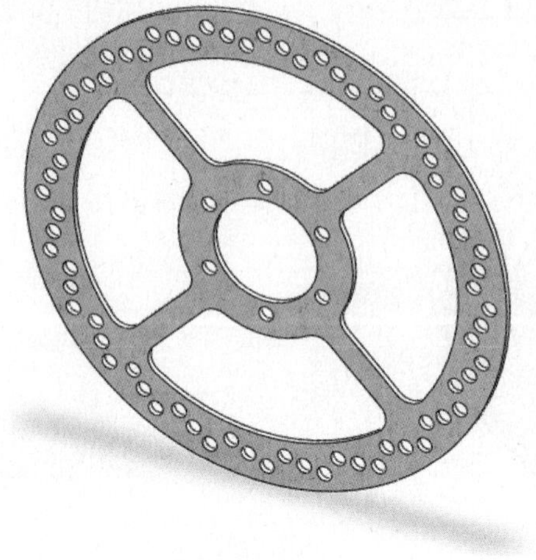

P2.8 Create an eDrawing of the brake rotor from **Problem P2.7**.

P2.9 Create a SOLIDWORKS drawing of the connecting rod described in **Problem P1.10**.

FIGURE P2.9

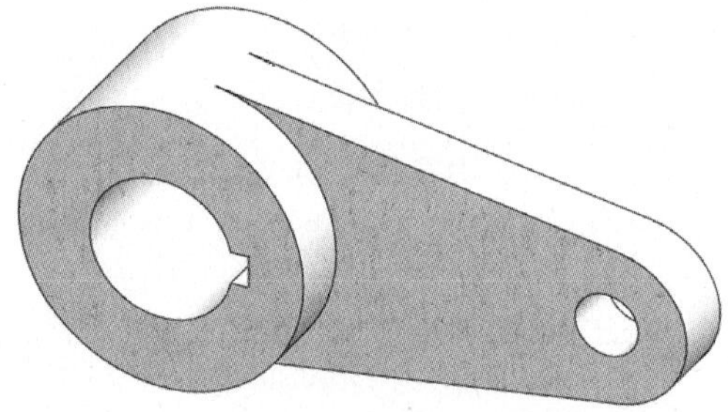

P2.10 Create an eDrawing of the connecting rod from **Problem P2.9**.

P2.11 **Problem P1.11** describes a required modification to the connecting rod of **Problem P1.10**. Using the eDrawing Markup Tool, add a markup to the eDrawing of **Problem P2.10** indicating to the designer that the weight reduction modification is necessary and recommending the area for weight removal.

CHAPTER 3

Additional Part Modeling Techniques

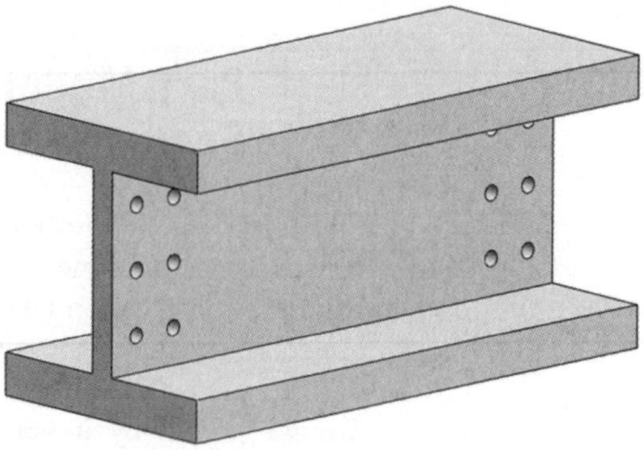

Introduction

The flange and pulley parts modeled in Chapter 1 required the use of a number of basic solid modeling construction techniques. In this chapter, some additional construction and dimensioning techniques will be explored. The parts that will be modeled in this chapter are a wide-flange beam segment, shown in **Figure 3.1**, and a wall-mounted bracket, shown in **Figure 3.2**.

FIGURE 3.1

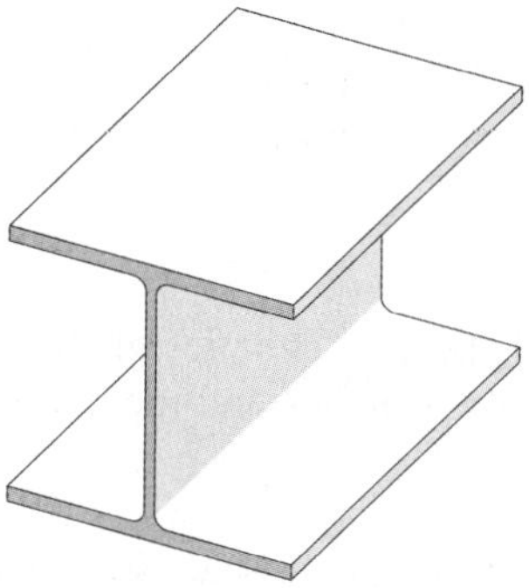

FIGURE 3.2

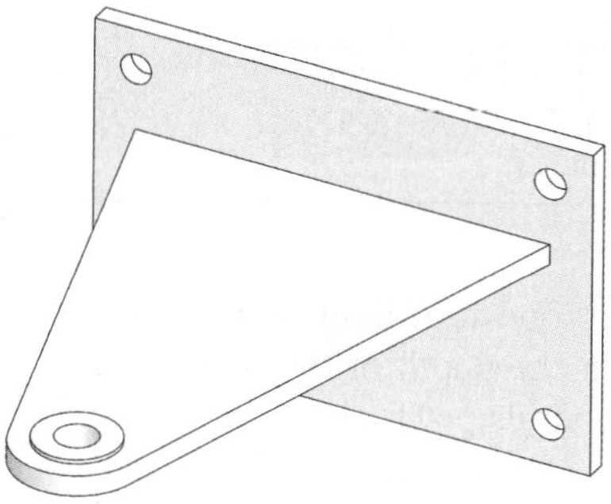

For each part, we will take advantage of the part's symmetry. A plane of symmetry exists if the half of the part on one side of the plane is a mirror image of the other half of the part. When a part contains a plane of symmetry, aligning that plane with one of the pre-defined SOLIDWORKS® planes (Front, Top, and Right) is good practice and makes many modeling and assembly operations easier.

Chapter Objectives

In this chapter, you will:

- use symmetry in part modeling,
- learn to create different configurations of a part,
- learn to use the Rib Tool,
- create a linear hole pattern,
- use the eDrawings® software to share a part model, and
- learn how to create an animation of a part.

SOLIDWORKS is a registered trademark of Dassault Systémes SolidWorks Corporation.
eDrawings is a trademark of Dassault Systémes SolidWorks Corporation.

3.1 Part Modeling Tutorial: Wide-Flange Beam Section

Structural steel members are made in standard shapes that are defined by the American Institute of Steel Construction (AISC). Wide-flange beams are commonly used members in steel construction. A wide-flange beam consists of two horizontal *flanges* joined by a vertical *web*. As shown in **Figure 3.3**, four dimensions are used to define the shape of the cross-section: the overall depth d, the flange width b, the thickness of the flanges t_f, and the thickness of the web t_w. (The edges between the web and flanges are filleted, but the fillet radius is allowed to vary somewhat and is not considered to be a standard dimension.)

FIGURE 3.3

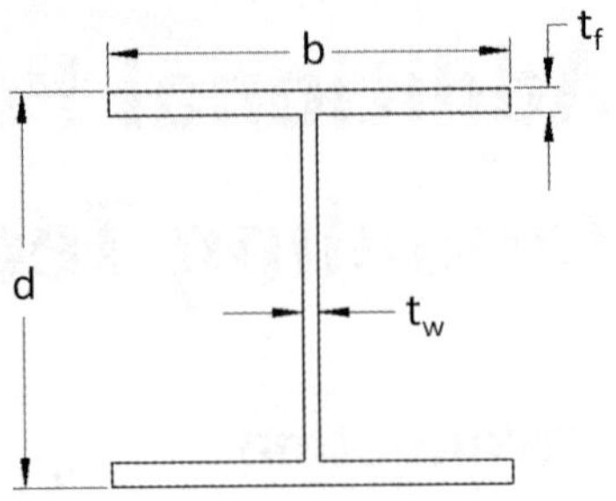

The dimensions of three wide-flange beam shapes are listed in **Table 3.1**.

TABLE 3.1 DIMENSIONS OF SOME WIDE-FLANGE BEAM SHAPES

Designation	Depth d, in	Flange Width b, in	Flange Thickness t_f, in	Web Thickness t_w, in	Area, in^2
W12 × 65	12.12	12.000	0.605	0.390	19.1
W10 × 45	10.10	8.020	0.620	0.350	13.3
W8 × 31	8.00	7.995	0.435	0.285	9.13

We will model a one-foot-long section of a W8×31 beam. The designation W8×31 means that the depth of the beam is approximately 8 inches and the beam weighs about 31 pounds per foot.

FIGURE 3.4

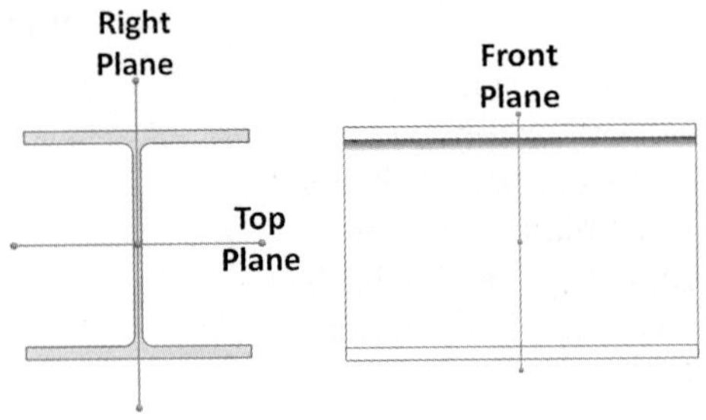

As we begin to model the beam section, we note that it is symmetric about three planes. We will create our model so that the three planes of symmetry coincide with the Front, Top, and Right planes of the model, as shown in the Front and Right Views of **Figure 3.4**.

Open a new part. In the FeatureManager, click on the Front Plane to select it as the initial sketch plane. From the Sketch group of the CommandManager, select the Corner Rectangle Tool, as shown in Figure 3.5. Sketch two rectangles, one above the origin and one below, as shown in Figure 3.6.

Select the Line Tool, and draw two vertical lines connecting the two rectangles, as shown in Figure 3.7. Make sure to snap the endpoints to the edges of the rectangles and not to any particular points (corners or midpoints). Select the Centerline Tool, as shown in Figure 3.8, and draw vertical and horizontal centerlines from the origin, as shown in Figure 3.9.

FIGURE 3.5

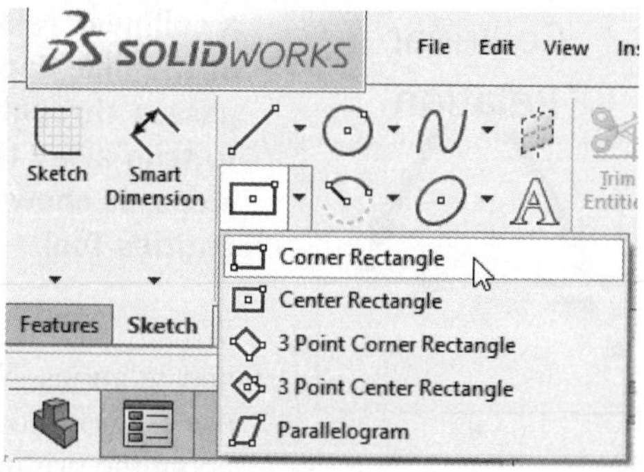

FIGURE 3.6

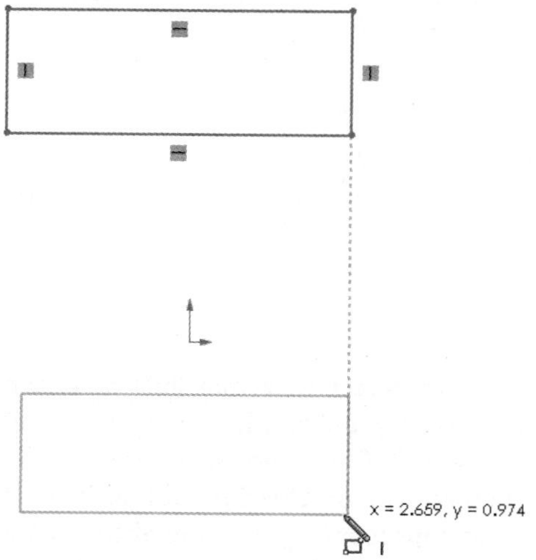

FIGURE 3.7

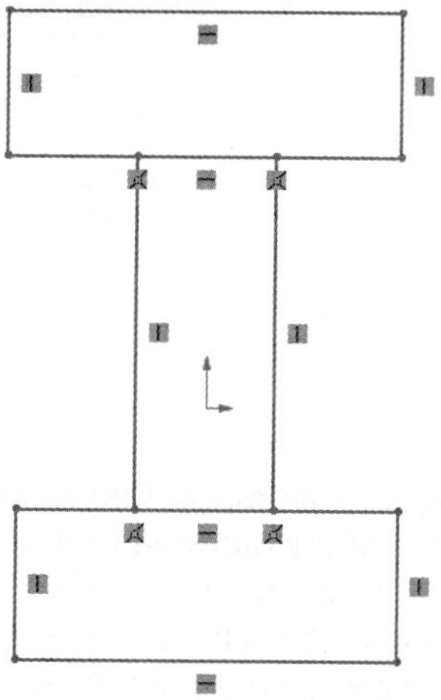

FIGURE 3.8

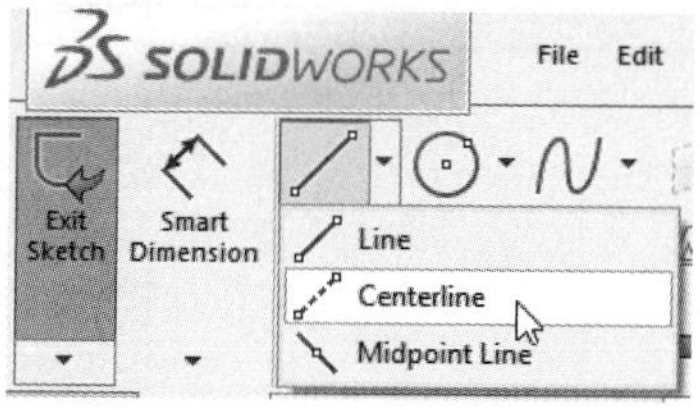

FIGURE 3.9

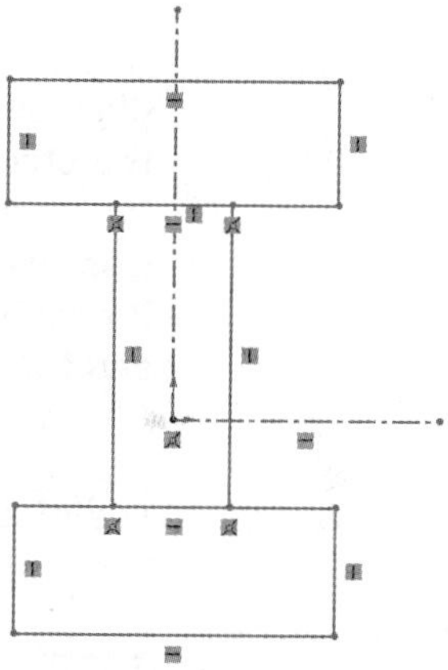

Notice that our sketch consists of three closed regions. For the pulley of Chapter 1, we saw that we can use a multiple-region sketch to create a 3-D shape by selecting the regions (contours) to be extruded or revolved. We will take a different approach here by trimming away the portions of the sketch not needed.

Select the Trim Entities Tool, as shown in Figure 3.10. In the PropertyManager, select Trim to closest, as shown in Figure 3.11. Move the cursor over the portion of the top line of the lower rectangle that lies between the two vertical lines, as shown in Figure 3.12.

FIGURE 3.10

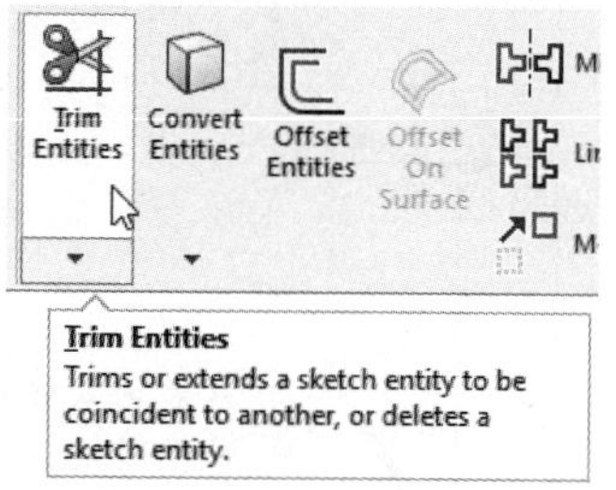

FIGURE 3.11

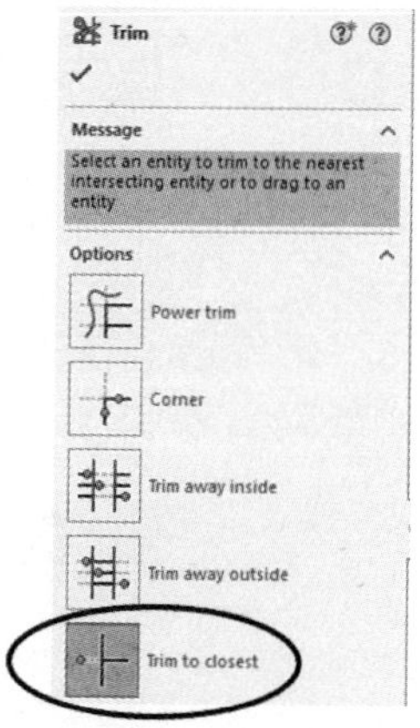

FIGURE 3.12

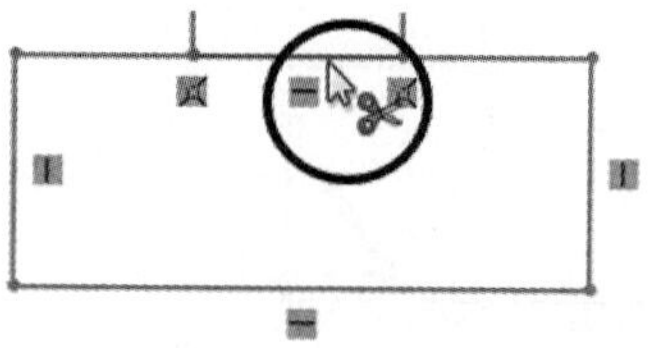

FIGURE 3.13

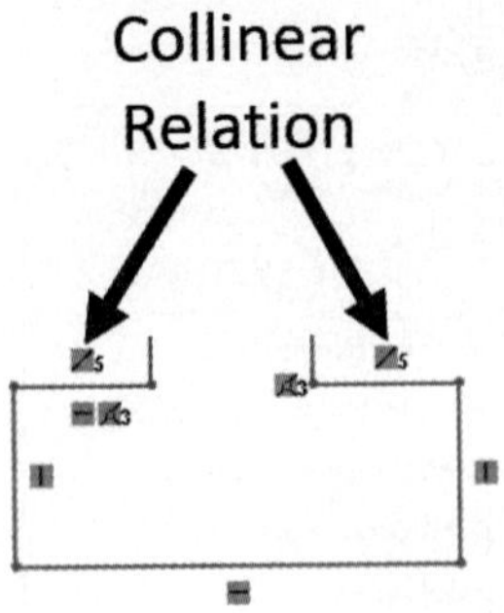

Click to trim away this portion of the line, as shown in Figure 3.13. Note that a collinear relation is automatically created between the remaining portions of the line. Repeat for the lower line of the upper rectangle. If the centerline passes through the section of the line to be trimmed, then it is necessary to trim away the sections of the line on either side of the centerline in two steps, as shown in Figures 3.14 and 3.15. Press the Esc key to turn off the Trim Entities Tool.

We are now ready to define the geometry of the beam section using dimensions and relations. We want to use only the four dimensions shown in **Table 3.1**, and use relations to achieve a fully defined sketch. To begin, we will center the section about the origin using symmetric relations.

FIGURE 3.14

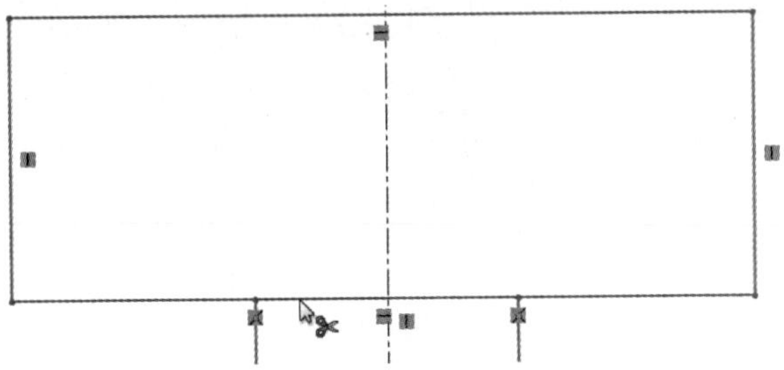

FIGURE 3.15

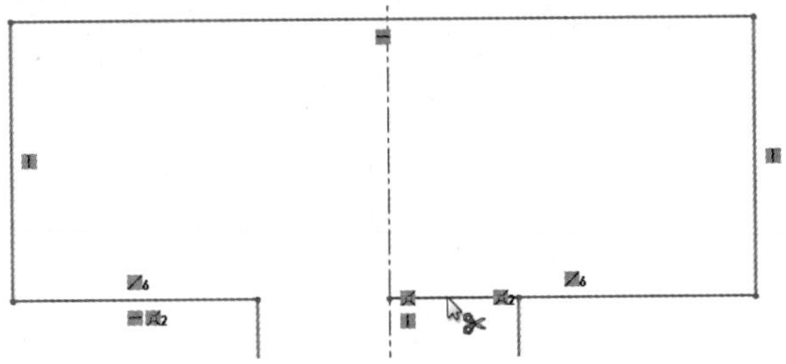

Click on the top horizontal line of the sketch to select it. While holding down the Ctrl key to allow multiple selections, click on the bottom horizontal line and the horizontal centerline, as shown in Figure 3.16. Click the Make Symmetric icon in the context toolbar, as shown in Figure 3.16, or the Symmetric icon in PropertyManager. Select the two vertical lines representing the web of the beam and the vertical centerline, as shown in Figure 3.17, and add a symmetric relation.

FIGURE 3.16

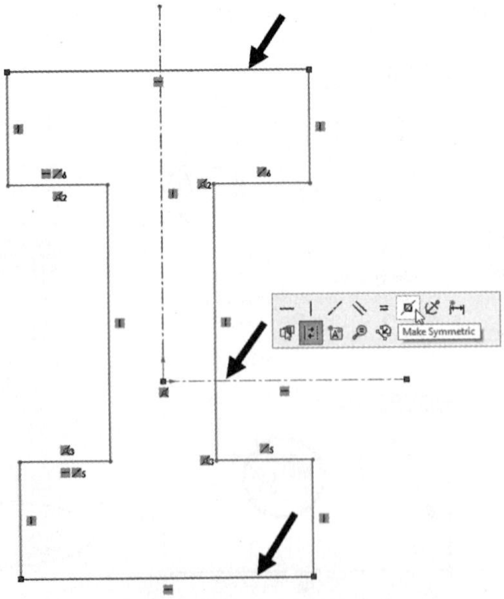

FIGURE 3.17

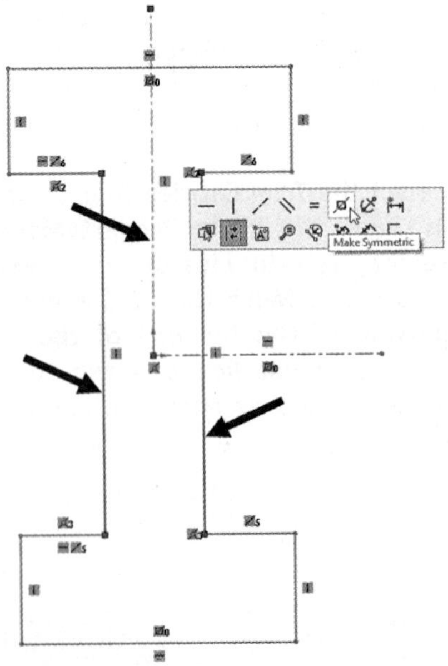

Select the Smart Dimension Tool, and add the four dimensions from Table 3.1, as shown in Figure 3.18. Click the Esc key to turn off the tool.

FIGURE 3.18

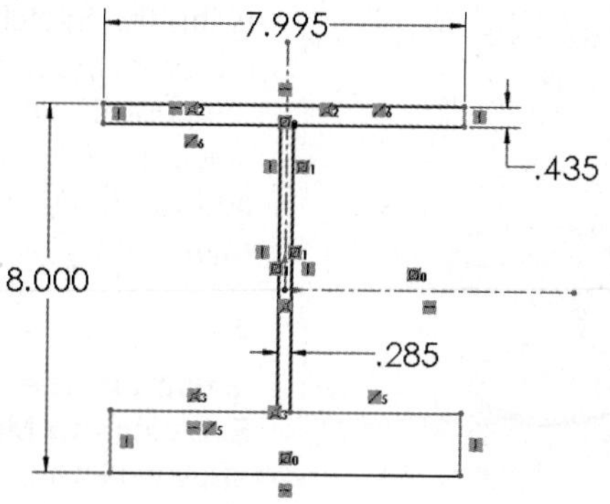

There is an almost limitless number of ways to add relations that will fully define the sketch, utilizing equal, symmetric, and/or collinear relations. The method presented here is probably the simplest, in that only two additional relations are required.

Select the four horizontal lines shown in Figure 3.19 and add an equal relation.

FIGURE 3.19

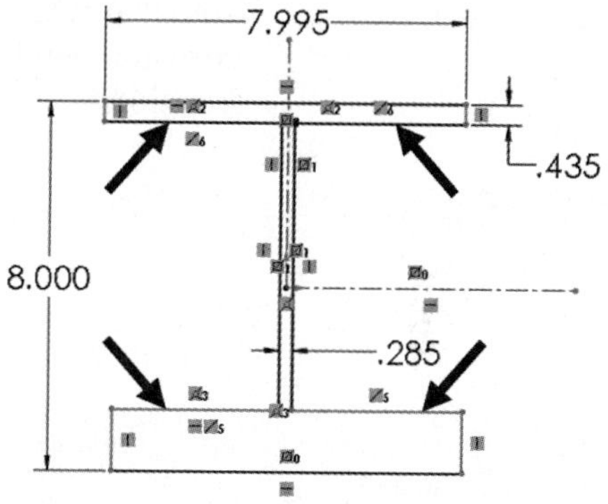

The result of this relation is shown in **Figure 3.20**. Note that the entire sketch is now symmetric about the vertical centerline. To determine what other relations are needed, it is helpful to see which entities are still blue, indicating that they are under defined. Another way to determine which entities are not fully defined is to click and drag a corner point. If the point does not move, then its position is defined. If the point does move, then the allowable movement is an indication of the relation needed to fully define the sketch. For example, if you click and drag the corner point shown in **Figure 3.21**, you will see that up-down movement of the point is allowed, indicating that the thickness of the bottom flange is not defined. You will also see that the right portion of the bottom flange moves along with the left portion as a result of the collinear relation that was applied automatically during the earlier trim operation. Finally, you will notice that left-to-right movement of the point is not possible, indicating that the point's position in that direction (its X-coordinate shown in the PropertyManager) is fixed. By moving this point, it is easy to conclude that a relation is needed to define the thickness of the bottom flange.

FIGURE 3.20

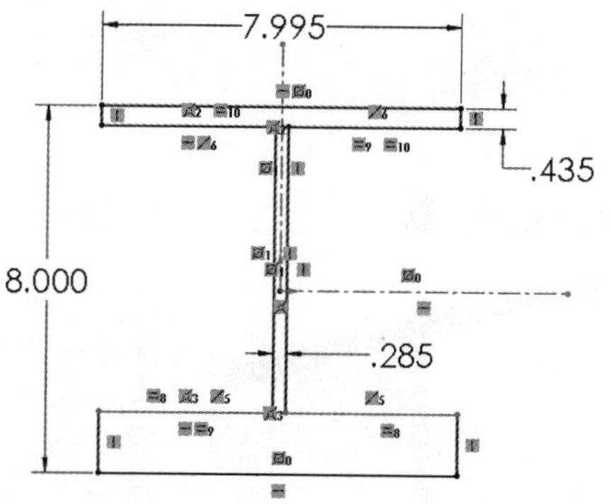

FIGURE 3.21

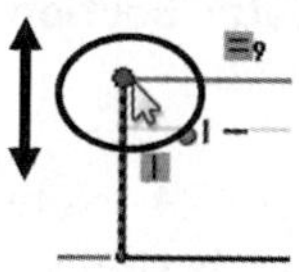

Select the two lines shown in Figure 3.22, and add an equal relation.

FIGURE 3.22

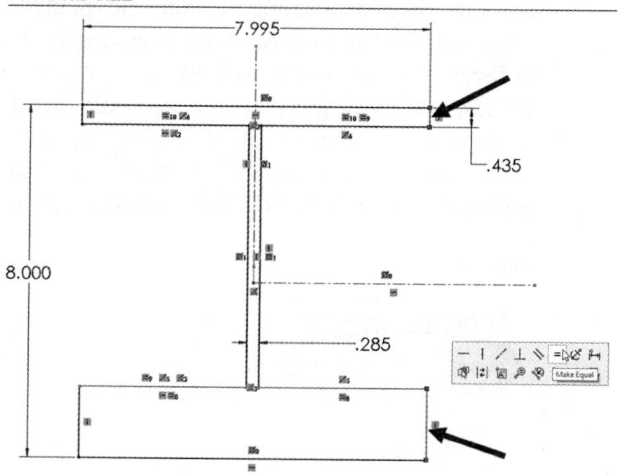

FIGURE 3.23

The sketch is now fully defined. Note that we could add the fillets to the sketch, using the Sketch Fillet Tool. However, doing so would change the lengths of the edges that join at the fillets, causing the equal relations applied to those lines to be temporarily lost. Also, our methodology is to keep sketches as simple as possible, since most errors occur in sketches. Therefore, we will add the fillets as a separate feature rather than defining them within the sketch.

FIGURE 3.24

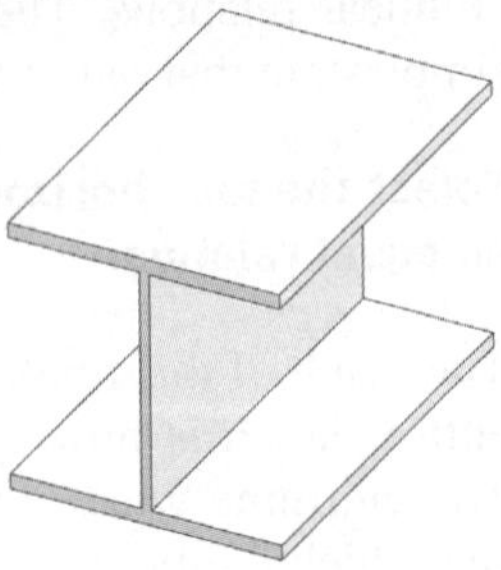

Select the Extruded Boss/Base Tool from the Features group of the CommandManager. Set the type of Extrusion to Mid Plane and the depth to 12 inches, as shown in Figure 3.23. Click the check mark to create the extrusion, which is shown in Figure 3.24.

Note that when a Mid Plane extrusion is specified, the extrusion depth is the total depth, not the depth in one direction.

FIGURE 3.25

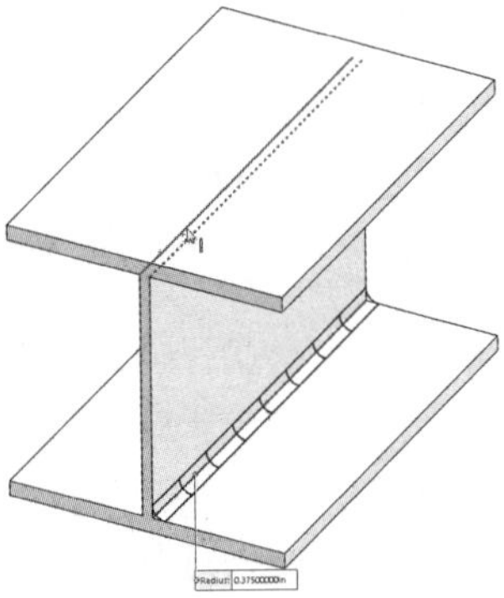

FIGURE 3.26

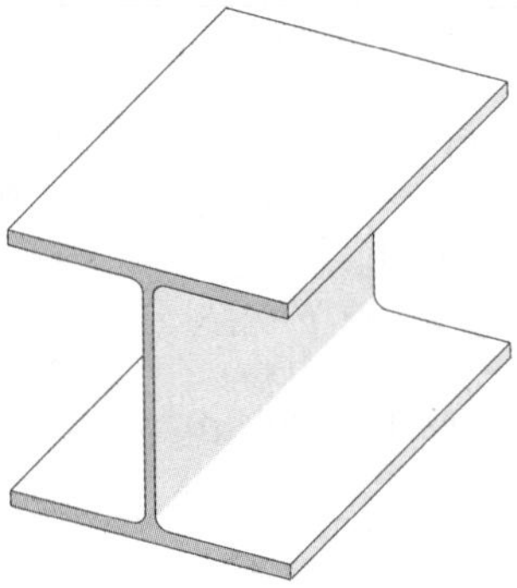

Select the Fillet Tool. Set the fillet radius to 0.375 inches, and select the four edges to be filleted. Note that the hidden edges can be selected "through" the part, as shown in Figure 3.25. Be careful to click on the correct edges. Click the check mark to apply the fillets, as shown in Figure 3.26.

The area of the cross-section can be found using the Section Properties Tool.

Click on the front surface of the part to select it, as shown in Figure 3.27. From the main menu, select Tools: Evaluate: Section Properties, as shown in Figure 3.28.

FIGURE 3.27

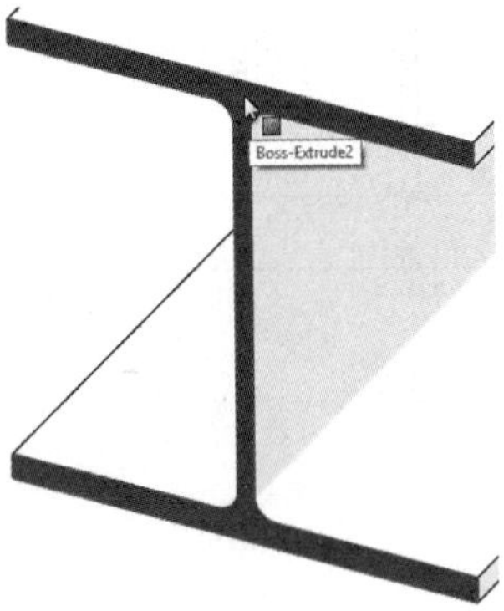

FIGURE 3.28

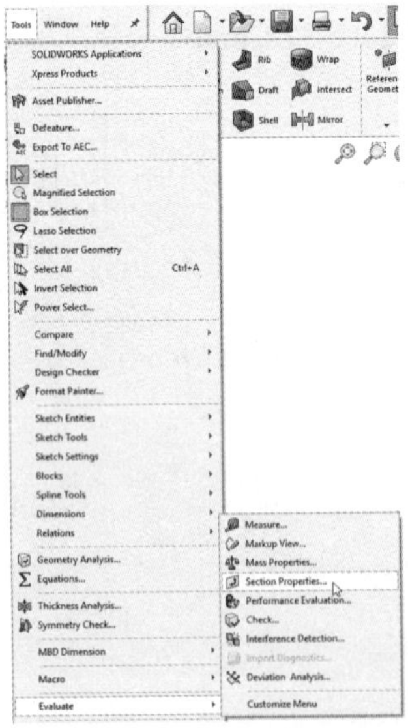

The results are shown in Figure 3.29. Note that the area of the section, 9.11 in^2, is slightly different from the 9.13 in^2 shown in Table 3.1. This slight difference is not unexpected, since the fillet radius that we selected may be different from the one used for the tabulated values. Other properties that are calculated include the location of the centroid (the geometric center) of the section and the moments of inertia of the section about

FIGURE 3.29

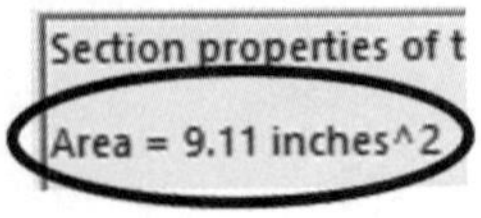

different axes. These moments of inertia are important for structural analysis and are measures of a beam's bending stiffness. Note that the default display of decimal places for these mass/section property calculations can be set under the Units entry on the Document Properties options tab.

We will now specify the part's material and check its weight.

Close the Section Properties window. Right-click on Material in the FeatureManager and select Edit Material, as shown in Figure 3.30. Locate ASTM A36 Steel in the list of steel materials and click to select it, as shown in Figure 3.31. Click Apply and Close. From the main menu, select Tools: Evaluate: Mass Properties, as shown in Figure 3.32.

FIGURE 3.30

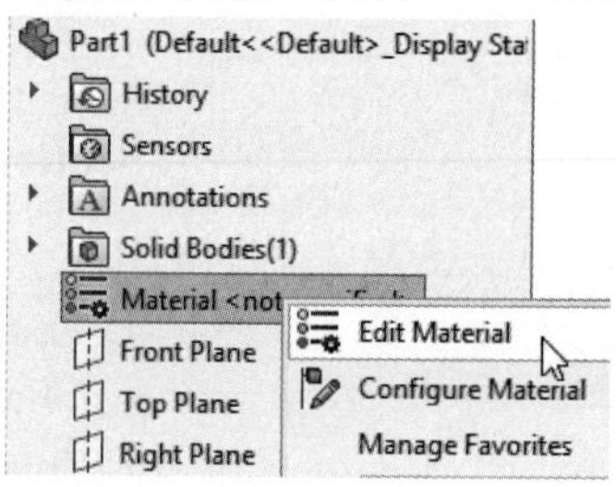

As shown in Figure 3.33, the weight is confirmed as 31 pounds for this one-foot-long beam section.

Close the Mass Properties window, and save the part with the name "W8×31 Beam Section."

FIGURE 3.31

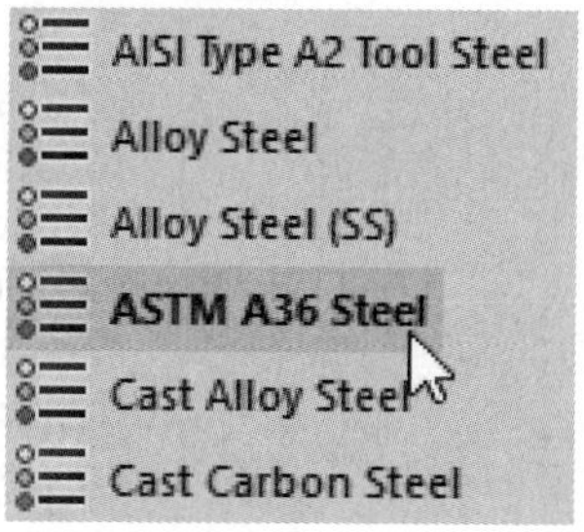

FIGURE 3.32

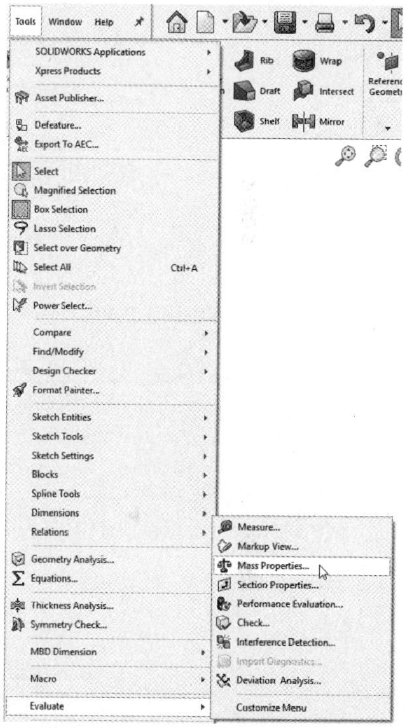

FIGURE 3.33

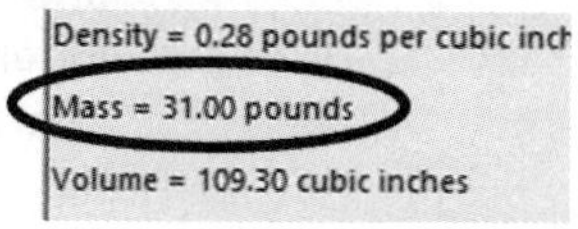

Because we used only the four dimensions of Table 3.1 to define the sketch of the section, it is easy to change this beam model into one of another standard shape. For example, we will create a model of a W10×45 section by modifying our W8×31 model.

Double-click Boss-Extrude1 in the FeatureManager, and its dimensions will be displayed, as shown in Figure 3.34. Drag the dimensions around the screen for better visibility, if desired. Double-click each dimension and change its value to that shown in Table 3.1 for the W10×45 shape, as shown in Figure 3.35. Click the Rebuild Tool, as shown in Figure 3.36.

FIGURE 3.34

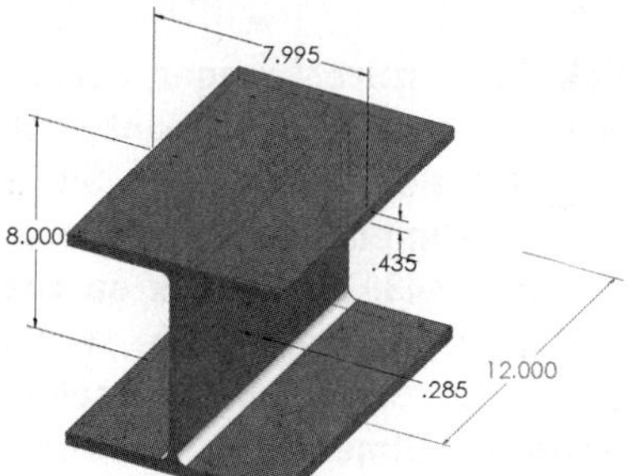

FIGURE 3.35

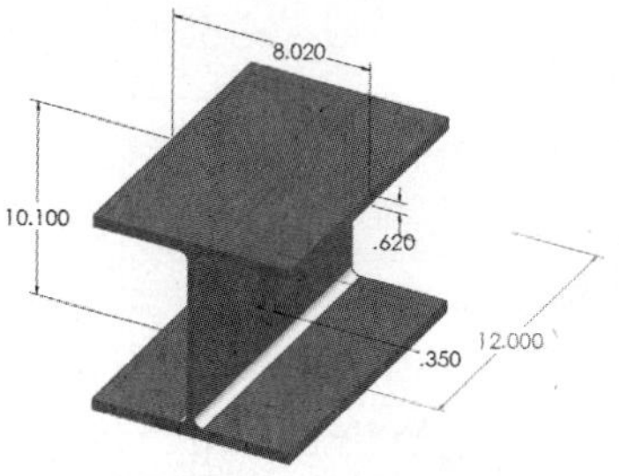

FIGURE 3.36

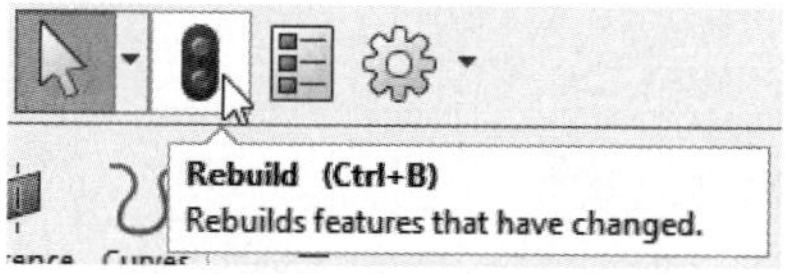

The new part is shown in **Figure 3.37**. If you check its mass properties, you will find that its weight is slightly less than 45 pounds (44.81 pounds). If the fillet radius is increased to 0.50 inches, then the weight is slightly more than 45 pounds (45.13 pounds).

FIGURE 3.37

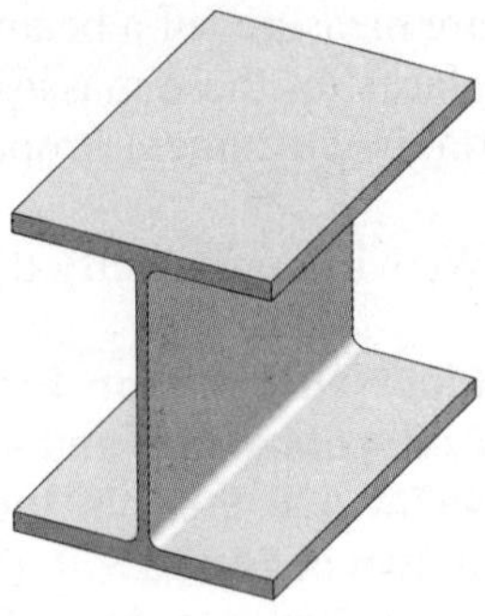

Save the modified part with the name "W10×45 Beam Section."

Using this technique, you can make a family of beam sections, each as a separate file with a unique name. Another approach is to use different *configurations* within a single file. The following steps will show you how to make a single file with two beam sections as configurations.

FIGURE 3.38

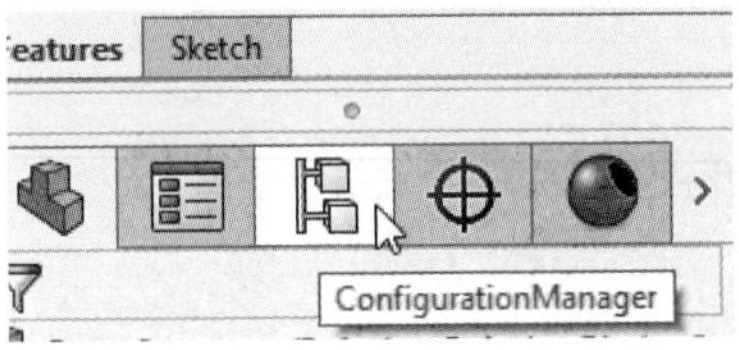

Choose File: Save As from the main menu. Save the file with the name "Beam Segments." Select the Configuration Manager, as shown in Figure 3.38. Right-click the Default configuration, and select Properties, as shown in Figure 3.39. As the configuration name, enter W10×45, as shown in Figure 3.40. Click the check mark.

FIGURE 3.39

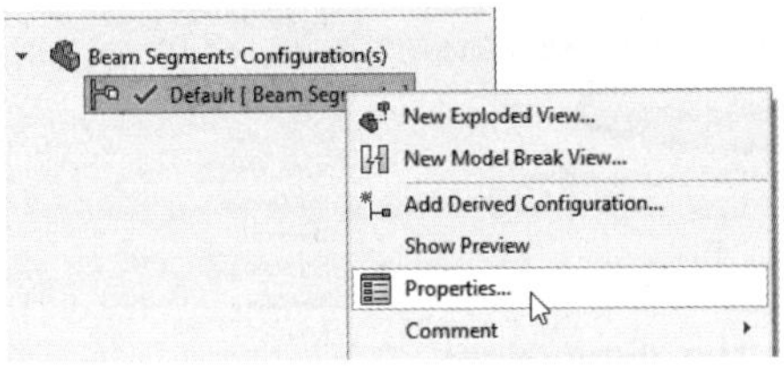

Right-click on Beam Segments Configuration(s) and select Add Configuration, as shown in Figure 3.41. Enter the name W8×31 for the new configuration and click the check mark.

There are now two configurations, but they are identical. The last configuration created, the W8×31, is now the selected configuration, and we can now edit this configuration.

FIGURE 3.40

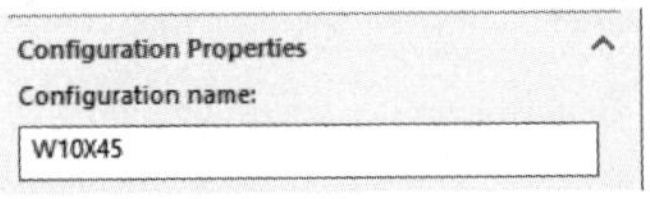

FIGURE 3.41

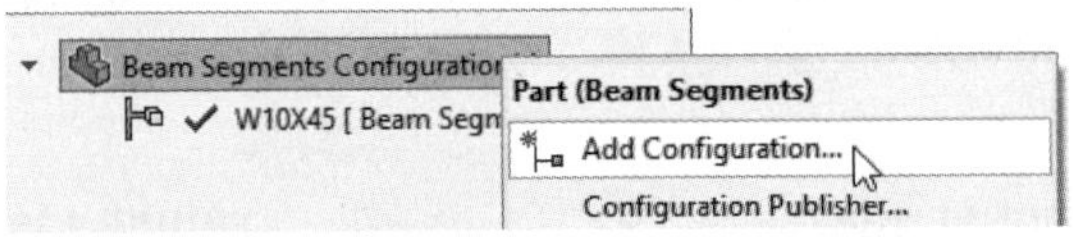

FIGURE 3.42

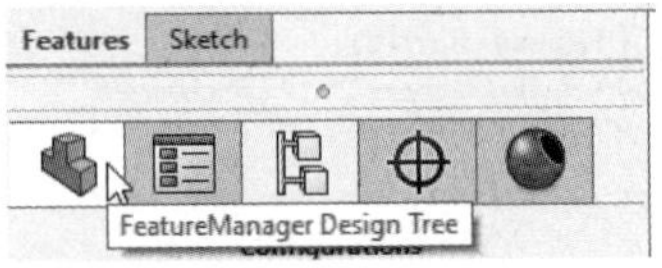

FIGURE 3.43

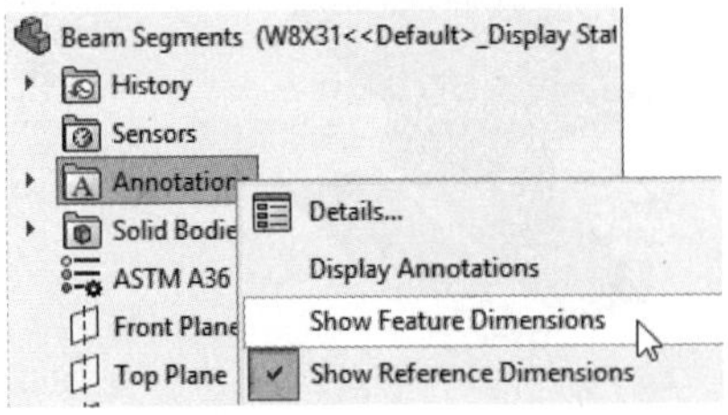

Click the FeatureManager icon, as shown in Figure 3.42. Right-click on Annotations and select Show Feature Dimensions, as shown in Figure 3.43. Right-click on the 8.020-inch flange width. From the menu that appears, select Configure Dimension, as shown

in Figure 3.44. The Modify Configurations Dialog Box will appear. Change the dimension associated with the W8×31 configuration to 7.995 inches, as shown in Figure 3.45, and click OK. Repeat for each of the other cross-sectional dimensions, changing the dimensions associated with the W8×31 to the values shown in Figure 3.34.

Now that there are two distinct configurations defined, you can switch from one to the other through the Configuration Manager.

Switch to the ConfigurationManager. Double-click the W10×45 configuration and notice that the dimensions will change to those of the selected configuration, as shown in Figure 3.46.

Add a third configuration to define the W12×65 section detailed in Table 3.1. Save the file.

In this example, we created the configurations manually. In Chapter 5, we will use a spreadsheet to define several different configurations of a cap screw.

FIGURE 3.44

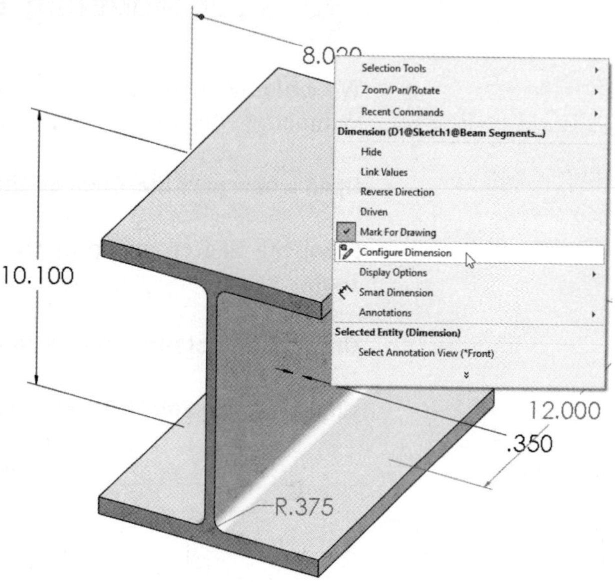

FIGURE 3.45

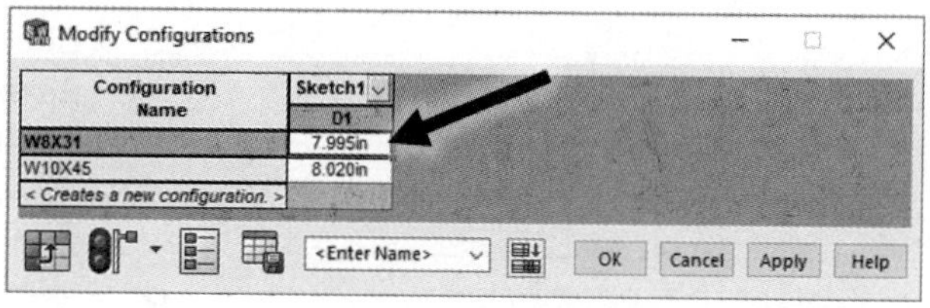

FIGURE 3.46

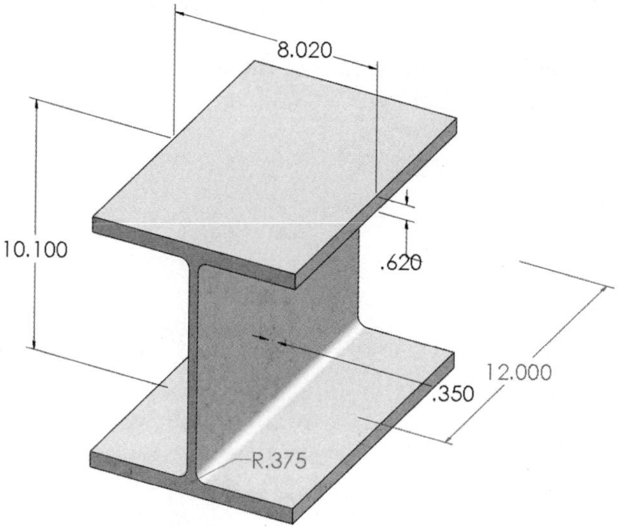

3.2 Part Modeling Tutorial: Bracket

We will now model the bracket shown in **Figure 3.2**. This part has a single plane of symmetry, which we will align with the Right Plane.

Open a new part file. Click on the Front Plane in the FeatureManager to select it.

From the Sketch group of the CommandManager, select the Center Rectangle Tool.

Drag out a rectangle centered at the origin.

Click on the Smart Dimension Tool, and dimension a horizontal side of the rectangle to 6 inches and a vertical side to 4 inches, as shown in Figure 3.47.

FIGURE 3.47

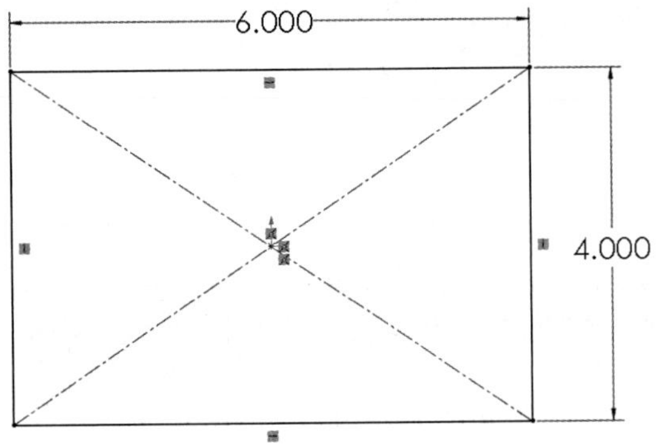

The rectangle is now both dimensioned and located, so the sketch is fully defined. By centering the rectangle at the origin, we will be placing the Right Plane so that it will be a plane of symmetry of the part.

Click on the Features tab and then the Extruded Boss/Base Tool. In the PropertyManager, enter 0.25 inches as the thickness, and click the check mark.

The horizontal boss of the bracket will now be constructed. The feature will be sketched in a plane 1.25 inches below the top surface of the base part. To accomplish this, we will first need to create the reference plane to be used for sketching.

FIGURE 3.48

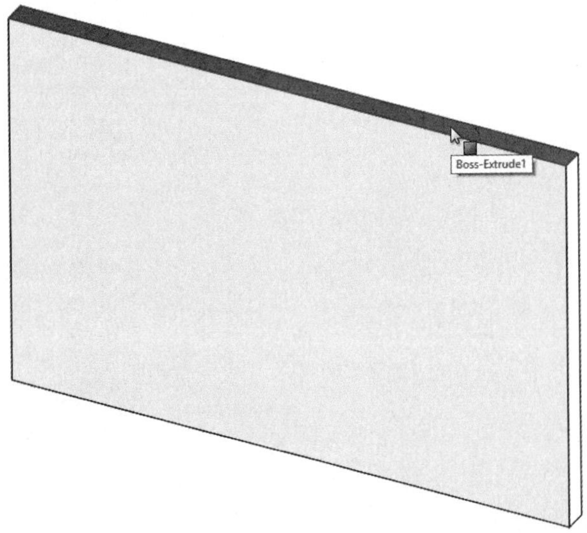

Select the top surface of the base part, as shown in Figure 3.48.

Be sure that the face is selected (a square symbol will appear next to the cursor prior to selection).

From the Features group of the CommandManager, select Reference Geometry: Plane, as shown in Figure 3.49. In the PropertyManager, click the Flip offset box to reverse the direction, so that the new plane will be below the top surface, as shown in the preview, and set the distance to 1.25 inches, as shown in Figure 3.50. Click the check mark to create the plane, which will be named Plane1.

FIGURE 3.49

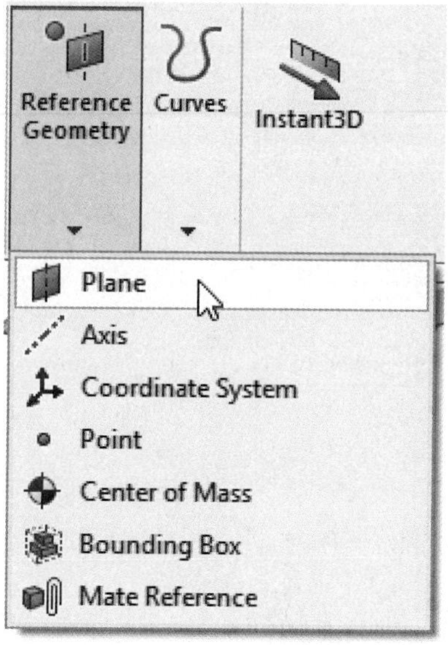

Note that the boundaries of the plane are similar in size and shape to those of the top surface. Of course, a plane extends infinitely, so these boundaries can be adjusted for better visibility.

Click and drag the move handles on Plane1 to change the plane's visible boundaries as desired, as shown in Figure 3.51.

The boss feature will now be sketched in this new plane. The symmetry of the boss about the Right Plane will be exploited in the creation of the sketch.

FIGURE 3.50

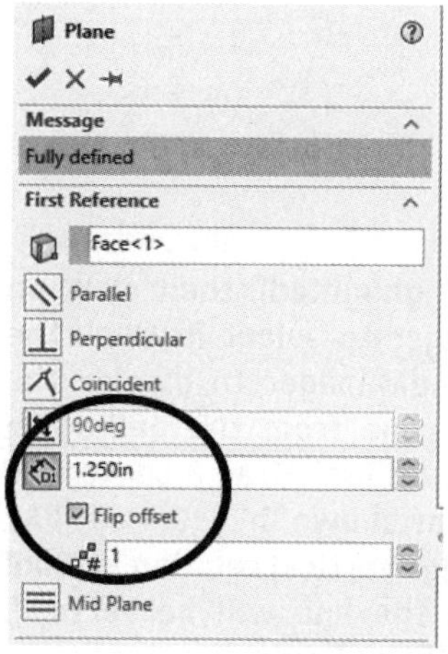

FIGURE 3.51

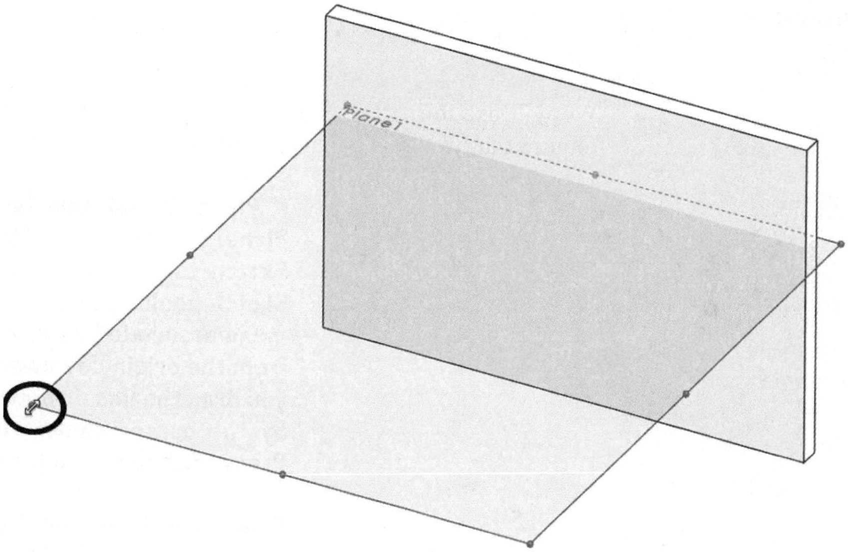

Switch to the Top View. Use the Zoom and Pan Tools if necessary to place the part toward the top of the screen, as shown in Figure 3.52.

FIGURE 3.52

We will use mirroring to assist in construction of the symmetric sketch. This will require that we draw a centerline first.

FIGURE 3.53

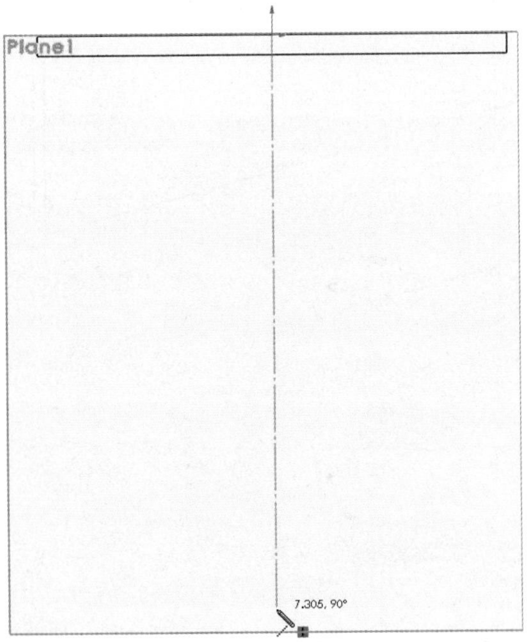

If Plane1 is not selected (highlighted), then click on Plane1 in the FeatureManager to select it. Click the Sketch tab of the CommandManager to display the Sketch tools. Select Centerline from the pull-down menu associated with the Line Tool. Drag a centerline from the origin downward, as shown in Figure 3.53. As you drag the line downward, a vertical relation symbol by the cursor shows that the line will be vertical. Press the Esc key to turn off the Centerline Tool.

Select the Line Tool from the Sketch group of the CommandManager. Move the cursor near the lower edge of the solid rectangle, as shown in Figure 3.54.

FIGURE 3.54

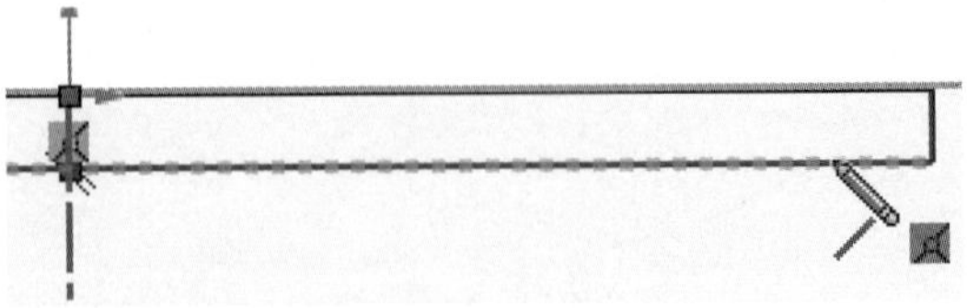

A coincident relation symbol will appear next to the cursor, indicating that you are snapping to the edge of the solid.

Click and hold, while dragging a line downward and at an angle, as shown in Figure 3.55.

Select the Tangent Arc Tool from the pull-down menu of arc tools, as shown in Figure 3.56. Move the cursor to the end of the line, snapping the start of the arc to this endpoint. Hold down the left mouse button and drag out an arc, with the endpoint of the arc snapped to the centerline, as shown in Figure 3.57. Press the Esc key to turn off the Tangent Arc Tool.

The arc that you drew will remain tangent to the first line, no matter how the dimensions of the sketch are changed.

We will now exploit the symmetry of the bracket, and construct a mirror image of our sketch about the centerline.

FIGURE 3.55

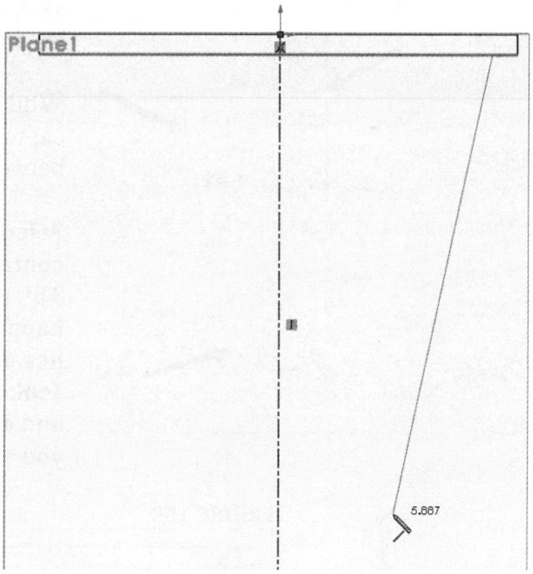

FIGURE 3.56

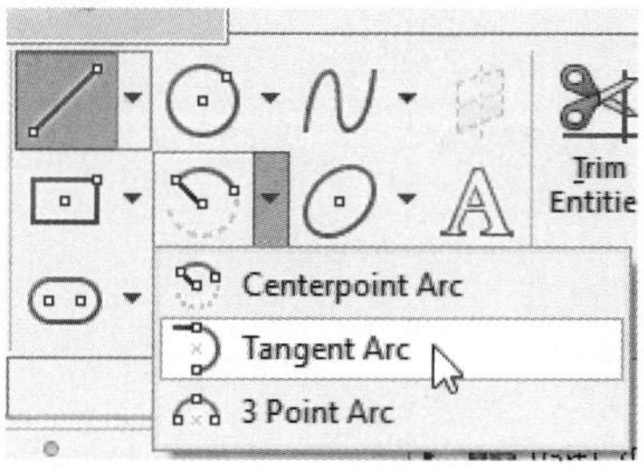

FIGURE 3.57

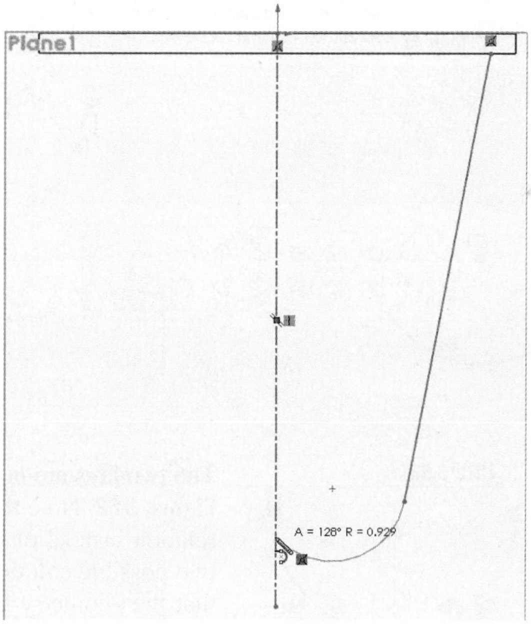

FIGURE 3.58

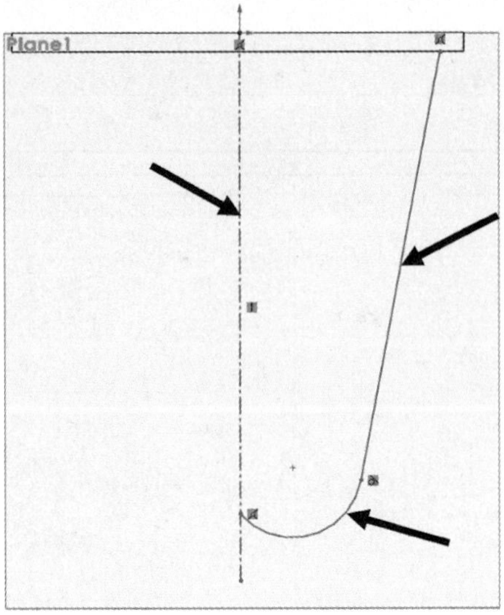

With the Ctrl key depressed (to allow for multiple selections), click to select the centerline, the solid line, and the tangent arc, as shown in Figure 3.58. Select Mirror Entities from the CommandManager, as shown in Figure 3.59. A symmetric sketch will be created about the selected centerline, as shown in Figure 3.60.

While this sketch does exhibit the required symmetry, it does not capture our design intent. We need to add an additional relation between the two tangent arcs.

Select the two arcs, and apply a Concentric relation from the context menu or from the PropertyManager, as shown in Figure 3.61. Note: There may be only a single arc in your sketch. If you happened to drag out the first tangent arc so that its center lies on the centerline, then when you used the Mirror Entities Tool, the mirrored arc and the original arc were concentric and automatically combined into a single arc. If this is the case, you can proceed to the next step.

FIGURE 3.59

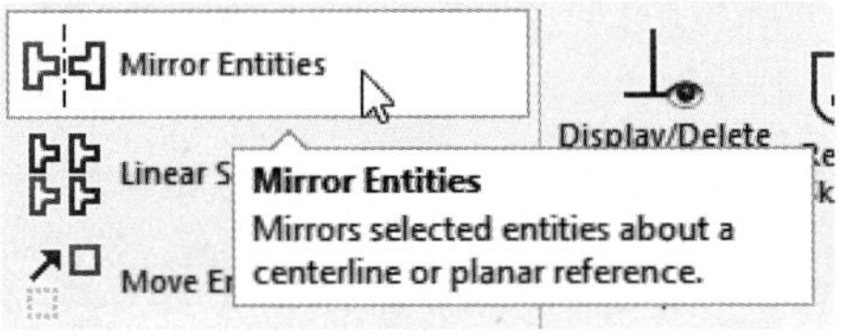

FIGURE 3.60

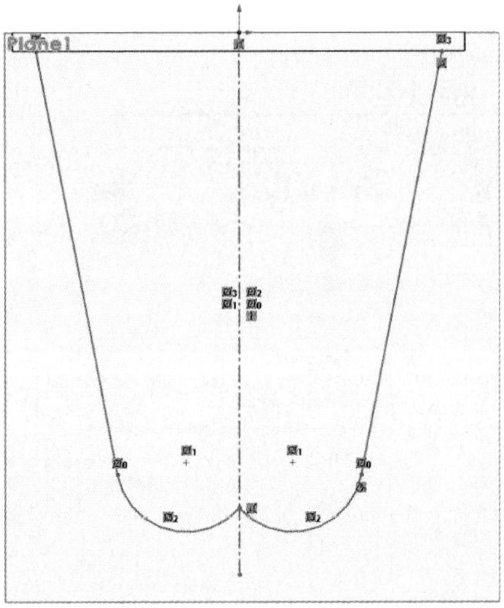

FIGURE 3.61

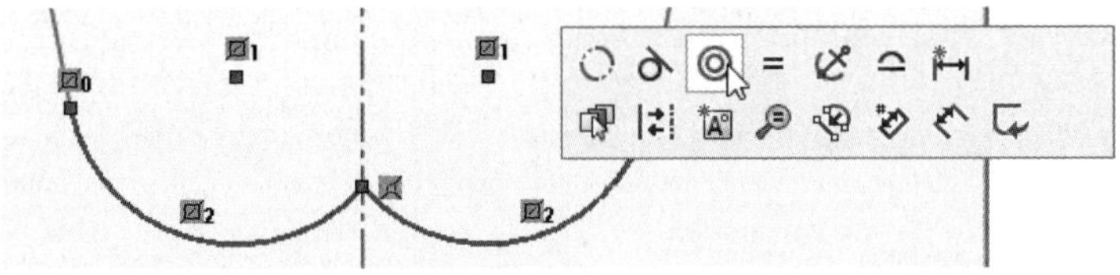

FIGURE 3.62

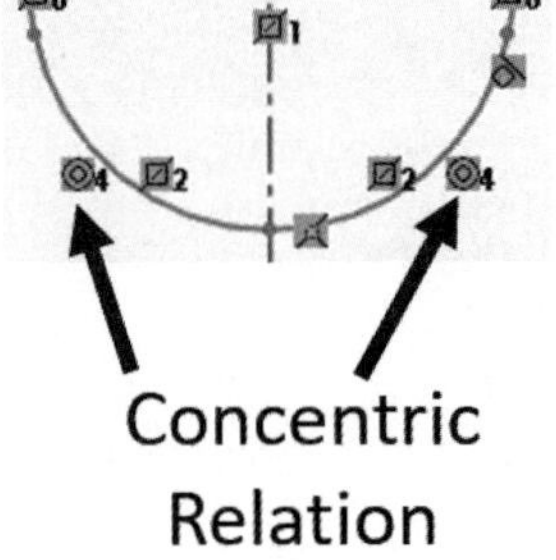

The two arcs are now tangent to each other, as shown in Figure 3.62. Note that we could have specified a tangent relation instead of the concentric relation, but there are two possible solutions to the tangent relation. To ensure that the geometry is the way we want it, the concentric relation provides only one solution.

Select the Line Tool, and finish the sketch with a horizontal line connecting the two free ends of the sketch, as shown in Figure 3.63. Select the Smart Dimension Tool, and dimension the sketch as shown in Figure 3.64. Add the 0.75-inch radius first.

FIGURE 3.63

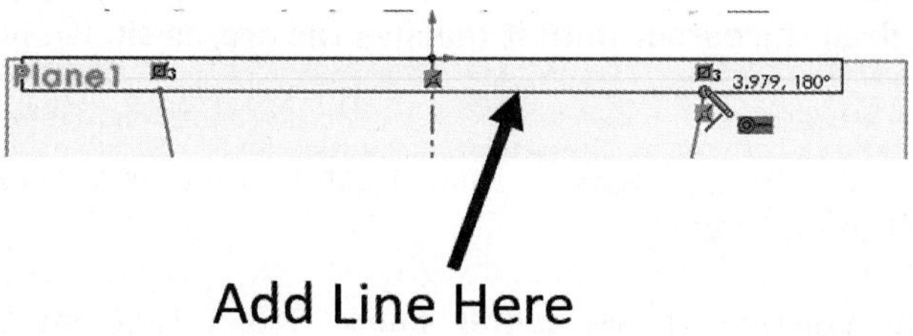

FIGURE 3.64

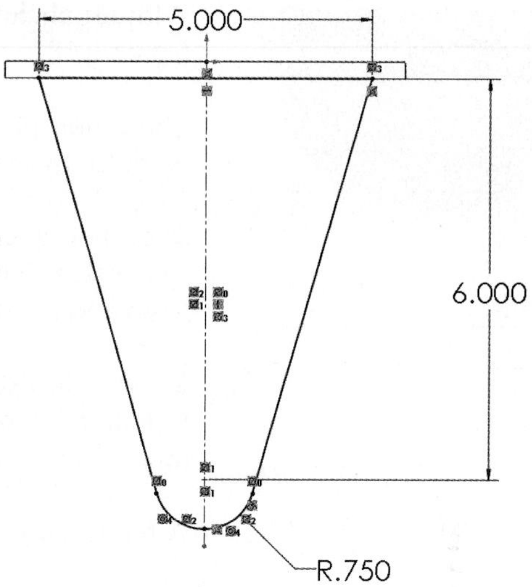

Note that the 6.00-inch dimension is to the center of the arcs.

Click on the Features tab of the CommandManager. Select the Extruded Boss/Base Tool. Click the Reverse Direction icon in the PropertyManager, as shown in Figure 3.65, and extrude the part downward 0.25 inches. Change to the Trimetric View to preview the extrusion, and click the check mark to complete the operation.

FIGURE 3.65

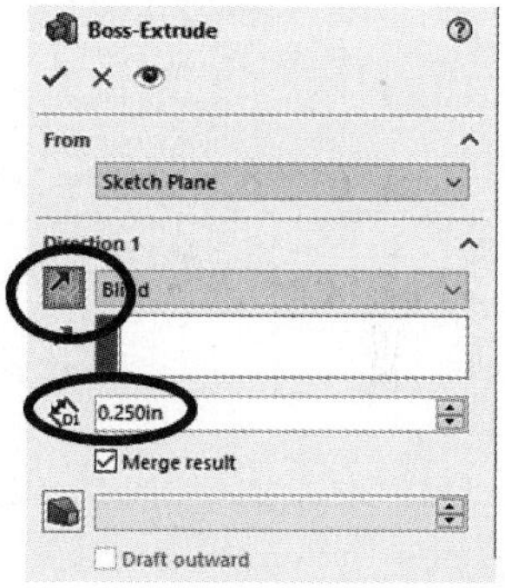

The bracket should appear as shown in Figure 3.66. Plane1 can be hidden by right-clicking on it in the FeatureManager and selecting Hide, or by turning off the display of all planes with the Hide/Show Items Tool of the Heads-Up View Toolbar.

Now we will add a raised cylindrical feature for reinforcement near what will become a mounting hole.

Select the top surface of the horizontal extruded boss, as shown in Figure 3.67. Select the Normal To View.

FIGURE 3.66

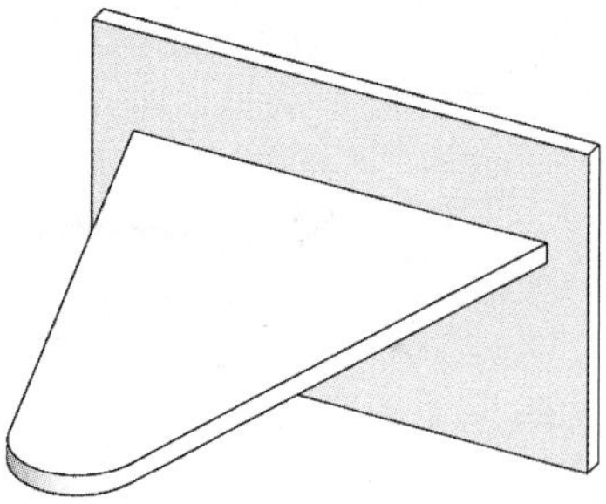

FIGURE 3.67

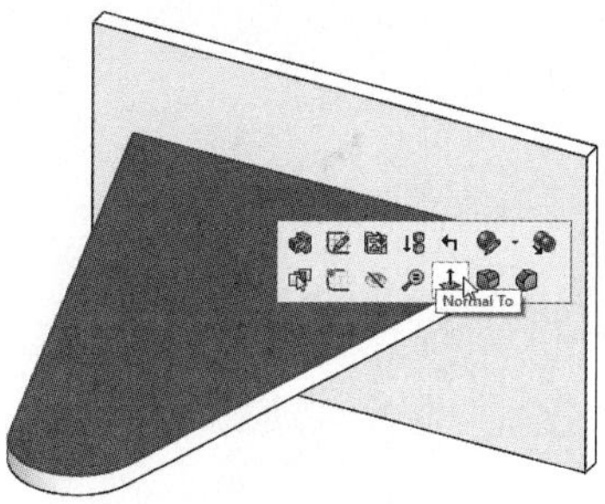

We want to locate the raised area concentric with the rounded tip of the boss. In order to establish this relation, we will first need to "wake up" the arc that makes up this rounded tip, allowing the center point to be used as a snap point in our sketch.

FIGURE 3.68

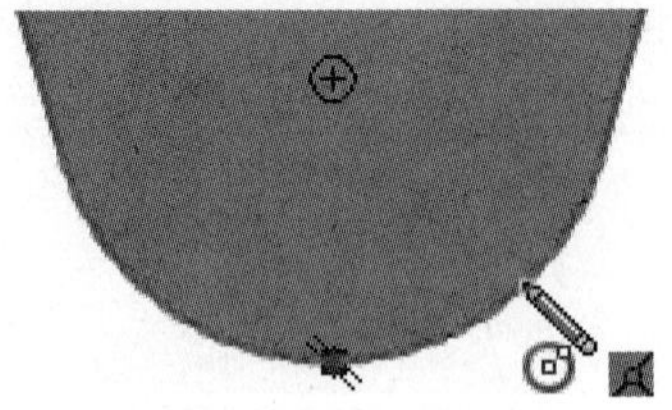

Click the Sketch tab of the CommandManager. Select the Circle Tool. Without clicking, drag the cursor until it touches the arc, as shown in Figure 3.68.

The center point of the arc now appears, and it can be used as a snap point for the center of the circle.

Drag out a circle, starting at the center mark. Select the Smart Dimension Tool, and dimension the circle to be 1 inch in diameter, as shown in Figure 3.69.

FIGURE 3.69

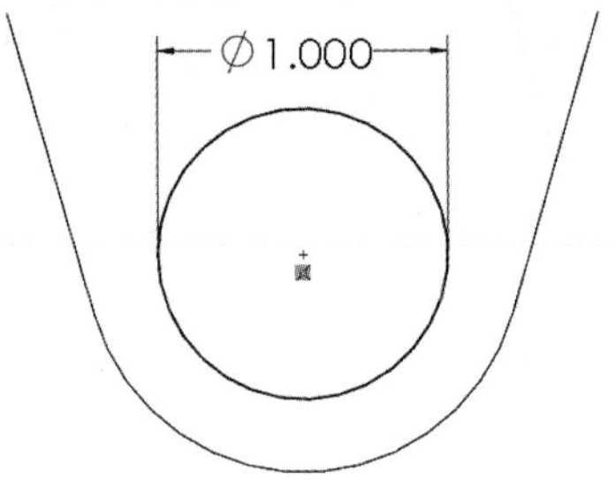

Select the Extruded Boss/Base Tool from the Features group of the CommandManager. Extrude the sketch upward for a distance of 0.050 inches, as shown in Figure 3.70.

A reinforcing rib will now be added to the part.

FIGURE 3.70

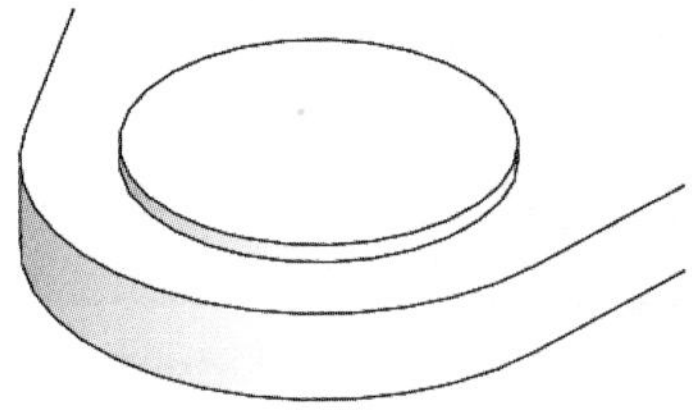

Select the Right Plane from the FeatureManager, and select the Right View. Select the Line Tool from the Sketch group of the CommandManager, and sketch the line shown in Figure 3.71. The endpoints of your line should snap to the edges of the solids, but not to any specific point on the edges. Recall that a coincident relation symbol next to the cursor indicates that you are snapping to the edge.

Add one horizontal and one vertical line coincident with the edges of the solid (snap to the endpoints of the first line and the intersection point of the edges), completing the profile of the rib, as shown in Figure 3.72.

FIGURE 3.71

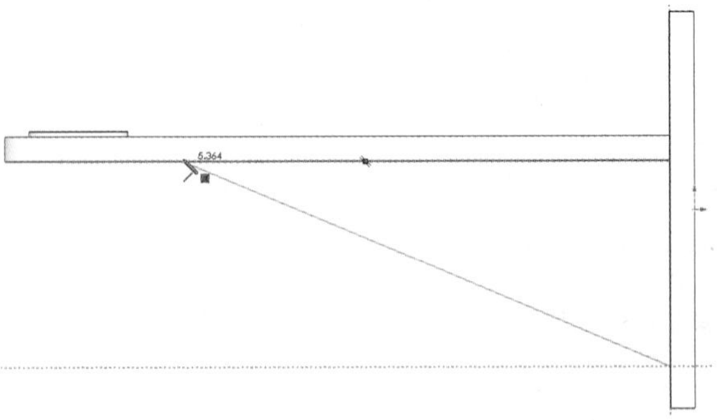

FIGURE 3.72

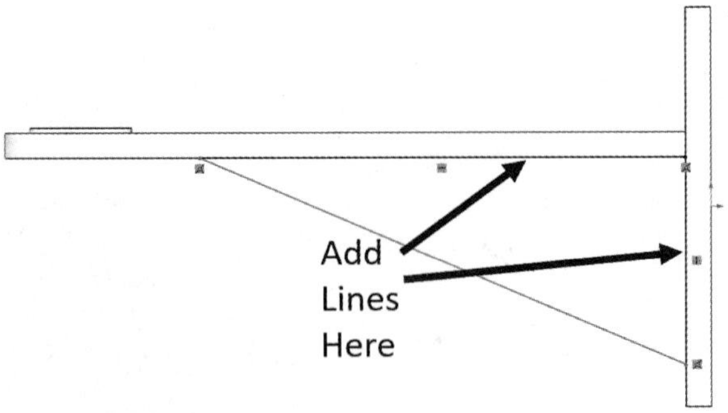

Select the Smart Dimension Tool, and dimension the sketch as shown in Figure 3.73.

FIGURE 3.73

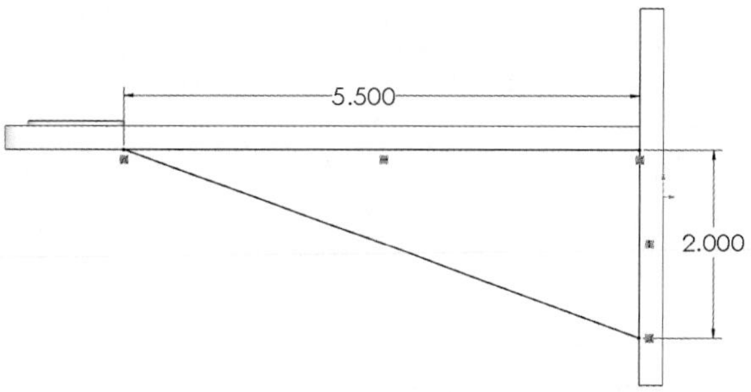

Select the Extruded Boss/Base Tool from the Features group of the CommandManager. In the PropertyManager, select Mid Plane as the type from the pull-down menu, and enter a thickness of 0.1875 inches, as shown in Figure 3.74. Click the check mark to complete the extrusion.

A front view of the part shows that the rib is symmetric about the Right Plane, as shown in Figure 3.75.

The next rib will be added in a slightly different manner, using the Rib Tool. But first, we must create a new plane for the rib. So far, we have used the Plane Tool from the Reference Geometry Tool of the CommandManager. Here, we will use a drag-and-drop technique that is similar to that used in other Windows applications.

FIGURE 3.74

FIGURE 3.75

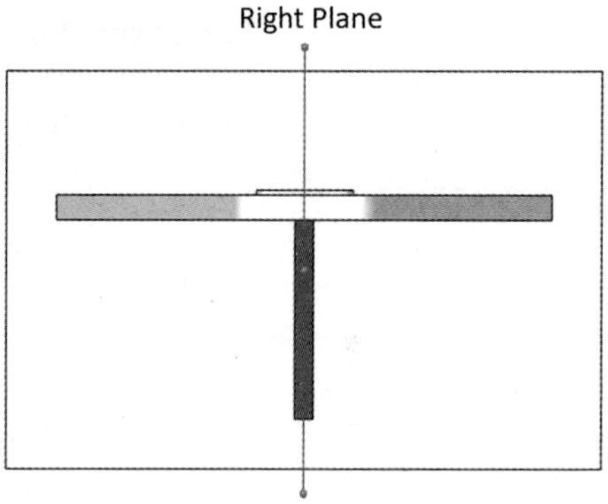

Select the Trimetric View. Click on the Right Plane in the FeatureManager to select it. Hold down the Ctrl key, and move the cursor over one of the lines defining the plane until the move arrows appear, as shown in Figure 3.76. Click and drag the mouse to the right. When the mouse button is released, a new plane is created, as shown in Figure 3.77. In the PropertyManager, set the offset distance to 1 inch, as shown in Figure 3.78. Click the check mark to complete the definition of the new plane, which will be labeled Plane2.

FIGURE 3.76

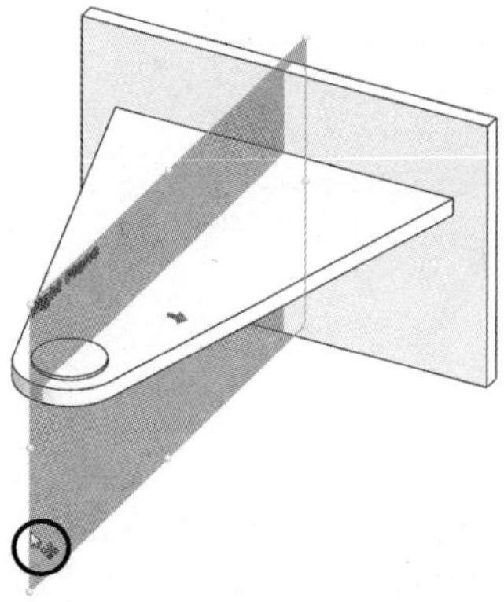

FIGURE 3.77

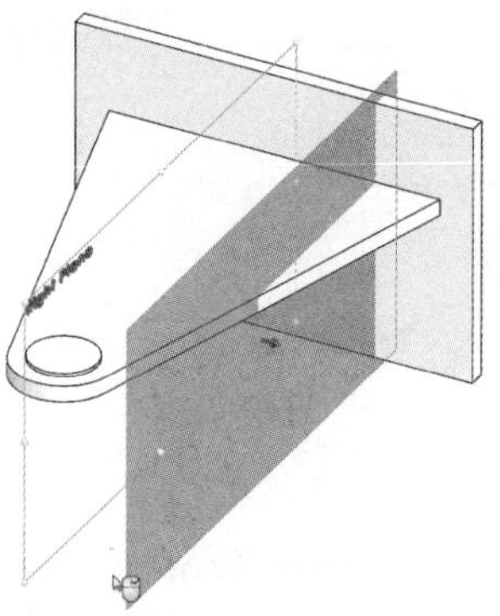

FIGURE 3.78

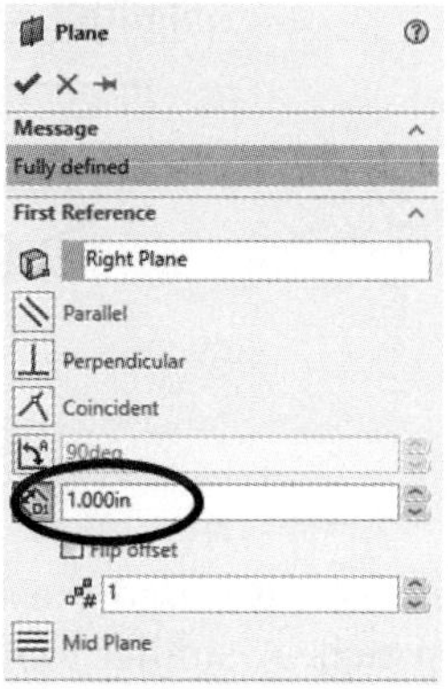

FIGURE 3.79

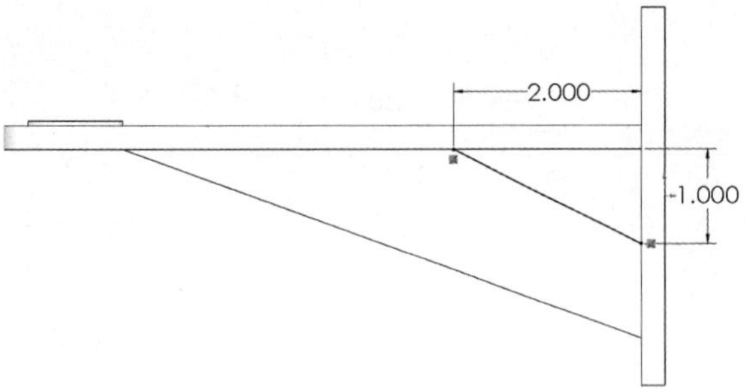

Switch to the Right View. With Plane2 selected, choose the Line Tool from the Sketch group of the CommandManager. Draw a diagonal line as shown in Figure 3.79, snapping the endpoints to the edges of the solids (but not to any specific points). Add the two dimensions shown. Do not add additional lines to close the sketch contour.

For the first rib, we added two lines to close the sketch and used an Extruded Boss command to complete the rib. A rib can be created from an open sketch using the Rib Tool.

Select the Rib Tool from the Features group of the CommandManager, as shown in Figure 3.80. In the FeatureManager, set the thickness to Both Sides and 0.100 inches, as shown in Figure 3.81.

FIGURE 3.80

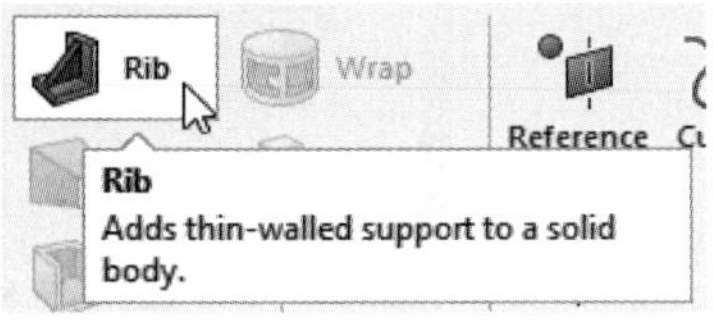

FIGURE 3.81

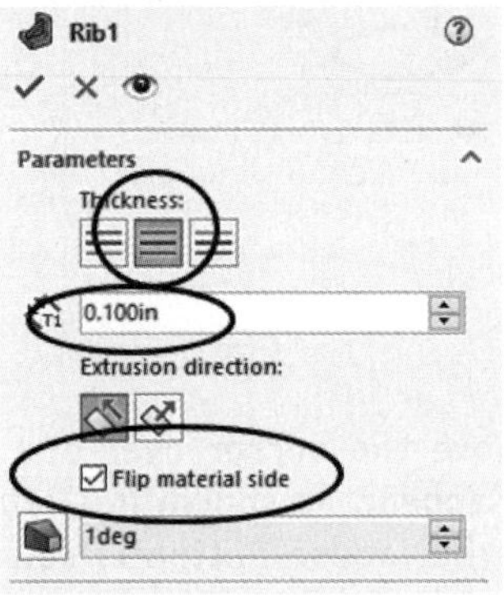

The thickness can be set to be offset from the plane in either direction, or so that the rib is symmetric about the plane (similar to the Mid Plane Extrusion used for the first rib). By default, the middle of the three buttons, designating a rib symmetric about the plane, is selected. For the extrusion direction, the rib can be defined as parallel to the plane of the sketch (default) or normal to the plane. Neither of these settings needs to be changed for our rib.

Check the "Flip material side" box, as shown in Figure 3.81, so that the arrow defining the rib direction points toward the part, as shown in Figure 3.82. Click the check mark to complete the rib, which is shown in Figure 3.83. Hide Plane2.

FIGURE 3.82

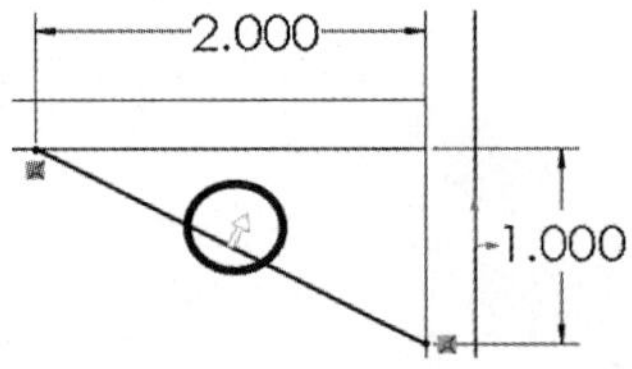

FIGURE 3.83

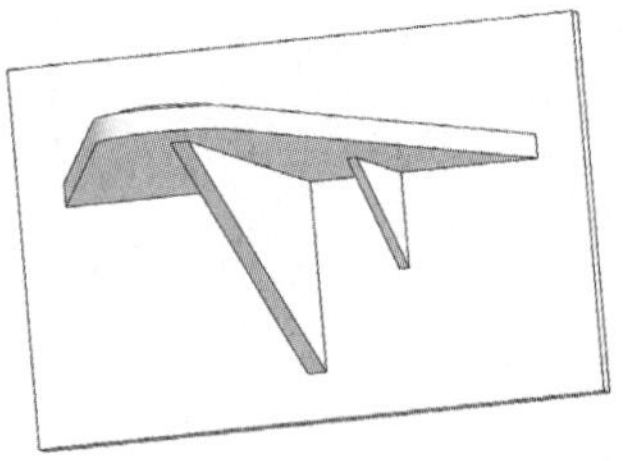

Earlier, we used the Mirror Entities Tool to create symmetric elements within a sketch. A similar tool in the Features group of the CommandManager, called the Mirror Tool, is used to create symmetric features in a part. We will use this tool to create the final rib.

Click on the Features Tab of the CommandManager. Select the Mirror Tool, as shown in Figure 3.84. In the PropertyManager, the Mirror Face/Plane box is active (highlighted).

FIGURE 3.84

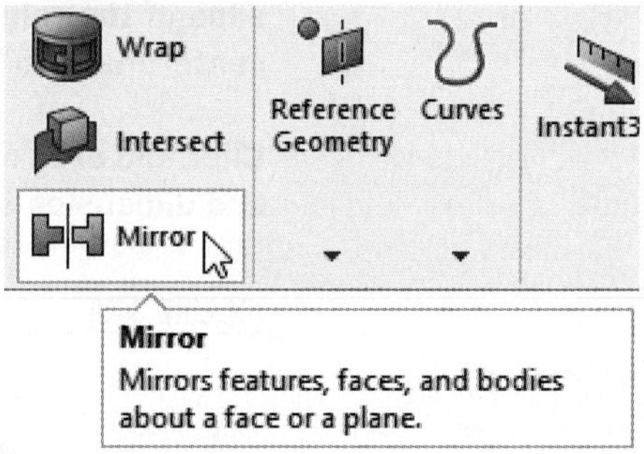

Click on the Right Plane in the FeatureManager, which has "flown out" to the right of the PropertyManager, as shown in Figure 3.85. It may be necessary to click on the arrow next to the part name to expand the FeatureManager before selection. The rib just created should be listed as the feature to mirror. If it is not, click that box to make it active and select the rib from the FeatureManager.

Click the check mark to complete the rib.

The three ribs are shown in Figure 3.86.

FIGURE 3.85

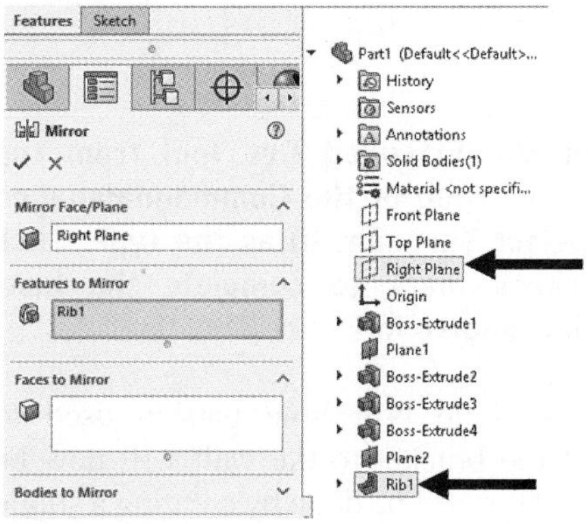

FIGURE 3.86

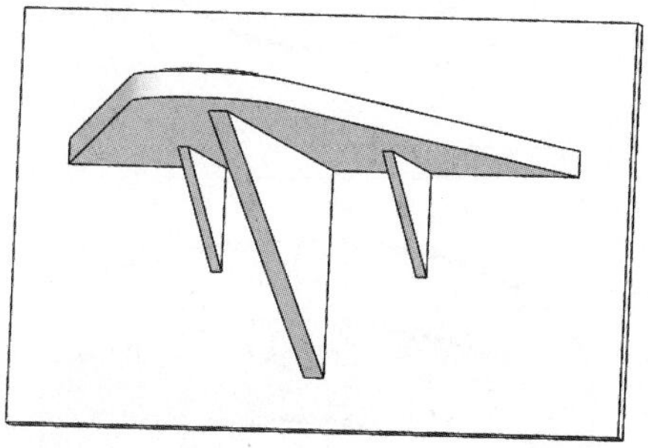

The holes will now be added to the part.

Select the top surface of the raised circle, as shown in Figure 3.87. Select the Circle Tool from the Sketch group of the CommandManager. Select the Top View.

As before, it is necessary to "wake up" a center mark at which the hole will be centered.

FIGURE 3.87

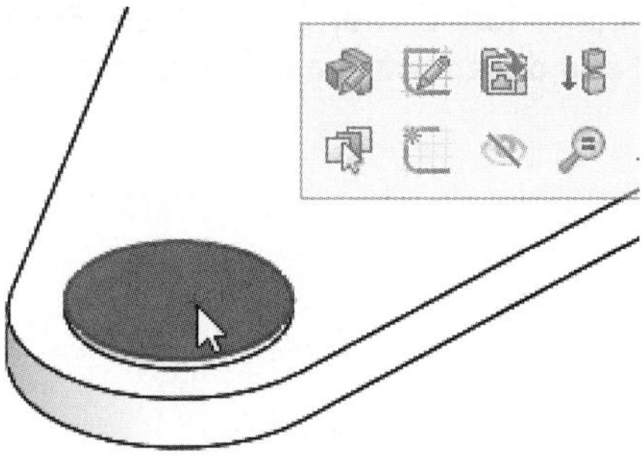

Move the cursor to the edge of the rounded end of the part, or the circular edge of the reinforced area, without clicking a mouse button. This will cause the center mark to be displayed, as shown in Figure 3.88.

Click and drag out a circle from the center mark. Select the Smart Dimension Tool, and dimension the diameter of the circle as 0.50 inches, as shown in Figure 3.89.

FIGURE 3.88

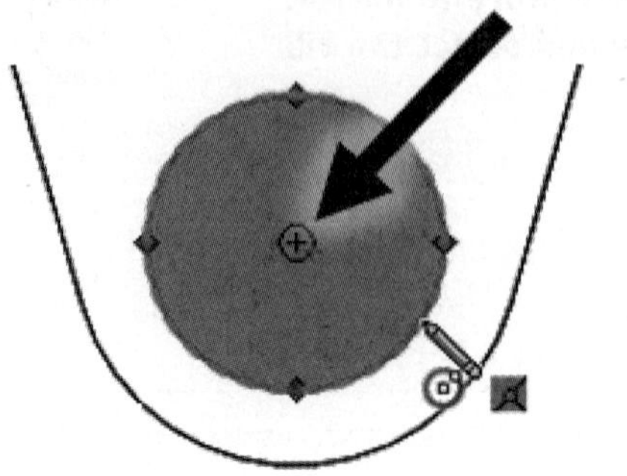

FIGURE 3.89

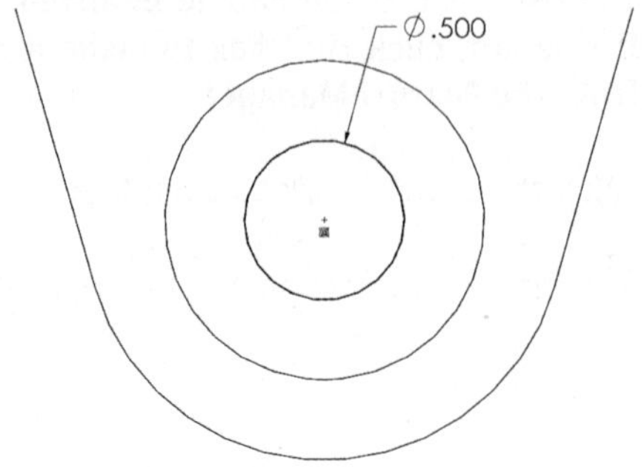

FIGURE 3.90

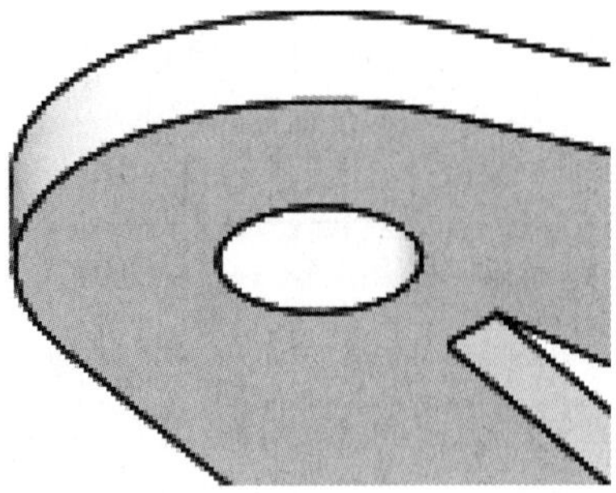

Select the Extruded Cut Tool from the Features group of the CommandManager, and select Through All as the type. Click the check mark to complete the hole, which is shown in Figure 3.90.

The four-hole bolt hole pattern used to mount the bracket to the wall will now be added. This will be done by creating a single hole, and then defining a *pattern* based on this hole. In Chapter 1, a *circular pattern* was used to create a bolt hole pattern in the flange. In this exercise, a new type of pattern known as a *linear pattern* will be introduced.

FIGURE 3.91

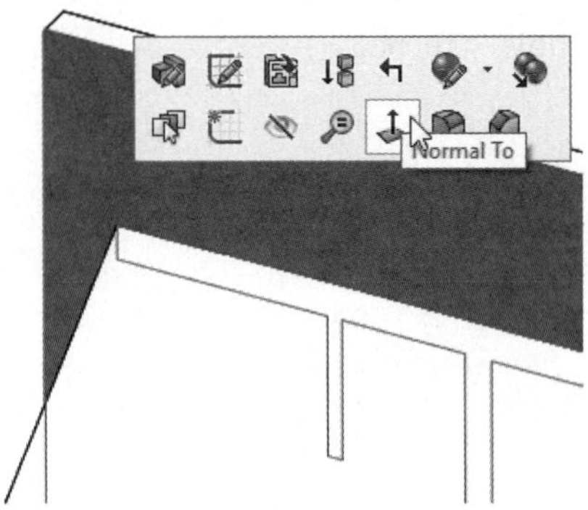

The first step will be the creation of a single bolt hole.

Select the large vertical face of the base part as shown in Figure 3.91 and select the Normal To View. Select the Circle Tool from the Sketch group of the CommandManager, and sketch a circle near the upper-left corner of the face. Select the Smart Dimension Tool and dimension the hole diameter as 0.375 inches.

Add linear dimensions between the circle and the edges, as shown in Figure 3.92.

FIGURE 3.92

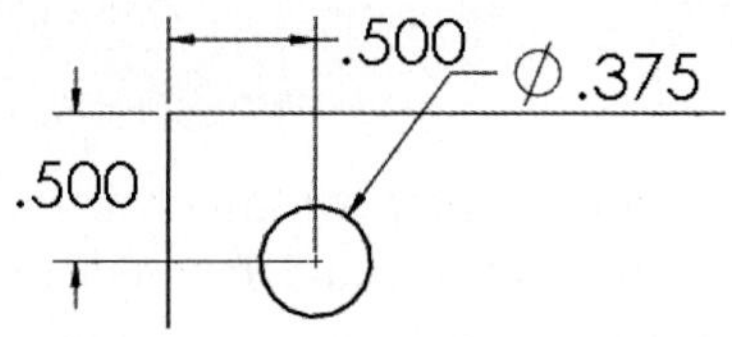

Select the Extruded Cut Tool from the Features group of the CommandManager, and select Through All as the type. Click the check mark to complete the hole, which is shown in Figure 3.93.

With this single hole defined, a linear hole pattern can now be created.

With the new hole selected, select the Linear Pattern tool from the Features group of the CommandManager, as shown in Figure 3.94.

FIGURE 3.93

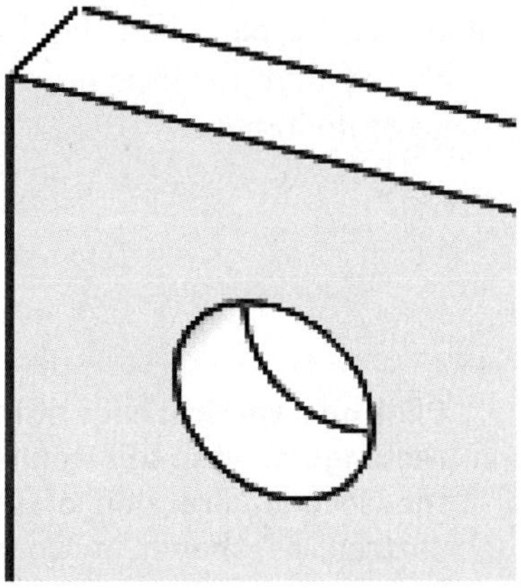

FIGURE 3.94

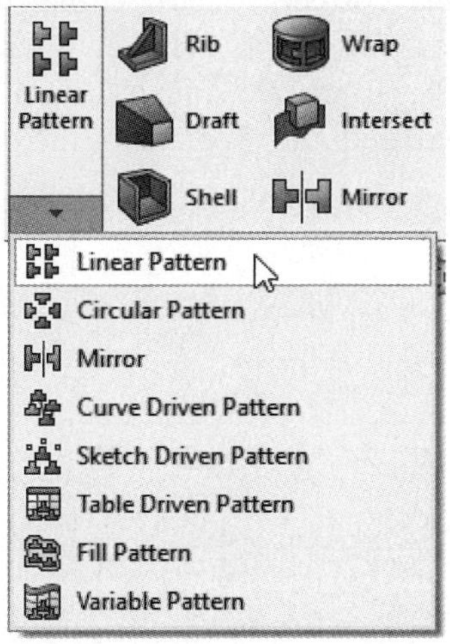

The feature Cut-Extrude2 (the hole just created) will be shown in the "Features to Pattern" box, indicating that this is the base feature for the linear pattern. The "Direction 1" box in the PropertyManager will be highlighted; this allows the first linear direction of the pattern to be established.

FIGURE 3.95

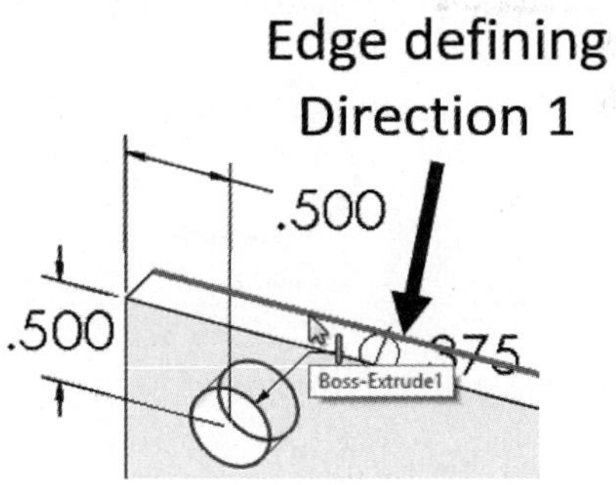

FIGURE 3.96

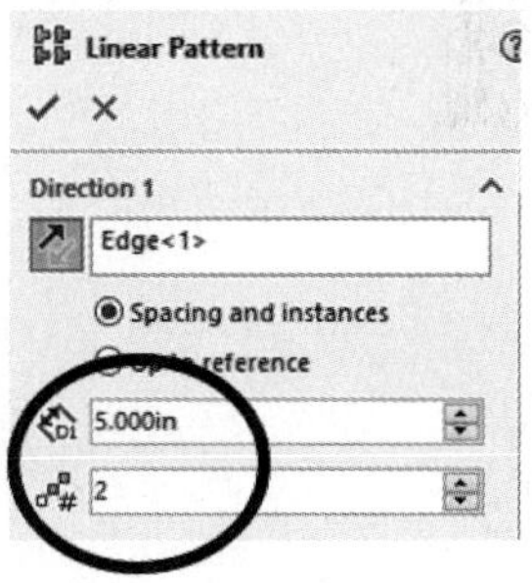

Click on the top horizontal edge of the rectangular base to establish the first direction, as shown in Figure 3.95. In the PropertyManager, set the Direction 1 distance to 5 inches and the number of instances to 2, as shown in Figure 3.96. The Direction 2 box in the PropertyManager will be highlighted.

DESIGN INTENT Symmetry in Modeling

In this chapter, we used mirroring techniques for sketch entities and features to define symmetric elements in a part. The four mounting holes in the bracket, which were created with a linear pattern, could have been defined with mirror commands. The advantage of doing so would be that if we change the dimensions of the vertical portion of the bracket (6 inches by 4 inches), the hole positions would be updated to the correct positions. The same associativity between the hole locations and the size of the bracket can be added if a linear pattern is used, but will require the addition of equations, which will be explained in Chapter 5.

Using symmetric elements in a part is good modeling practice. If a part contains planes of symmetry, plan your model to take advantage of those planes. In the bracket model, we centered the initial sketch about the origin. This placed the part so that the Right Plane coincided with the symmetry plane of the entire part, and the Top Plane coincided with an additional symmetry plane of the back plate of the bracket.

FIGURE 3.97

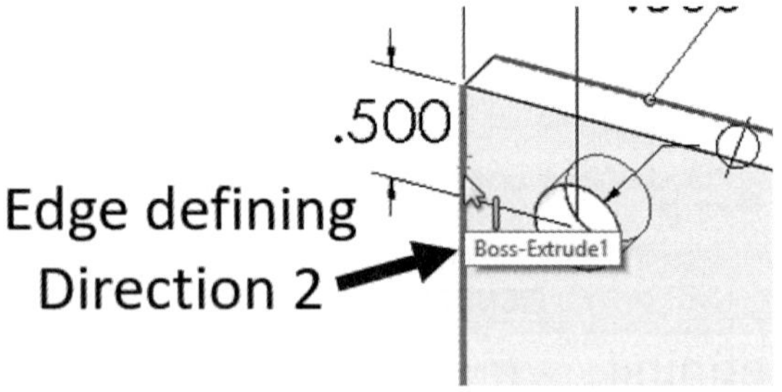

FIGURE 3.98

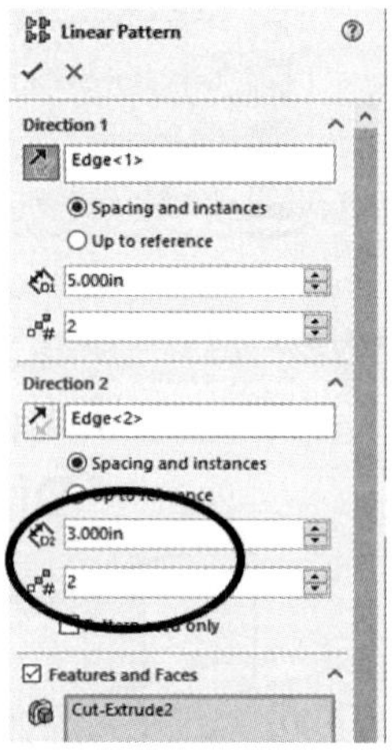

Click on a vertical edge of the rectangular base to establish the second direction of the pattern, as shown in Figure 3.97.

In the PropertyManager, set the Direction 2 distance to be 3 inches and the number of instances to be 2, as shown in Figure 3.98.

If the preview of the patterned holes shows that the pattern is created in the wrong direction, click on the arrow at the end of the edge defining the pattern direction, as shown in Figure 3.99.

FIGURE 3.99

VIDEO EXAMPLE 3

In this chapter, we created the mounting holes in the bracket using a linear hole pattern. Alternatively, the mounting holes could have been created by:

- creating one hole feature, and using feature mirroring to create the four holes, or,
- sketching one circular hole section, and using sketch mirroring to create a sketch of the four hole sections before extruding all four holes in a single cut.

These two alternatives are demonstrated in a video at the book website, *www.mhhe.com/howard2019.*

When the preview of the pattern appears correct, as shown in Figure 3.100, click the check mark to complete the pattern.

FIGURE 3.100

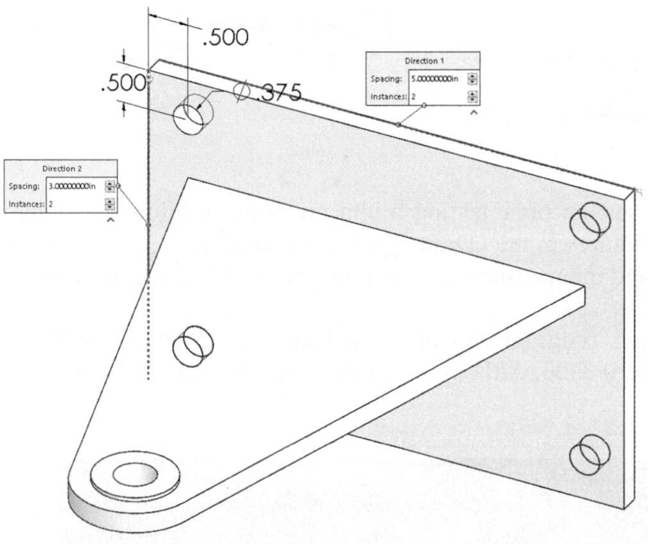

The completed part is shown in Figure 3.101.

FIGURE 3.101

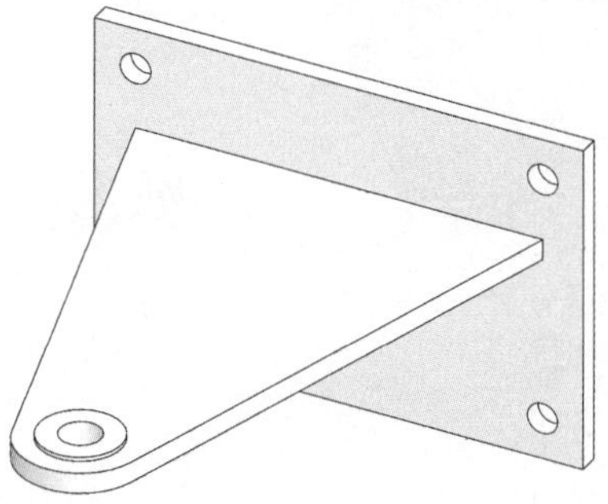

From the FeatureManager, right-click Material and select Edit Material. Select ABS from the Plastics group, as shown in Figure 3.102. Click Apply, and then Close.

FIGURE 3.102

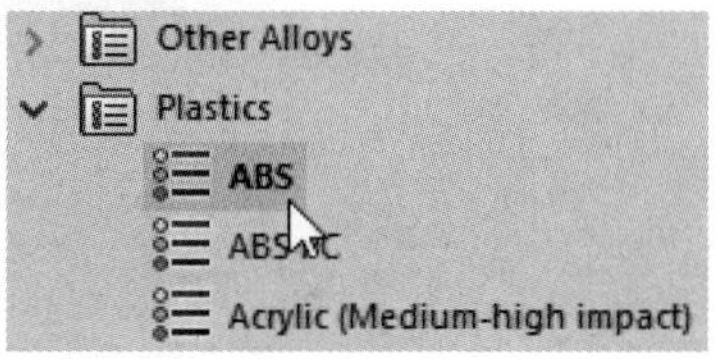

Save the part file.

3.3 Sharing and Displaying the Solid Model

FIGURE 3.103

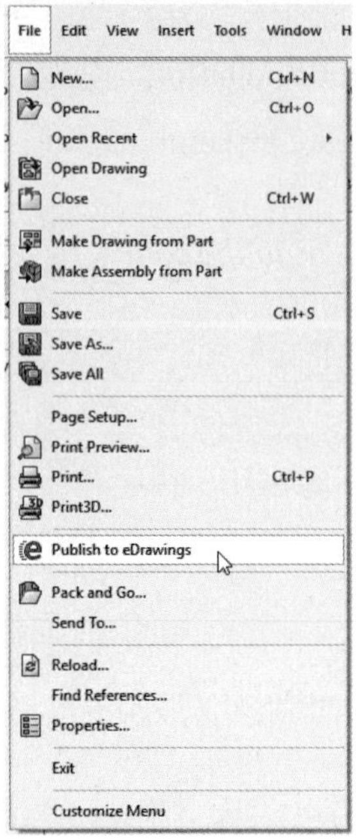

In Chapter 2, we created an eDrawing of a 2-D drawing. You can also create an eDrawing of a part. An eDrawing can be viewed by anyone who downloads the free eDrawing viewer. The small file sizes of eDrawings make them easy to share via e-mail.

Open the bracket created in the last section. From the main menu, select File: Publish to eDrawings, as shown in Figure 3.103. In the dialog box that opens, click Options, as shown in Figure 3.104. Check the box in System Options labeled "OK to measure this eDrawings file," as shown in Figure 3.105. Click OK to close the System Options menu, and OK to close the dialog box.

FIGURE 3.104

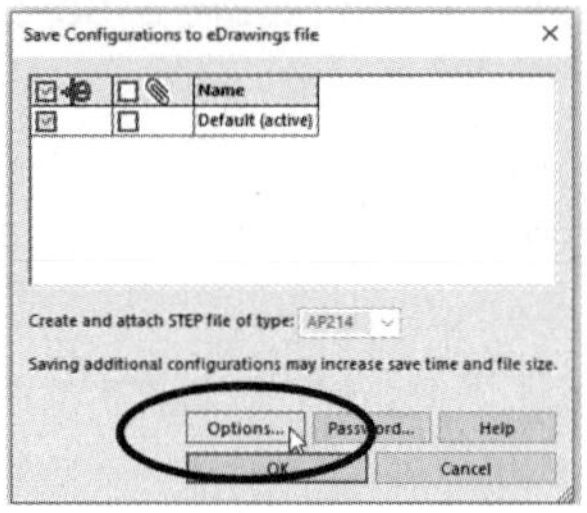

FIGURE 3.105

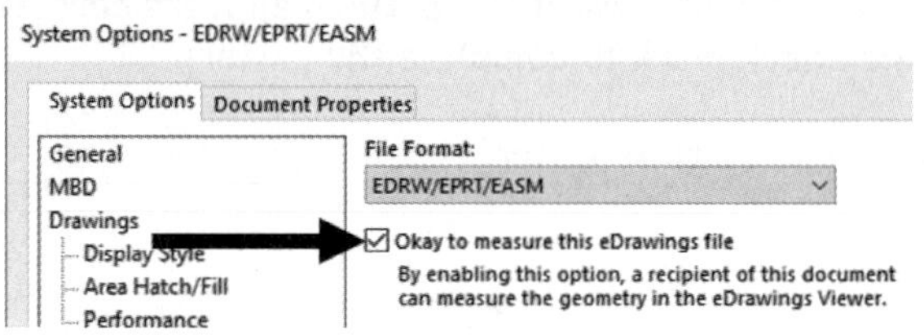

In certain cases, an organization might not want to allow customers or suppliers to measure features in the eDrawings file. In such cases, the option box can be left unchecked, and the measurement tool will be disabled in eDrawings.

The eDrawings program will open, and the bracket model will be displayed, as shown in **Figure 3.106**. Although the eDrawings model does not contain the history

FIGURE 3.106

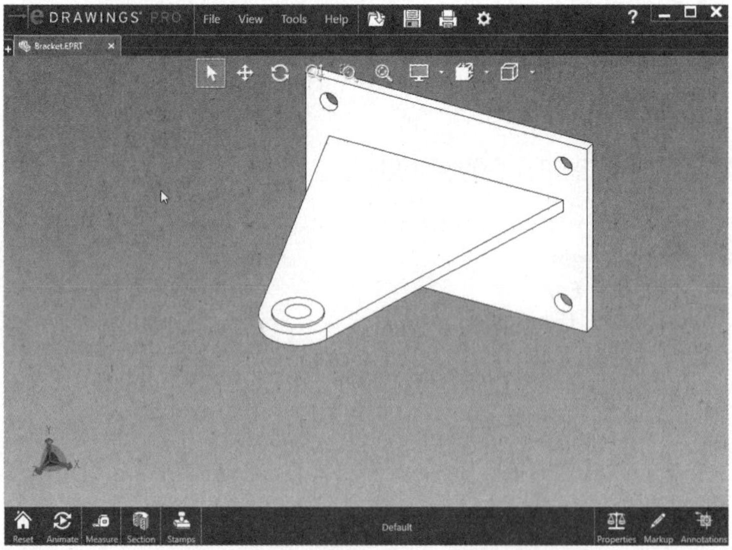

of the steps that were used to create it, and cannot be edited, it does contain the geometric data sufficient to view the model from various angles and view a cross-section, and it also contains the mass properties data.

Click the Animate button, and the Next button will appear. Click the Next button several times, as shown in Figure 3.107.

FIGURE 3.107

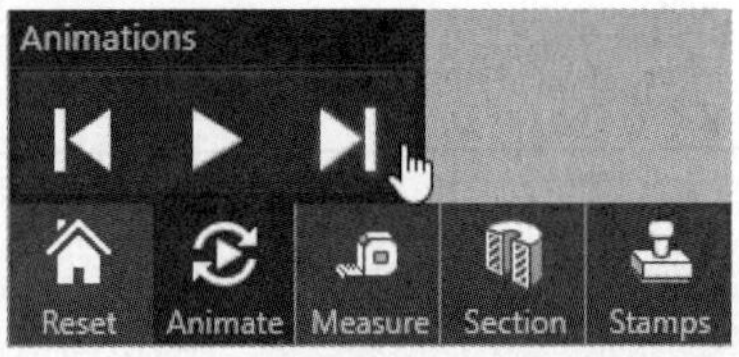

The Model View will shift from the Default View from the SOLIDWORKS file (trimetric) to isometric, and then to each of the principal views. The Play button will automatically rotate the part through the various views until the Stop button is pressed. The Reset button returns the view to the default. You can zoom in and out with tools similar to those of SOLIDWORKS.

Select the Reset button, and then the Section Tool, as shown in Figure 3.108. From the pop-up buttons that appear, choose the YZ Plane as the section plane.

FIGURE 3.108

Note that you can click and drag the plane through the part, as shown in **Figure 3.109**.

FIGURE 3.109

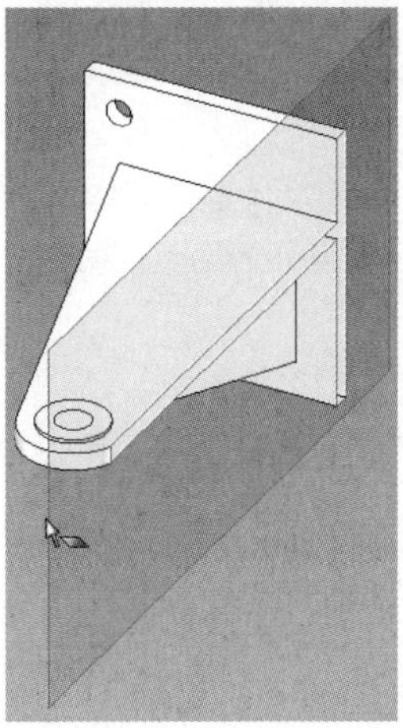

Click the Section Tool again to turn it off, and select the Properties Tool, as shown in Figure 3.110.

The mass, volume, and surface area are displayed, as shown in **Figure 3.111**. Note that you can select different units and more decimal places if desired.

FIGURE 3.110

FIGURE 3.111

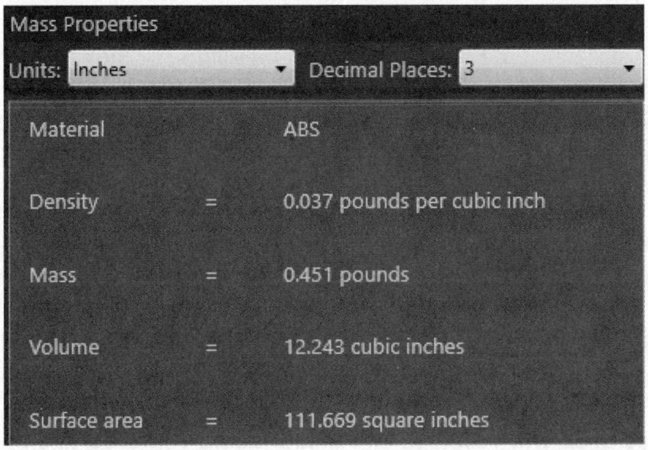

FIGURE 3.112

Click on the Measure Tool, as shown in Figure 3.112. Click on an edge, and the parameters of that edge (diameter or length) are displayed. If two entities are selected, then the dimensions between the two entities are displayed. Figure 3.113 shows two circular edges selected. The distance between the centers of the circles is displayed, along with the differences in x, y, and z coordinates. When you click in the space around the part, the selections are cleared. Experiment with the Measure Tool, and then select Save and save the file to the desired location.

FIGURE 3.113

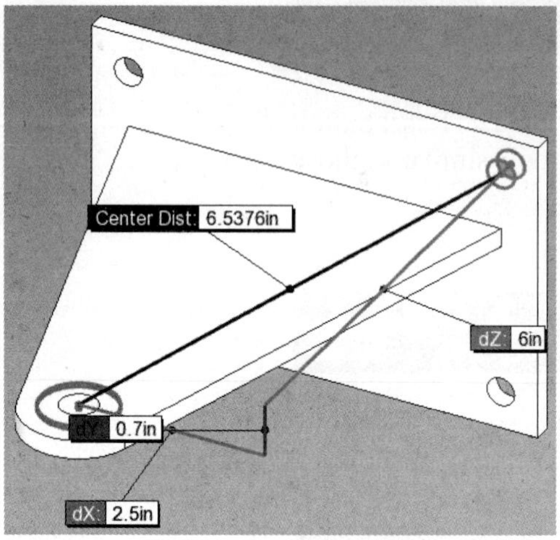

Note that the file size is about the same as that of a small word-processing file. This makes the file ideal for sharing via e-mail. The Send Tool from the Main Menu allows the eDrawing file to be sent by e-mail to a recipient. Options for sending eDrawings files are described in Chapter 2.

Close eDrawings, and return to SOLIDWORKS.

Often, you will want to display your solid model in a text document or in a presentation. The easiest way to do this is with screen captures of the model.

From the main menu, select View: Screen Capture: Image Capture, as shown in Figure 3.114. Open a Word document, and select Paste.

FIGURE 3.114

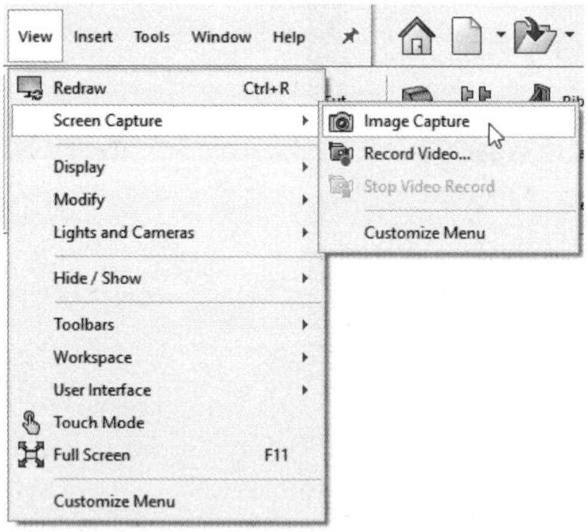

The image is copied to the Windows clipboard and then pasted into the Word (or PowerPoint) document. While the Prnt Scrn key performs a similar function, an advantage of SOLIDWORKS Image Capture is that only the graphics area is captured. With the Prnt Scrn key, the entire screen is captured and must be cropped if only the model is desired in the image.

The menu also includes a Record Video option, which records the screen as the model is rotated on the screen. However, we can create animations with the MotionManager, which gives us better control over the manipulation of the model.

Select the Motion Study 1 tab at the bottom of the screen, as shown in Figure 3.115. (If the tab is not visible, right-click in the CommandManager and check MotionManager.)

FIGURE 3.115

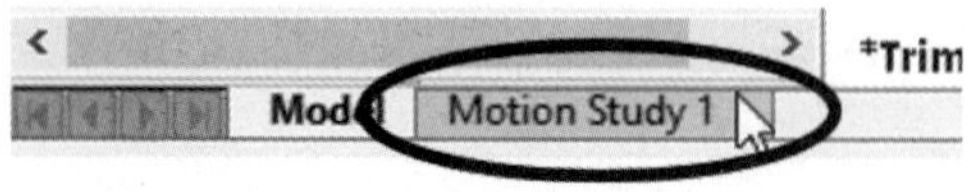

A timeline is displayed. We will set the view orientations of the model that we desire along the timeline, using *keys*.

Select the Trimetric View. Zoom and pan the model as desired. Since the recorded video will not show the FeatureManager, placing the model slightly to the left of center in the graphics area will result in the model being centered left-to-right in the video. Right-click the diamond-shaped key beside Orientation and

Camera Views, and select Replace Key, as shown in Figure 3.116. Move the cursor to the 2-second mark on the timeline, to the right of the Orientation and Camera Views entry. Right-click and select Place Key, as shown in Figure 3.117.

FIGURE 3.116

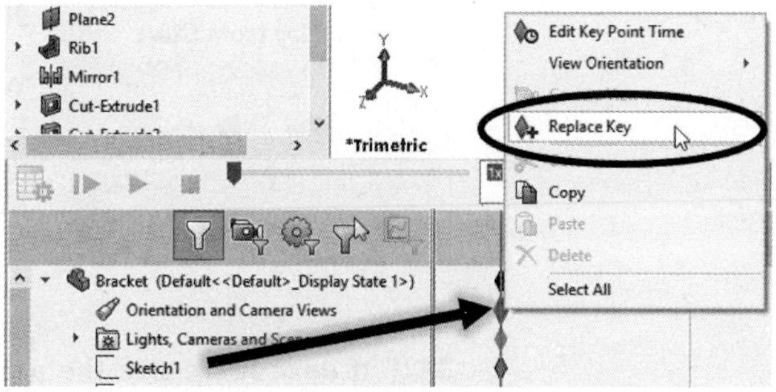

The purpose of placing this second key is to display the model in the selected orientation for two seconds before beginning to rotate it. Since the ribs are best seen from the Bottom View, we will now rotate the model to that view.

Select the Bottom View. Zoom and pan the model as desired, again placing it to slightly left of center in the graphics window. Move the cursor to the 4-second mark, and to the right of Orientation and Camera Views, right-click and select Place Key. Repeat for the 6-second mark.

FIGURE 3.117

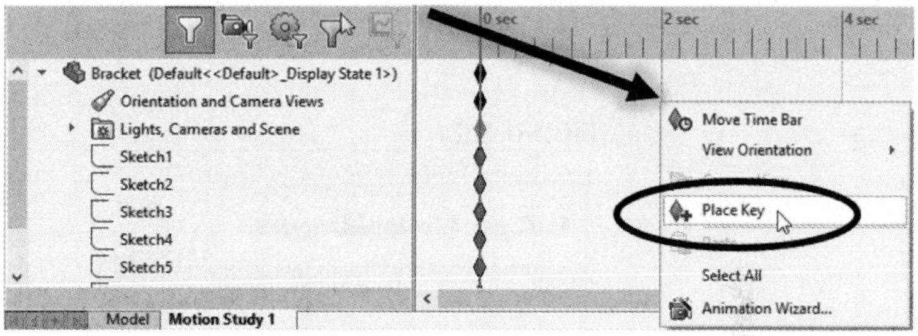

The timeline will now show a line between the keys at 2 and 4 seconds, as shown in Figure 3.118. This indicates that the views are different at these two points, and so the model will be rotated into the new view over this 2-second interval.

FIGURE 3.118

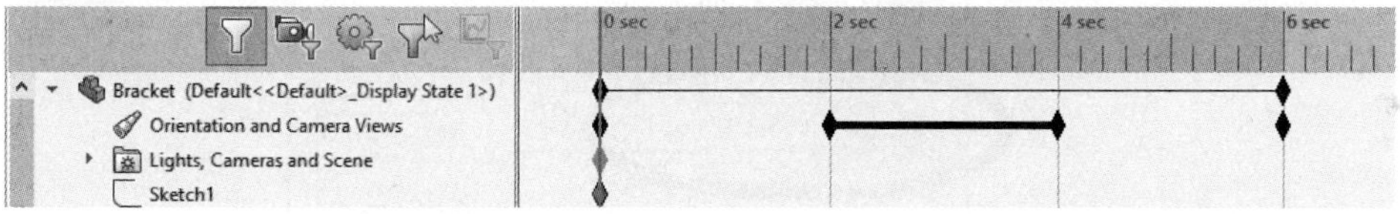

Select the Right View and zoom and pan the model as desired. Place keys at 8 and 10 seconds. Select the Trimetric View, zoom and pan as desired, and place keys at 12 and 14 seconds.

The finished timeline is shown in Figure 3.119. Note that individual keys can be deleted or replaced as necessary by right-clicking on that key. To start over, you can right-click a key, choose Select All, and press the Delete key.

FIGURE 3.119

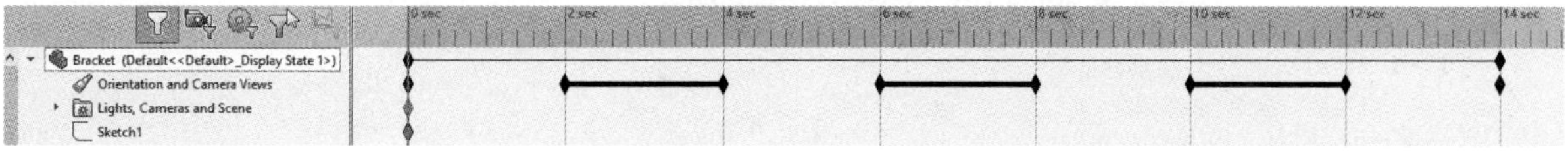

FIGURE 3.120

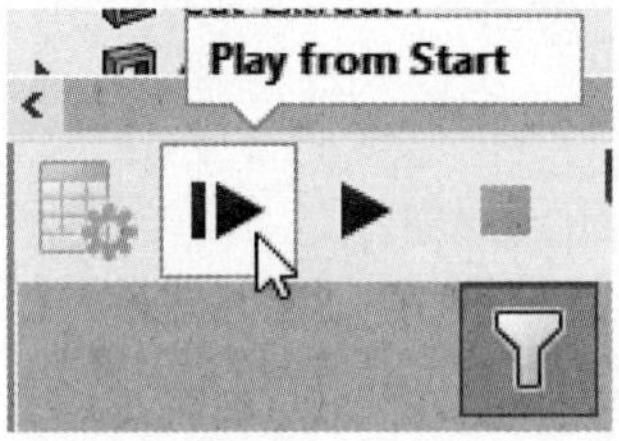

To play the animation, select Play from Start, as shown in Figure 3.120.

We will now save the animation as a video file. First, we will minimize the MotionManager so that the aspect ratio of the graphics area is better for the finished video.

Collapse the MotionManager by selecting the arrow at the right edge of the screen, as shown in Figure 3.121. If desired, re-size the part window to the desired aspect ratio. Select the Save Animation Tool, as shown in Figure 3.122. Select the file name and directory, and accept the default settings. At the Video Compression screen, choose the Microsoft Video 1 compressor and uncheck the Key Frames box, as shown in Figure 3.123.

FIGURE 3.121

FIGURE 3.122

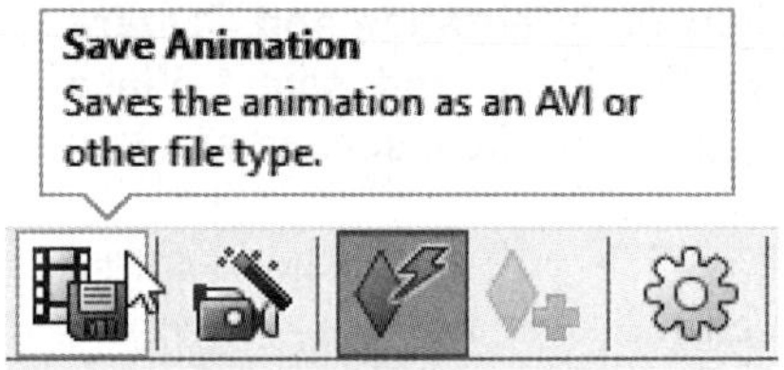

FIGURE 3.123

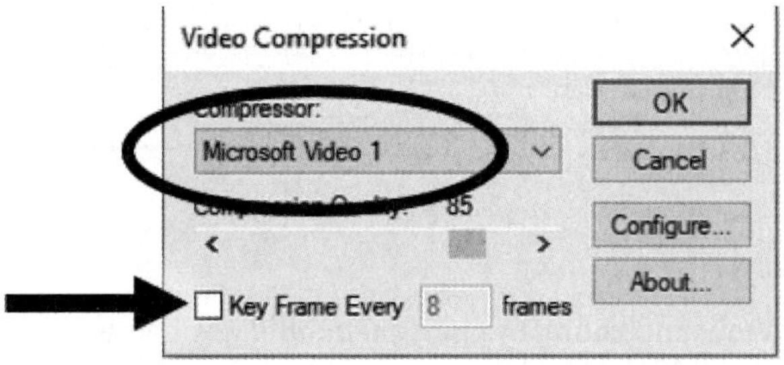

The animation will run on the screen as the file is created. When the animation is completed, the saved video file can be viewed from a standard player. Videos can also be embedded into PowerPoint presentations. This allows a model to be displayed and rotated during the presentation without having to open and run SOLIDWORKS on the presentation computer.

Save and close the SOLIDWORKS file.

PROBLEMS

P3.1 Modify the wide flange beam model created in Section 3.1 to create a one-foot segment of a W12×65 beam, using the dimensions from **Table 3.1**.

P3.2 The cross-section of a W14×370 Wide-Flange Beam is shown in **Figure P3.2A**. Create a model of a 36-inch long segment of a W14×370 beam by creating the sketch shown in **Figure P3.2B** and mirroring it about the two centerlines. (Note: This must be done in two steps. Select the items to be mirrored and one of the centerlines, and then select the Mirror Entities Tool. Then select all entities to be mirrored and the other centerline and select the Mirror Entities Tool. Only one centerline can be selected for each mirror operation.) Extrude the sketch to yield the beam section shown in **Figure P3.2C**. Set the material type to Plain Carbon Steel and find the weight of the beam segment.

(Answer: 1101 lb)

FIGURE P3.2A

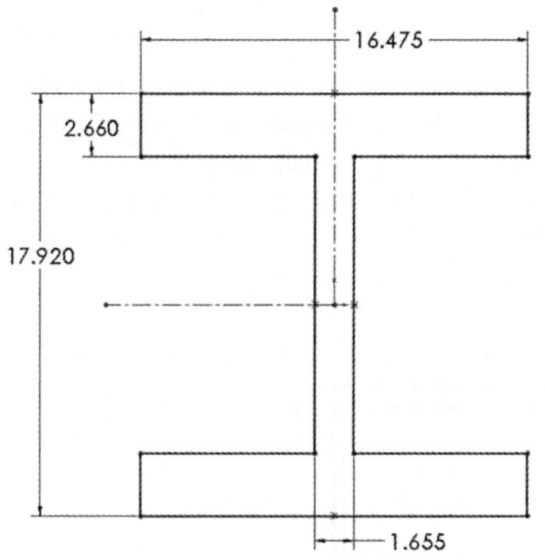

FIGURE P3.2B

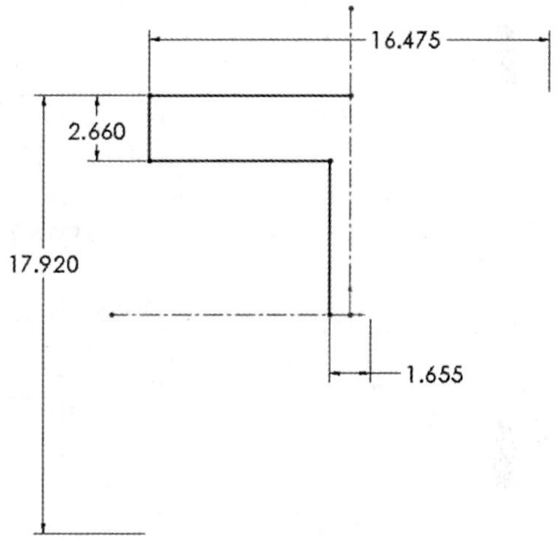

FIGURE P3.2C

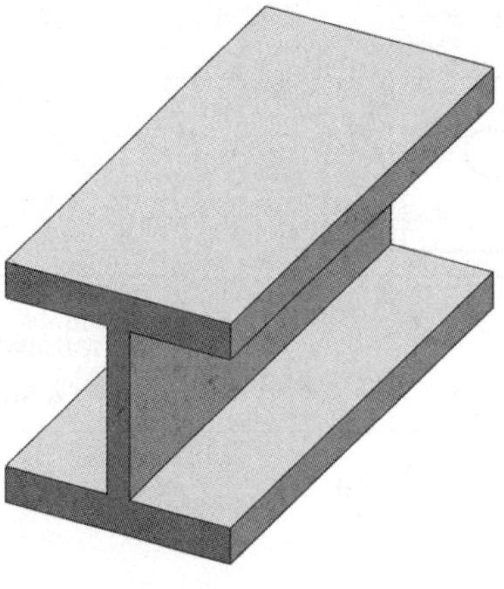

P3.3 Modify the wide-flange beam segment of **Problem P3.2** by adding six 1-inch-diameter holes to each end of the beam, as shown in **Figure P3.3A**. The locations of the holes are shown in **Figure P3.3B**. Create a single hole, and use a linear pattern to place the other five holes in one end. Then use a mirror command to place the holes in the other end.

FIGURE P3.3A

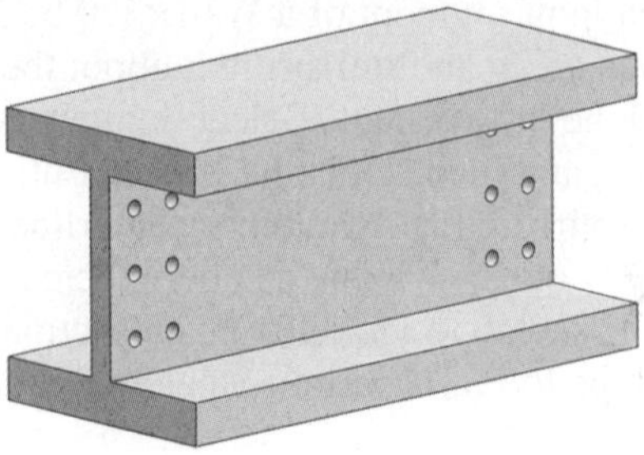

FIGURE P3.3B

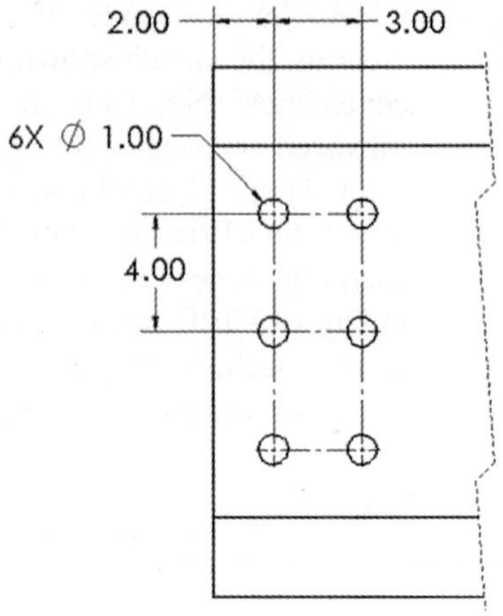

P3.4 Create a part model of the plastic bracket shown in **Figure P3.4A** and detailed in **Figure P3.4B**. Use symmetry in your model so that if you change the width of the part from 2 to 3 inches, the rib and hole placements remain symmetric, as shown in **Figure P3.4C**.

FIGURE P3.4A

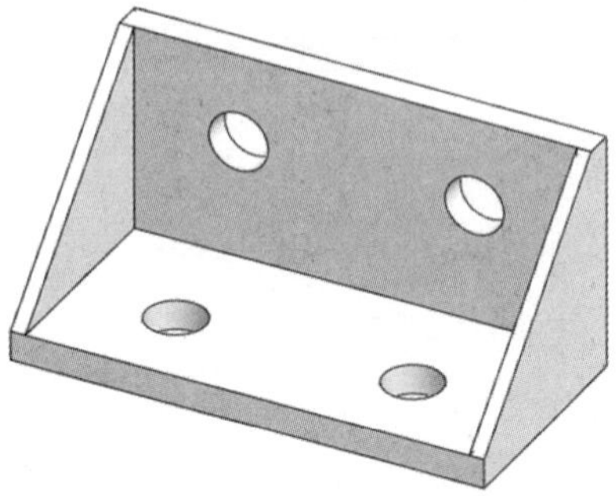

FIGURE P3.4B

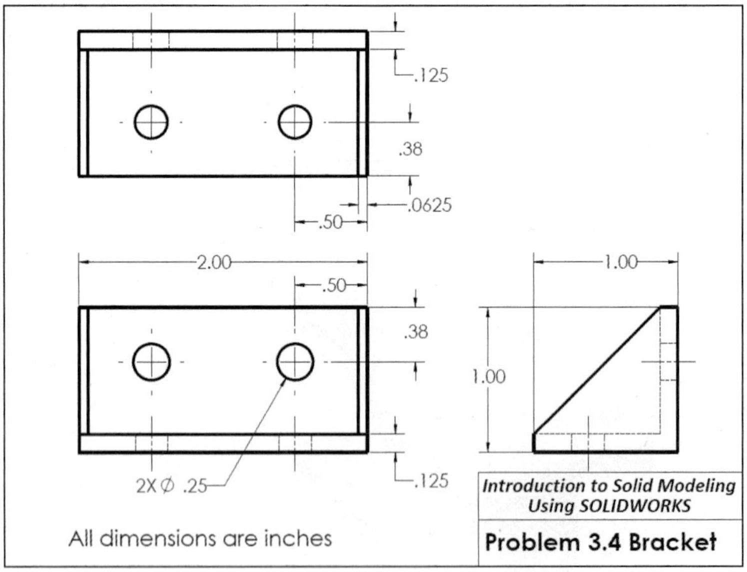

FIGURE P3.4C

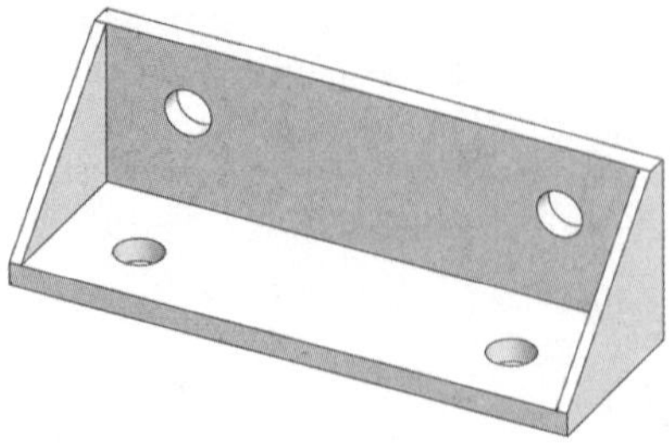

P3.5 Create a model of the perforated board shown in **Figure P3.5A** by the following procedure:

a. Begin with a 4-inch by 2-inch sketch, centered at the origin (**Figure P3.5B**).

b. Extrude the sketch 0.1 inch.

c. Create and locate a square thru-hole, as shown in **Figure P3.5C**.

d. Use a linear pattern to create an evenly spaced 40×20 grid of holes.

FIGURE P3.5A

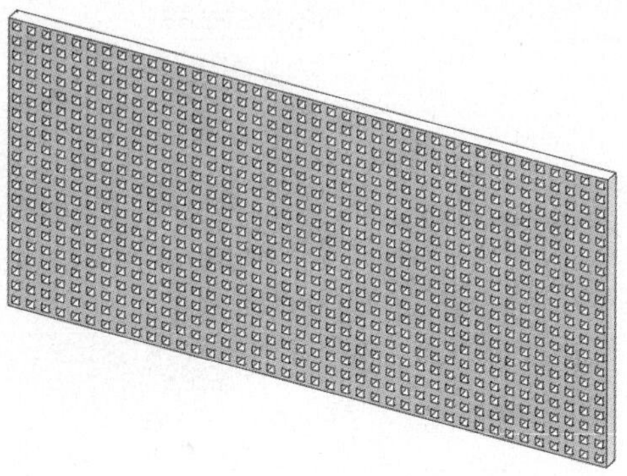

FIGURE P3.5B

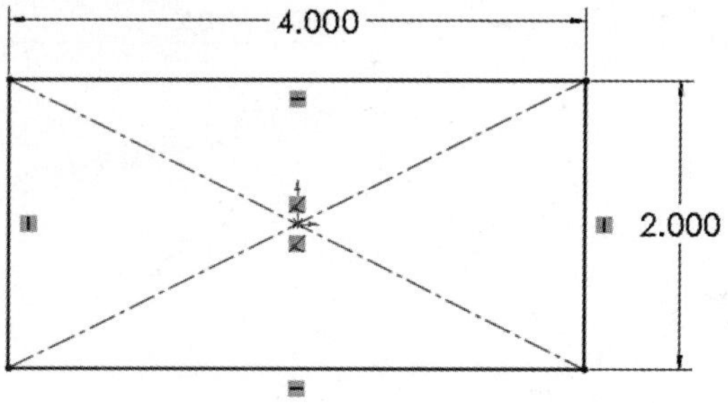

FIGURE P3.5C

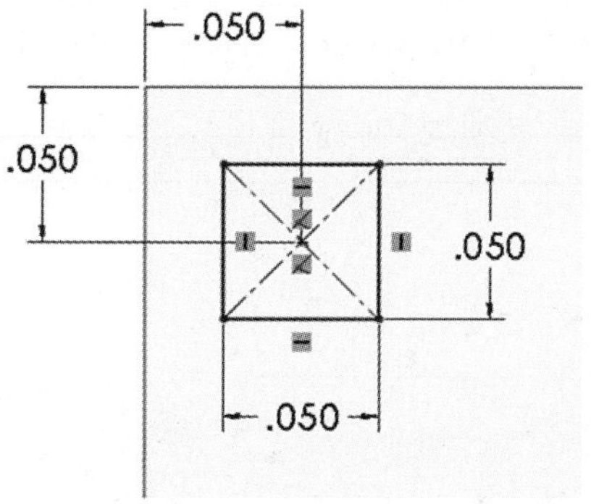

P3.6 Create a solid model of a hacksaw blade (as shown in **Figure P3.6A**, with a close-up view of the teeth shown in **Figure P3.6B**), using a linear patterned cut to create the saw teeth. Follow the procedure outlined below.

a. Begin by creating a sawblade "blank," using the dimensions shown in **Figure P3.6C** and extruding the shape to a 0.02-inch depth.

b. Sketch a single tooth profile, and extrude a cut through the blank (see **Figure P3.6D**).

c. Create a linear pattern to copy the tooth profile the length of the sawblade.

FIGURE P3.6A

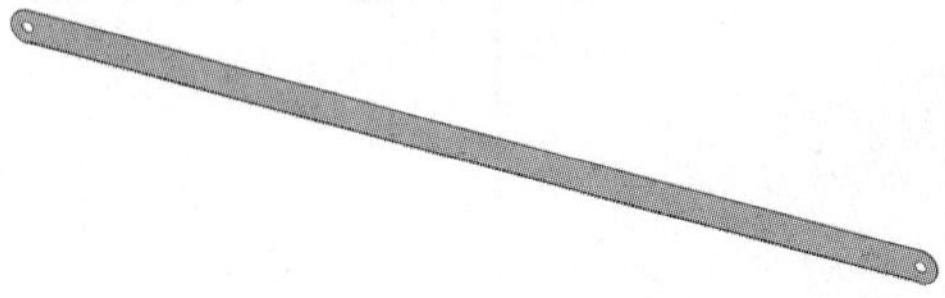

FIGURE P3.6B

FIGURE P3.6C

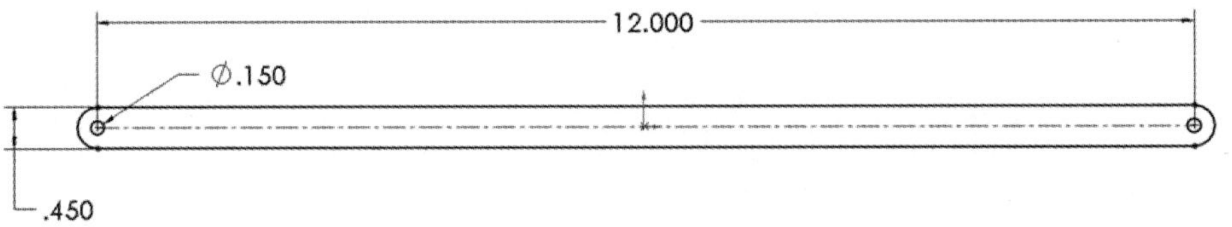

FIGURE P3.6D

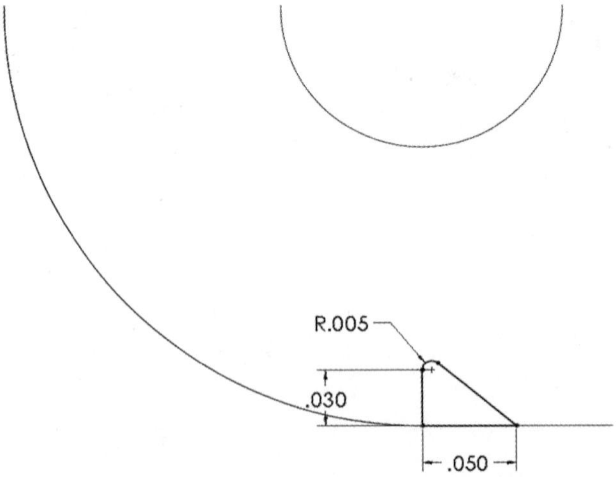

P3.7 Create a solid model of the muffin pan shown in **Figure P3.7A** and detailed in **Figure P3.7B**. The entire pan is 0.050 inches thick. Each of the 12 muffin wells is 3 inches in diameter at the top, is 1.5 inches deep, and has tapered sides, as shown in **Figure P3.7C**. The entire top surface and the insides of the muffin wells are to be coated with a nonstick material. Find the surface area that will be coated. (Hint: The Mass Properties Tool gives you the surface area of the entire part, but to get the surface area for a single surface or a specific group of surfaces, select the surface(s), using the Ctrl key to select multiple items, and select Tools: Measure from the main menu.)

FIGURE P3.7A

FIGURE P3.7B

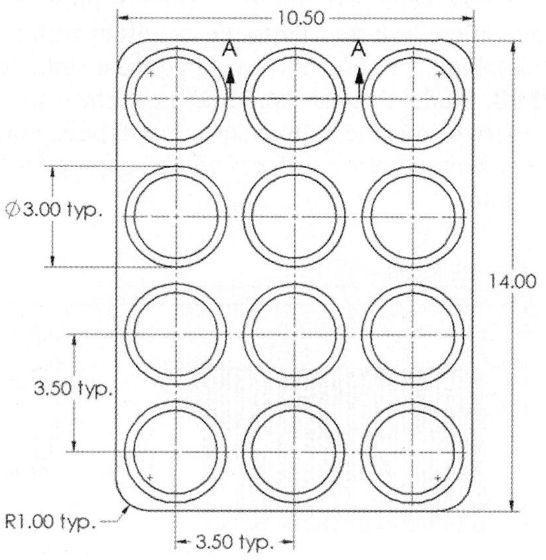

FIGURE P3.7C

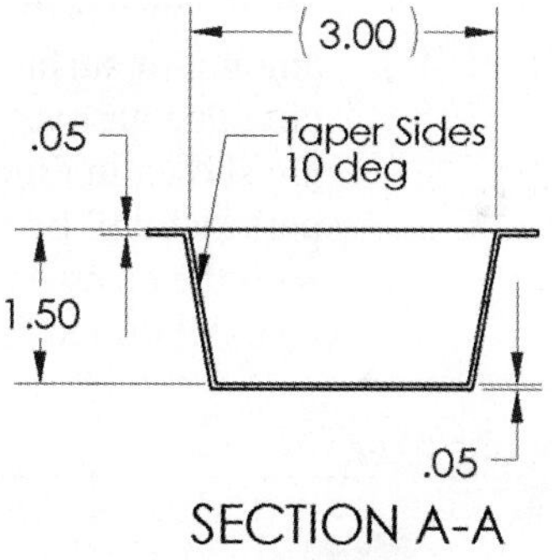

P3.8 Create a solid model of the gear rack shown in **Figure P3.8A** using the following procedure:

a. Begin creating a "gear blank" by extruding a 0.75 × 0.75-inch cross section to an overall depth of 11.16 inches.

b. At the left end of the gear blank, create the sketch shown in **Figure P3.8B** (dimensions in inches), and extrude a cut to begin the gear tooth definition.

c. Add 0.005-inch fillets at the root of the tooth.

d. Create a linear pattern of both the extruded cut and the fillets to create the final gear profile. The tooth-to-tooth spacing should be 0.215 inches.

FIGURE P3.8A

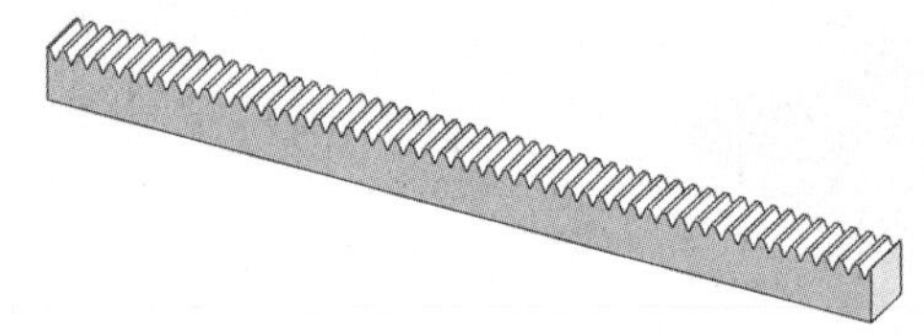

FIGURE P3.8B

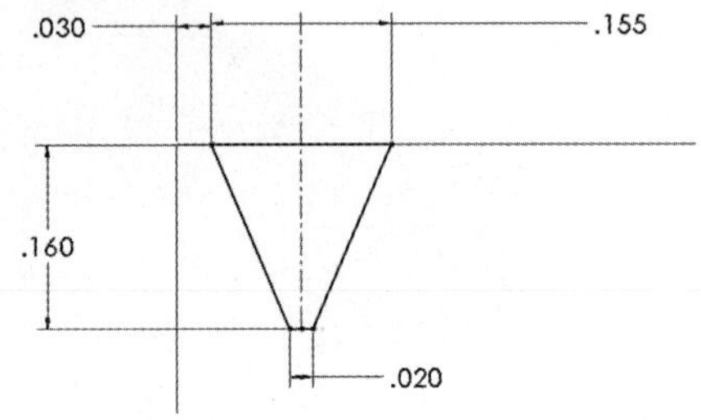

P3.9 Many electronic devices use a heat sink to conduct heat away from a component that may be damaged by high temperatures. A simple heat sink often consists of a conductive metal part with fins. The fins provide a large amount of surface area to allow heat transfer to the air (often with a fan to circulate air over the fins). **Figure P3.9A** shows a copper heat sink; details are shown in **Figure P3.9B**. Model this part, and find the surface area of the part that will be exposed to the air (the bottom surface will be in contact with the electronic component; all other surfaces will be exposed to the air). All dimensions are mm.

FIGURE P3.9A

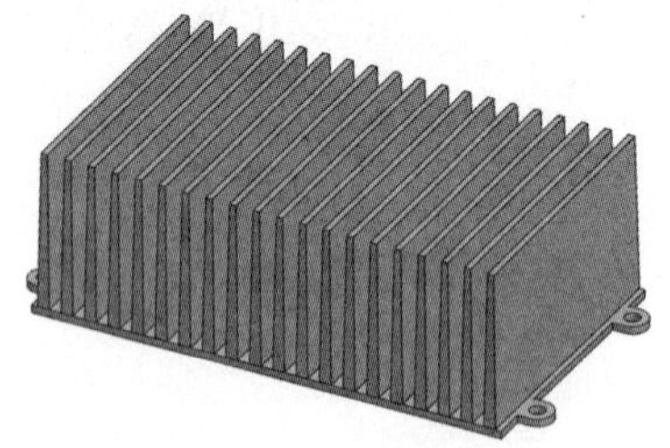

FIGURE P3.9B

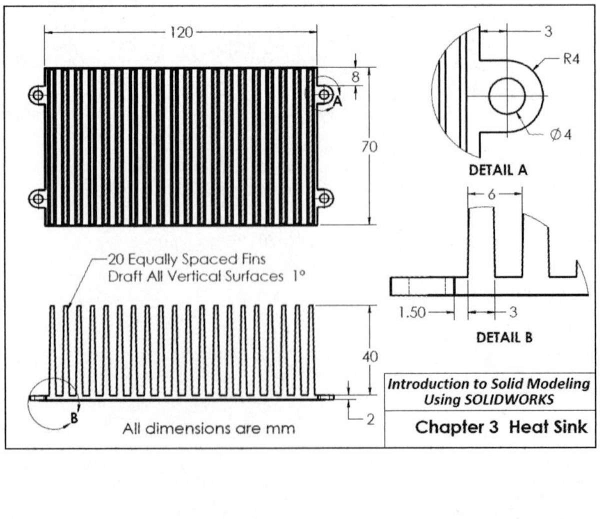

P3.10 Design a wall grate to cover the 12″ × 4″ heating duct shown in **Figure P3.10A**. The grate must attach to the wall with six #10 screws, and must have between 9 in^2 and 10 in^2 of open area through which air will pass. In your design, use linear patterns to define the hole pattern and the grate pattern. The grate must be symmetric for ease of mounting. An example is shown in **Figure P3.10B**.

FIGURE P3.10A

FIGURE P3.10B

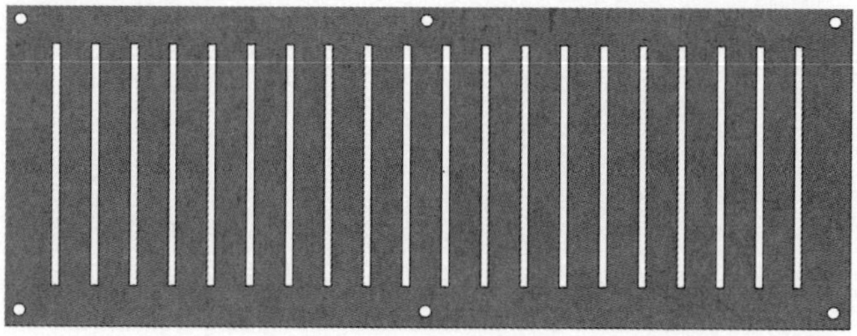

P3.10 Design a wall plate to cover the 12 × 8 in. heating duct shown in Figure P3.10A. The plate must mount to the wall with six #10 screws, and must have between 8 in² and 10 in² of openings through which air will pass. Be creative in using pattern features to define the hole pattern and the grate pattern. The plate should be symmetric for ease of mounting. An example is shown in Figure P3.10B.

FIGURE P3.10A

FIGURE P3.10B

CHAPTER 4

Advanced Part Modeling

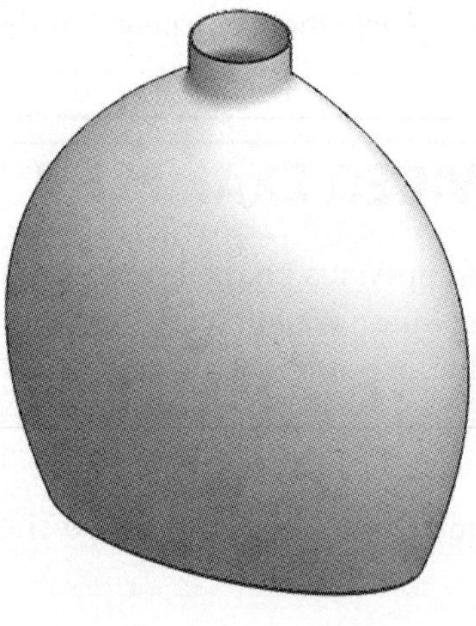

Introduction

The parts that we have made so far have been made primarily with extruded and revolved bases, bosses, and cuts. In this chapter, we will introduce several other tools for creating and modifying parts, including the *Loft, Shell,* and *Sweep* Tools.

Chapter Objectives

In this chapter, you will:

- create a lofted feature from multiple sketches,
- use the Shell Tool to create a thin-wall part,
- learn how to change the order in which features are created and modified,
- create raised letters on a part,
- learn how to create a 3-D sketch, and
- create a feature by sweeping a cross-section around a planar or 3-D path.

4.1 A Lofted and Shelled Part

In this exercise, we will construct the business card holder shown in **Figure 4.1**. Note that the top of the part is rounded, while the bottom of the part is rectangular. These dissimilar shapes will be joined into a solid with the Loft Tool. Also, notice that the part is not solid, but rather is hollow underneath, as the view in **Figure 4.2** shows. The Shell Tool allows this type of construction to be easily modeled.

FIGURE 4.1

FIGURE 4.2

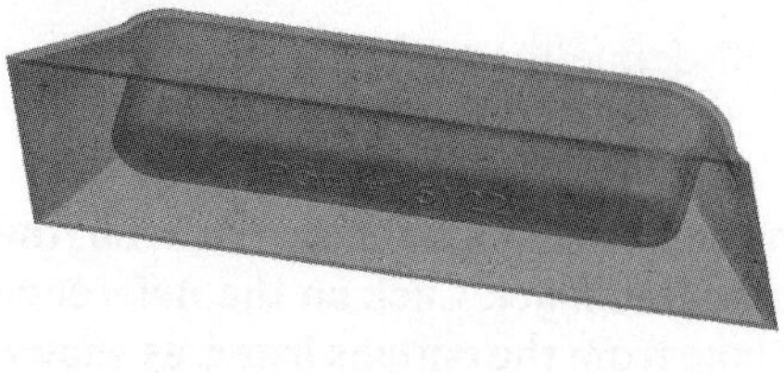

VIDEO EXAMPLE 4

In previous chapters, we created solid features by extruding or revolving a 2-D sketch. In this chapter, we introduce two methods of creating solid features from multiple sketches: lofted features between two or more sketches, and swept features made by sweeping one sketch (the profile) around a second sketch (the path).

The video at *www.mhhe.com/howard2019* demonstrates the creation of these parts using lofts and sweeps, and also shows the use of the Shell Tool for creating thin-wall parts.

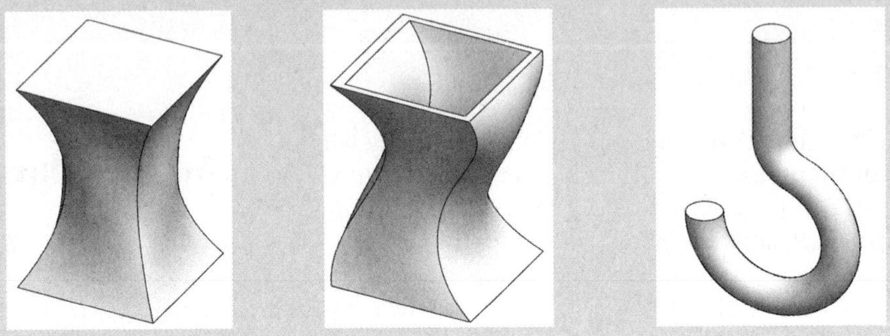

FIGURE 4.3

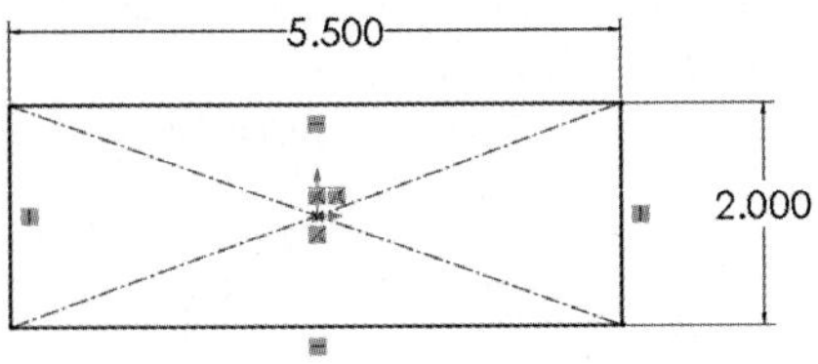

Open a new part. Select the Top Plane. Select the Center Rectangle Tool from the Sketch group of the CommandManager. Drag out a rectangle, centered at the origin.

Select the Smart Dimension Tool, and dimension the rectangle to be 5.50 inches by 2.00 inches. The resulting sketch is shown in Figure 4.3.

FIGURE 4.4

In previous tutorials, upon completing a sketch we have then used a Features tool to convert the sketch into a 3-D object. However, a lofted feature requires at least two sketches. Therefore, we will close this sketch and begin the second sketch.

Close the sketch by clicking on the icon indicated in Figure 4.4, in the upper-right corner of the screen.

FIGURE 4.5

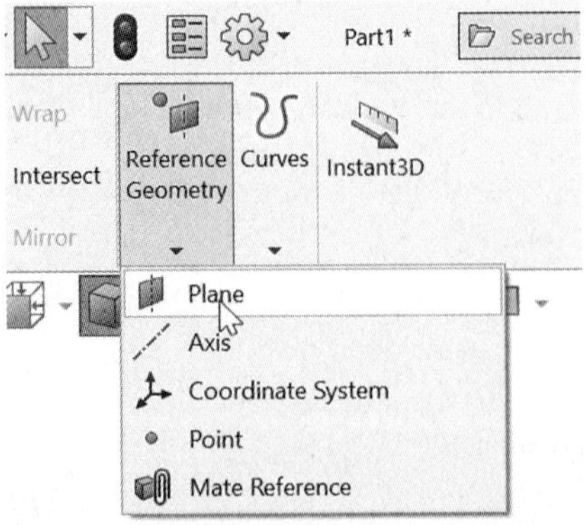

You can also close a sketch by choosing Exit Sketch from the Sketch group of the CommandManager.

The second sketch, which will define the top of the part, will be created in a new plane.

Select the Top Plane from the FeatureManager. Click on the Features group of the CommandManager. Click on the Reference Geometry Tool, and choose Plane from the options listed, as shown in Figure 4.5. In the PropertyManager, set the offset distance as 1 inch.

Make sure that the new plane is above the Top Plane, as shown in Figure 4.6 (check the Flip box to change the direction, if necessary), and click the check mark to create the new plane.

FIGURE 4.6

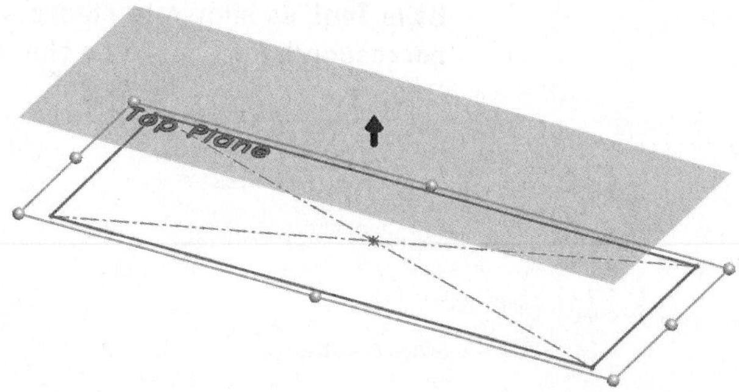

With the new plane selected (Plane1), select the Centerpoint Straight Slot Tool from the Sketch group of the CommandManager, as shown in Figure 4.7. In the PropertyManager, check the box labeled "Add dimensions," as shown in Figure 4.8.

Switch to the Top View. Click on the origin, and then drag the cursor to the right, as shown in Figure 4.9. Click to define the endpoint of the centerline of the slot geometry, and then drag the cursor upwards, as shown in Figure 4.10. Make sure that you do not drag the cursor onto the construction line of the first sketch; doing so will create an unwanted relation. Click to complete the slot, and dimensions will be added automatically. Press Esc to end the Slot Command. Double-click each dimension and set its value as shown in Figure 4.11. Close the sketch. Hide Plane1 by right-clicking its name in the FeatureManager and selecting Hide.

Both of the sketches required for the lofted feature are now in place.

FIGURE 4.7

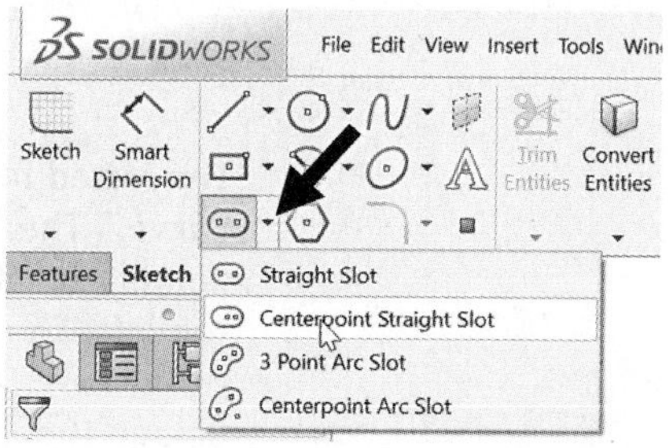

FIGURE 4.8

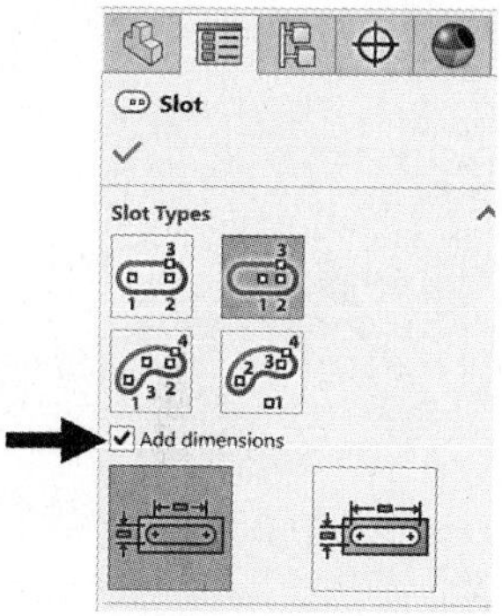

FIGURE 4.9

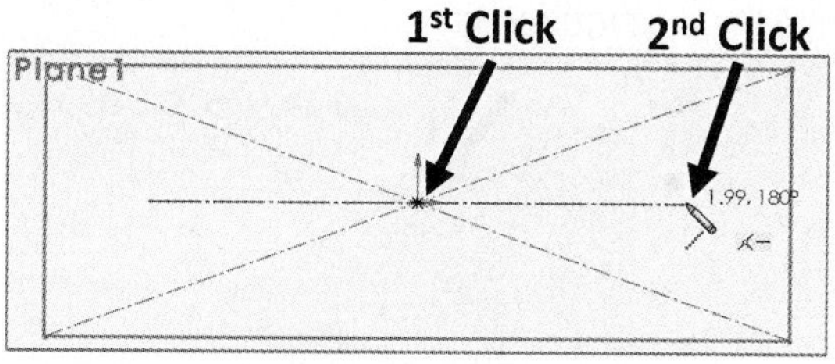

FIGURE 4.10

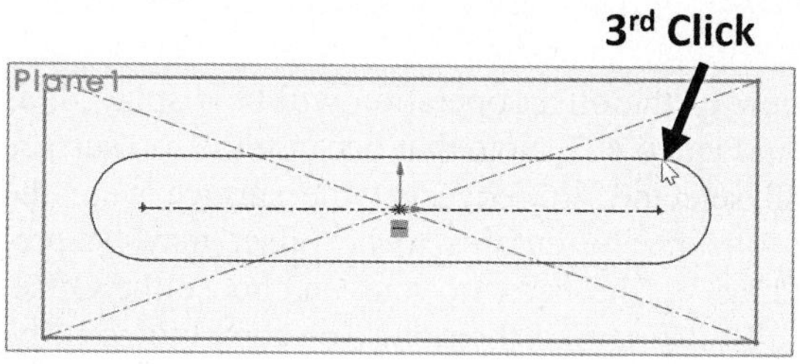

FIGURE 4.11

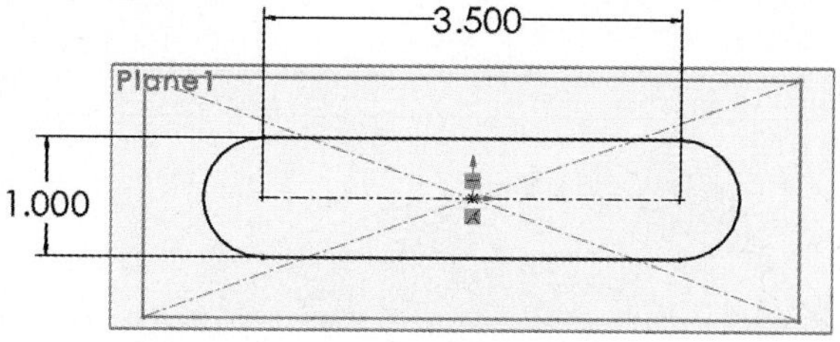

Click on the Features tab of the CommandManager. Select the Lofted Boss/Base Tool, as shown in Figure 4.12. Now click on each of the two sketches, near corresponding corners, as shown in Figure 4.13.

FIGURE 4.12

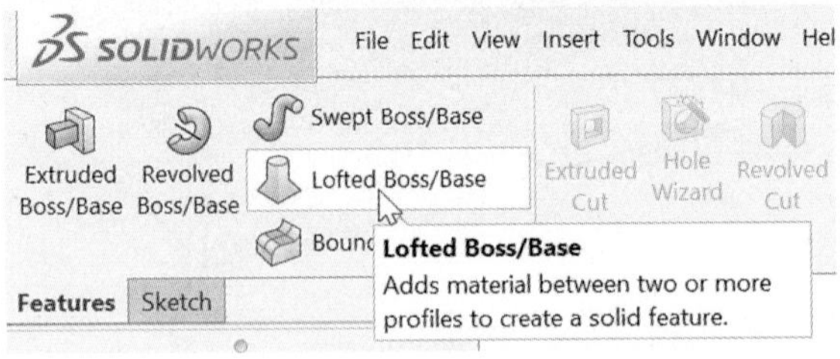

FIGURE 4.13

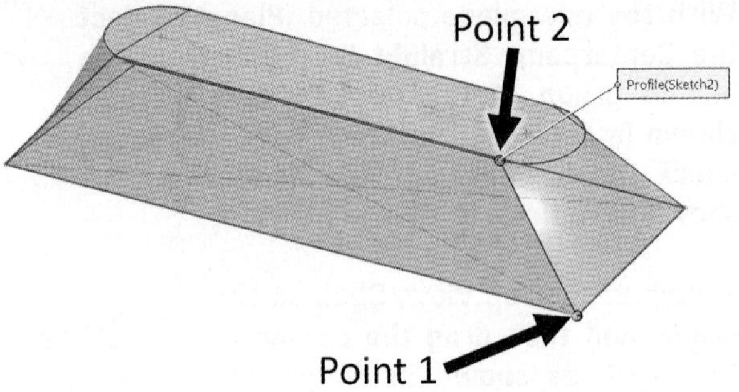

FIGURE 4.14

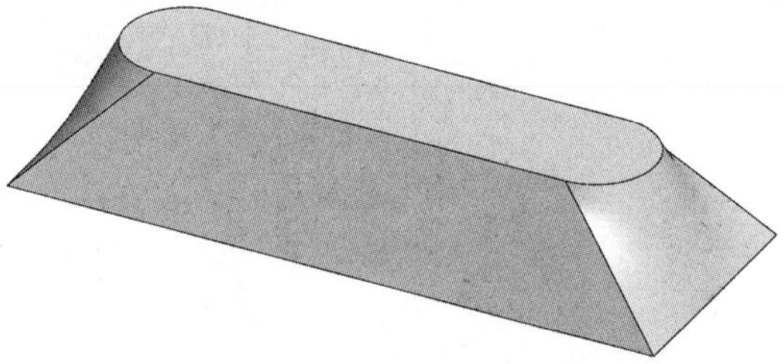

The lofted feature will be created based on a "guide curve." The guide curve is created based on the points selected. Click the check mark to complete the loft. The resulting part is shown in Figure 4.14.

Select the top surface of the part. Select Sketch from the pop-up toolbar, as shown in Figure 4.15, to open a sketch on the selected surface. Select the Offset Entities Tool, as shown in Figure 4.16.

FIGURE 4.15

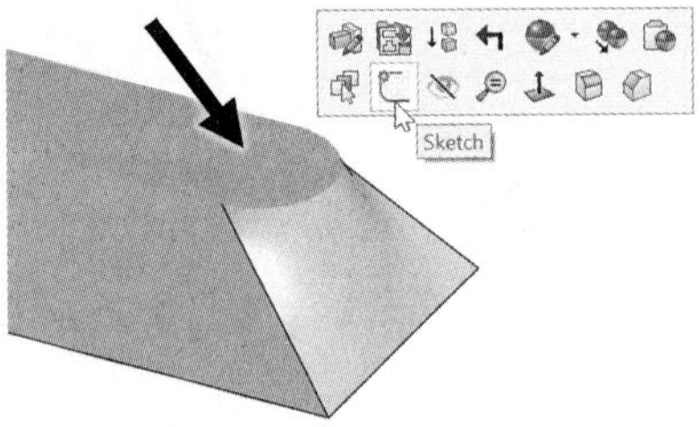

FIGURE 4.16

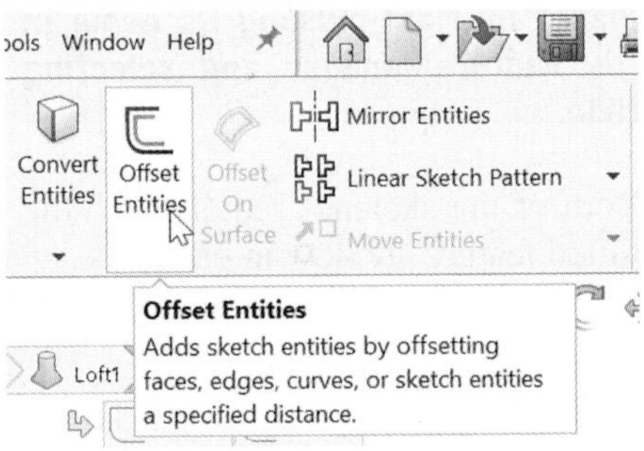

FIGURE 4.17

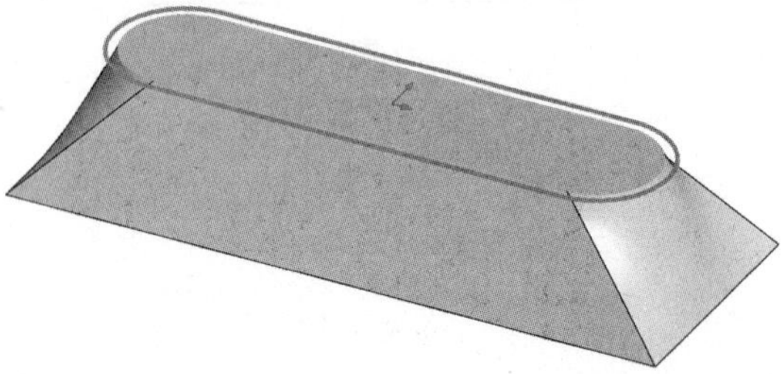

A preview of the offset operation will be displayed, as shown in **Figure 4.17**. Note that because the top surface was still selected, the edges of the surface were the entities offset. The entities to be offset may be preselected, as we did here, or you can choose the Offset Entities Tool first and then click on the entities to be offset.

In the PropertyManager, set the offset distance as 0.20 inches. Check the Reverse box, as shown in Figure 4.18, so that the offset entities are inside of the edges of the face. Click on the check mark to finish.

The finished sketch is shown in Figure 4.19. Note that the 0.20-inch offset distance is shown as a dimension that can be edited by double-clicking it. Also note that unlike most tools in the Sketch group, the Offset Entities Tool cannot be activated by simply selecting the sketch plane and then clicking on the tool. Instead, a sketch must be opened before selecting the Offset Entities Tool. The Convert Entities Tool is similar in this regard; a sketch must be opened before accessing the tool.

FIGURE 4.18

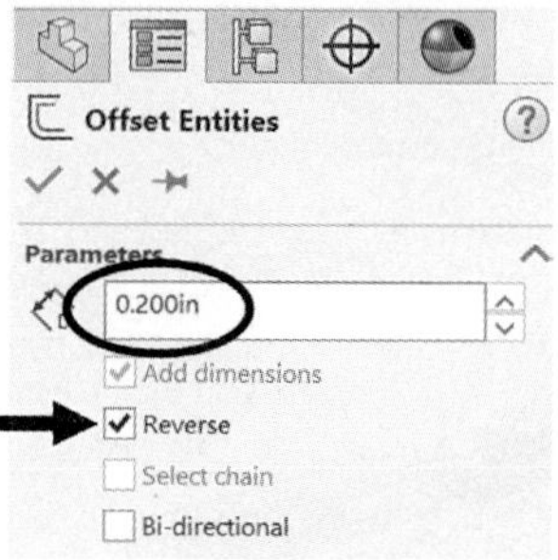

FIGURE 4.19

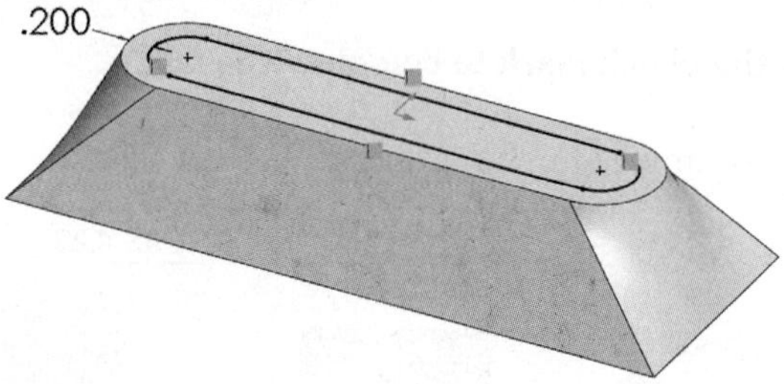

Before cutting the shape of the sketch into the part, we need to consider the design intent. We want this cut to be blind, so that the cards sit on the bottom of the hole, but we want the cut depth to vary with the overall height of the card holder. Therefore, neither a Blind nor Through All type of cut will meet our design intent. Rather, we will specify the depth of cut so that the bottom of the cut is a fixed distance above the bottom of the card holder.

Select the Extruded Cut Tool from the Features group of the CommandManager. Switch to the Trimetric View. In the PropertyManager, set the type of extrusion to Offset From Surface and the offset distance to .125 inches, as shown in Figure 4.20. As the surface, we want to select the bottom face of the part.

To select the bottom surface, we could switch to Bottom View or rotate the model until the bottom surface is visible, but in the steps that follow, we will learn a handy technique for selecting a nonvisible surface *without* rotating the model from the Trimetric View.

FIGURE 4.20

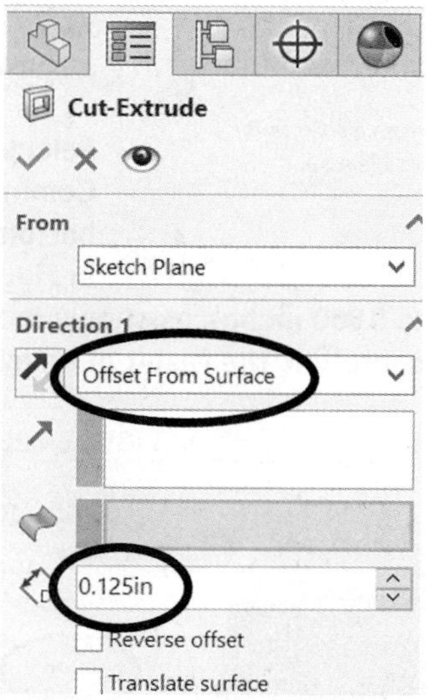

FIGURE 4.21

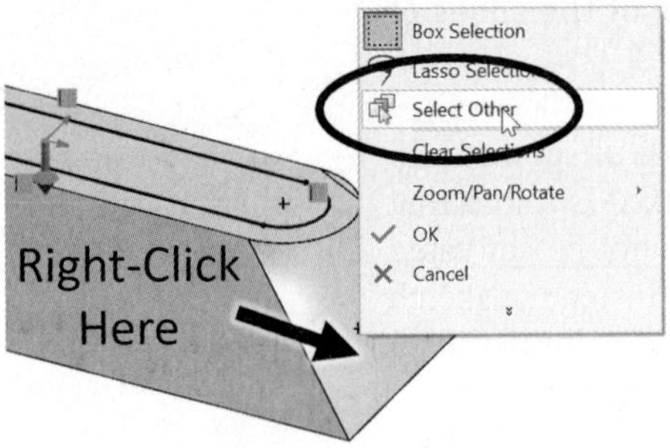

Position the cursor above the bottom face. Do not click the left mouse button; doing so would select the visible outer surface. Right-click, and choose Select Other from the menu, as shown in Figure 4.21. (Note: The menu will have many more items than the ones shown in Figure 4.21; most have been hidden for clarity.)

With the bottom face highlighted, as shown in Figure 4.22, click the left mouse button.

If there were other possible selections available, they could be selected from the pop-up menu that is shown in Figure 4.22.

Click the check mark to complete the cut.

The resulting geometry is shown in Figure 4.23.

FIGURE 4.22

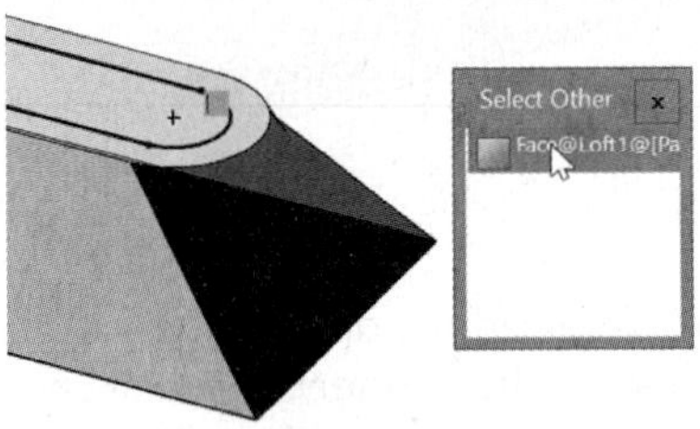

FIGURE 4.23

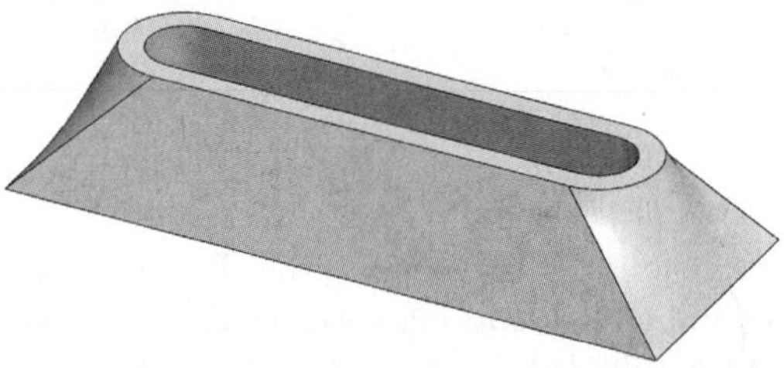

FIGURE 4.24

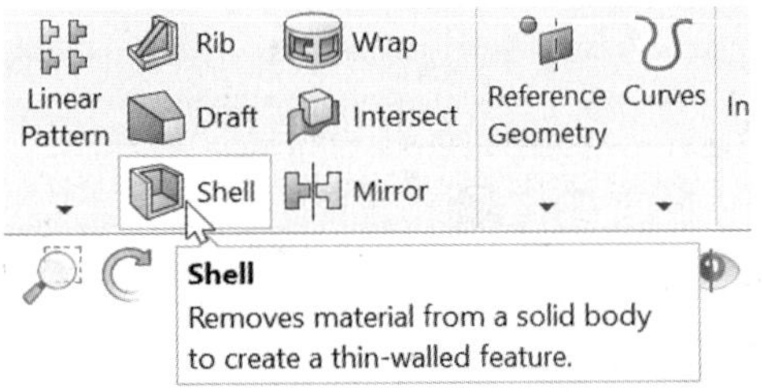

We will now introduce the Shell command. This command is often used with molded plastic parts to create thin-wall geometries. Since the part we are modeling does not need to have any significant strength, making the part solid would be a waste of material. We will make the wall thickness of the part a constant 0.060 inches.

Select the Shell Tool from the Features group of the CommandManager, as shown in Figure 4.24. Select the bottom face of the part as the face to be removed, as shown in Figure 4.25. In the PropertyManager, set the wall thickness to 0.060 inches, as shown in Figure 4.26. Click the check mark to complete the shell operation, the result of which is shown in Figure 4.27.

FIGURE 4.25

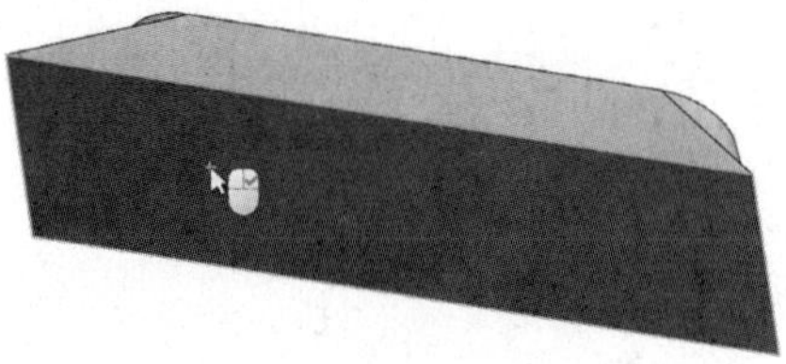

FIGURE 4.26

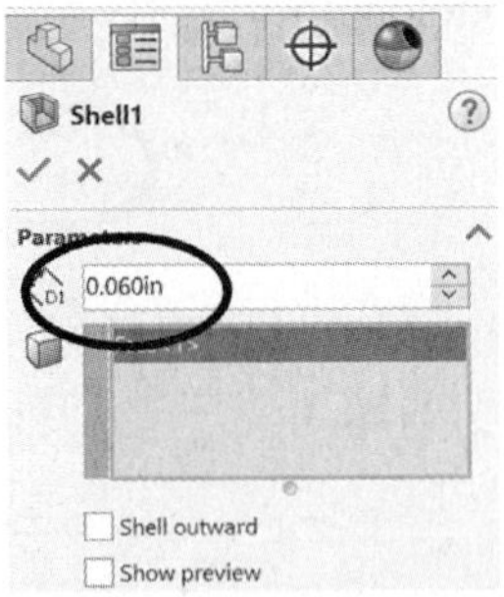

FIGURE 4.27

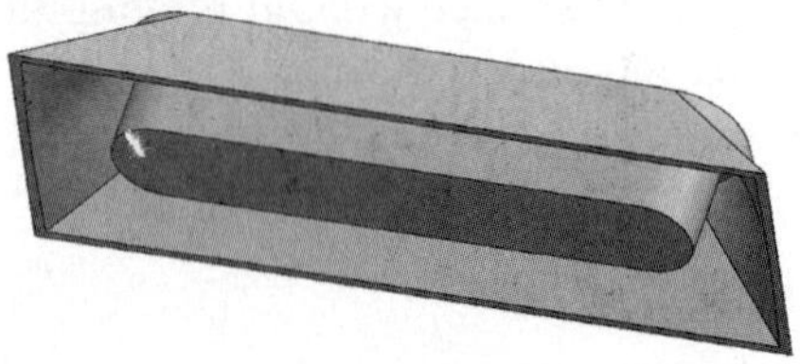

Select the Fillet Tool from the Features group of the CommandManager, as shown in Figure 4.28. In the PropertyManager, set the fillet radius of 0.10 inches. Select the top face, as shown in Figure 4.29, and click the check mark to apply the fillet.

FIGURE 4.28

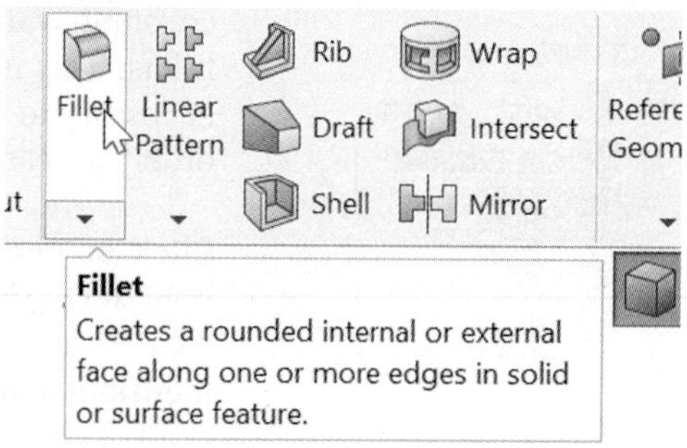

Note that fillets are usually applied to edges, not faces. Selecting a face causes all of the edges of that face to be filleted, as shown in Figure 4.30.

Select the Fillet Tool again. Select one of the edges at the bottom of the cavity as shown in Figure 4.31 (as long as the "Tangent propagation" box is checked, then the fillet will be extended completely around the bottom edge). Set the radius as 0.050 inches, and click the check mark to apply the fillet.

FIGURE 4.29

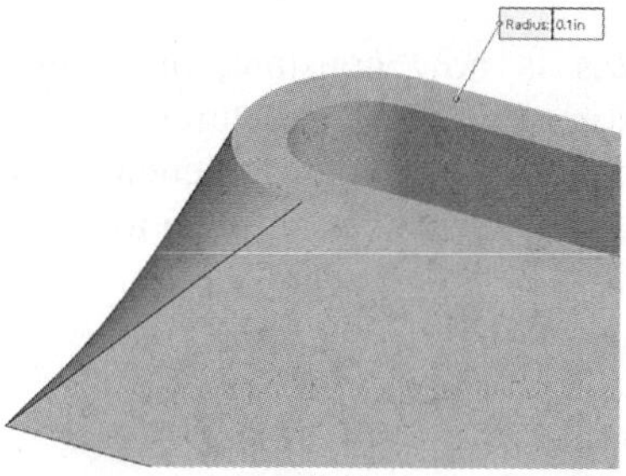

FIGURE 4.30

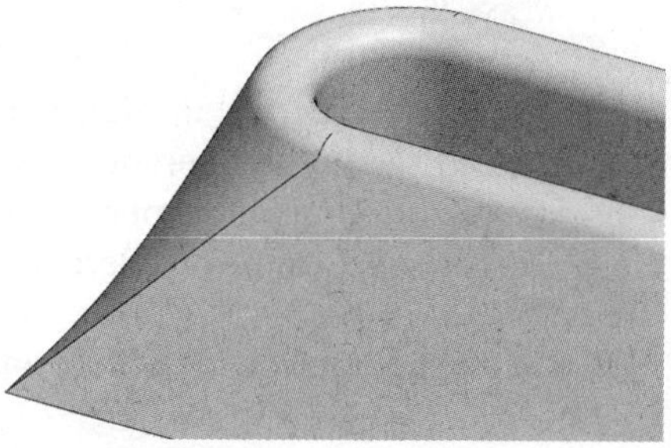

FIGURE 4.31

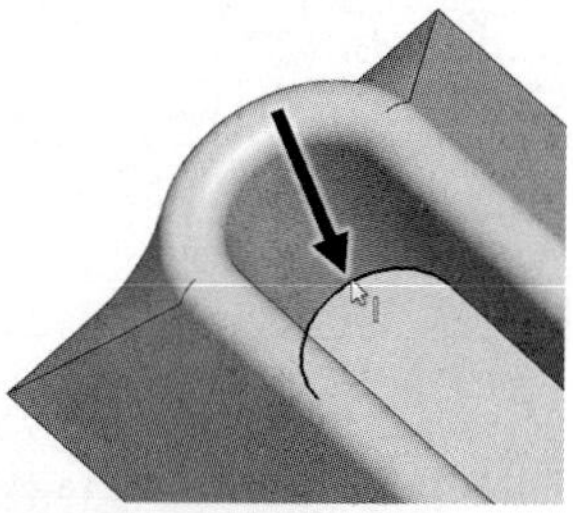

Although our part appears to be finished, there is a problem that may not be evident from examining the part from standard views. We will use a section view to get a better look at the problem areas.

FIGURE 4.32

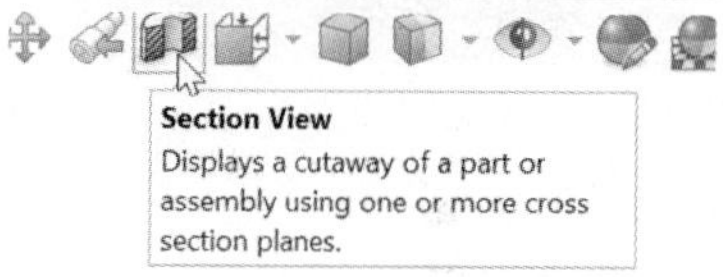

Select the Front Plane from the FeatureManager. Click on the Section View Tool in the Heads-Up Toolbar, as shown in Figure 4.32. Click the check mark, and the cross-section of the part is displayed, as shown in a Front View in Figure 4.33.

FIGURE 4.33

Note that we pre-selected the Front Plane as the section plane, but either of the other principal planes (Top or Right) can be selected from the PropertyManager. Also, the section plane can be offset from the selected principal plane by entering an offset distance in the PropertyManager or by dragging the arrow that appears when the Section View Tool is selected.

Zooming in on the filleted edges shows the problem. Because the filleting was performed after the shell operation, the wall is thinner than desired at the upper corners, as shown in Figure 4.34. (Similarly, the wall is thicker than desired in the lower corners.)

FIGURE 4.34

FIGURE 4.35

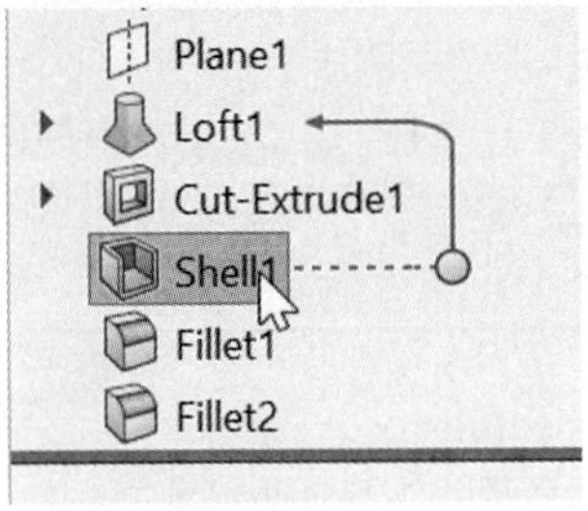

To correct this problem, we could fillet the sharp corners to maintain a constant wall thickness. We could also shell the part after creating the fillets. This is the easier way to produce the constant wall thickness. It is not necessary to delete the shell and fillet operations and redo them in the proper order; we can simply reorder them in the FeatureManager.

Click on the Shell in the FeatureManager, as shown in Figure 4.35, and hold down the left mouse button. Note the arrow showing that the fillet is related to the lofted feature. This is an example of the dynamic references mentioned in Chapter 1.

FIGURE 4.36

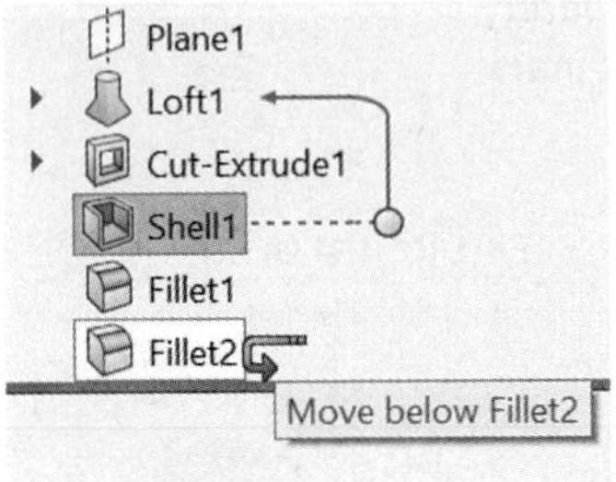

FIGURE 4.37

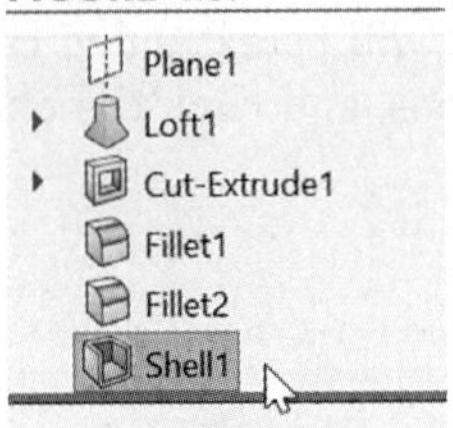

Drag the cursor down until the arrow and the pop-up message indicate that the Shell will be moved after the second fillet, as shown in Figure 4.36.

Release the mouse button, and the features are reordered, as shown in Figure 4.37.

The wall thickness is now constant, as shown in **Figure 4.38**. Note that not all features can be reordered, as some operations will be based on geometries created by prior operations. The display of the dynamic references helps to identify which features can be reordered.

FIGURE 4.38

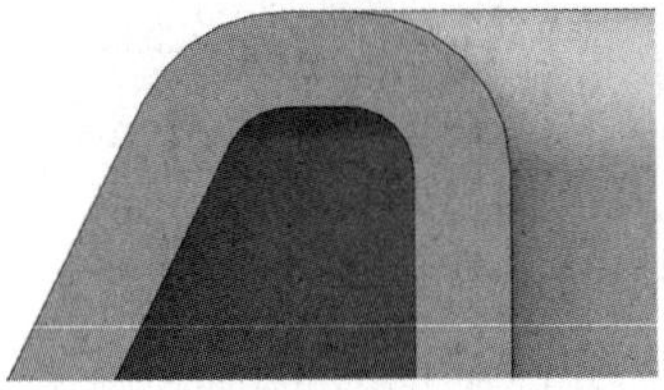

Click on the Section View Tool again.

This will toggle off the display of the section view.

Text can be added to a part as a sketch entity and then extruded into raised or embossed letters. We will add a part number in raised letters to the bottom of the part.

FIGURE 4.39

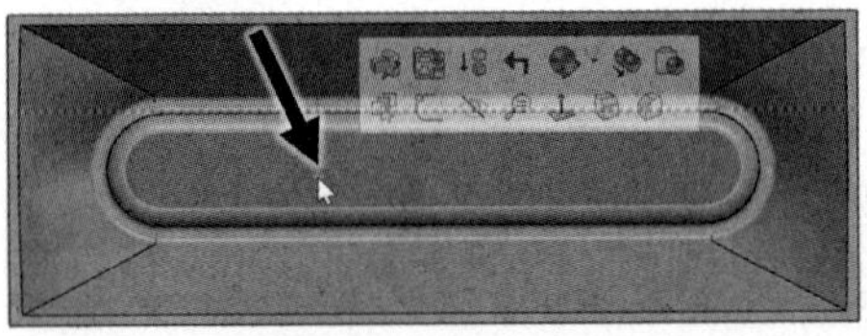

Switch to Bottom View and select the flat surface as shown in Figure 4.39.

Select the Text Tool from the Sketch group of the CommandManager, as shown in Figure 4.40. In the PropertyManager, type in the text as shown in Figure 4.41.

By default, the font specified in the Options will be used. To use another font, click on the "Use document font" box to uncheck it, which will allow you to select the Font button and edit the type and size of the font.

FIGURE 4.40

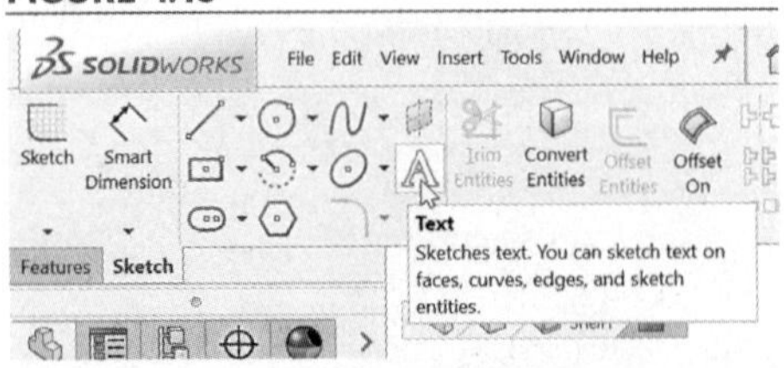

FIGURE 4.41

Click the check mark in the PropertyManager to close the text box. The text can now be moved by clicking and dragging it around the sketch. Center the text on the face, as shown in Figure 4.42.

FIGURE 4.42

If necessary, dimensions can be added to the marker at the lower left of the text to positively locate the text. However, for this example, locating the text approximately in the center of the face is adequate.

Click on the Features tab of the CommandManager, and select the Extruded Boss/Base Tool.

Extrude the sketch 0.020 inches out from the part, as shown in Figure 4.43.

FIGURE 4.43

This card holder may be made from a translucent plastic. The SOLIDWORKS® program allows parts to be displayed in the color desired, and also for optical properties such as transparency and shininess to be set.

Select the part name from the FeatureManager (so that the entire part is selected), and select Edit Appearance from the Heads-Up Toolbar, as shown in Figure 4.44.

FIGURE 4.44

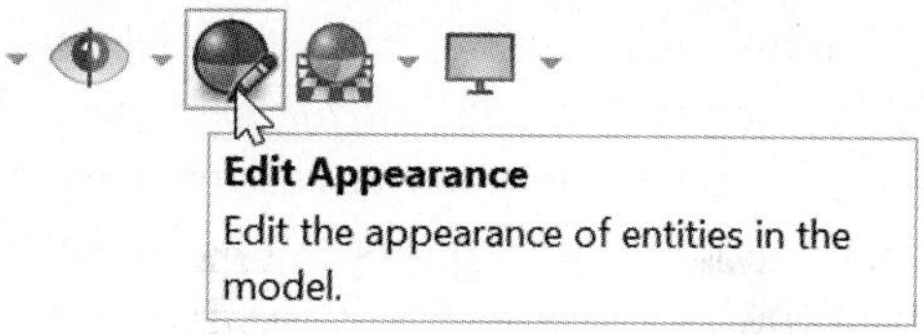

In the PropertyManager, under the Basic option, select a gray color, as shown in Figure 4.45. (You may also choose a color and texture from the Task Pane, if you prefer.) Select the Advanced option, the Illumination tab, and move the Transparent amount slider bar toward the right, as shown in Figure 4.46. Click the check mark to apply the desired color and properties.

FIGURE 4.45

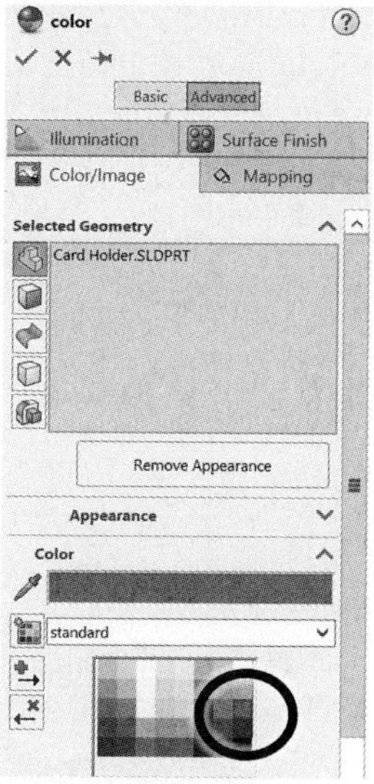

FIGURE 4.46

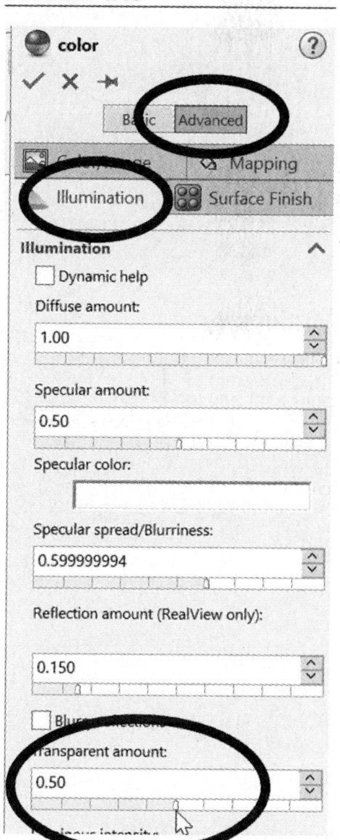

The translucent part will look more realistic without the edges displayed, so select Shaded (without edges) as the display mode, as shown in Figure 4.47.

The translucent part is shown in Figure 4.48.

Save this part with the name "Card Holder" for use in future exercises.

FIGURE 4.47

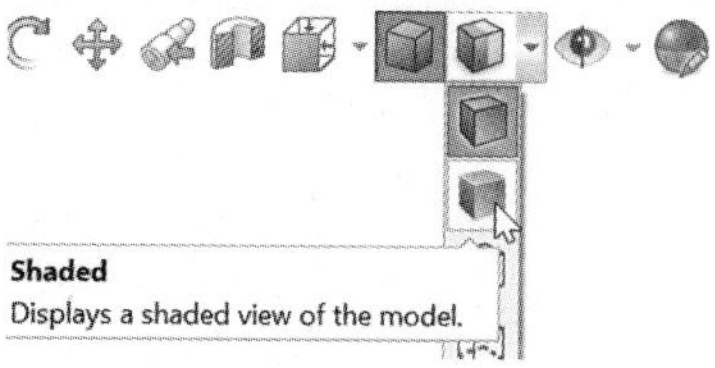

FIGURE 4.48

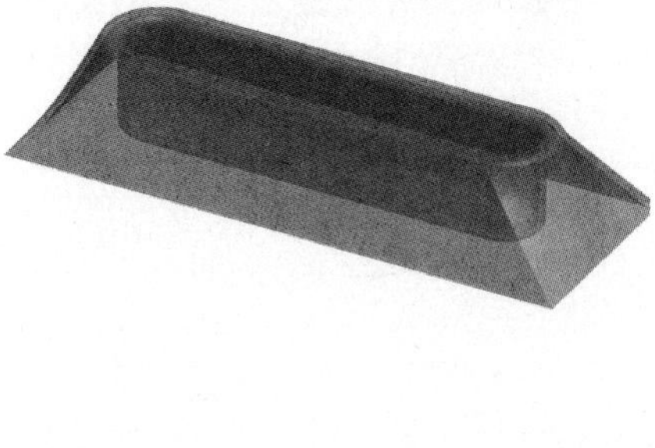

FUTURE STUDY

Industrial Design

In this chapter, we created a part by using the Loft command to smoothly blend two shapes. Engineers in many industries are sometimes reluctant to use this type of construction, since the resulting surfaces are difficult to define mathematically. This makes the geometry difficult to specify on engineering drawings and often impossible to make with traditional manufacturing methods.

Complex geometries have always been an important tool for *industrial designers.* Industrial designers are important members of product development teams, but perform a different task than do design engineers. The Industrial Designers Society of America defines the role of the industrial designer as:

> The industrial designer's unique contribution places emphasis on those aspects of the product or system that relate most directly to human characteristics, needs and interests. This contribution requires specialized understanding of visual, tactile, safety and convenience criteria, with concern for the user. (www.idsa.org)

The image that many associate with industrial designers is that of an artist working on a clay model of an automobile, creating the shapes that would eventually be seen on the showroom floor. The clay model would eventually be digitized by measuring thousands of points on the surface in order to define the shape for the tooling used to stamp the sheet metal body parts. Industrial designers now create many of their models with the "virtual clay" of solid modeling software, computer-controlled machining centers, and rapid prototyping machines.

The roles of design engineer and industrial designer have begun to overlap as they have access to a similar set of product development tools. However, engineers should recognize and utilize the unique capabilities of industrial designers within the product design process.

FIGURE 4.49

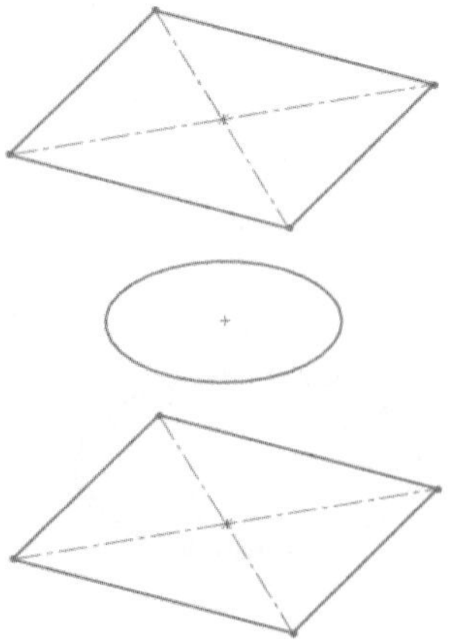

FIGURE 4.50

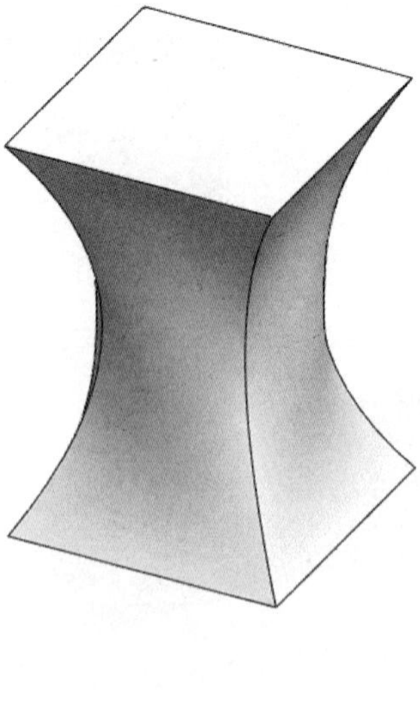

Note that more than two sketches can be used to create a lofted feature. Consider the three sketches shown in **Figure 4.49**.

The lofted solid created from these sketches is shown in **Figure 4.50**.

Also, an additional sketch defining a *guide curve* can be introduced, allowing more control over the loft. **Figure 4.51** shows a guide curve added to the three previous sketches. The resulting solid is shown in **Figure 4.52**.

FIGURE 4.51

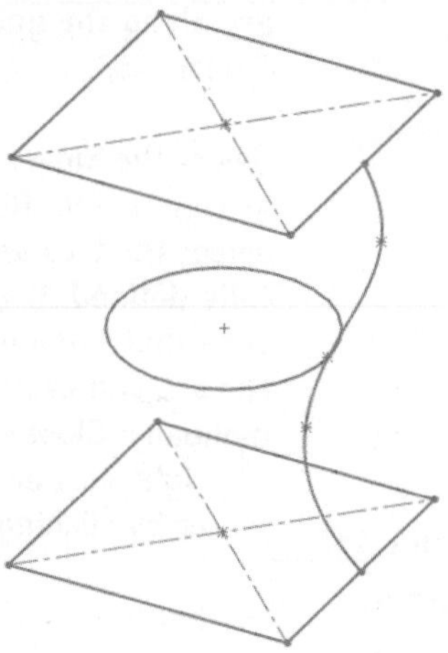

FIGURE 4.52

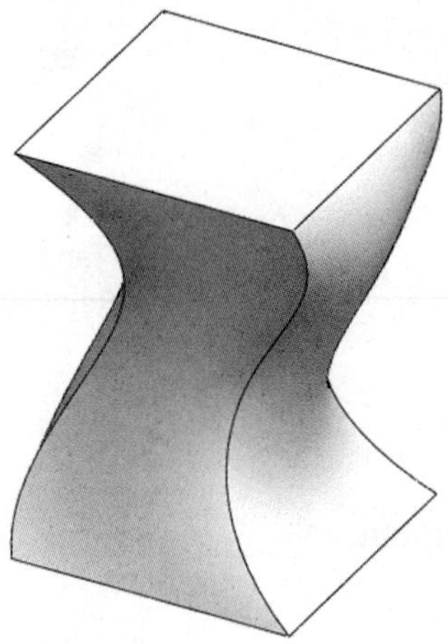

4.2 Parts Created with Swept Geometry

In this section, we will learn how to create a solid by "sweeping" a cross-section along a path. We will start with a simple part in which the path is planar. Later, we will use a helix curve to form a helical spring. In the next section, we will introduce the 3-D sketch, which will be used to define a sweep path in 3-D space.

The first part that we will create is a bent tubing section. We begin by creating the sweep path. The geometry of the sweep path, which defines the centerline of the tubing, is shown in **Figure 4.53**.

Open a new part. Select the Top Plane from the FeatureManager. Choose the Line Tool from the Sketch group of the CommandManager, and draw a vertical line beginning at the origin. Choose the Tangent Arc Tool, and drag out an arc from the endpoint on the line, as shown in Figure 4.54.

FIGURE 4.53

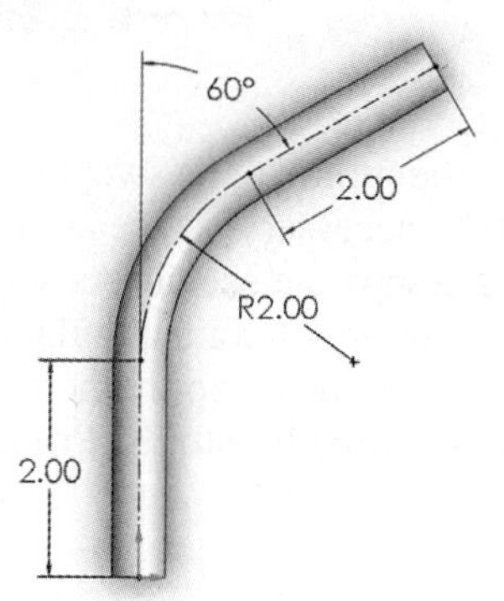

FIGURE 4.54

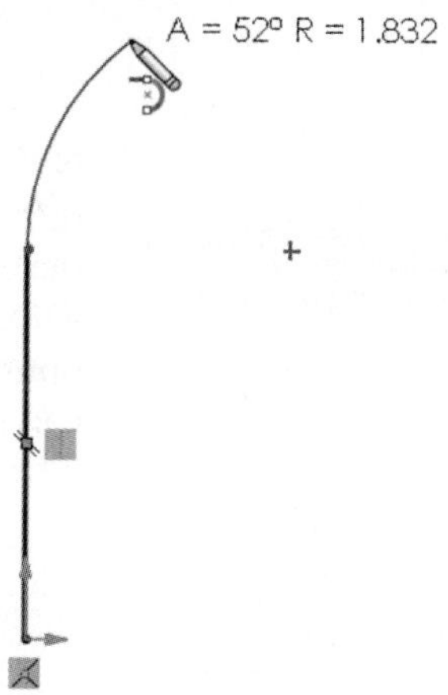

FIGURE 4.55

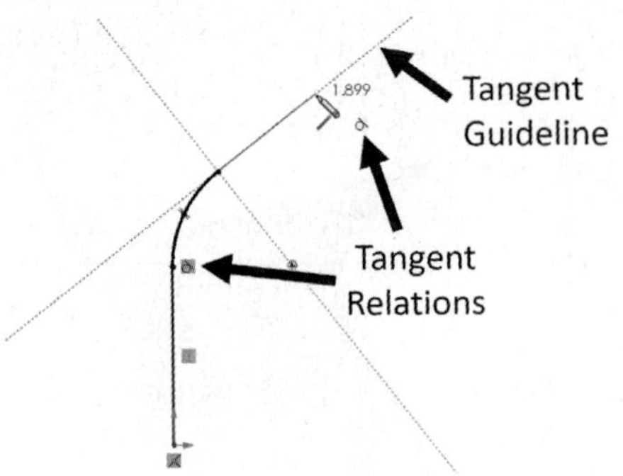

Choose the Line Tool. Drag out a line from the endpoint of the arc along the guideline that is tangent to the arc, as shown in Figure 4.55.

Select the Smart Dimension Tool, and add the dimensions shown in Figure 4.56. Recall that to create an angular dimension, you select the two lines that define the angle. The sketch should be fully defined. If the sketch is not fully defined, check to make sure that tangent relations were added automatically between the arc and each of the lines. If either relation is missing, add it manually. Close the sketch by clicking on the icon in the upper right corner of the graphics area, as shown in Figure 4.57, or by clicking the Exit Sketch Tool in the CommandManager.

FIGURE 4.56

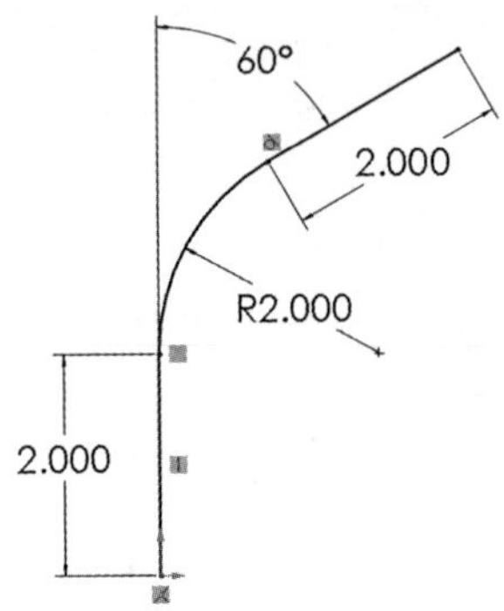

FIGURE 4.57

Switch to the Front View. Select the Front Plane from the FeatureManager. Select the Circle Tool, and drag out two circles, both centered at the origin. Select the Smart Dimension Tool, and add diameter dimensions of 0.40 and 0.50 inches, as shown in Figure 4.58. Close the sketch.

Click on the Features tab of the CommandManager. Select the Swept Boss/Base Tool, as shown in Figure 4.59. In the PropertyManager, select the sketch containing the two circles (Sketch2) as the Profile, the cross section to be swept, and select the first sketch defining the geometry of the tubing centerline (Sketch1) as the Path of the sweep, as shown in Figure 4.60. Each sketch can be selected by clicking in the graphics area on any entity contained within the sketch. A preview of the swept geometry is shown in Figure 4.61. Click the check mark to complete the operation.

FIGURE 4.58

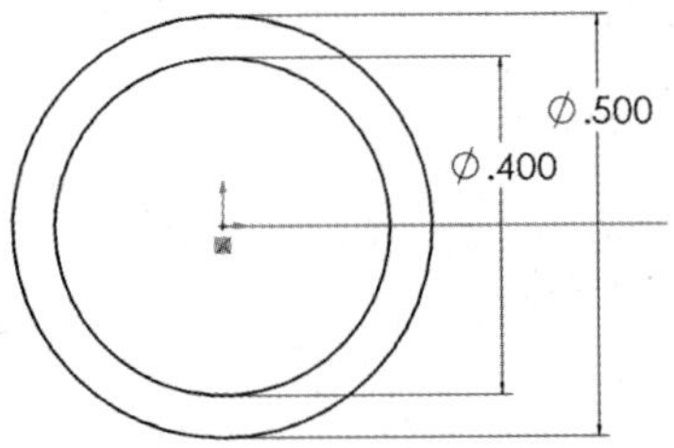

FIGURE 4.59

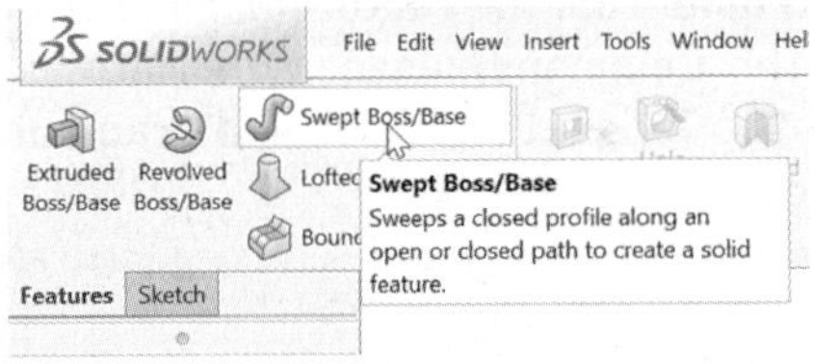

FIGURE 4.60

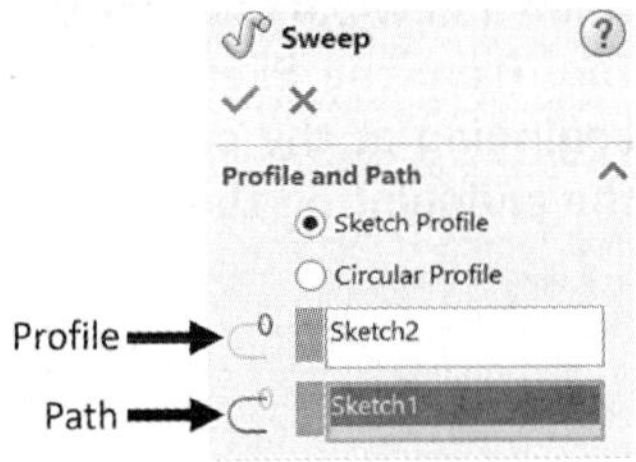

FIGURE 4.61

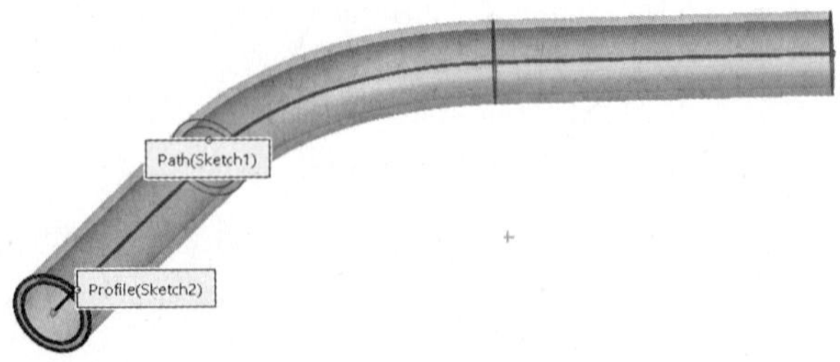

Note that if the profile to be swept is a solid circle, then it is not necessary to create and dimension the circle in a separate sketch. As shown in **Figure 4.60**, there is an option for a circular profile. If this option is chosen, then the diameter of the circle can be entered directly into the PropertyManager.

The completed part is shown in **Figure 4.62**.

FIGURE 4.62

In the next exercise, we will use a more complex sweep path, a helix, to create a helical spring.

Open a new part. Select the Front Plane from the FeatureManager. Select the Circle Tool from the Sketch group of the CommandManager. Draw a circle centered at the origin. Select the Smart Dimension Tool, and dimension the circle's diameter as 2 inches, as shown in Figure 4.63.

FIGURE 4.63

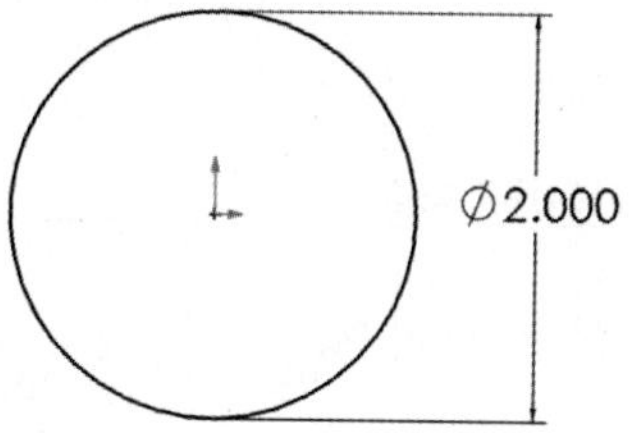

From the main menu, select Insert: Curve: Helix/Spiral, as shown in Figure 4.64.

A helix can be defined by specifying any two of the following three quantities:

1. The height, or overall length of the helix,
2. The pitch, the distance between similar points on successive turns of the helix, and
3. Revolutions, the total number of complete turns of the helix.

FIGURE 4.64

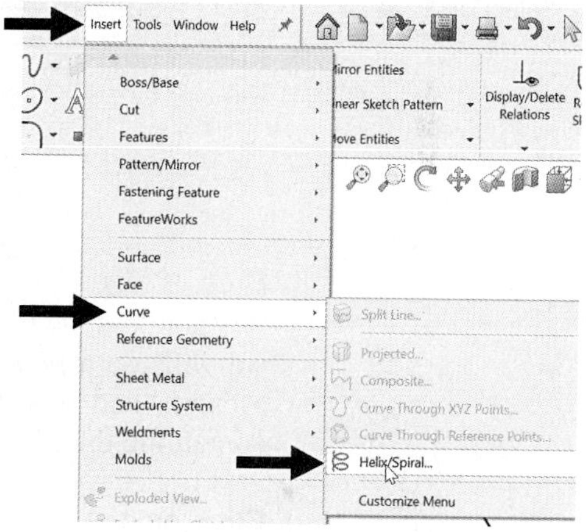

We will define the height and the number of revolutions and allow the pitch to be calculated.

Select Height and Revolution in the "Defined By" box. Set the height to 6 inches and the number of revolutions to 8. Set the starting angle to 135 degrees, as shown in Figure 4.65.

A preview of the helix geometry is shown in **Figure 4.66**. The start angle is not critical here. If the start angle is set to a multiple of 90 degrees, then we can define our profile sketch in either the Right or the Top Plane. We have chosen to use an angle value which is not a 90-degree multiple to illustrate the procedure for creating a plane at the end of a path sketch.

FIGURE 4.65

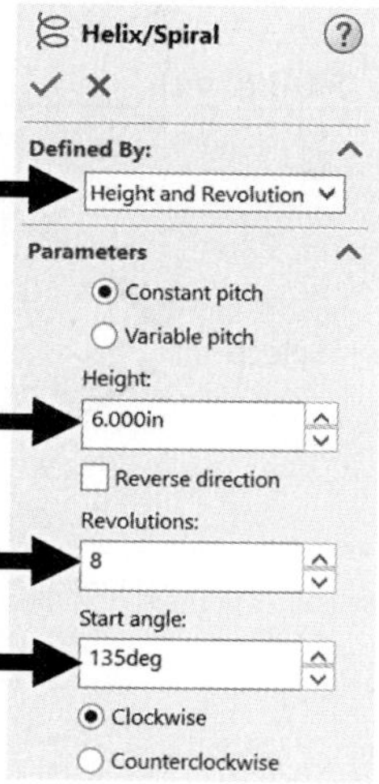

FIGURE 4.66

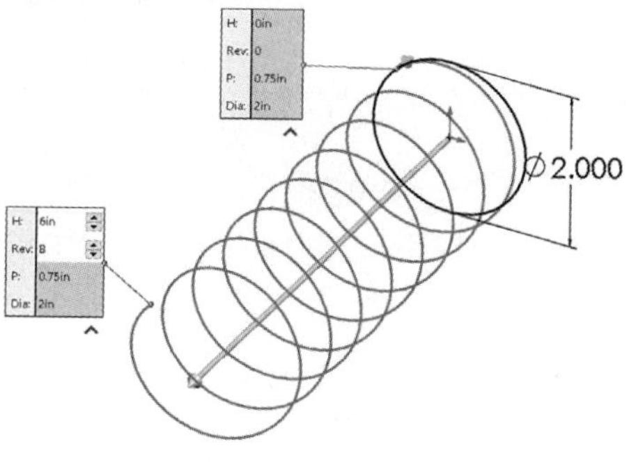

Click the check mark to accept the helix definition and close the sketch. Press the Esc key to clear the selection of the sketch.

We will now create a new plane at the end of the helix, perpendicular to the helix at that point.

FIGURE 4.67

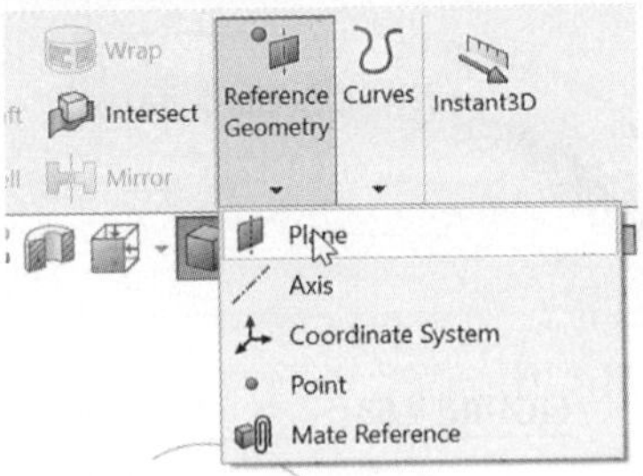

Click on the Features tab of the CommandManager. Click on the Reference Geometry Tool, and choose Plane, as shown in Figure 4.67. Click once on the helix curve, and then click on the endpoint of the curve, as shown in Figure 4.68. Note that by selecting a curve and an endpoint, the plane created will be through the endpoint and perpendicular to the curve, as shown in Figure 4.69. Click the check mark to create the plane.

Click the Sketch tab of the CommandManager. With the new plane selected, choose the Circle Tool. Drag out a circle, as shown in Figure 4.70. Click the check mark to complete the circle.

FIGURE 4.68

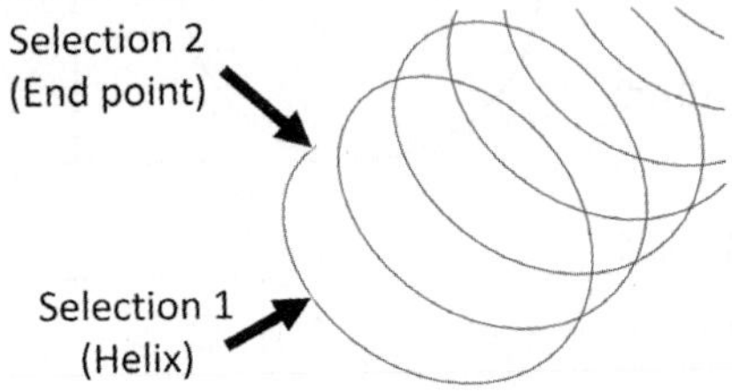

FIGURE 4.69

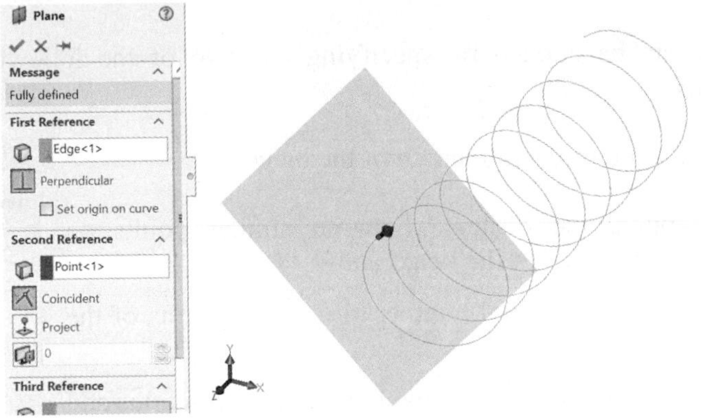

It is not possible to snap the center of the circle to the endpoint of the helix. Rather, we will use a *pierce* relation. The pierce relation is defined between a point and a curve, and sets the point at the location where the curve "pierces" the sketch containing the point.

Click on the center point of the circle to select it. While holding down the Ctrl key, select the helix near the endpoint, as shown in Figure 4.71. In the PropertyManager, click on Pierce, as shown in Figure 4.72. Click the check mark

FIGURE 4.70

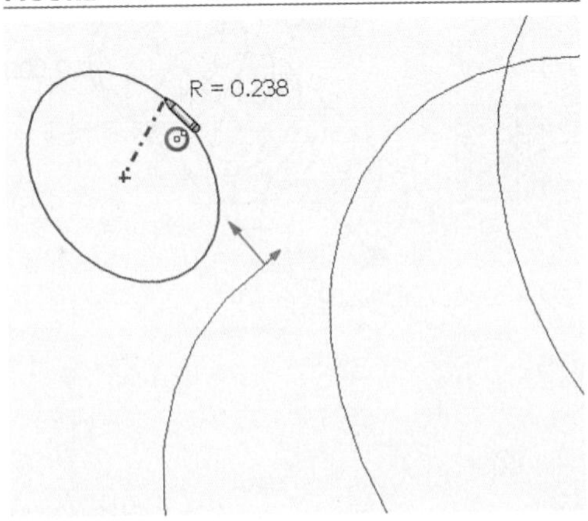

FIGURE 4.71

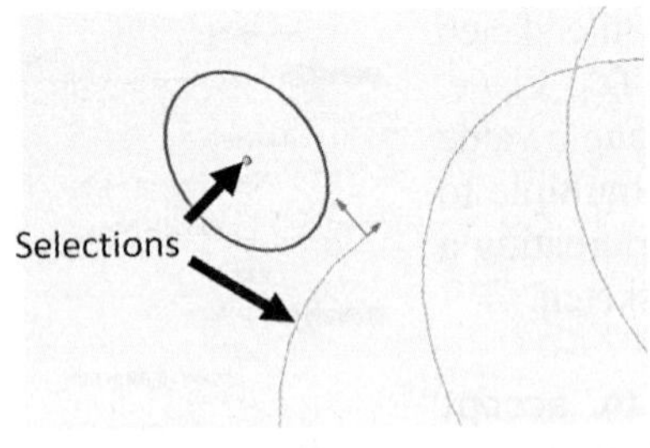

FIGURE 4.72

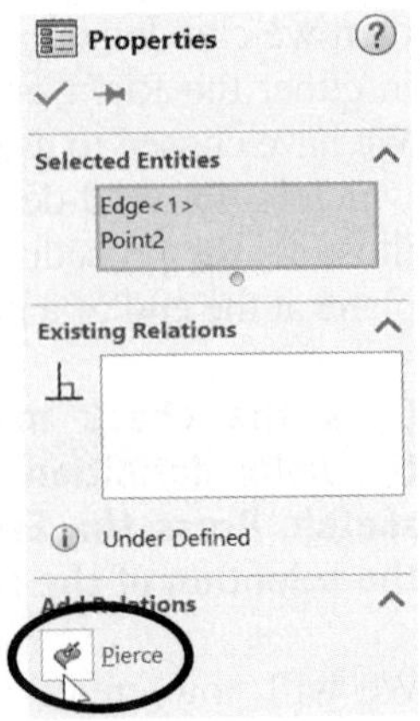

FIGURE 4.73

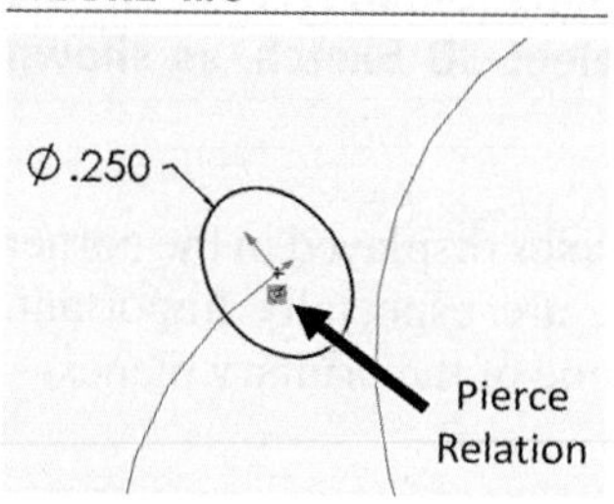

to add the relation. Choose the Smart Dimension Tool, and add a 0.25-inch-diameter dimension to the circle. The sketch should now be fully defined, as shown in Figure 4.73. Close the sketch.

Click on the Features tab of the CommandManager. Choose the Swept Boss/Base Tool. In the Property-Manager, choose the sketch containing the 0.25-inch-diameter circle as the Profile, and the helix as the Path, as shown in Figure 4.74. Click the check mark to complete the sweep operation. Hide both the plane and the helix.

FIGURE 4.74

The completed spring is shown in the Isometric View in Figure 4.75. Note that instead of creating the sketch profile, we could have simply selected the Circular Profile Option, as shown in Figure 4.74. We chose our method to illustrate the use of the Pierce Relation, as well as the creation of a plane at the end of the path sketch. In this exercise, we used a curve in 3-D space as the sweep path. This curve was created from a 2-D sketch (a circle). In the next section, we will use the more general 3-D Sketch Tool to define the sweep path in three-dimensional space.

FIGURE 4.75

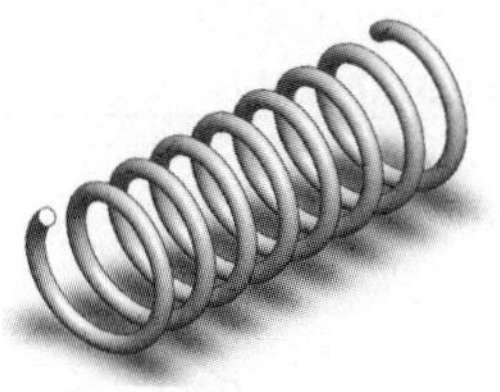

4.3 A Part Created with a 3-D Sketch as the Sweep Path

The use of a 3-D sketch as a sweep path allows more complex parts to be created. As the name implies, a 3-D sketch contains entities in 3-D space, whereas typical sketches contain entities that exist in a plane. Not all sketch entities are available in a 3-D sketch.

FIGURE 4.76

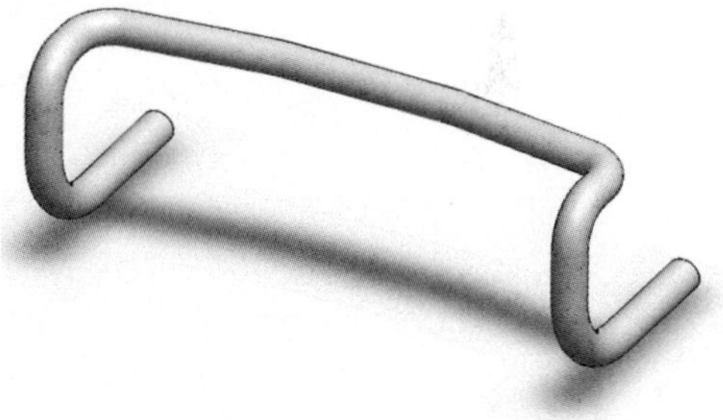

In this exercise, we will model the handlebars shown in Figure 4.76. Since the handlebars are defined in metric units, we need to set the units accordingly.

Open a new part. In the Status Bar at the bottom of the screen, click on the unit system (IPS) and select MMGS as the new unit system, as shown in Figure 4.77. Click on the unit system again and select Edit Document Units. Set the number of decimal places for length measurements to None.

FIGURE 4.77

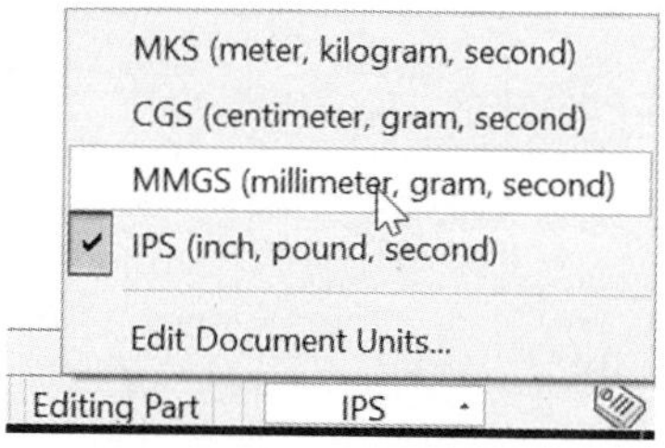

Since most of the dimensions in this part will be in integer values of millimeters, setting the number of decimal places to zero is a good choice. (It is standard convention to show integer values of mm without a decimal point.)

FIGURE 4.78

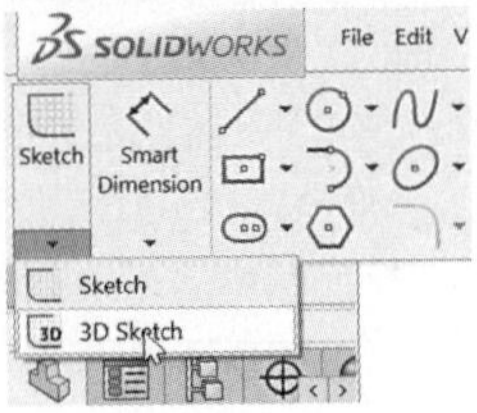

FIGURE 4.79

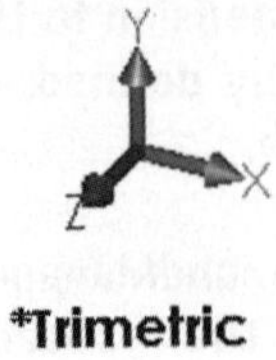

From the Sketch group of the CommandManager, click the arrow under the Sketch Tool and select 3D Sketch, as shown in Figure 4.78.

When working with 3-D sketches, the axes displayed in the corner of the screen, shown in Figure 4.79, are especially important. When we draw lines, we will do so in one of the primary planes—XY, YZ, or ZX.

FIGURE 4.80

Select the Line Tool.

Note that beside the cursor, the plane that the line will be drawn in is displayed. We want our first two lines to be sketched in the ZX plane, so we will change this orientation before proceeding.

FIGURE 4.81

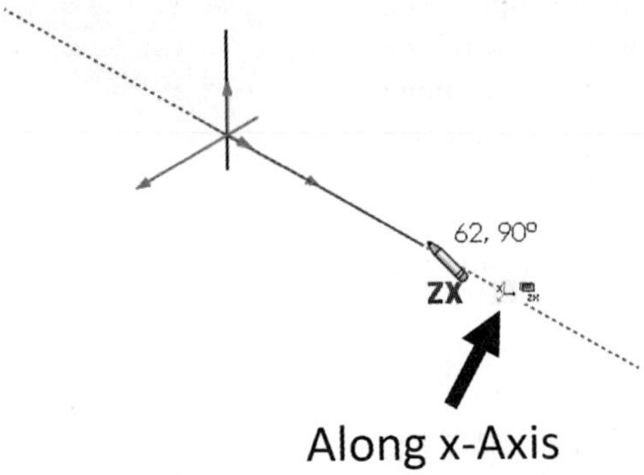

Press the Tab key, which causes the sketch plane to cycle between the three principal planes. Stop when the plane selected is ZX, as shown in Figure 4.80.

Switch to the Isometric View. Drag out a line from the origin along the X axis, as shown in Figure 4.81. Make sure that the "Along X-axis" relation icon is displayed before releasing the mouse button.

From the endpoint of the first line, drag out a diagonal line, as shown in Figure 4.82.

FIGURE 4.82

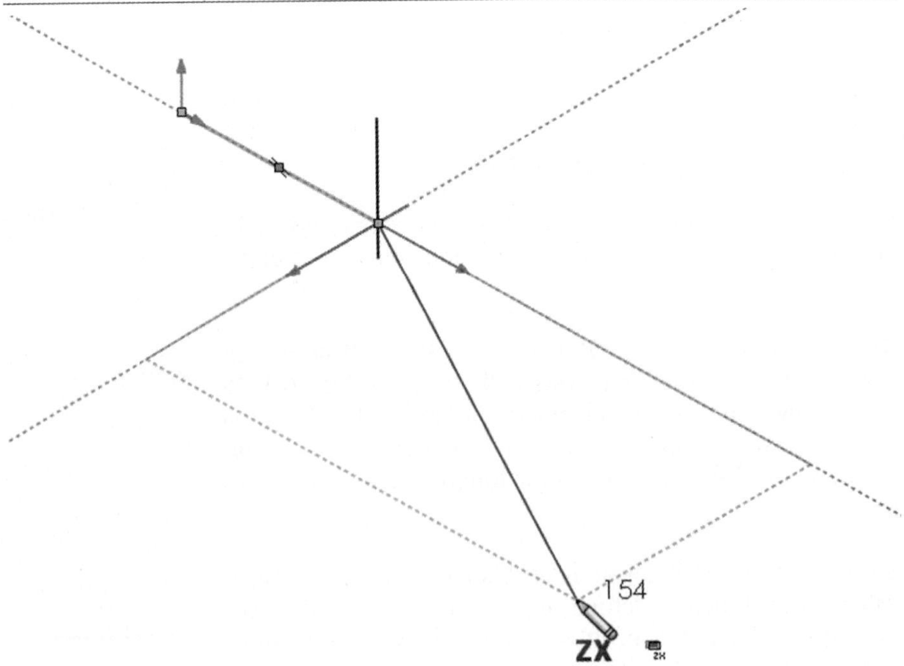

When dimensioning a 3-D sketch, not all of the options available with 2-D sketches can be used. For example, if we try to add a dimension from the origin to the endpoint of the second line, only the straight-line distance between the two points can be displayed. Similarly, when adding a dimension to a line, the resulting dimension will define the length of the line, but the options available in a 2-D sketch of placing the dimension to show a horizontal or vertical dimension are not available. If we want to dimension the x and z distances from the origin to the point, then the use of a centerline is necessary.

Note that sketch relations in a 3-D sketch are not shown by default, as they are in a 2-D sketch. Sketch relations can be shown by selecting View: Hide/Show: Sketch Relations. For clarity, sketch relations are not shown on the 3-D sketch figures in this section.

Select the Centerline Tool from the Line Tool pull-down menu. Make sure that the ZX designation still shows beside the cursor. Drag out a centerline from the origin along the Z axis, as shown in Figure 4.83.

Select the Smart Dimension Tool. Click once on the first solid line drawn, and place a 50-mm dimension, as shown in Figure 4.84.

FIGURE 4.83

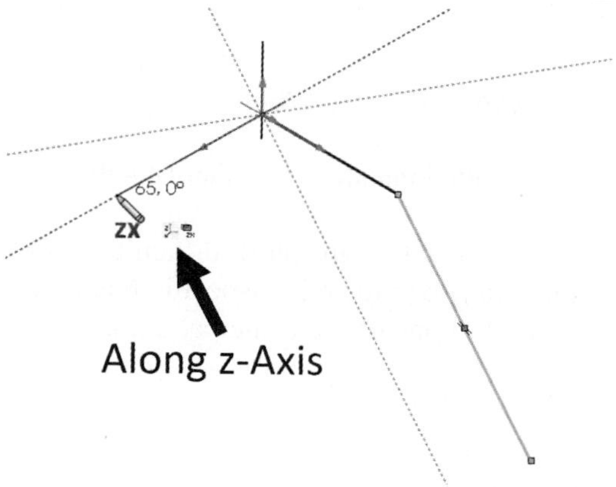

FIGURE 4.84

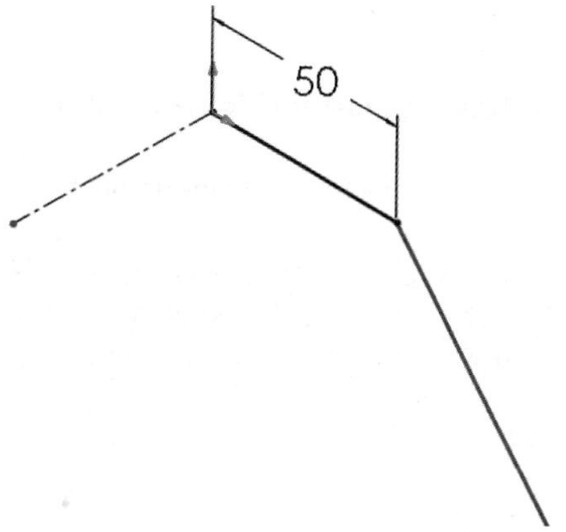

Add other dimensions as shown in Figure 4.85. To add the 210-mm dimension, click on the centerline and then on the endpoint of the diagonal line. To add the 20-mm dimension, click on the 50-mm-long line and then on the endpoint of the diagonal line. Note that when dimensioning in the isometric view, it may be helpful to orient the dimensions by selecting the Options Tool: System Options: Display/ Selection and checking the box labeled "Display dimensions flat to screen."

The other lines of the sketch will be drawn in the YZ plane.

FIGURE 4.85

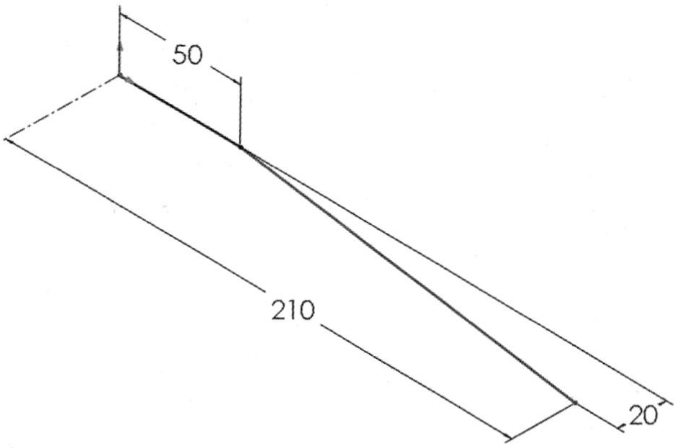

Select the Line Tool. Press the Tab key until YZ is shown as the drawing plane, as shown in Figure 4.86.

FIGURE 4.86

Drag out a line from the last endpoint of the second solid line in the Z-direction, as shown in Figure 4.87.

Drag a line downward (in the –Y direction), as shown in Figure 4.88.

FIGURE 4.87

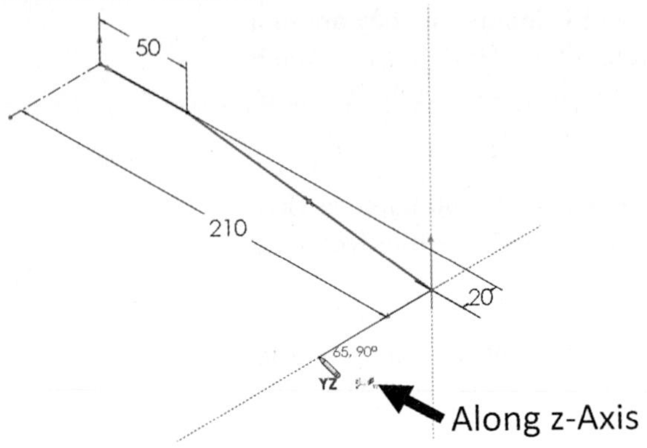

FIGURE 4.88

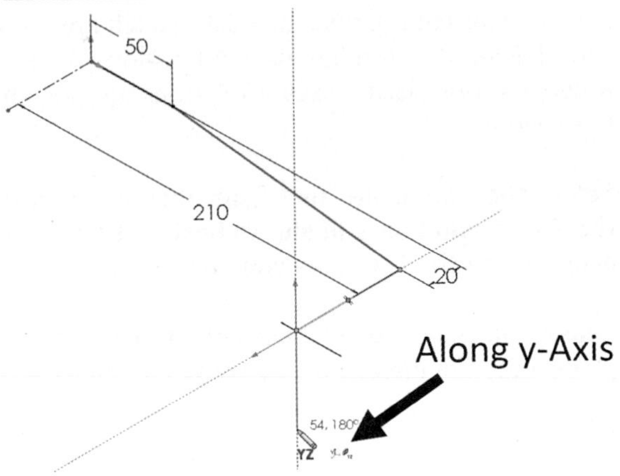

Drag the last line in the –Z direction, as shown in Figure 4.89.

Select the Smart Dimension Tool. Add the dimensions shown in Figure 4.90.

Note that the sketch is not fully defined. 3-D sketches are more difficult to fully define than are 2-D sketches. Although a line may be drawn in a specific plane, it is not *locked* into that plane unless it is aligned with one of the principal axes.

FIGURE 4.89

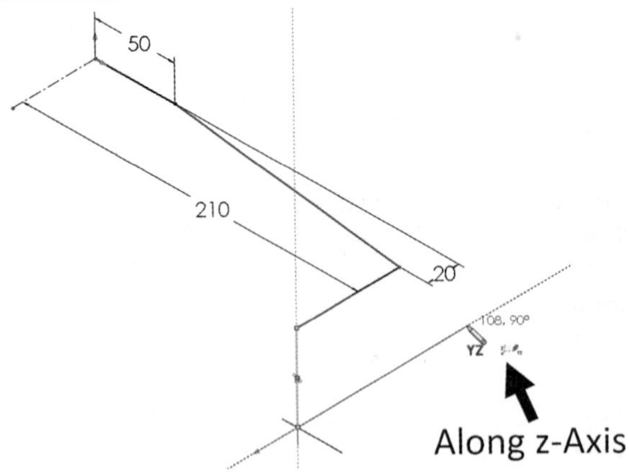

FIGURE 4.90

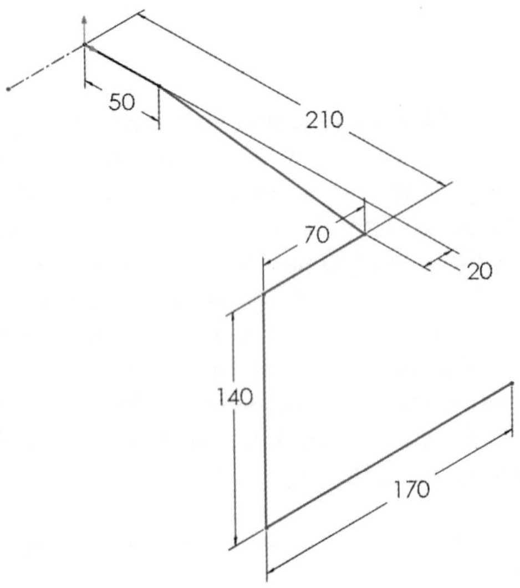

We will now add fillets to the sharp corners of our sketch. The fillet radii to be added are shown in **Figure 4.91**.

FIGURE 4.91

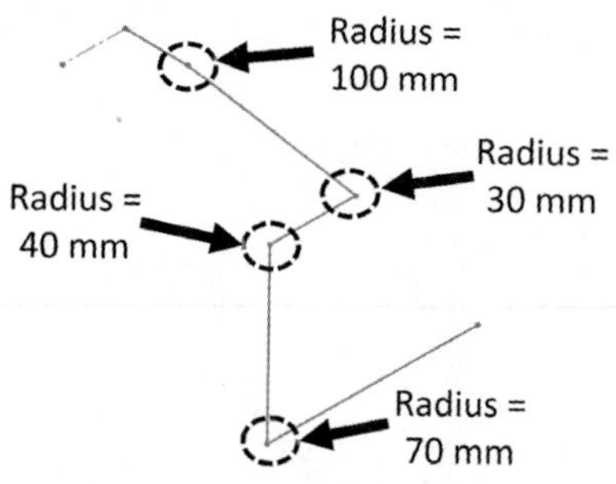

Select the Sketch Fillet Tool, as shown in Figure 4.92. In the PropertyManager, make sure that the pushpin icon (Keep Visible) is pushed "in," so that the Sketch Fillet Tool remains open for multiple fillets, as shown in Figure 4.93. Make sure that the box labeled "Keep constrained corners" is checked; otherwise, the dimensions to the corners will be lost when the corner is filleted. Leave the box labeled "Dimension each fillet" unchecked; this option only applies when you have multiple fillets with the same radius. Set the radius to 100 mm, and select the first corner to be filleted, as shown in Figure 4.94. Click the check mark to apply the fillet. Change the fillet radius to 30 mm, select the next corner to be filleted, and click the check mark. Repeat for the other two fillets, and then click the check mark again to close the Sketch Fillet Tool. The finished sketch is shown in Figure 4.95.

FIGURE 4.92

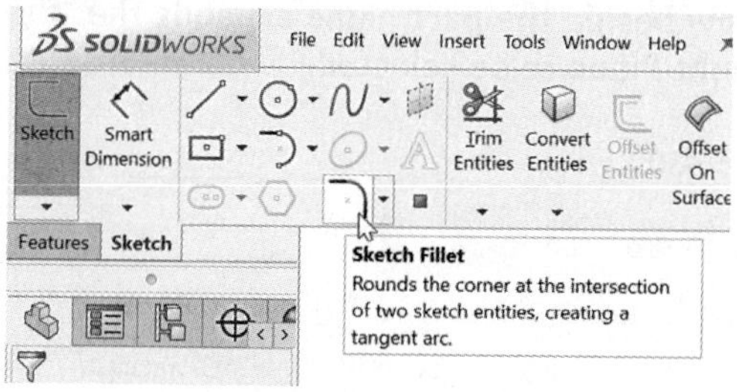

FIGURE 4.93

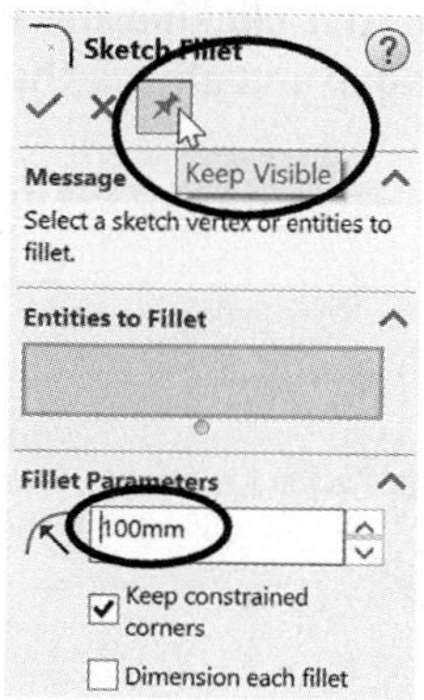

FIGURE 4.94

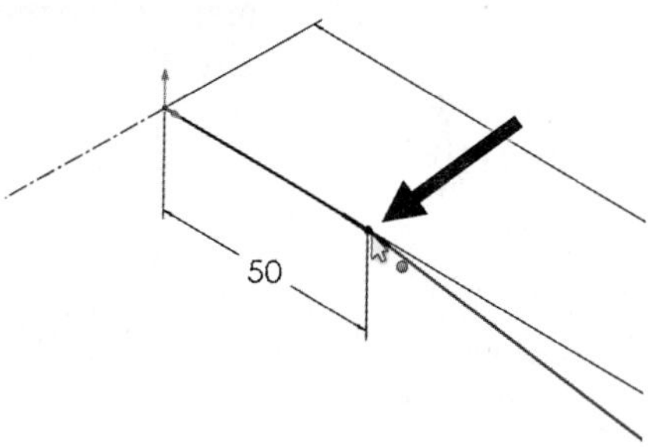

Close the sketch. Recall that the path and profile for a swept feature must be defined in separate sketches.

Select the Right Plane from the FeatureManager. Select the Circle Tool. Drag out two circles from the origin, as shown in Figure 4.96.

FIGURE 4.95

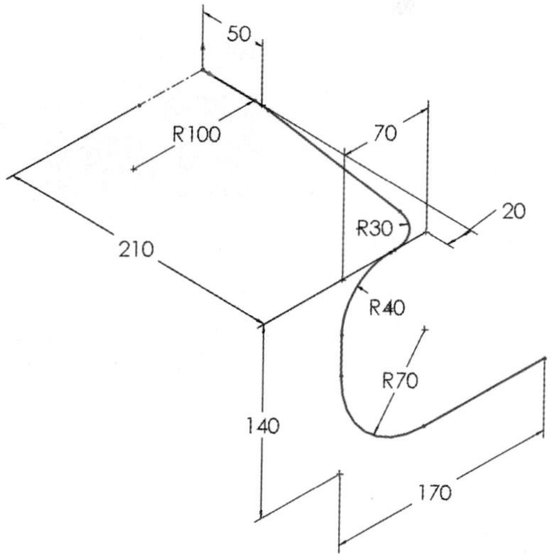

FIGURE 4.96

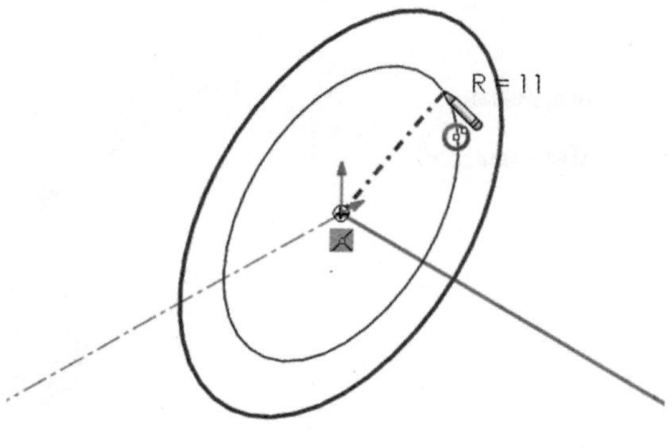

FIGURE 4.97

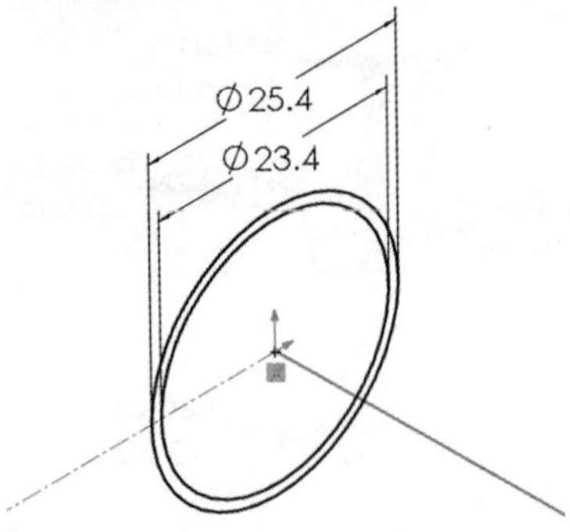

Add diameter dimensions of 25.4 mm and 23.4 mm, as shown in Figure 4.97. The decimal value can be shown by selecting each dimension and setting the unit precision in the PropertyManager.

(Although most bicycle components are specified in metric units, a 1-inch outer handlebar diameter, 25.4 mm, is a standard size.)

Close the sketch.

Click on the Features tab of the CommandManager. Select the Swept Boss/Base Tool. In the PropertyManager, select the 2-D sketch just completed as the profile and the 3-D sketch as the path, as shown in Figure 4.98.

Click the check mark to complete the sweep. The result is shown in Figure 4.99.

Select the Mirror Tool, as shown in Figure 4.100. In the PropertyManager, select the Right Plane as the mirror plane, as shown in Figure 4.101, and click the check mark. (Remember that clicking the arrow beside the part name expands the "fly-out" FeatureManager and allows the Right Plane to be selected.)

FIGURE 4.98

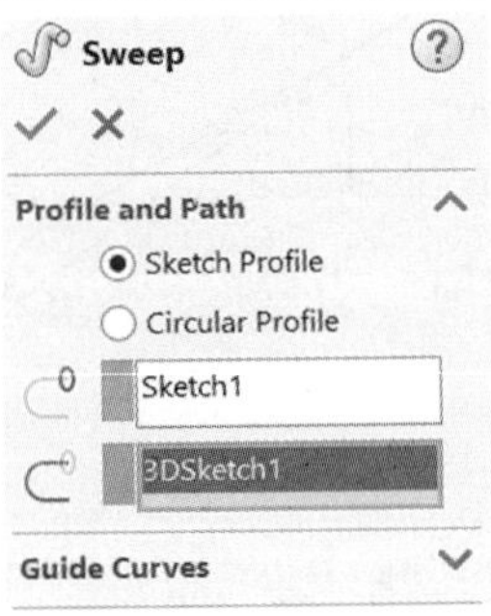

FIGURE 4.99

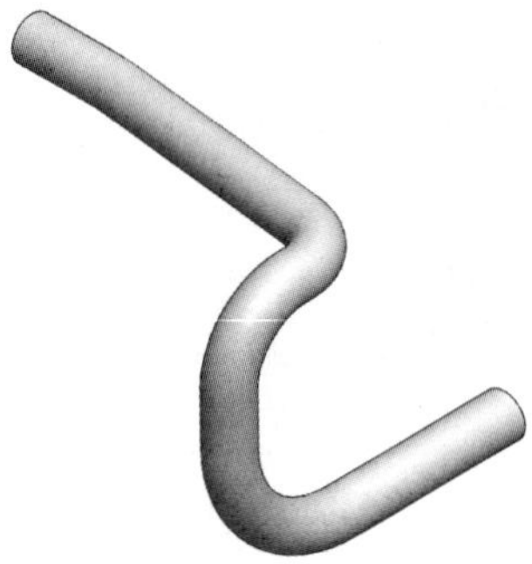

FIGURE 4.100

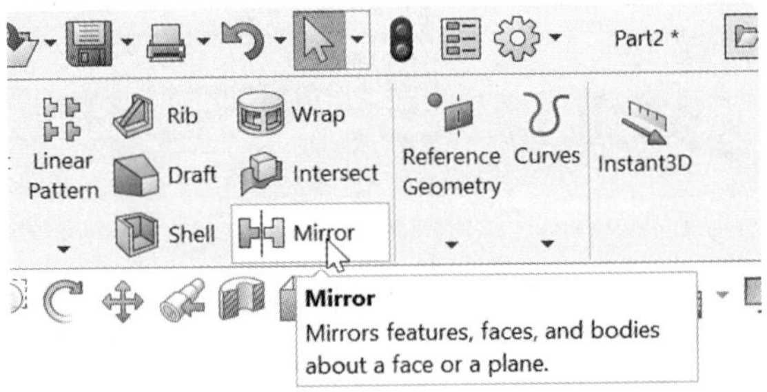

FIGURE 4.101

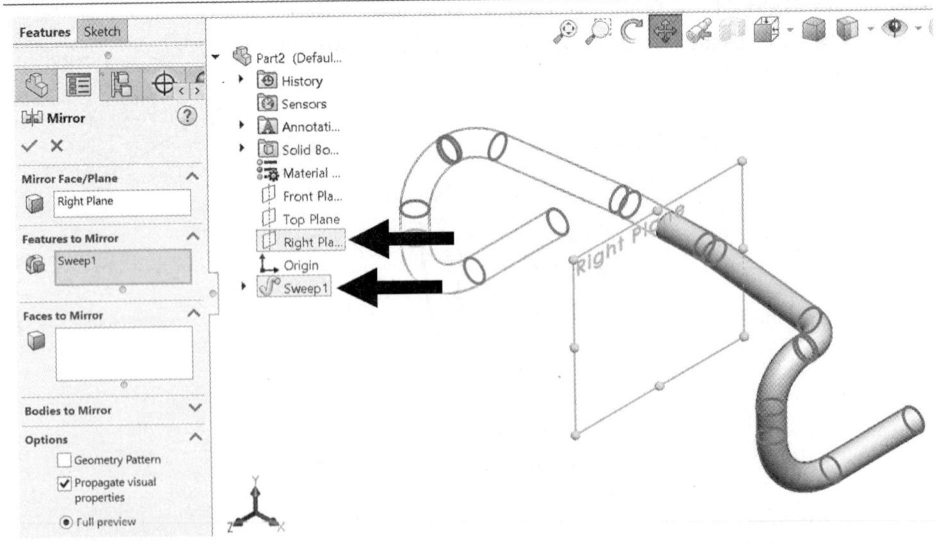

The completed handlebars are shown in the Trimetric View in **Figure 4.102**.

FIGURE 4.102

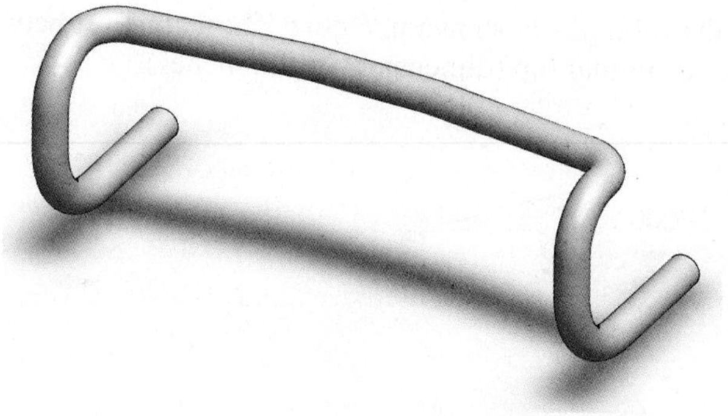

We will now determine the mass of the handlebars, which will be made from 6061 aluminum.

Right-click on Material in the FeatureManager. Select Edit Material and select 6061 Alloy from the list of aluminum alloys. Click Apply and Close.

Select Tools: Evaluate: Mass Properties from the Main Menu.

The mass of the handlebars is calculated as about 221 grams, as shown in **Figure 4.103**.

Save the part file as "Handlebars."

FIGURE 4.103

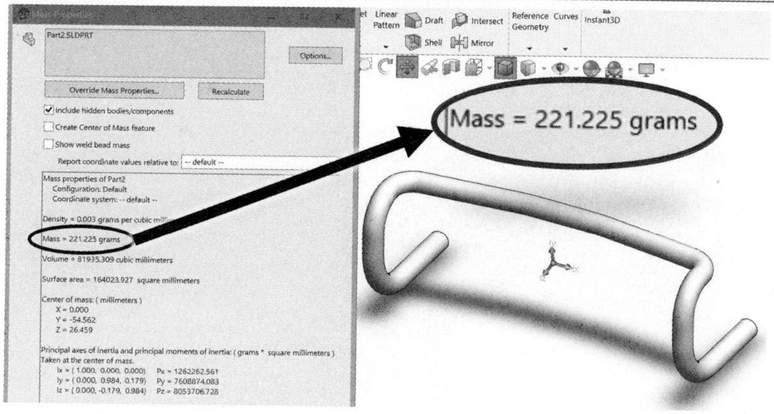

PROBLEMS

P4.1 Create the solid object shown in **Figure P4.1** with a loft between a square base and a circular top (dimensions are in inches).

FIGURE P4.1

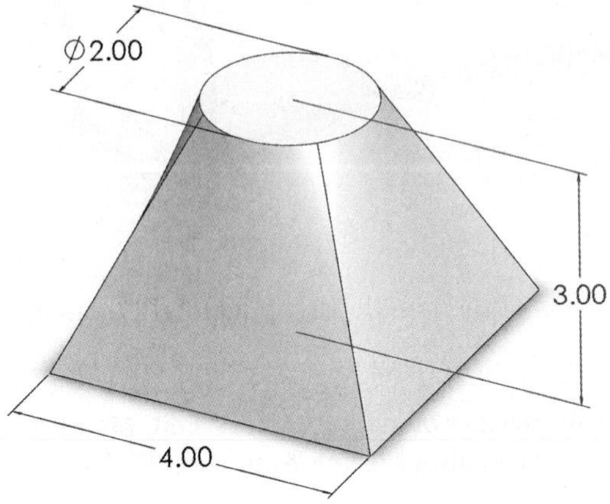

P4.2 Create the part shown in **Figure P4.2A**. Create the sketch shown in **Figure P4.2B** in the Top Plane. Create a new plane 4 inches above the Top Plane, and create the second sketch consisting of the two arcs and two lines indicated in **Figure P4.2C**, snapping to corresponding points of the first sketch. Make sure that both sketches are fully defined, or the loft operation may not work properly. Create a loft between the two sketches to finish the part. All dimensions are in inches.

FIGURE P4.2A

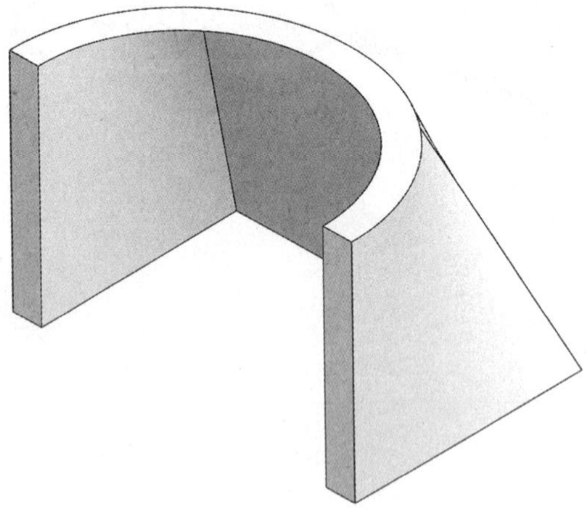

FIGURE P4.2B

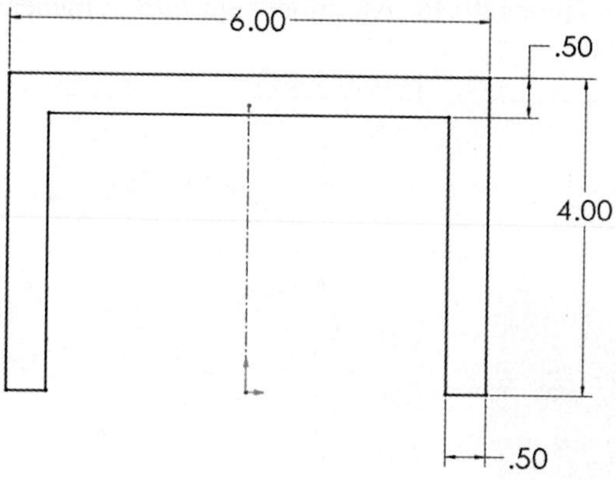

FIGURE P4.2C

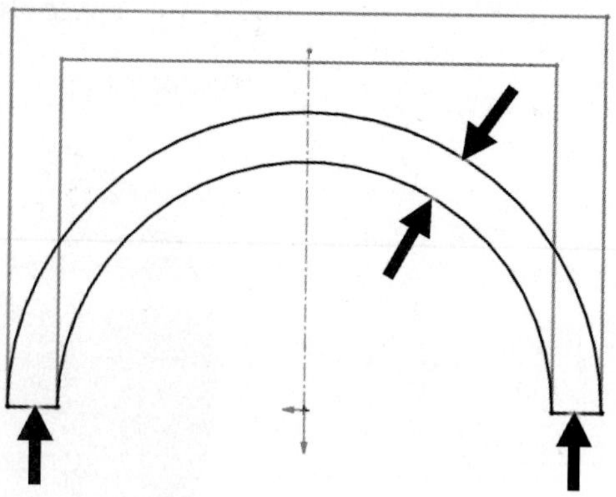

P4.3 Create the part shown here, with a circular cross section of 1-inch diameter. All dimensions are in inches.

FIGURE P4.3A

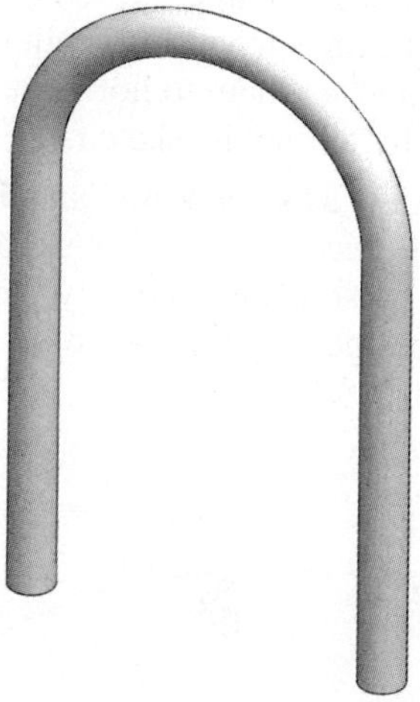

FIGURE P4.3B

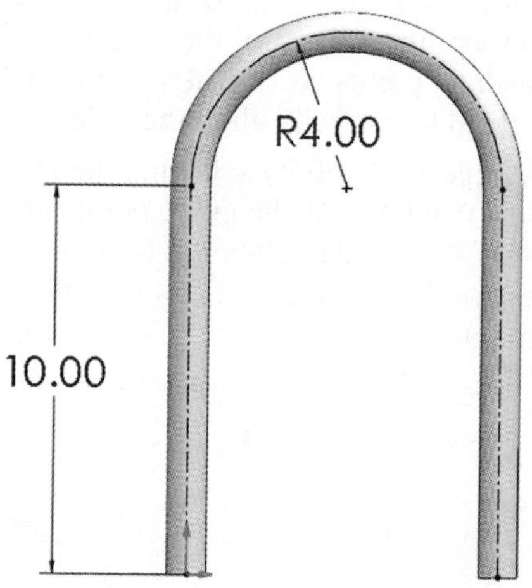

P4.4 Create the shape shown in **Figure P4.4A** as a loft defined by the two ellipses and one circle shown in **Figure P4.4B**. All dimensions are in inches.

FIGURE P4.4A

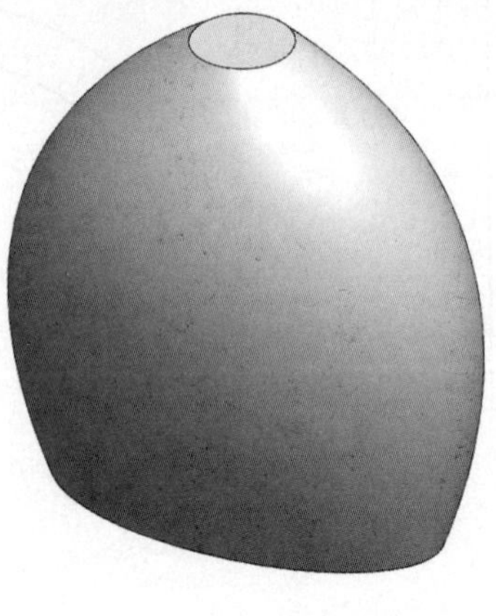

FIGURE P4.4B

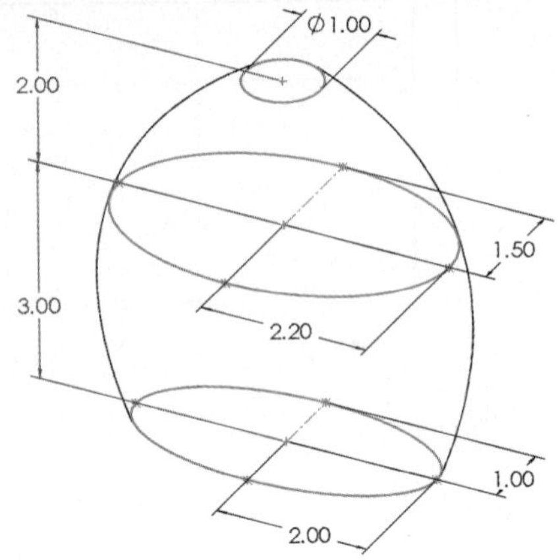

To create an ellipse, select the Ellipse Tool from the Sketch group of the CommandManager. Drag out a circle from the origin, as shown in **Figure P4.4C**, and then click and drag a point on the edge of the circle to "flatten" it into an ellipse, as shown in **Figure P4.4D**. Quadrant points are created at four locations on the ellipse; use these points to define the semimajor and semiminor axes of the ellipse. Add centerlines from the origin to the quadrant points, as shown in **Figure P4.4E**, and set them to horizontal and vertical to correctly align the ellipse and fully define the sketch.

Complete the loft by selecting the guide points as shown in **Figure P4.4F**. Note that because the guide points are not vertices, you cannot snap to them. The guide points can be aligned to a reasonable degree by viewing the preview from the Top View and aligning the guide points to be along a straight line projection, as shown in **Figure P4.4G**.

FIGURE P4.4C

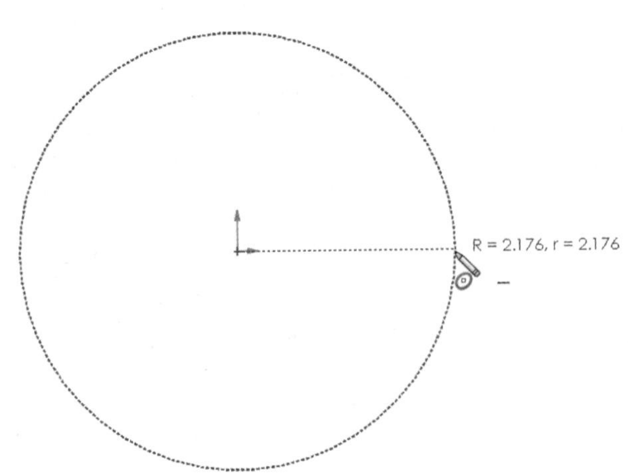

FIGURE P4.4D

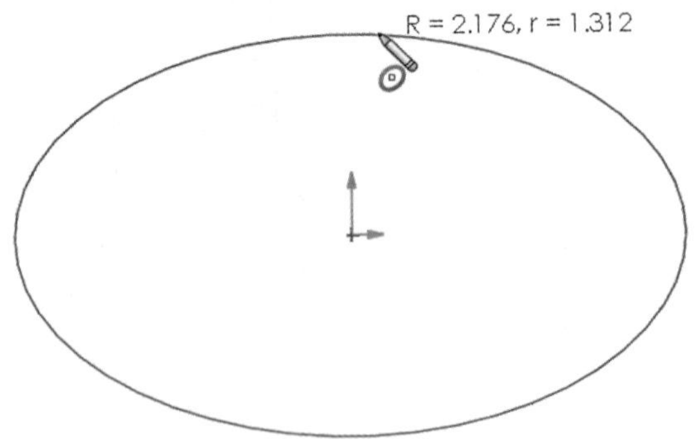

FIGURE P4.4E

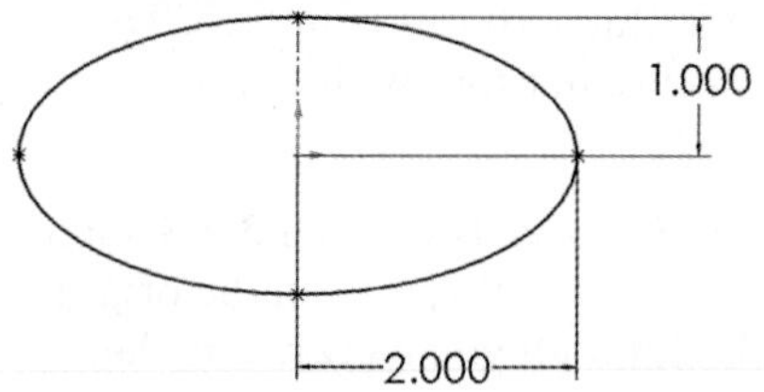

FIGURE P4.4F

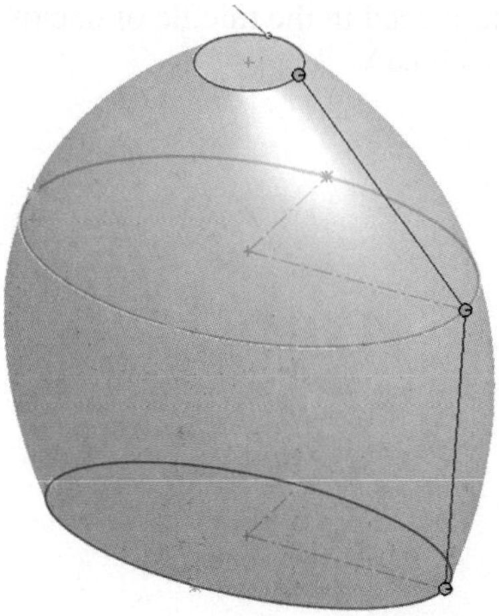

FIGURE P4.4G

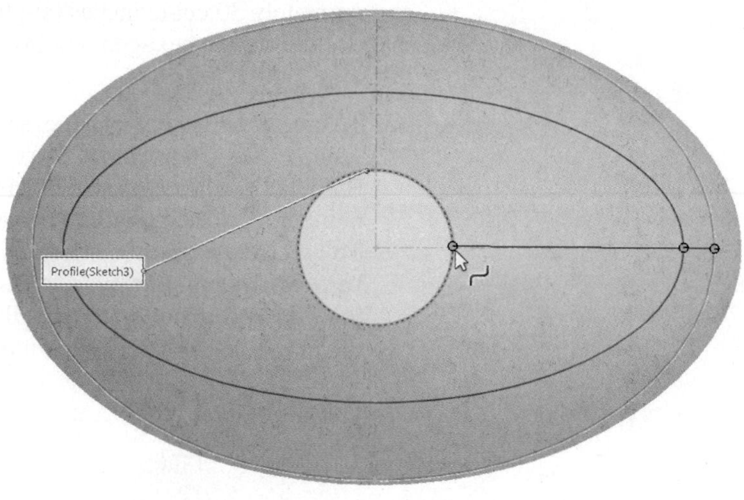

P4.5 Turn the shape created in **Problem P4.4** into the bottle shown in **Figure P4.5A**. Extrude a circular neck, as shown in **Figure P4.5B**, add a 0.25-inch fillet to the neck-to-body junction, and use the Shell Tool to hollow the bottle, leaving a 0.020-inch wall thickness. What is the volume of the space within the bottle? (Hint: To find the volume, move the rollback bar to just before the shell command, as shown in **Figure P4.5C** and find the volume of the solid before shelling. Then move the rollback bar past the shell, and find the volume of the bottle itself. The difference in the two volumes is the volume contained within the bottle.)

FIGURE P4.5A

FIGURE P4.5B

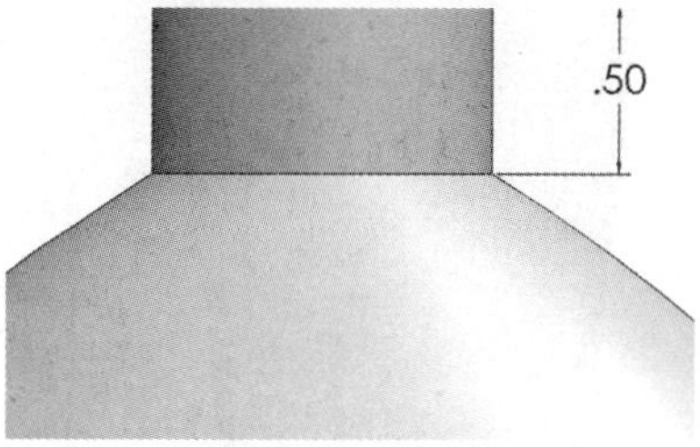

FIGURE P4.5C

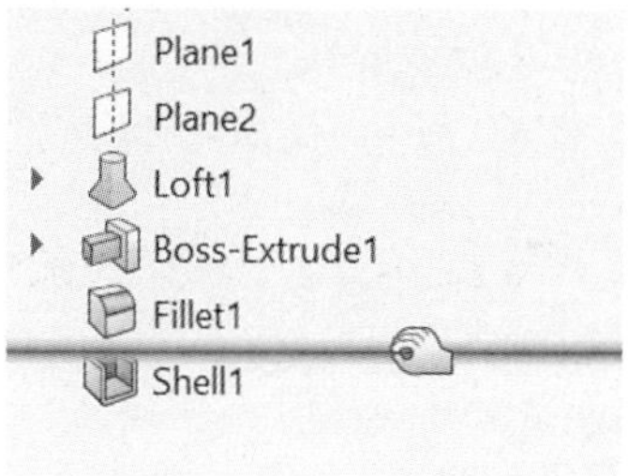

P4.6 Design a mug as shown in **Figure P4.6A**. The mug should hold approximately 30 cubic inches (slightly more than 16 fluid ounces). The mug should have a hexagonal bottom and a circular top, with a wall thickness of 0.125 inches.

- To create the hexagon shape, select the Polygon Tool from the Sketch group of the CommandManager and drag out a shape from the origin. By default, a hexagon will be created, although you can change the number of sides in the PropertyManager to create other polygons. To precisely orient the hexagon, select a side and add a horizontal or vertical relation. For this exercise, adding a vertical relation to one of the sides will result in the handle being placed in the middle of one of the mug's segments, as shown in **Figure P4.6A**.

FIGURE P4.6A

- After shelling the mug's body, create the handle by sketching a sweep path and a circular profile using the Swept Boss Tool. To make sure that the mug's body and the handle interface with no gaps, extend the sweep path well into the mug's body, as shown in **Figure P4.6B**. Trim away the portions of the handles inside the body by opening a sketch on each handle end, using the Convert Entities Tool to create a circle

FIGURE P4.6B

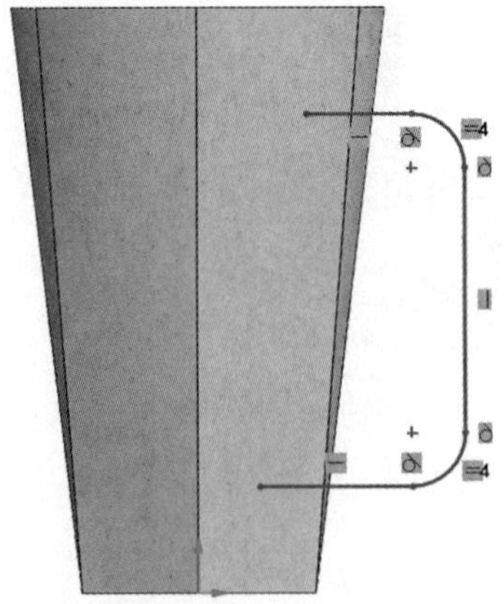

in the sketch plane, and then using the Extruded Cut Tool with a type of Up to Surface to remove the excess portion of the handle. Select the inner surface of the mug as the surface defining the end of the cut, as shown in **Figure P4.6C**. (This step can be avoided if you shell the mug after creating the handle, but that causes the handle to be shelled as well.)

FIGURE P4.6C

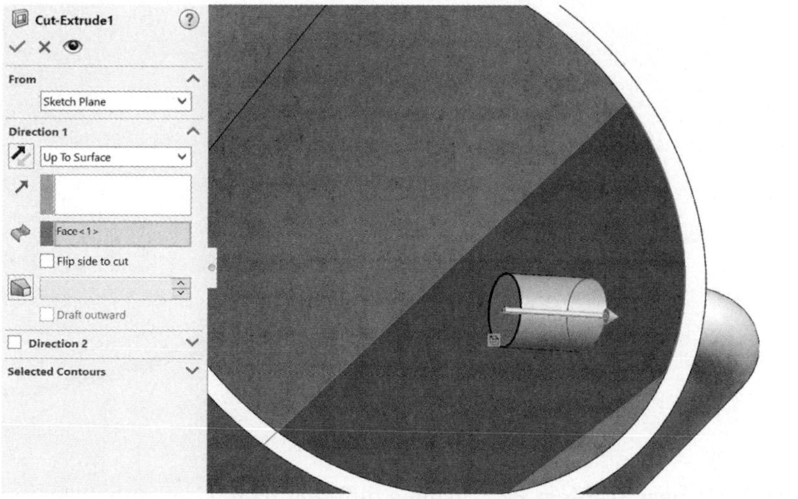

- Add fillets at the intersections of the handle and the body.
- If your SOLIDWORKS license includes the add-in PhotoView 360, you can experiment with creating a photorealistic rendering of the mug using different materials and backgrounds, as shown in **Figure P4.6D**. To see if you have access to the add-in, select Tools: Add-Ins and look for PhotoView 360 in the list. A tutorial for using PhotoView 360 is available from the book's website: *www.mhhe.com/howard2019*.

FIGURE P4.6D

P4.7 Model a bent tube with 0.5-inch outer diameter and 0.375-inch inner diameter to join the ends of the two tubes shown in **Figure P4.7A**. Follow the route shown by the centerlines in **Figure P4.7A**, and add 1-inch radius fillets to the corners.

The completed model is shown in **Figure P4.7B**.

FIGURE P4.7A

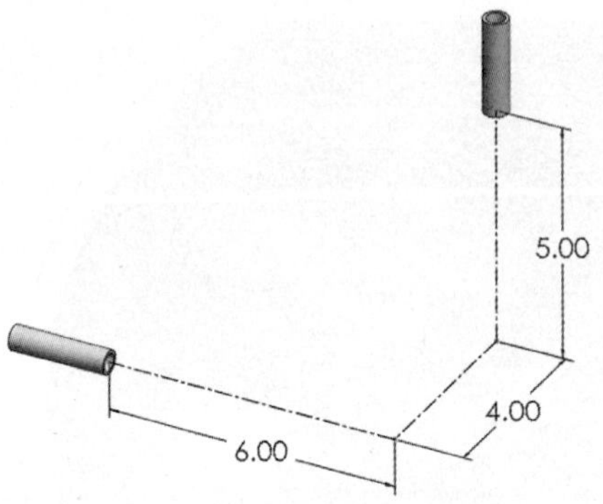

FIGURE P4.7B

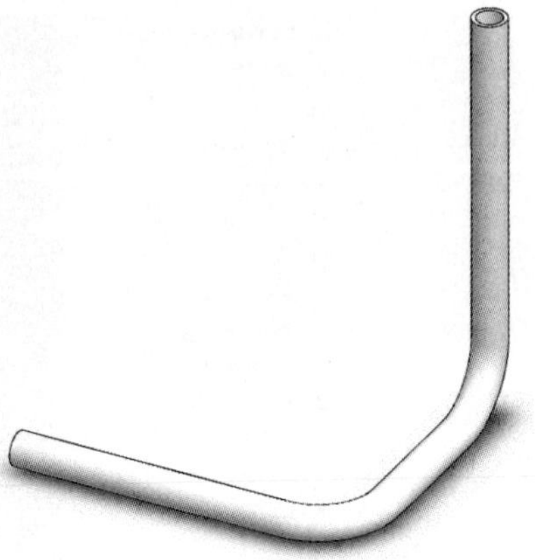

P4.8 Repeat **Problem P4.7**, using a different path. Use a tube geometry that minimizes the tube length while maintaining a straight section at each end, as shown in **Figure P4.8**.

FIGURE P4.8

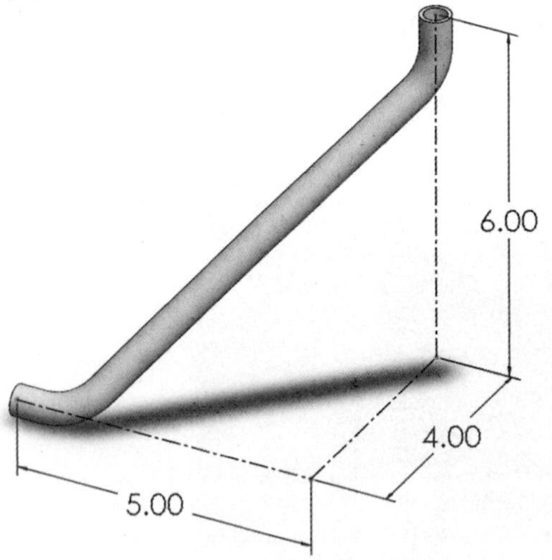

P4.9 Screw threads on parts are not usually modeled. Instead, cosmetic threads are added to represent the threads. We will learn how to apply cosmetic threads in Chapter 5. In some cases, however, a threaded part needs to be prototyped, so the threads must be modeled. **Figure P4.9A** shows the drive screw from a table vise. The threads on a drive screw are typically Acme threads, a trapezoidal-profile thread. The thread shape can be seen in the cross-section of the screw of **Figure P4.9B**. Model the screw, following these steps:

FIGURE P4.9A

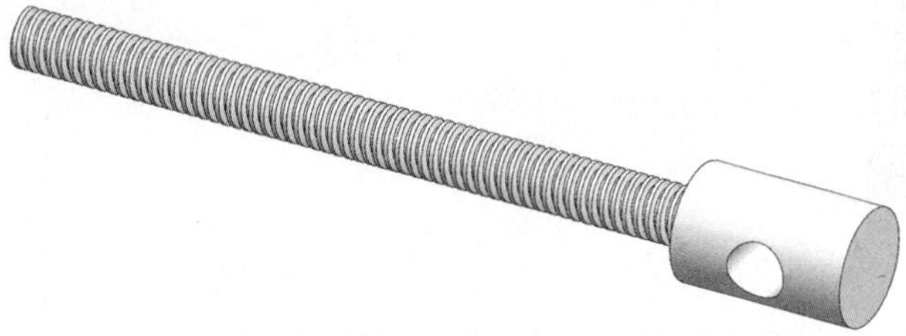

FIGURE P4.9B

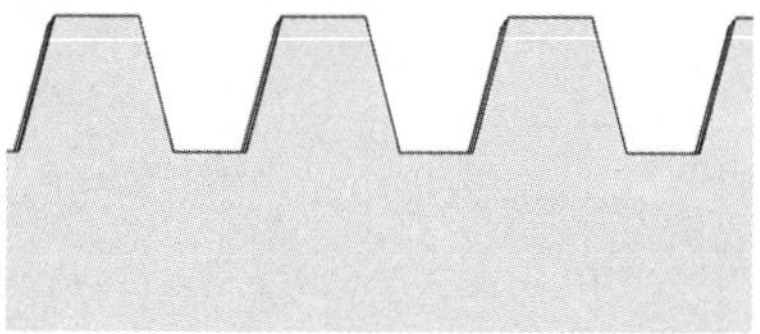

1. Create a cylinder 0.5 inches in diameter and 6 inches in length.
2. Create a new plane 0.1 inches offset from one end of the cylinder.
3. Sketch a 0.5-inch diameter circle in the new plane, and insert a clockwise helix with a pitch of 0.1 inches (10 threads per inch) and a height of 6.2 inches, as shown in **Figure P4.9C**.

FIGURE P4.9C

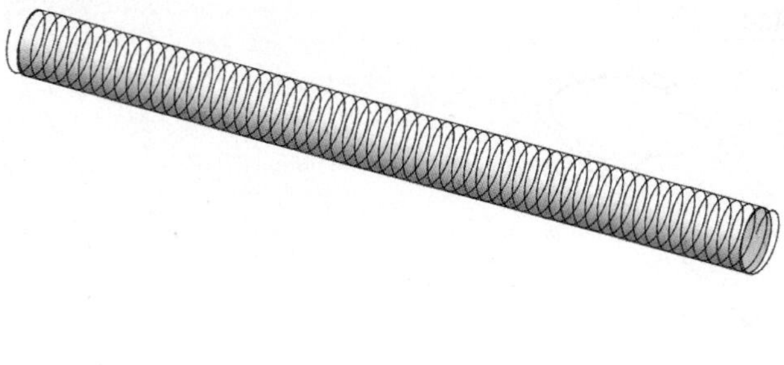

4. Open a new sketch in a plane normal to the helix at its endpoint. Sketch the profile as shown in **Figure P4.9D**, and close the sketch.

FIGURE P4.9D

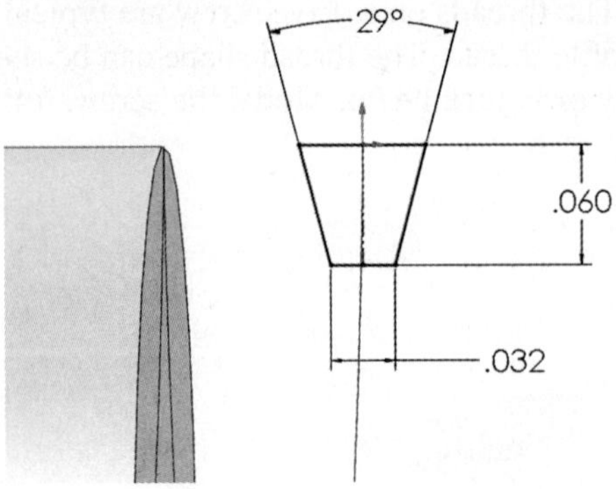

5. Select the Swept Cut Tool. Sweep the profile along the helix to cut the threads.
6. Add the head of the screw, using the dimensions shown in **Figure P4.9E**.

FIGURE P4.9E

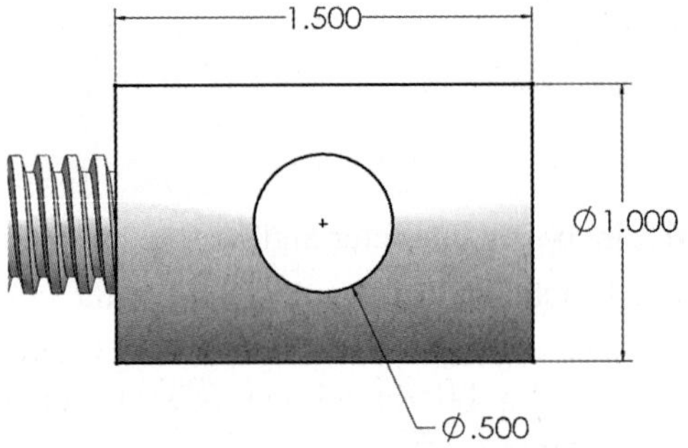

P4.10 Measure a standard paper clip, and use your measurements to create a solid model similar to the one shown in **Figure P4.10**.

FIGURE P4.10

CHAPTER 5

Parametric Modeling Techniques

Introduction

One of the tremendous advantages of solid modeling software is the ability to represent geometric relationships in a parametric form. In a parametric model, the description of the model contains both a geometric description (shape, orientation, etc.) and specific numerical parameters (dimensions, number of instances, etc.). Consider, for example, a simple parameterized solid model of a cylinder; the definition of the model contains both a geometric description defining the primitive shape (cylinder), and two numeric parameters (diameter and length) to fully define the model. A whole "family" of cylindrical parts of different diameters and lengths could be modeled with a single parameterized model, with only a numeric table of values used to differentiate the various parts in the family.

The Linear Pattern Tool used to define the hole pattern in Chapter 3 is another good example of a parameterized model; the geometric information used to describe a hole pattern includes:

- the shape of the hole (circular)
- the type of extruded cut ("through all")

The numeric parameters that must also be defined include:

- the vectors (axes) that define the two directions of the pattern
- the two "repeat" dimensions along the two directions
- the number of instances of the holes in each direction

Changes can be readily made to the parameters to update the model, without requiring any changes to the geometric information.

This concept of using numeric parameters to drive a solid model can be exploited in design. One powerful tool that can be employed is the development of mathematical relationships (equations) that relate the values of two or more parameters in a model.

Chapter Objectives

In this chapter, you will:

- use the Draft Tool to add draft to an existing solid part,
- learn how to use a shared sketch to create more than one feature,
- develop equations to embed parameterized relationships into a solid model,
- use a design table to create a family of similar parts, and
- create a multiple-configuration part drawing.

In our hole pattern, for example, we might want to restrict our model so that the number of instances of holes in one direction is always the same as the number of instances of holes in the other direction. This is accomplished by relating the two parameters together with an equation, for example:

(# of instances along Direction 2) = (# of instances along Direction 1)

Once this parameterized relationship is established, any change made to the *independent parameter* (the number of instances of holes along Direction 1) would automatically update the value of the *dependent parameter* (the number of instances of holes along Direction 2), and the model would be updated accordingly. This allows the engineer to embed design intelligence and "rules-of-thumb" directly into solid models.

In this chapter, an example of a mechanical part that contains embedded equations to drive specific parameters will be presented. In addition, the tutorial will introduce some new modeling operations that have not yet been utilized in the preceding chapters. A second example will be presented where a spreadsheet-based *design table* is used to drive the dimensions of a model, producing an entire family of parts.

5.1 Modeling Tutorial: Molded Flange

In this exercise, the flange shown in **Figure 5.1** will be created. Since this is to be a molded part, draft will be added to the appropriate surfaces.

FIGURE 5.1

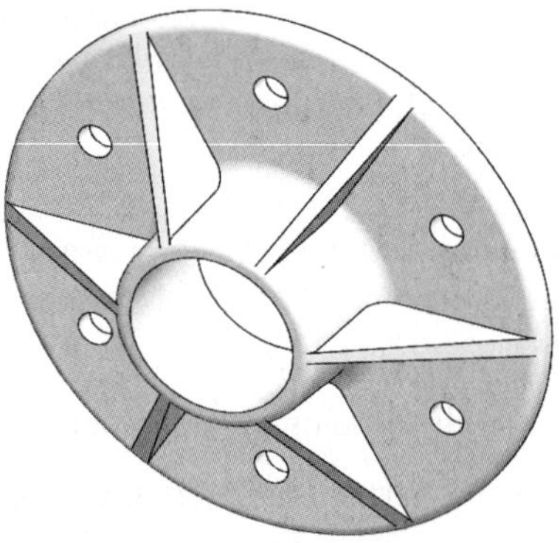

Begin by starting a new part, and selecting the Front Plane from the FeatureManager. Select the Circle Tool from the Sketch group of the CommandManager. Drag out a circle centered at the origin. Select the Smart Dimension Tool and add a 6-inch diameter dimension, as shown in Figure 5.2.

Select the Extruded Boss/Base Tool from the Features group of the CommandManager. Extrude the sketch outward to a thickness of 0.25 inches.

FIGURE 5.2

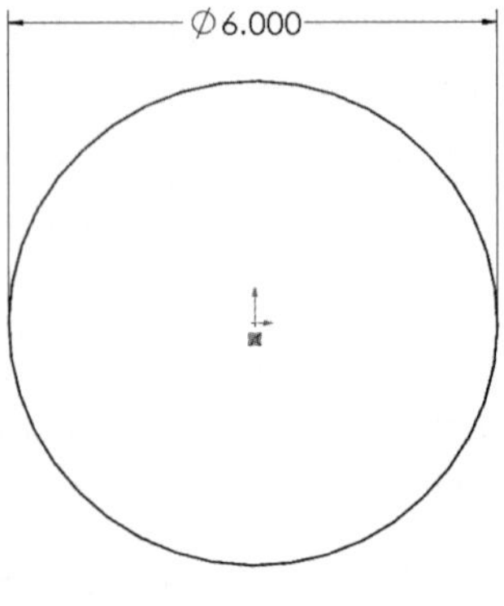

FIGURE 5.3

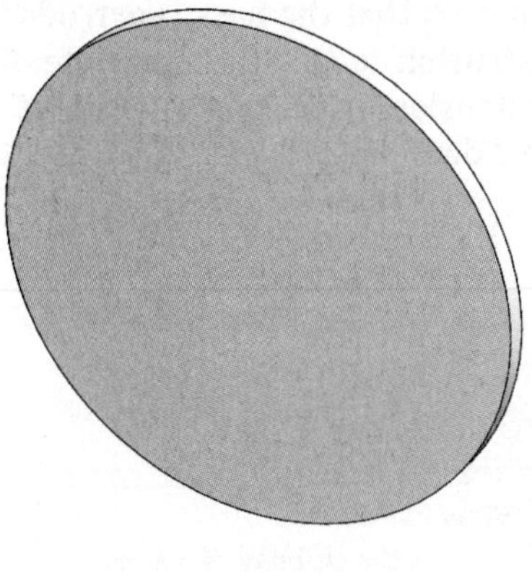

The extruded solid is shown in Figure 5.3.

Since the next feature will be drafted, we will define its dimensions in a new plane and extrude its sketch back toward the rest of the part. Therefore, we need to create a new construction plane offset from the Front Plane.

Begin by selecting the Front Plane from the FeatureManager.

The selected plane will be highlighted.

FIGURE 5.4

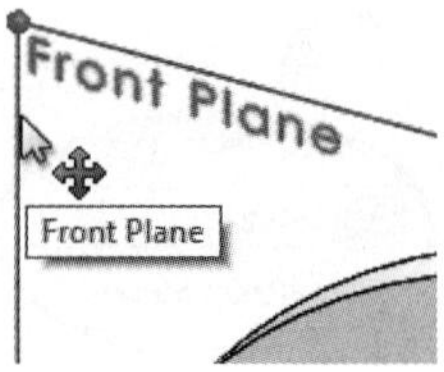

While holding down the Ctrl key, click and drag the plane outward. Make sure that the move arrows, as shown in Figure 5.4, appear before clicking. Release the mouse button, and the new plane is previewed, as shown in Figure 5.5.

FIGURE 5.5

In the PropertyManager, set the offset distance to 1.5 inches, as shown in Figure 5.6, and click the check mark to complete the operation.

The new plane will be labeled Plane1 in the FeatureManager. A cylindrical boss will now be extruded with a draft from this new plane back toward the base feature.

FIGURE 5.6

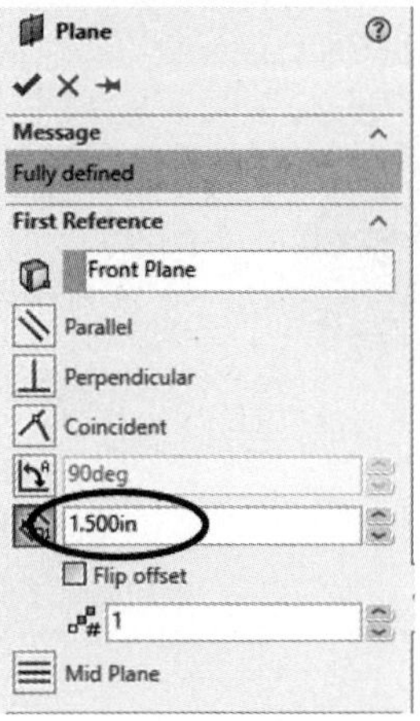

Click on the boundary of the new plane (Plane1) and choose the Normal to View from the Context Toolbar, as shown in Figure 5.7, or from the Heads-Up View Toolbar. Select the Circle Tool from the Sketch group of the CommandManager. Drag out a circle from the origin. Select the Smart Dimension Tool and dimension the circle's diameter as 2.2 inches, as shown in Figure 5.8.

FIGURE 5.7

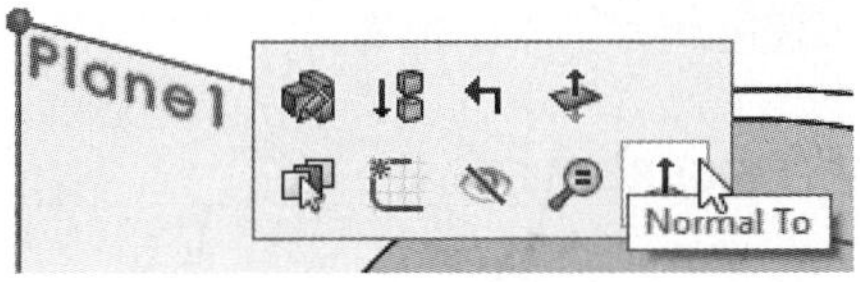

FIGURE 5.8

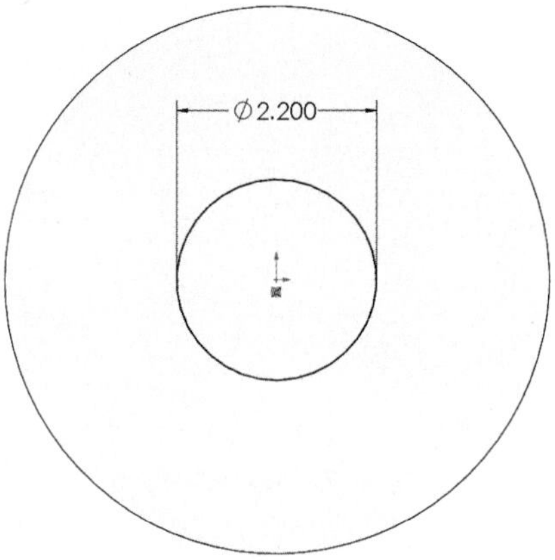

FIGURE 5.9

Select the Extruded Boss/Base Tool from the Features group of the CommandManager, and set the direction such that the boss is extruded toward the base. Change the type of extrusion to Up To Next. (Note: If Up To Next is not available, then the extrusion direction is incorrect.) Turn the draft on. Set the draft angle to 3 degrees and check the "Draft outward" box, as shown in Figure 5.9. Click on the check mark to complete the extrusion. From the Hide/Show Items menu of the Heads-Up View Toolbar, select View Planes, as shown in Figure 5.10. This will toggle off the display of planes.

FIGURE 5.10

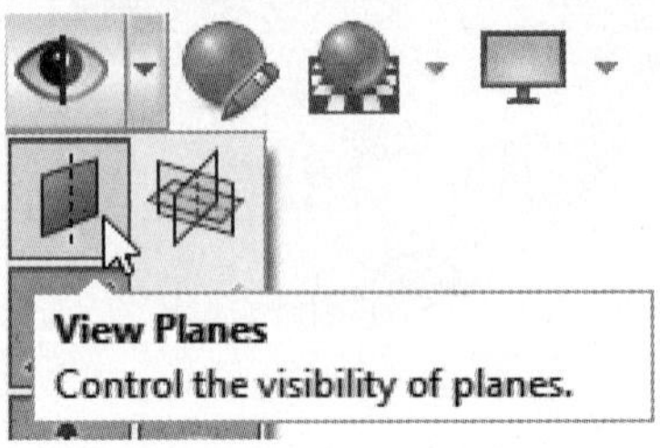

The result of the drafted extrusion is shown in **Figure 5.11**.

Select the face shown in Figure 5.12, and select the Normal To View. Select the Circle Tool from the Sketch group of the CommandManager. Drag out a circle centered at the origin. Select the Smart Dimension Tool. Dimension the circle's diameter at 1.8 inches, as shown in Figure 5.13.

FIGURE 5.11

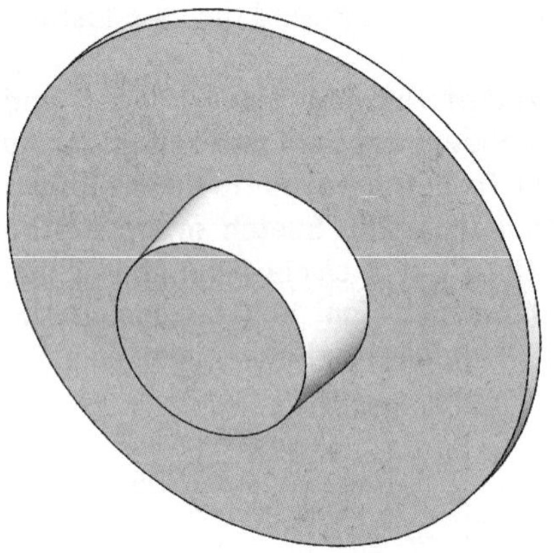

FIGURE 5.12

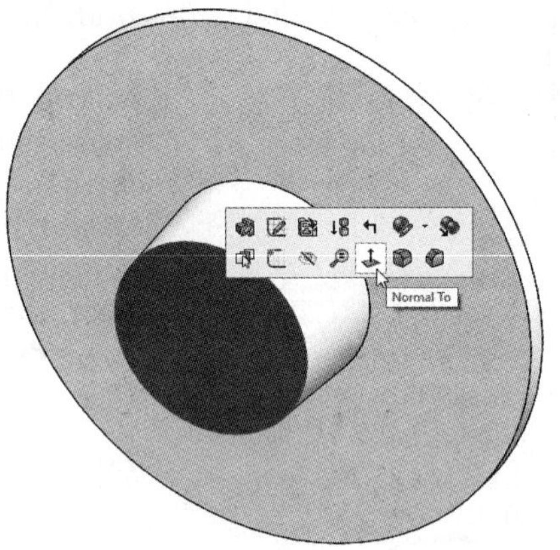

FIGURE 5.13

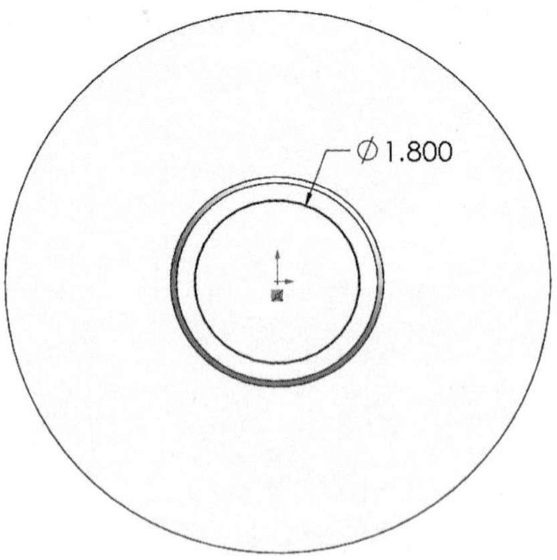

Select the Extruded Cut Tool from the Features group of the CommandManager. Select Through All as the type, turn the draft on, and set the draft angle to 2 degrees. Make sure that the "Draft outward" box is checked, as shown in Figure 5.14.

FIGURE 5.14

Click the check mark to apply the cut, which is shown in Figure 5.15.

To see the effect of the draft on the hole, viewing the model with a section view is helpful.

Select the Section View Tool from the Heads-Up View Toolbar, as shown in Figure 5.16. In the PropertyManager, select the middle button, representing the Top Plane, as shown in Figure 5.17.

Reverse the direction of the section by clicking on the icon next to the plane selection box.

FIGURE 5.15

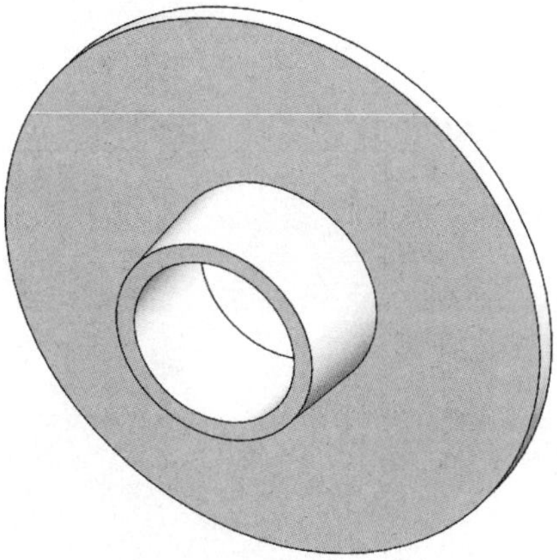

FIGURE 5.16

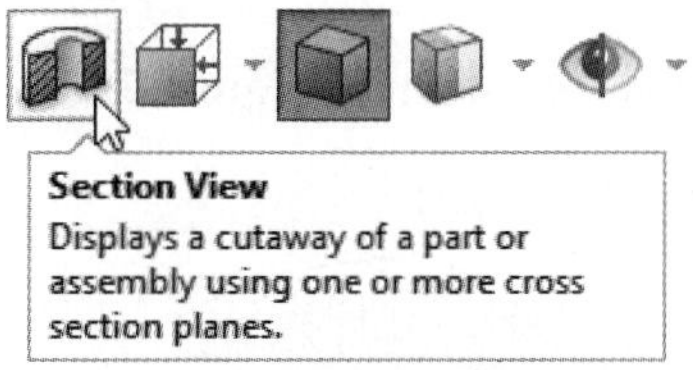

FIGURE 5.17

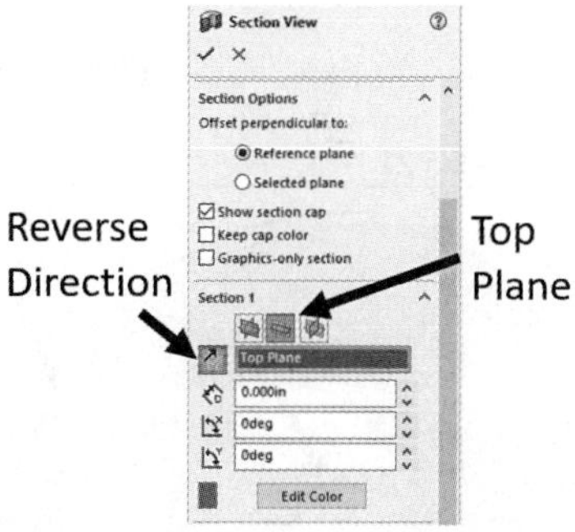

Click the check mark to apply the settings for the section view. Select the Bottom View from the Heads-Up View Toolbar, as shown in Figure 5.18.

FIGURE 5.18

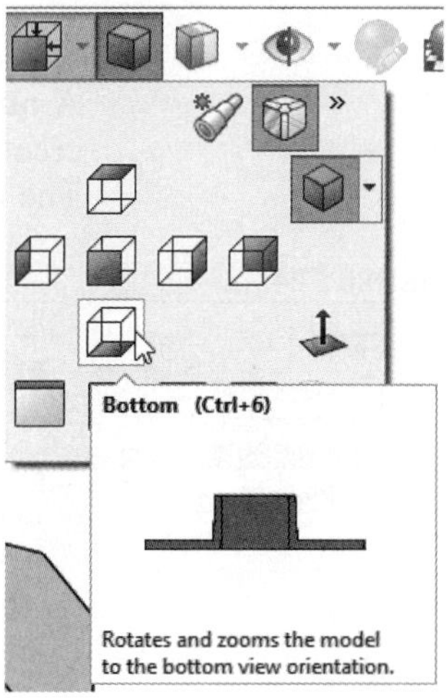

The cross section of the part, as shown in **Figure 5.19**, clearly shows the drafted center hole. The section view will be displayed until turned off with the Section View Tool.

FIGURE 5.19

FIGURE 5.20

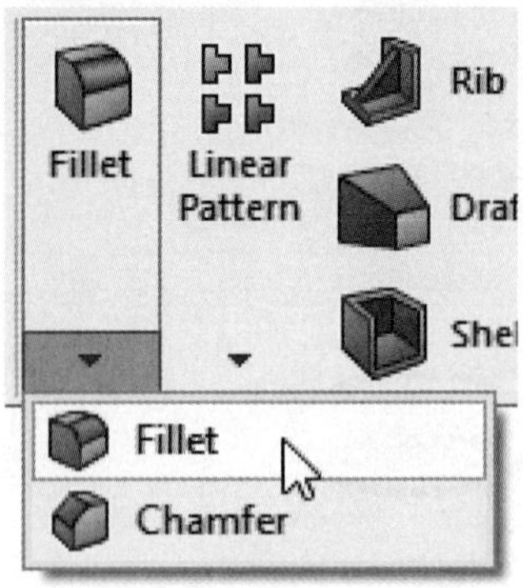

Click the Section View Tool to return to the display of the entire model.

Before adding the ribs and holes, we will fillet several of the sharp edges of the part.

Start by selecting the Fillet Tool from the Features group of the CommandManager, as shown in Figure 5.20.

Since two of the edges are to have the same radius, we can add these fillets at the same time.

Set the radius to 0.125 inches, and select the two edges shown in Figure 5.21.

Be sure to select the edge and not the face. If a face is selected, then all of the edges of that face will be filleted.

Click the check mark to apply the fillets. Using a similar procedure, fillet the edge shown in Figure 5.22 with a fillet radius of 0.25 inches.

The filleted part is shown in **Figure 5.23**.

FIGURE 5.21

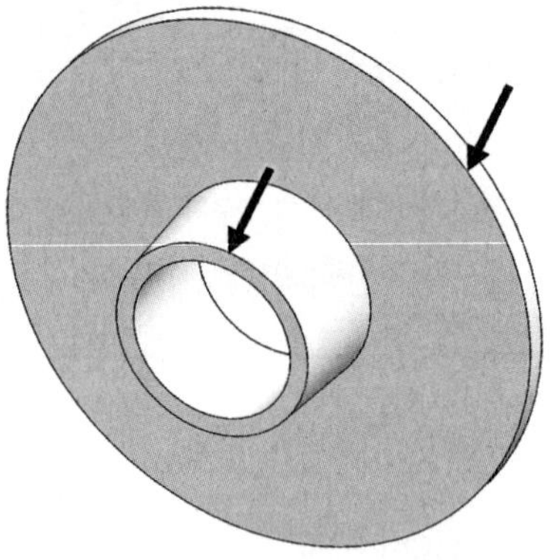

FIGURE 5.22

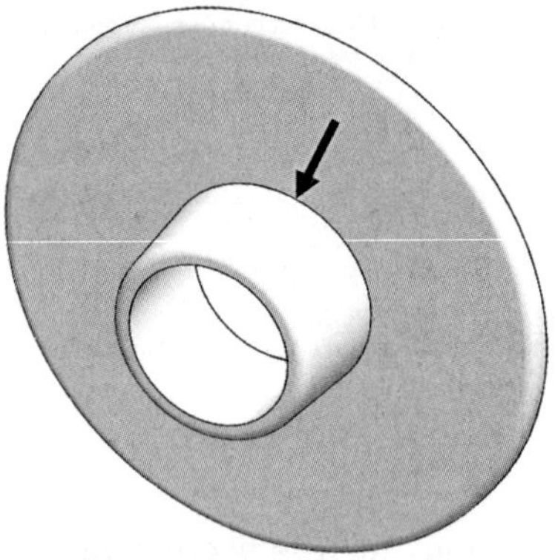

FIGURE 5.23

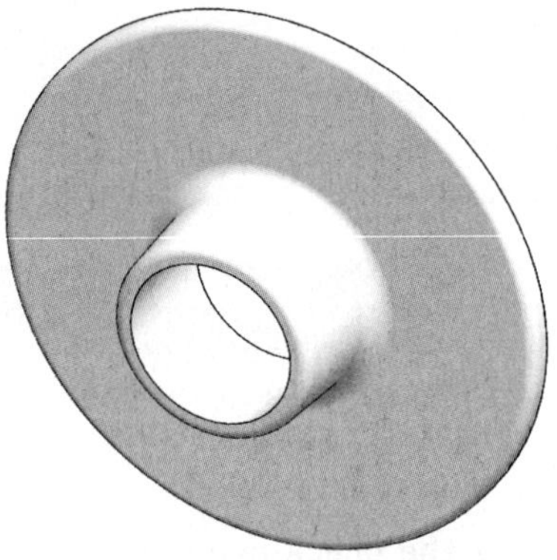

A number of stiffening ribs will now be added to the part. This will be done by creating a single rib and duplicating it a number of times using a circular pattern. The first rib in the pattern will now be created.

FIGURE 5.24

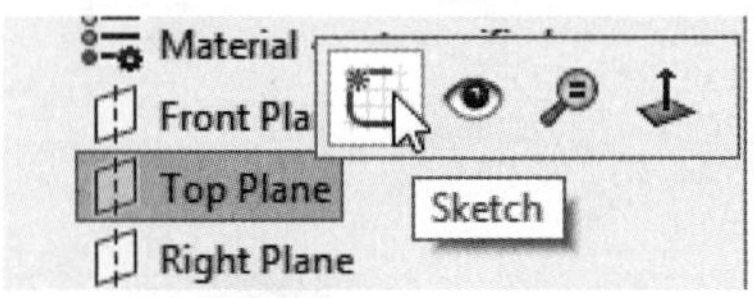

Select the Top Plane from the FeatureManager. Select the Sketch Tool from the context toolbar, as shown in Figure 5.24, or from the Sketch group of the CommandManager. Switch to the Bottom View.

Note that most of the time when we begin a sketch, we simply choose a sketch tool. However, for the step we are about to perform, it is necessary to open the sketch before choosing the desired tool (Convert Entities).

Choose the Wireframe display mode with hidden lines visible, as shown in Figure 5.25. Zoom in on the right side of the part.

FIGURE 5.25

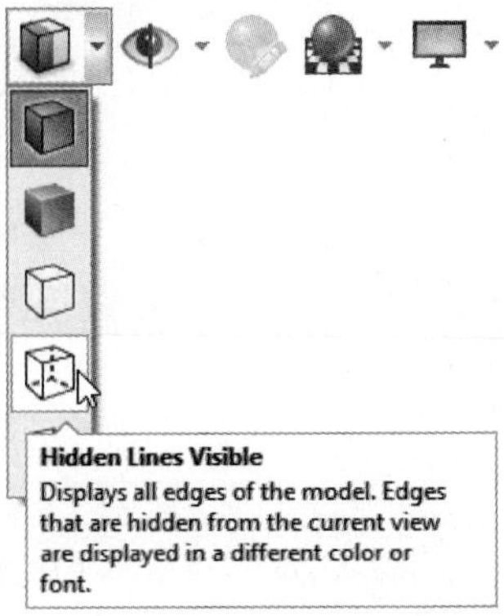

Since the rib will blend into the fillet radii, we need to select the intersections of the radii with the plane. While the edge of the solid part appears to be an arc when viewed from this perspective, there is no physical sketch entity associated with it; however, we can construct an arc coincident with this projected solid edge by using the Convert Entities Tool.

Move the cursor to the intersection of the top fillet and the plane. When you have positioned the cursor at the appropriate location, the silhouette symbol will appear (Figure 5.26). Click to select this silhouette.

FIGURE 5.26

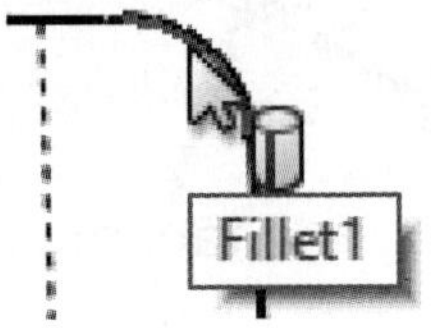

While holding down the Ctrl key, locate and select the silhouette of the edge of the other fillet, as shown in Figure 5.27.

FIGURE 5.27

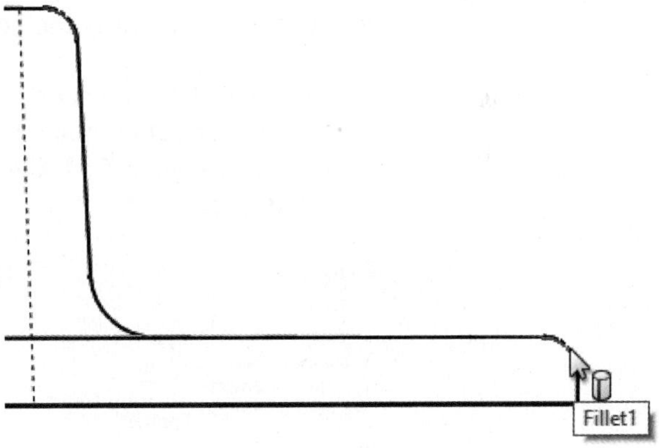

With the silhouettes selected, click the Convert Entities Tool, as shown in Figure 5.28.

This will add to your sketch arcs that are coincident with the silhouettes of the fillets, as shown in Figure 5.29.

Select the Line Tool. Move the cursor over the top arc so that the coincident relation icon appears, as shown in Figure 5.30, indicating that the endpoint of the line will snap to the arc. Do not snap to the midpoint of the arc.

FIGURE 5.28

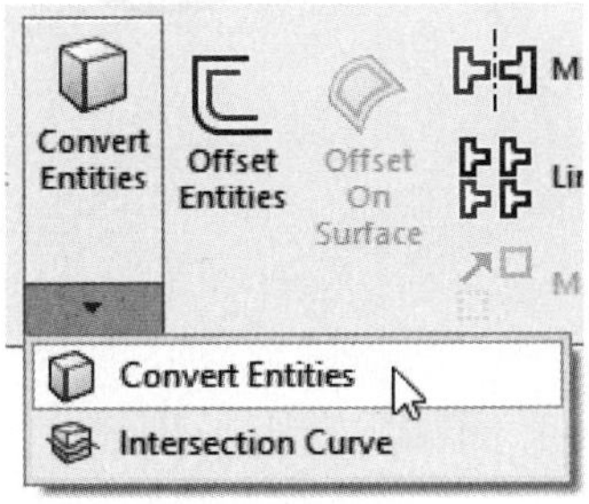

FIGURE 5.29

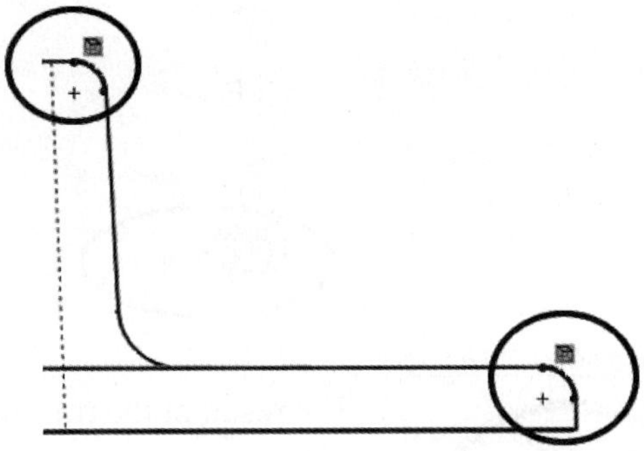

FIGURE 5.30

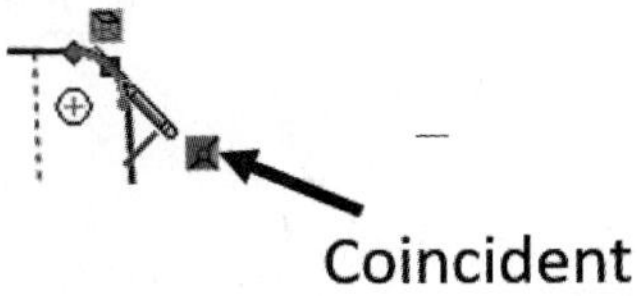

FIGURE 5.31

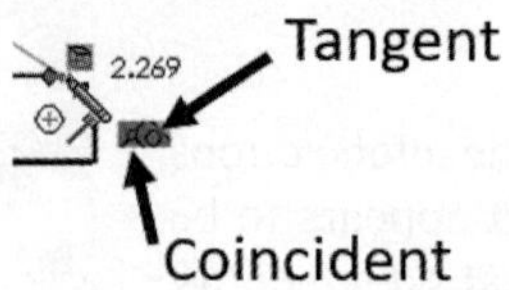

Click and drag a line to the other arc. Place the endpoint so that the coincident and tangent relation icons appear, as shown in Figure 5.31. Press Esc to turn off the Line Tool.

FIGURE 5.32

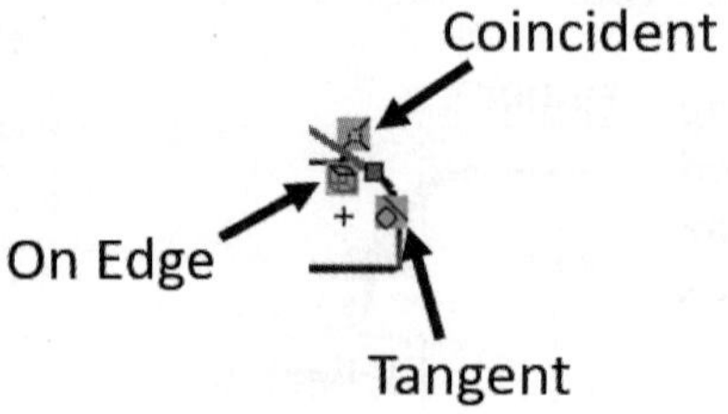

Zoom in on the fillet at the top of the part, as shown in Figure 5.32. A tangent relation should have been added automatically to the arc and the line. If the relation did not add automatically, then add it manually by selecting both entities (use the Ctrl key to select more than one entity) and adding a tangent relation in the PropertyManager. Note that there are also relation icons for On Edge (the arc is on the silhouette edge of the fillet) and Coincident (the end of the line lies on the arc).

We will now trim away the portions of the arcs that are not needed in the sketch.

Zoom in on the intersection of one of the arcs and the line. Select the Trim Entities Tool, as shown in Figure 5.33. Choose "Trim to closest" as the trim option, as shown in Figure 5.34. Click to trim the portion of the arc to be removed, as shown in Figure 5.35.

FIGURE 5.33

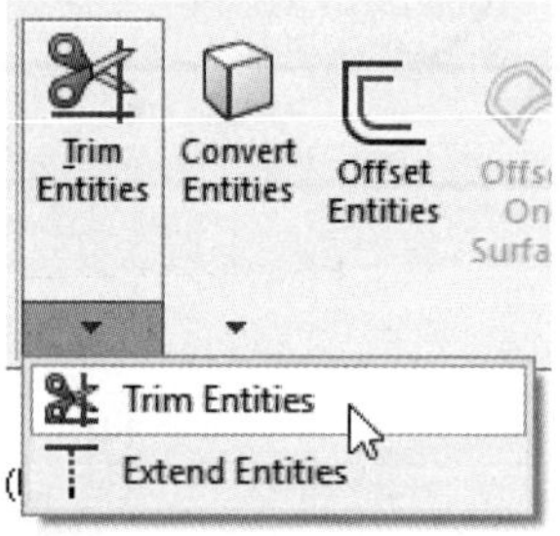

FIGURE 5.34

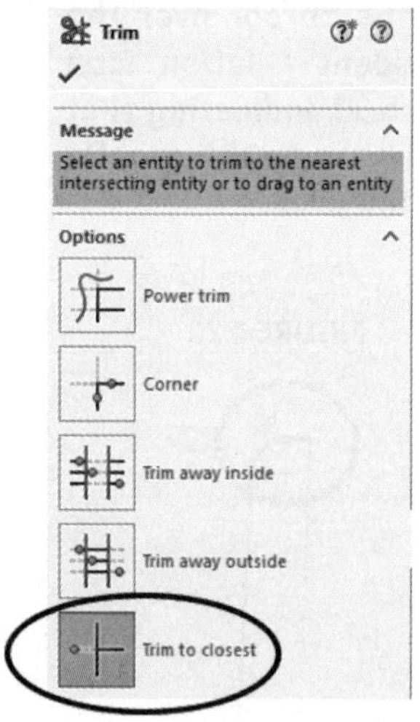

FIGURE 5.35

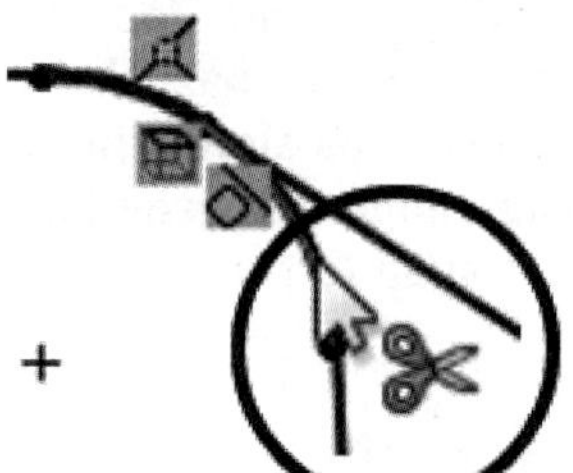

FIGURE 5.36

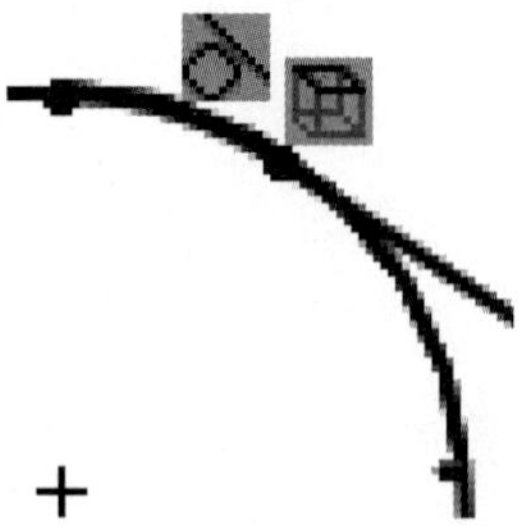

FIGURE 5.37

The result of the trimming operation is shown in **Figure 5.36**.

Repeat this process on the other arc/line intersection, as shown in Figure 5.37.

Add the two lines indicated in Figure 5.38 to close the sketch contour.

FIGURE 5.38

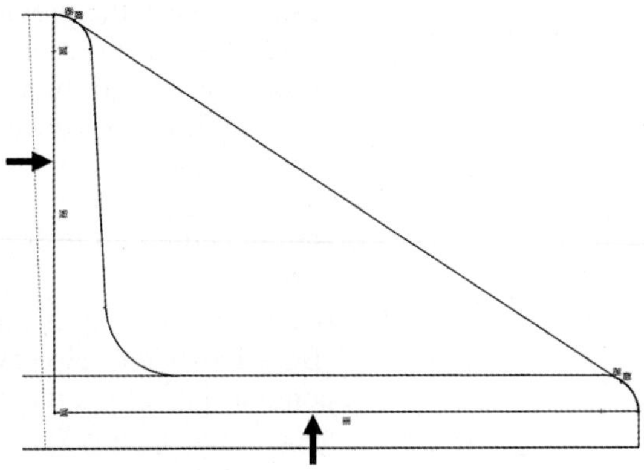

If the lines are horizontal and vertical, then the sketch should be fully defined.

Select the Extruded Boss/Base Tool from the Features group of the CommandManager. Extrude the rib as a Mid Plane Extrusion with a thickness of 0.125 inches, as shown in Figure 5.39.

The result is shown in Figure 5.40.

FIGURE 5.39

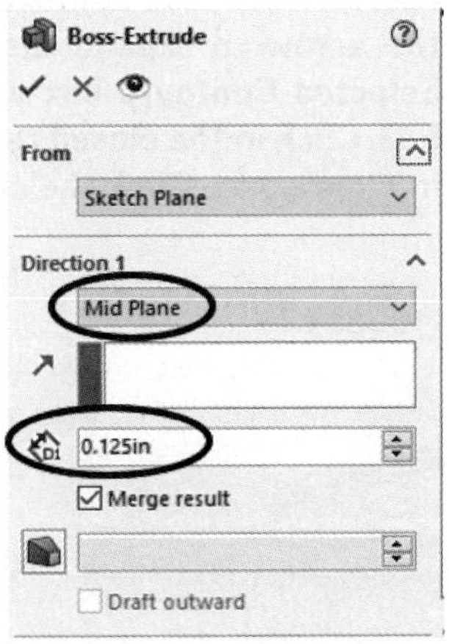

FIGURE 5.40

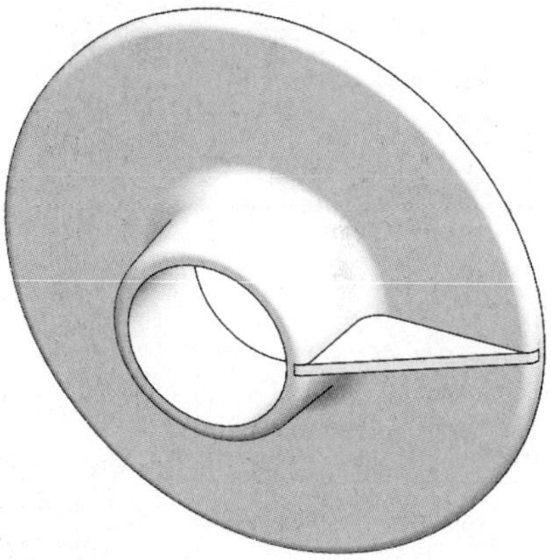

While the geometry of the rib appears to be satisfactory, there are two areas where the faces do not blend smoothly. These areas are the intersections between the rib and the fillets of the flange. The intersection at the lower fillet is shown in Figure 5.41; a similar area exists at the intersection with the upper fillet. We have a flat part (the rib) mating with a curved surface (the fillet). While the contour of the rib matches that of the fillet perfectly at the mid-surface of the rib, as the rib is extruded outward its surface is slightly higher than the fillet's. Although the difference is small, this type of mismatch can result in errors when exporting the geometry to a 3-D printer, tool path program, or finite element analysis program.

FIGURE 5.41

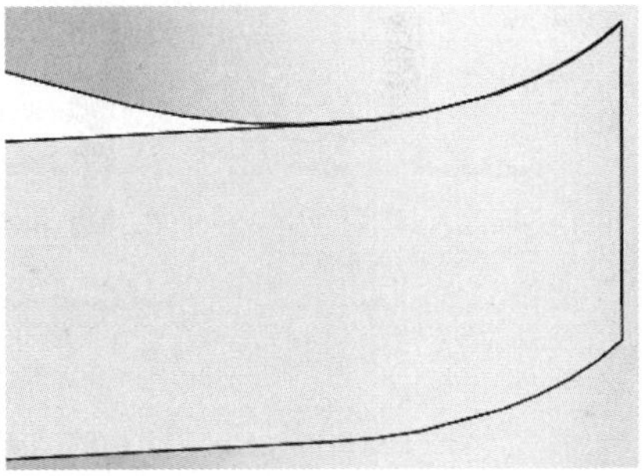

One solution to this mismatch would be to replace the extruded rib with one that is created by revolving the cross section about the center axis of the flange through a small angle. However, this would result in a rib that is much thicker toward the outer edge of the flange. We could also try adding some small fillets to smooth the mismatched surfaces. A better approach is to use a revolved cut to remove the portions of the rib that extend above the fillets. Think of this operation as creating a cutting tool that precisely matches the profile of the mid-surface of the rib, and is swept around the center axis of the flange.

One way to match the rib profile is to open a new sketch and repeat the steps of identifying the silhouettes of the fillets and converting them to arcs, adding a line tangent to both arcs, and trimming away the excess portions, as we did previously. We cannot do this in the same location as the previous sketch, however, because the rib geometry has eliminated the silhouettes of the fillets. We could recreate the sketch steps in a new sketch in another location, say in the Right Plane, but it is easier to reuse the previous sketch. Another option would be to copy the previous sketch and edit it to create the cutter geometry, but the copied sketch would lose associativity with the rest of the flange geometry and would therefore not rebuild correctly if a change were made (for example, if the offset distance of Plane1 were to be changed).

Before adding on to the previous sketch, we need to make a small change to the definition of the rib. Since the sketch (Sketch4 in the FeatureManager) has only one closed contour, it was not necessary to select the contour when defining the rib (Boss-Extrude3). However, when we edit the sketch, we will be adding a second contour. Therefore, there will be an ambiguity in the definition of the rib unless we specify the contour to be used.

FIGURE 5.42

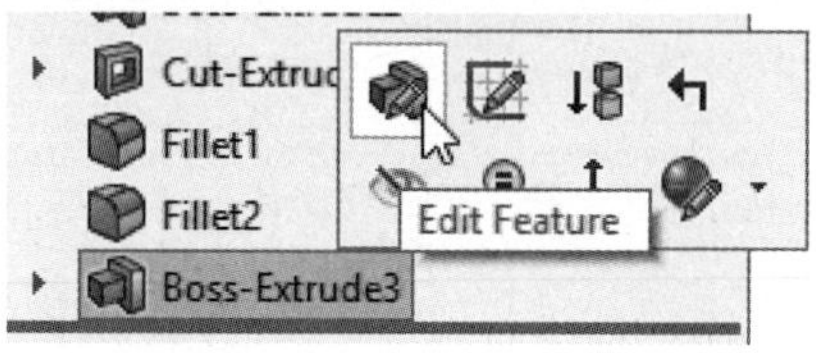

Select Boss-Extrude3 in the FeatureManager, and select Edit Feature from the context toolbar, as shown in Figure 5.42. Click the arrow to expand the box labeled "Selected Contours." The Selected Contours box will be highlighted, as shown in Figure 5.43. Click in the closed region of the sketch to select it, as shown in Figure 5.44. Click the check mark.

FIGURE 5.43

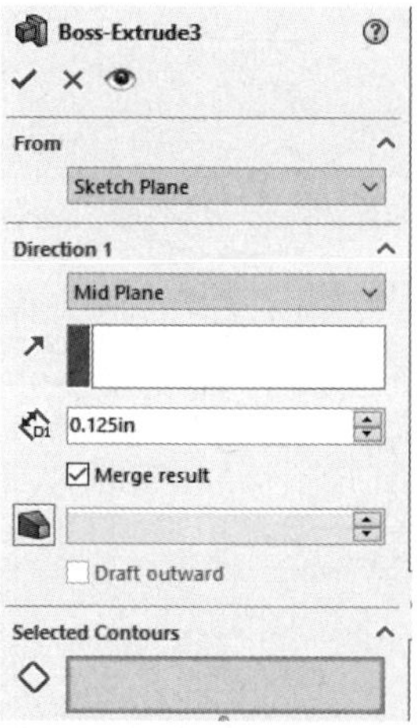

FIGURE 5.44

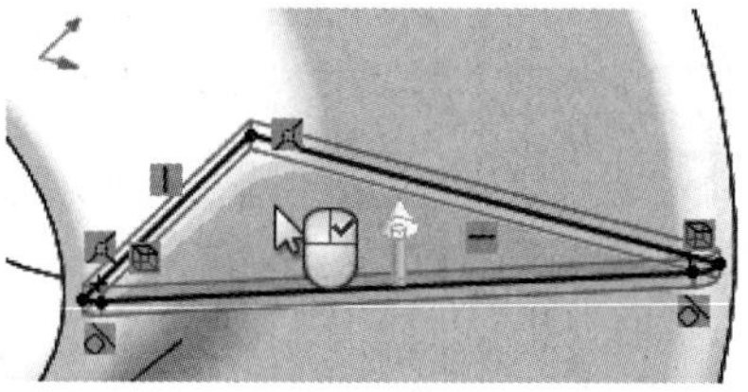

If the step above is not performed, then the addition to the sketch that follows will result in an error rebuilding the rib. The error can be fixed by editing the rib and selecting the sketch contour, but it is easier to preclude the error by selecting the contour before editing the sketch.

FIGURE 5.45

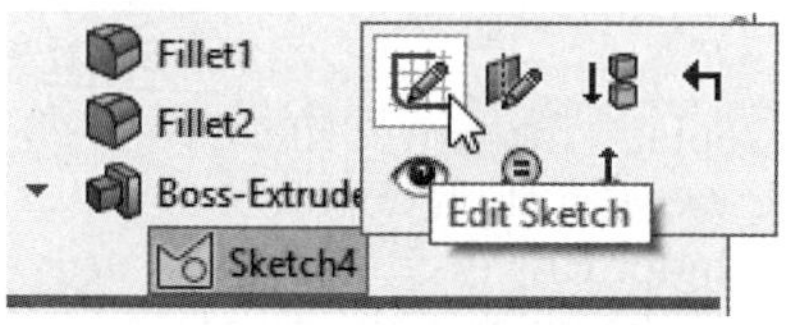

Click the arrow next to Boss-Extrude3 to expand it. Click on the rib sketch (Sketch4) to select it, and select Edit Sketch from the context toolbar, as shown in Figure 5.45. Switch to the Bottom View. Add the four horizontal and vertical lines indicated in Figure 5.46, and the two dimensions, which will fully define the sketch.

FIGURE 5.46

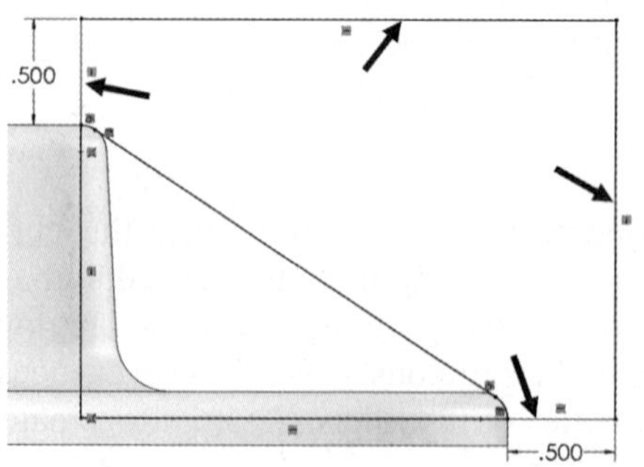

The 0.5-inch dimensions are arbitrary. However, by placing these dimensions, we are assured that the cutter's contour will always be larger than the flange, even if the flange dimensions change.

Rebuild the part. Click on Sketch4 again to select it, and choose the Revolved Cut Tool from the Features group of the CommandManager. Click to highlight the Selected Contours box. Select the upper portion of the sketch as the contour to be revolved, as shown in Figure 5.47.

FIGURE 5.47

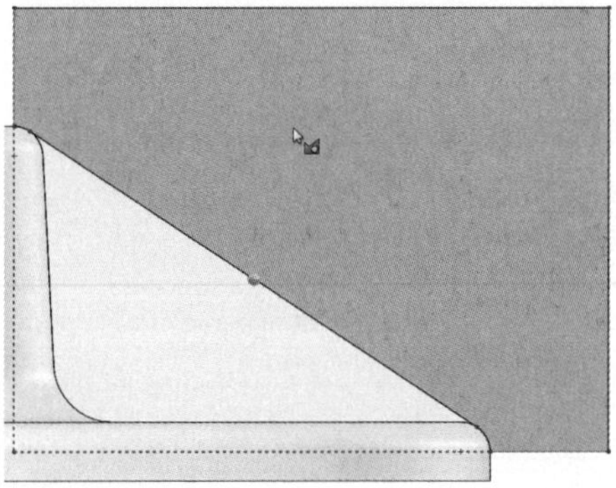

We need to define the axis of the revolution. We could add a centerline to the sketch, but instead we will use the center axis of the flange. Every cylindrical feature has an associated axis, called a temporary axis. By default, these axes are not visible.

From the Heads-Up Toolbar, click the Hide/Show Items icon and click View Temporary Axes, as shown in Figure 5.48. Click in the Axis of Revolution box in the PropertyManager to highlight it, and then click on the axis at the center of the flange, as shown in Figure 5.49. Click the check mark to complete the cut.

FIGURE 5.48

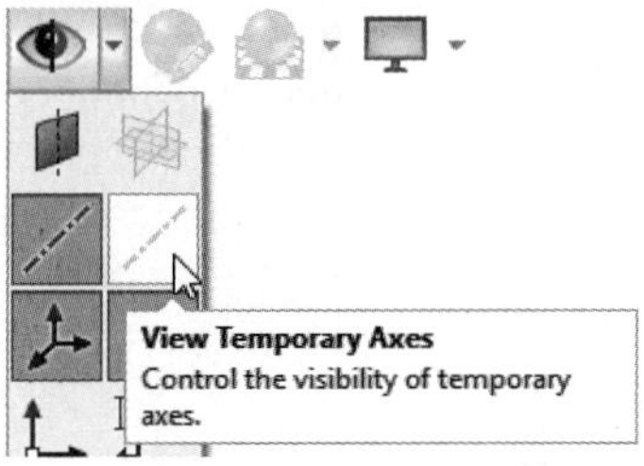

The rib and the fillets now blend smoothly, as shown in **Figure 5.50**. (Note that there is a tangent edge between the rib and fillet. If the display of tangent edges is turned off, then no edge will be visible. If the tangent edge is visible, select Options: System Options: Display and check Removed for Part/Assembly tangent edge display, as we did in Chapter 1.)

FIGURE 5.49

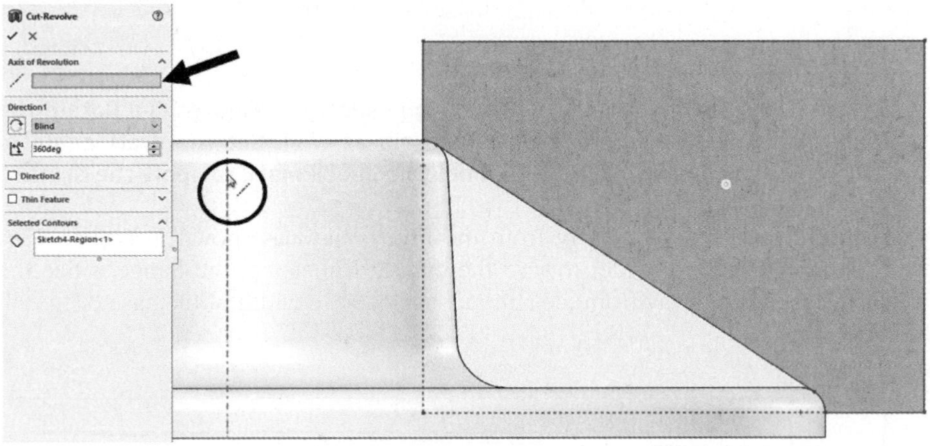

Note that Sketch4 is shown in the FeatureManager under both the rib (Boss-Extrude3) and the cut (Cut-Revolve1). The hand under each sketch icon, as shown in **Figure 5.51**, indicates that the sketch is shared with another feature.

FIGURE 5.50

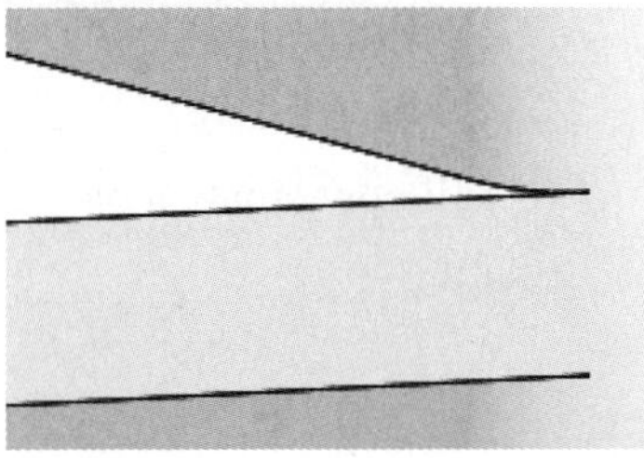

FIGURE 5.51

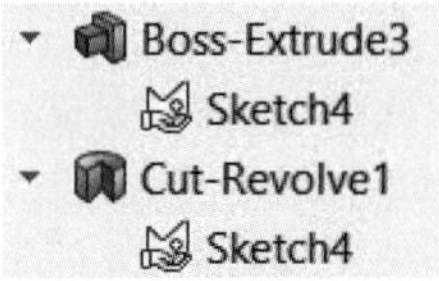

FIGURE 5.52

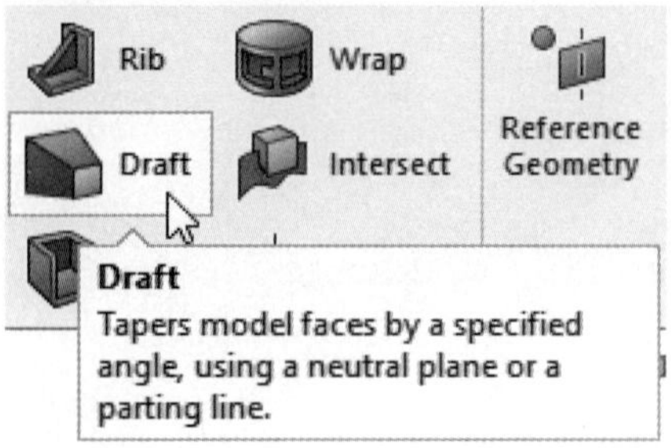

We will now add draft to the rib. Unlike the boss feature and the extruded cut, where the draft was specified as part of the extrusion step, the necessary draft on the rib is not in the same direction as the extrusion. Therefore, draft must be added as a secondary operation.

Select the Draft Tool from the Features group of the CommandManager, as shown in Figure 5.52. Make sure the Manual tab is selected. For the Neutral Plane, which defines the Direction of Pull, select the flat surface shown in Figure 5.53 (make sure not to select the curved fillet surface).

Note the arrow pointing away from the flange body. This is the direction in which the feature will get smaller as the draft is applied (i.e., the direction in which a mold half would be pulled). Since this is the correct direction for the draft, we do not need to change it.

FIGURE 5.53

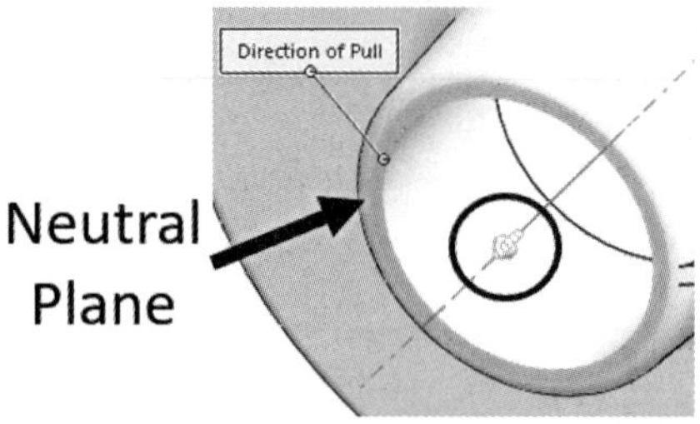

FIGURE 5.54

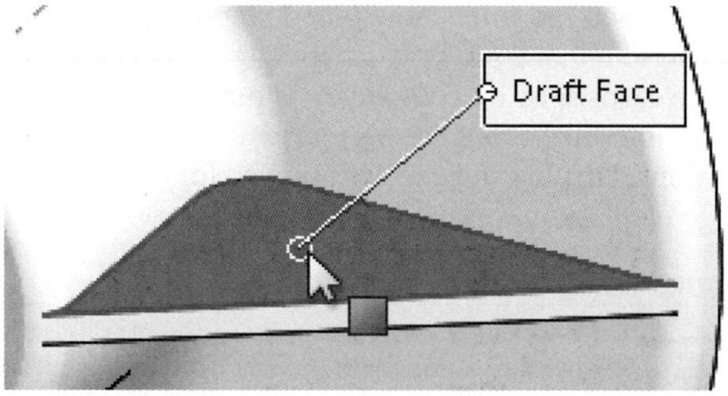

FIGURE 5.55

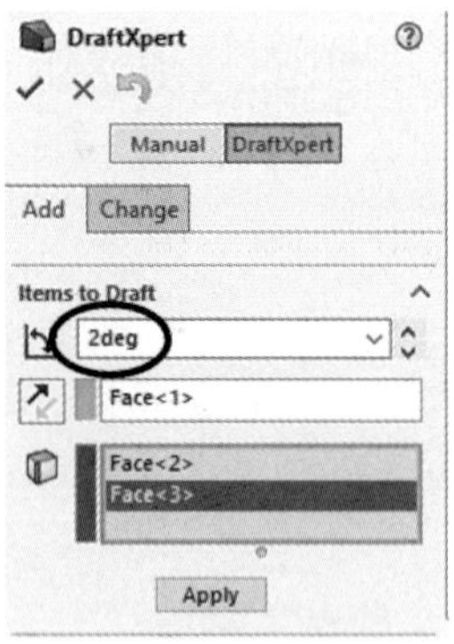

For the Faces to Draft, click on the surface shown in Figure 5.54. Rotate the flange and select the other side of the rib as well. Set the draft angle to 2 degrees, as shown in Figure 5.55. Click the check mark to apply the draft.

The draft can be seen clearly from the Front View, as shown in **Figure 5.56**. The top of the rib is wider toward the outer perimeter of the flange, since the surface is farther away from the neutral surface. The width of the base of the rib is constant.

FIGURE 5.56

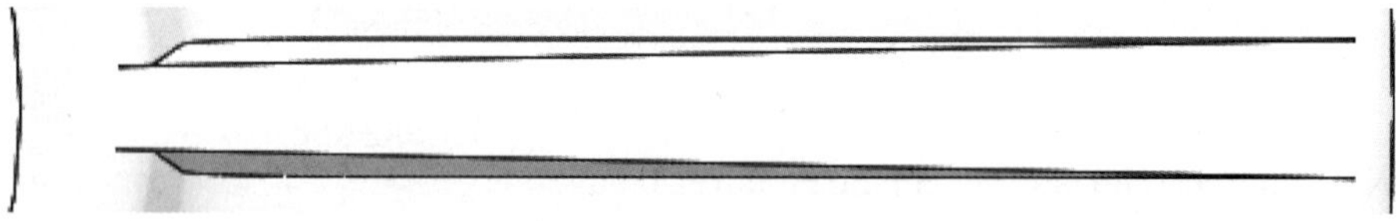

We will now create a pattern to add the other ribs.

Select the Circular Pattern Tool from the pull-down menu under the Linear Pattern Tool. Expand the fly-out FeatureManager by clicking the arrow beside the part name. For the features to pattern select the rib (Boss-Extrude3), the cut (Cut-Revolve1), and the draft, as shown in Figure 5.57. Click in the Pattern Axis box to highlight it, and select the temporary axis at the center of the part, as shown in Figure 5.58. Select Equal spacing, set the number of instances to 6, and click the check mark. Turn off the display of the temporary axes from the Hide/Show Items Tool of the Heads-Up Toolbar.

FIGURE 5.57

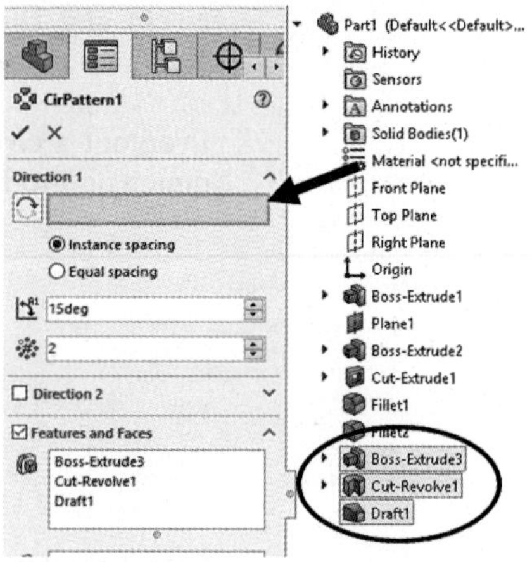

The rib pattern is shown in **Figure 5.59**.

The holes will now be added.

Select the face shown in Figure 5.60, and change to the Normal To View.

FIGURE 5.58

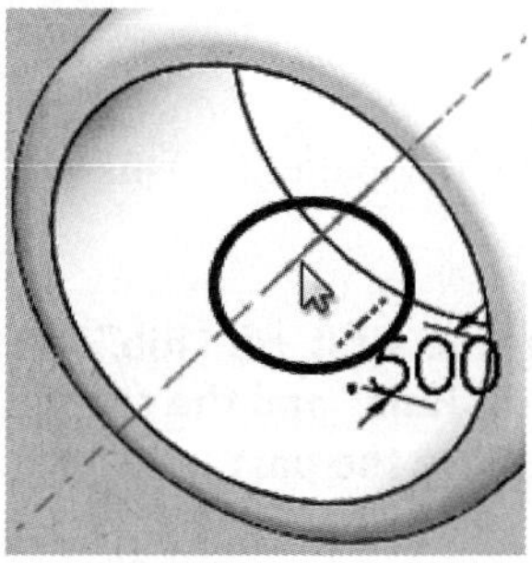

FIGURE 5.59

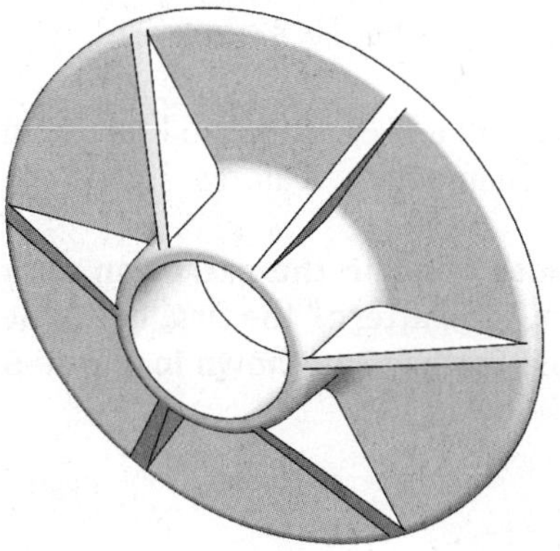

FIGURE 5.60

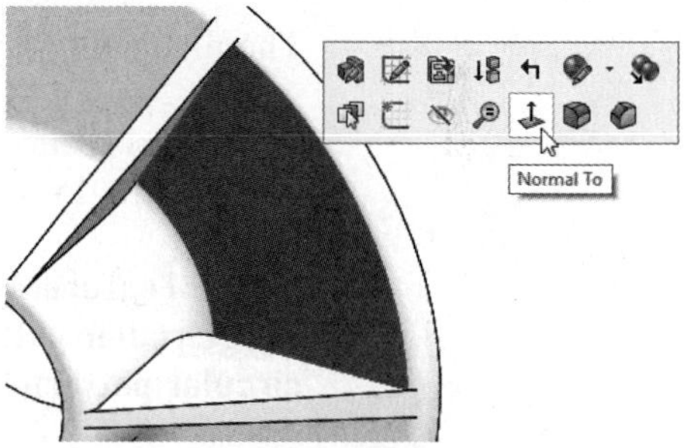

Select the Circle Tool from the Sketch group of the CommandManager. Draw a circle centered at the origin. Check the "For construction" box in the PropertyManager. Select the Smart Dimension Tool, and dimension the circle diameter as 4.5 inches.

Select the Centerline Tool and draw two centerlines, both originating from the origin. One of the centerlines should be horizontal and the other diagonal, as shown in Figure 5.61. Select the Smart Dimension Tool and add a 30-degree angular dimension between the centerlines.

FIGURE 5.61

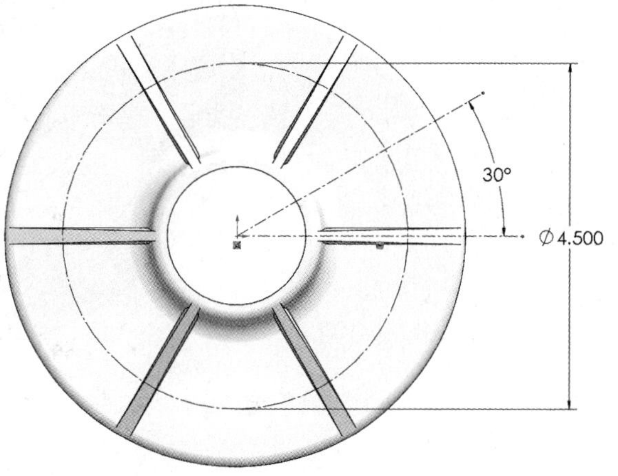

Select the Circle Tool. Move the cursor to the intersection of the construction circle and the diagonal centerline, so that the intersection icon appears, as shown in Figure 5.62.

Drag out a circle. Select the Smart Dimension Tool and add a 3/8-inch diameter dimension to the circle, as shown in Figure 5.63.

FIGURE 5.62

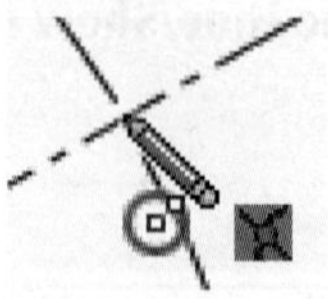

FIGURE 5.63

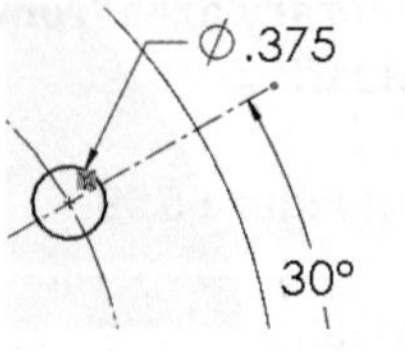

Select the Extruded Cut Tool from the Features group of the CommandManager. Set the type as Through All, and click the check mark to create the first hole.

Select the Circular Pattern Tool. Create a six-hole pattern.

The finished part is shown in **Figure 5.64**.

In the next section, we will add equations to link the rib and hole patterns. This will be easier to do if we rename these features.

In the FeatureManager, rename the extrusion defining the first rib "Rib," the first circular pattern "Rib Pattern," the first bolt hole "Bolt Hole," and the second circular pattern "Hole Pattern," as shown in Figure 5.65. Save the part.

FIGURE 5.64

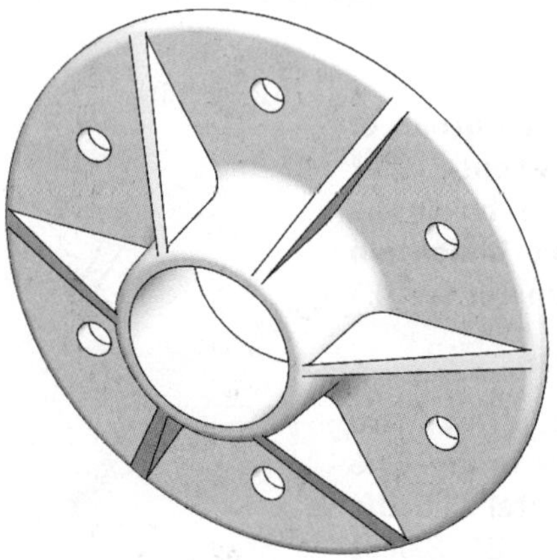

FIGURE 5.65

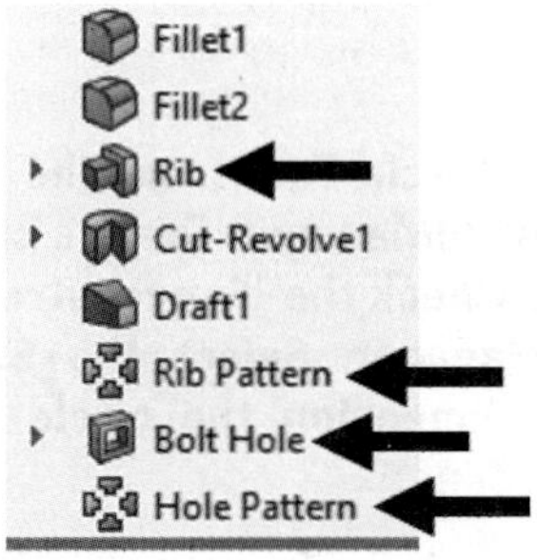

5.2 Creation of Parametric Equations

The use of parametric equations embedded in models to control dimensions is a powerful engineering tool. These equations can be used to embed design intelligence directly into solid models. In this section, the flange created in Section 5.1 will be modified to include parametric equations relating the hole pattern to the rib pattern.

With the flange model open, select Tools: Equations from the main menu, as shown in Figure 5.66.

FIGURE 5.66

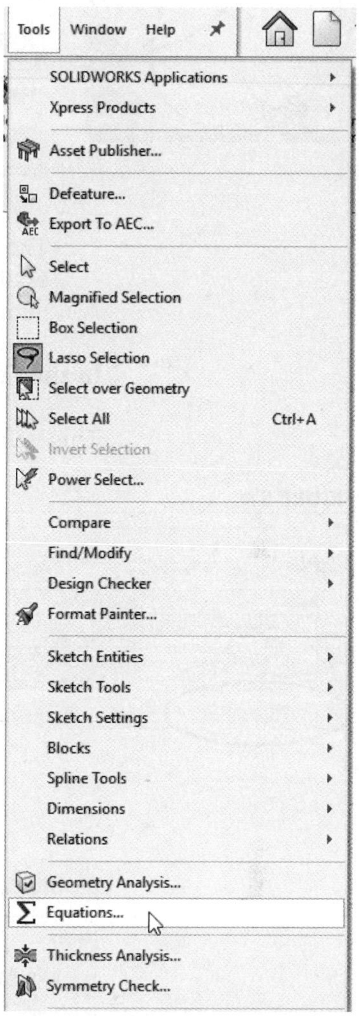

The Equations, Global Variables, and Dimensions dialog box will appear. In the model, the number of holes in the Hole Pattern feature will be controlled by the number of ribs in the Rib Pattern feature. An equation will be written to establish this parametric relationship. Equations can be established relating model parameters to one another, or relating model parameters to user-defined variables. In this tutorial, since the model is relatively complicated, and since two different equations controlling model parameters will eventually be developed involving the number of ribs in the Rib Pattern, we will use the technique that involves the creation of a user-defined variable. We will therefore begin by defining a *global variable* that contains the number of ribs.

Double-click on the Rib Pattern in the FeatureManager.

This will display the parameters (6 ribs, 360° spacing) in the model window.

With the Equation View button selected, click on the cell in the "Name" column labeled *Add global variable*, as shown in Figure 5.67. In the cell, type the variable name Ribs. Click in the cell in the "Value/ Equation" column next to it, as shown in Figure 5.68.

FIGURE 5.67

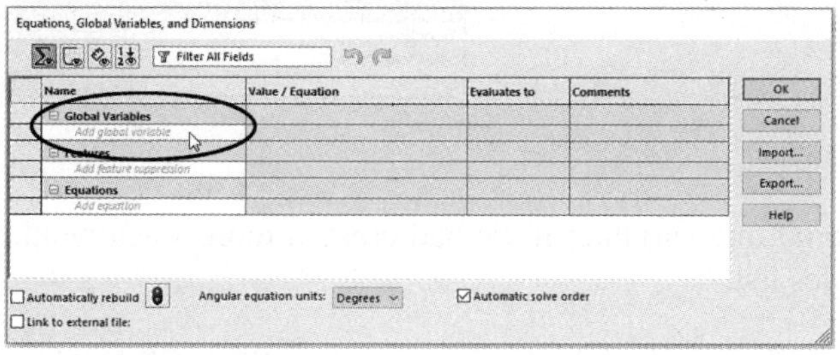

FIGURE 5.68

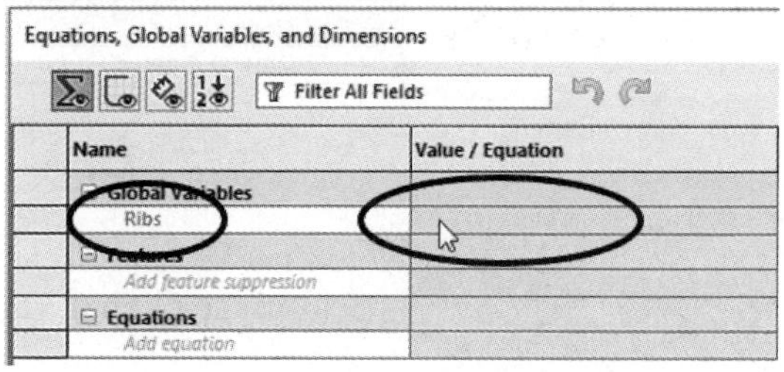

FIGURE 5.69

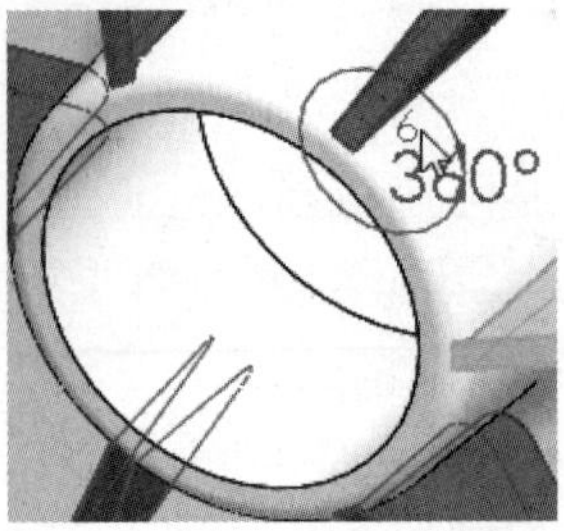

Click on the "6" parameter (denoting the number of ribs) in the modeling window, as shown in Figure 5.69 (you may need to zoom in and/or rotate to find it). Click the green check mark, and the value of the parameter (6) will be associated with the variable Ribs and added to the "Evaluates to" column, as shown in Figure 5.70.

FIGURE 5.70

Name	Value / Equation	Evaluates to
⊟ Global Variables		
"Ribs"	= "D1@Rib Pattern"	6

The variable Ribs can now be used as a driving value in an equation. We will establish an equation that ties the number of holes in the hole pattern to the number of ribs in the rib pattern, which has been given the variable name Ribs.

In the FeatureManager, click on the Hole Pattern entry.

This will display the parameters (6 holes, 360° spacing) in the model window.

FIGURE 5.71

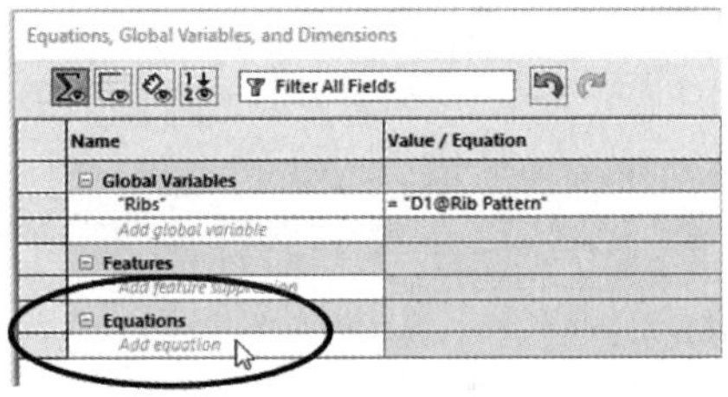

Click in the cell in the "Name" column labeled *Add equation,* as shown in Figure 5.71. In the model window, select the parameter "6."

The parameter, called by the default name "D1@Hole Pattern," will be added to the cell, as shown in **Figure 5.72**. The variable becomes the left-hand side of the equation.

In the Value/Equation column, select "Global Variables" from the menu that appears. Select the variable Ribs from the list, as shown in Figure 5.73.

FIGURE 5.72

Name
⊟ Global Variables
"Ribs"
Add global variable
⊟ Features
Add feature suppression
⊟ Equations
"D1@Hole Pattern"

FIGURE 5.73

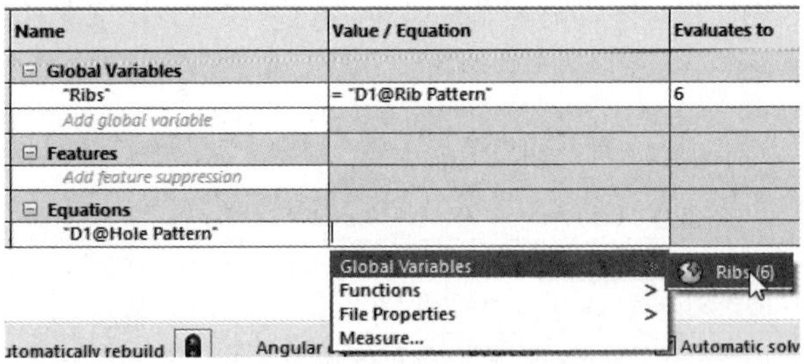

Note that we have created only one variable; if we had created more, each would appear in the Global Variables list.

Click the check mark to complete and evaluate the equation. Click OK to close the dialog box.

This equation establishes a parametric relation that ties the number of holes in the hole pattern directly to the number of ribs in the rib pattern. We will test the equation at this point.

Double-click the Rib Pattern to display its parameters, and double-click on the value "6" (the number of ribs in the pattern) and change it to "3." Click the check mark, and then click the Rebuild Tool to rebuild the model.

The model will be rebuilt, with the Rib Pattern modified and the driven Hole Pattern modified as well, as shown in **Figure 5.74**.

FIGURE 5.74

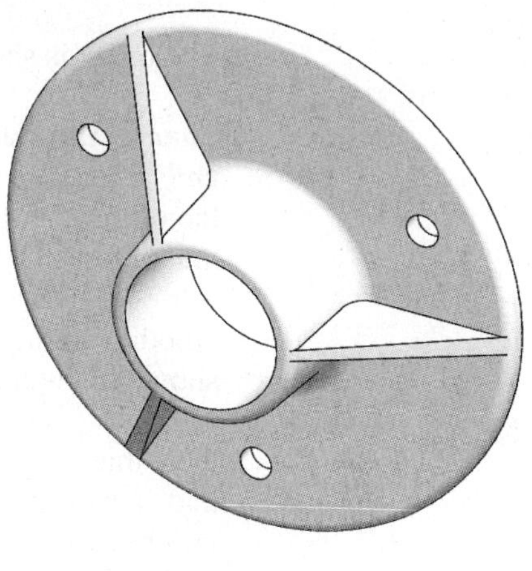

While the number of holes is correctly tied to the number of ribs, our design intent may not be satisfied by this model. Note that the angular location of the holes was set at 30 degrees from the center of a rib; this centered the hole between two ribs when there were six ribs in the pattern, but it no longer provides for centering of the holes when the number of ribs is modified. We will create a new parametric equation to establish the relationship required by our design intent; the equation will drive the angular dimension of the holes so that they are centered between the ribs.

From the main menu, select Tools: Equations. In the dialog box, click the *Add Equation* cell. Click on the Bolt Hole in the FeatureManager to display the dimensions associated with the bolt hole.

The angular spacing between ribs is 360 degrees divided by the number of ribs. The first hole is located at one-half of this value away from the first rib. Therefore, the equation required to set the angular dimension in degrees to the desired value will be:

Angular Dimension = (1/2) (360/Number of Ribs) or

Angular Dimension = 180/Number of Ribs

With the hole dimensions still displayed, select the angular dimension locating the hole, as shown in Figure 5.75.

FIGURE 5.75

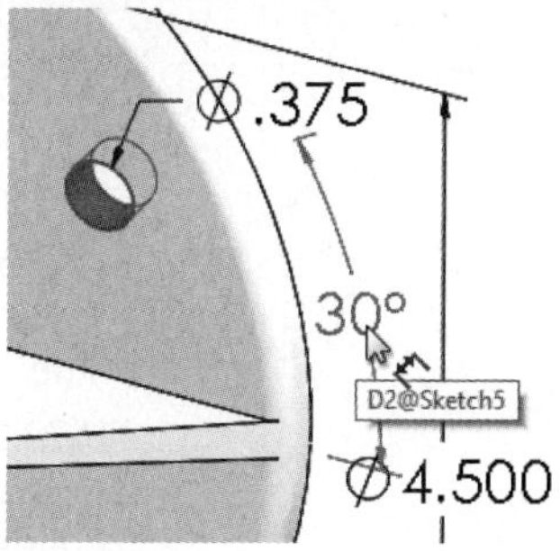

In the "Value/Equation" column, type 180/"Ribs," as shown in Figure 5.76. Click the check mark.

FIGURE 5.76

⊟ Equations		
"D1@Hole Pattern"	= "Ribs"	3
"D2@Sketch5"	=180/"Ribs"	✓

The "Evaluates to" column confirms that the dimension will equal 60 degrees with the current number of ribs (3), as shown in **Figure 5.77**.

FIGURE 5.77

⊟ Equations		
"D1@Hole Pattern"	= "Ribs"	3
"D2@Sketch5"	= 180 / "Ribs"	60deg

Click OK to close the dialog box.

The result is shown in **Figure 5.78**.

FIGURE 5.78

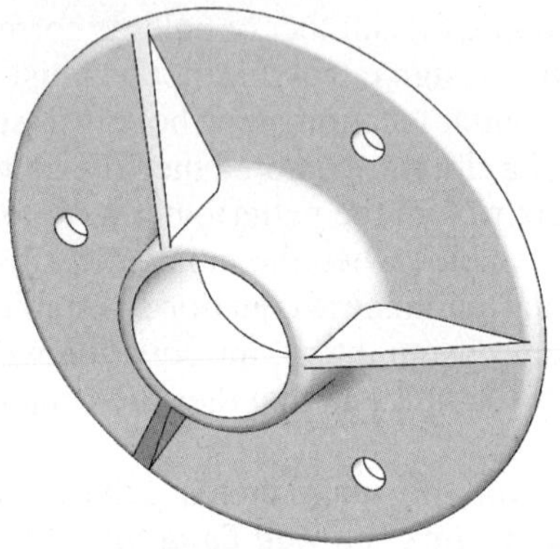

Change the number of ribs to other values and check to see that the number and locations of the holes change in a consistent manner.

Another example, with five ribs and holes, is shown in **Figure 5.79**.

FIGURE 5.79

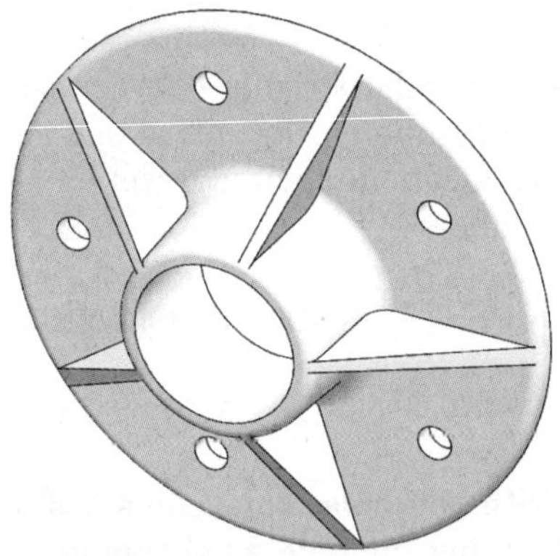

Note that there is a difference between *driving* and *driven* parameters. In this example, the number of ribs is a free design choice, and is therefore a driving parameter. As such, it appears only on the right side of the equal sign in parametric equations. Conversely, the number of holes and the angular hole location are driven parameters; they are determined by the choice of the number of ribs, and appear on the left side of the equal sign in parametric equations. Since their values are set by the values established by the driving parameters, driven parameters cannot be modified directly in the model.

Perform the following demonstration to verify this.

Double-click on the Hole Pattern in the FeatureManager to display the associated parameters in the model window. Double-click on number of holes in the pattern, to try to change the value.

The "Σ" symbol displayed beside the value (see **Figure 5.80**) indicates that the dimension is controlled by an equation and cannot be changed.

FIGURE 5.80

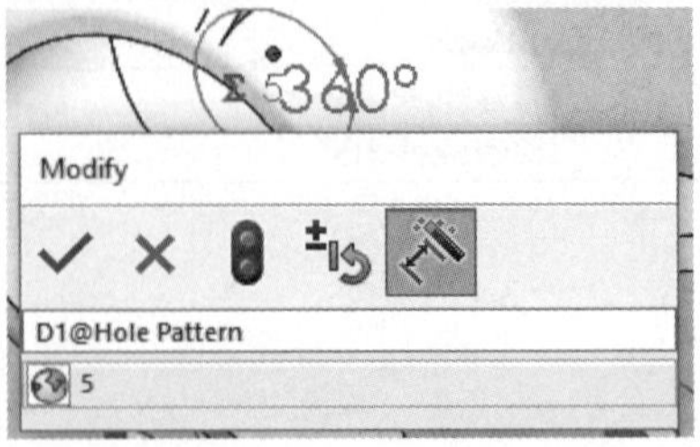

Click the check mark to close the Modify box. Reset the number of ribs and holes to six, rebuild the model, and save the part file.

5.3 Modeling Tutorial: Cap Screw with Design Table

Note: This tutorial requires Microsoft Excel to be installed on your computer.

In this section, we will create a family of similar parts. Many parts are defined this way, especially common parts such as fasteners, washers, seal rings, and so on. Rather than creating separate model files and drawings for every different part, a single part with multiple configurations is made. A single drawing can be made to define the parameters of all of the different configurations. The specifications of the dimensions that define each configuration are contained in a spreadsheet called a *design table*.

FIGURE 5.81

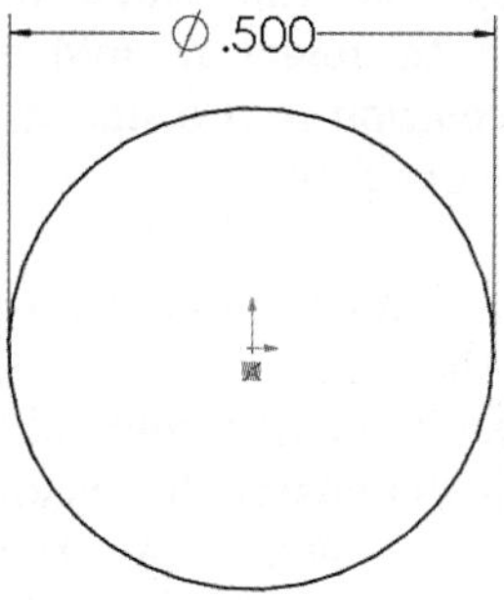

Open a new part. Select the Right Plane from the FeatureManager, and select the Circle Tool from the Sketch group of the CommandManager. Draw a circle centered at the origin. Select the Smart Dimension Tool, and dimension the diameter of the circle as 0.50 inches, as shown in Figure 5.81.

Select the Extruded Boss/Base Tool from the Features group of the CommandManager, and extrude the circle 1.50 inches, in the direction shown in Figure 5.82.

Select the face shown in Figure 5.83, and select the Normal To View.

FIGURE 5.82

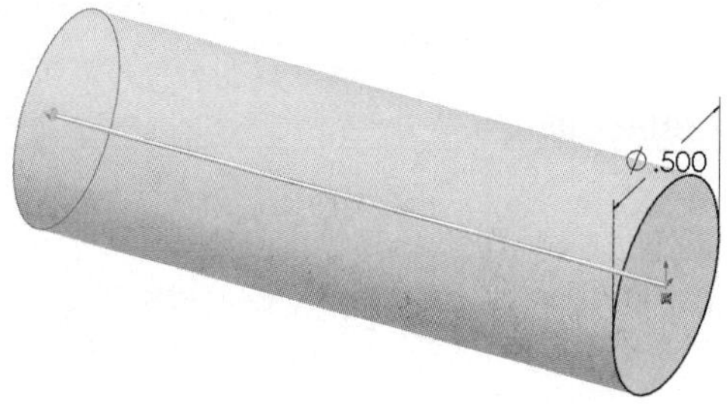

Select the Circle Tool from the Sketch group of the CommandManager. Draw a circle centered at the origin. Select the Smart Dimension Tool and dimension the diameter of the circle as 0.75 inches, as shown in Figure 5.84.

FIGURE 5.83

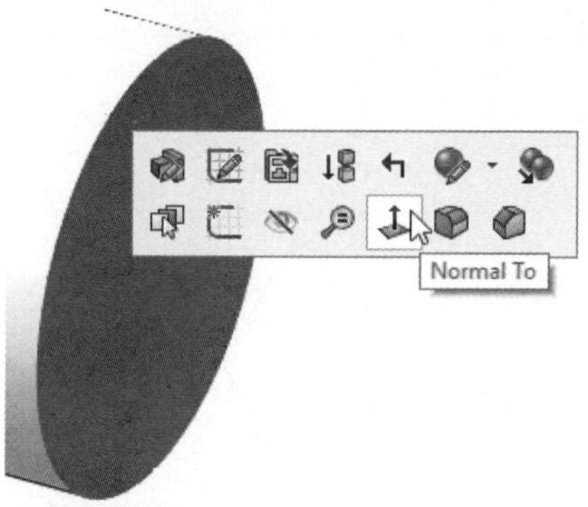

FIGURE 5.84

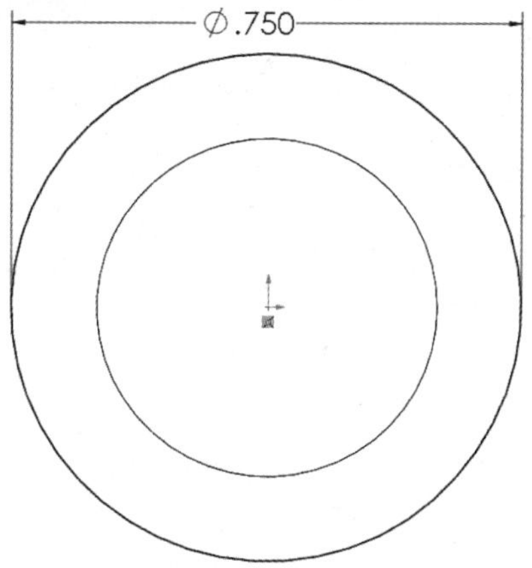

FIGURE 5.85

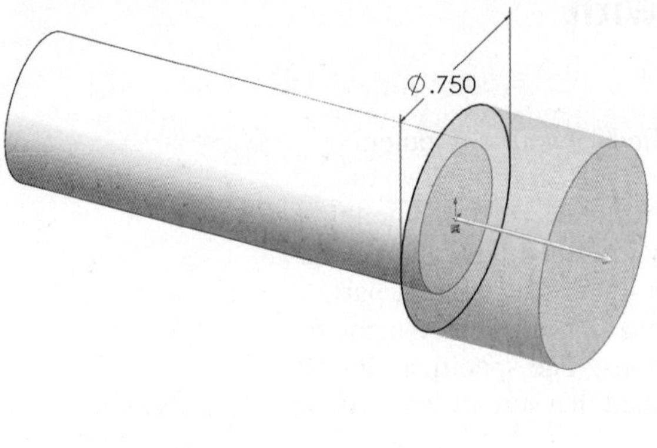

Select the Extruded Boss/Base Tool from the Features group of the CommandManager, and extrude the circle 0.50 inches to form the head of the screw, as shown in Figure 5.85.

Select the top of the screw head, as shown in Figure 5.86. Choose the Normal To View.

Select the Polygon Tool from the Sketch group of the CommandManager, as shown in Figure 5.87. Drag out a polygon from the origin, as shown in Figure 5.88. By default, the number of sides is six (a hexagon).

FIGURE 5.86

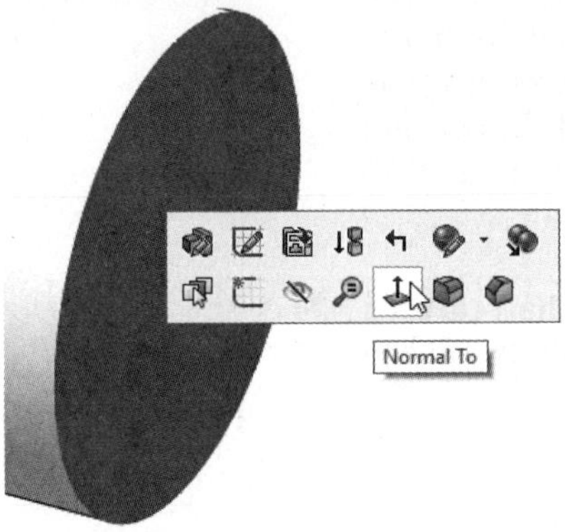

FIGURE 5.87

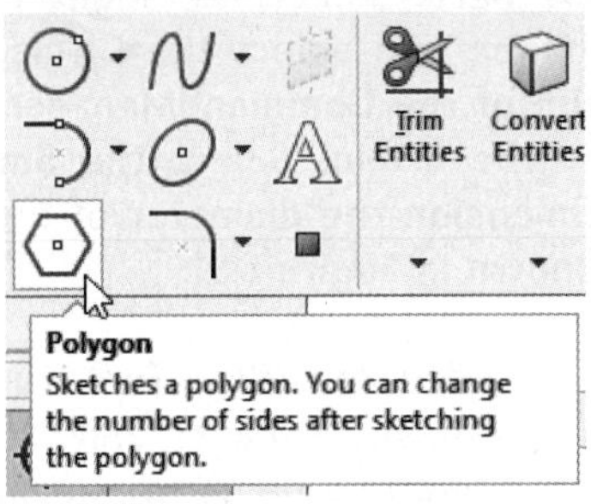

Deselect the Polygon Tool, and click on one of the sides of the hexagon to select it. Add a Horizontal relation to the side, as shown in Figure 5.89.

Select the Smart Dimension Tool from the Sketch group of the CommandManager, and add a 0.375-inch dimension between two opposite sides of the hexagon, as shown in Figure 5.90.

The sketch is now fully defined.

FIGURE 5.88

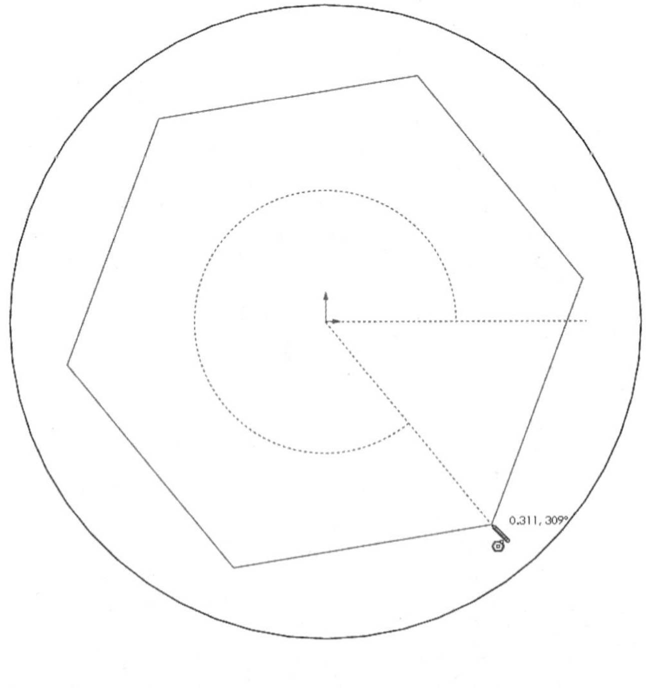

FIGURE 5.89

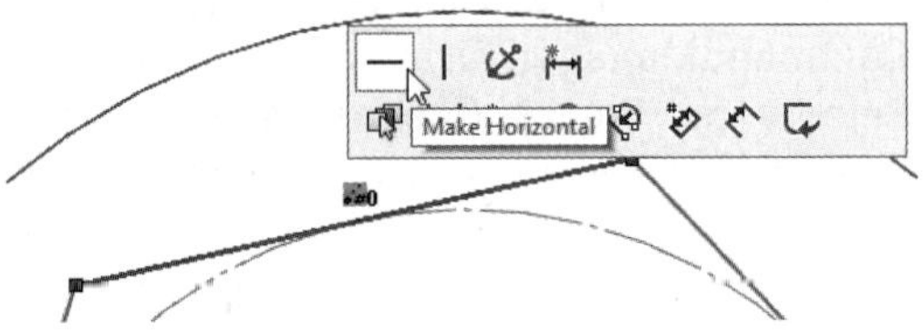

FIGURE 5.90

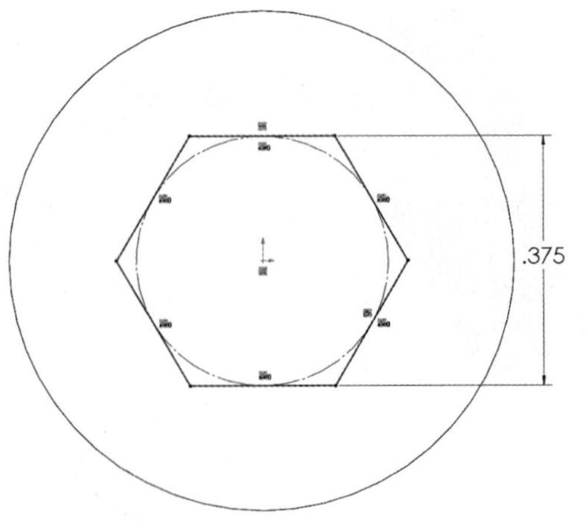

Select the Extruded Cut Tool from the Features group of the CommandManager, and extrude a cut 0.245 inches deep to form the hex socket in the head, as shown in Figure 5.91.

FIGURE 5.91

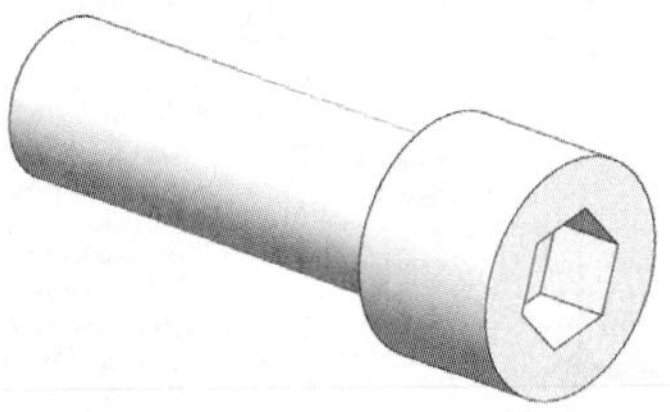

While screw threads can be modeled as solid features with the Thread Tool, this is rarely done, since it can slow down the performance of your computer. Also, since they are standardized, the thread profiles are not required in order to specify them on a drawing. Rather, *cosmetic threads*, graphical features used to show the threaded regions, are much more commonly used. Cosmetic threads can be added to any cylindrical feature.

Select the edge where the threads will begin, as shown in Figure 5.92.

FIGURE 5.92

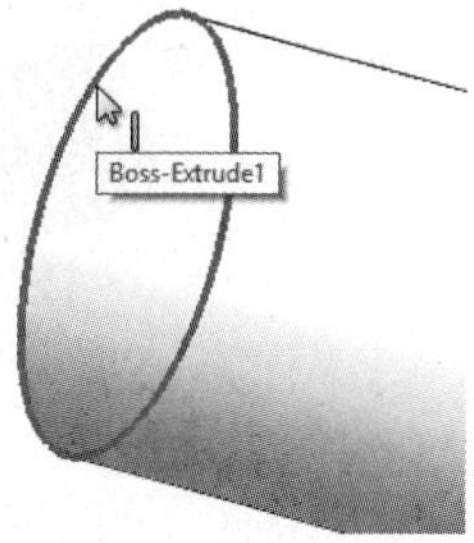

Select Insert: Annotations: Cosmetic Thread from the main menu, as shown in Figure 5.93.

FIGURE 5.93

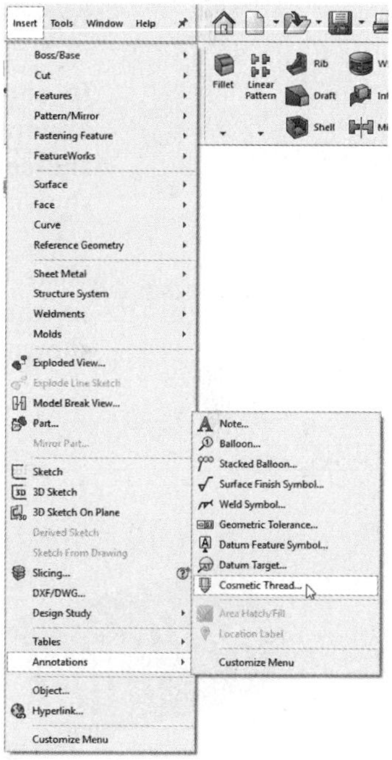

Set the thread length as 1.00 inch, the minor diameter as 0.40 inches, and other options as shown in Figure 5.94. Click the check mark to add the thread. Switch to the Front View. The resulting thread display is shown in Figure 5.95.

The minor diameter is the diameter at the "root" of the thread. The actual minor diameter value of a UNC (unified series coarse) 1/2-inch thread is 0.408 inches. Since the minor diameter of the cosmetic thread is only for display purposes, using an approximate value is acceptable.

FIGURE 5.94

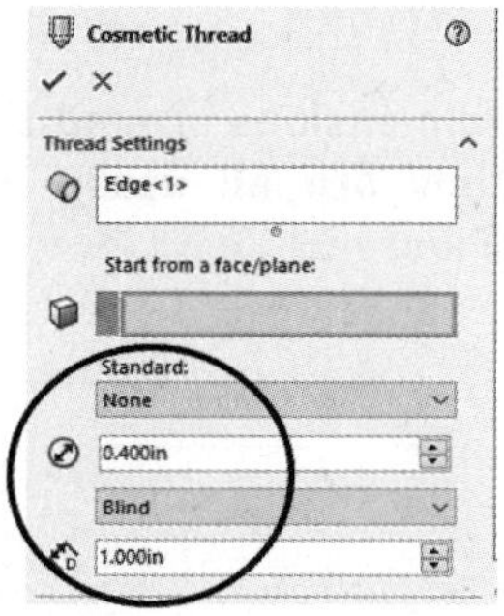

If preferred, the cosmetic thread can be displayed with a shaded thread pattern on the surface rather than with the dashed lines shown in **Figure 5.95**. If you want to change the display mode of the cosmetic thread, choose the Options Tool, and under the Document Properties tab, click on Detailing. Click on the "Shaded cosmetic threads" box, as shown in

FIGURE 5.95

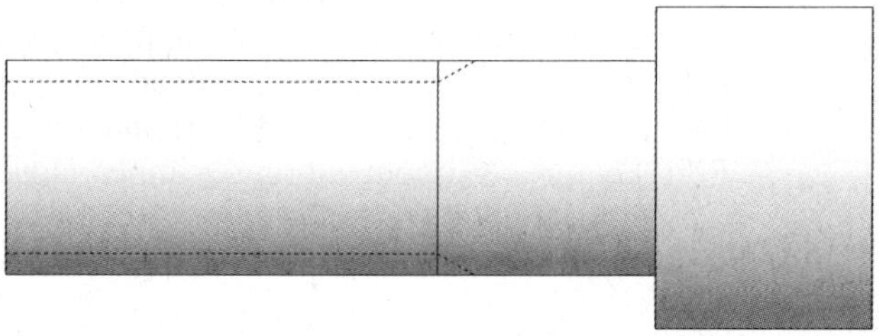

FIGURE 5.96

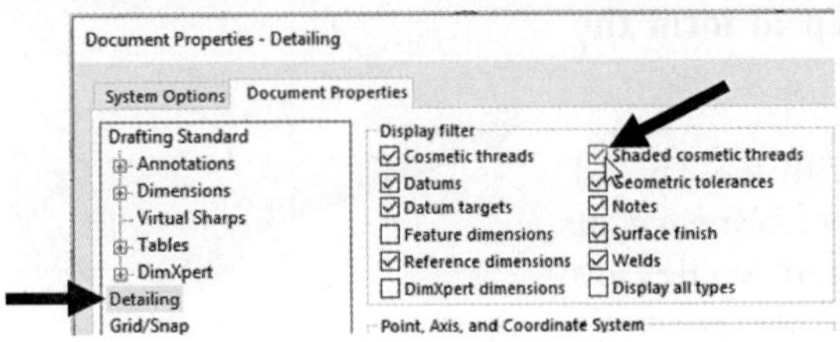

Figure 5.96. The resulting display is shown in **Figure 5.97**.

The definition of the thread appears in the FeatureManager, attached to the cylindrical feature that is to be threaded, as shown in **Figure 5.98**.

Save the part file as "Cap Screw."

FIGURE 5.97

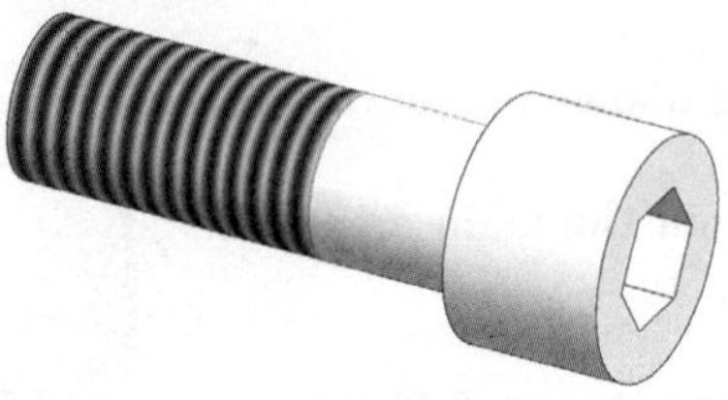

FIGURE 5.98

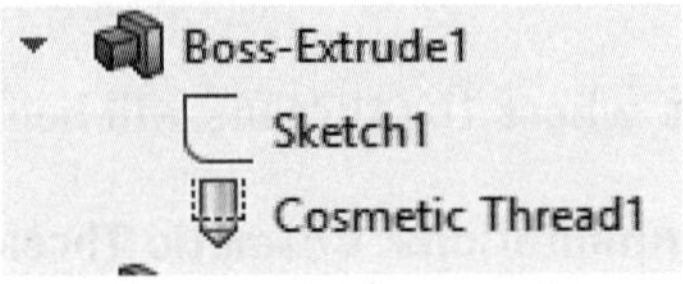

We have now defined one configuration of the cap screw. To define more configurations, we are going to create a design table.

We will need to access all of the dimensions used to create the part.

To show the dimensions, right-click on Annotations in the FeatureManager, and select Show Feature Dimensions, as shown in Figure 5.99.

Move the dimensions around the screen so that they are all easily visible, as shown in Figure 5.100.

FIGURE 5.99

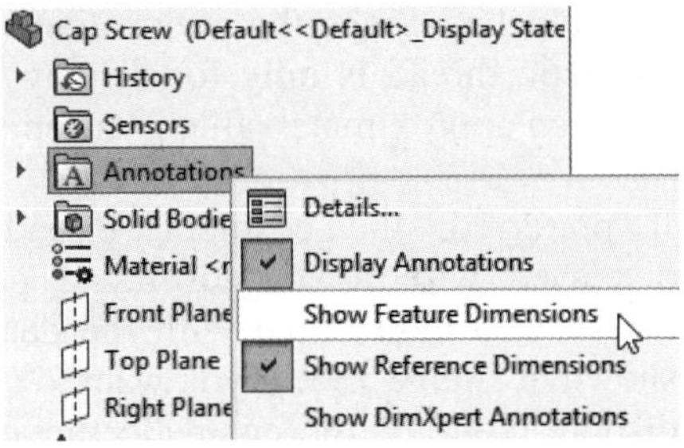

FIGURE 5.100

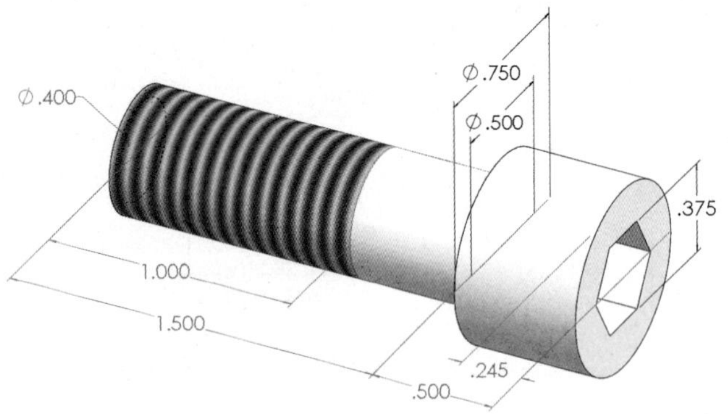

If desired, the display of dimensions can be changed so that the text is always oriented relative to the screen by clicking the Options tool, selecting System Options: Display and checking the box labeled "Display dimensions flat to screen." Note that if the dimensions of the cosmetic thread do not appear, you should locate the cosmetic thread in the FeatureManager, right-click on it, and select Edit Feature. Without editing, simply close the PropertyManager, and the dimensions should appear.

From the main menu, select Insert: Tables: Design Table, as shown in Figure 5.101.

FIGURE 5.101

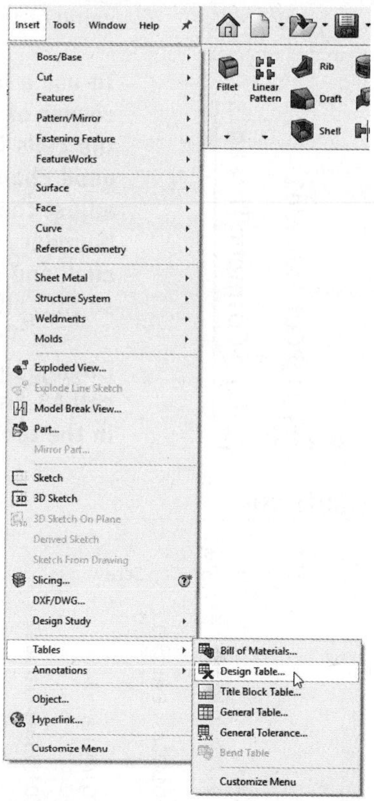

In the PropertyManager, leave the Source option as Auto-create. In the Edit Control box, choose the second option, as shown in Figure 5.102, so that dimensions entered in the table cannot be changed outside of the table. Clear all of the options for adding new rows and columns, and click the check mark to begin creating the table.

FIGURE 5.102

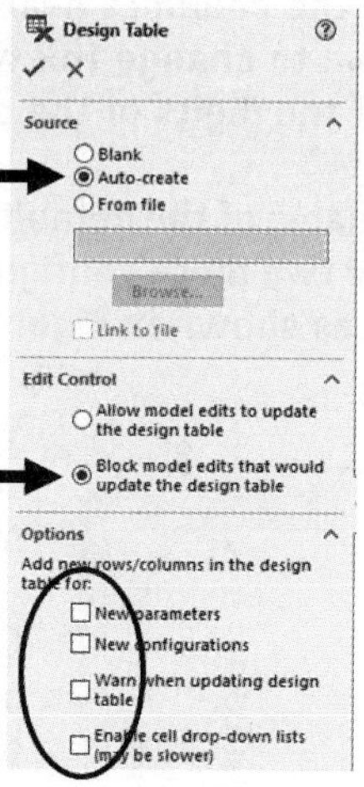

A dialog box will open, prompting you to select the dimensions that will be included in the table, as shown in Figure 5.103. Select all of them by clicking on the first and, while holding the Shift key, clicking on the last. Click OK.

A window containing a Microsoft Excel spreadsheet will open, as shown in **Figure 5.104**. Note that the Excel tools are available at the top of the screen whenever the table is open. Be careful not to click in the white space of the graphics window while the table is open; doing so will close the table. If you close the table accidentally, you can re-open it by clicking on the ConfigurationManager tab above the FeatureManager, right-clicking the table name in the ConfigurationManager, and selecting Edit Table.

FIGURE 5.103

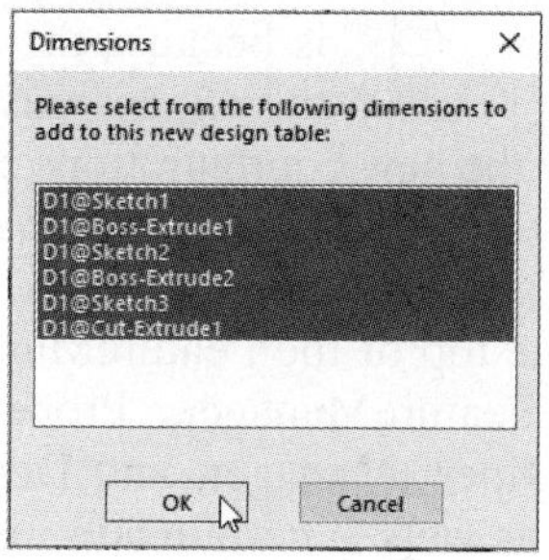

FIGURE 5.104

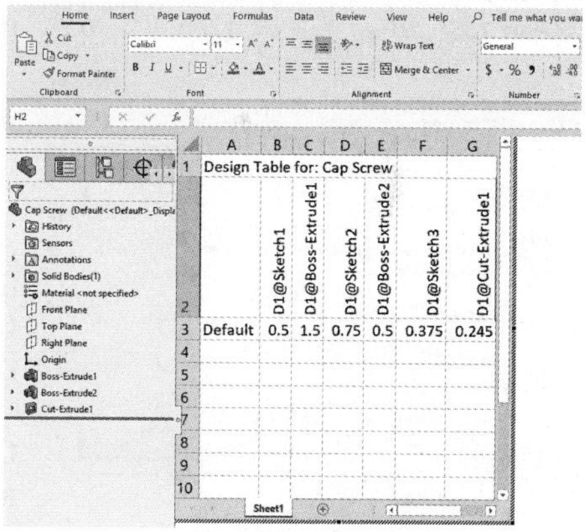

Note that row 2 of the table contains the SOLIDWORKS® name of each dimension, and row 3 contains the current values of the dimensions as the "default" configuration. The only dimension that we will need to add manually is the thread length, since dimensions associated with cosmetic threads are not automatically added to design tables.

Click in cell H2 to select it, as shown in Figure 5.105. Double-click the dimension defining the length of the cosmetic threads, as shown in Figure 5.106.

FIGURE 5.105

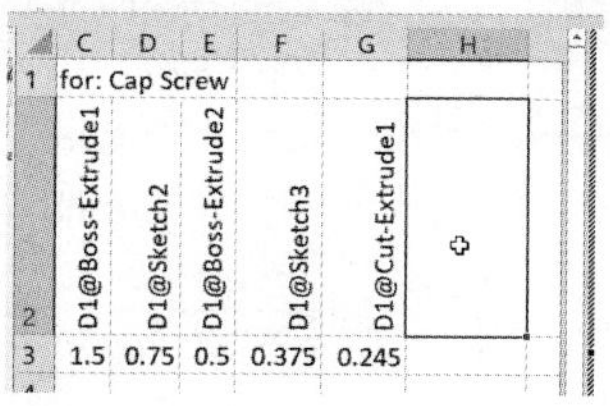

FIGURE 5.106

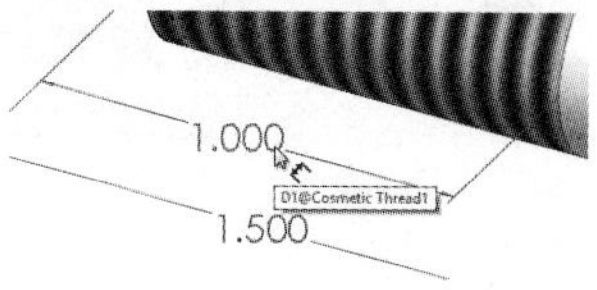

FIGURE 5.107

G	H
D1@Cut-Extrude1	D1@Cosmetic Thread1
0.245	1

The name of the dimension will be placed in cell H2. The dimension's current value (1 inch) will be placed in cell H3, as shown in **Figure 5.107**.

To use a larger font size, click the blank area in the upper left corner of the table, as shown in Figure 5.108. This selects all of the cells in the table. Choose a new font size (14–16 point is a good choice) from the Home group of the Excel tool ribbon. To adjust the width of each column, move the cursor to the right boundary of the column's header, as shown in Figure 5.109, and click and drag to change the width. Change the column widths so that all of the digits of the dimensions are displayed.

FIGURE 5.108

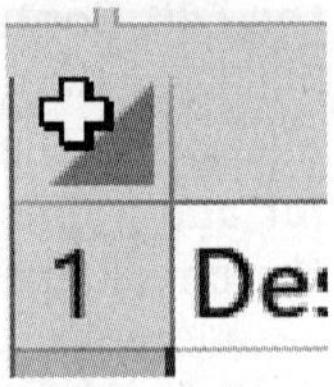

FIGURE 5.109

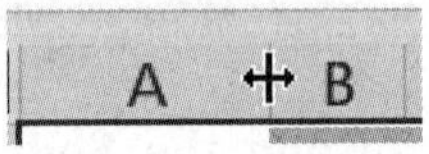

Change the name of the default configuration to "Part 101" in cell A3. Enter two more configurations and their dimensions in the table, as shown in Figure 5.110.

FIGURE 5.110

	A	B	C	D	E	F	G	H
1	Design Table for: Cap Screw							
2		D1@Sketch1	D1@Boss-Extrude1	D1@Sketch2	D1@Boss-Extrude2	D1@Sketch3	D1@Cut-Extrude1	D1@Cosmetic Thread1
3	Part 101	0.5	1.5	0.75	0.5	0.375	0.245	1
4	Part 102	0.75	2	1.125	0.75	0.625	0.37	1.5
5	Part 103	1	3	1.5	1	0.75	0.495	2

When the table is closed, the program will search column A (below row 2) for configuration names, much like a lookup table in Excel.

Click in the white space in the modeling window, outside of the spreadsheet window. This will cause the design table to close, and a message will be displayed that indicates the new configurations have been created, as shown in Figure 5.111. Click OK.

Note that the part has not changed on the screen. That is because the default configuration is the one with the dimensions used to model the part originally. To view the new configurations created from the design table, we need to use the ConfigurationManager.

FIGURE 5.111

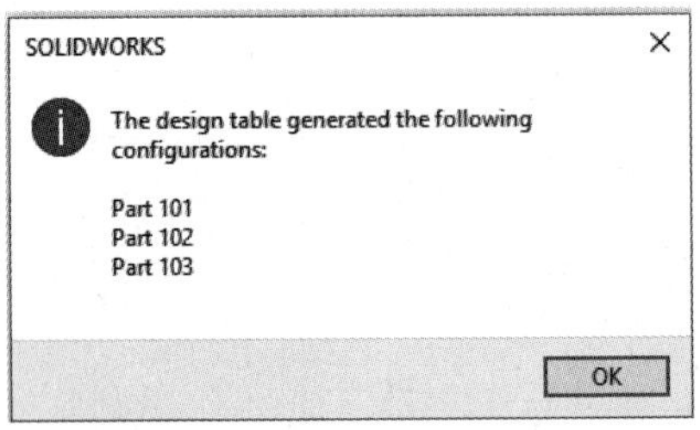

At the top of the FeatureManager, there are tabs corresponding to the FeatureManager, PropertyManager, ConfigurationManager, DimXpert Manager, and DisplayManager. (There could be other tabs as well, if certain add-ins are present.)

Click on the icon representing the ConfigurationManager, as shown in Figure 5.112.

FIGURE 5.112

FIGURE 5.113

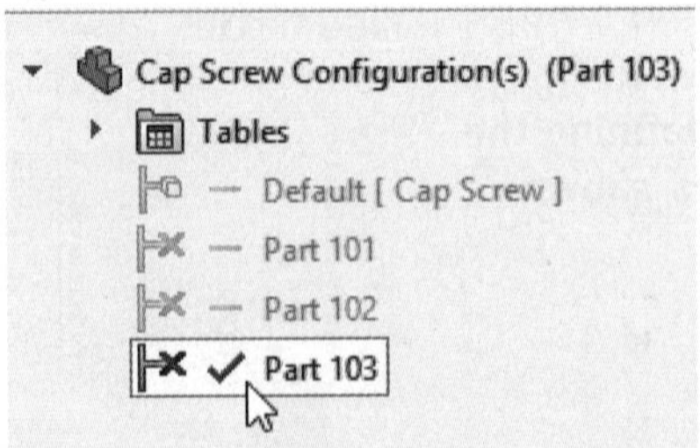

In the ConfigurationManager, double-click on Part 103, as shown in Figure 5.113.

The part is rebuilt to the dimensions specified in the design table for Part 103, as shown in **Figure 5.114**. Dimensions controlled by the design table may appear in a different color on your screen. Note that when you move a dimension, you will receive a message that the dimension's value cannot be edited, as shown in **Figure 5.115**.

FIGURE 5.114

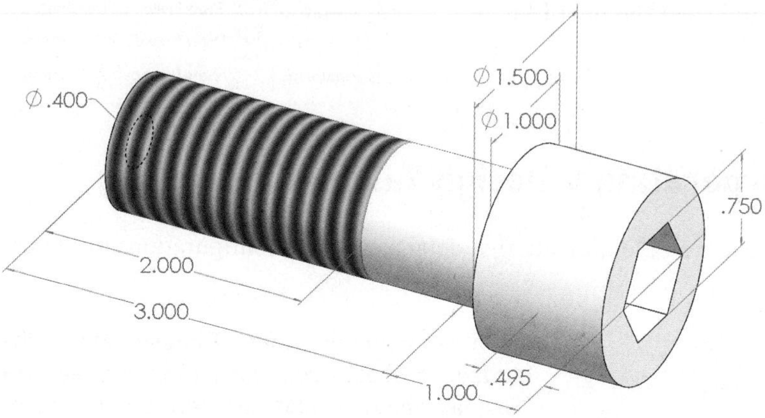

One dimension that was left unchanged is the minor thread diameter of the cosmetic thread. Since this is not a dimension used to define the part, we did not include it in the design table. However, we would like to have the display of the cosmetic thread look reasonable on the screen. We can add an equation to control this dimension. In the previous section, we established equations that related model parameters to a user-defined variable. In this model, we will establish an equation directly between two model parameters.

FIGURE 5.115

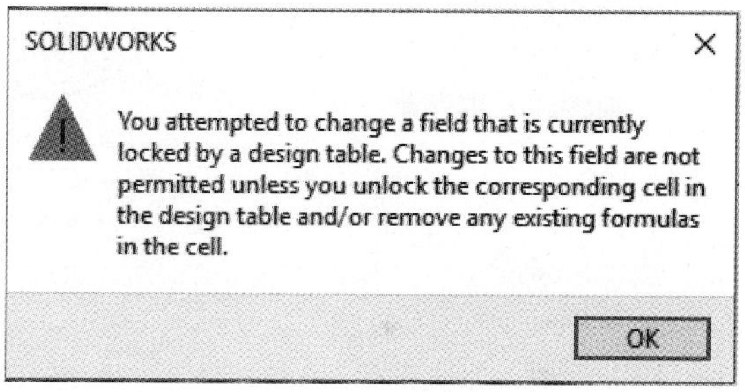

From the main menu, select Tools: Equations.

Click the Add equation cell to create a new equation. Click on the dimension representing the minor thread diameter (0.400). In the Value/Equation column, type "0.8*". Click on the dimension representing the diameter of the shank (the 1.00-inch diameter). The equation should appear as shown in Figure 5.116.

Click the check mark to evaluate the equation, and OK to close the dialog box. Click the diameter dimension to update it.

FIGURE 5.116

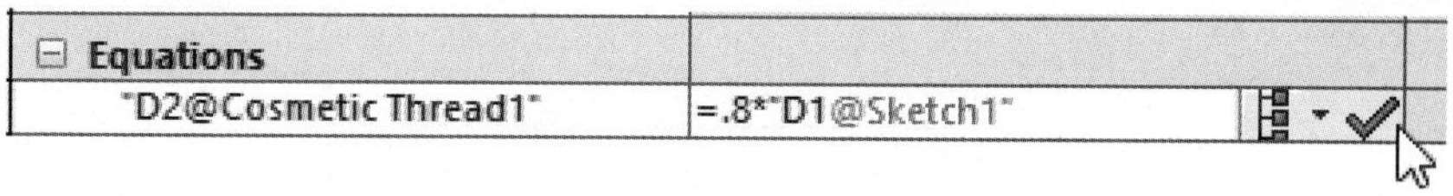

The updated thread diameter is shown in **Figure 5.117**.

FIGURE 5.117

Click on the FeatureManager tab to return to the FeatureManager. Right-click Annotations in the FeatureManager, and select Show Feature Dimensions, as shown in Figure 5.118, to turn off the display of the dimensions on the screen. Change back to the Part 101 configuration, and save the part file.

FIGURE 5.118

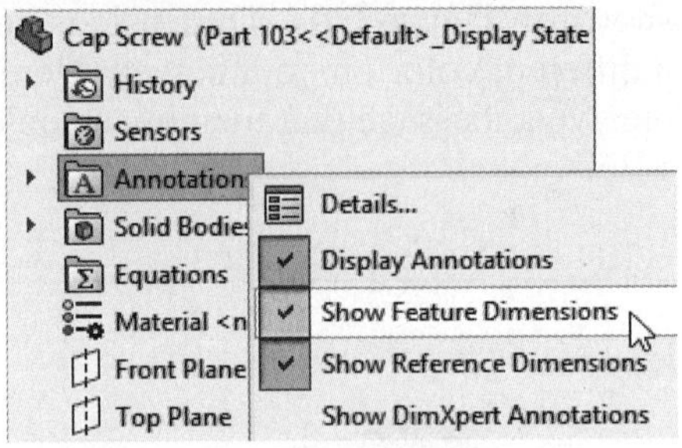

5.4 Incorporating a Design Table in a Drawing

We now will make a single drawing that details all three configurations of the cap screw.

FIGURE 5.119

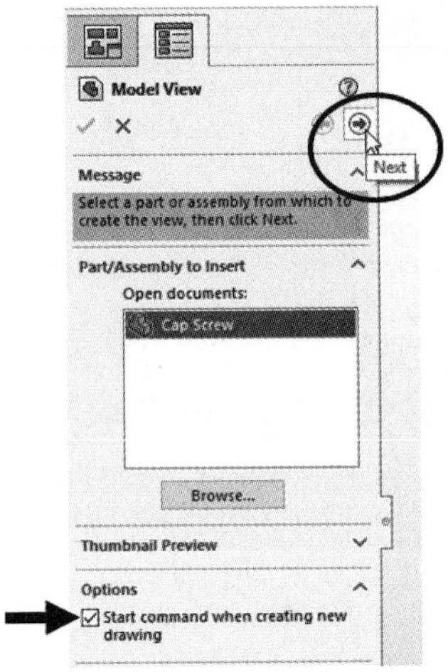

FIGURE 5.120

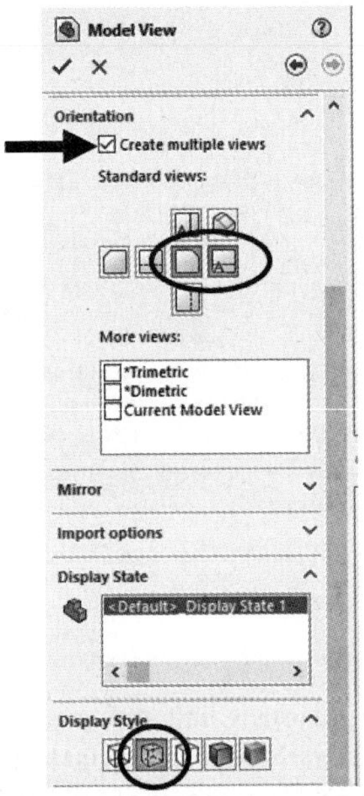

Open a new drawing. Choose an A-size landscape sheet size, and either uncheck the "Display sheet format" box for a plain sheet, or select the sheet format that you created in Chapter 2.

If the Model View command does not start automatically, then select the Model View Tool from the Drawings group of the CommandManager. If desired, check the box labeled "Start command when creating new drawing," as shown in Figure 5.119. Click on the Cap Screw in the Open documents box, or browse to find it. Click the Next arrow in the PropertyManager.

Select "Create multiple views," and select the Front and Right Views. Choose the wireframe display style with the hidden lines visible, as shown in Figure 5.120.

Click the check mark to create the views, which are shown in Figure 5.121.

FIGURE 5.121

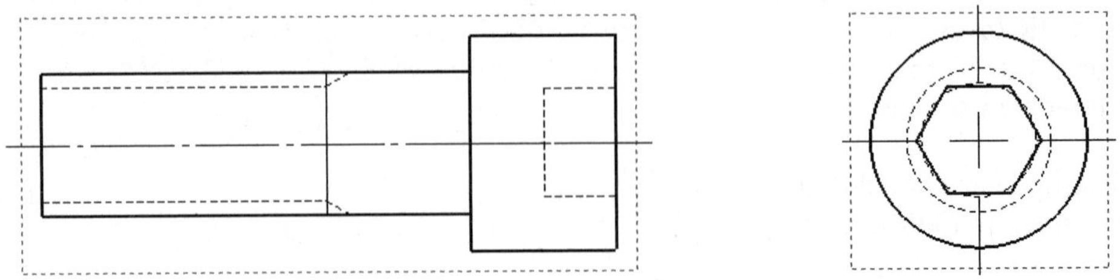

The model views that are created reflect the configuration that is current when the drawing is created. The configuration used for any model view can be changed by clicking on that view, and changing the Reference Configuration in the Drawing View PropertyManager. For this tutorial, the configuration used is not important.

With one of the drawing views selected, select Design Table from the Tables Tool in the Annotation group of the CommandManager, as shown in Figure 5.122.

FIGURE 5.122

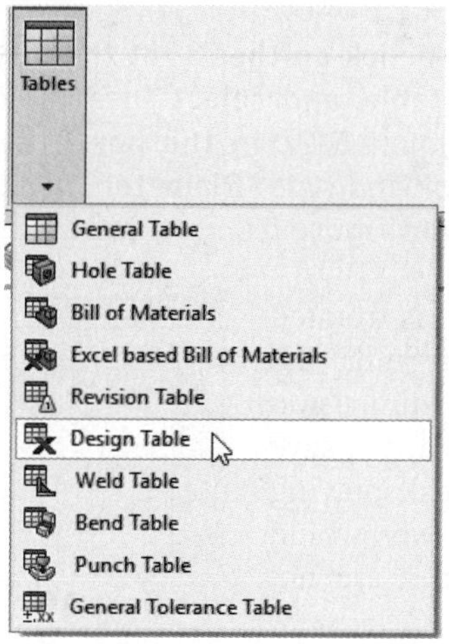

FIGURE 5.123

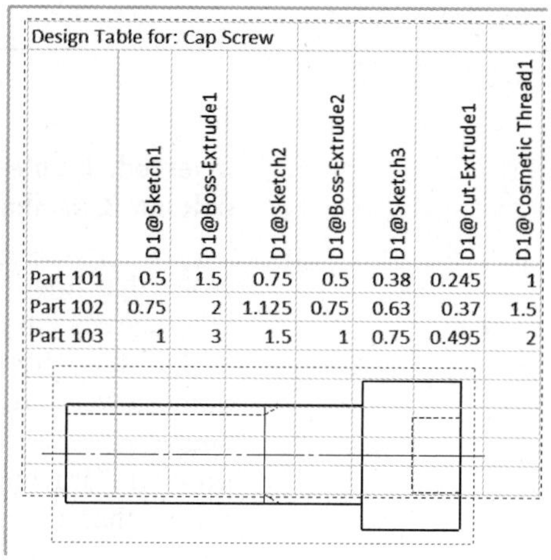

Design Table for: Cap Screw							
	D1@Sketch1	D1@Boss-Extrude1	D1@Sketch2	D1@Boss-Extrude2	D1@Sketch3	D1@Cut-Extrude1	D1@Cosmetic Thread1
Part 101	0.5	1.5	0.75	0.5	0.38	0.245	1
Part 102	0.75	2	1.125	0.75	0.63	0.37	1.5
Part 103	1	3	1.5	1	0.75	0.495	2

The design table is inserted into the drawing, as shown in Figure 5.123. The design table can be located on the drawing sheet by clicking and dragging it. Note that the format of the table is not appropriate for a drawing; there are blank rows and columns visible, dimensions are shown to only the number of decimal places we input, and so on. We will now edit the table to improve its appearance.

Double-click on the design table, and you will be taken back to the part screen, with the spreadsheet open in a window, as shown in Figure 5.124.

FIGURE 5.124

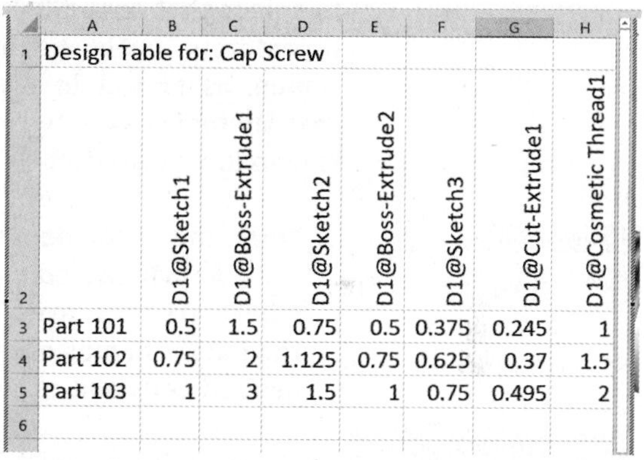

	A	B	C	D	E	F	G	H
1	Design Table for: Cap Screw							
2		D1@Sketch1	D1@Boss-Extrude1	D1@Sketch2	D1@Boss-Extrude2	D1@Sketch3	D1@Cut-Extrude1	D1@Cosmetic Thread1
3	Part 101	0.5	1.5	0.75	0.5	0.375	0.245	1
4	Part 102	0.75	2	1.125	0.75	0.625	0.37	1.5
5	Part 103	1	3	1.5	1	0.75	0.495	2
6								

Remember that the design table is a Microsoft Excel spreadsheet, and we will be editing it the same way we would edit any other Excel spreadsheet. To begin, the first two rows do not need to be displayed on the drawing. We cannot delete them, since they contain information necessary to the design table, but we can hide them.

Click and hold the mouse button on the number 1 on the left side of the spreadsheet, drag the cursor down onto the 2, and release. Rows 1 and 2 will be highlighted, as shown in Figure 5.125. Right-click, and select Hide.

FIGURE 5.125

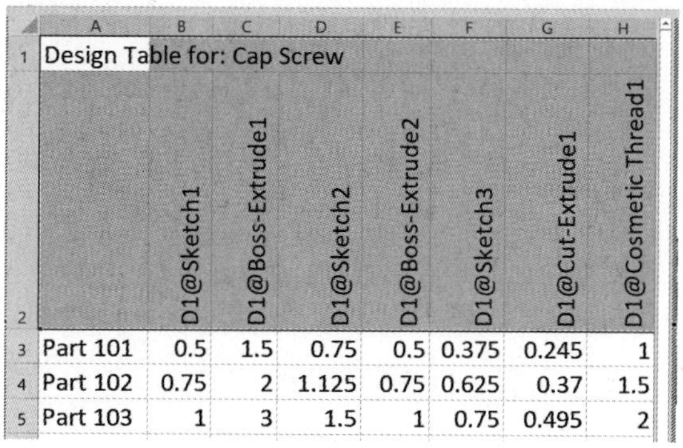

	A	B	C	D	E	F	G	H
1	Design Table for: Cap Screw							
2		D1@Sketch1	D1@Boss-Extrude1	D1@Sketch2	D1@Boss-Extrude2	D1@Sketch3	D1@Cut-Extrude1	D1@Cosmetic Thread1
3	Part 101	0.5	1.5	0.75	0.5	0.375	0.245	1
4	Part 102	0.75	2	1.125	0.75	0.625	0.37	1.5
5	Part 103	1	3	1.5	1	0.75	0.495	2

FIGURE 5.126

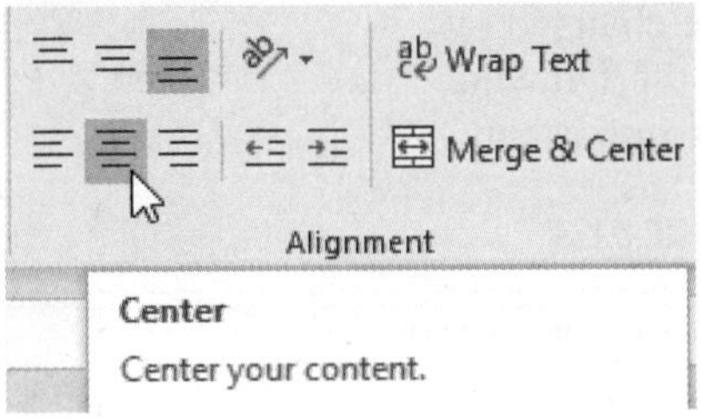

Select Columns B–H. Click the Center icon on the Excel menu, as shown in Figure 5.126.

We will now add headers to our table. We hid the row containing the SOLIDWORKS names of the dimensions, and we will now add a row that contains more descriptive names.

Right-click on the "3" at the left side of the table, and select Insert, as shown in Figure 5.127. In the new row that is inserted, double-click and type "Diameter" in the cell in column B, as shown in Figure 5.128.

FIGURE 5.127

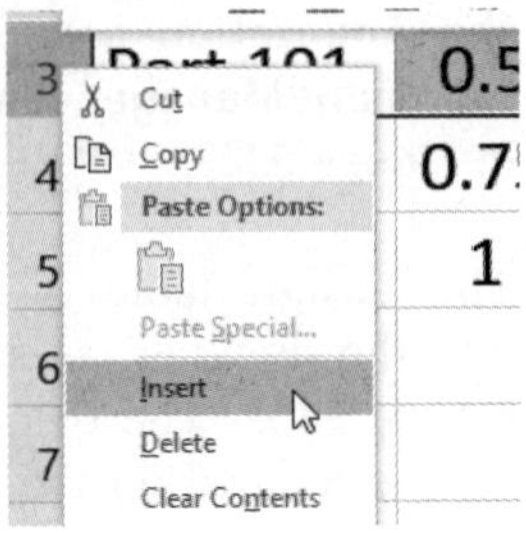

Note that the text is rotated 90 degrees. By default, the SOLIDWORKS dimension names were rotated to save space. However, for displaying on our drawing, we would prefer that the names are not rotated.

FIGURE 5.128

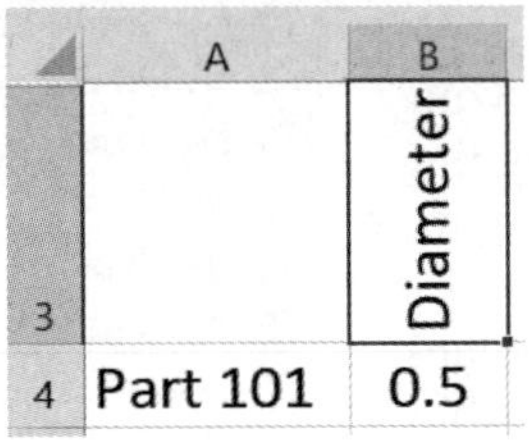

Click on the "3" on the left side of the table to select that row. Click the arrow beside the Orientation Tool in the Home group of the Excel ribbon, as shown in Figure 5.129. The Rotate Text Up tool is selected; click to de-select it, as shown in Figure 5.130.

FIGURE 5.129

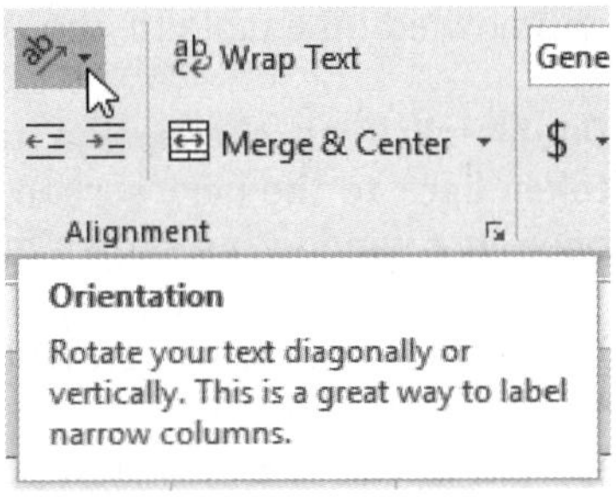

FIGURE 5.130

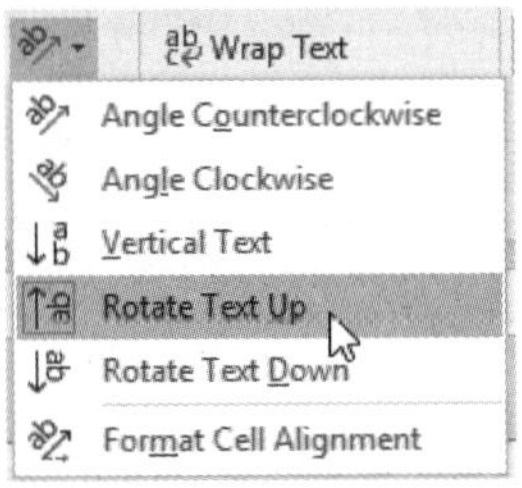

Enter the other names, as shown in Figure 5.131. Adjust the column widths so that the labels are displayed completely. To show the longer names on two lines, highlight the desired cells and choose the Wrap Text Tool, as shown in Figure 5.132.

Be sure to leave cell A3 blank. Recall that column A is reserved for configuration names.

FIGURE 5.131

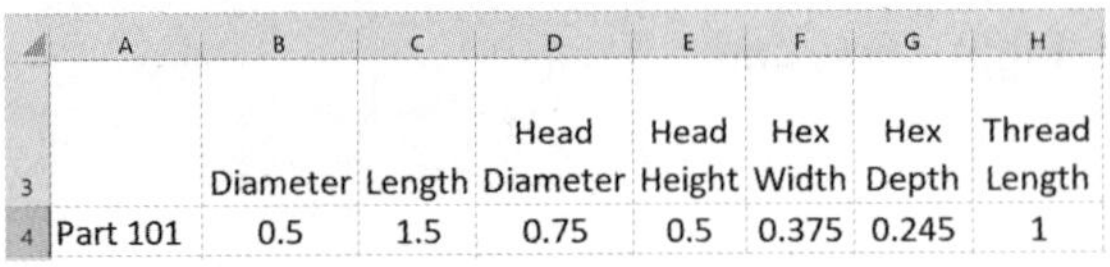

	A	B	C	D	E	F	G	H
3		Diameter	Length	Head Diameter	Head Height	Hex Width	Hex Depth	Thread Length
4	Part 101	0.5	1.5	0.75	0.5	0.375	0.245	1

FIGURE 5.132

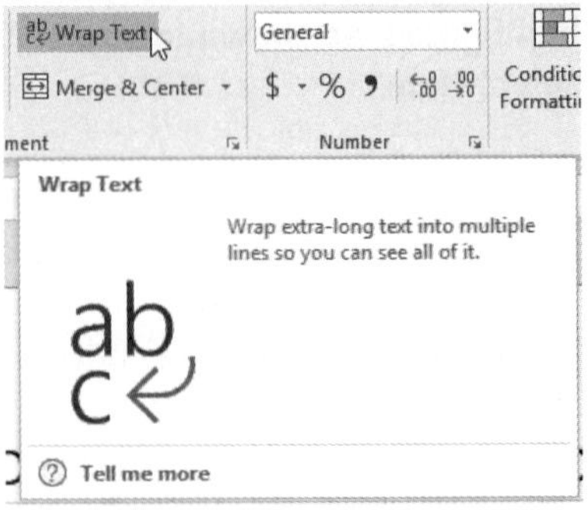

Highlight all of the cells containing numerical values. Click the Increase Decimal Tool, shown in Figure 5.133, until all values are shown to three decimal places.

FIGURE 5.133

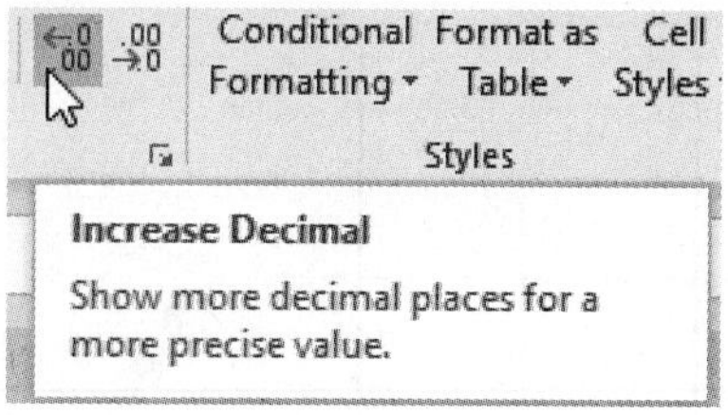

Click and drag the "handle" at the lower right corner of the table until all blank rows and columns are hidden, as shown in Figure 5.134. Click in the white space of the graphics area to close the table. From the menu, select Window and choose the name of the drawing.

FIGURE 5.134

	A	B	C	D	E	F	G	H
3		Diameter	Length	Head Diameter	Head Height	Hex Width	Hex Depth	Thread Length
4	Part 101	0.500	1.500	0.750	0.500	0.375	0.245	1.000
5	Part 102	0.750	2.000	1.125	0.750	0.625	0.370	1.500
6	Part 103	1.000	3.000	1.500	1.000	0.750	0.495	2.000

Sheet1

Click on the Rebuild Tool to update the design table in the drawing, which now appears as shown in Figure 5.135.

FIGURE 5.135

	Diameter	Length	Head Diameter	Head Height	Hex Width	Hex Depth	Thread Length
Part 101	0.500	1.500	0.750	0.500	0.375	0.245	1.000
Part 102	0.750	2.000	1.125	0.750	0.625	0.370	1.500
Part 103	1.000	3.000	1.500	1.000	0.750	0.495	2.000

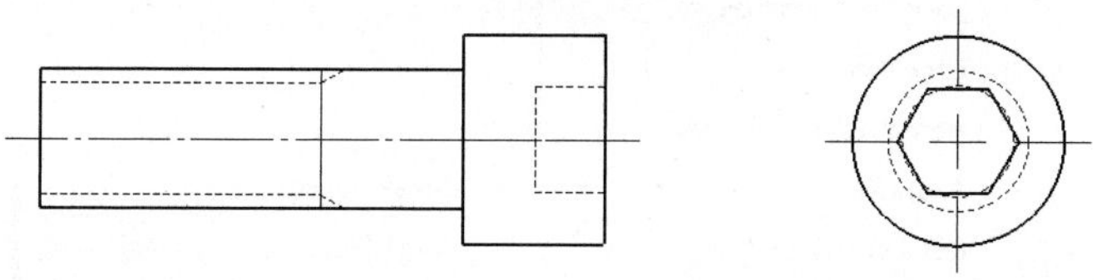

Before adding dimensions to the drawing, we need to add a view. Since the depth of the hex cavity needs to be dimensioned, we will add a section view of the head. This will let us avoid dimensioning a hidden feature. It is good practice to refrain from using hidden lines for dimensioning.

FIGURE 5.136

Zoom in on the head of the screw in the Front View. Choose the Line Tool from the Sketch group of the CommandManager, as shown in Figure 5.136. Hold the cursor momentarily at the midpoint of the top edge of the head to "wake up" this feature as shown in Figure 5.137. Move the cursor to the

FIGURE 5.137

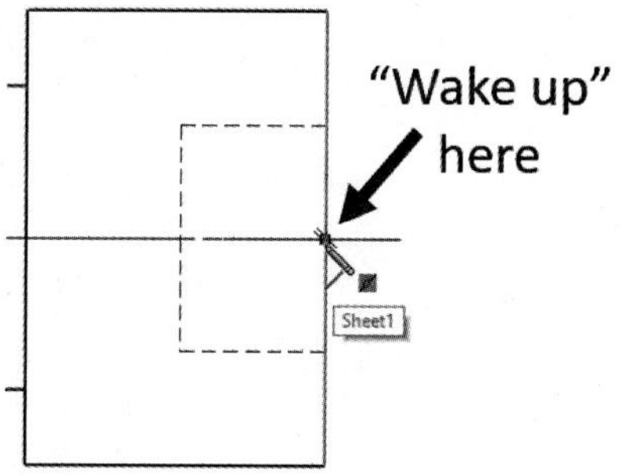

FIGURE 5.138

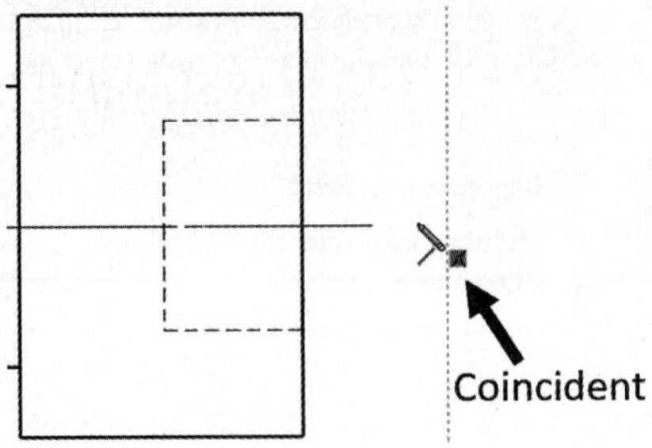

FIGURE 5.139

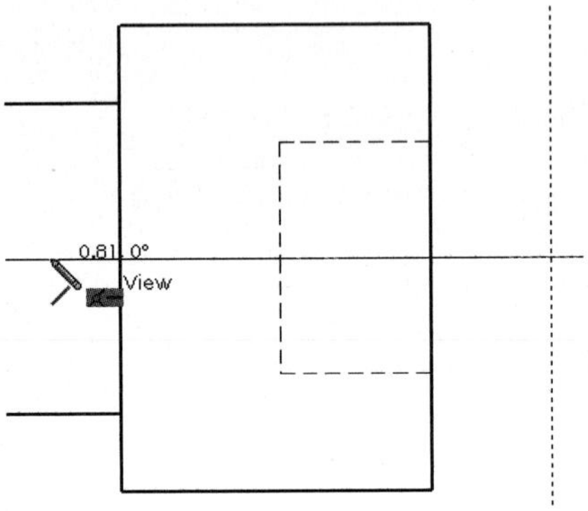

right, as shown in Figure 5.138. The coincident relation icon will appear, indicating that the cursor is aligned horizontally with the midpoint of the top edge of the head. Click and drag a horizontal line through the head, as shown in Figure 5.139. Note the coincident and horizontal relation icons. Click the check mark, and hit Esc to close the Line Tool.

Select the horizontal line, and create a partial section view by choosing the Section View Tool from the View Layout group of the CommandManager, as shown in Figure 5.140. Drag the section view away from the Front View and click to place it. Right-click, and choose Alignment: Break Alignment, as shown in Figure 5.141.

FIGURE 5.140

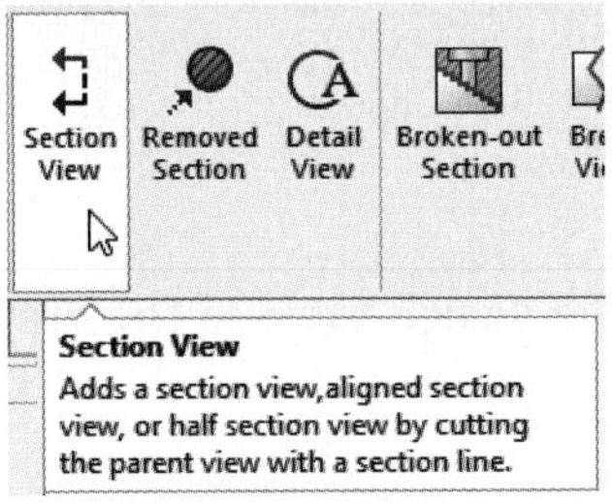

You will now be able to click and drag the section view to any location on the sheet, as shown in Figure 5.142. Change the font on the section label if desired.

FIGURE 5.141

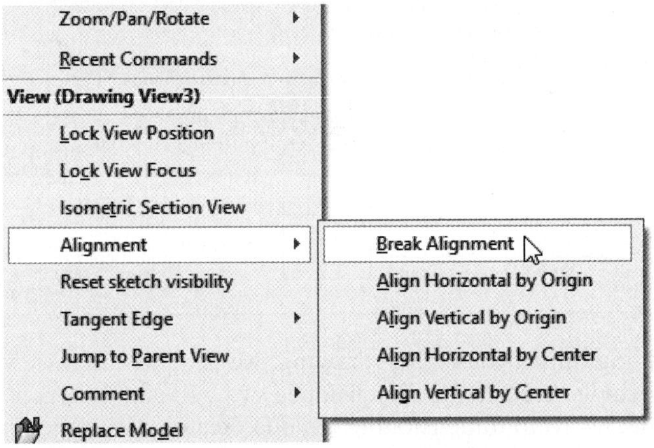

FIGURE 5.142

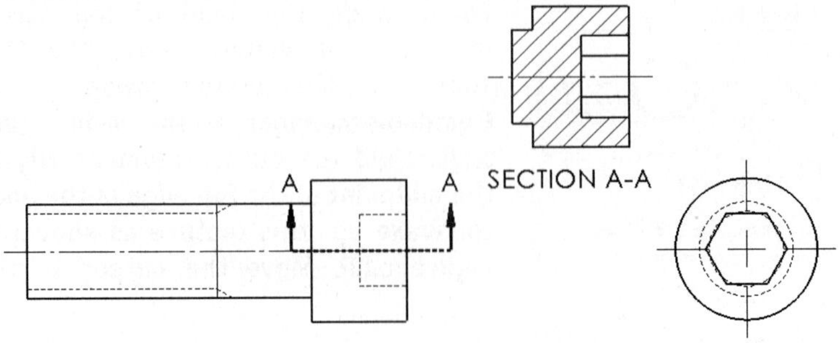

We will now add dimensions to the drawing. Since we want most of the dimensions to appear in the Front View, we will import dimensions into that view first.

FIGURE 5.143

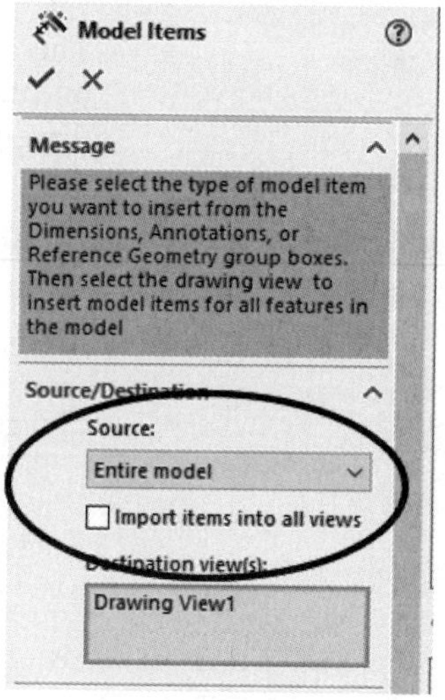

Select the Front View. Select the Model Items Tool from the Annotations group of the CommandManager. In the PropertyManager, select Entire Model as the source, as shown in Figure 5.143. Make sure that the box labeled "Import items into all views" is unchecked. Click the check mark to apply the dimensions.

Repeat for the Right View and then for the section view, so that all dimensions are imported, as shown in Figure 5.144.

FIGURE 5.144

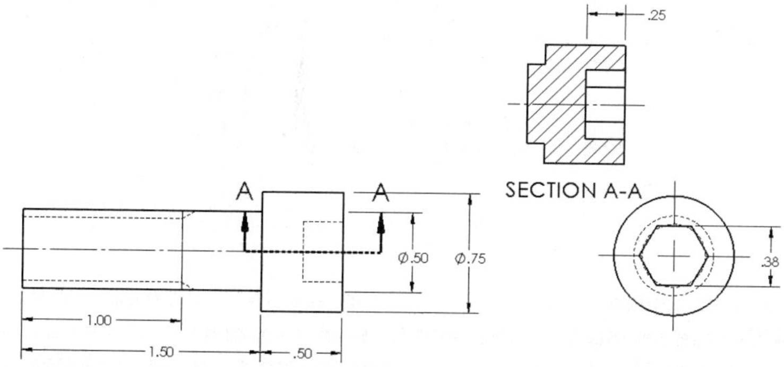

The dimensions shown are for the Part 101 configuration, which was the configuration selected when the part file was saved. If you saved the part with another configuration selected, then the dimensions will be different. This is not a problem, as we will be replacing the numerical values with the dimension names.

The dimensions defined by the design table may be shown in a different color than black. (The default color is magenta.) While this may be desirable in the part file, for the drawing we would prefer that all dimensions be displayed in black.

Select the Options Tool. Under the System Options tab, select Colors and scroll down the list of entities to find "Dimensions, Controlled by Design Table," as shown in Figure 5.145. Select Edit, and set the color to black.

FIGURE 5.145

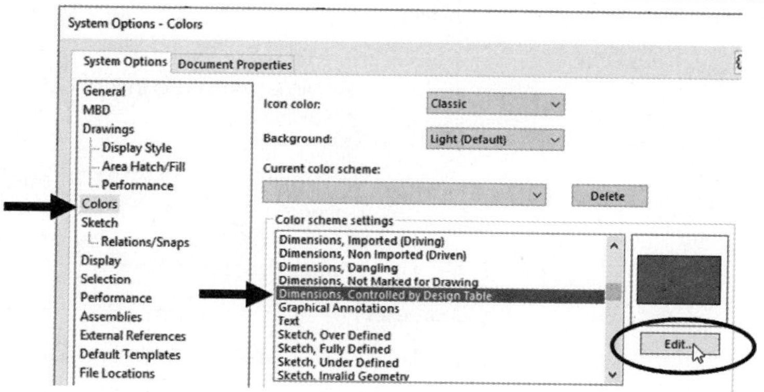

Move the dimensions and/or views on the screen so that they are all visible.

The dimensions displayed are those for one configuration. To make this drawing one that defines the dimensions for all configurations, we need to display the dimension names in the drawing views.

FIGURE 5.146

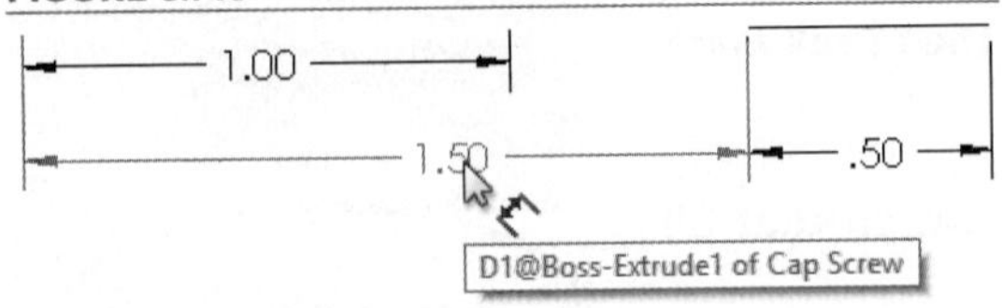

Click on the dimension defining the length, as shown in Figure 5.146. In the PropertyManager, replace the "<DIM>" in the Dimension Text box and type in "Length," as shown in Figure 5.147. A message will appear, warning you that any tolerances specified for this dimension will not be displayed (see Figure 5.148). Check the "Don't show again" box, and click Yes.

FIGURE 5.147

FIGURE 5.148

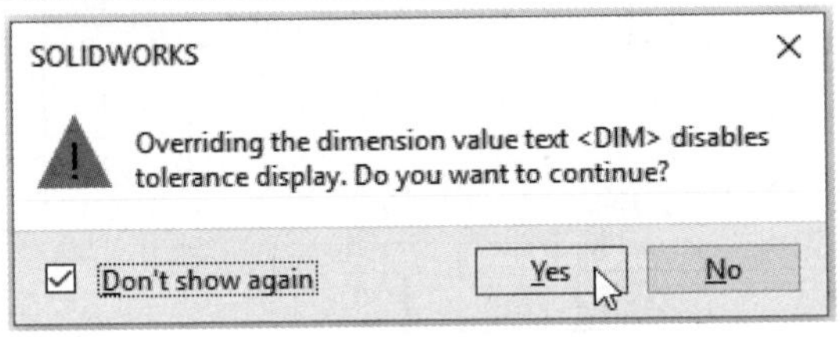

Repeat for the other dimensions. Add a note calling out UNC-2A threads (UNC = Unified Standard Coarse series, 2 = tolerance level, A = external threads) and a note specifying inches as the dimensions. The completed drawing is shown in Figure 5.149.

FIGURE 5.149

	Diameter	Length	Head Diameter	Head Height	Hex Width	Hex Depth	Thread Length
Part 101	0.500	1.500	0.750	0.500	0.375	0.245	1.000
Part 102	0.750	2.000	1.125	0.750	0.625	0.370	1.500
Part 103	1.000	3.000	1.500	1.000	0.750	0.495	2.000

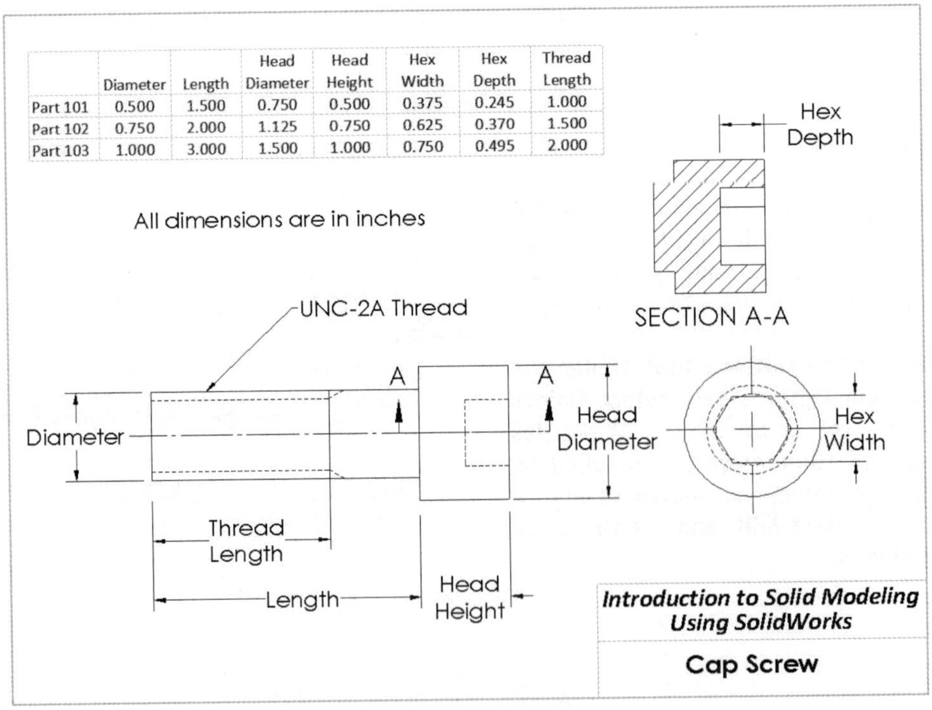

PROBLEMS

P5.1 Consider the flange model developed in Chapter 1. Create a "blank" for this part (without the holes), as shown in **Figure P5.1A**, using the following procedure:

a. Sketch a 5.5-inch diameter circle on the Top Plane, and extrude it upward 2.25 inches to create a solid cylinder (**Figure P5.1B**).

b. In the Front Plane, sketch a "cutting tool" to create the flange "blank" from the solid (**Figure P5.1C**).

c. Use the Revolved Cut command to create the blank.

FIGURE P5.1A

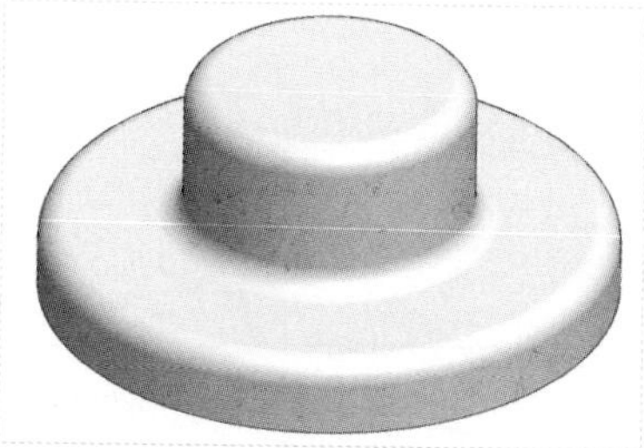

FIGURE P5.1B

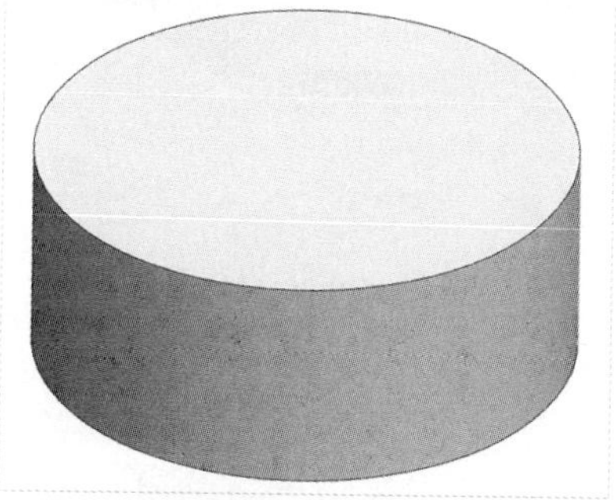

FIGURE P5.1C

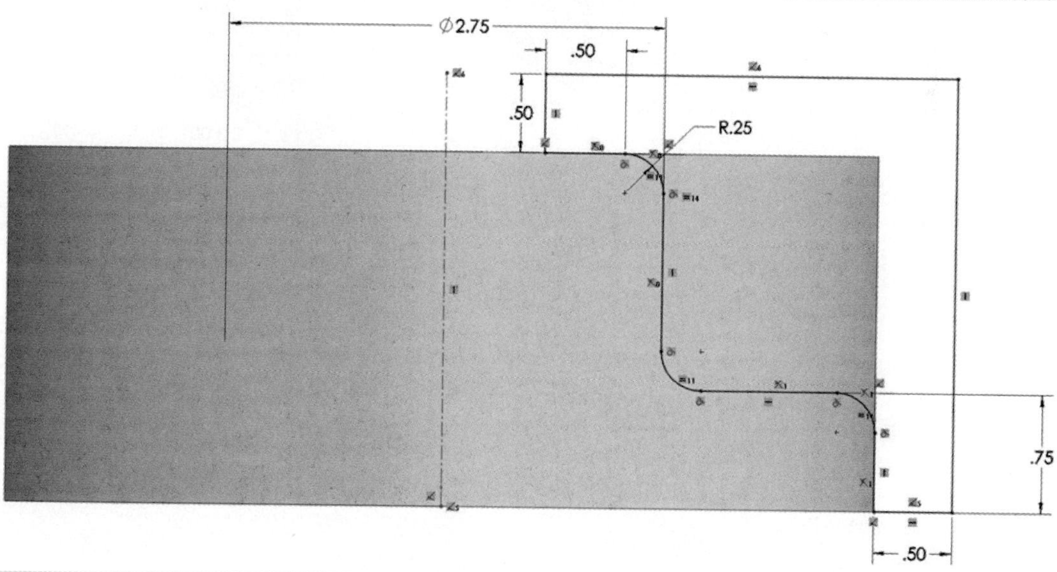

P5.2 Create a solid box, as shown in **Figure P5.2A**. Using equations, set the depth $d = 2w$, and the height $h = 3w$, where the width w is the driving dimension. Show the box for a few different values of the driving dimension, as shown in **Figure P5.2B**.

FIGURE P5.2A

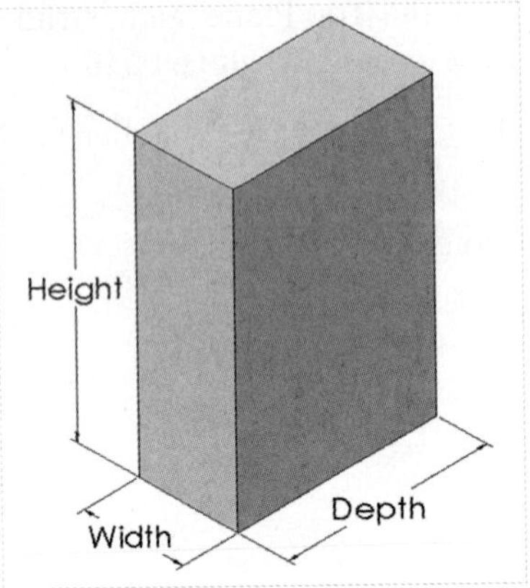

FIGURE P5.2B

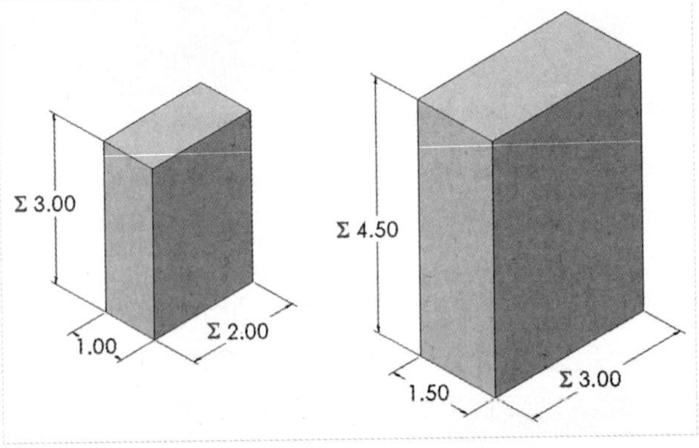

P5.3 Model the bracket shown in **Figure P5.3A**. Use only the dimensions shown in **Figure P5.3B**; use relations and symmetry as required so that these dimensions completely define the part. (The fillet radius, 0.125 inches, is the same for the three fillets.)

Add equations to the model so that these relationships exist between the dimensions:

1. Height = 0.40 * Width
2. Leg Depth = 0.50 * Height
3. Hole Spacing = Width – 2 inches
4. Hole Location = 0.50 * (Height – 0.25 inches)
5. Slot Width = 0.50 * Hole Spacing
6. Slot Location = 0.50 * Leg Depth

Check to see that the equations work for Width values from 3 to 8 inches.

FIGURE P5.3A

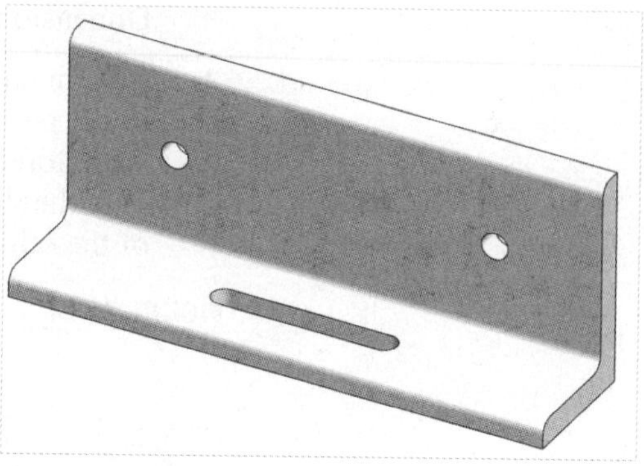

Note: The slot can be created using the Slot Tool from the Sketch group of the CommandManager.

FIGURE P5.3B

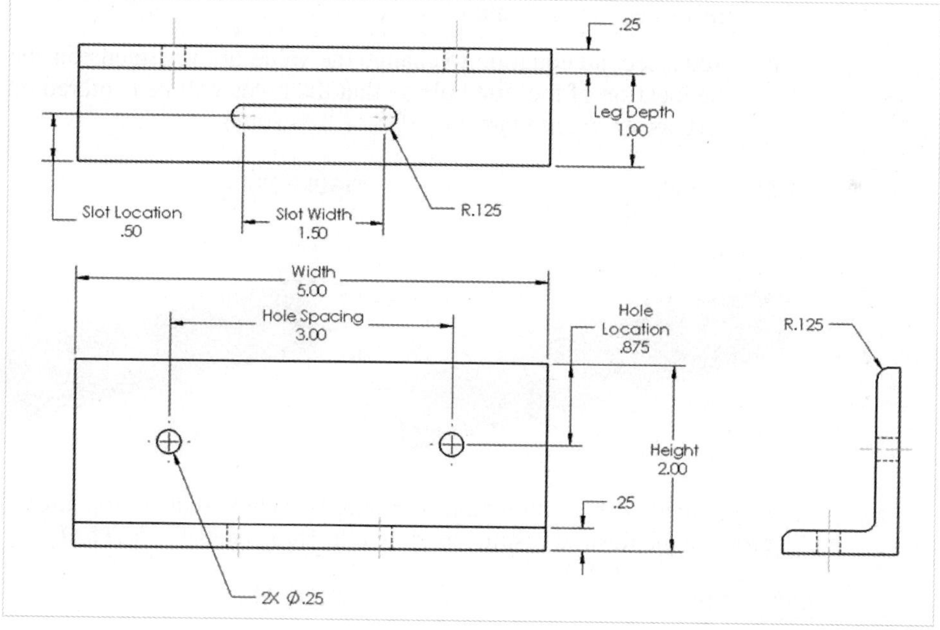

P5.4 In this exercise, you will model a part in which an integer design parameter (number of holes) is controlled by an equation.

a. Model the part shown in **Figure P5.4A**. Use a linear pattern to place the second hole. Show all of the dimensions by right-clicking on Annotations in the FeatureManager and clicking on "Show Feature Dimensions."

b. Add an equation so that the number of holes is equal to the length of the part (4 inches in the current configuration) divided by two. Therefore, there will be one hole for every 2 inches of length. Show only the dimensions shown in **Figure P5.4B** by right-clicking on each of the other dimensions and selecting "Hide."

FIGURE P5.4A

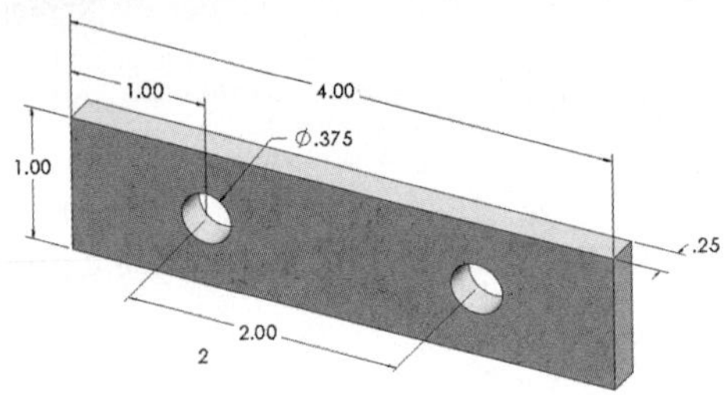

FIGURE P5.4B

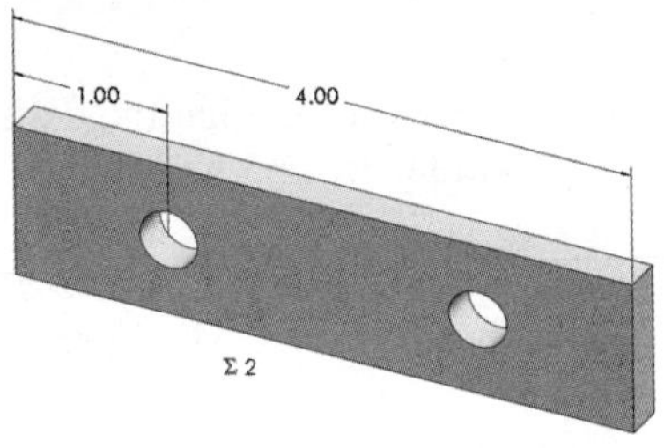

c. Change the length of the part to 5 inches, as shown in **Figure P5.4C**. Note that the program has rounded the value of the equation (2.5) to the closest integer value (3).

d. Add a second equation to change the value of the dimension specifying the location of the first hole so that the holes will be centered on the part, as shown in **Figure P5.4D**.

FIGURE P5.4C

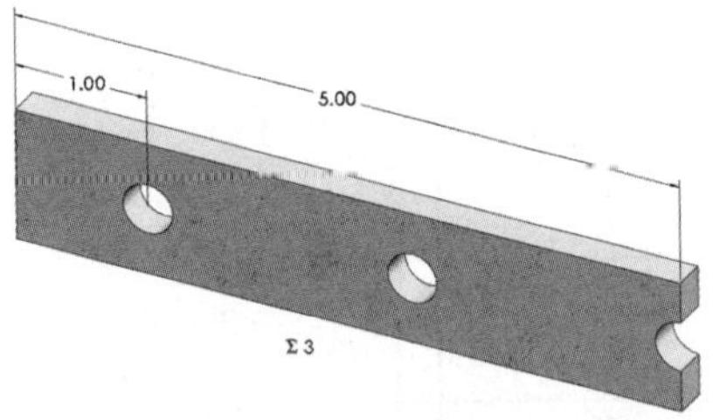

FIGURE P5.4D

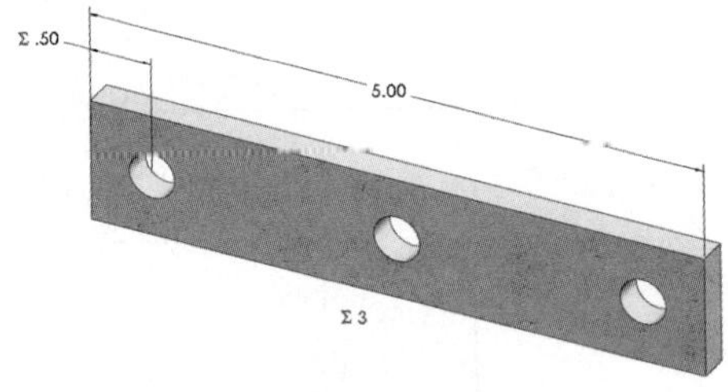

Experiment with several values of length to show that the equations produce the desired results, as shown in **Figures P5.4E** and **P5.4F**.

FIGURE P5.4E

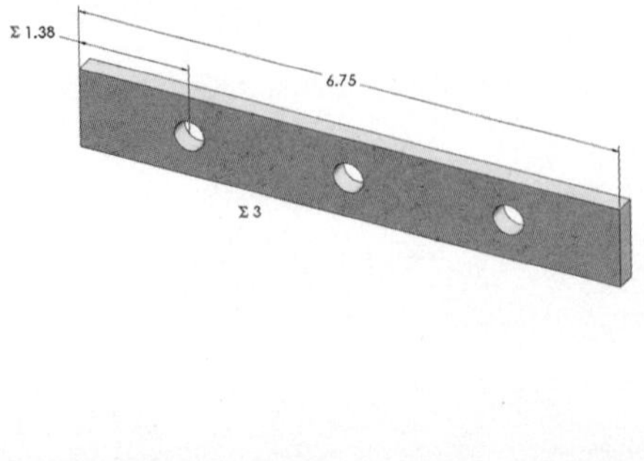

FIGURE P5.4F

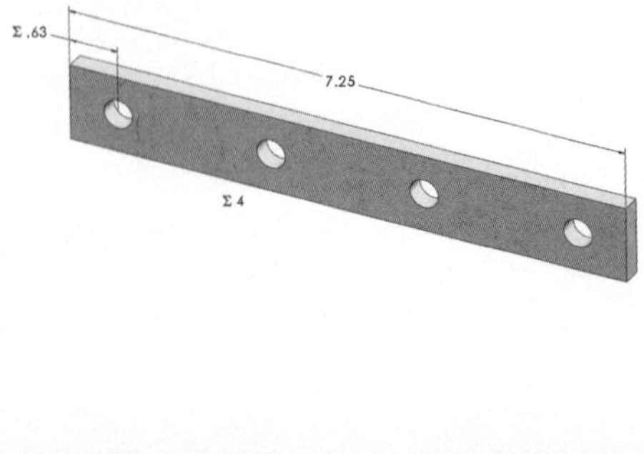

P5.5 Revisit the perforated board created in **Problem P3.5**. Add equations to resize the overall board dimensions based on the number of holes needed in the board (using the same hole size and spacing used in the original problem). Show that the equations work by generating new boards for the following cases:

a. A 70 × 50 grid of holes

b. A 80 × 30 grid of holes

c. A 45 × 15 grid of holes

P5.6 Revisit the hacksaw blade created in **Problem P3.6**. Add equations to calculate the proper tooth spacing and number of teeth in the linear pattern based on the blade length (the 12-inch dimension in **Figure P3.6C**) and the tooth length (the 0.050-inch dimension in **Figure P3.6D**). Show that the equation works by creating blades for the following three cases:

a. 18-inch length, 0.050-inch tooth length

b. 15-inch blade length, 0.060-inch tooth length

c. 15.5-inch length, 0.060 inch-tooth length

P5.7 Consider the heat sink from **Problem P3.9**, which is shown in **Figure P5.7A**. Make a copy of this part, and add equations as follows (refer to **Figure P5.7B** on the following page for the dimension names):

FIGURE P5.7A

1. The depth = one-half of the width, plus 10 mm.
2. The fin height = one-third of the width.
3. The number of fins = the width divided by 6 mm.
4. The offset distance to the first fin is such that the fins are placed symmetrically of the part.

Show that the equations work correctly for width values from 60 to 180 mm.

FIGURE P5.7B

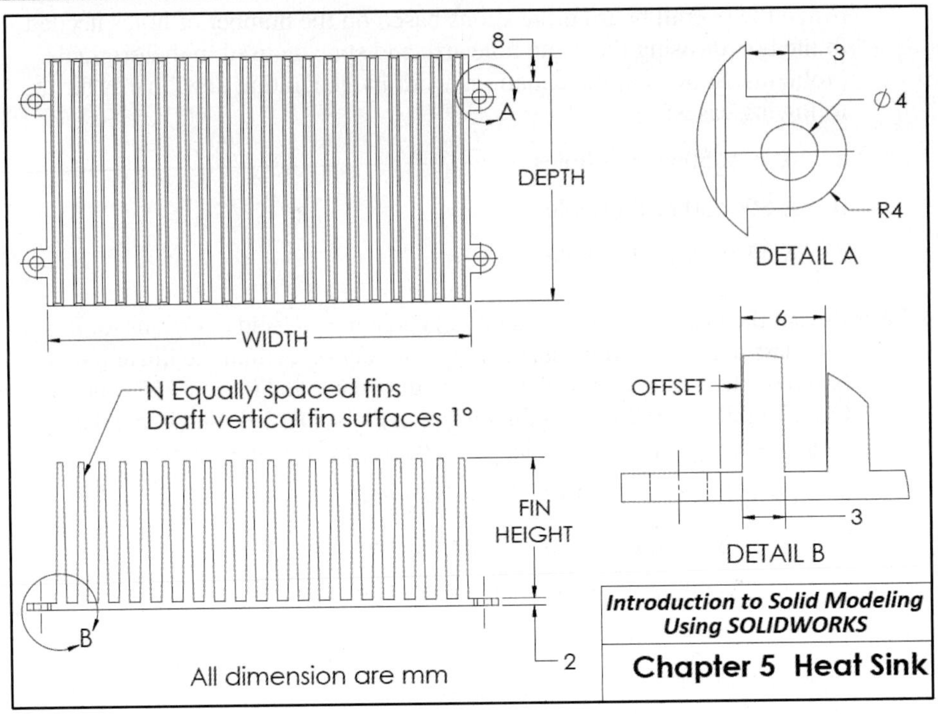

P5.8 Use a design table to create the W18 series of wide-flange shape beams, according to **Figure P5.8** and **Table P5.8**. The model of the beam should be extruded to 36 inches in all configurations.

FIGURE P5.8

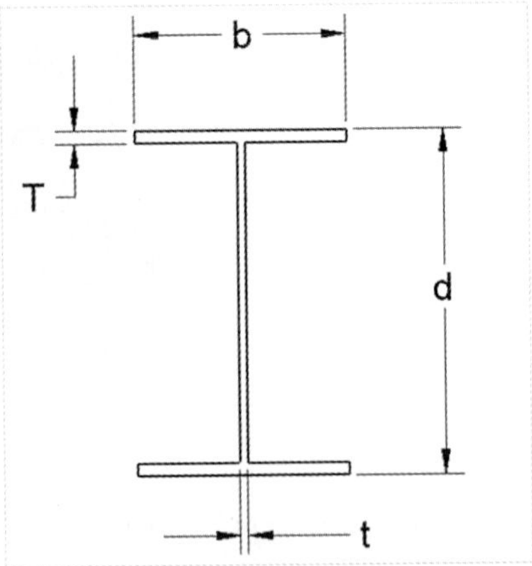

TABLE P5.8

Designation	Depth d	Flange width b	Flange thickness T	Web thickness t
W18 × 106	18.73	11.200	0.940	0.590
W18 × 76	18.21	11.035	0.680	0.425
W18 × 50	17.99	7.495	0.570	0.355
W18 × 35	17.70	6.000	0.425	0.300

Note: All dimensions in inches.

P5.9 Create a multiconfiguration drawing of the model created in **Figure P5.8**.

P5.10 Make a copy of the flange created in Chapter 1, which is shown in **Figure P5.10A** with some of the dimensions hidden. Create three different configurations of the flange (the flange as created in Chapter 1 will be the first configuration), using both equations and a design table, as detailed below.

FIGURE P5.10A

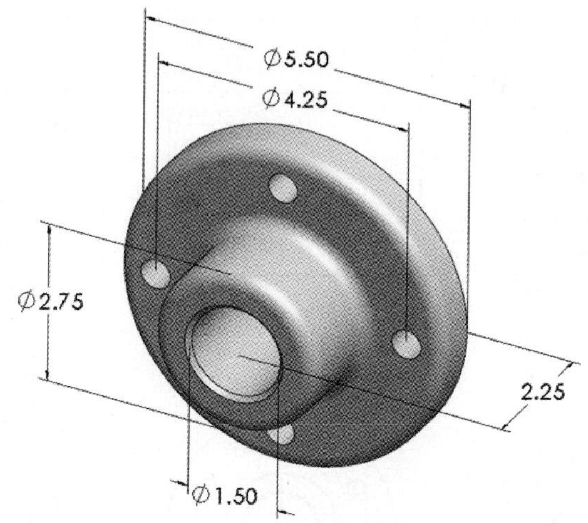

a. Add two equations:

   (1) The boss diameter (2.75 in.) is equal to one-half of the flange diameter (5.50 in.).

   (2) The center hole diameter (1.50 in.) is equal to the boss diameter minus 1.25 inches.

b. Create a design table to define the dimensions of two additional configurations, as specified in **Table P5.10**.

c. Make a 2-D drawing showing the three configurations, with only the dimensions that change shown, as in **Figure P5.10B**. This type of drawing is used often in product literature and catalogs to illustrate the relative sizes of different parts.

TABLE P5.10

Flange	Diameter	Height	Bolt circle diameter	Number of bolt holes
Part 1	5.5	2.25	4.25	4
Part 2	7.0	3.00	5.25	6
Part 3	8.0	3.50	6.25	8

Note: All dimensions in inches.

Insert three Front Views and three Top Views in the drawing. Right-click on each view and select Properties, and select the appropriate named configuration for each view. Hide unwanted dimensions by selecting View: Hide/Show Annotations from the main menu and selecting the dimensions to be hidden. Press the Esc key to return to the normal drawing mode.

To precisely align the views, select a view, right-click, and choose Align: Align Horizontal by Origin, and select another view to align to. Repeat until all views are aligned.

FIGURE P5.10B

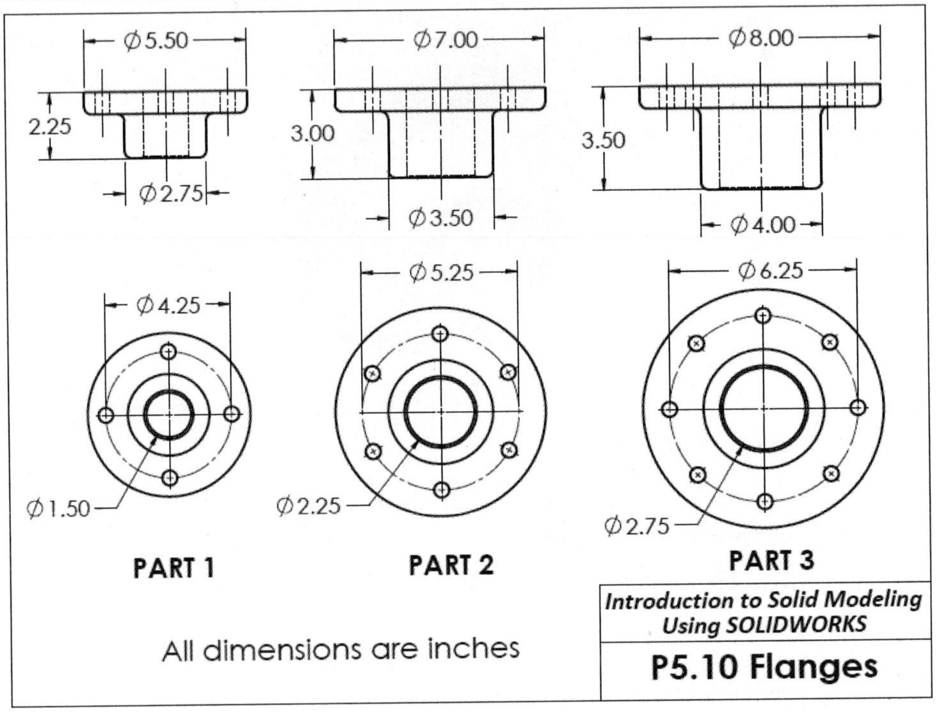

P5.11 An O-ring is an elastomeric seal in the shape of a torus, as shown in **Figure P5.11A**. Standard sizes of O-rings used in the United States are defined by a Society of Automotive Engineers (SAE) specification. Create a solid model of an O-Ring, with the six configurations as detailed in **Table P5.11**. (Note that while it is normally good practice to dimension to the centers of circular features rather than to their edges, O-rings are defined by their inner diameters. This allows a design engineer to determine how much the seal will have to stretch to fit into a groove of a specified size.) Make a multi-configuration drawing, as shown in **Figure P5.11B**.

FIGURE P5.11A

When adding the inner diameter dimension to the sketch defining the part geometry, note that the Smart Dimension Tool attempts to define the dimension to the center of the circle representing the O-ring cross section. To dimension to the edge of the circle, use the Point Tool to add a point to a quadrant point of the circle, and dimension to that point.

TABLE P5.11

Part Number	Inner Diameter	Thickness
125	1.299 ± .020	.103 ± .003
142	2.362 ± .020	.103 ± .003
235	3.109 ± .024	.139 ± .004
254	5.484 ± .035	.139 ± .004
350	4.600 ± .030	.210 ± .005
362	6.225 ± .035	.210 ± .005

While the tolerance values cannot be used to change values within the part (even if you define the tolerance associated with a dimension on the part file, only the nominal value appears in the design table), they can be added in separate columns in the design table. To add the plus/minus symbol, select Insert: Symbol from the Excel tools.

FIGURE P5.11B

	Inner Diameter	Thickness	Diameter Tolerance	Thickness Tolerance
125	1.299	0.103	±.020	±.003
142	2.362	0.103	±.020	±.003
235	3.109	0.139	±.024	±.004
254	5.484	0.139	±.035	±.004
350	4.600	0.210	±.030	±.005
362	6.225	0.210	±.035	±.005

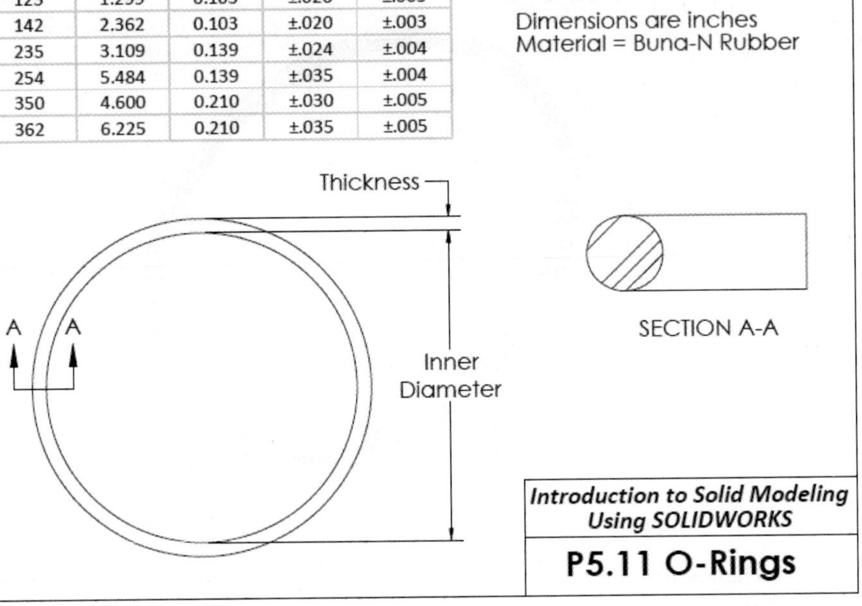

P5.12 Revisit the grate design problem from Chapter 3, **Problem P3.10**. Design two alternative configurations of the grate by varying the slot sizes. The original design will be considered the Medium Capacity grate. The other two will be a Low Capacity model, with an open area 15% less than the Medium Capacity grate, and a High Capacity model, with an open area 15% more than the Medium Capacity grate. Show the three designs in a multiple-configuration drawing incorporating a design table.

CHAPTER 6

Creation of Assembly Models

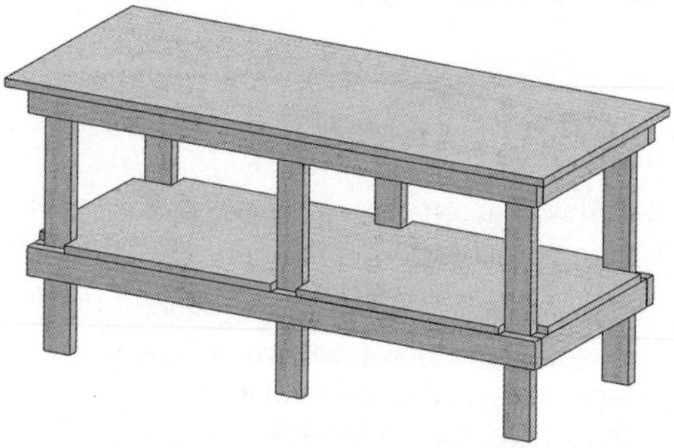

Introduction

In the preceding chapters, the development of solid models of parts was covered in detail. In this chapter, methods for combining such part models into complex, interconnected solid models will be described. These types of models, composed of interconnected part models, are called *assembly models*.

The assembly that will be constructed in this chapter is a model of a hinged door, as shown in **Figure 6.1**.

The chapter will begin with a tutorial describing the construction of the part models of the components used in this assembly. After the models are constructed, they will be interconnected into an assembly model. The assembly model will be used to demonstrate the creation of an exploded configuration of the model.

FIGURE 6.1

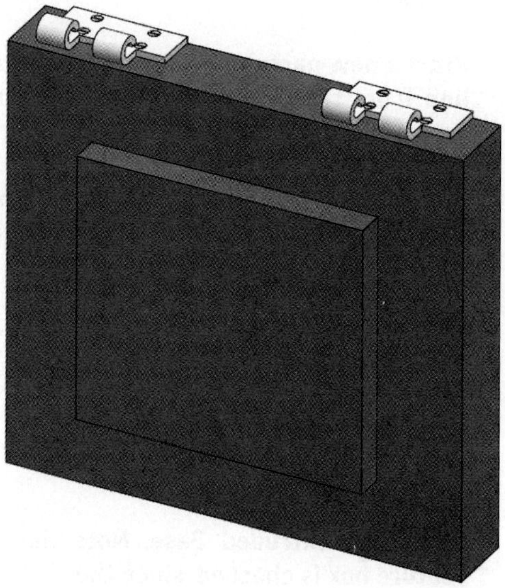

Chapter Objectives

In this chapter, you will:

- create holes using the Hole Wizard,
- learn to import part models into an assembly,
- use assembly mates to define how components fit together in an assembly,
- add assembly-level features to a model,
- create an exploded configuration of an assembly model, and
- create an animation of an exploded assembly.

SOLIDWORKS is a registered trademark of Dassault Systémes SolidWorks Corporation.

VIDEO EXAMPLE 5

In this chapter, we will build assembly models out of pre-existing parts. An analogy for assembly modeling is building structures from blocks.

A video demonstrating the creation of a simple assembly from prefabricated blocks, as well as the solid models of the blocks, can be found at the book website, *www.mhhe.com/howard2019.*

6.1 Creating the Part Models

Before an assembly can be created, the parts to be assembled must be modeled. In this first step, solid models of the hinge and door components will be created.

The first model that we will create is the hinge component.

Start a new part model, and sketch a horizontal line from the origin and a tangent arc in the Front Plane, as shown in Figure 6.2.

FIGURE 6.2

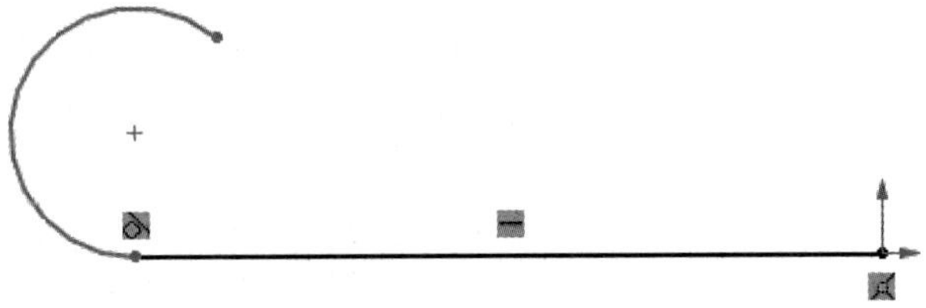

Add a horizontal relation between the center point and endpoint of the arc, and dimension the sketch as shown in Figure 6.3.

FIGURE 6.3

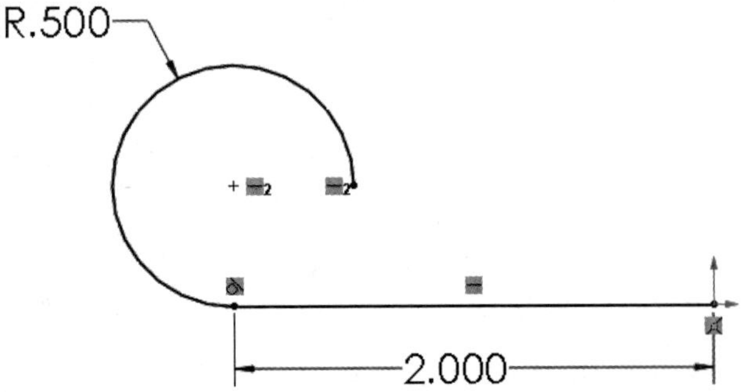

The sketch should now be fully defined. While in previous exercises we have used sketches with closed contours to make extrusions, in this case we will use the open-contour sketch to create a *thin-feature* extrusion.

Create an Extruded Base. Note that the Thin Feature box is checked, since the sketch contour is open. Set the extrusion type to Mid Plane, the

extrusion length to 4 inches, and the Thin Feature thickness to 0.25 inches, as shown in Figure 6.4. Set the Thin Feature direction such that the 0.5-inch radius applies to the exterior of the hinge, as shown in the preview in Figure 6.5.

The completed extrusion is shown in **Figure 6.6.**

FIGURE 6.4

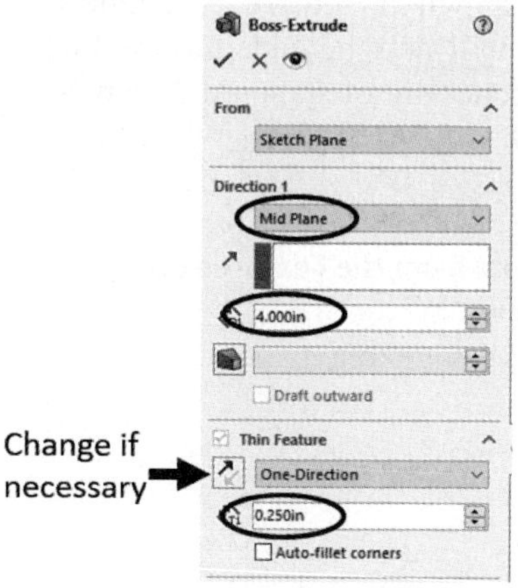

FIGURE 6.5

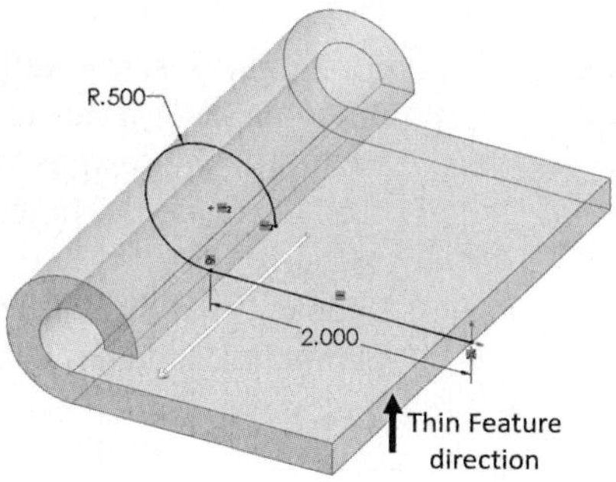

FIGURE 6.6

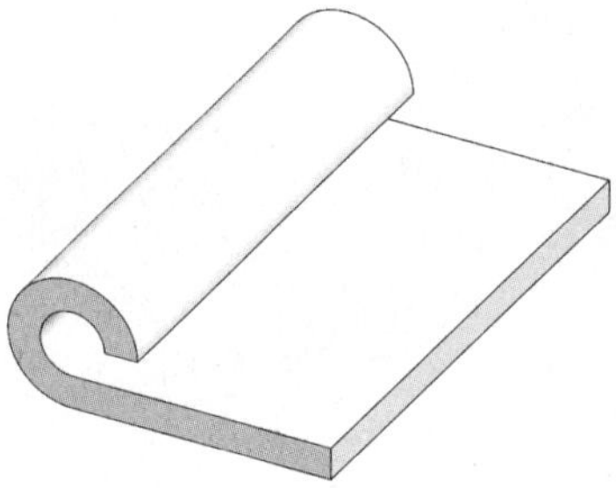

Select the top horizontal surface of the hinge, as shown in Figure 6.7, for the next sketch. Choose the Normal To View.

Sketch the 1-inch square area to be cut away from the basic hinge shape, as shown in Figure 6.8. Use snaps and/or relations to align the edges of the square with the edges of the hinge, so that the sketch is fully defined with only the single dimension shown.

FIGURE 6.7

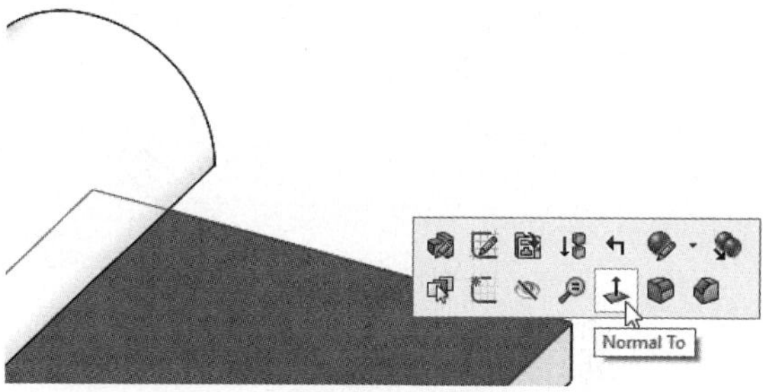

FIGURE 6.8

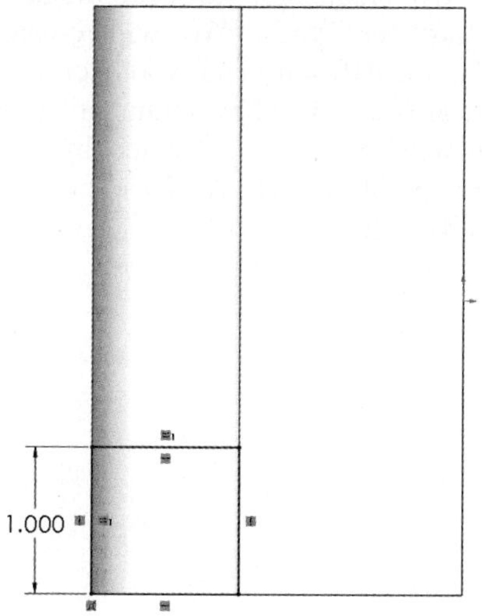

Use the Extruded Cut Tool to cut Through All - Both, as shown in Figure 6.9.

FIGURE 6.9

FIGURE 6.10

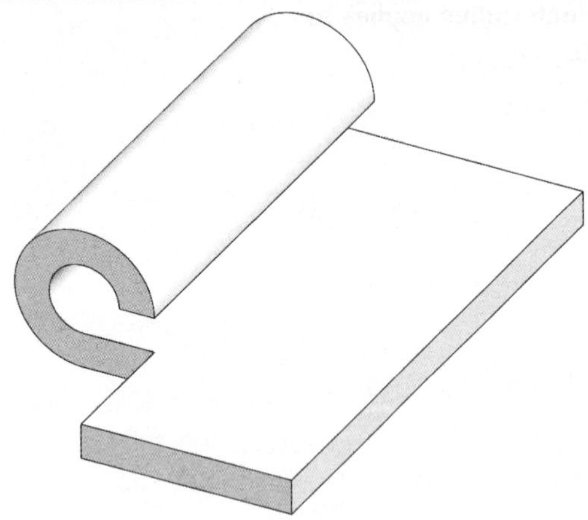

The result of the cut is shown in Figure 6.10.

Select Cut-Extrude1 from the FeatureManager, and define a linear pattern to create another instance of this feature 2 inches along the horizontal edge of the hinge, as shown in the preview in Figure 6.11.

The completed pattern is shown in Figure 6.12. A pattern of countersunk screw holes will now be added to the hinge. Since fastener holes are generally of standard dimensions, an intelligent design tool known as the Hole Wizard will be used to create the holes.

Choose the Hole Wizard Tool from the Features group of the CommandManager, as shown in Figure 6.13.

FIGURE 6.11

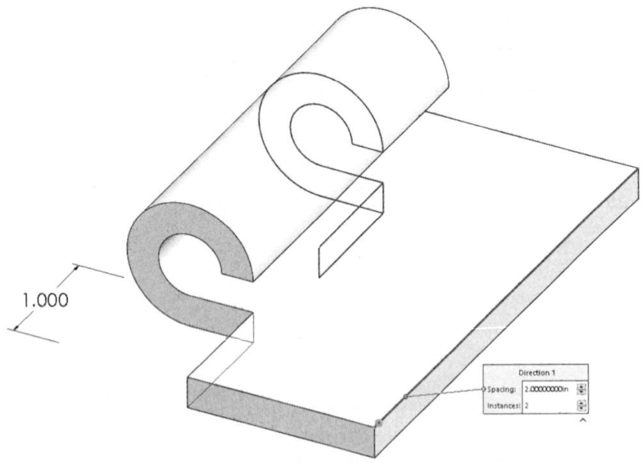

FIGURE 6.12

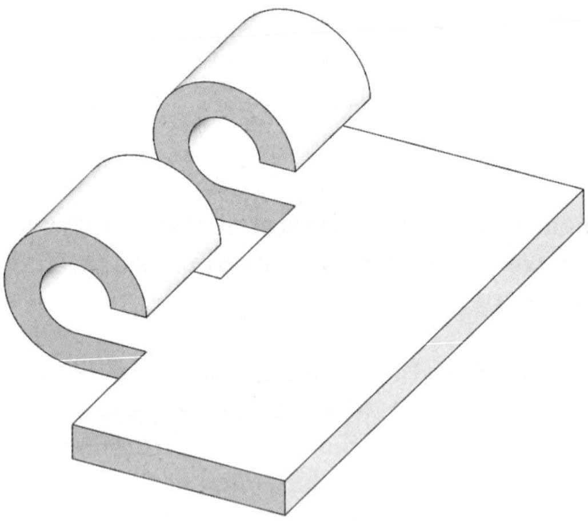

FIGURE 6.13

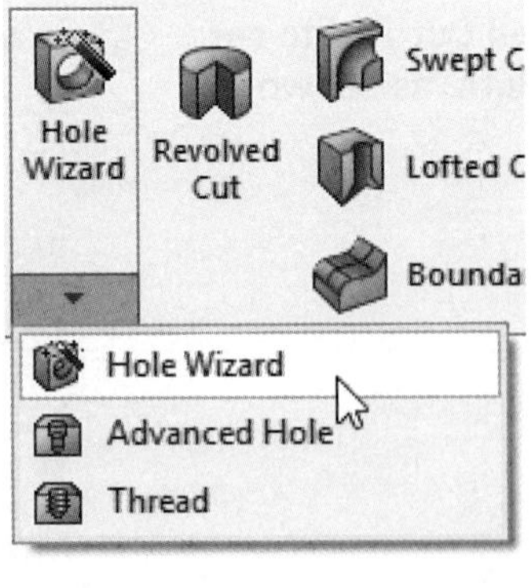

The Hole Wizard dialog box appears. The Hole Wizard can be used to create holes to accommodate most standard fastener types. We will create countersunk holes for #10 flat head wood screws. (Screw diameters are designated by a number from 1 to 12, after which they are designated by the diameter as a fraction of an inch. A #10 screw has a diameter of 0.190 inches.)

In the Hole Specification PropertyManager under the Type tab, click to set the hole specification to Countersink. Set the Standard to Ansi Inch, the Type to Flat Head Screw (82), the size to #10, and the End Condition to Through All, as shown in Figure 6.14. Click the Positions tab in the PropertyManager to initiate the Hole Position PropertyManager.

FIGURE 6.14

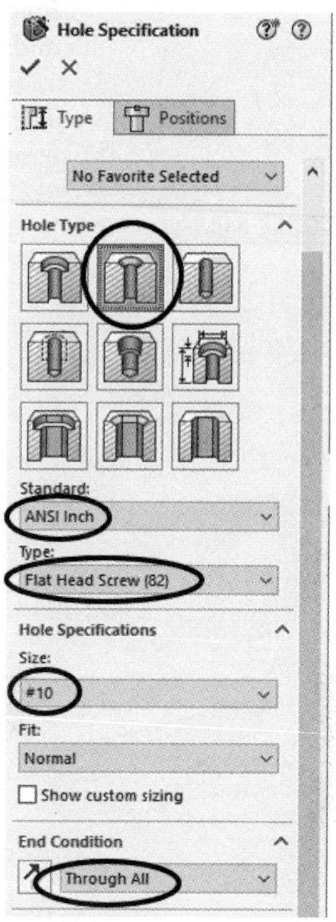

We are now prompted to enter the hole location, as shown in Figure 6.15. We will create a single hole, and replicate it using a linear pattern.

Change to the Top View. Click on the flat surface, then place the center of the hole by clicking in the approximate location shown in Figure 6.16.

FIGURE 6.15

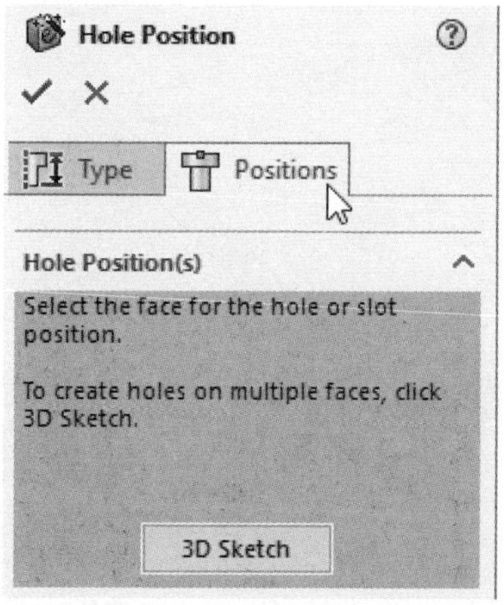

FIGURE 6.16

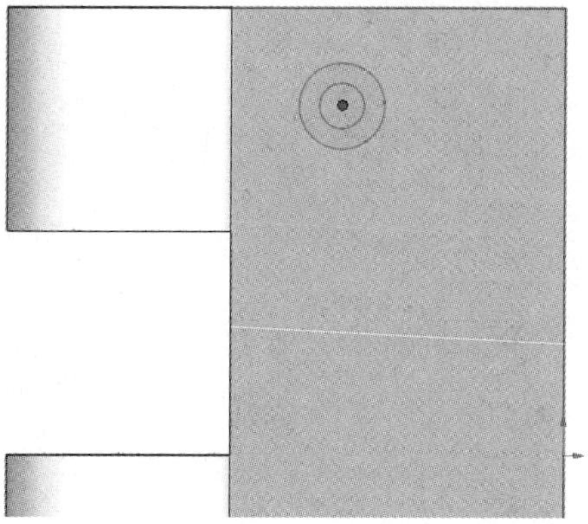

Select the Smart Dimension Tool, and add dimensions to the hole location as shown in Figure 6.17.

Click the check mark to create the hole, which is shown in Figure 6.18.

FIGURE 6.17

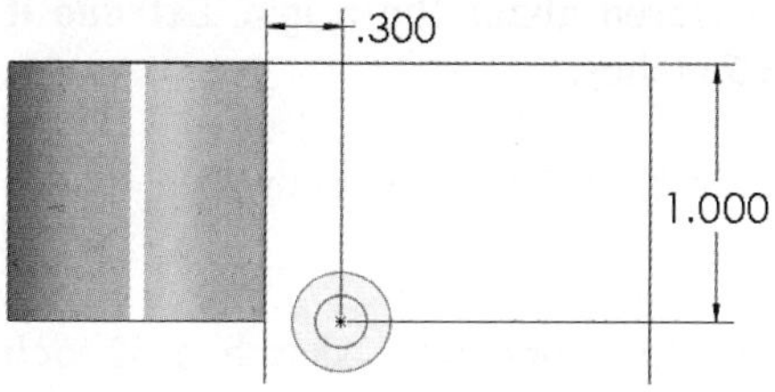

FIGURE 6.18

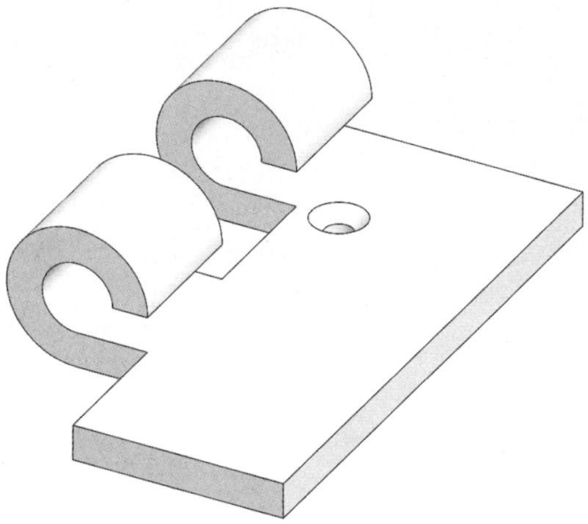

Select the new hole from the FeatureManager, and create a linear pattern with four total instances of the hole, spaced 2 inches along the long side of the hinge and 0.9 inches along the short side, as shown in the preview in Figure 6.19.

The final model of the hinge is shown in **Figure 6.20**. Since it will be used in a later assembly it must be saved.

FIGURE 6.19

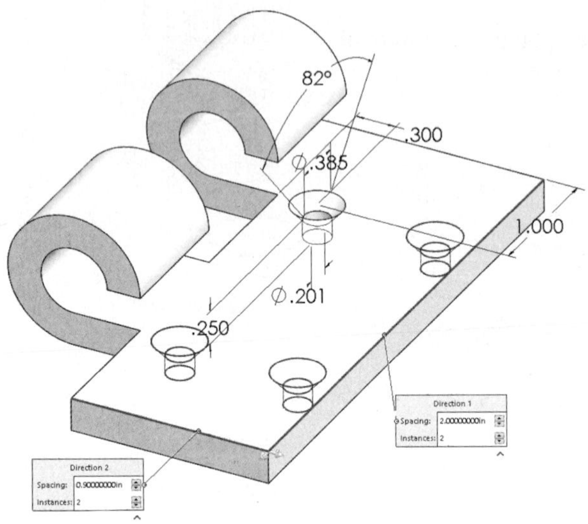

FIGURE 6.20

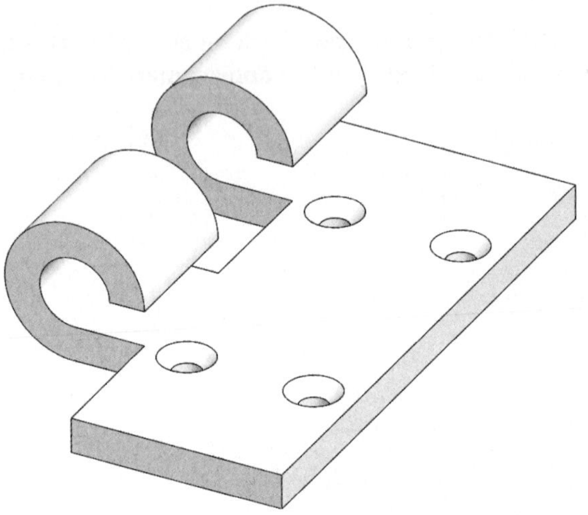

FIGURE 6.21

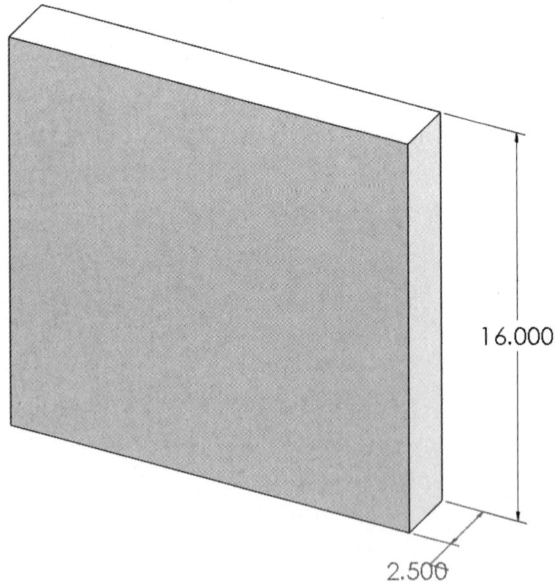

Save this using the file name "Hinge," and close the file.

Now, the second major component will be created.

Open a new part, and sketch a 16-inch by 16-inch square in the Front Plane, centered about the origin. Extrude it 2.5 inches.

This base feature is shown in **Figure 6.21**.

On the front face, sketch a 10-inch square centered about the origin. Extrude it 1 inch from the face.

This completes the second component, as shown in **Figure 6.22**. When we assemble the parts later, it will be helpful if they are different colors.

If desired, press the Esc key to clear any selections, and select the Edit Appearance Tool from the Heads-Up View Toolbar, as shown in Figure 6.23. Pick the new color for the hatch, and click the check mark to close the Color PropertyManager.

Save it using the file name "Hatch." Close the file.

FIGURE 6.22

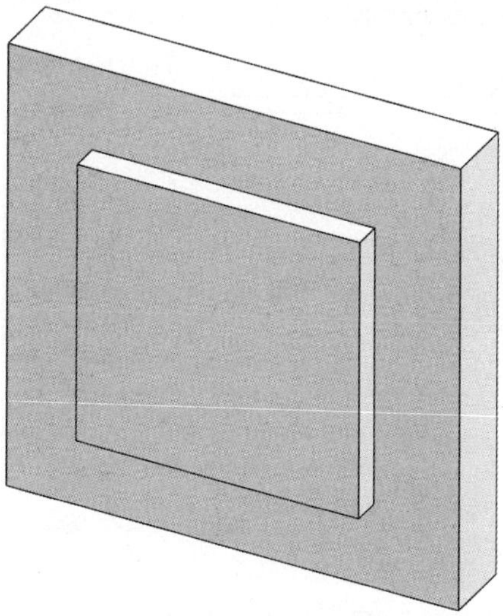

FIGURE 6.23

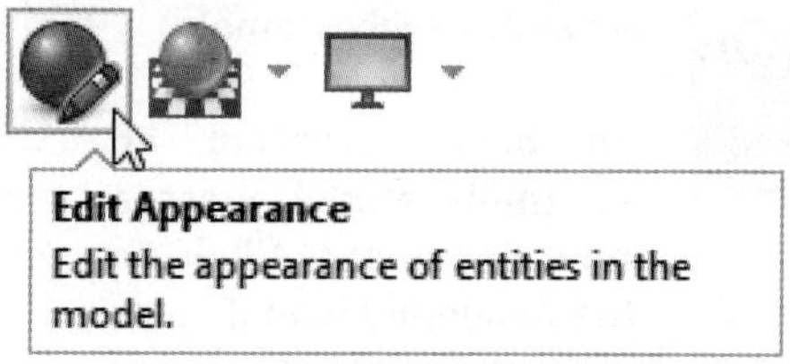

These parts will now be used to create a model of the door assembly.

6.2 Creating an Assembly of Parts

The features of the software that we will employ are the *assembly* capabilities. *Assemblies* are complex solid models that are made up of simpler part models, with specifically defined geometric relationships between the parts. The SOLIDWORKS® program provides us with the ability to relate surfaces and other geometric features of one part to those of another part. For example, we could:

- *Define two flat surfaces as* ***coincident****:* This places the two flat surfaces in the same plane.
- *Define two flat surfaces as* ***parallel****.*
- *Define two flat surfaces a preset* ***distance*** *apart:* This makes the two surfaces parallel, with a specified distance between them.
- *Define two lines or planes as* ***perpendicular*** *to one another.*
- *Define two lines or planes at a preset* ***angle*** *to one another.*

- *Define a cylindrical feature as* ***concentric*** *with another cylindrical feature:* This aligns the axes of two cylindrical features.
- *Define a cylindrical feature as* ***tangent*** *to a line or plane.*

These geometric relationships are called *mates*. There are many other geometric relationships that can be accommodated as well.

FIGURE 6.24

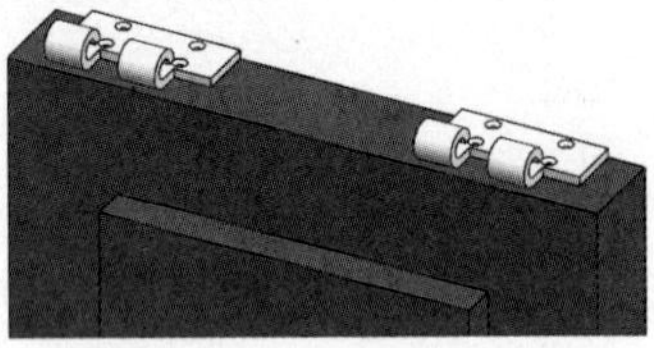

In this section, a tutorial will be presented in which we will create a simple assembly by attaching a set of hinges to the hatch component, as shown in Figure 6.24. This assembly will be used in the following chapter as a small assembly (subassembly) within a larger assembly.

To begin creating an assembly model, start a new document. Click on the Assembly icon (Figure 6.25), and click OK.

FIGURE 6.25

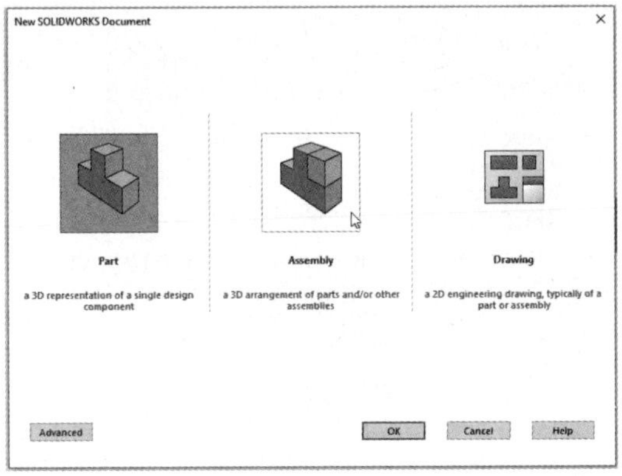

A new assembly window will be created. For the purposes of this assembly, the main "base" part will be the hatch. We will begin by importing this component into the assembly.

FIGURE 6.26

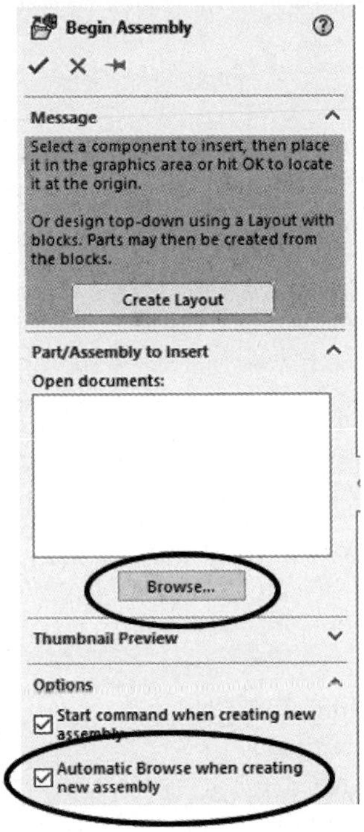

The PropertyManager will open, as shown in Figure 6.26. If the Automatic Browse option is checked in the PropertyManager, then the box shown in Figure 6.27 will also open. If not, then click on Browse. (Note: If the PropertyManager does not appear when you start a new assembly, select Insert: Component: Existing Part/Assembly from the main menu.)

Browse to the location where the hatch file was saved, as shown in Figure 6.27, and open it. If necessary, change the file type option to Part (*.prt, *.sldprt) to find the part files.

FIGURE 6.27

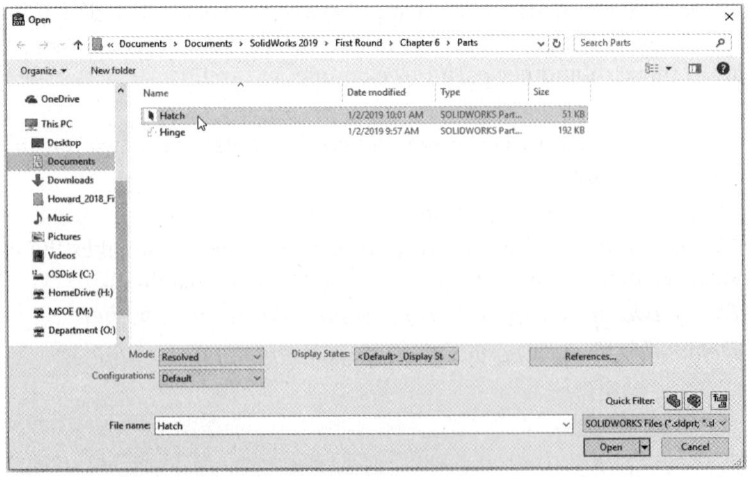

DESIGN INTENT **Planning an Assembly Model**

When we imported the first part into our assembly, we were careful to locate the part at the origin of the assembly space. More precisely, we made the origin of our part model coincident with the origin of the assembly space, and therefore also made the Front, Top, and Right planes of our part model coincident with the corresponding planes in the assembly space. This is not strictly necessary; we could have located the origin of the part model anywhere in the assembly space. However, by taking advantage of the default Front, Top, and Right planes (as well as the origin), we can use these as references for the addition of new features (such as holes, bosses, etc.) at the assembly level (as we will do later in the chapter), as well as in construction of assembly drawings (as we will do in Chapter 8). Judicious choice of the location of the first part we bring into an assembly can simplify subsequent tasks, if we anticipate our future use of the assembly model.

As prompted by the message shown in Figure 6.28, click the check mark (the OK button) to insert the hatch part at the origin.

FIGURE 6.28

This will import the hatch model into the assembly, with the origin and default planes of the hatch coincident with the origin and default planes of the assembly window.

From the Heads-Up View Toolbar, click the Apply Scene Tool and select Plain White as the background, as shown in Figure 6.29. (See appendix A for instructions on the creation of an assembly template.)

FIGURE 6.29

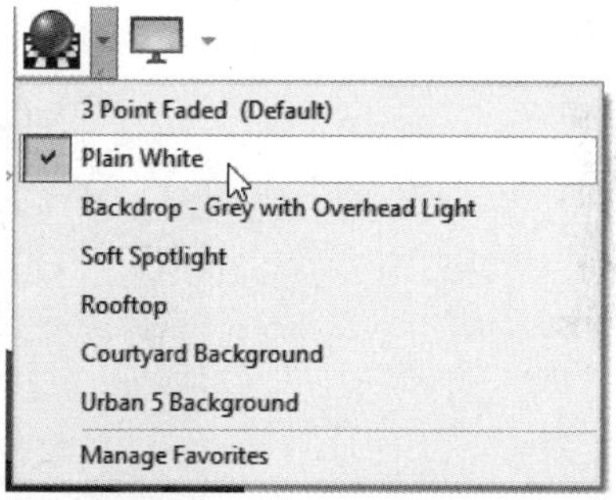

Note that the name of the component (Hatch) now appears in the FeatureManager (**Figure 6.30**). The designation (f) means that the component is "fixed"; it is fully constrained in the assembly window, and cannot be moved or rotated. Note the Mates group at the bottom of the FeatureManager. As we define the geometric relations between components, these relations will be stored under this Mates group.

FIGURE 6.30

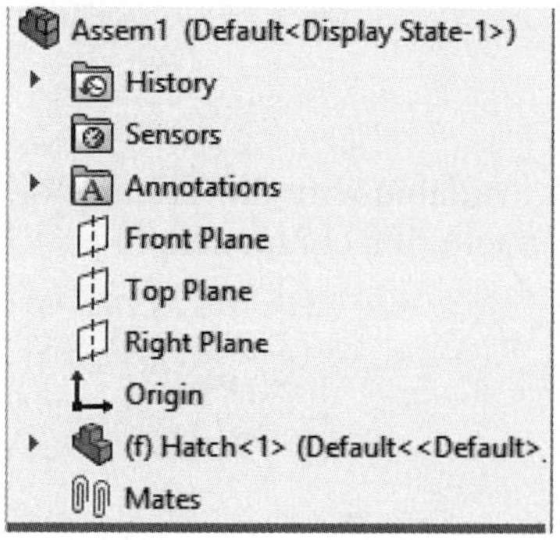

The hinge component will now be brought into the assembly. Note that the CommandManager has an Assembly group containing many of the tools we will use in assembling the components.

Select the Insert Components Tool from the Assembly group of the CommandManager, as shown in Figure 6.31. Browse to find the hinge file. Click Open, and move the hinge

FIGURE 6.31

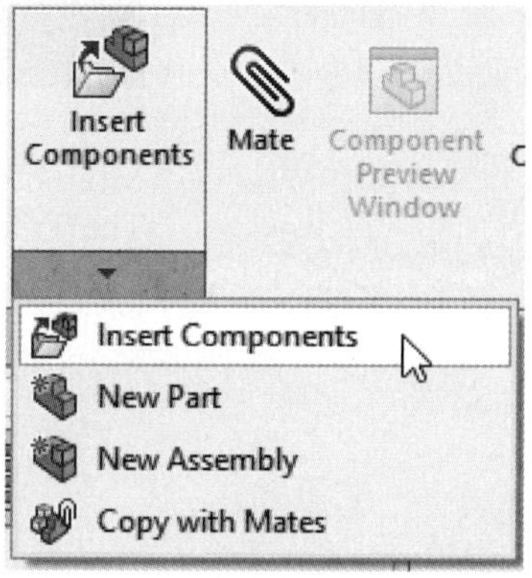

FIGURE 6.32

to the approximate position shown in Figure 6.32. Click to place the hinge.

This will import the hinge component, with its origin located at the point selected. The exact position and orientation are not important at this point, since the Mate Tool will be used to establish the position and orientation with respect to the base part.

Note that the name of the hinge component now appears in the FeatureManager, with the (–) designation preceding it. This designation indicates that the component is "floating," and can be moved or rotated (as its degrees of freedom allow).

Although the CommandManager contains Move Component and Rotate Component Tools, parts can also be moved or rotated directly with click-and-drag operations.

Click on the hinge with the left mouse button, and, while holding the button down, drag the hinge to a new position, as shown in Figure 6.33. Hit Esc to exit this mode.

Click on the hinge with the right mouse button and, while holding the button down, rotate the hinge to a new orientation, as shown in Figure 6.34. Press the Esc key to exit this mode.

FIGURE 6.33

FIGURE 6.34

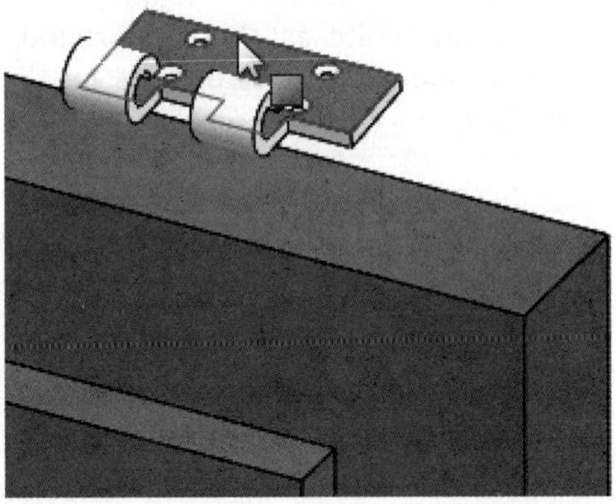

More control over the move and rotate commands is available with the Triad Tool. The Triad Tool can be activated from the right-click menu of a component.

FIGURE 6.35

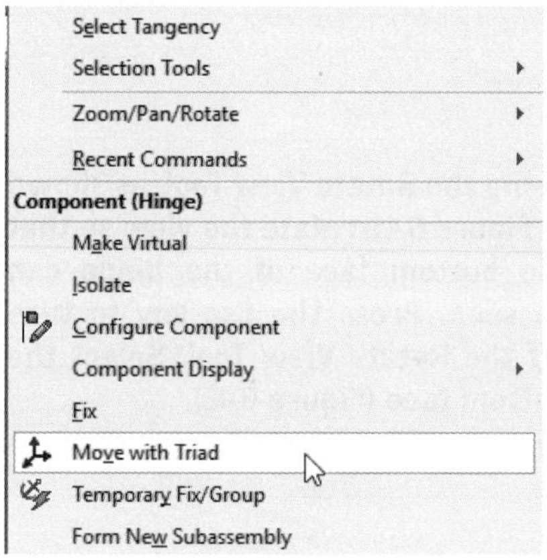

Right-click on the hinge and select Move with Triad from the menu, as shown in Figure 6.35.

The Triad Tool is shown in **Figure 6.36**. The arrows correspond to the principal axes of the assembly. Click and drag on one of the arrows, as shown in **Figure 6.37**, to translate the part in that direction. Click and drag on one of the circles, as shown in **Figure 6.38**, to rotate the part within the plane defined by that circle.

FIGURE 6.36

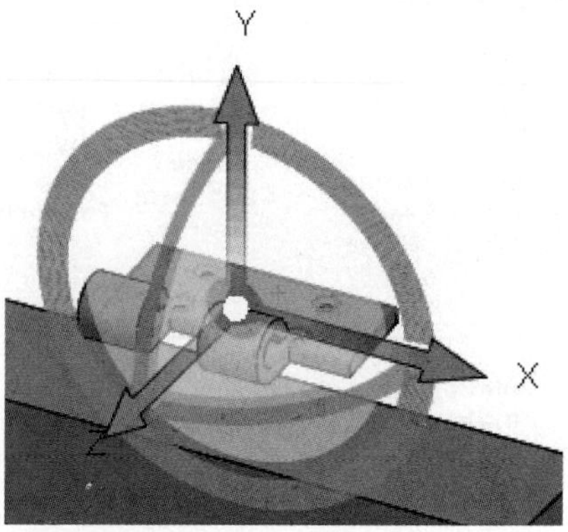

FIGURE 6.37

FIGURE 6.38

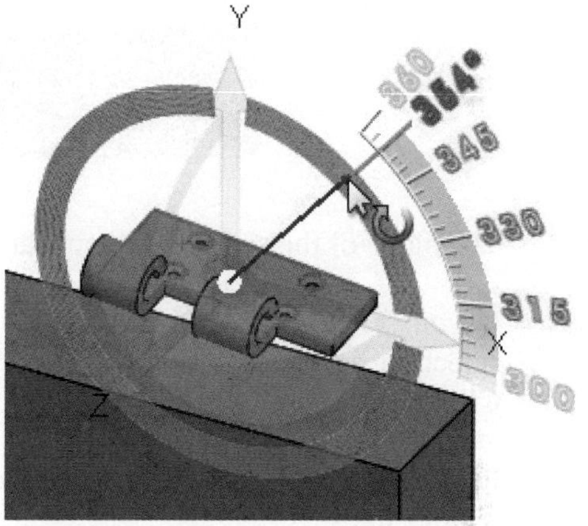

Experiment with moving and rotating the hinge component with the Triad Tool. Place the hinge in the approximate position and orientation shown in Figure 6.39. Click in the white space around the hinge to turn off the Triad Tool.

FIGURE 6.39

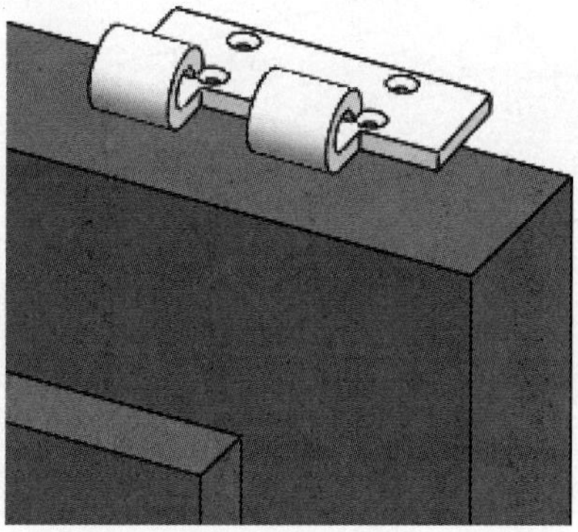

While it is not strictly necessary to place the component in its approximate position and orientation prior to defining mate instructions, doing so can remove ambiguity and simplify the establishment of the mates.

The Mate Tool will now be used to establish the first geometric relationship between the hinge and the hatch.

Click the Mate Tool from the Assembly group of the CommandManager, as shown in Figure 6.40.

This brings up the Mate PropertyManager.

FIGURE 6.40

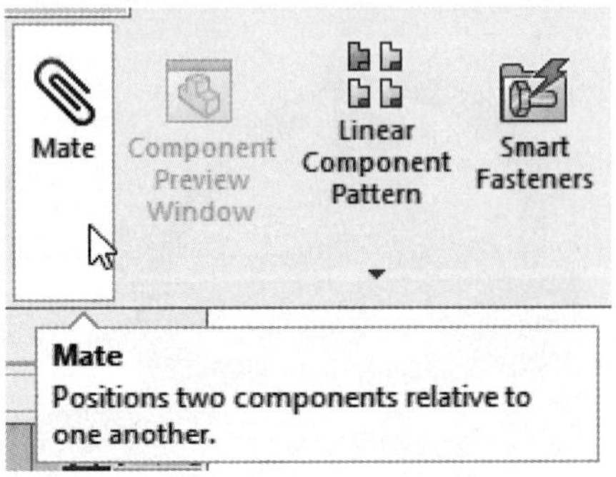

FIGURE 6.41

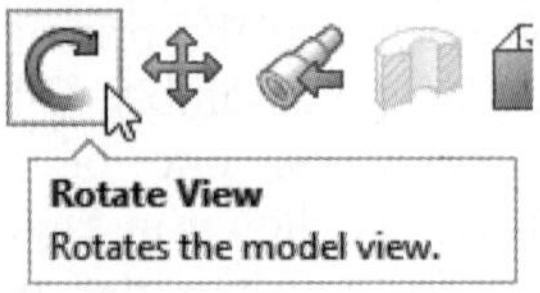

Using the Rotate View Tool, as shown in Figure 6.41, rotate the view so that the bottom face of the hinge can be seen. Press the Esc key to turn off the Rotate View Tool. Select the bottom face (Figure 6.42).

FIGURE 6.42

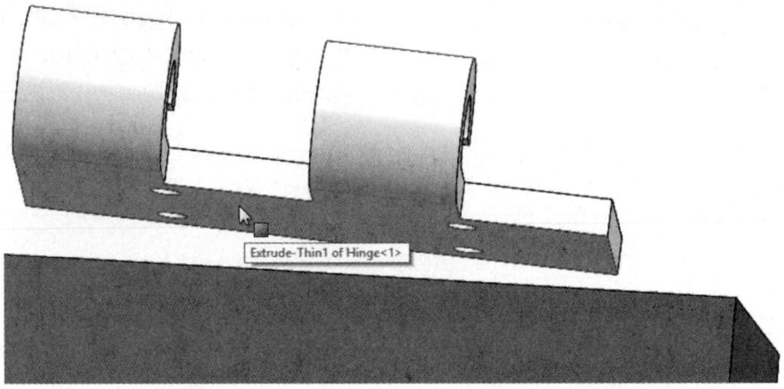

The name of the surface will appear in the highlighted area of the Mate dialog box.

If you select something incorrectly during a mate operation, you can clear the selection box at any time by right-clicking in the graphics area and selecting the Clear Selections option.

Return to a Trimetric View, and select the top face of the hatch component (Figure 6.43).

FIGURE 6.43

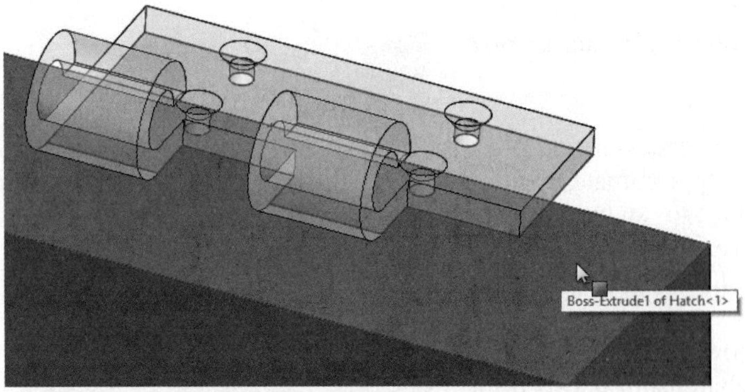

In the Mate dialog box, the selected faces are shown, and a list of possible mates is shown (**Figure 6.44**). By default, a coincident mate is selected when two flat surfaces are selected. This means that the two selected faces will be coplanar. The hinge will move to satisfy the selected mate configuration, as shown in **Figure 6.45**.

FIGURE 6.44

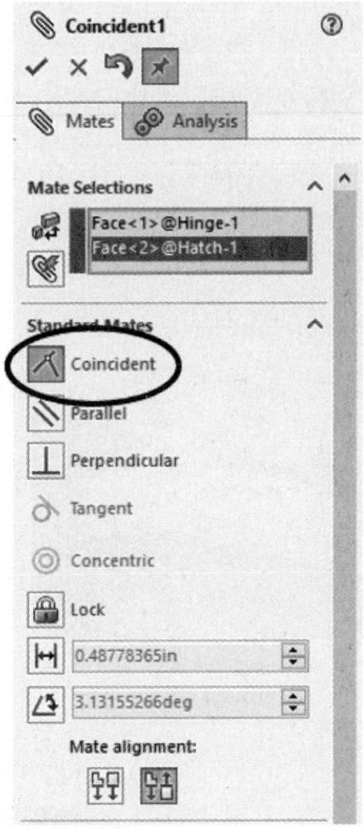

FIGURE 6.45

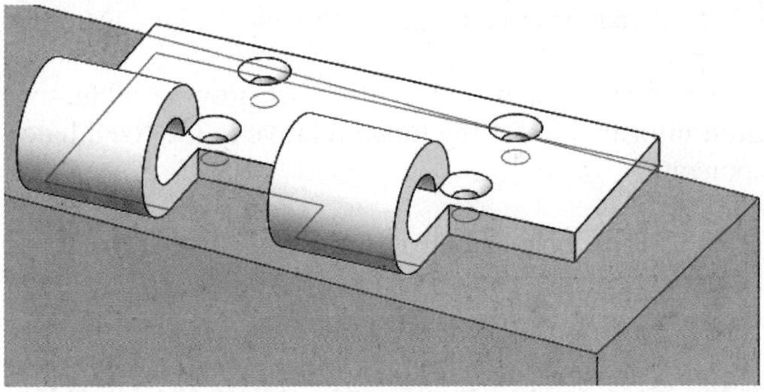

The Mate Alignment Tools in the dialog box are important because there is often more than one configuration that meets the specification of the selected mate. For example, the hinge could be upside-down and the selected mate could still be satisfied. An advantage of placing and orienting a component before applying mates is that the default alignments of the mates are usually correct. However, we will illustrate the use of the Mate Alignment Tools before applying the mate to the hatch and hinge.

FIGURE 6.46

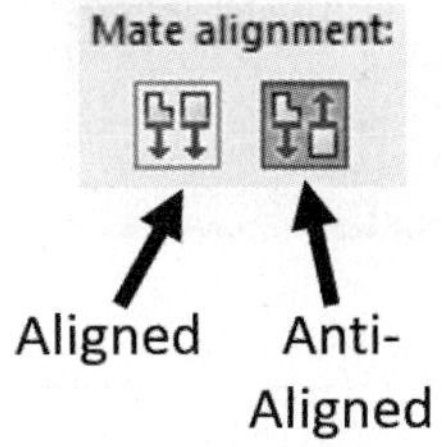

Toggle between the Aligned and Anti-Aligned Tools, as shown in Figure 6.46. Note that the hinge is flipped, as shown in Figure 6.47. Choose the tool which results in the proper alignment.

To better view the effect of the mate, switch to the Right View.

FIGURE 6.47

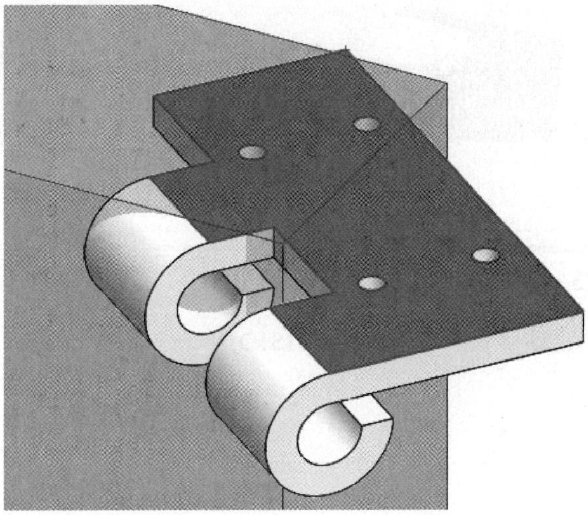

FIGURE 6.48

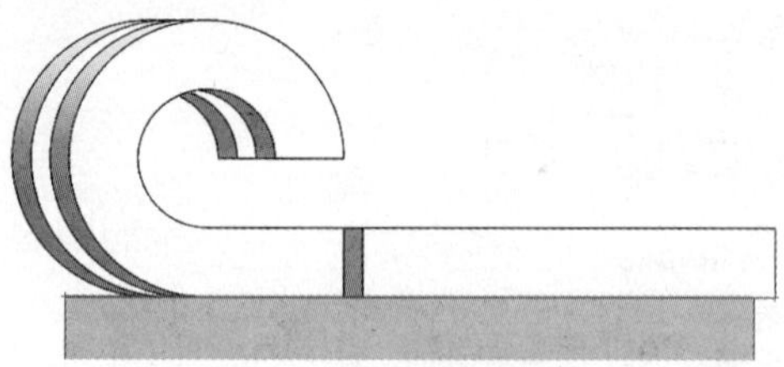

As shown in **Figure 6.48**, the bottom of the hinge and the top of the hatch are coplanar.

Click the check mark in the Mate dialog box or the pop-up box to apply the mate.

Note that the hinge component can be moved with respect to the fixed hatch component, but that the two mated faces remain coplanar. That is the geometric effect of the mate.

FIGURE 6.49

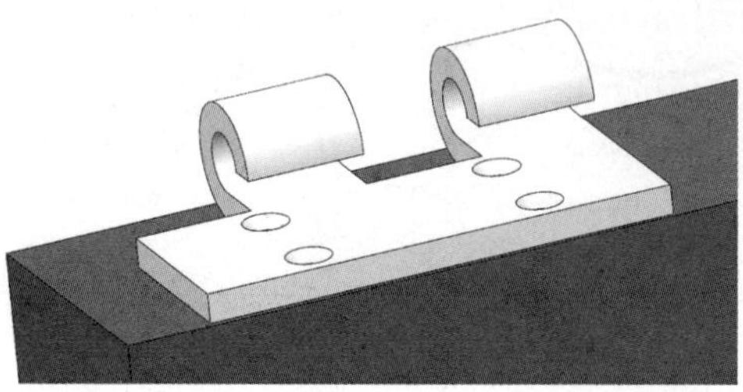

A second mate will now be added to provide additional location information for the hinge relative to the fixed hatch component.

Rotate the view orientation so that the back face of the hatch is visible. Click and drag the hinge to the approximate position shown in Figure 6.49.

FIGURE 6.50

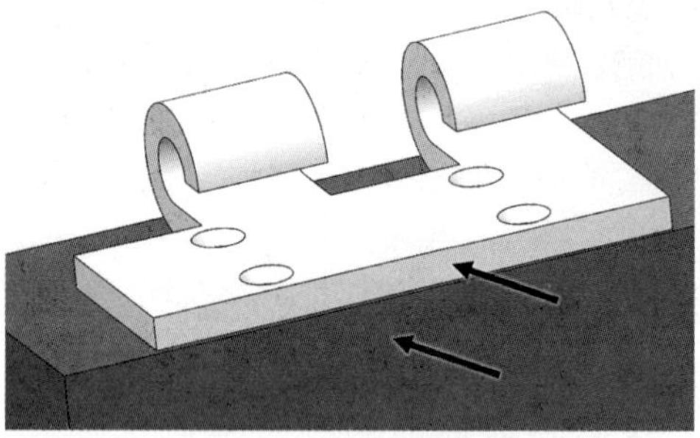

With the Mate dialog box still open (select the Mate Tool if you closed it accidentally), select the two faces shown in Figure 6.50. Click the check mark to apply this second coincident mate.

The result of the mate is shown in **Figure 6.51**.

Select the two faces shown in Figure 6.52.

In this case, we do not want to apply the default coincident mate. Rather, we want these faces to be a specific distance apart.

FIGURE 6.51

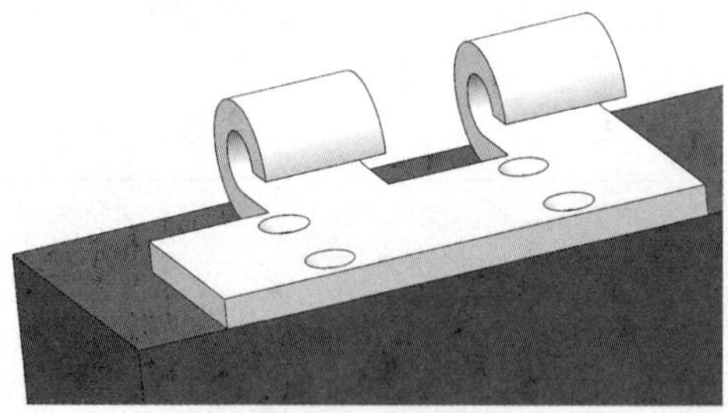

FIGURE 6.52

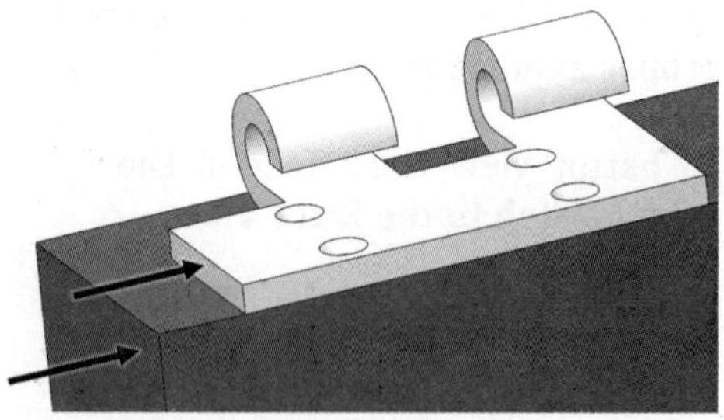

FIGURE 6.53

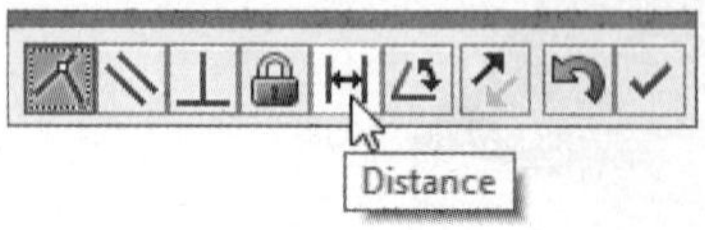

FIGURE 6.54

Select a Distance mate from the Mate dialog box or the pop-up box, as shown in Figure 6.53. Set the distance to 1 inch, as shown in Figure 6.54.

A distance mate has two possible configurations. If you click the Flip Dimension button, then the other configuration is selected, as shown in **Figure 6.55**.

FIGURE 6.55

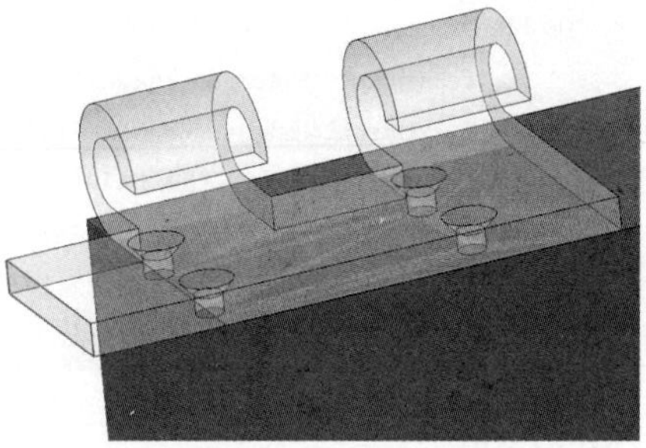

With the mate defined correctly, as in Figure 6.56, click the check mark to apply the mate.

Click Esc to close the Mate dialog box. Try to move the hinge by clicking and dragging it.

A message appears, as shown in **Figure 6.57**, stating that the part cannot be moved. The three mates that we have applied have completely defined its position and orientation relative to the fixed hatch.

FIGURE 6.56

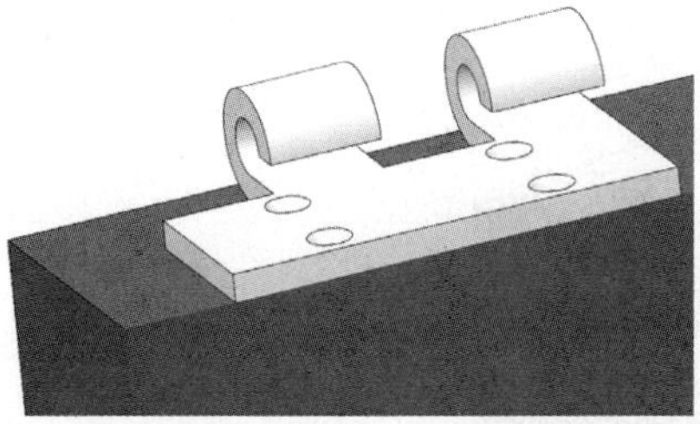

FIGURE 6.57

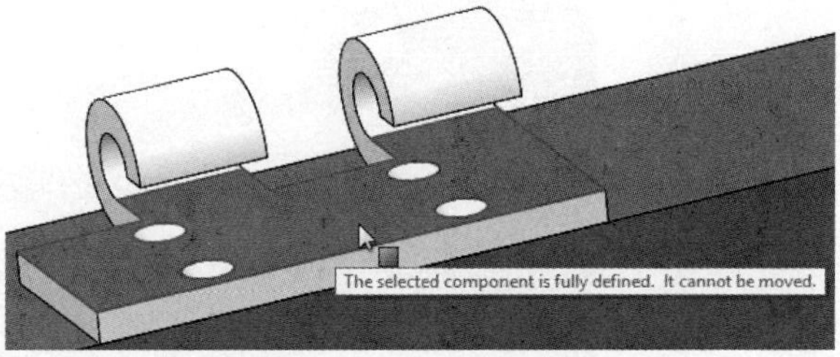

A second hinge will now be added to the assembly. Since the second hinge will be identical to the first, no new component needs to be created. A second instance of the first hinge can simply be added to the assembly.

Return to the Trimetric View. Select the Insert Component Tool, and browse to the Hinge file. Place the hinge into the assembly window, as shown in Figure 6.58.

Using a procedure similar to the one outlined previously in this chapter, add two coincident mates and a distance mate to fully constrain the second hinge in the position shown in Figure 6.59.

FIGURE 6.58

FIGURE 6.59

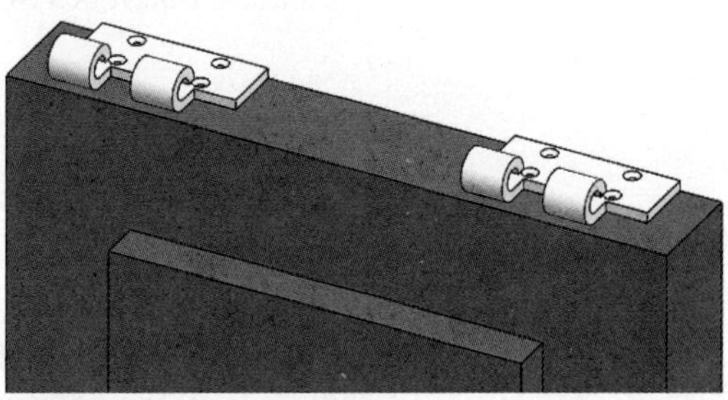

Once defined, mates can be easily modified. For instance, assume that a redesign specified that the location of the second hinge should be 4 inches from the mated edge rather than 1 inch.

FIGURE 6.60

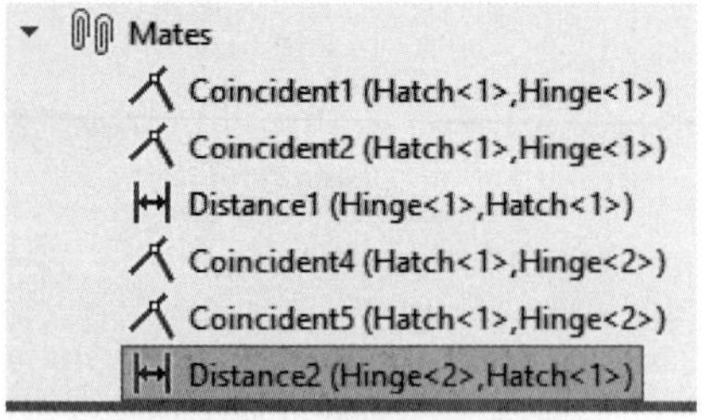

Click the arrow next to the Mates entry in the FeatureManager. Locate the distance mate associated with the second hinge (called Distance2), and right-click on the mate name, as shown in Figure 6.60. Select the Edit Feature option.

The PropertyManager associated with this mate will reopen, allowing for editing of the mate.

Change the value of the distance to 4.00 inches, as shown in Figure 6.61. Accept the change by clicking the check mark.

FIGURE 6.61

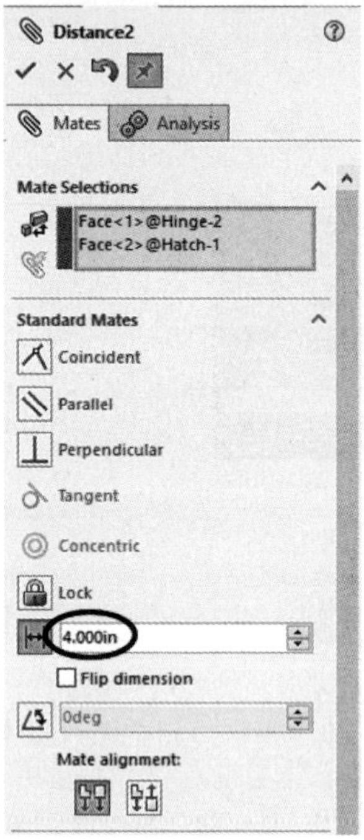

The resulting position of the second hinge is shown in **Figure 6.62**.

FIGURE 6.62

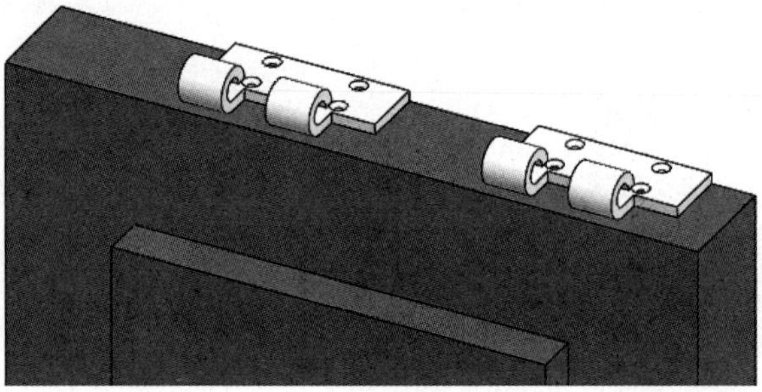

In this way, a parameter of the mate can be readily redefined. Since we will want to continue with the 1-inch value rather than the modified 4-inch value, we will revert to the previous value.

Press the Esc key to close the Mate dialog box. Click the Undo Tool, shown in Figure 6.63, to revert to the 1-inch dimension. Save the assembly file using the name "Door," and close the file.

FIGURE 6.63

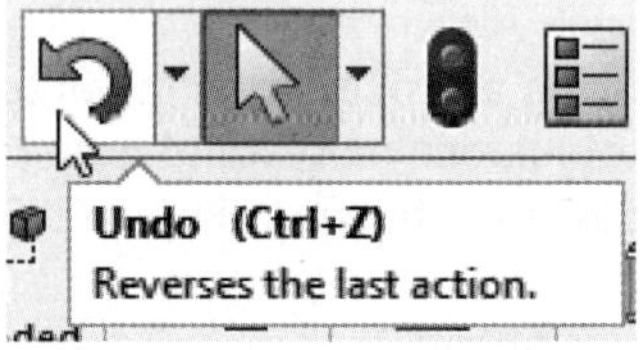

Note that the new file has the extension .SLDASM, indicating that it is a SOLIDWORKS assembly file.

6.3 Adding Features at the Assembly Level

Assembly models are created from pre-existing part files. However, it is sometimes desirable to make modifications to the parts during the creation of assemblies. In this section, the addition of features at the assembly level will be described.

In our example, based on our design intent, we would like to produce a hole in the hatch component corresponding to each of the four countersunk holes in the hinge component. Therefore, we will use the existing holes in the hinge to establish relations that precisely locate our new holes to match the holes in the hinge. The first step will be to create points on the top surface of the hatch where the holes will be added.

FIGURE 6.64

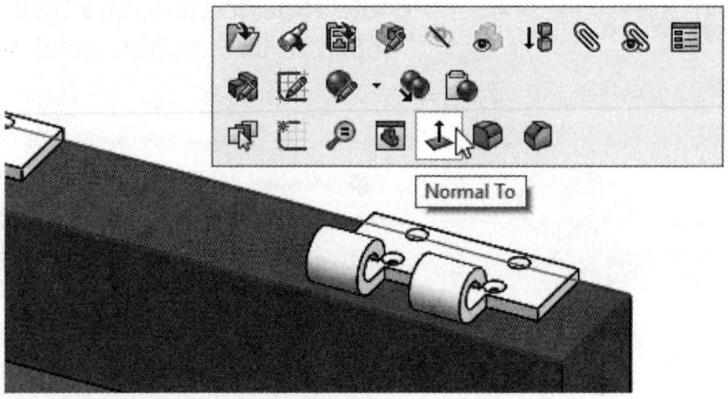

In the Door assembly, select the top surface of the hatch and choose the Normal To View, as shown in Figure 6.64. Zoom in on the right-hand hinge. Select the Point Tool from the Sketch group of the CommandManager, as shown in Figure 6.65 (if you receive a warning, you can ignore it). Add points at the center of each of the holes in the hinge, as shown in Figure 6.66. (It may be necessary to hold the cursor momentarily over the perimeter of the hole to "wake up" its center mark.) Exit the sketch, as shown in Figure 6.67.

FIGURE 6.65

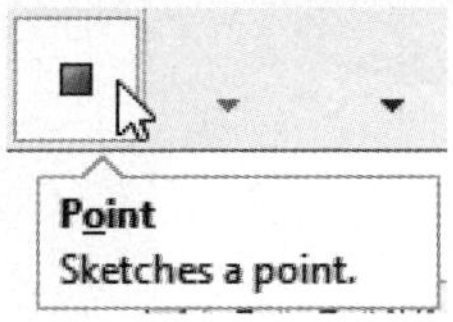

FIGURE 6.66

It will be easier to add the holes in the proper position if the hinge is hidden. Otherwise, the Hole Wizard may not find the proper surface to define the hole.

Select the first instance of the hinge from the FeatureManager. From the pop-up menu, select Hide Components, and shown in Figure 6.68.

FIGURE 6.67

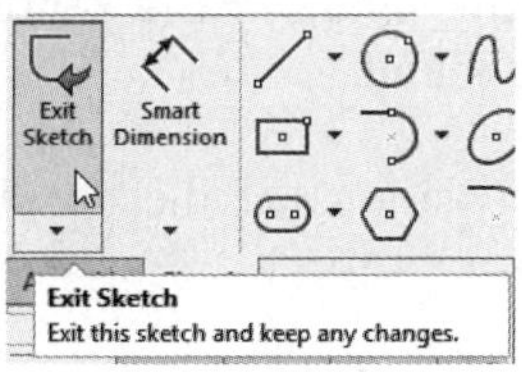

FIGURE 6.68

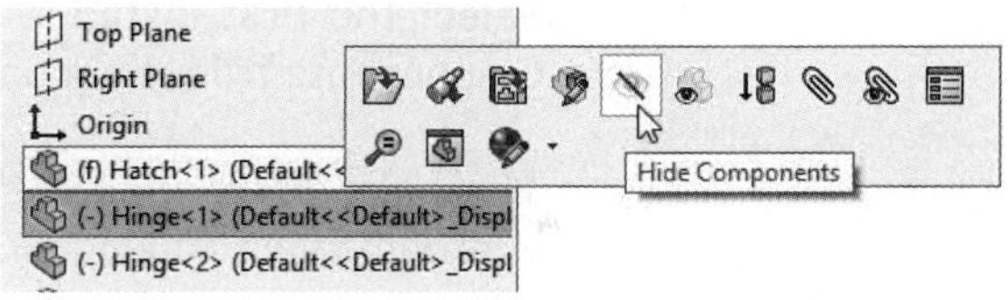

The hinge is now hidden from view, as shown in Figure 6.69. The points of the sketch just created should be visible. If they are not, select View Sketches from the Hide/Show Items menu of the Heads-Up View Toolbar.

Select the Assembly group of the CommandManager. Click on the Assembly Features Tool, which reveals a menu of assembly-level features, as shown in Figure 6.70. Select the Hole Wizard.

FIGURE 6.69

FIGURE 6.70

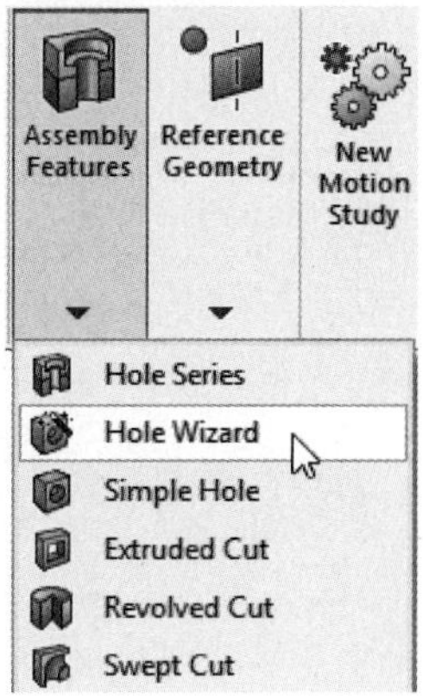

Four *pilot holes* will be added to the hatch, matching the positions of the holes in the hinges. (Note: the holes in the hatch are drilled undersized to allow the screws to thread into the hatch.)

FIGURE 6.71

In the Hole Specification PropertyManager, with the Type tab selected, set the Hole Type to Hole, the Size to 7/64 (inches), the End Condition to Blind and the depth to 0.75 inches, as shown in Figure 6.71.

FIGURE 6.72

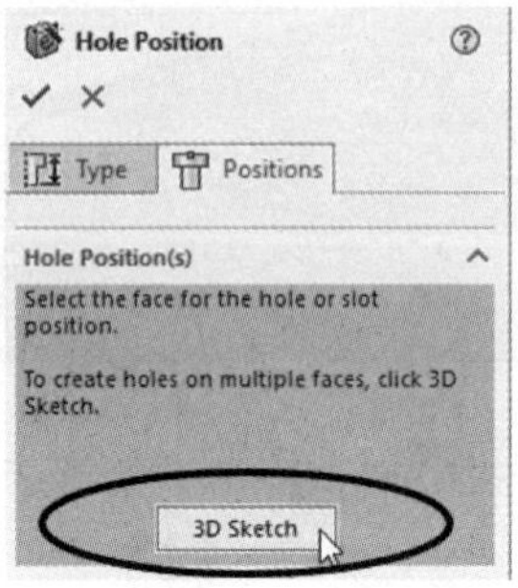

Click on the Positions tab in the PropertyManager, and the Hole Position PropertyManager will prompt you for the position of the holes. Either select the top hatch surface, or click the 3D Sketch button, as shown in Figure 6.72. Click on each of the points created earlier, as shown in Figure 6.73, and click the check mark to create the holes.

FIGURE 6.73

Select the first instance of the hinge in the FeatureManager, and select Show Components from the pop-up menu.

The holes added match the positions of the holes in the hinge, as shown in a wireframe view in Figure 6.74.

Repeat the entire operation on the other hinge.

Hide the point sketches by right-clicking on each in the FeatureManager and selecting Hide from the menu.

Save the changes to the assembly.

FIGURE 6.74

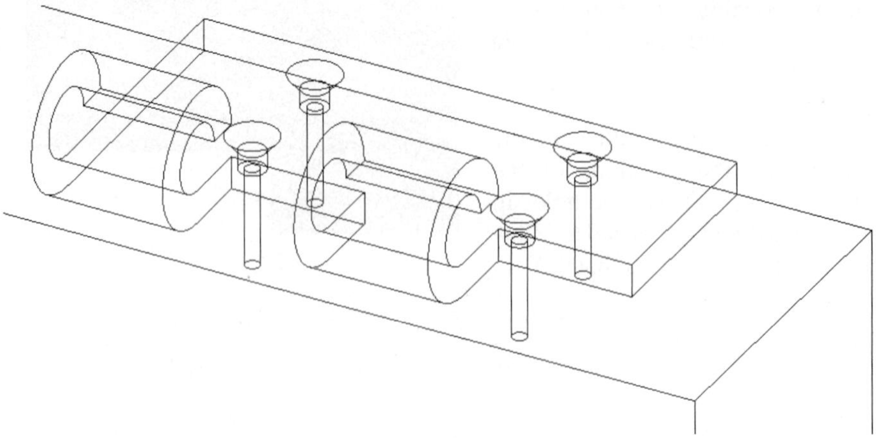

6.4 Adding Fasteners to the Assembly

FIGURE 6.75

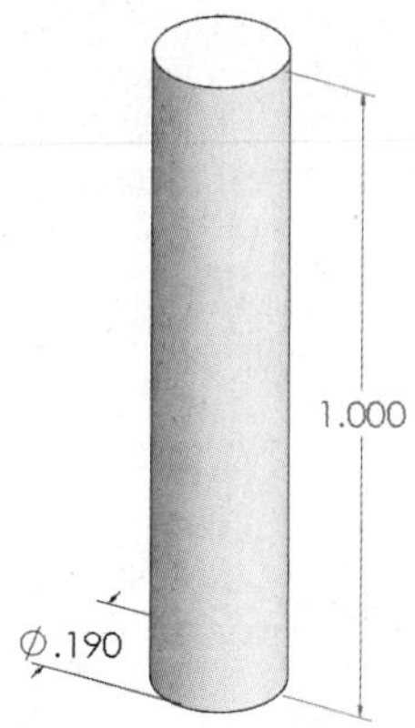

We will now add wood screws to the assembly model. Commercial and educational licenses of SOLIDWORKS have an add-in feature called SmartFasteners that can be used to intelligently insert appropriate mechanical fasteners into an assembly. Some limited-license SOLIDWORKS products do not contain this feature, so we will create a wood screw part model for this exercise.

Open a new part. In the Top Plane, sketch and dimension a 0.190-inch diameter circle, centered at the origin. Extrude the circle upward to a height of 1.00 inches, as shown in Figure 6.75.

Switch to the Front View. In the Front Plane, sketch and dimension the three lines and vertical centerline shown in Figure 6.76. Be sure to add the 0.365-inch dimension as a diameter by clicking on the corner point and the centerline, and dragging the dimension to the left of the centerline. Select the Revolved Boss/Base Tool from the Features group of the CommandManager, as shown in Figure 6.77. Click the check mark to accept the 360-degree default revolution.

FIGURE 6.76

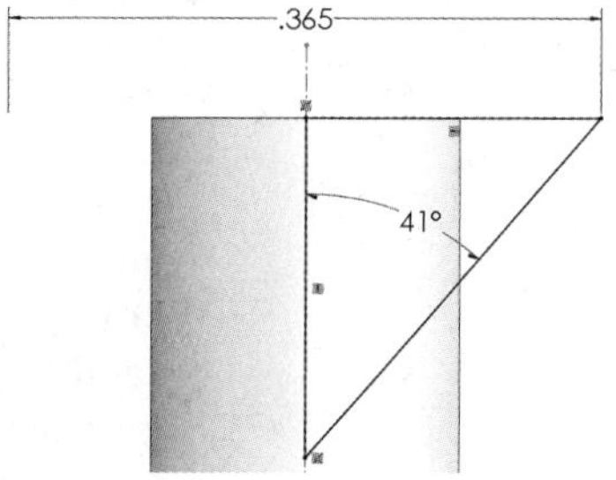

FIGURE 6.77

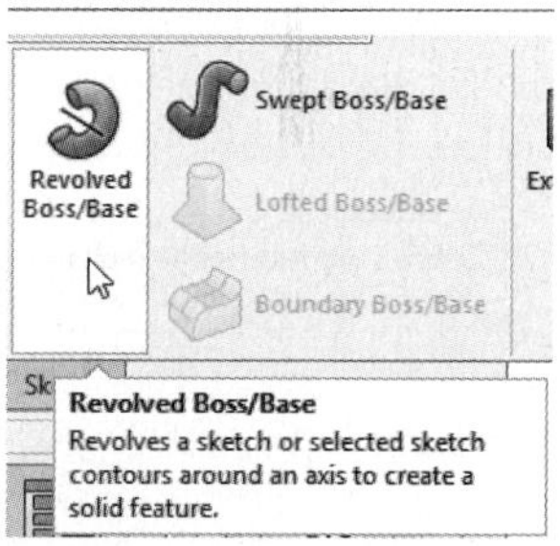

DESIGN INTENT Part-Level and Assembly-Level Features

Since the holes that we are adding in the tutorial affect only the hatch part, we could have added them to the part file. Instead, we have added them at the assembly level, and the holes do not appear in the part model (verify this by opening the hatch file after adding one or more of the holes to the assembly). There are two types of features: *part-level* features, and *assembly-level* features. Where we apply the features should reflect the actual manufacturing and assembly processes. For example, if the holes are predrilled into the hatch before the assembly with the hinges, then the holes should be added to the part model and are considered part-level features. If the holes are added after placing the hinge on the hatch and using the hinge's holes to locate the drilled holes in the hatch, then these are assembly-level features.

The difference between part-level and assembly-level features is especially important when creating detailed drawings. In this example, the part drawings would contain all information needed to manufacture and/or inspect the hinge and hatch parts. The assembly drawing would show which parts make up the assembly (in a *Bill of Materials*), and would contain only the dimensions necessary to assemble them. In this case, the dimensions to locate the hinges and the hole definitions would be defined on the assembly drawing (we will learn how to make an assembly drawing in Chapter 8).

FIGURE 6.78

The revolved screw head is shown in Figure 6.78.

At the bottom of the screw, sketch and dimension the lines and vertical centerline shown in Figure 6.79 in the Front Plane.

FIGURE 6.79

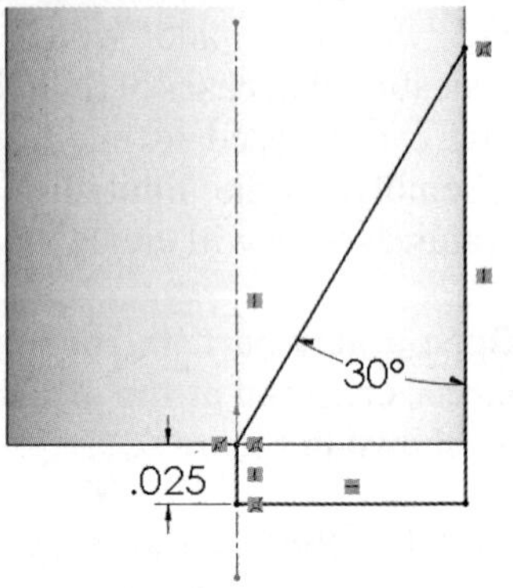

Note that the "cutting tool" profile extends below the bottom of the part. An error is encountered when the profile comes to a single point on the centerline.

FIGURE 6.80

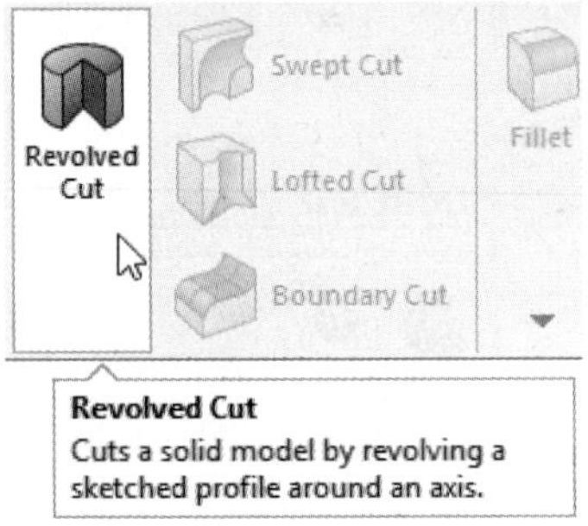

Select the Revolved Cut Tool from the Features group of the CommandManager, as shown in Figure 6.80. Click the check mark to accept the 360-degree default revolution.

FIGURE 6.81

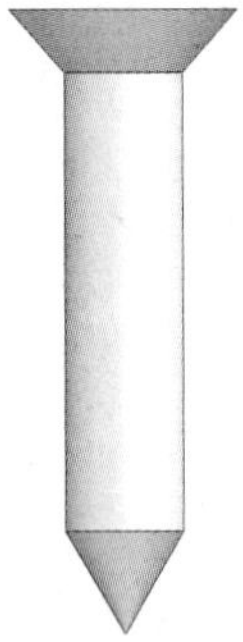

The screw body is shown in Figure 6.81.

Open a sketch on the screw head and sketch a rectangle centered at the origin. The rectangle should extend beyond the edges of the screw by 0.05 inches (a construction line tangent to the circle will be helpful in establishing this dimension), and should be 0.055 inches tall, as shown in Figure 6.82. Extrude a cut 0.040 inches deep to create the slot in the screw head, as shown in Figure 6.83. Save this part file as "Screw."

FIGURE 6.82

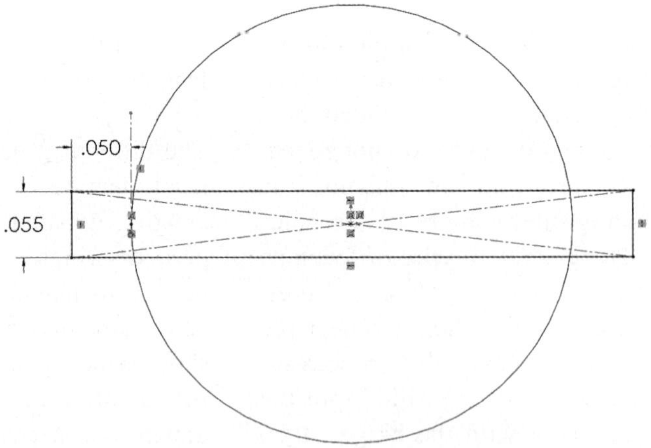

FIGURE 6.83

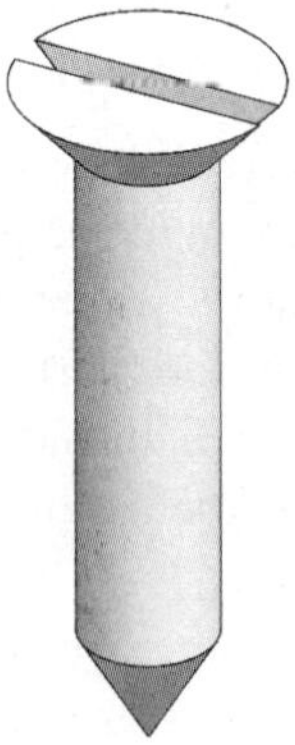

FIGURE 6.84

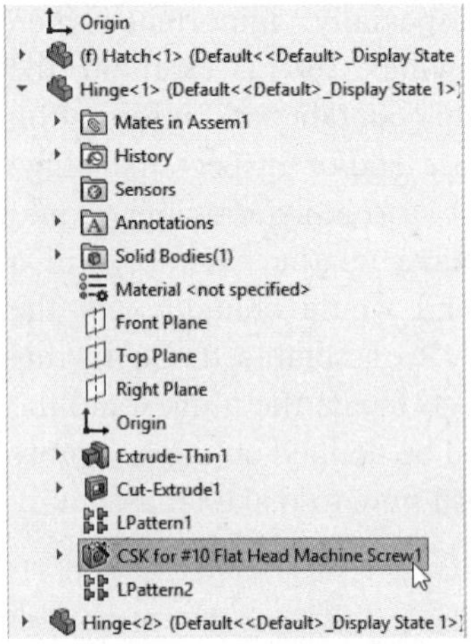

Switch back to the door assembly. Expand the definition of the first hinge in the FeatureManager, and click on the first hole created in the hinge, as shown in Figure 6.84. This will select and highlight the hole. Note the position of this first hole, as it will be used in the next assembly step.

By placing our first screw in the first hole created, we can use the pattern that was previously defined for the holes to place the other screws, rather than placing them one at a time or defining a new pattern.

Insert a screw into the assembly, placing it approximately in the position shown in Figure 6.85.

FIGURE 6.85

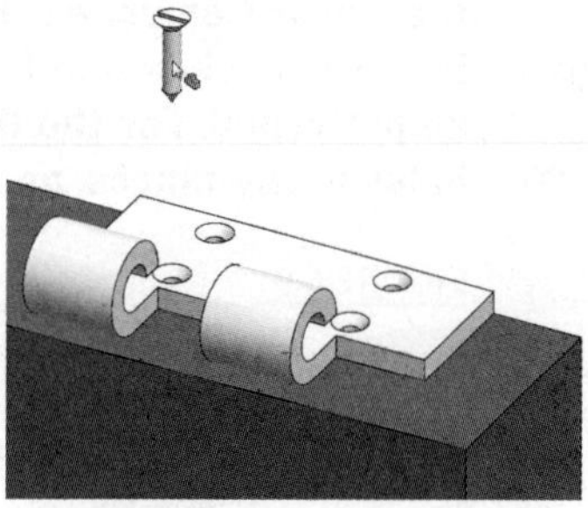

Select the Mate Tool. Select the conical surface of the screw head, as shown in Figure 6.86. Select the conical face of the first hole in the hinge, as shown in Figure 6.87.

FIGURE 6.86

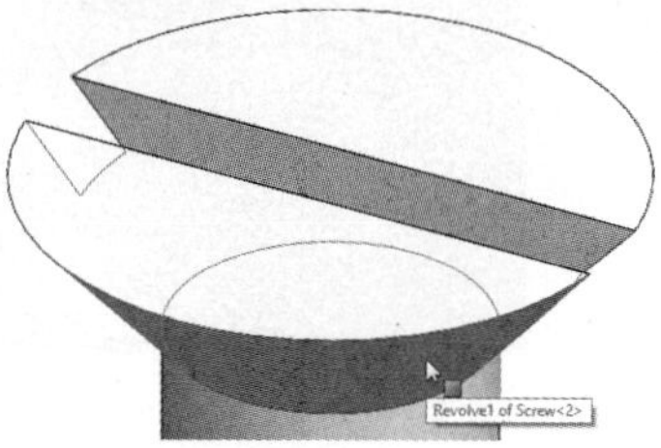

Adding a coincident mate to the two conical surfaces will completely locate the screw in the hole.

In the Mate PropertyManager, set the type of mate to Coincident, as shown in Figure 6.88. Click the check mark to apply the mate, and click the check mark again to close the Mate PropertyManager.

FIGURE 6.87

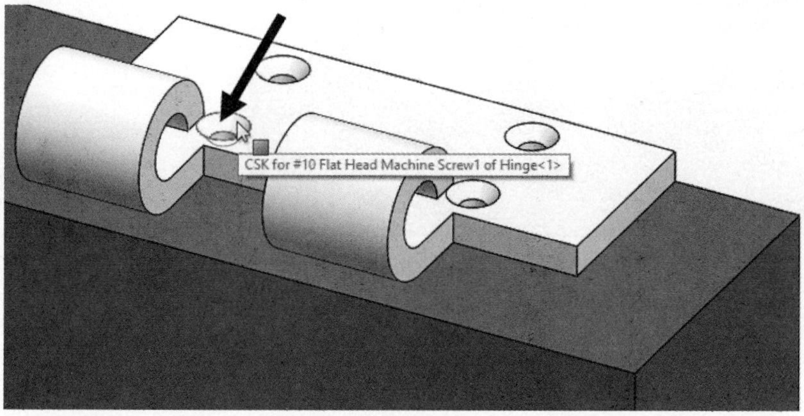

FIGURE 6.88

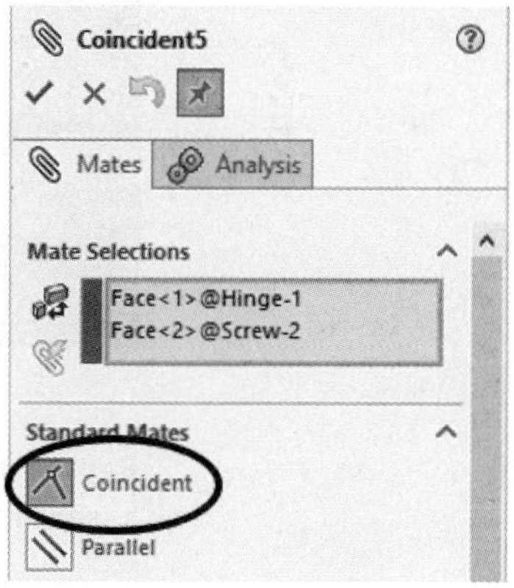

The screw is shown in its final position in Figure 6.89. Rather than add the remaining screws individually, we will create a pattern.

FIGURE 6.89

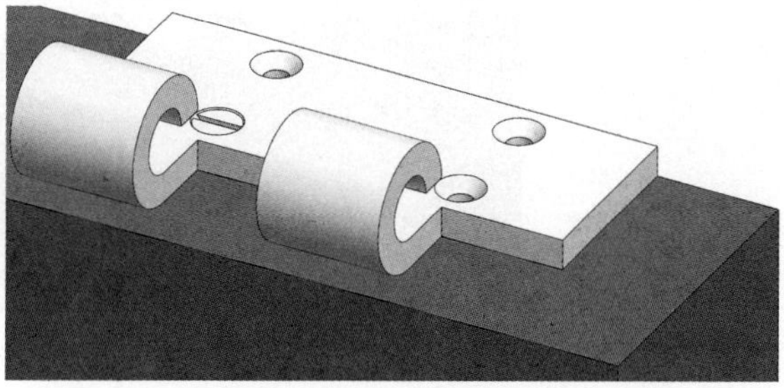

Since the pattern of the screws will follow that of the holes in the hinges, we will use that pattern to create a *feature-driven* pattern.

FIGURE 6.90

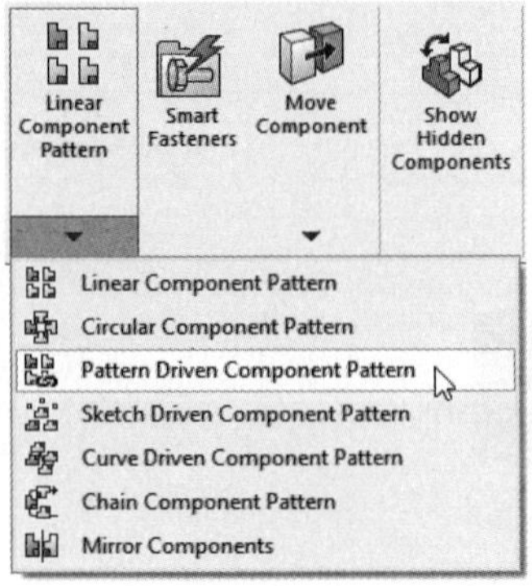

From the Assembly group of the CommandManager, click the arrow under the Linear Component Pattern Tool and select Pattern Driven Component Pattern, as shown in Figure 6.90. Select the screw as the component to be patterned. For the Driving Feature or Component, click on one of the holes in the hinges, as shown in Figure 6.91.

FIGURE 6.91

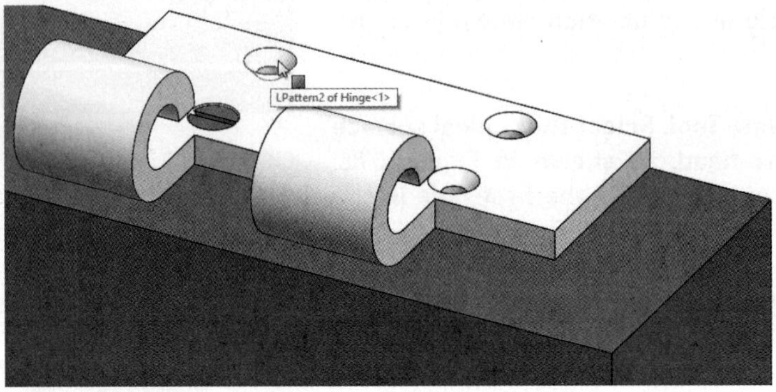

Click the check mark to complete the pattern, which is shown in Figure 6.92.

FIGURE 6.92

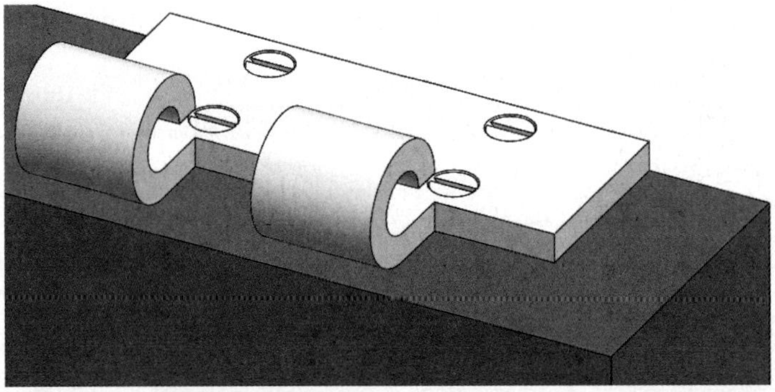

Repeat for the other hinge, as shown in Figure 6.93. Save the changes to the assembly.

FIGURE 6.93

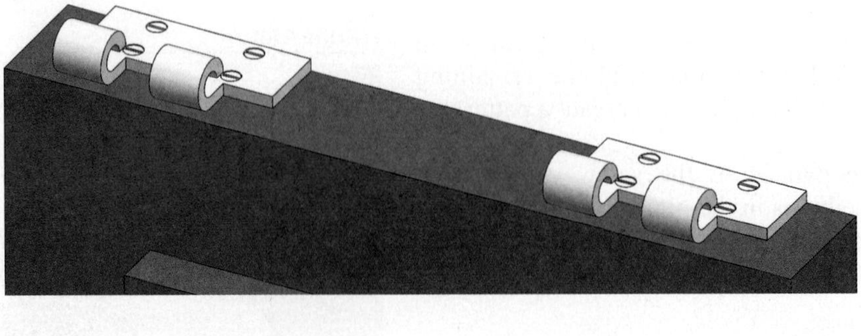

6.5 Creating an Exploded View

Exploded views are often used to visualize assemblies. In this section, an exploded view of the door assembly will be created.

Select the Exploded View Tool from the Assembly group of the CommandManager, as shown in Figure 6.94.

FIGURE 6.94

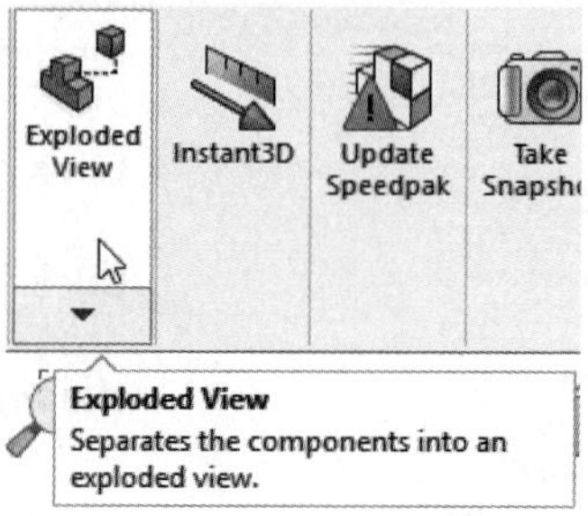

Click on each of the screws to select them. If you accidentally select one of the hinges, click on it again to cancel its selection.

These components will appear in the Settings box in the PropertyManager. A manipulator handle will appear in the model window, as shown in **Figure 6.95**. This allows for "drag and drop" explosion of assembly components.

FIGURE 6.95

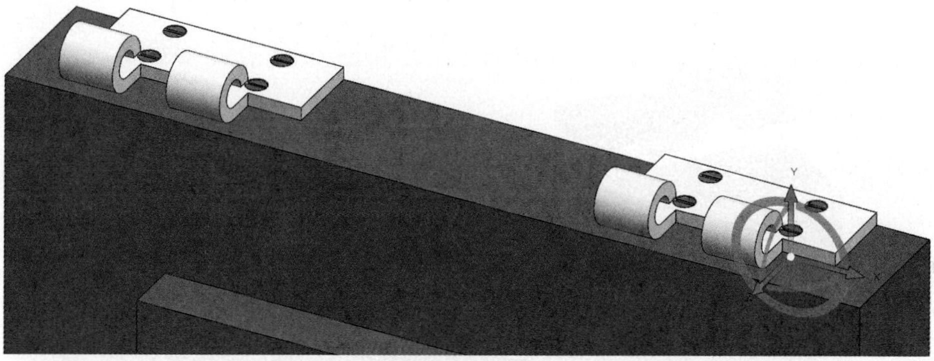

Click and hold on the manipulator handle that points in the Y direction, and drag the fasteners up to the desired location, as shown in Figure 6.96. Release the mouse button to place the components.

FIGURE 6.96

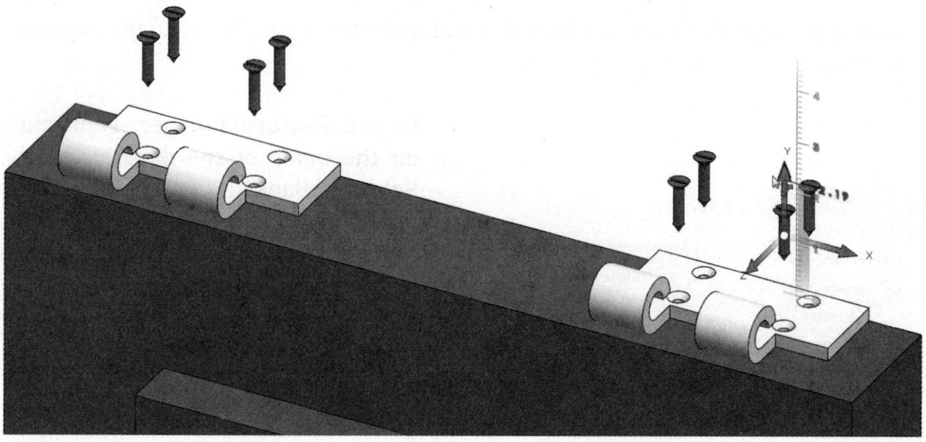

FIGURE 6.97

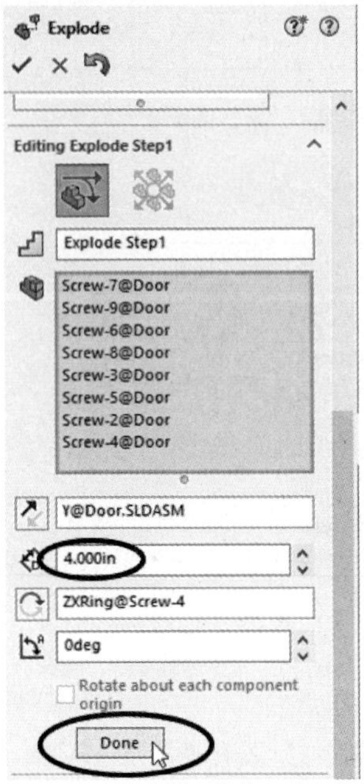

Note that this explosion step, denoted Explode Step1, now appears in the Explode Steps box of the PropertyManager. We can make modifications to this step by using the PropertyManager entries.

Change the distance to 4 inches, as shown in Figure 6.97. Click Done to end Explode Step1.

The Explode PropertyManager will remain open on the screen. The selections for the second explode step can now be made.

Click on each of the hinges to select them.

Click and hold on the manipulator handle that points in the Y direction, and drag the hinges up to the desired location, as shown in Figure 6.98. Release the mouse button to place the components.

FIGURE 6.98

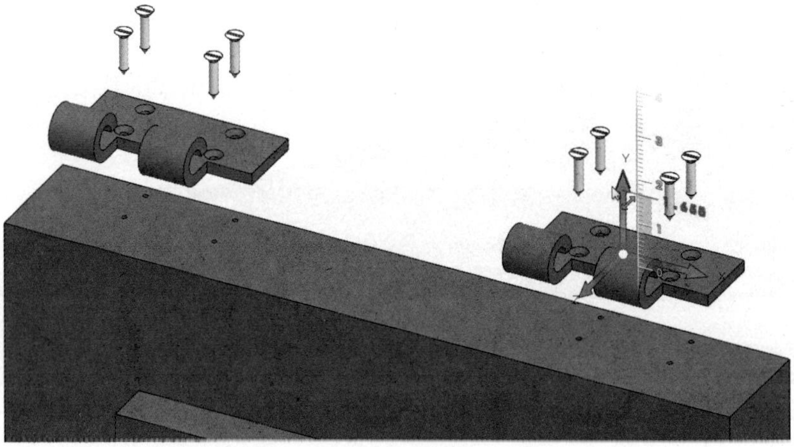

Change the distance to 2 inches. Click Done to end the explode step.

Click the check mark to complete the exploded view. Save the modified assembly.

Now that the exploded view has been defined, you may toggle between the exploded and collapsed views at any time.

FIGURE 6.99

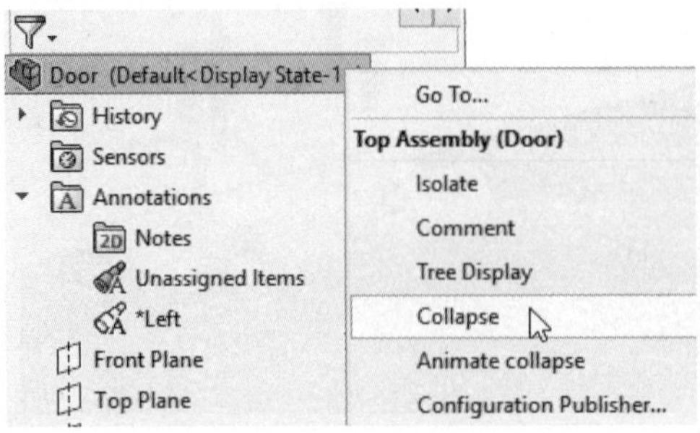

In the FeatureManager, right-click on the name of the door assembly. Select Collapse (Figure 6.99).

The exploded view will be toggled to the collapsed view.

DESIGN INTENT **Manufacturing Considerations**

The placement of the holes in the hatch and the hinges can be accomplished in one of three ways, and each way simulates a different approach to the manufacturing of the door assembly. First, the holes could be added to the hatch at the part level, just as they are in the hinge part. In this method, at the final assembly step, the hatch and the hinges would be received with the holes in place. The hinges would be aligned with the holes in the hatch, and the fasteners inserted. In this case, the holes would be classified as part-level features. In the second method, neither the hinge nor the hatch would have holes before arriving at the assembly step, where the holes would be classified as assembly-level features. In the third method, the hatch would be received without holes, and the hinges received with the holes pre-drilled. The holes in the hatch would be drilled to match the holes in the hinges. This is the method modeled in the tutorial. In this case, we have combined a part-level feature (the holes in the hinges) with assembly-level features (the holes in the hatch). Although the final assembly would look the same regardless of which method is chosen, the definition of where the holes are added can be an important consideration in the actual manufacturing process, and the method used to create the solid model should represent the actual manufacturing steps.

Right-click on the name again, and select Explode, as shown in Figure 6.100.

FIGURE 6.100

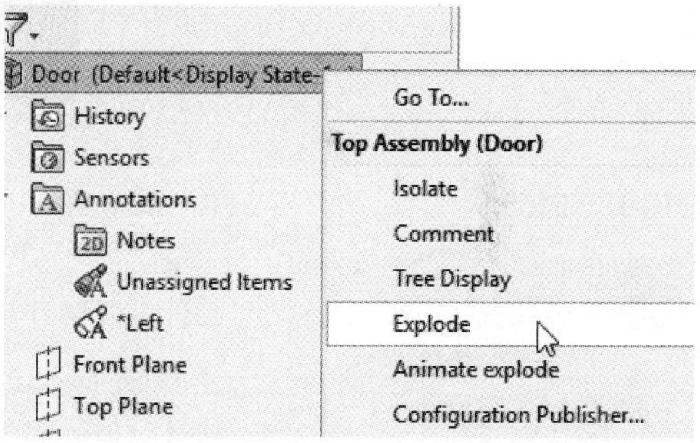

This will toggle the display back to the exploded state.

Sketch lines can be added to the exploded view to show how the parts fit together. Although they are not really necessary for a simple assembly such as our door, they can be very helpful in more complex assemblies.

Select the Explode Line Sketch Tool from the Assembly group of the CommandManager, as shown in Figure 6.101.

A 3-D sketch will be opened. The Route Line PropertyManager will open, as shown in **Figure 6.102**.

FIGURE 6.101

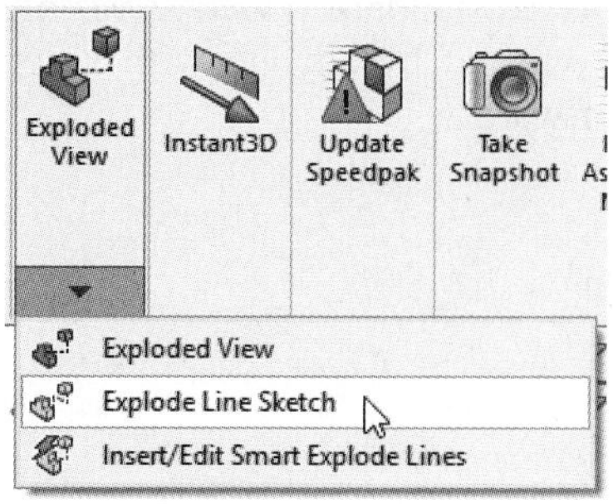

FIGURE 6.102

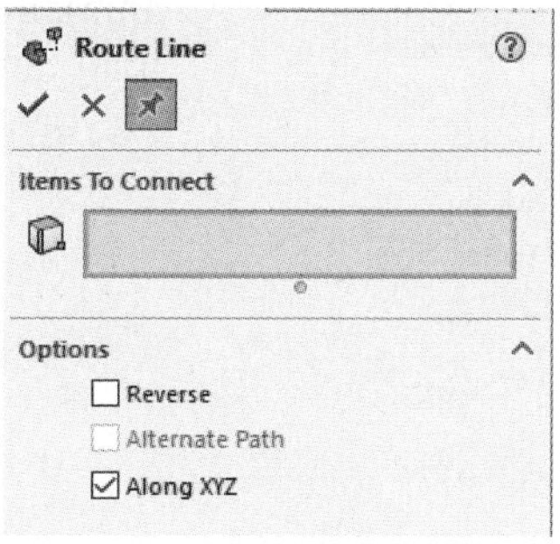

FIGURE 6.103

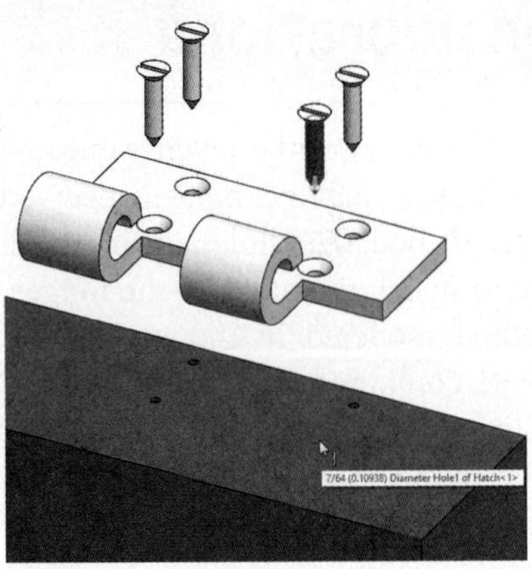

Click on the lower half of the cylindrical surface of one of the fasteners. Then click on the edge of the corresponding hole in the hatch, as shown in Figure 6.103. Click the check mark to create the sketch line.

Repeat for the other fasteners, and close the sketch to complete the operation.

The completed explode lines are shown in **Figure 6.104**. Clicking the ConfigurationManager tab shows the 3-D sketch containing the explode lines with the associated exploded configuration, as shown in **Figure 6.105**.

FIGURE 6.104

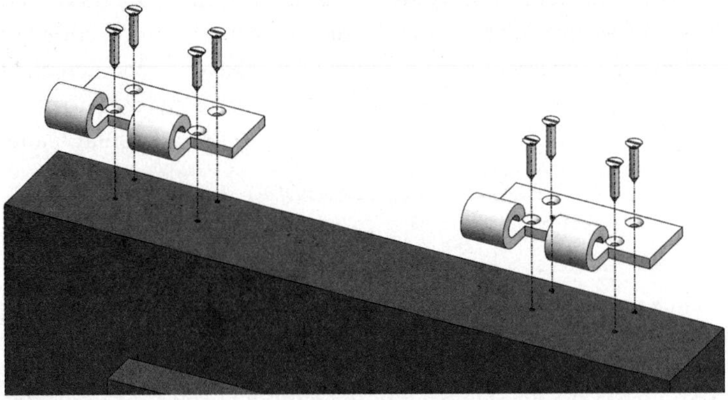

FIGURE 6.105

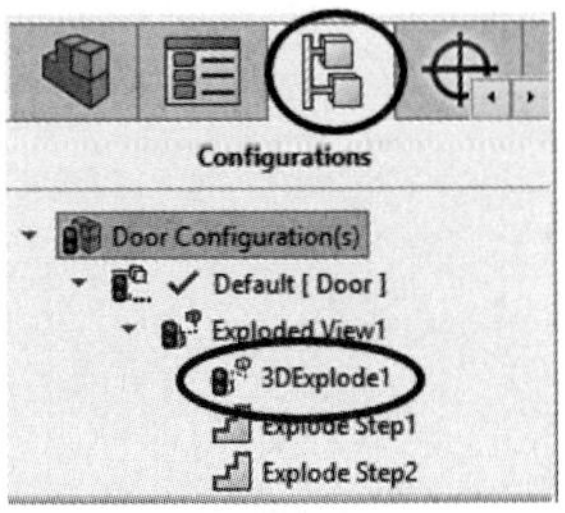

FIGURE 6.106

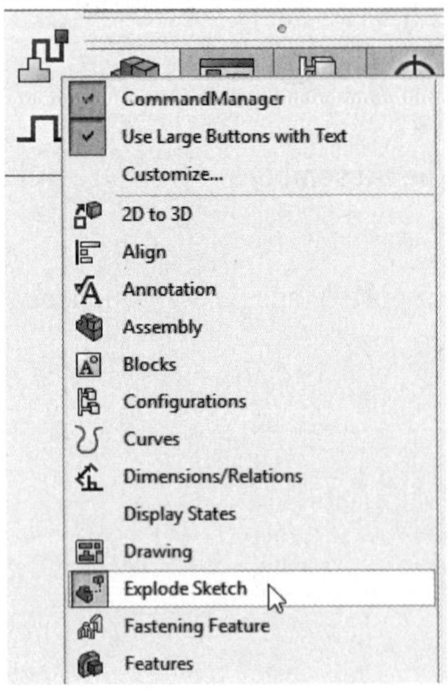

Right-click on the CommandManager, and clear the entry for Explode Sketch to close the Route Line Tool, as shown in Figure 6.106.

While this exploded configuration offers a clear picture of how the components are assembled together, we may wish to produce an animation file to demonstrate the explosion step-by-step. Not only can we easily produce this animation using SOLIDWORKS, we can export the animation as a standard video AVI file that can be viewed by anyone with video playback software.

Switch back to the FeatureManager. Right-click on the Door and select Animate Collapse, as shown in Figure 6.107.

FIGURE 6.107

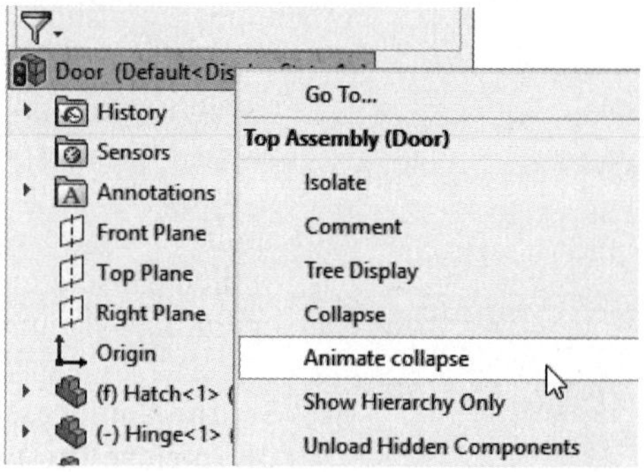

The Animation Controller will initiate (shown in Figure 6.108), and the animation of the collapse/explosion will begin. The Animation Controller contains buttons to control the play, playback mode, and play speed of the on-screen animation.

Experiment with the Play/Pause, and Playback Mode buttons. Make sure the Playback Mode is set to Normal, as shown in Figure 6.109.

FIGURE 6.108

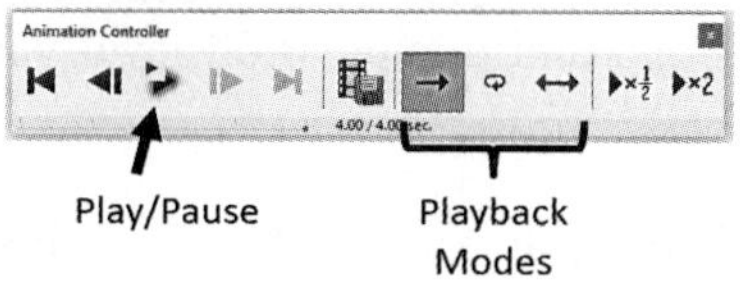

FIGURE 6.109

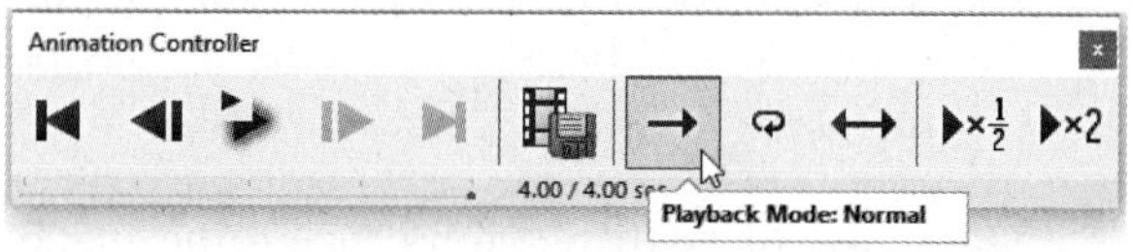

In order to view this animation using the Animation Controller, access to the assembly model, all part models used in the assembly, and the SOLIDWORKS software package are required. To allow someone without access to the models and software to view this animation, we can save it as a video AVI file.

In the Animation Controller, click the Save Animation button, as shown in Figure 6.110.

FIGURE 6.110

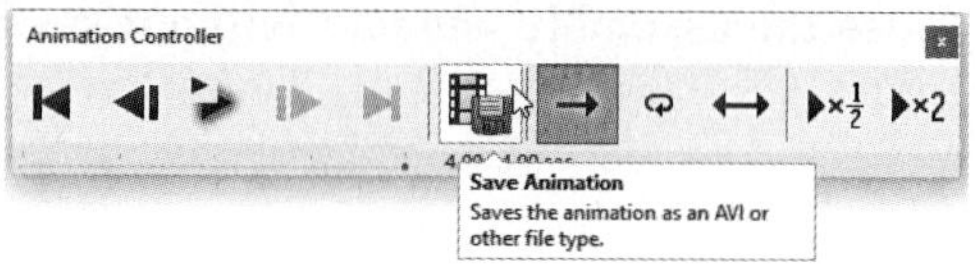

The Save Animation to File dialog box will appear, as shown in Figure 6.111. This allows us to select the appropriate directory and filename for a video file. It also allows us to set some parameters that control file size and image quality, such as the number of frames per second to be saved when we create our video file. For this application, we will accept the default values.

Browse to the appropriate file location, and click Save.

FIGURE 6.111

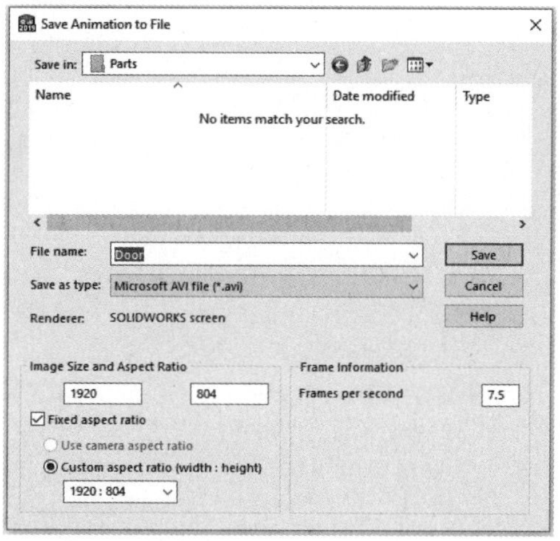

The Video Compression dialog box will appear, as shown in **Figure 6.112**. This dialog box allows us to select the video resolution and other parameters of the video file. Higher-resolution images will result in larger file sizes. For this application, we will not use Key Frames (which cause animation errors with some operating systems).

FIGURE 6.112

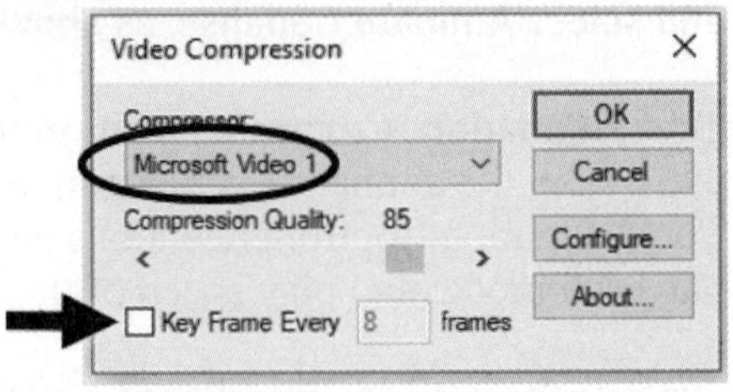

Set the Compressor to Microsoft Video 1, and clear the Key Frame checkbox. Click OK to close the Video Compression dialog box.

The animation will be replayed in the graphics window. While this is occurring, the video file will be written at the frames per second and compression level we specified.

The video file will be stored using the name and location we specified. In this case, the file will be name Door.avi. This file is a stand-alone video file; it is not linked in any way to your SOLIDWORKS part or assembly files. As such, it can be viewed by anyone with standard video playback software, such as Windows Media Player or RealPlayer. Do note, however, that all associativity with the model is lost; subsequent changes to the assembly in SOLIDWORKS will not affect this video file. If changes are desired, a new video file must be created.

Close the assembly window, without saving the last changes made.

PROBLEMS

P6.1 Create a 6.5-inch × 2.5-inch × 2-inch block. Using the Hole Wizard, add a hole counterbored for a 1/2-inch socket head cap screw with a depth of 2 inches (see **Figure P6.1A**). Use a linear pattern to create a pattern of four evenly spaced holes in the block with a distance of 1.5 inches between hole centers. The resulting block is shown in **Figure P6.1B**, and a section view is shown in **Figure P6.1C**.

FIGURE P6.1A

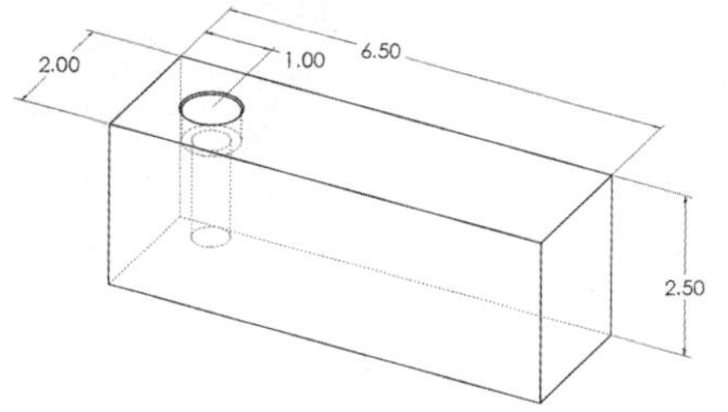

FIGURE P6.1B

FIGURE P6.1C

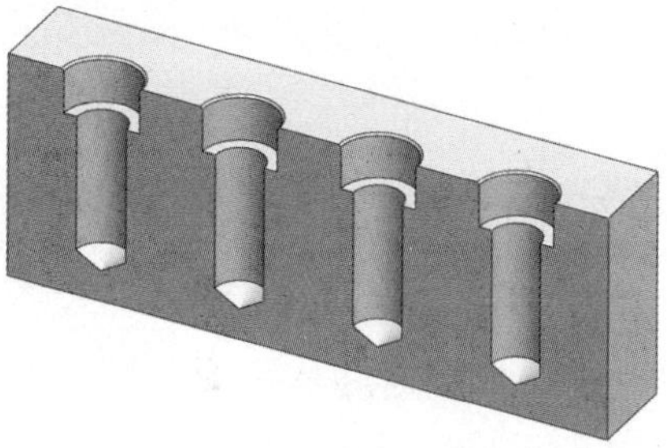

P6.2 Import the part model created in **Problem P6.1** into a new assembly. Import a socket head cap screw from Chapter 5 into the assembly. Right-click on the cap screw and select properties. Select the 101 configuration. Add mates to locate the cap screw, and define a linear pattern to place cap screws in the other holes.

FIGURE P6.2

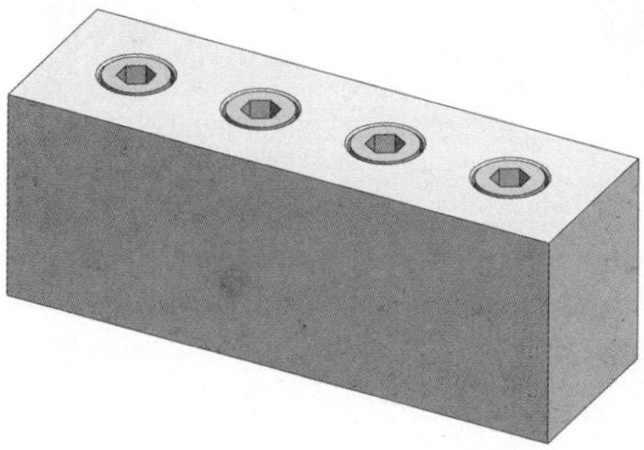

P6.3 Create solid models of a 5-ft. long, 1.5-in. diameter pole (**Figure P6.3A**), and a 4-in. diameter sphere with a 1.5-in. diameter hole in it (**Figure P6.3B** and **Figure P6.3C**). Using these models and the flange model created in Chapter 1 (**Figure P6.3D**), create an assembly model of a flagpole (**Figure P6.3E**).

FIGURE P6.3A

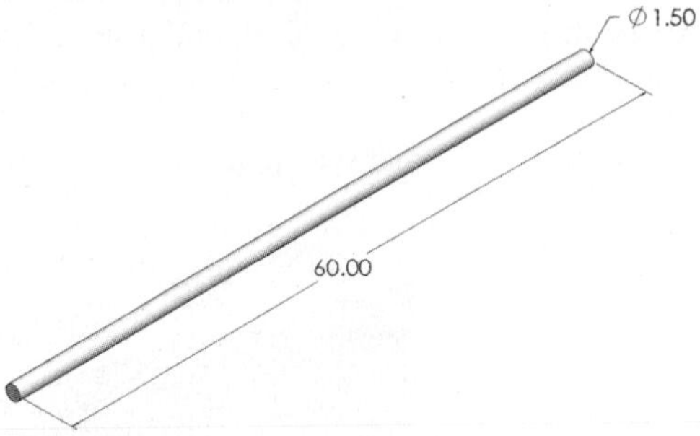

FIGURE P6.3B

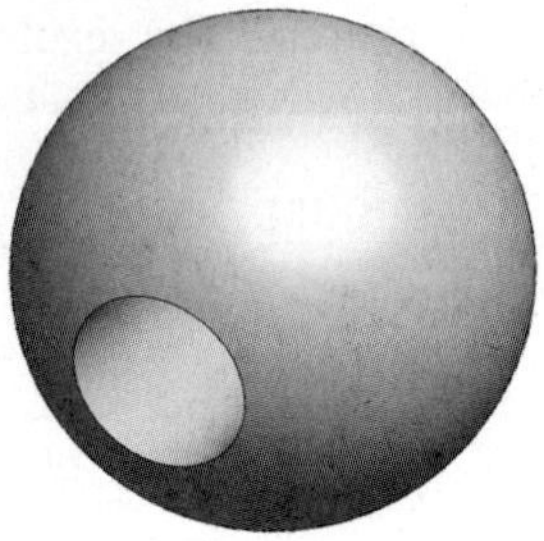

FIGURE P6.3C

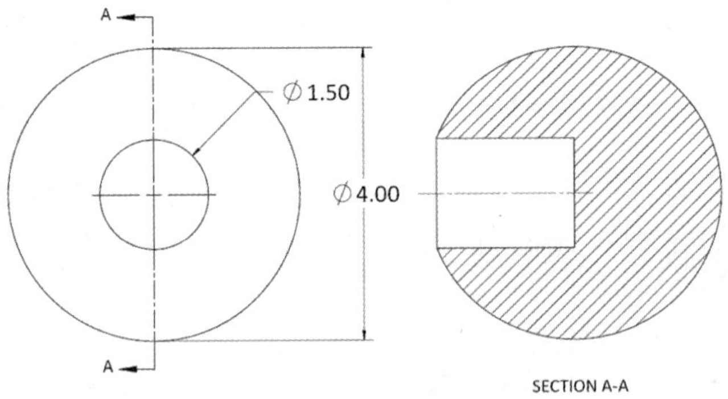

FIGURE P6.3D

FIGURE P6.3E

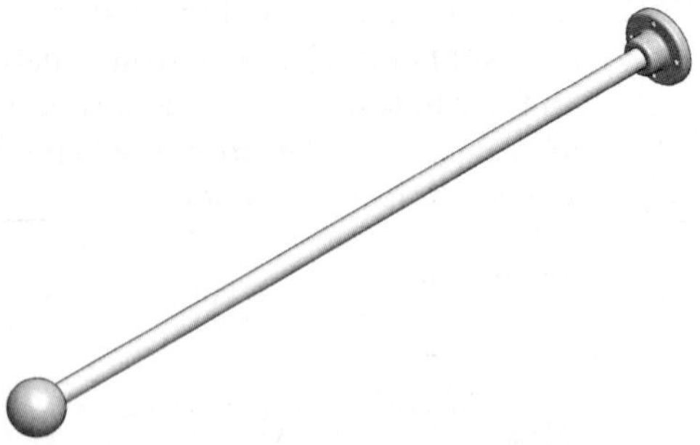

P6.4 Create solid models of the shaft segment shown in **Figure P6.4A** and **Figure P6.4B**, and the key shown in **Figure P6.4C**. Create an assembly with these two parts and the pulley model described in Chapter 1. The finished assembly is shown in exploded state in **Figure P6.4D** and in normal state in **Figure P6.4E**.

FIGURE P6.4A

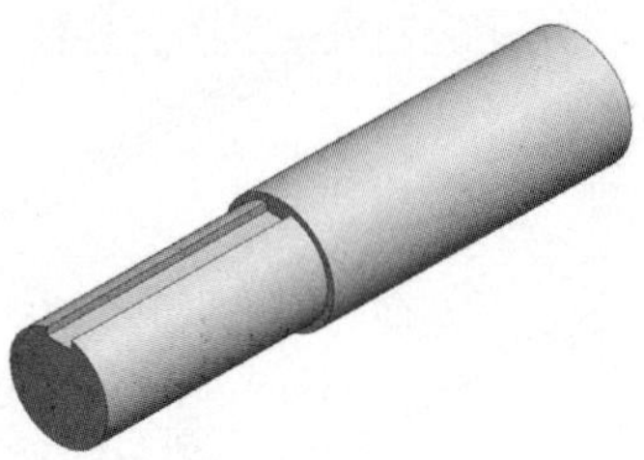

FIGURE P6.4B

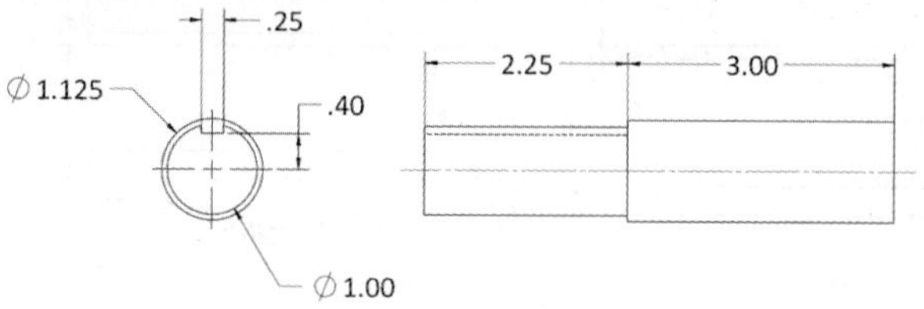

FIGURE P6.4C

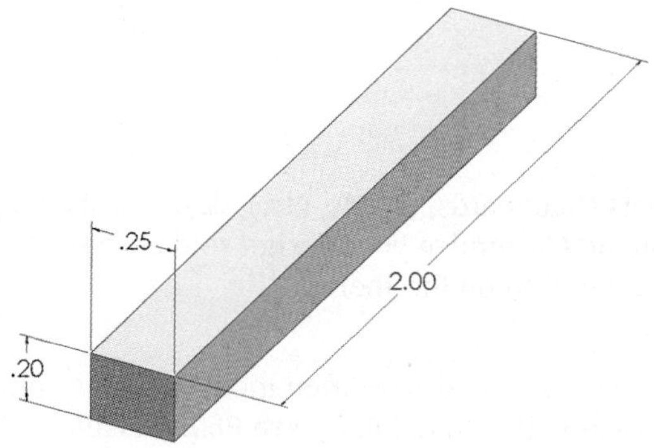

FIGURE P6.4D

FIGURE P6.4E

P6.5 Create models of the 2 × 4 members shown in **Figure P6.5A**. (Note that 2 × 4s have actual finished dimensions of 1.5 × 3.5 inches.) Create an assembly from these 2 × 4s as shown in **Figure P6.5B**. Consider creating one or more subassemblies to reduce the total number of mates required.

FIGURE P6.5A

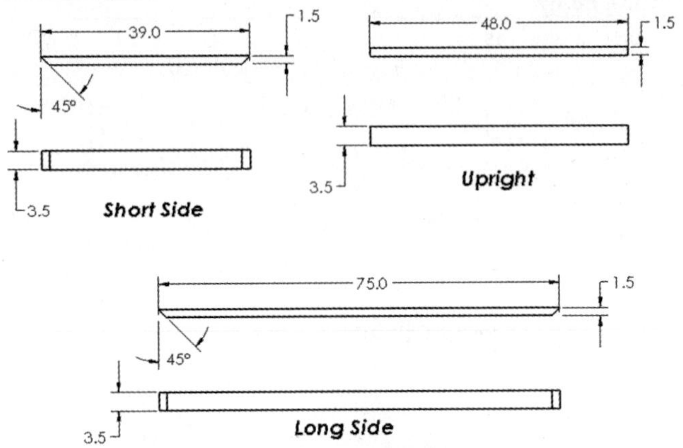

FIGURE P6.5B

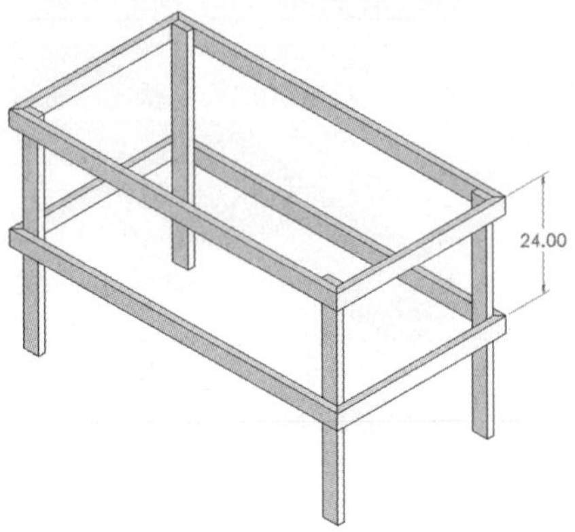

P6.6 Using simple components (flat boards, 2 × 4s, etc.), design a shelving unit for your room or apartment. Customize your design so that some items that you own (TV, stereo, etc.) will fit on the shelves.

P6.7 The atomic crystal structures of metals are often modeled using spheres to represent atoms. The structure illustrated in **Figure P6.7A** is called a body-centered cubic structure. Create a model of this crystal structure as follows:

FIGURE P6.7A

a. Create a 2-inch diameter sphere part model.
b. Open a new assembly.
c. Open a 3-D sketch in the assembly.
d. Add lines to create a box shape, as shown in **Figure P6.7B**. Add relations so that all lines are along the x, y, and z axes. Add an Equal relation to three of the lines so that the sketch defines a cube.
e. Add a centerline connecting two opposite corners of the cube. Add a 4-inch dimension of this line, which will ensure that the atoms on the corners will touch the center atom. Add a point to the midpoint of the centerline, as shown in **Figure P6.7C**.
f. Close the sketch.

FIGURE P6.7B

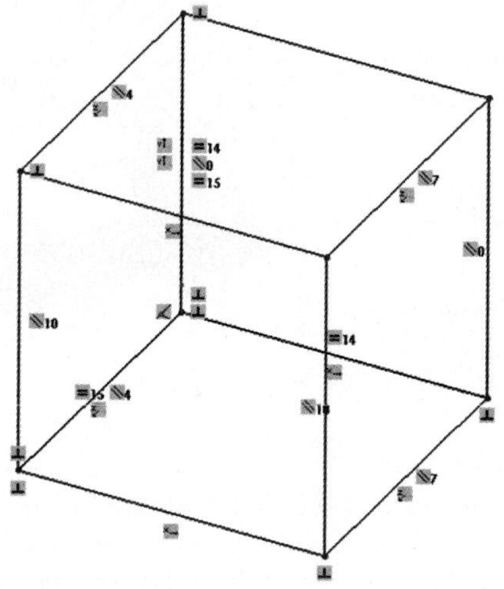

FIGURE P6.7C

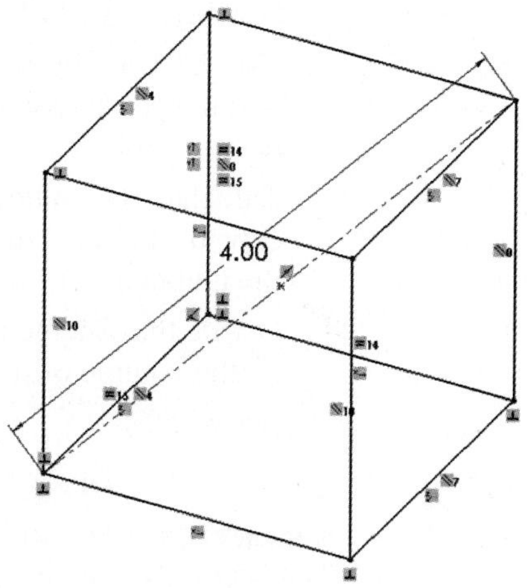

g. Insert nine atoms into the assembly. Turn on the display of the origins.
h. Add mates between the origin of each of the atoms and a corner or the center point of the 3-D sketch, as shown in **Figure P6.7D**.
i. Turn off the display of the origins, and hide the 3-D sketch.

FIGURE P6.7D

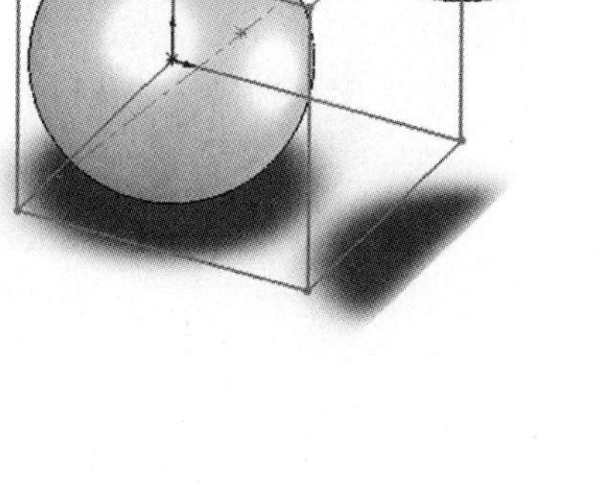

P6.8 The *packing factor* of a crystal structure quantifies how tightly packed the atoms of the structure are, and is defined as the volume of atoms contained within a *unit cell,* or basic building block of the crystal structure, divided by the total volume available within the cell. Find the packing factor of the body-centered cubic structure by following this procedure:

a. Cut away parts of the model created in **Problem 6.7** to create a unit cell, as shown in **Figure P6.8**. (Note: Extruded cuts can be made in assemblies using a similar tool to that used in parts. The Extruded Cut Tool for assemblies can be accessed from the Assembly Features Tool of the Assembly Group.)

b. Use the Mass Properties Tool to find the volume of atoms within the unit cell.

c. Calculate the volume of a solid cube of the same dimensions as the unit cell.

d. Divide the volume of the atoms by the volume of the solid cube.

(Answer: 68%)

FIGURE P6.8

P6.9 Repeat **Problem 6.7** for the *face-centered cubic* structure in **Figure P6.9A**. The 3-D sketch defining the atom positions is shown in **Figure P6.9B**. Note that a point must be added at the center of each of the six faces of the cube.

FIGURE P6.9A

FIGURE P6.9B

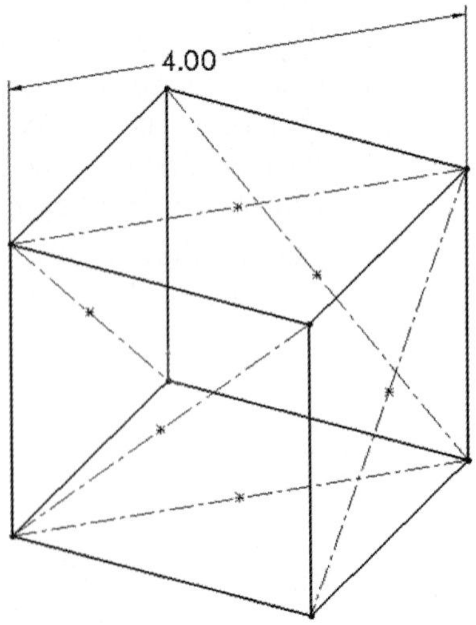

P6.10 Find the packing factor of the face-centered cubic structure model in **Problem 6.9**. The unit cell is shown in **Figure P6.10**.

(Answer: 74%)

FIGURE P6.10

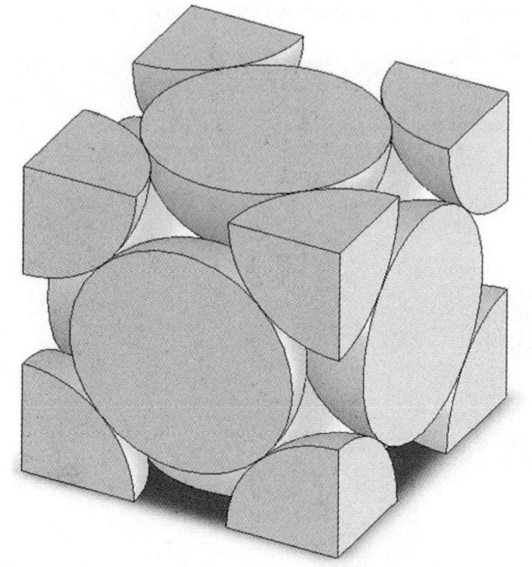

P6.11 Create an assembly model of a split hub clamp, as shown in **Figure P6.11A**. To do this, create the part shown in **Figure P6.11B**. Assemble two instances of this part along with two appropriately sized socket head cap screws to complete the assembly.

FIGURE P6.11A

FIGURE P6.11B

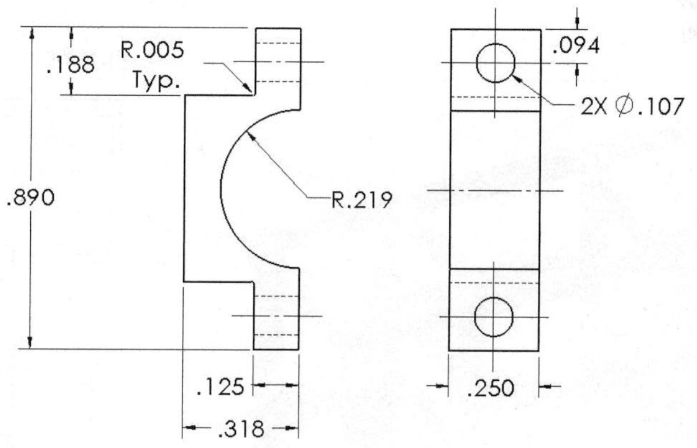

P6.12 Model the workbench shown in **Figure Problem 6.12**. The bench is constructed from 2×4 members (actual dimensions 1.5 by 3.5 inches) and 1-inch-thick plywood for the top and shelf. The top is 72 inches wide by 30 inches deep, and is 34 inches from the floor. The top has a 2-inch overhang on the front and both sides (but not the back). Set the material to Pine for all members, and use the Mass Properties Tool to estimate the weight of the assembled bench.

FIGURE P6.12

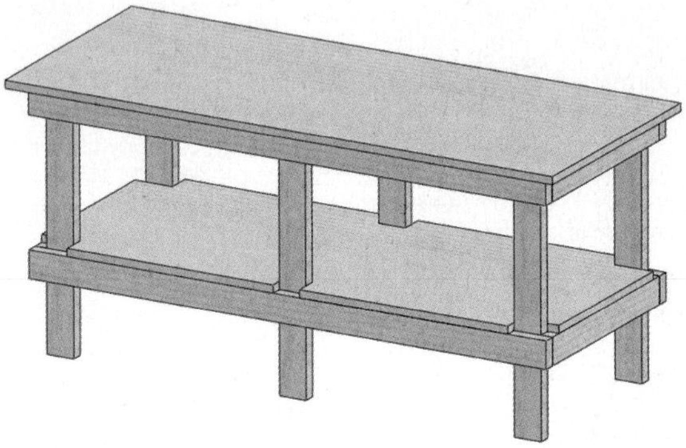

P6.13 The frame shown in **Figure P6.13A** is made up of wood 2×4 members (actual cross section dimensions are 1.5 by 3.5 inches). Create a single 2×4 part with multiple configurations that are 21.5, 27, 30, 36, 40, and 72 inches long. Create the frame using these configurations of the 2×4 and the dimensions shown in **Figure P6.13B**. The frame is symmetric from left to right and from front to back.

FIGURE P6.13A

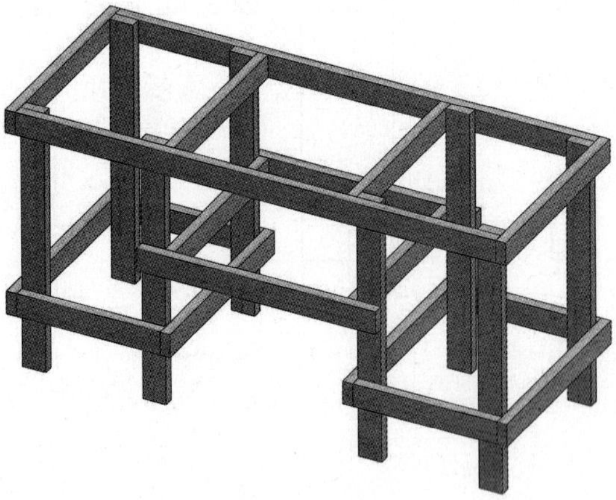

FIGURE P6.13B

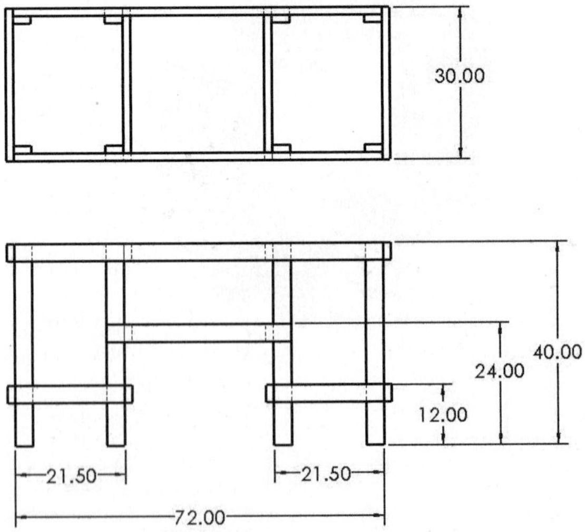

P6.14 Create an exploded view of the assembly created in **Problem P6.2**.

FIGURE P6.14

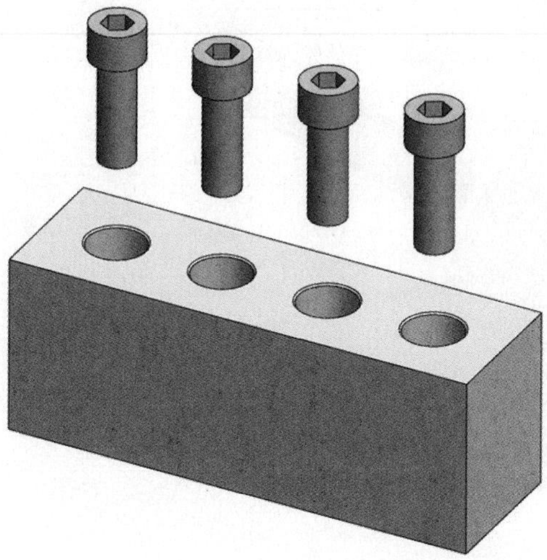

P6.15 Create an exploded view of the flagpole assembly model created in **Problem P6.3**.

FIGURE P6.15

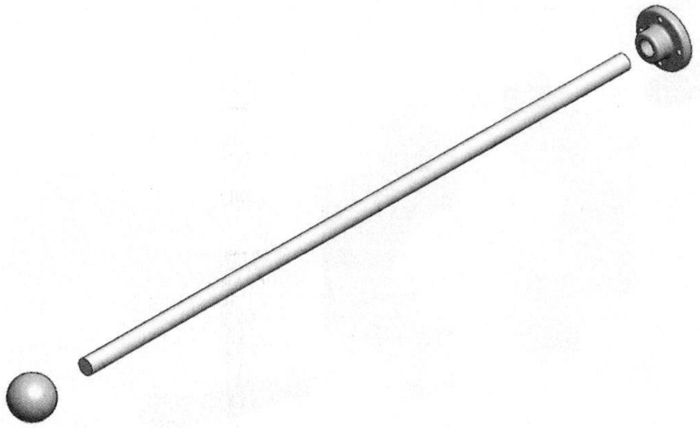

P6.16 Create an exploded view of the pulley assembly model created in **Problem P6.4**.

FIGURE P6.16

P6.17 Create and export an animation (*.avi file) of the explosion from **Problem P6.16**.

P6.18 Create an exploded view of the split hub clamp assembly model created in **Problem P6.11**.

FIGURE P6.18

P6.19 Create and export an animation (*.avi file) of the explosion from Problem **P6.18**.

CHAPTER 7

Advanced Assembly Operations

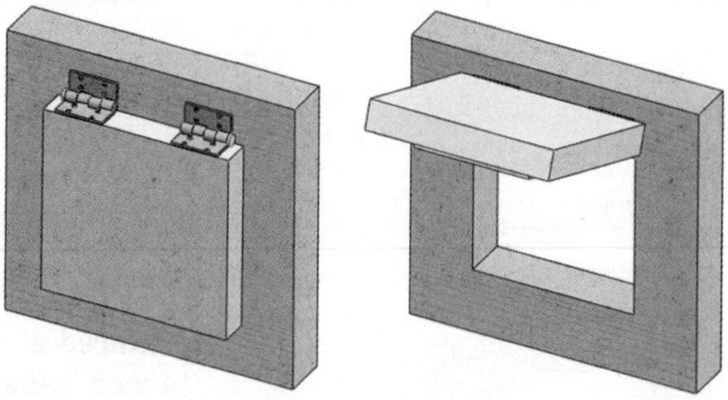

Introduction

In Chapter 6, the basic operations required to create a SOLIDWORKS® assembly were introduced. In this chapter, some advanced modeling, visualization, and analysis operations will be developed. The new assembly model that will be created is shown in **Figure 7.1**. It includes the door assembly created in Chapter 6, as well as additional part models.

FIGURE 7.1

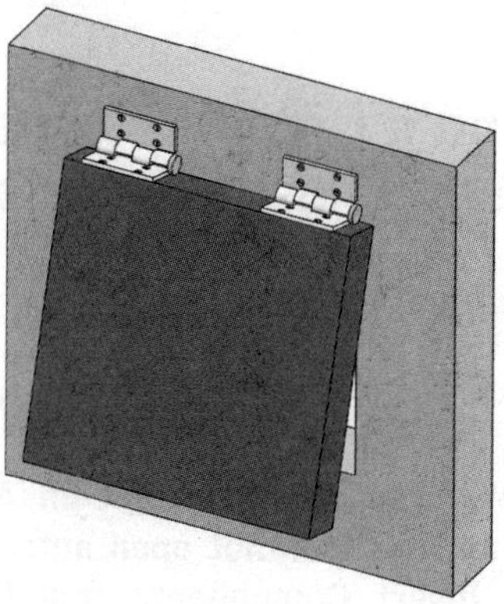

Chapter Objectives

In this chapter, you will:

- use an assembly model as a subassembly in a more complex assembly model,
- create an assembly model that allows for motion between components, and
- use the interference detection and collision detection features to analyze the assembly model.

7.1 Creating the Part Models

Two additional parts must be created for the new assembly model.

The first model that we will create is the hinge pin.

Open a new part, and sketch a 0.5-inch diameter circle in the Right Plane, centered at the origin. Extrude it 4 inches. Add a 1-inch diameter cap on the pin, extruding it away from the pin body 0.25 inches.

The pin is shown in **Figure 7.2**. Using a new color for this component will aid in visualization of the final assembly.

FIGURE 7.2

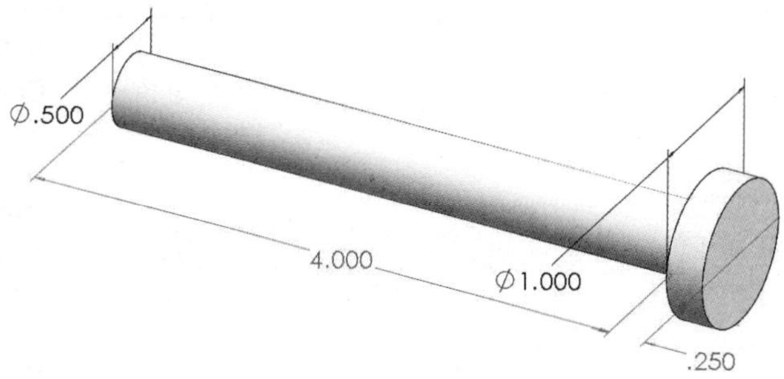

SOLIDWORKS is a registered trademark of Dassault Systémes SolidWorks Corporation.

Change the color of the part, if desired, and save the part in a file named "Pin." Close the file.

FIGURE 7.3

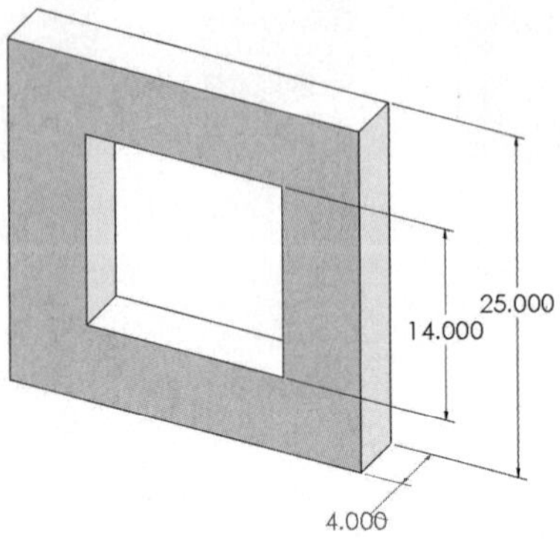

The final component needed for the assembly is a frame.

Open a new part, and in the Front Plane sketch a 25-inch square centered at the origin. Extrude it to a depth of 4 inches. Extrude a 14-inch square cut (centered on the front face of the component), yielding the part shown in Figure 7.3.

Change the color, if desired. Save this part in a file named "Frame," and close it.

These parts, along with the assembly model from Chapter 6, will now be used to create a model of the hatch assembly.

7.2 Creating a Complex Assembly of Subassemblies and Parts

FIGURE 7.4

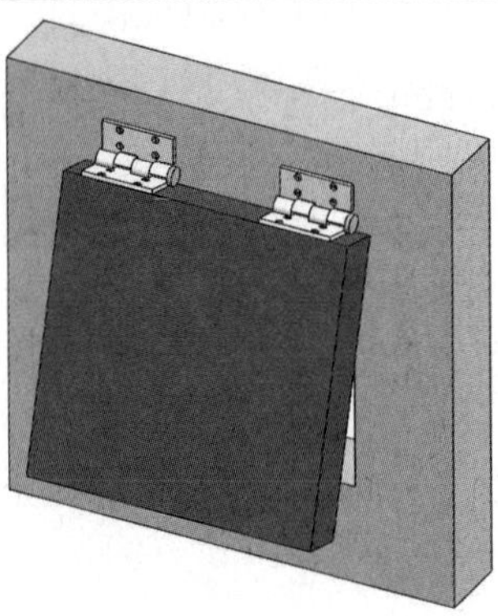

In this section, we will create a new assembly using the door assembly created in the previous chapter as a subassembly. The assembly will be a model of a hinged hatch, shown in **Figure 7.4**.

Open a new assembly. If the Begin Assembly PropertyManager does not open automatically, select the Insert Components Tool from the Assembly group of the CommandManager. Browse to the Frame file, and click the check mark to place it at the origin, as shown in Figure 7.5.

FIGURE 7.5

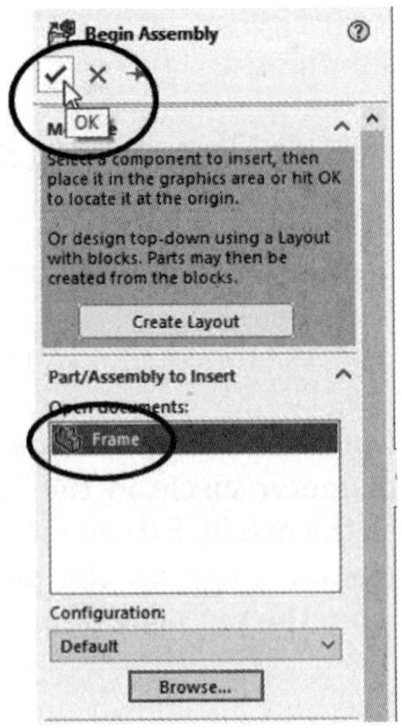

In the next step, the door assembly created in the previous section will be used as a subassembly within the new assembly. An assembly can be inserted just like a part.

FIGURE 7.6

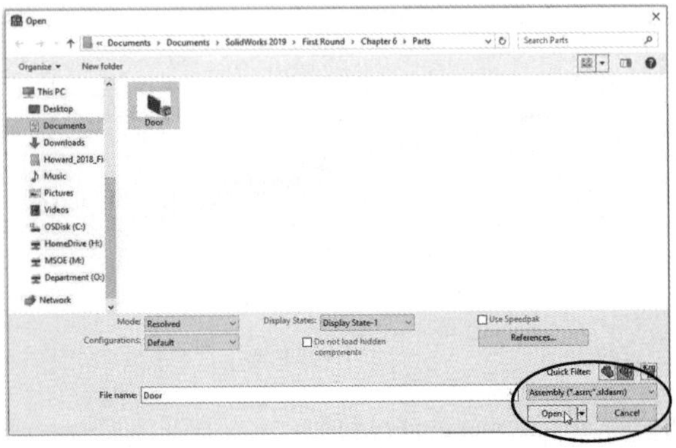

Select the Insert Components Tool, and browse for a file. In the dialog box, change the file type option to Assembly files, or All Files, and locate the file Door.SLDASM. Select it, and click Open, as shown in Figure 7.6.

Drop the subassembly into the assembly window in the approximate position shown in Figure 7.7 by clicking.

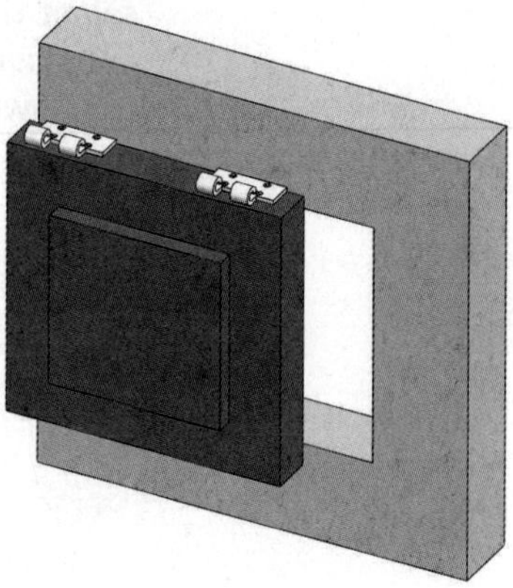
FIGURE 7.7

While the door is an assembly and not a part, it can be handled (mated, rotated, moved, etc.) just like a part component in the assembly window.

Right-click and drag on the door subassembly to rotate it into the approximate orientation shown in Figure 7.8, with the curved portion of the hinges toward the frame.

In our previous mates, we have established relationships between surfaces of the components to be mated. While this is often the procedure we will use, it is sometimes preferable to use the default (Front, Top, or Right) planes, the origins, or other reference geometries in our mates. In centering the Door subassembly with respect to the frame, we will establish a coincident mate between the Right Plane used in the Door subassembly model and the Right Plane used in the Frame part model.

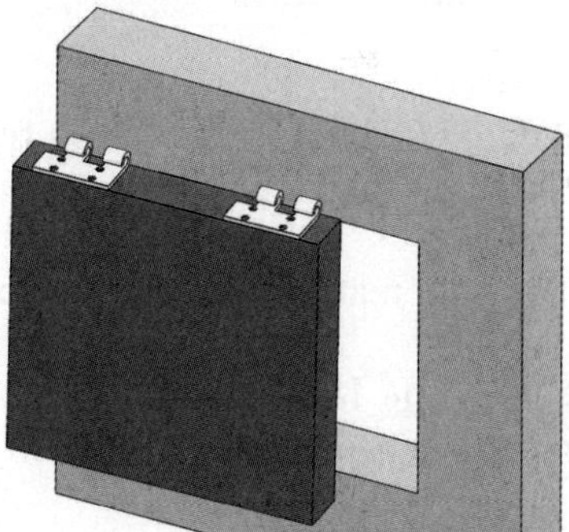
FIGURE 7.8

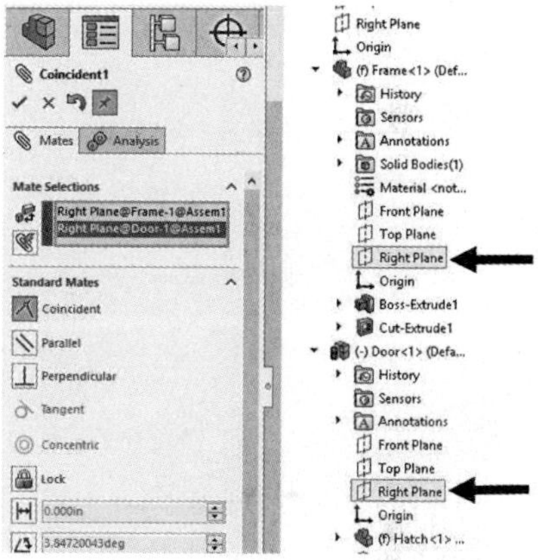

FIGURE 7.9

Select the Mate Tool. Expand the FeatureManager "flyout" by clicking the arrow, and expand the entries for Frame and Door. Select the Right Plane from each, and establish a coincident relation between them, as shown in Figure 7.9. Click the check mark twice to apply the mate and close the Mate Tool.

Instances of the hinge parts will now be added to the frame.

Select the Insert Components Tool. Browse for a file, changing the file type option back to part files. Select the hinge part. Drop it into the assembly, and rotate and move the hinge to place it into the approximate location and orientation shown in Figure 7.10.

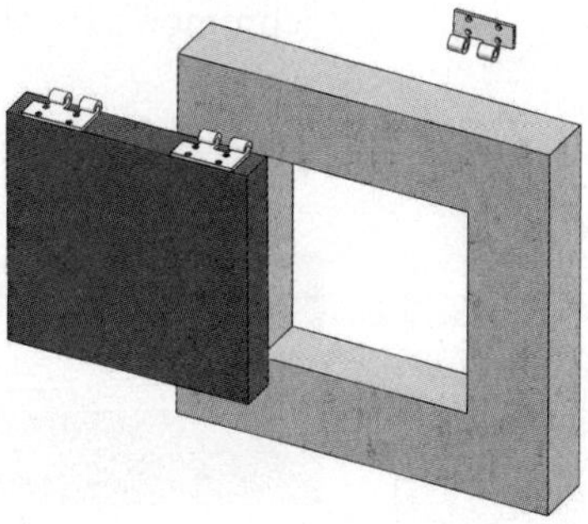
FIGURE 7.10

Click on the Mate Tool to begin a mate. Instead of rotating the view orientation to select the bottom face of the hinge, move the cursor over the hinge, right-click, and choose Select Other, as shown in Figure 7.11. With the back face highlighted, as shown in Figure 7.12, click the left mouse button.

FIGURE 7.11

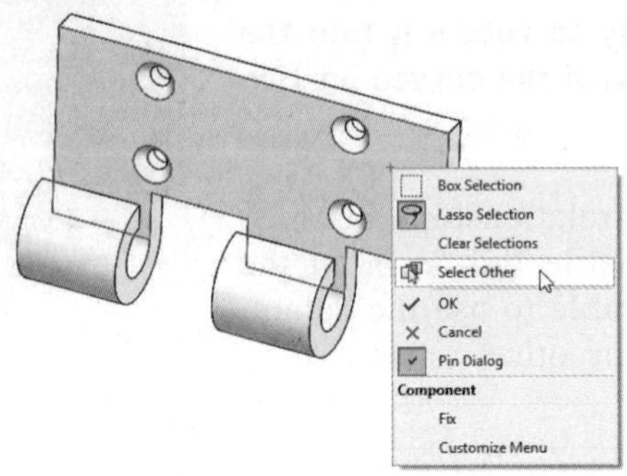

FIGURE 7.12

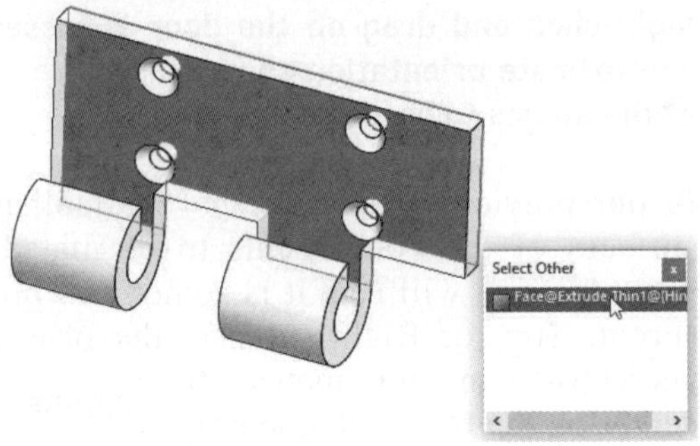

FIGURE 7.13

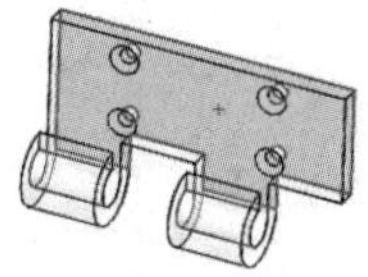

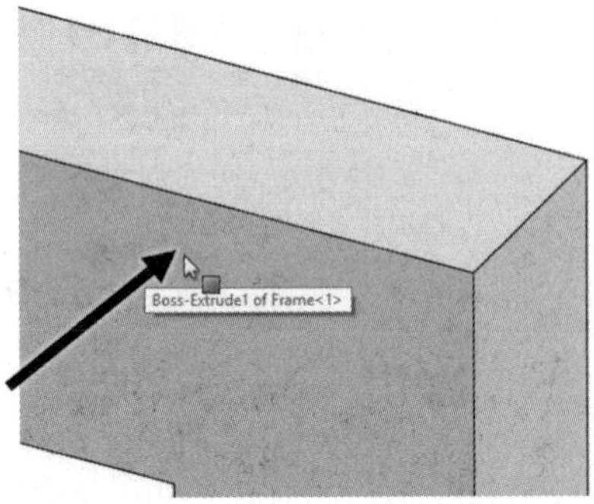

Select the front face of the frame, as shown in Figure 7.13. Click the check mark to apply a coincident mate.

The next mate will align the faces of the two hinges to be engaged.

Select the face shown in Figure 7.14 on the hinge that was just mated to the frame.

Also select the corresponding face on the hinge mounted to the hatch, as shown in Figure 7.15. Click the check mark to apply a coincident mate.

The effect of the mate can be most easily seen from the Front View, as shown in Figure 7.16.

These two faces now lie in the same plane, but additional mates are required to fully constrain the desired relationship between these components. The hinges will now be brought together to share a common axis of rotation.

FIGURE 7.14

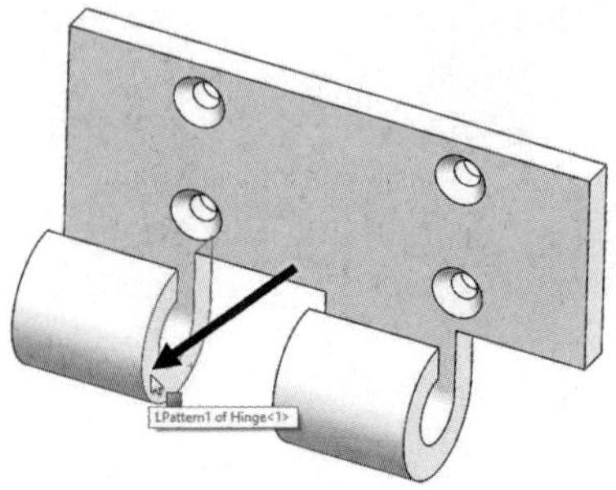

FIGURE 7.15

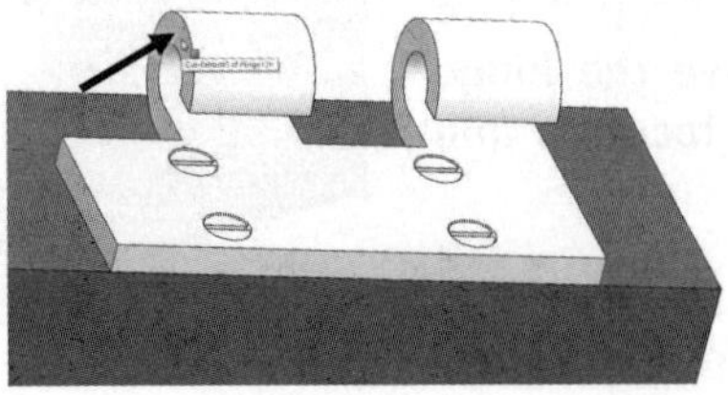

FIGURE 7.16

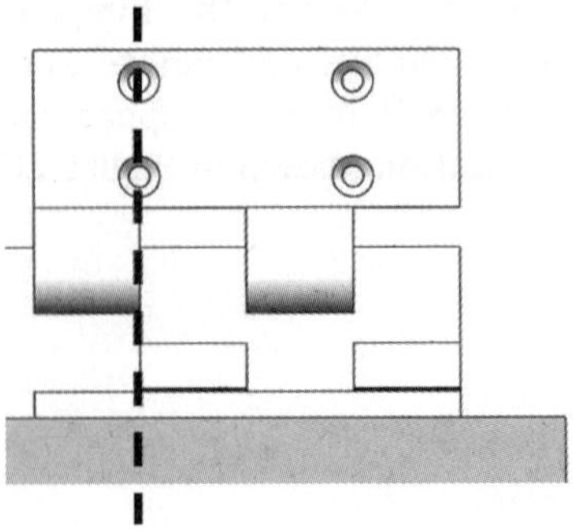

Select the cylindrical faces of each of the two hinges to be mated, as shown in Figure 7.17.

FIGURE 7.17

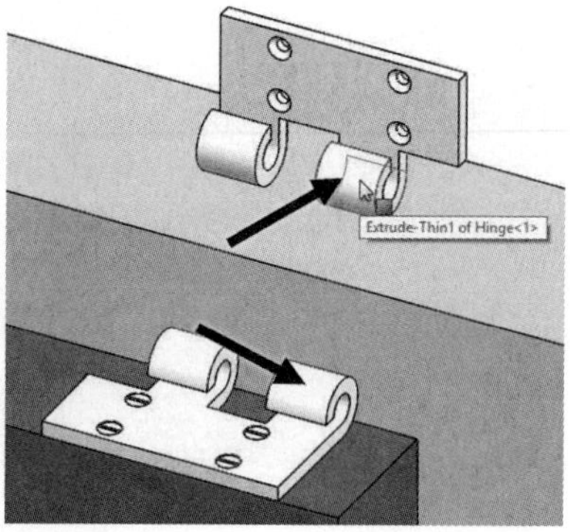

By default, a concentric mate will be previewed between the two cylindrical faces, as shown in **Figure 7.18**.

Click the check mark to apply the mate.

The final mate required will locate the hinge on the frame.

If necessary, click and drag the hatch downward until the top of the hinge is below the top of the frame.

Select the top surfaces of the hinge and frame, as shown in Figure 7.19, and apply a 2-inch distance mate. Close the Mate PropertyManager.

FIGURE 7.18

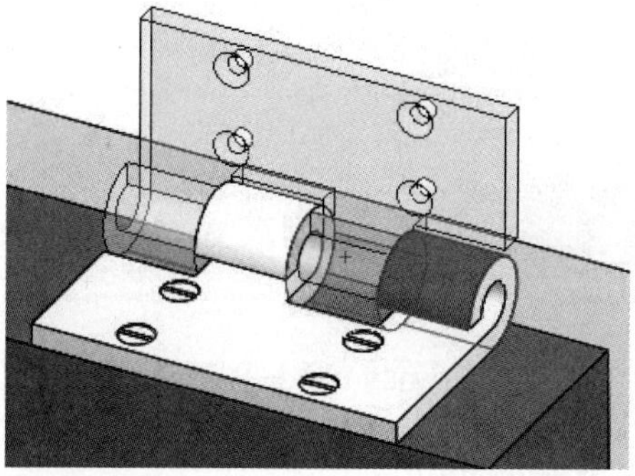

FIGURE 7.19

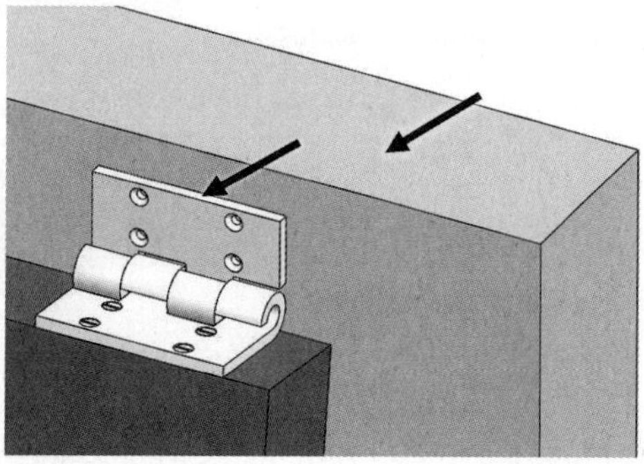

The hinge mate is now fully defined. We can see the effect of the mates by attempting to move the hatch.

Click and drag on the hatch, as shown in Figure 7.20.

FIGURE 7.20

FIGURE 7.21

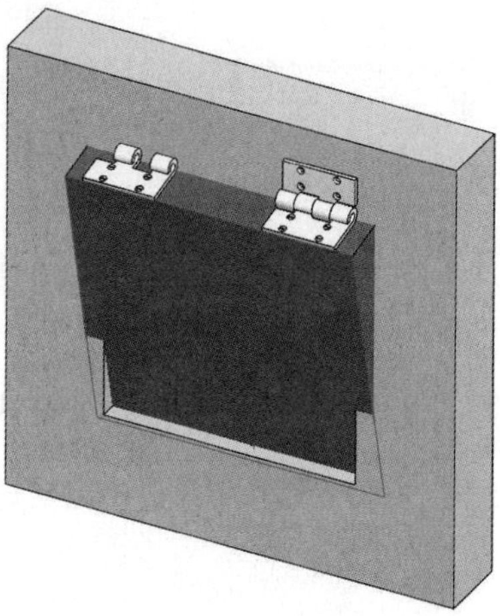

Note that the hatch has a degree of freedom about the hinge, but is constrained against all other types of motion.

Note also that the hatch can apparently rotate through the frame, as shown in **Figure 7.21**. This is due to the fact that only geometric relations between the entities have been defined, but no true physical characteristics have been imparted to the objects. (We will learn how to use collision detection to limit the motion of the model in the next section.)

FIGURE 7.22

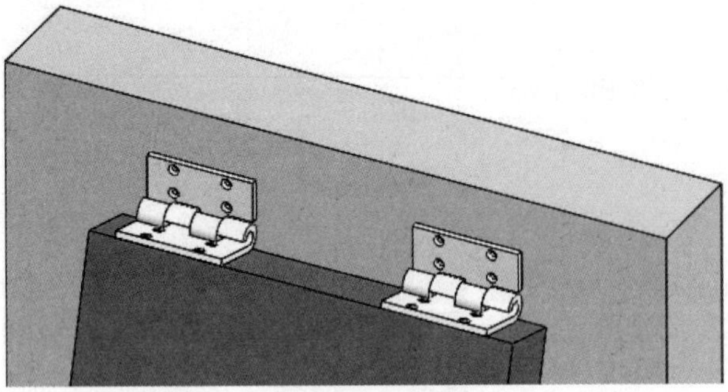

Insert a second hinge part, and add mates to place it as shown in Figure 7.22.

We will now add pins to the assembly.

Choose the Insert Components Tool. Close the browsing dialog box if it opens. Click on the pushpin icon, as shown in Figure 7.23. This will enable the insertion of multiple parts. Browse to find the pin, and click two locations to place pins in the assembly, as shown in Figure 7.24. Click the check mark to close the Insert Component PropertyManager.

FIGURE 7.23

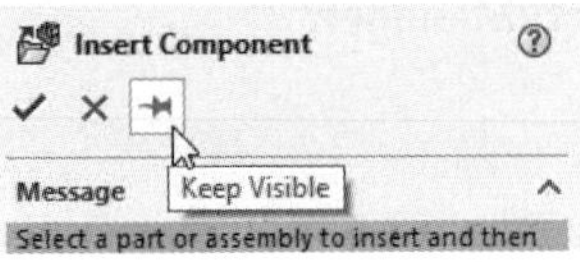

FIGURE 7.24

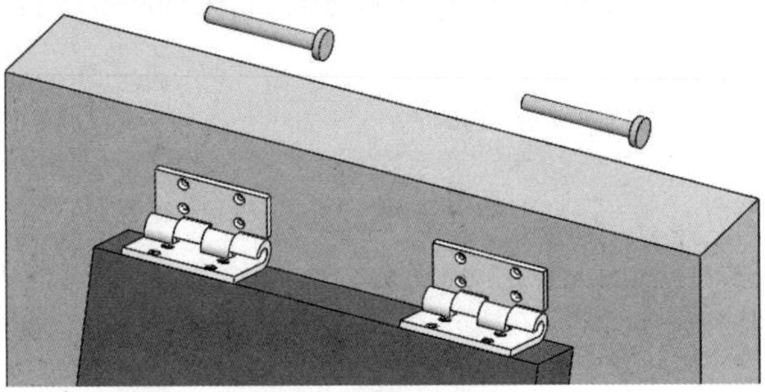

To aid in the selection of small details, a *filter* is sometimes used. When a filter is active, only certain entities can be selected. We are adding mates between faces, so a filter that allows only faces to be selected will be helpful.

FIGURE 7.25

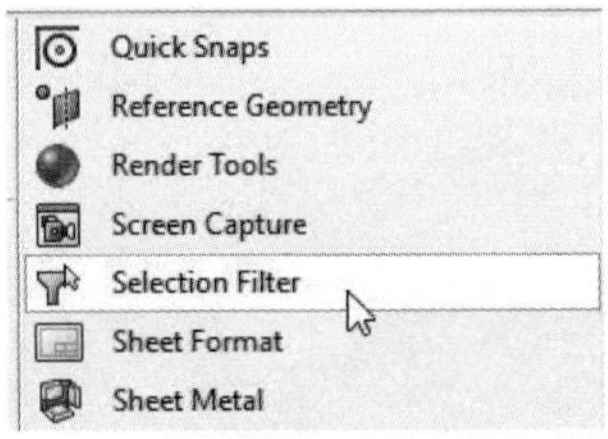

Right-click on the menu bar or CommandManager. From the list of available toolbars, select Selection Filter, as shown in Figure 7.25.

FIGURE 7.26

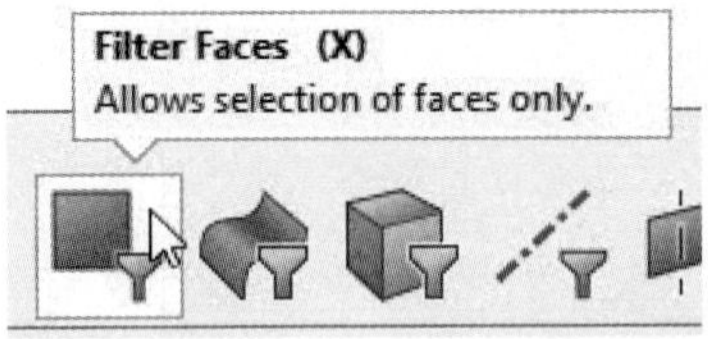

In the Selection Filter toolbar, select the Filter Faces Tool, as shown in Figure 7.26.

Whenever a filter is active, a filter icon appears beside the cursor, as shown in Figure 7.27.

FIGURE 7.27

Select the Mate Tool, and select the cylindrical face of one of the pins, as shown in Figure 7.28. (Note that with the filter set, you cannot select an edge.)

FIGURE 7.28

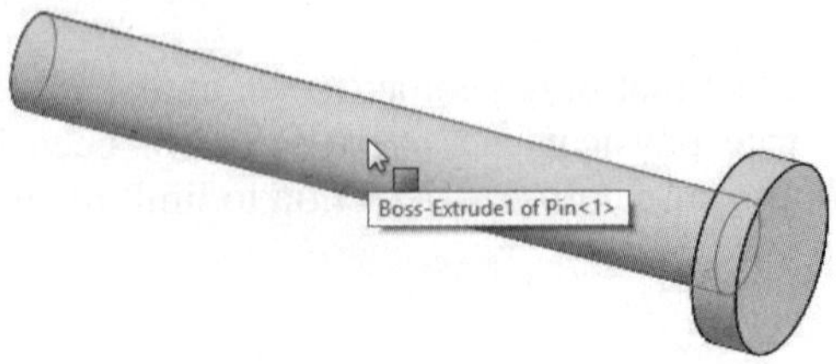

Select one of the cylindrical faces of the corresponding hinge, as shown in Figure 7.29. (Either the outer or the inner face may be selected; since they are concentric, the resulting mate will be the same.) Establish a concentric mate.

FIGURE 7.29

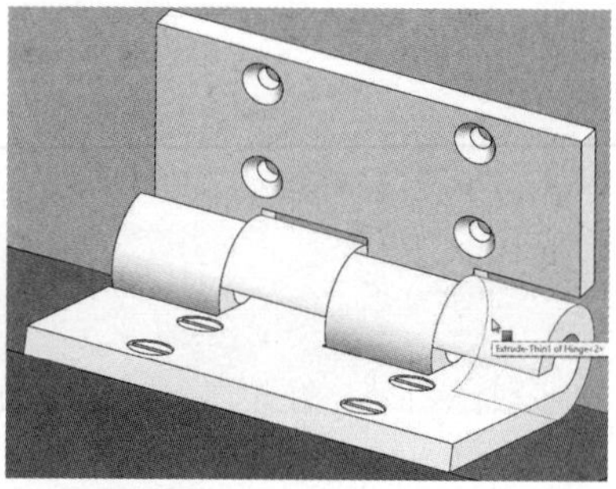

The pin and the hinge will align concentrically in the preview of the mate. If necessary, change the alignment (from anti-aligned to aligned, or vice versa) of the mate to orient the head of the pin appropriately, as shown in Figure 7.30. Click the check mark to apply the mate. If necessary, drag the pin to the approximate location shown.

FIGURE 7.30

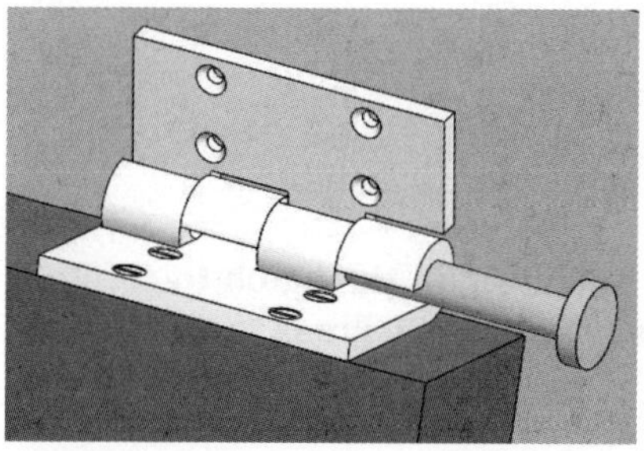

Select the face on the hinge shown in Figure 7.31, and then the underside of the head of the pin, as shown in Figure 7.32.

Click on the check mark to apply the mate, the result of which is shown in Figure 7.33. Close the Mate PropertyManager.

Using this procedure, duplicate the assembly of the second pin.

FIGURE 7.31

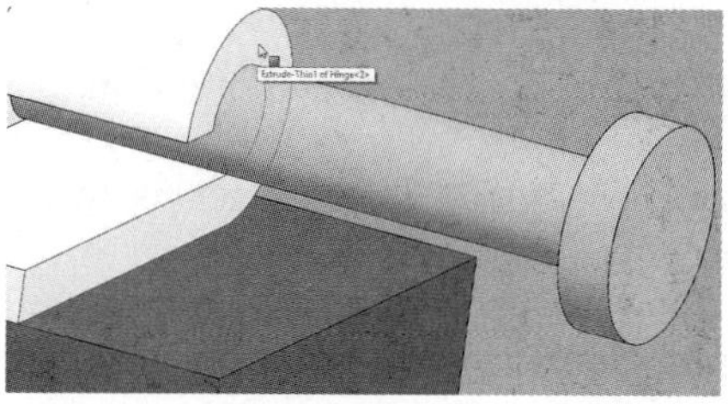

FIGURE 7.32

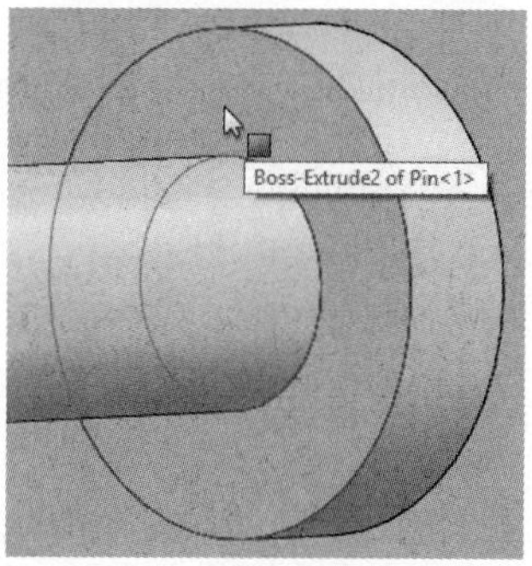

FIGURE 7.33

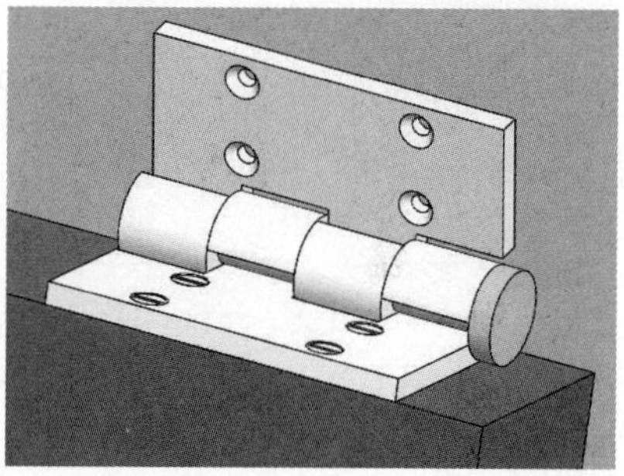

The assembly is now complete, as shown in **Figure 7.34**. Since we no longer need the filters, it is a good idea to clear them.

Select the Clear All Filters Tool, as shown in Figure 7.35. Right-click on the menu bar, and de-select the Selection Filter tool.

The Filter Faces Tool is very useful when working with mates. Because it is used so often, there is a keyboard shortcut that can be used to activate it. Pressing the X key will toggle the Filter Faces Tool on or off. The E and V keys can be used to activate the Filter Edges Tool and Filter Vertices Tool in a similar manner, and the F6 key turns off all filters.

FIGURE 7.34

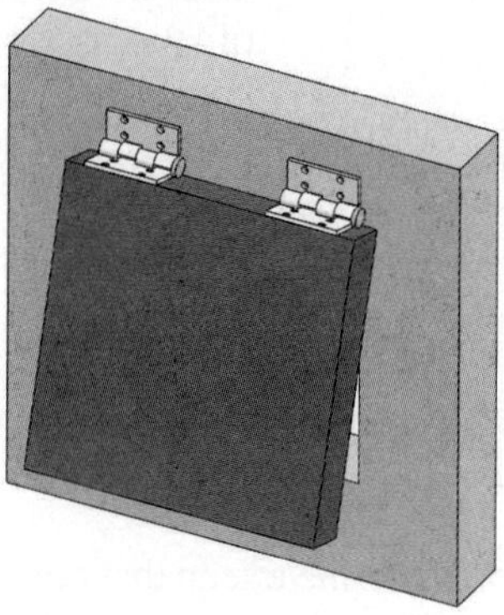

FIGURE 7.35

FIGURE 7.36

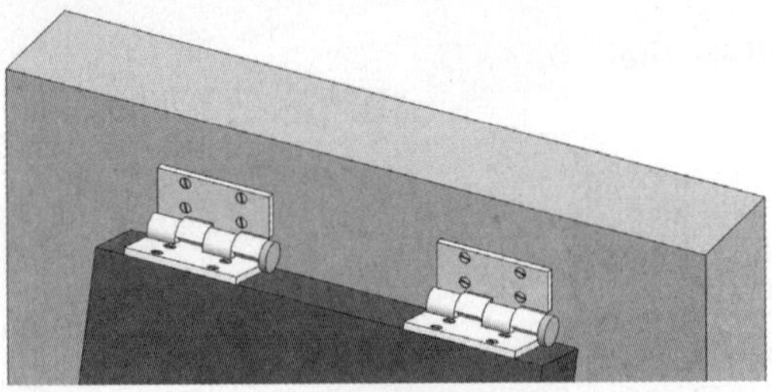

Using the procedures from Sections 6.3 and 6.4, add holes and fasteners to finish the assembly, as shown in Figure 7.36.

Save the assembly in a file entitled "Hatch Assembly."

7.3 Detecting Interferences and Collisions

In the assembly mode, the SOLIDWORKS program has the ability to check for interferences between components. It does this by determining locations where solid volumes overlap. This tutorial will demonstrate these capabilities.

FIGURE 7.37

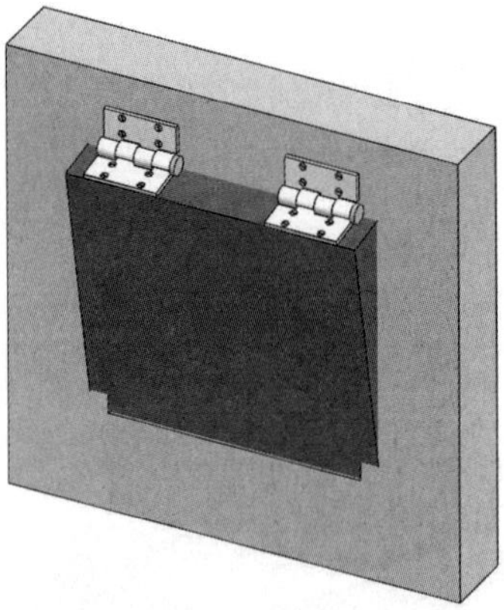

Open the Hatch Assembly file. Click and drag on the hatch to rotate it into the approximate position shown in Figure 7.37. Press the Esc key to deselect the hatch.

In doing this, we have intentionally introduced an interference between the door and the frame.

From the main menu, select Tools: Evaluate: Interference Detection.

The PropertyManager shows the parts for which interference is to be analyzed. By default, the entire assembly is selected.

Click the Calculate button to commence interference detection, as shown in Figure 7.38.

FIGURE 7.38

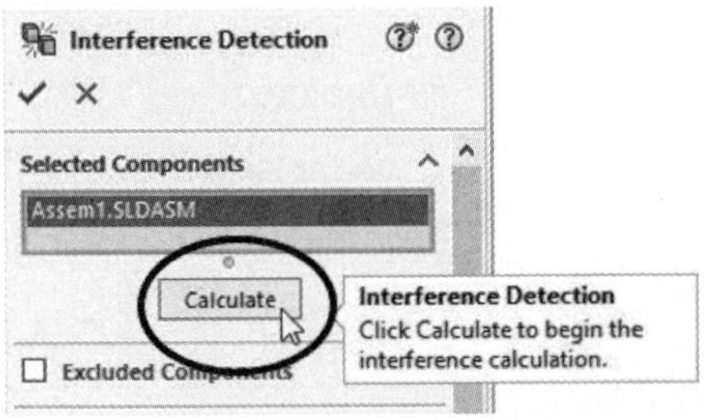

In the PropertyManager, the interference between the frame and hatch is identified, as shown in Figure 7.39. Note that there are seventeen regions of interference detected. In addition to the overlap of the door and the frame, the sixteen screws in their undersized holes are detected. The Ignore button can be used to eliminate the intentional screw interferences from the list.

FIGURE 7.39

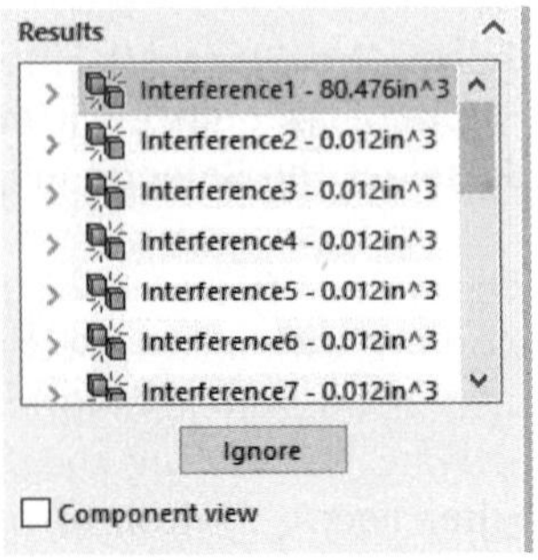

Click the X in the PropertyManager to end interference detection.

Although we have seen that components in an assembly can be moved by simply clicking and dragging them, the Move Component Tool allows collision detection to be incorporated into part movements.

Click and drag the hatch so that it no longer overlaps the frame.

FIGURE 7.40

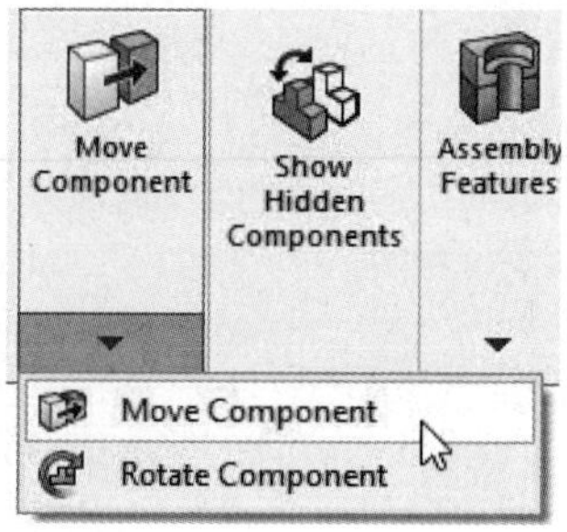

Select the Move Component Tool from the Assembly group of the CommandManager, as shown in Figure 7.40. In the PropertyManager, check "Collision Detection," "Stop at collision," and "Dragged part only." Under Advanced Options, select "Highlight faces" and "Sound," as shown in Figure 7.41.

FIGURE 7.41

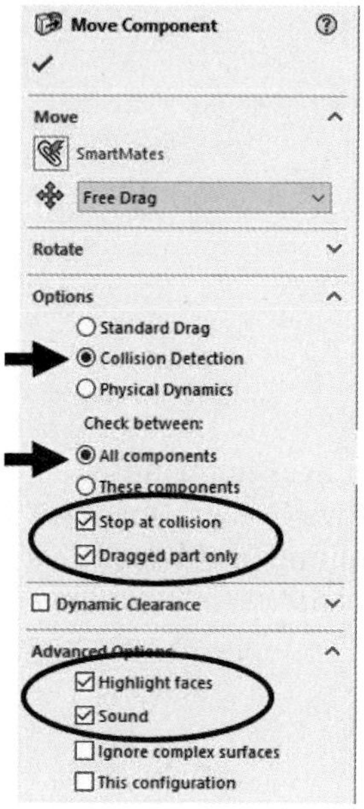

Move the hatch toward the frame until it stops, as shown in Figure 7.42.

Note that when the hatch and frame faces touch, the faces are highlighted and a tone signifies a collision.

Now drag the hatch upward as far as possible, as shown in Figure 7.43.

FIGURE 7.42

FIGURE 7.43

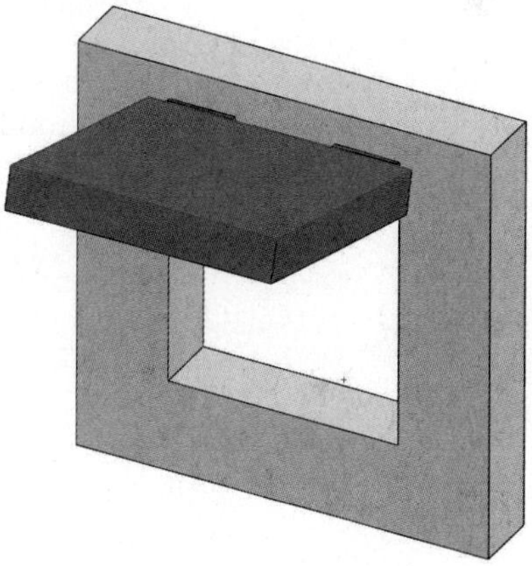

The movement stops when the hinges collide, as shown in the Right View of Figure 7.44.

Click the check mark or hit the Esc key to deselect the Move Component Tool, and close the file.

FIGURE 7.44

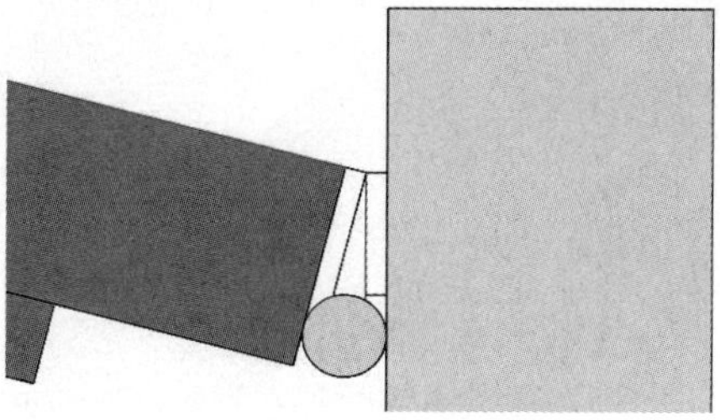

PROBLEMS

P7.1 Using the parts created in Chapter 6, create a working assembly model of a single hinge (**Figure P7.1**).

FIGURE P7.1

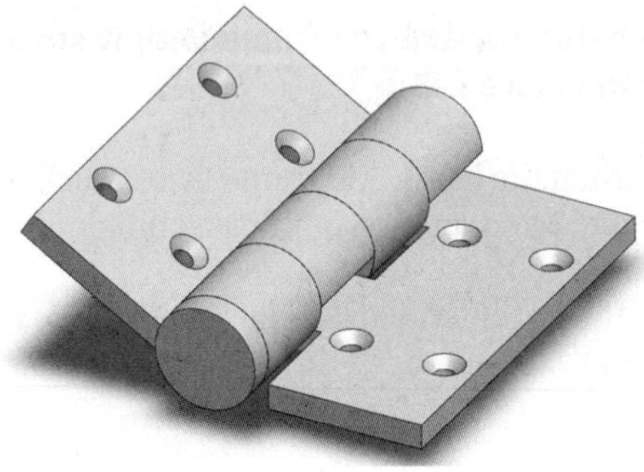

P7.2 Add an angular dimension between the faces of the hinges in the hinge assembly model, as shown in **Figure P7.2**. Use the Move Component Tool with collision detection to move the hinge into its limiting positions. What is the total angle through which the hinge can be rotated?

(Answer: 284 degrees)

FIGURE P7.2

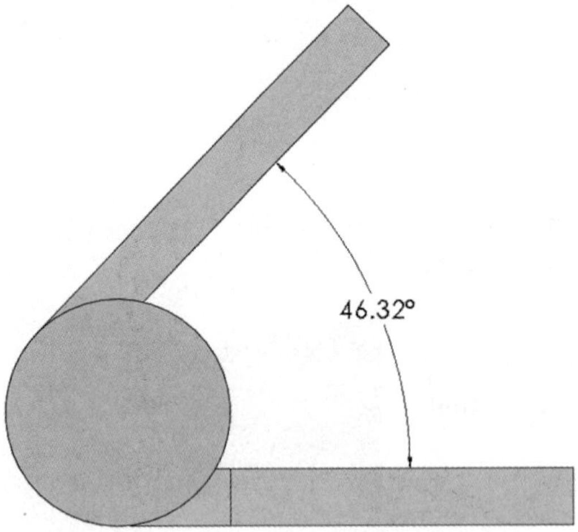

P7.3 Open the flagpole assembly created in **Problem P6.3**. Modify the diameter of the pole component to 2.00 inches, and rebuild the assembly. Use the Interference Detection Tool to locate the interferences between the components in the rebuilt assembly.

P7.4 Create the assembly model of the cylindrical roller bearing shown in **Figure P7.4A**. To do this, perform the following steps:

FIGURE P7.4A

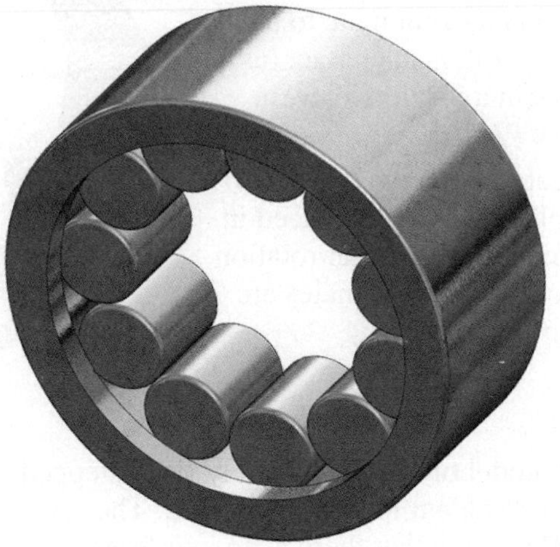

a. Create the bearing race by revolving the cross section shown in **Figure P7.4B** (dimensions in inches), and adding .02″ radius fillets to external edges.

b. Create a roller by modeling a 0.4″ long × 0.32″ diameter cylinder, and adding .02″ fillets to the edges.

c. Insert the race into a new assembly model and insert a single instance of the roller. Use one coincident and one tangent mate to add the first rolling element, as shown in a Section View in **Figure P7.4C**. It should be centered in the race.

FIGURE P7.4B

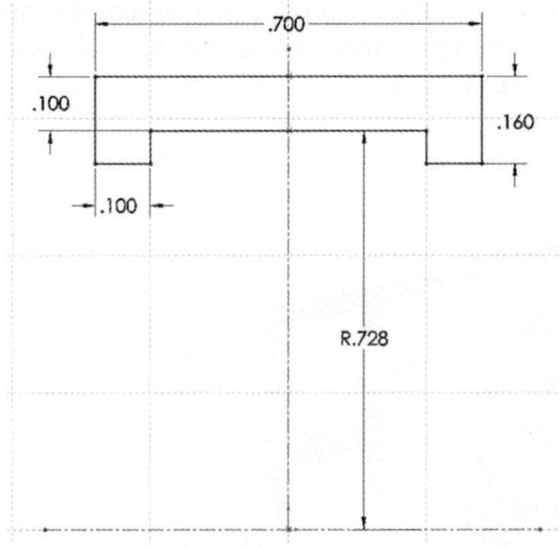

FIGURE P7.4C

d. Insert a second instance of the roller. Use one coincident mate and one tangent mate to place the roller in the race and a second tangent mate to place the two rollers in contact (as shown in a Section View in **Figure P7.4D**).

e. Repeat the previous step until all rollers have been placed in the model. Check that rotation can occur after the mates are complete.

FIGURE P7.4D

P7.5 Create a model of an appropriately sized stepped shaft, and assemble the bearing of **Problem P7.4** onto the shaft. The assembly should be similar to **Figure P7.5**. Show that there is no interference in the assembly.

FIGURE P7.5

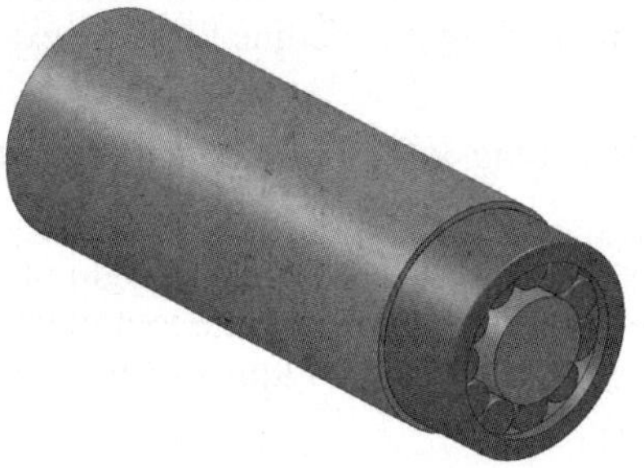

P7.6 Revisit the assembly in **Problem P6.13**. Build a new assembly model by creating subassemblies for the front and back sections, as shown in the exploded view in **Figure P7.6**.

FIGURE P7.6

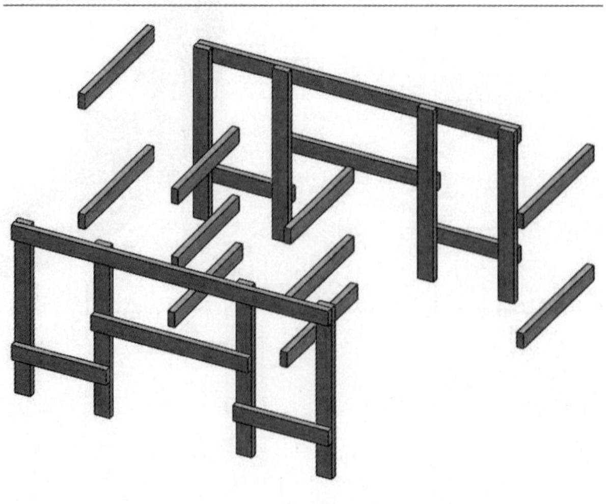

CHAPTER 8

Assembly Drawings

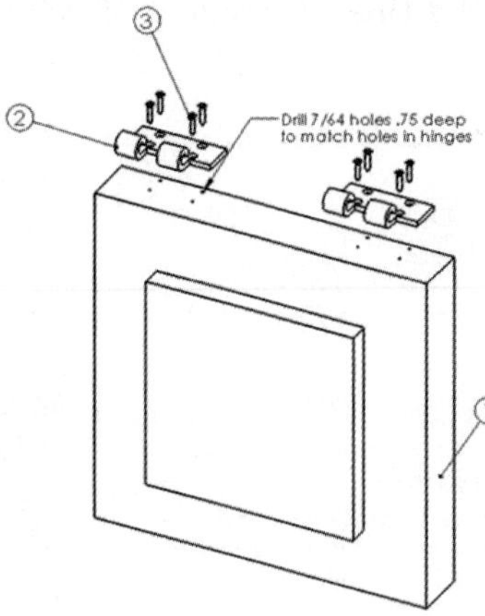

Introduction

In Chapters 6 and 7, the development of solid models of assemblies from SOLIDWORKS® part files was described. In this chapter, the documentation of these assemblies through the use of 2-D assembly drawings will be introduced.

> **Chapter Objectives**
>
> In this chapter, you will:
>
> - create a 2-D assembly drawing,
> - incorporate an exploded view into an assembly drawing, and
> - generate a Bill of Materials.

8.1 Creating an Assembly Drawing

In this section, an assembly drawing of the door assembly created in Chapter 6 will be produced.

Open the assembly file "Door.SLDASM." Make sure that the assembly is in the "collapsed" configuration. Click the arrow next to the New Document Tool, and select Make Drawing from Part/Assembly, as shown in Figure 8.1.

Select the sheet format that you created in Chapter 2, as shown in Figure 8.2, if desired. If you prefer a blank drawing sheet, select A-Landscape and clear the "Display sheet format" box.

FIGURE 8.1

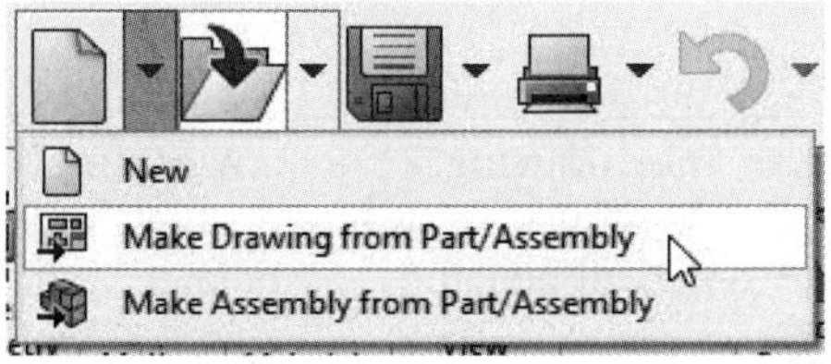

FIGURE 8.2

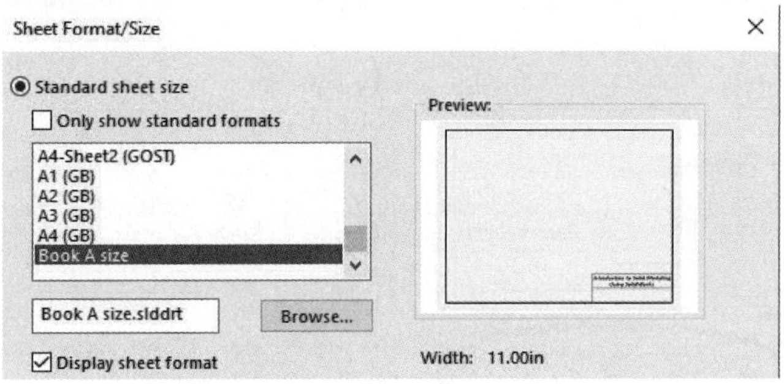

SOLIDWORKS is a registered trademark of Dassault Systémes SolidWorks Corporation.

FIGURE 8.3

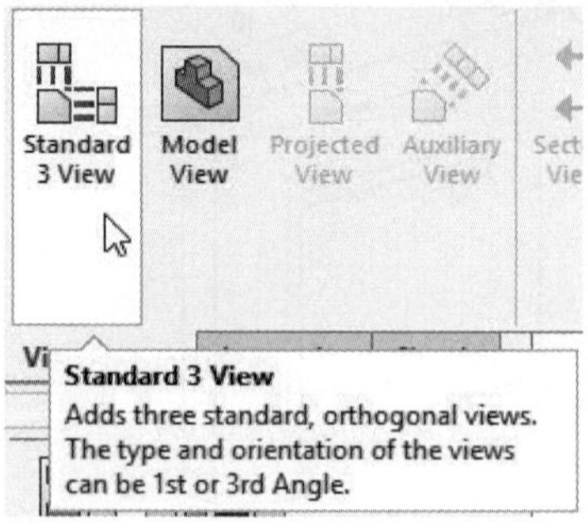

Select the Standard 3 View Tool from the View Layout group of the CommandManager, as shown in Figure 8.3. In the PropertyManager click on the assembly name ("Door") to select it. Click the check mark.

Three standard drawing views will be displayed, as shown in **Figure 8.4**.

If your drawing views display hidden lines, click on the Front View and select the Hidden Lines Removed Style from the Display Style menu of the Heads-Up View Toolbar, as shown in Figure 8.5. Also, click on each of the centerlines, as shown in Figure 8.6, and delete them. Repeat this in all views.

FIGURE 8.4

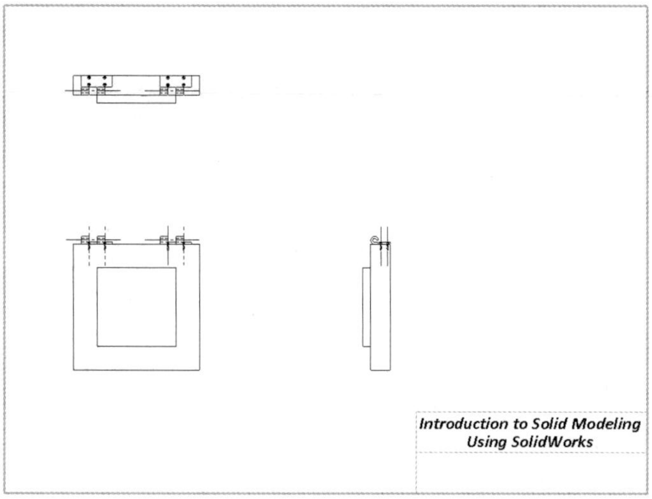

FIGURE 8.5

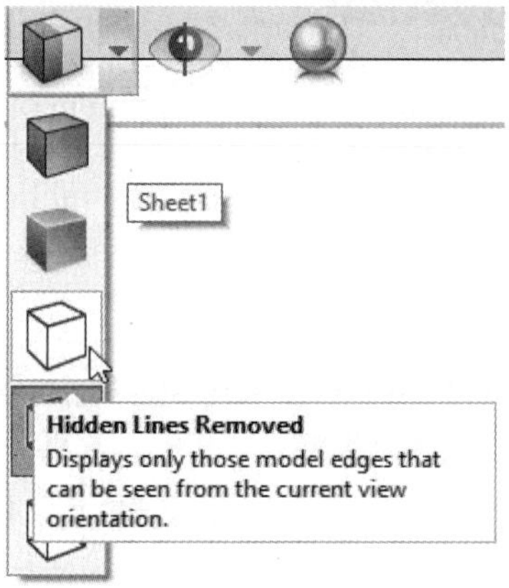

FIGURE 8.6

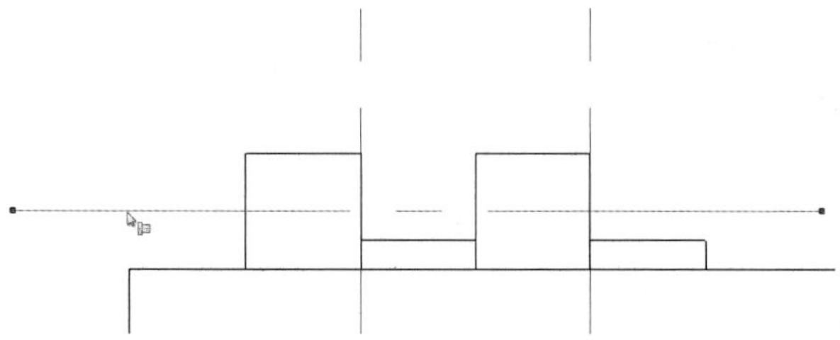

Since an assembly drawing's purpose is to show how components are joined, hidden edges and centerlines are not usually displayed.

Right-click in the blank area of the drawing, expand the menu by clicking the double-arrows if necessary, and select Properties. Change the scale to 1:5, as shown in Figure 8.7.

FIGURE 8.7

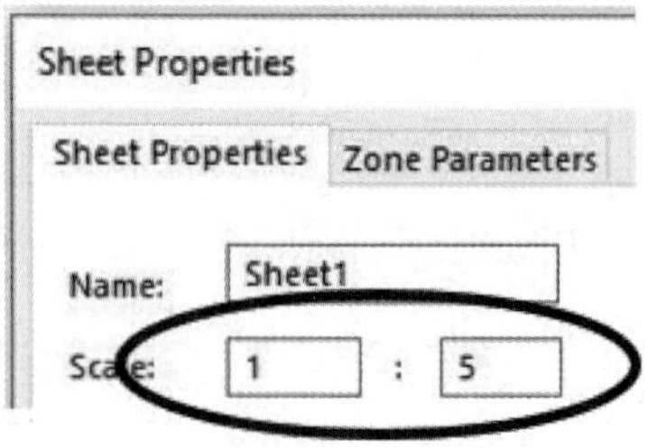

FIGURE 8.8

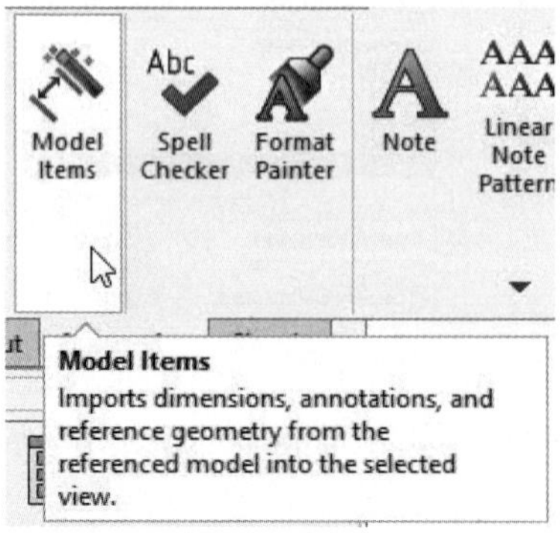

We will now add dimensions to the drawing. For an assembly drawing, we want to show only the dimensions associated with assembly-level features and operations.

Select the Top View. From the Annotation group of the CommandManager, select Model Items, as shown in Figure 8.8.

In the Model Items PropertyManager, select "Only assembly" from the Source menu, as shown in Figure 8.9. Click the check mark to complete the operation.

FIGURE 8.9

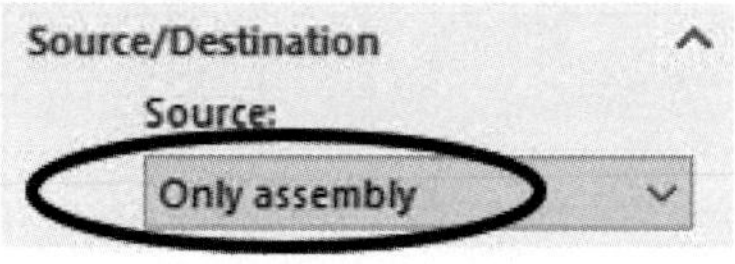

The dimensions related to the placement of the hinges are imported. Drag them into the positions shown in **Figure 8.10**.

FIGURE 8.10

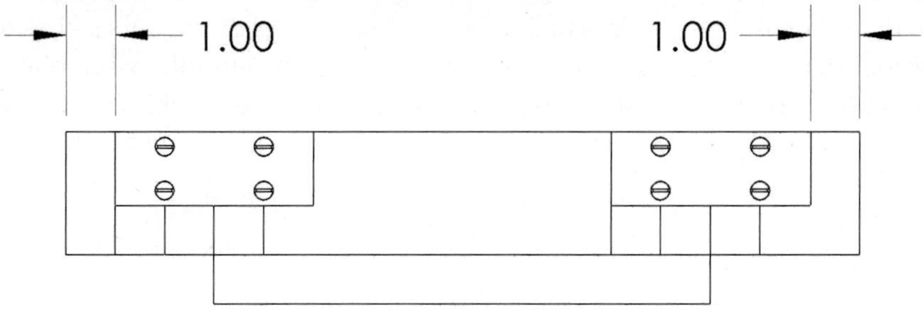

Since the two hinges were placed in the assembly separately, two separate 1-inch dimensions are imported. Note that if we wanted to add an equation to the assembly model, then only one of the dimensions would be needed.

8.2 Adding an Exploded View

In Chapter 6, an exploded view of the door assembly was created. This exploded view can be easily added to the drawing.

Select the Model View Tool from the View Layout group of the CommandManager, as shown in Figure 8.11.

The door assembly should be selected by default in the PropertyManager. If it is not, select it. Click the Next arrow, as shown in Figure 8.12.

FIGURE 8.11

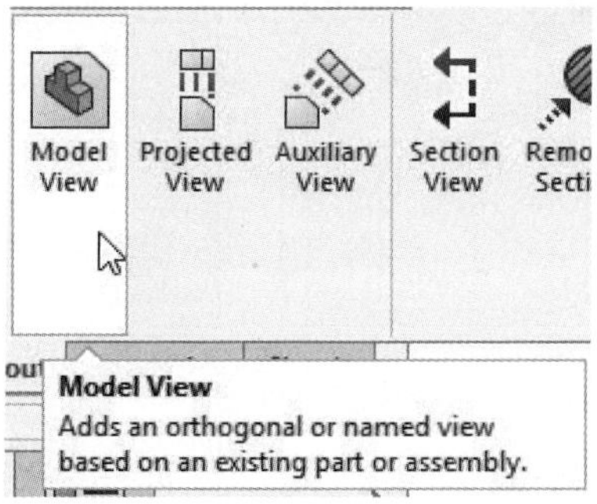

FIGURE 8.12

DESIGN INTENT Assembly-Level Dimensions

When importing dimensions into our assembly drawing, we selected the option labeled "Only assembly," as opposed to "Entire model." As a result, we imported dimensions associated only with assembly-level features. If the other option had been chosen, then the dimensions defining the components would have been imported as well. These dimensions could be edited, resulting in changes to the part files.

Including component dimensions in an assembly drawing is not usually recommended. One reason is that the components are defined in separate drawings, so adding the dimensions to the assembly drawings is redundant. Another reason is that a component may be used in multiple assemblies, so editing a component at the assembly level may produce unexpected changes to other assemblies.

FIGURE 8.13

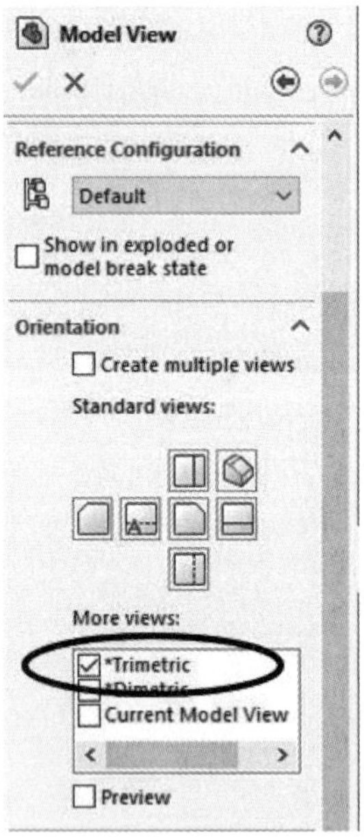

The PropertyManager will now contain a list of available orientations associated with the door assembly.

Insert a single Trimetric View, as shown in Figure 8.13. Scroll down in the PropertyManager, select Hidden Lines Removed as the Display Style, and check the "Use sheet scale" option, as shown in Figure 8.14.

FIGURE 8.14

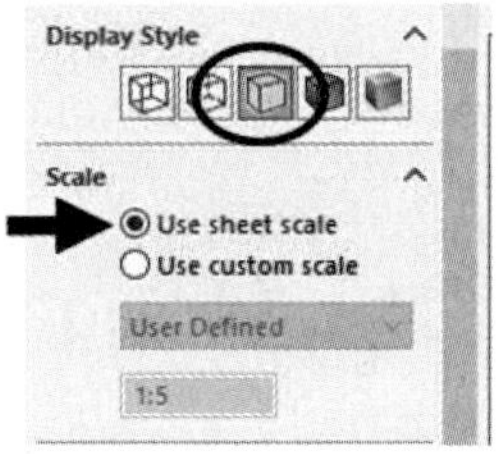

Click on the location in the drawing window where the Trimetric View will appear, as shown in Figure 8.15.
Click and delete each of the centerlines in the new view if necessary.

Right-click on the Trimetric View, and select Properties from the menu.

FIGURE 8.15

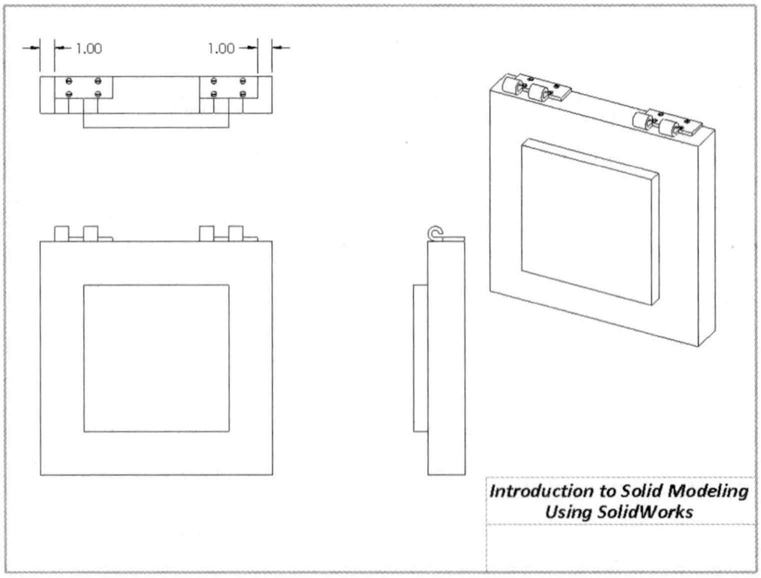

The Drawing View Properties dialog box will appear.

Check the "Show in exploded or model break state" box to change the currently selected view to an exploded view (Figure 8.16), and click OK.

FIGURE 8.16

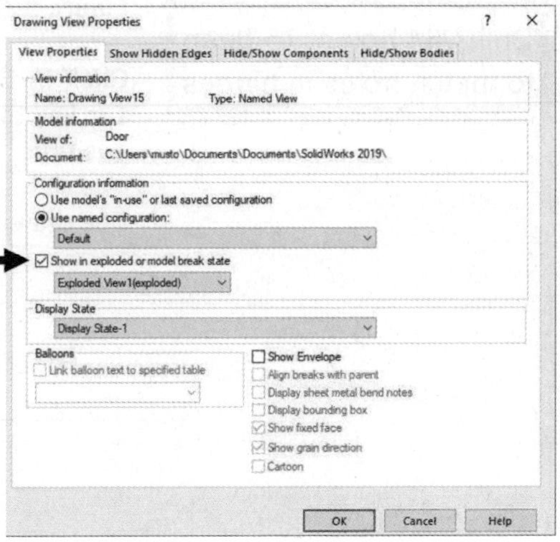

The exploded view will now appear in the drawing window, as shown in Figure 8.17.

FIGURE 8.17

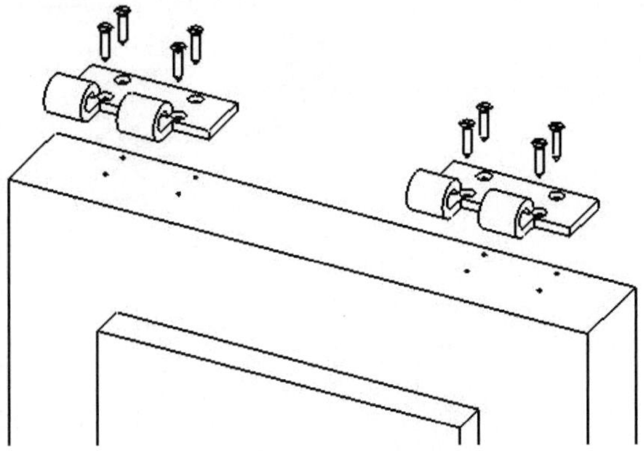

We will now add a note regarding the placement of the holes in the hatch.

FIGURE 8.18

Select the Note Tool from the Annotation group of the CommandManager, as shown in Figure 8.18.

Click on the edge of one of the holes in the exploded view. This will create a leader for the note, as shown in Figure 8.19.

Click on the approximate location of the note, as shown in Figure 8.20.

FIGURE 8.19

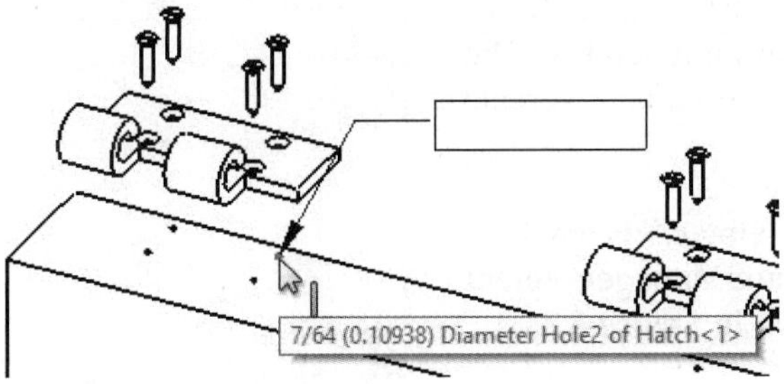

FIGURE 8.20

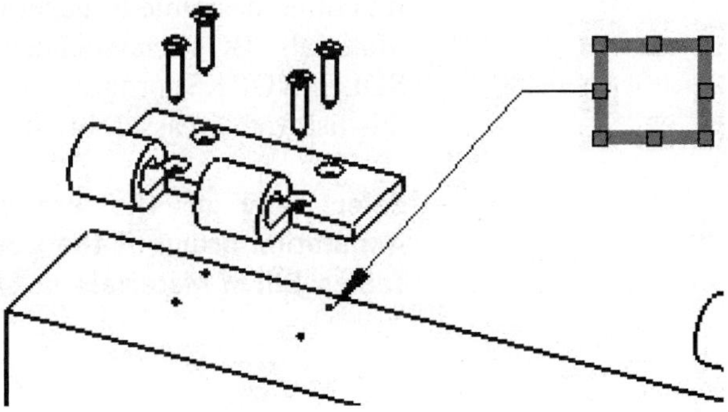

FIGURE 8.21

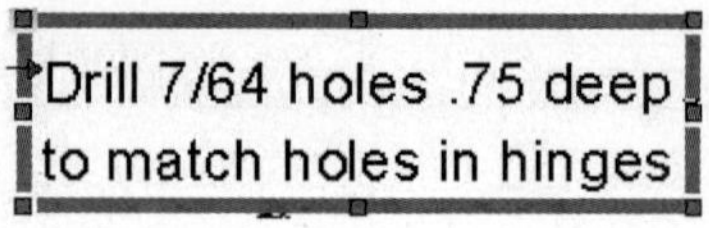

If desired, change the font type and size. Enter the text shown in Figure 8.21.

Click in the drawing window to place the note, and then press the Esc key to end the Note command. Click and drag the note to its final position.

The drawing is shown in **Figure 8.22**.

FIGURE 8.22

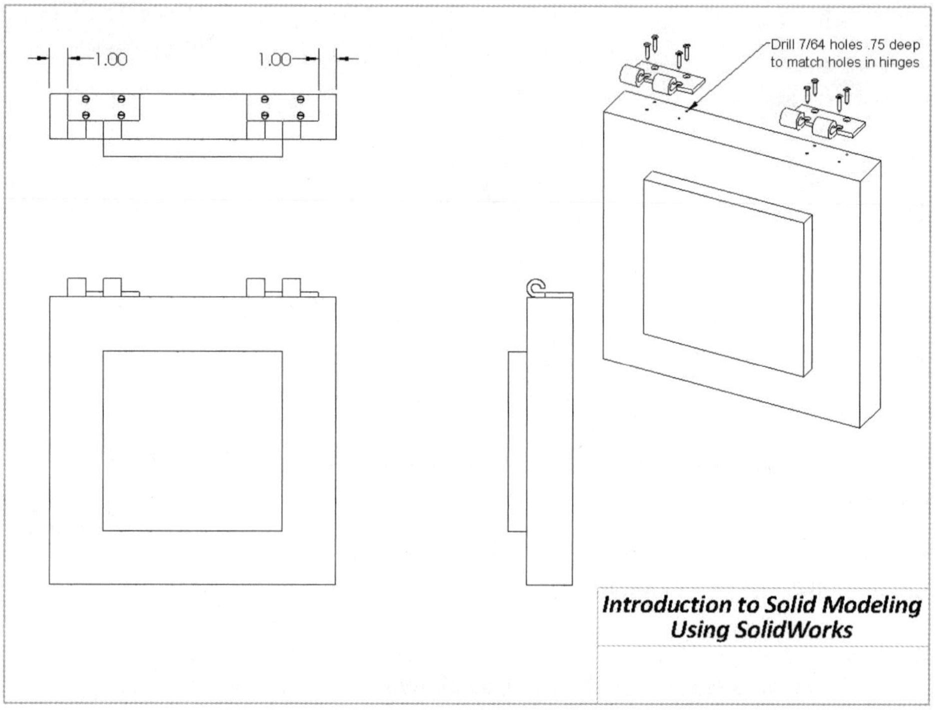

8.3 Creating a Bill of Materials

It is often desirable to generate a parts list, or Bill of Materials (BOM), associated with an assembly. The SOLIDWORKS program can automatically create this list from an assembly file.

Select any of the drawing views. From the Annotation group of the CommandManager, select Tables: Bill of Materials, as shown in Figure 8.23.

FIGURE 8.23

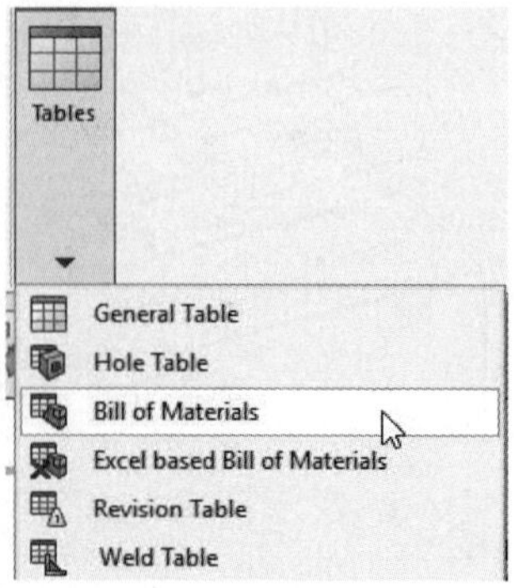

Accept the default selections listed in the PropertyManager, and click the check mark (Figure 8.24).

FIGURE 8.24

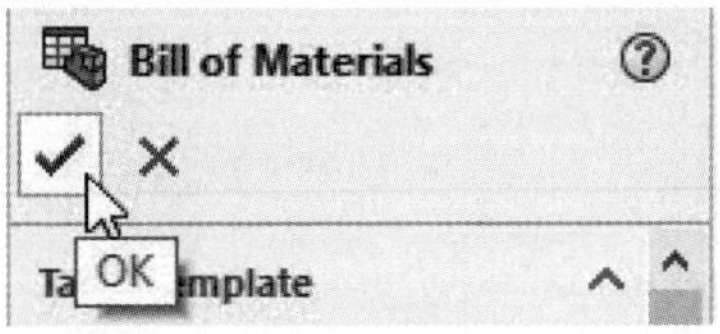

Click anywhere in the drawing to place the Bill of Materials, as shown in Figure 8.25.

FIGURE 8.25

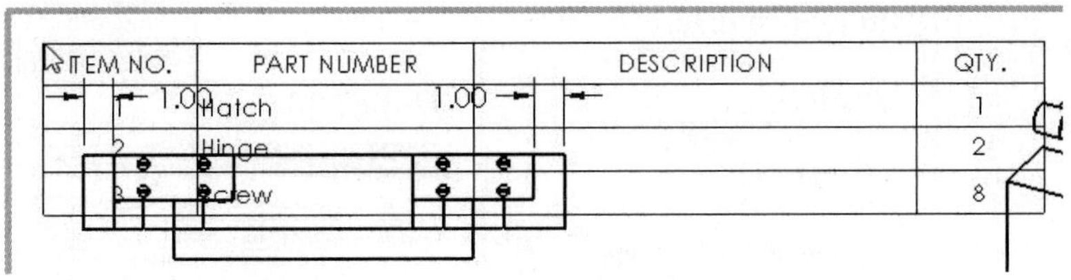

The table can be edited much like an Excel spreadsheet. The formatting of your Bill of Materials may differ slightly, but the following steps in general can be used to control appearance.

Click anywhere in the table to open it for editing.

We will now make changes to the appearance and formatting of the table.

FIGURE 8.26

Click on the arrows in the upper-left corner of the table. This causes the entire table to be selected. Click the "Use document font" icon, as shown in Figure 8.26, to override the default font. Select a new font type, if desired, and a larger size, as shown in Figure 8.27.

Change the width of any column by clicking and dragging on the right boundary of the column, as shown in Figure 8.28.

FIGURE 8.27

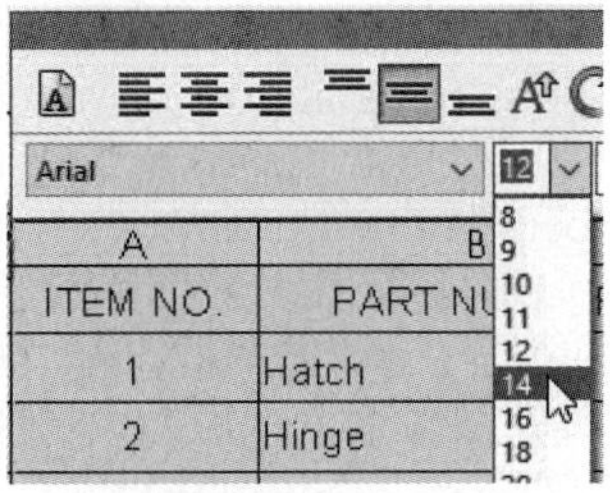

FIGURE 8.28

Columns that are not needed, such as the "Description" column in this example, can be deleted.

Click on the top of column C to select the entire column. Right-click in the selected column, and choose Delete: Column from the menu, as shown in Figure 8.29.

FIGURE 8.29

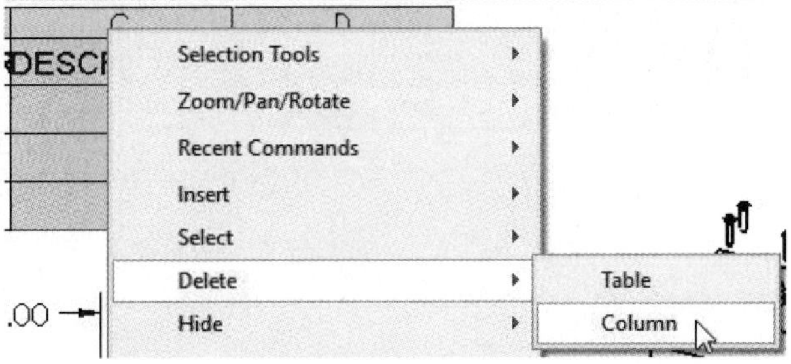

Individual row heights can be changed in a manner similar to column widths; however, we usually want the rows to be the same height. Therefore, we will specify the height for all rows.

FIGURE 8.30

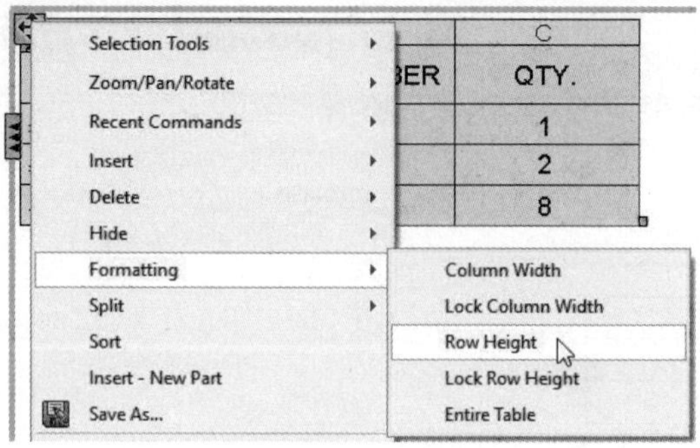

Right-click the arrows in the upper-left corner of the table. Select Formatting: Row Height, as shown in Figure 8.30. Set the row height to 0.30 inches, as shown in Figure 8.31.

Click on the arrows in the upper-left corner to select the entire table. Click the Center Align tool, as shown in Figure 8.32.

FIGURE 8.31

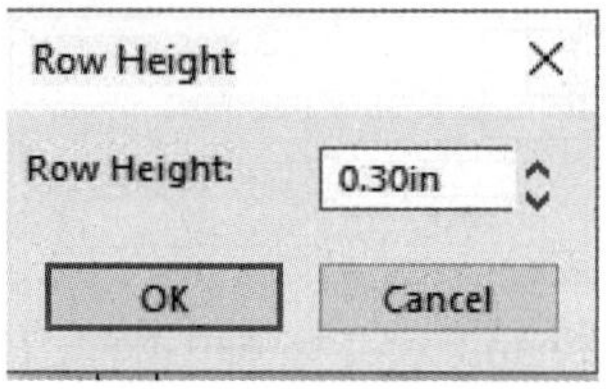

FIGURE 8.32

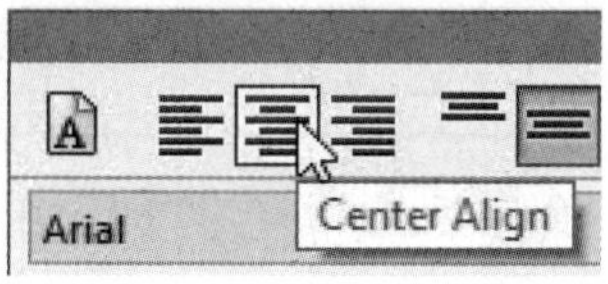

By default, one of the columns is labeled "PART NUMBER." This column contains the file name for each component, as most companies store part files by number rather than the descriptive names that we use in this book. We can easily change the column heading to reflect our method of naming parts.

FIGURE 8.33

Double-click the cell containing the label "PART NUMBER." Change the text to "PART NAME," as shown in Figure 8.33. Click outside the table to accept the change.

Click and drag the move icon in the upper left corner of the Bill of Materials to the position desired.

Our drawing is almost complete, but the item numbers in the Bill of Materials are not linked with the components in the drawing. We will add "balloons" with part numbers to the drawing.

FIGURE 8.34

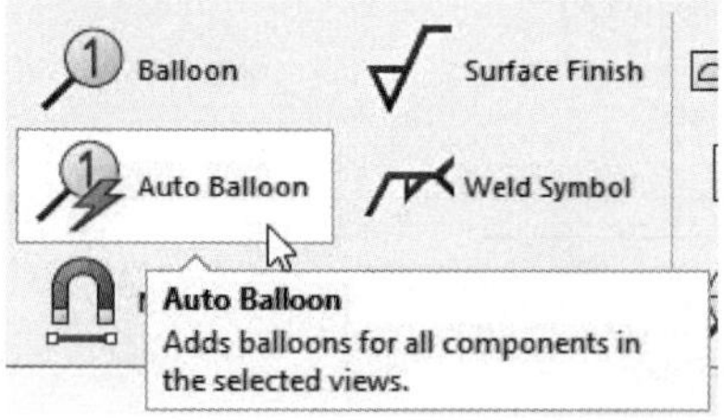

Select the exploded Trimetric View. Select the AutoBalloon Tool from the Annotation group of the CommandManager, as shown in Figure 8.34.

Balloons will be added to the view, as shown in Figure 8.35. The appearance of the balloons can be changed from the PropertyManager.

FIGURE 8.35

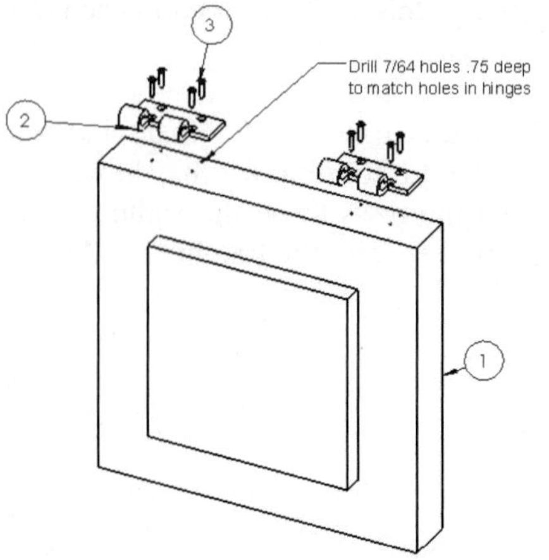

In the PropertyManager, select 1 Character from the Size pull-down menu, as shown in Figure 8.36. Click the check mark to close the PropertyManager. If desired, the balloon font size can be changed by clicking the Options tool, then selecting Document Properties: Annotations: Balloons.

FIGURE 8.36

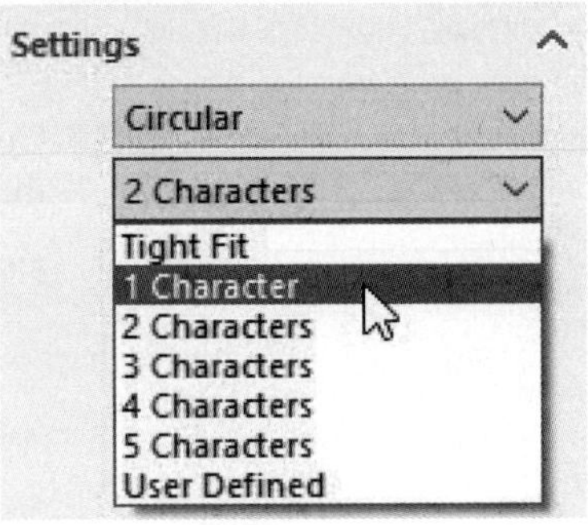

Click and drag the part numbers and views to position the balloons as desired.

You can also move the arrow end of the leader. Note that if the leader is attached to an edge or a point, an arrow appears. If the leader is attached to a face, a dot appears at the end of the leader.

Move the drawing views to the desired locations, and add the note and title with the Note Tool from the Annotation group of the CommandManager. The completed drawing is shown in Figure 8.37.

Save the drawing with the file name "Door," and close it.

Note that the drawing file has the extension .SLDDRW.

FIGURE 8.37

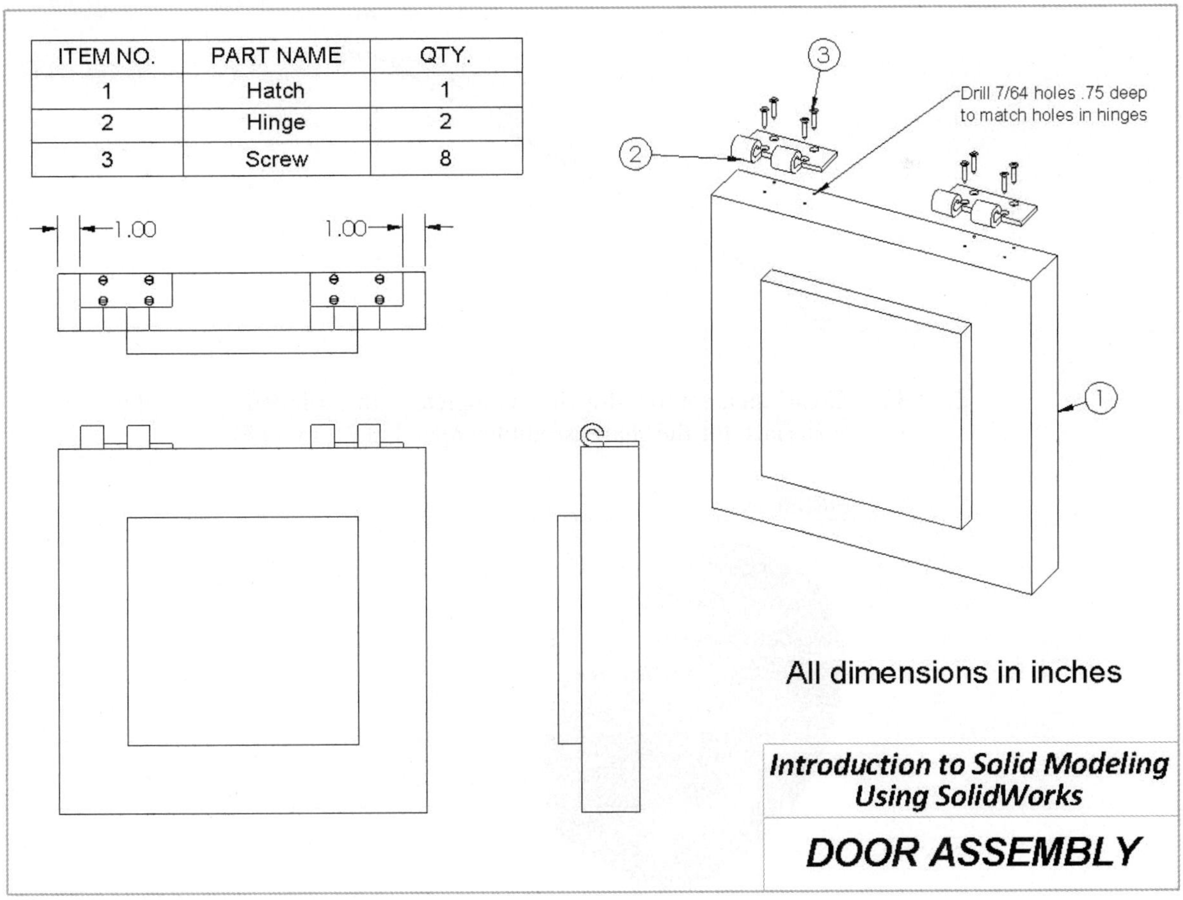

PROBLEMS

P8.1 Create an assembly drawing, complete with exploded view and bill of materials, for the hinge assembly created in **Problem P7.1**.

FIGURE P8.1

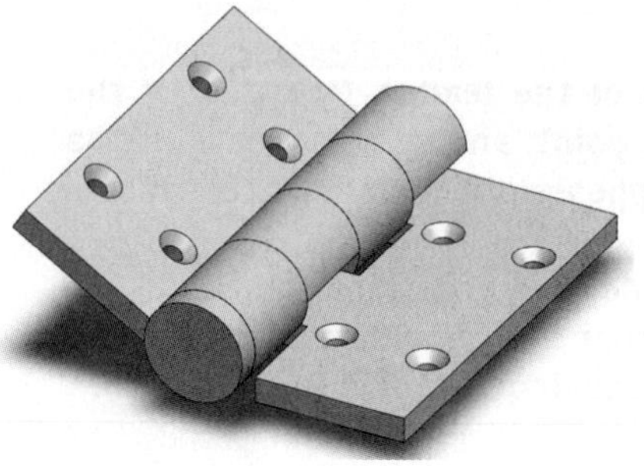

P8.2 Create an assembly drawing, complete with exploded view and bill of materials, for the flagpole assembly created in **Problem P6.3**.

FIGURE P8.2

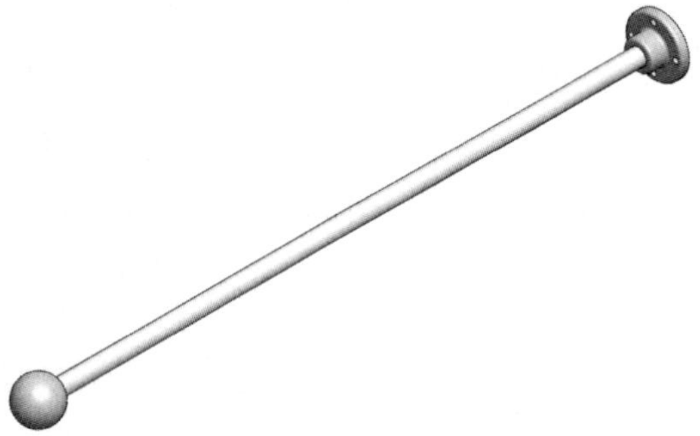

P8.3 Create an assembly drawing, complete with exploded view and bill of materials, for the shaft assembly created in **Problem P6.4**.

FIGURE P8.3

P8.4 Create an assembly drawing, complete with bill of materials, for the frame assembly created in **Problem P6.5**.

FIGURE P8.4

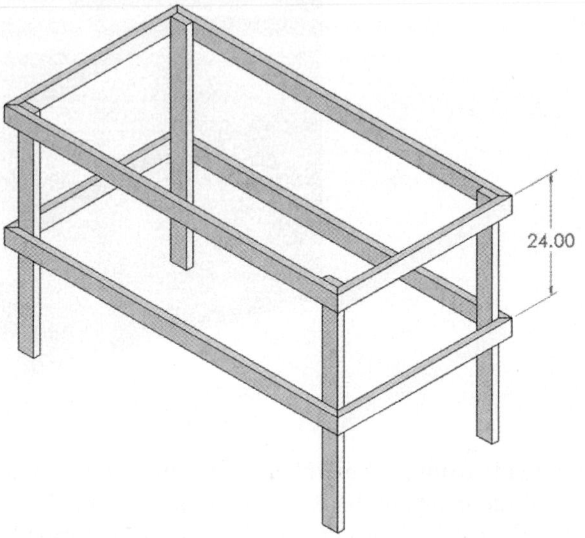

P8.5 Create an assembly drawing, complete with exploded view and bill of materials, for the split hub clamp model created in **Problem P6.11**.

FIGURE P8.5

P8.6 Create an assembly drawing, complete with exploded view and bill of materials, for the workbench model created in **Problem P6.12**.

FIGURE P8.6

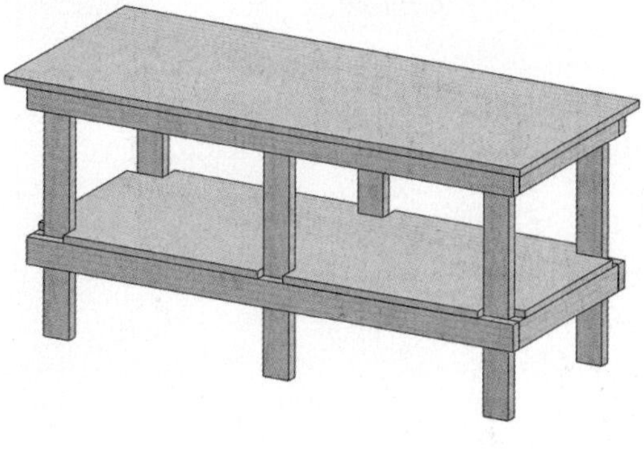

P8.7 Create an assembly drawing of the working hatch assembly created in Chapter 7.

FIGURE P8.7

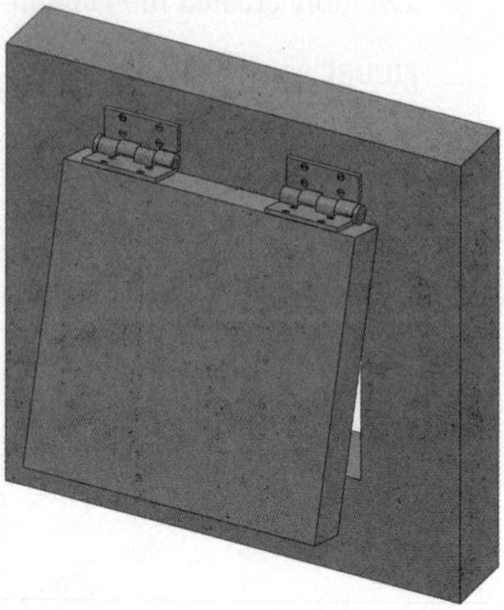

P8.8 Consider the frame made from 2×4 members of **Problem P7.6**. Create assembly drawings of the front and back subassembly and the final assembly. Add bills of materials showing the lengths of each member. To do this, edit the 2×4 part file. For each configuration, select Properties and enter the length in the Description box, as shown in **Figure P8.8A**. Check the Use in bill of materials box. The descriptions can then be shown in the drawing's bill of materials, as shown in **Figure P8.8B**.

FIGURE P8.8A

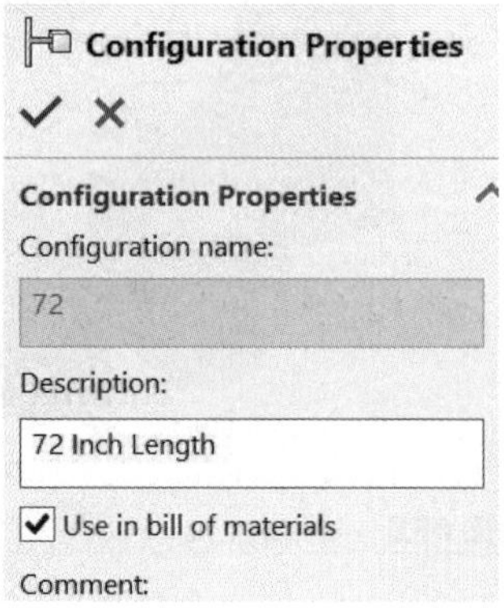

FIGURE P8.8B

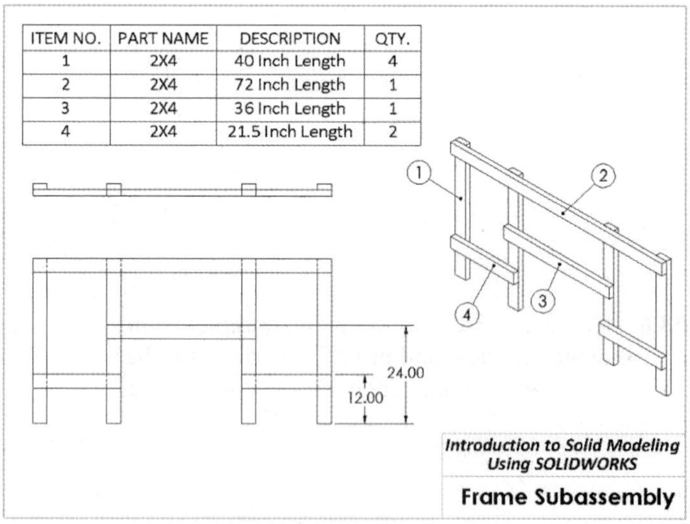

ITEM NO.	PART NAME	DESCRIPTION	QTY.
1	2X4	40 Inch Length	4
2	2X4	72 Inch Length	1
3	2X4	36 Inch Length	1
4	2X4	21.5 Inch Length	2

PART TWO

Applications of SOLIDWORKS®

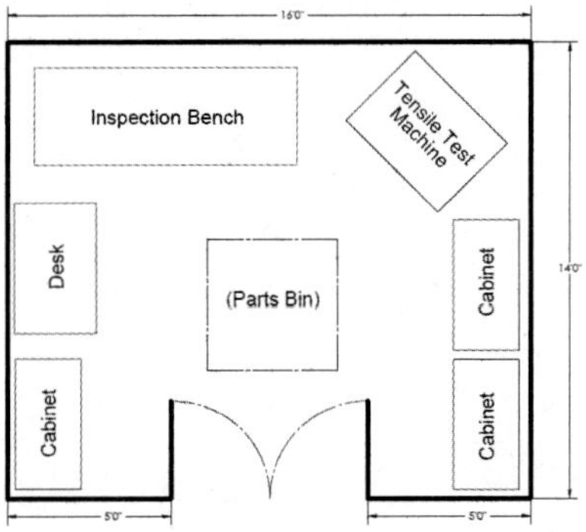

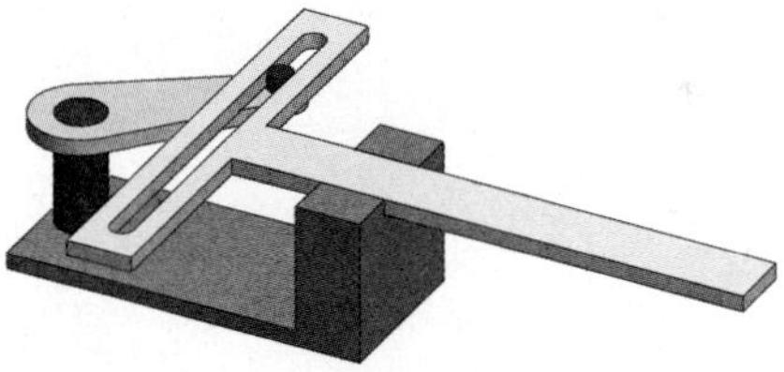

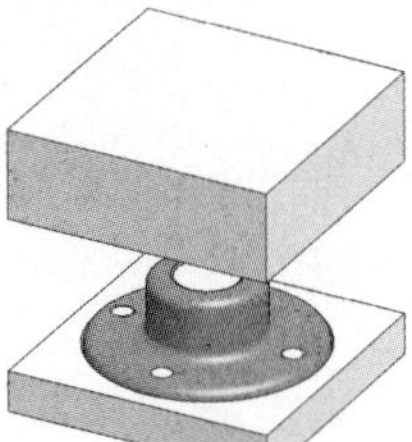

CHAPTER 9

Generation of 2-D Layouts

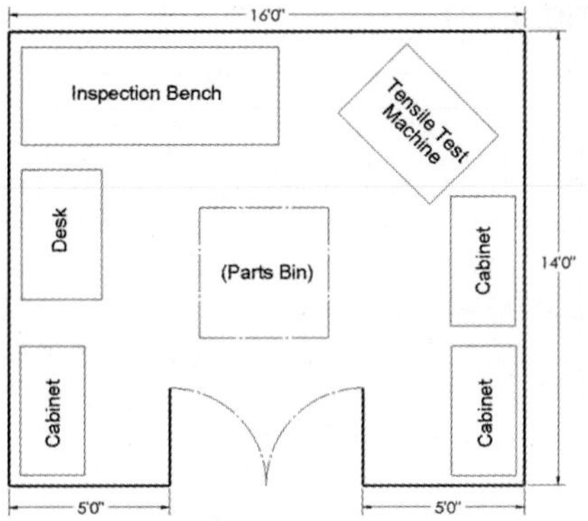

Introduction

Earlier we learned to create 2-D drawings from solid parts and assemblies. For some tasks, however, working directly in the 2-D environment is preferred. For example, floor plans and site drawings, plant equipment layouts, and electrical schematic drawings are usually created in 2-D. The SOLIDWORKS® program can be used for these applications, and the ease of changing dimensions allows multiple configurations to be quickly evaluated.

Chapter Objectives

In this chapter, you will:

- choose and set the scale for a 2-D layout,
- use a unit system of feet and inches,
- create a floor plan and learn how to modify the layout to evaluate alternate configurations,
- expand your knowledge of using relations with 2-D entities, and
- calculate the areas, centroid locations, and moments of inertias of 2-D areas.

9.1 A Simple Floor Plan Layout

In this exercise, we will prepare a layout drawing of a simple quality assurance lab for a manufacturing shop. The following items need to be placed in the lab:

- A tensile test machine, with a rectangular "footprint" of 4 feet wide by 3 feet deep
- An inspection bench, 8 feet by 3 feet
- A desk, 4 feet by 30 inches
- Three cabinets for storing measuring tools and fixtures, each 4 feet wide by 2 feet deep

SOLIDWORKS is a registered trademark of Dassault Systémes SolidWorks Corporation.

FIGURE 9.1

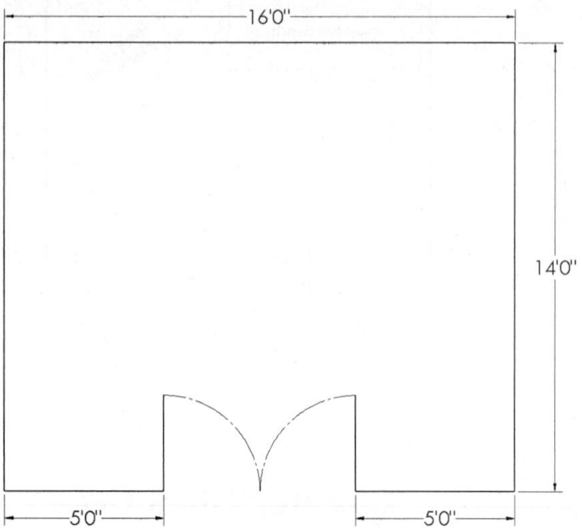

Parts will be brought in bins that are 4 feet square, so there will need to be room for one parts bin. The room that has been proposed for the lab is shown in Figure 9.1. We would like to prepare a layout drawing to show how the equipment should be placed in the room.

Begin by opening a new drawing. When prompted to select a sheet format, pick A-Landscape as the paper size, and make sure the "Display sheet format" box is unchecked, as shown in Figure 9.2. Click OK. If the Model View dialog appears in the PropertyManager, click the x to end the Model View command.

When we import parts into drawings, an appropriate scale is automatically set. When preparing a 2-D layout, we must specify the scale. To make the room fill most of the paper, we can set the scale at 1 inch equals 2 feet. Thus, the 14-foot dimension will appear as 7 inches on the drawing.

Right-click anywhere in the drawing area, and select Properties. Set the scale to 1:24, as shown in Figure 9.3. Click the Apply Changes button.

FIGURE 9.2

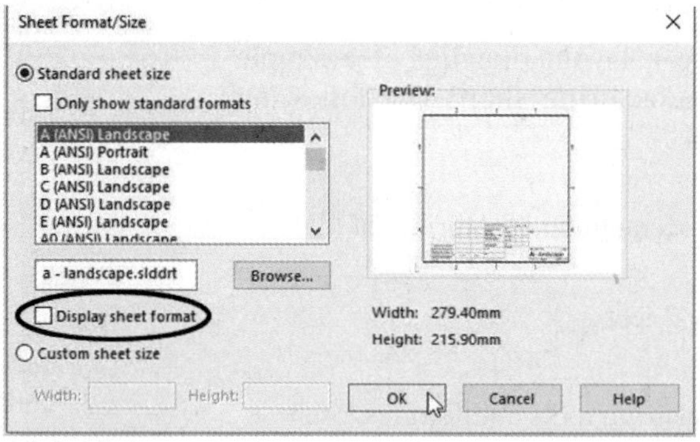

FIGURE 9.3

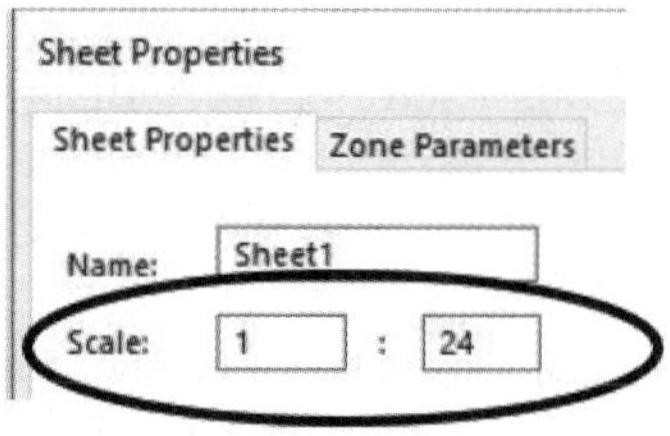

We can use units of feet, inches, or mixed feet and inches for this layout. We will use mixed feet and inches.

Click on the units designation on the Status Bar, and select Edit Document Units. Select Custom as the unit system, and feet & inches from the pull-down menu of length units. Set the decimal places to None, as shown in Figure 9.4. Click OK.

FIGURE 9.4

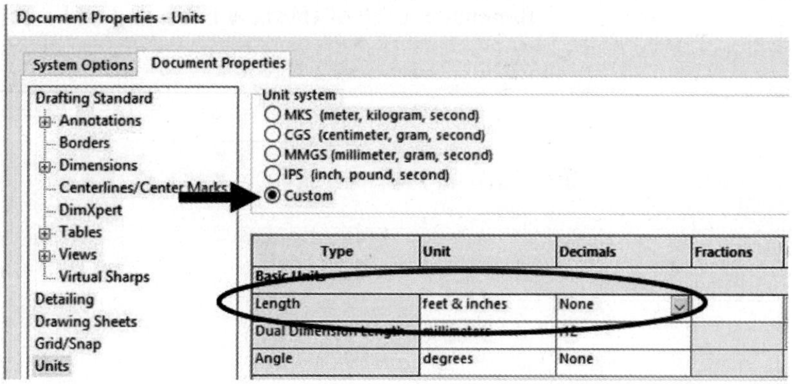

Select the Corner Rectangle Tool from the Sketch group of the CommandManager, and drag out a rectangle. Dimension the sides of the rectangle, as shown in Figure 9.5. When entering the dimensions as feet, include the foot symbol (') after the number. Otherwise, the dimension units default to inches.

If sketch relation icons do not appear on the drawing, select View: Hide/Show: Sketch Relations from the main menu.

With the Dimension Tool turned off, click and drag the upper-left corner of the rectangle until the shape is approximately centered on the sheet. Select Fix in the PropertyManager to fix the point on the sheet, as shown in Figure 9.6.

FIGURE 9.5

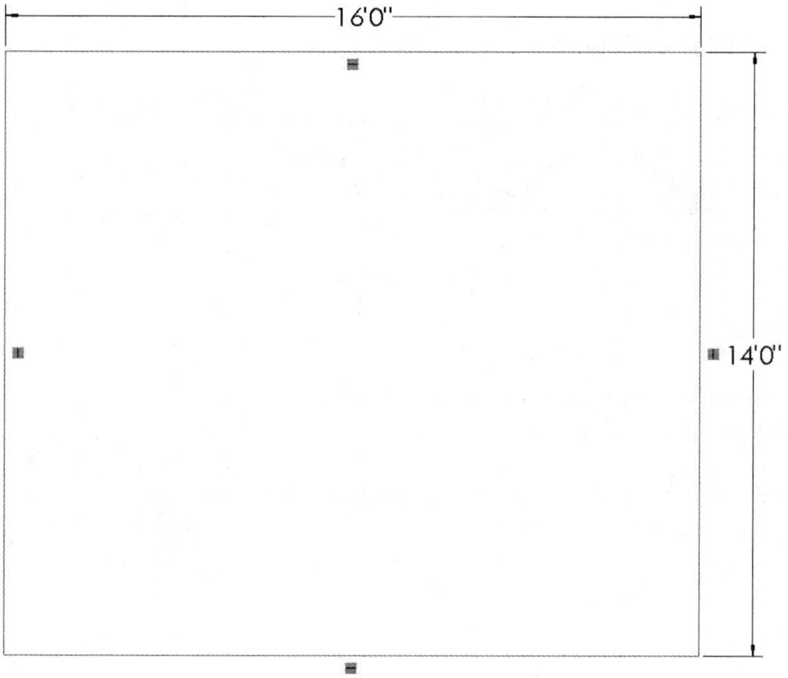

FIGURE 9.6

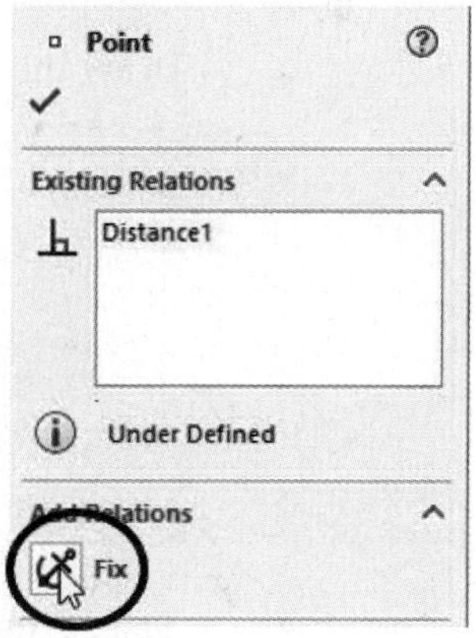

Delete the bottom line and replace it with the two lines shown in Figure 9.7. Dimension each of the new lines, and add a collinear relation to the lines.

FIGURE 9.7

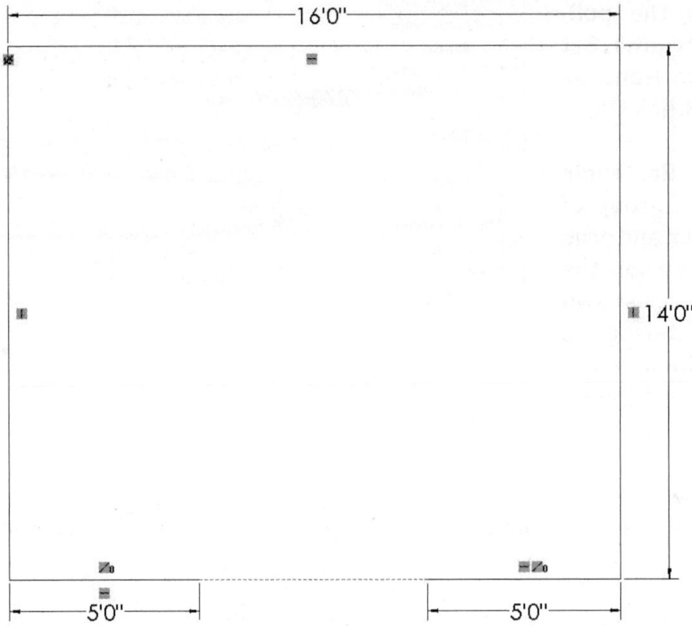

The drawing should be fully defined.

There are several ways that the doors and the arcs that represent their swing paths can be drawn and dimensioned. The method here uses relations to define the geometry.

Draw the vertical lines as shown in Figure 9.8.

FIGURE 9.8

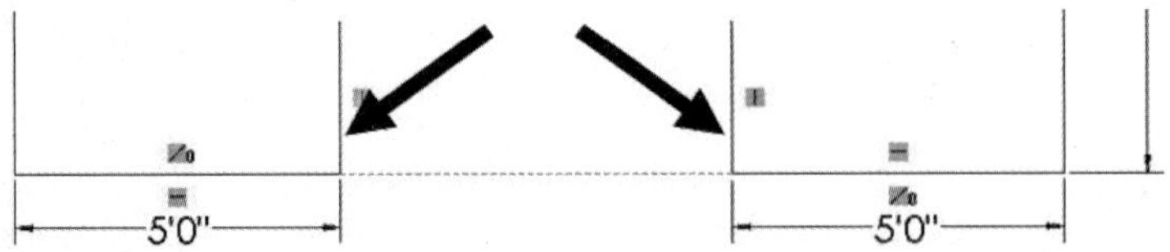

Select the Centerpoint Arc Tool from the Sketch group of the CommandManager, as shown in Figure 9.9.

This tool allows you to construct an arc by picking the center of the arc and then the two endpoints.

FIGURE 9.9

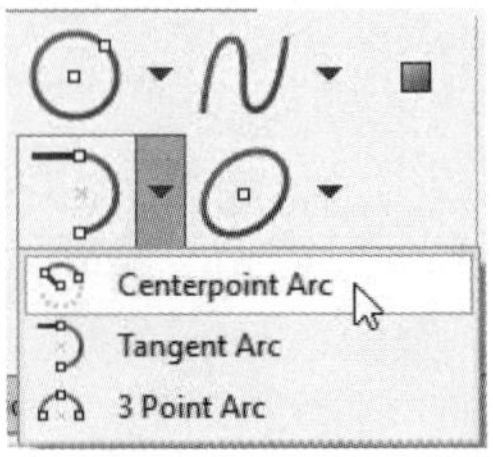

Move the cursor directly over the intersection of one of the vertical lines and the adjacent horizontal line, as shown in Figure 9.10. Click once on the intersection to set the center point of the arc. Click again on the other end of the vertical line, as shown in Figure 9.11, to set the starting point of the arc. Drag an arc 90 degrees, as shown in Figure 9.12, and click to finish the construction of the arc.

FIGURE 9.10

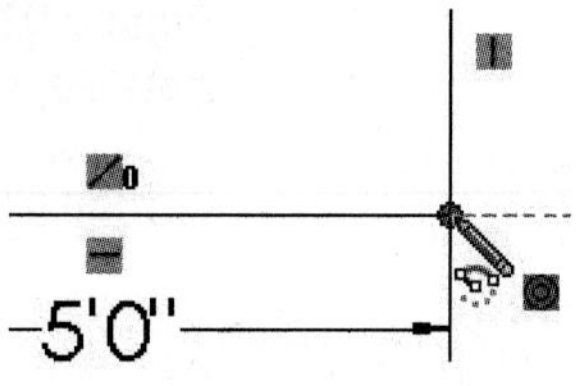

FIGURE 9.11

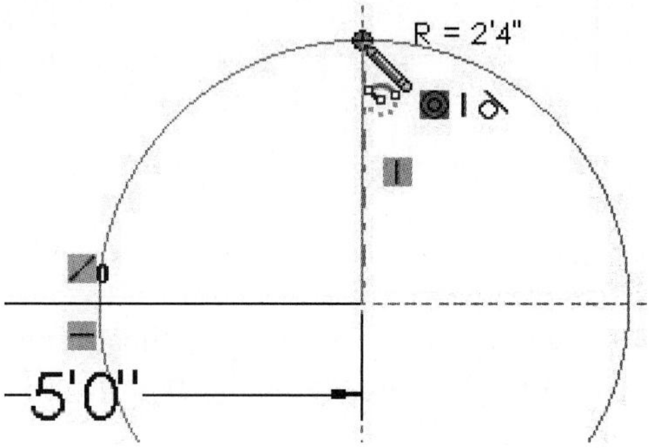

FIGURE 9.12

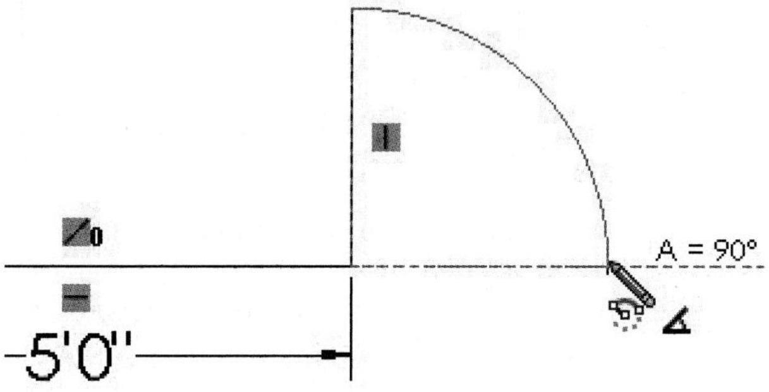

Check the "For construction" box in the PropertyManager.

Repeat for the other door, as shown in Figure 9.13.

With the Ctrl key depressed, select the endpoints of the arcs, as shown in Figure 9.14.

FIGURE 9.13

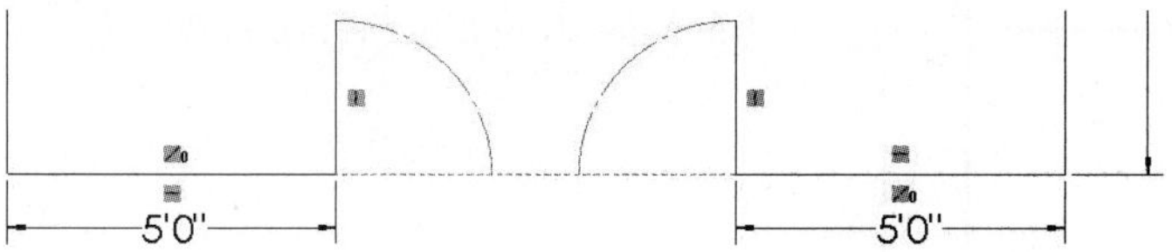

FIGURE 9.14

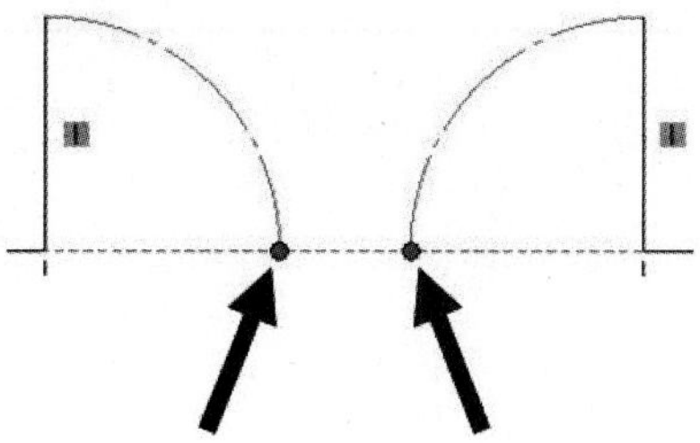

In the PropertyManager, click Merge, which joins the two points, as shown in Figure 9.15. Select both arcs, and add Equal and Tangent relations. The arcs will now appear as shown in Figure 9.16.

FIGURE 9.15

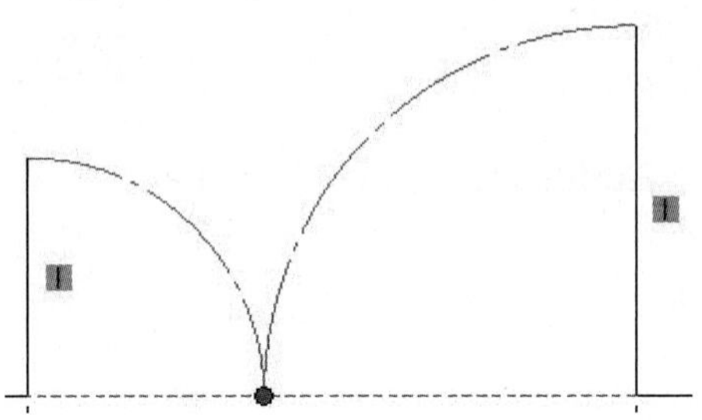

FIGURE 9.16

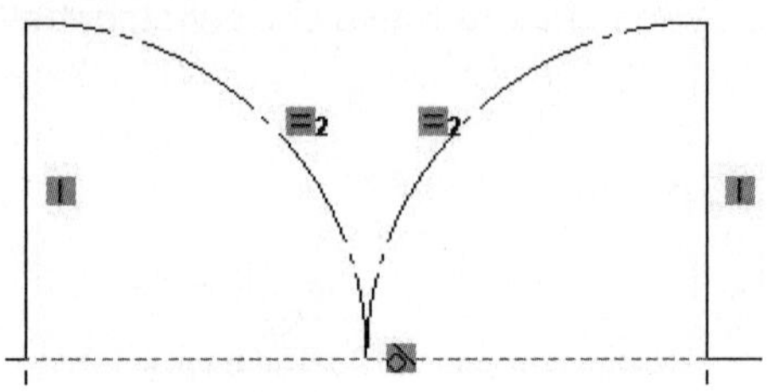

The drawing should once again be fully defined.

In order to make the outer walls and doors stand out, we will make those lines thicker. For this, we will use the Line Format toolbar, as shown in Figure 9.17.

FIGURE 9.17

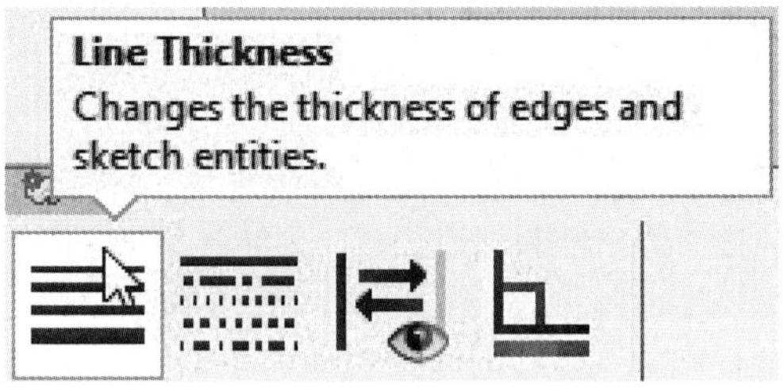

If the Line Format toolbar is not shown, right-click in the Menu Bar or CommandManager. Click on Line Format to add the toolbar.

Select the lines representing the walls and doors. In the Line Format toolbar, select the Line Thickness Tool, as shown in Figure 9.17, and pick a thicker line than the default. Click the check mark in the PropertyManager to apply the selected line thickness.

FIGURE 9.18

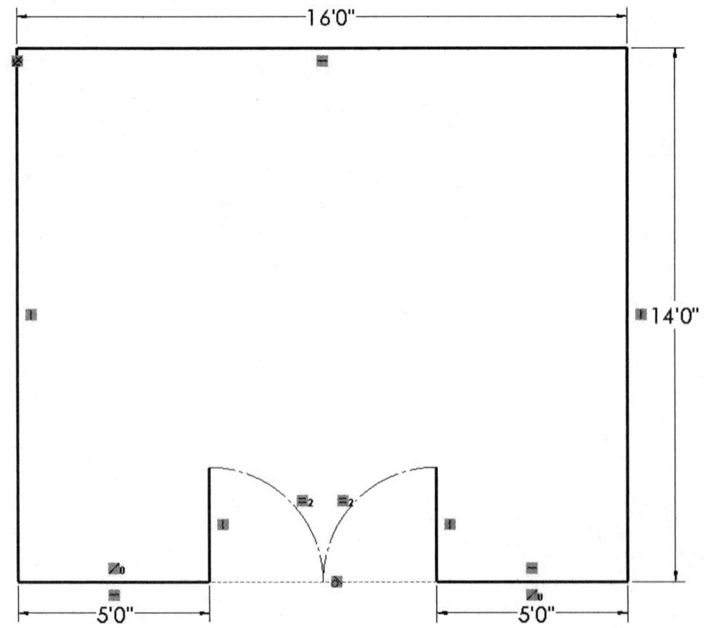

The walls and doors now stand out from the dimension and construction lines, as shown in Figure 9.18.

We will now place the tensile test machine in the room.

Draw a rectangle somewhere in the room, and dimension it as 4 feet by 3 feet, as shown in Figure 9.19.

FIGURE 9.19

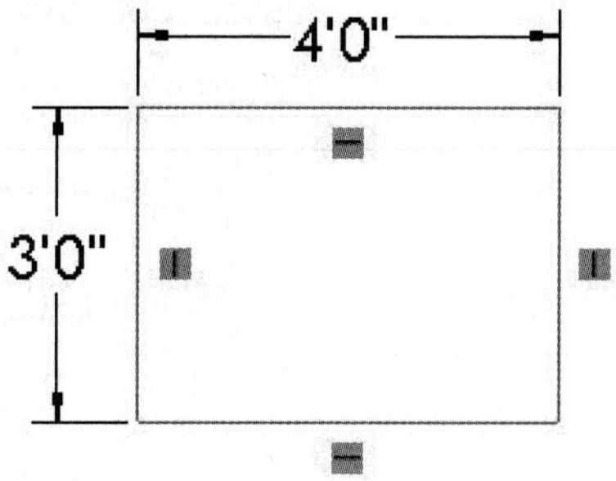

The dimensions that we added are not necessary to show in the drawing, but are required to set the size of the machine's footprint. We can hide the dimensions so they do not show, but still control the size of the rectangle.

Turn off the Smart Dimension Tool. Right-click on one of the dimensions and select Hide, as shown in Figure 9.20. Repeat for the other dimension.

FIGURE 9.20

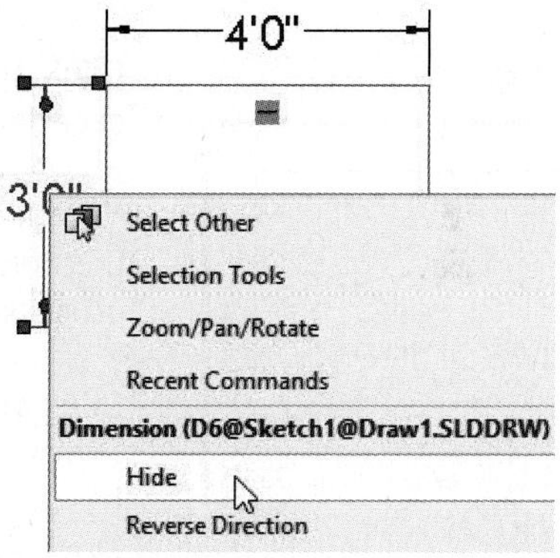

The rectangle is fully defined except for its position on the drawing. By clicking and dragging on one of the corners, you can move it.

Click and drag a corner of the rectangle until it is placed near the upper-right corner of the room, as shown in Figure 9.21.

Note: if you drag a corner until it contacts another point or a line, as shown in Figure 9.22, then a relation is automatically created with that entity, and you will be unable to move the rectangle away from the entity. If you desire to do so, then click on the coincident icon and delete it.

While it is not necessary to fix the rectangle in place, doing so is a good idea in that it will make the drawing fully defined if all other required dimensions and relations are present. If you fix the position of the rectangle and the status bar reports that the drawing is still underdefined, then you will probably want to check existing dimensions and relations to see why.

FIGURE 9.21

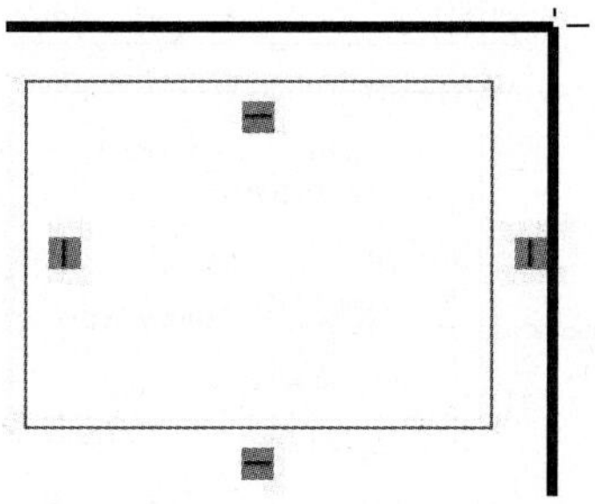

FIGURE 9.22

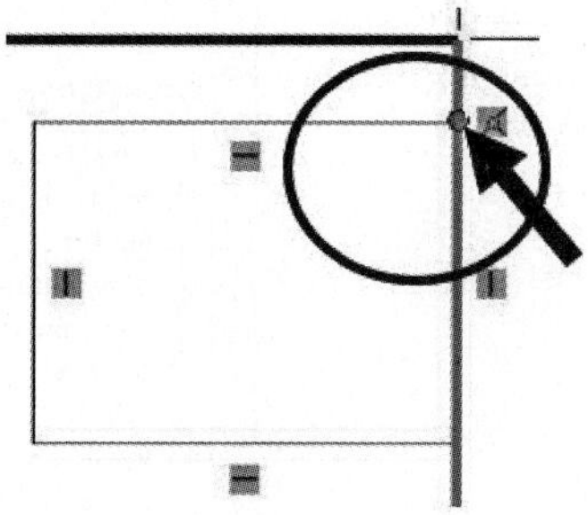

Click on a corner of the rectangle to select that point and display its properties in the PropertyManager. Click Fix to fix that point, and close the PropertyManager. Press Esc to deselect the point.

The drawing should now be fully defined. We will now label the rectangle as the location of the tensile test machine.

Click on the Annotation tab of the CommandManager. Select the Note Tool, as shown in Figure 9.23. Click at the approximate location of the note, near the new rectangle.

FIGURE 9.23

Set the desired font (Arial), size (20 point), and alignment (centered), as shown in Figure 9.24. Type "Tensile Test Machine" in the text box.

FIGURE 9.24

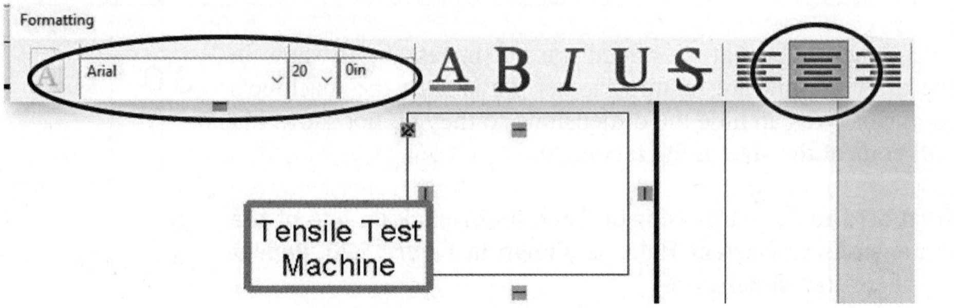

FIGURE 9.25

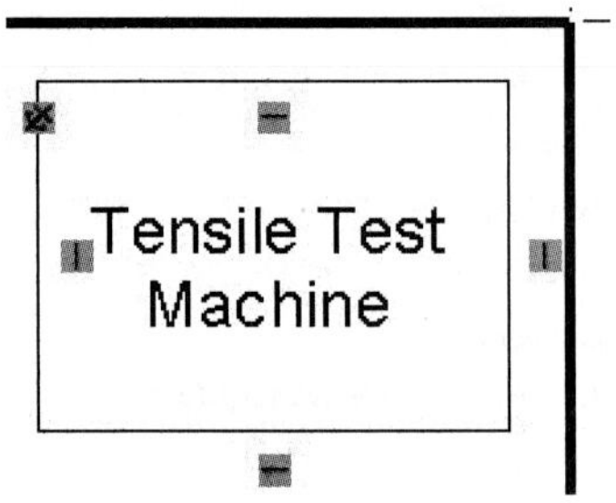

Click anywhere outside of the note box, and then Esc to end the note creation.

(If you want to place the same note in another location, then you can click the note down in multiple locations before using the Esc key to end the process.)

Click and drag the note to its final position, as shown in Figure 9.25.

If you want the note font changed for the entire drawing, do so from Document Properties: Annotations, as shown in Figure 9.26.

FIGURE 9.26

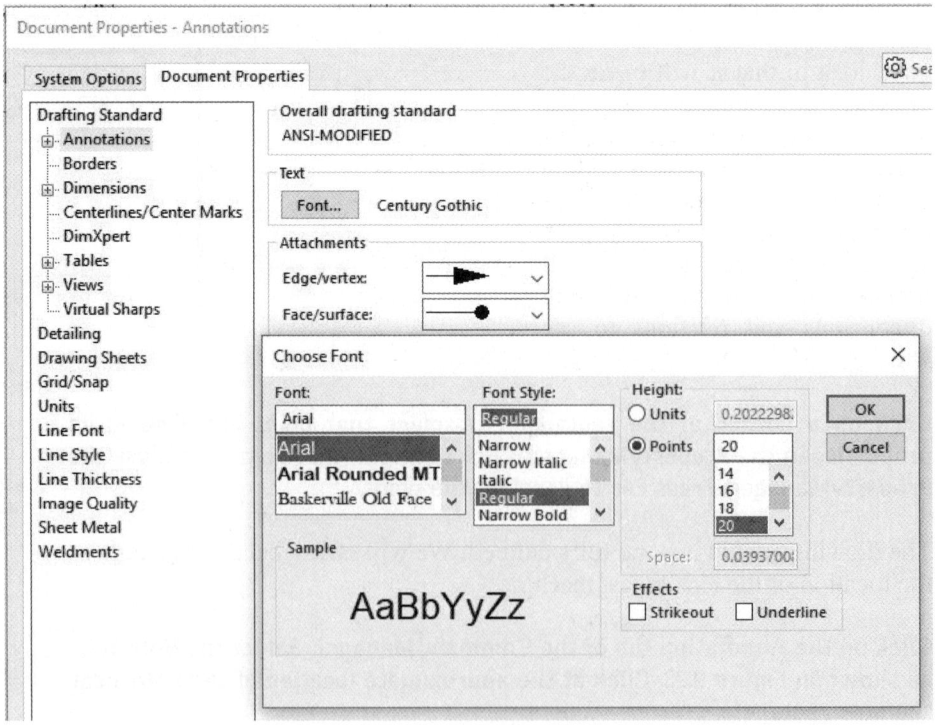

Add the first of the cabinets by drawing and dimensioning the rectangle as shown in Figure 9.27. Hide the dimensions.

FIGURE 9.27

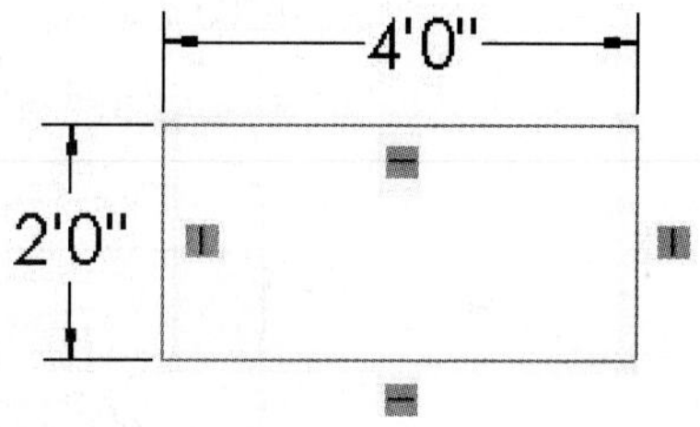

After drawing the cabinet and moving its location around, you may decide that turning the cabinet 90 degrees will allow it to fit into the space better. The easiest way to do this is simply to switch the values of the dimensions. This will require showing and editing the currently hidden dimensions.

Select View: Hide/Show: Annotations from the main menu, as shown in Figure 9.28. The hidden dimensions will be shown in gray. Click on each dimension that you want to show, as shown in Figure 9.29. Select View: Hide/Show: Annotations again or Esc to return to the editing mode. Change the cabinet dimensions as shown in Figure 9.30. Hide the dimensions again.

FIGURE 9.28

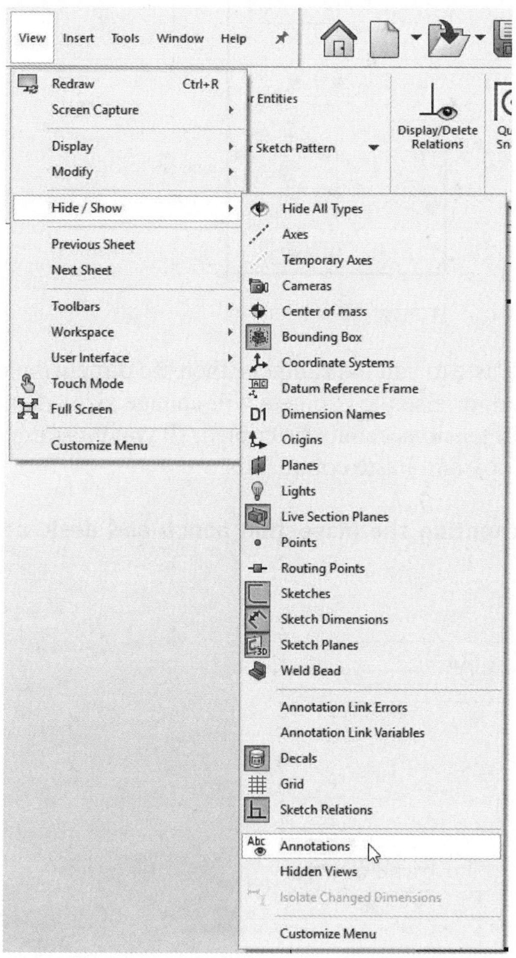

Add a note identifying the cabinet. Rotate the text 90 degrees by clearing the "Use document font" box and changing the angle in the PropertyManager, as shown in Figure 9.31. Click the check mark.

FIGURE 9.29

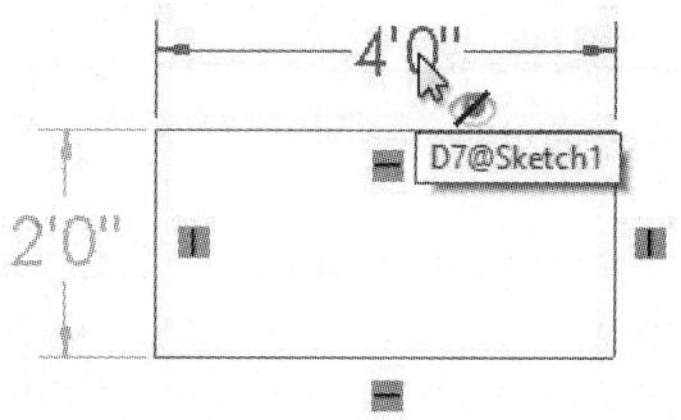

FIGURE 9.30

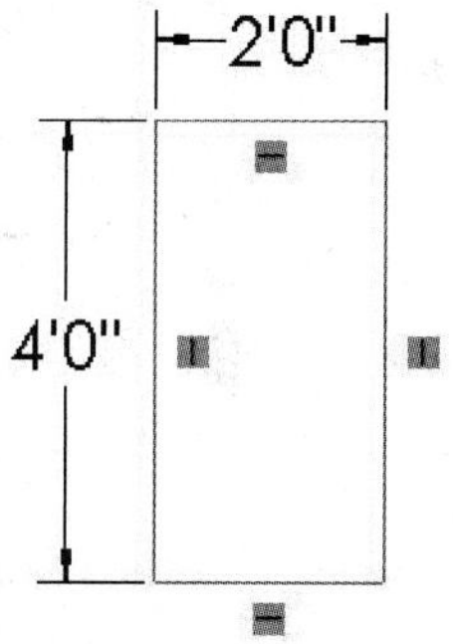

FIGURE 9.31

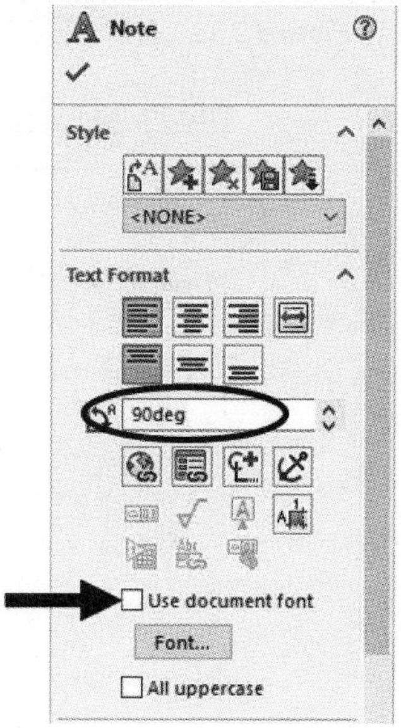

Add the other two cabinets. Dimension, locate, and label them, as shown in Figure 9.32.

FIGURE 9.32

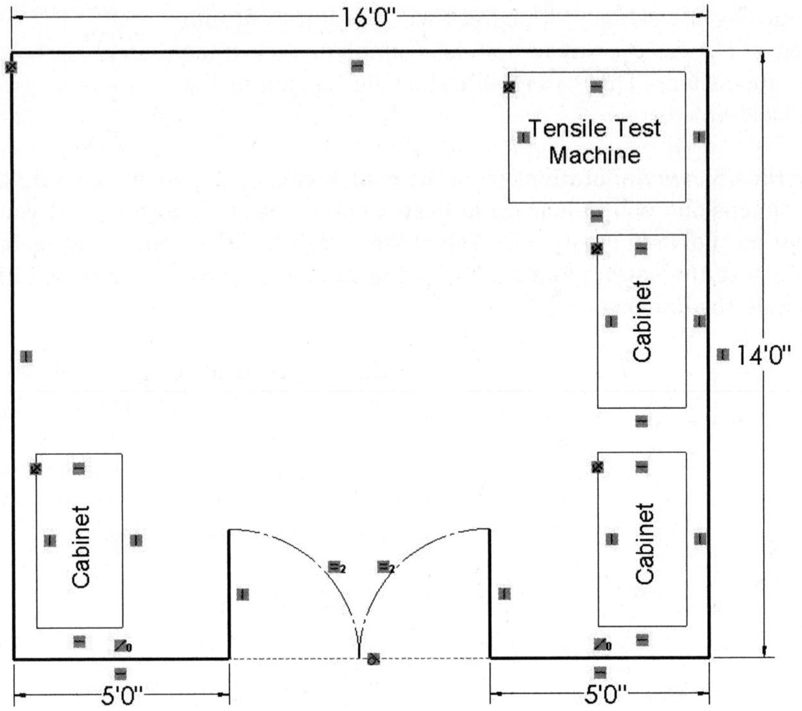

Note that if you want to use Copy and Paste to add the cabinets, then the dimensions defining their sizes must also be copied, or else the cabinets will change size when dragged into a new position. Hidden dimensions cannot be copied, so you must first unhide the dimensions before using Copy and Paste commands.

Add and dimension rectangles representing the inspection bench and desk, as shown in Figure 9.33.

FIGURE 9.33

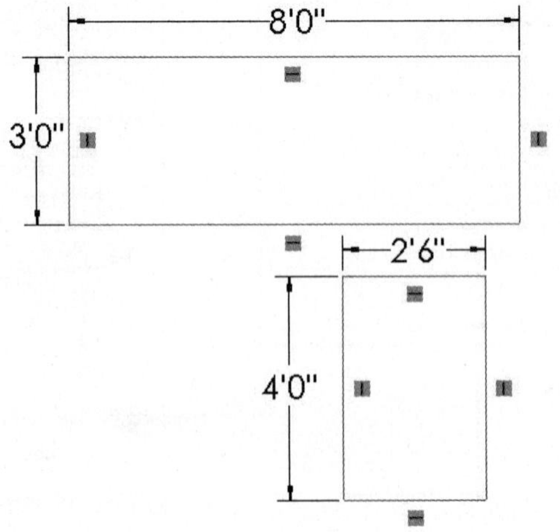

Label the inspection bench and desk. Add and label the 4-foot-square parts bin, with the lines representing it shown as construction lines. Fix all of the entities in position so that the drawing is fully defined. Hide the dimensions defining the bench, desk, and parts bin. The finished drawing is shown in Figure 9.34. Note that the sketch relation icons are not shown when the drawing is printed.

FIGURE 9.34

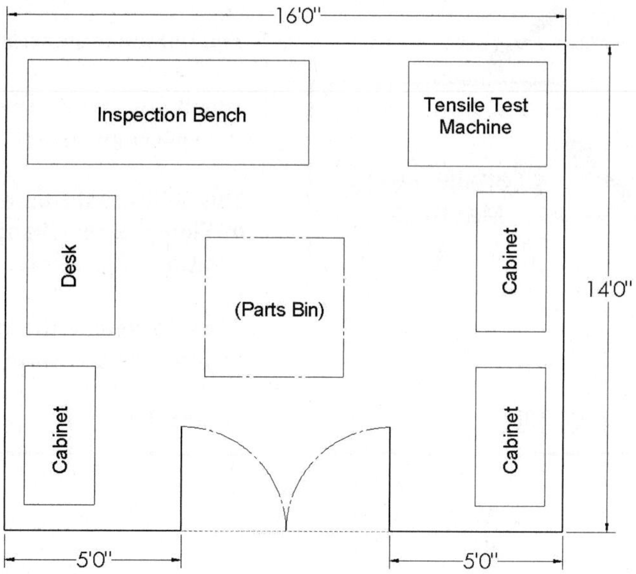

Suppose that we wish to rotate the tensile test machine 45 degrees to give the operator better access to the machine. To do so, it will be necessary to delete some of the relations created automatically when the rectangle was drawn, and to add some new relations to maintain the rectangular shape and orient the rectangle properly.

Click on the Fixed icon, as shown in Figure 9.35. Delete it. Delete the horizontal and vertical relations from the four lines defining the machine position. Add parallel relations between each of the pairs of opposite sides, as shown in Figure 9.36. Select two adjacent sides, and add a perpendicular relation, as shown in Figure 9.37.

FIGURE 9.35

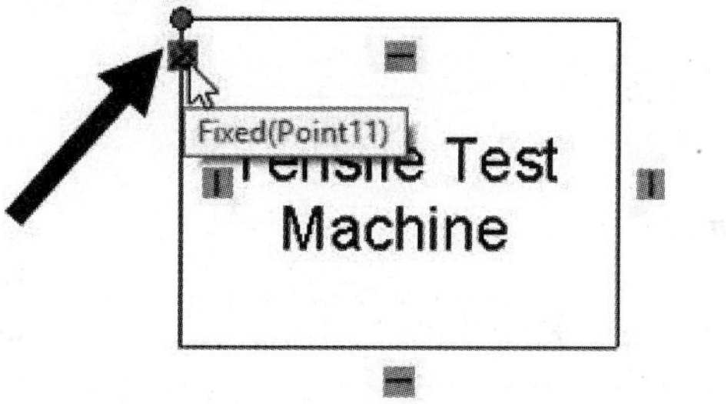

FIGURE 9.36

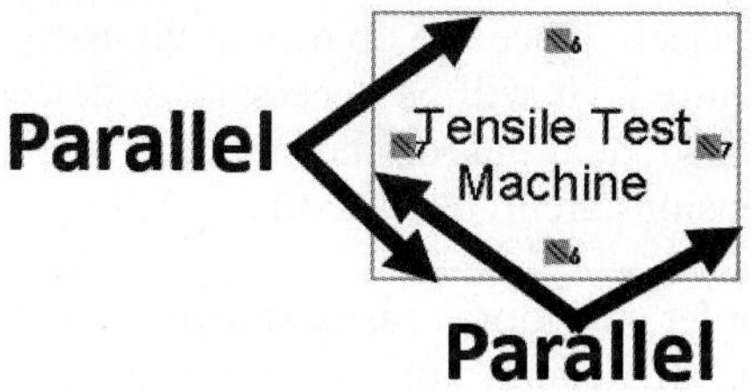

FIGURE 9.37

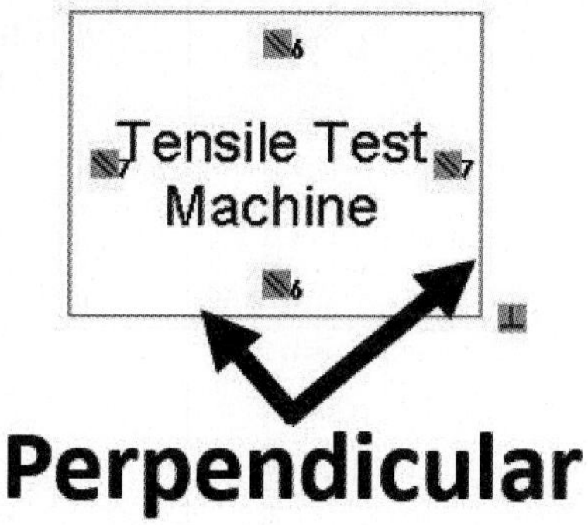

FIGURE 9.38

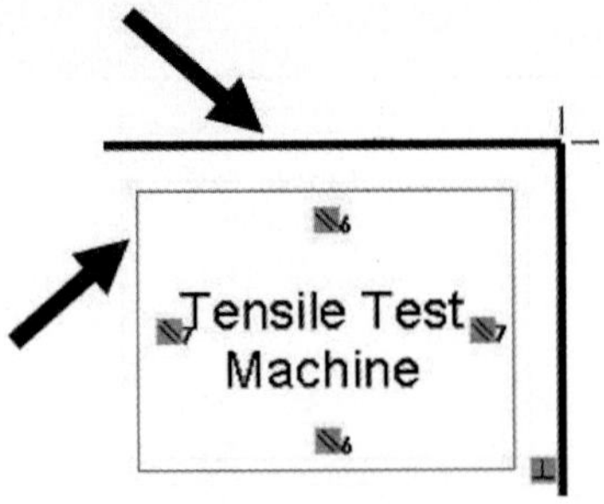

The addition of these relations defines the shape as a rectangle, but the removal of the horizontal and vertical relations allows it to be rotated. The hidden dimensions still apply to the rectangle, and fix its size.

Select the Smart Dimension Tool. Select one of the vertical lines of the rectangle and then a horizontal wall, as shown in Figure 9.38.

This will create an angular dimension between the lines, as shown in Figure 9.39. Change the dimensions to 135 degrees, reflecting a rotation of 45 degrees, as shown in Figure 9.40.

Click to select the note. Rotate the text labeling the machine −45 degrees, as shown in Figure 9.41.

FIGURE 9.39

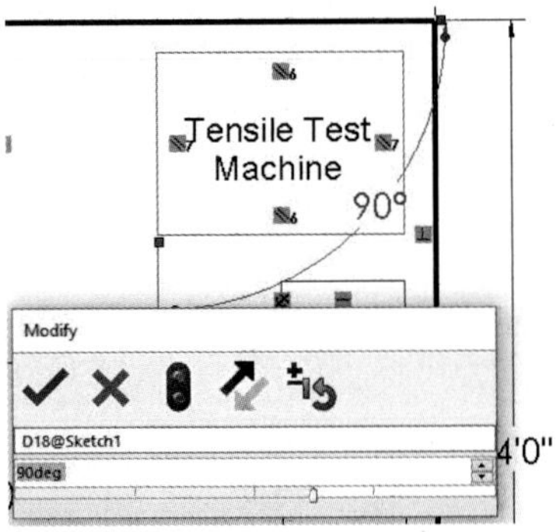

FIGURE 9.40

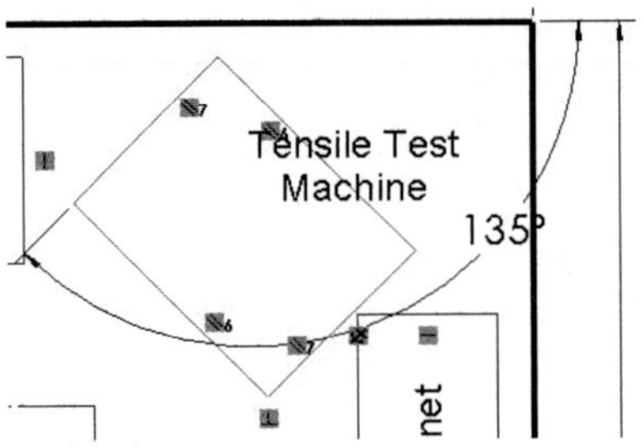

FIGURE 9.41

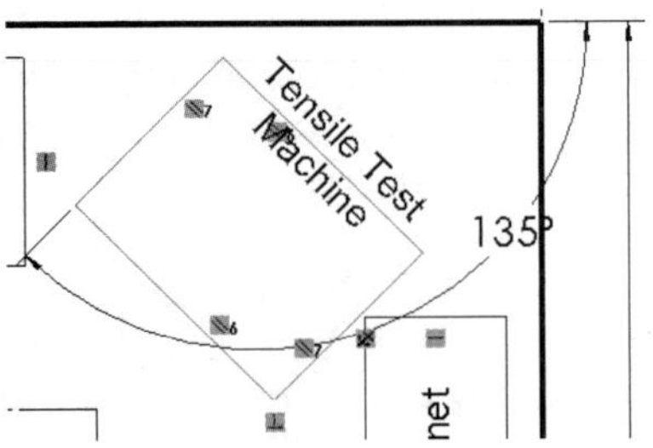

FIGURE 9.42

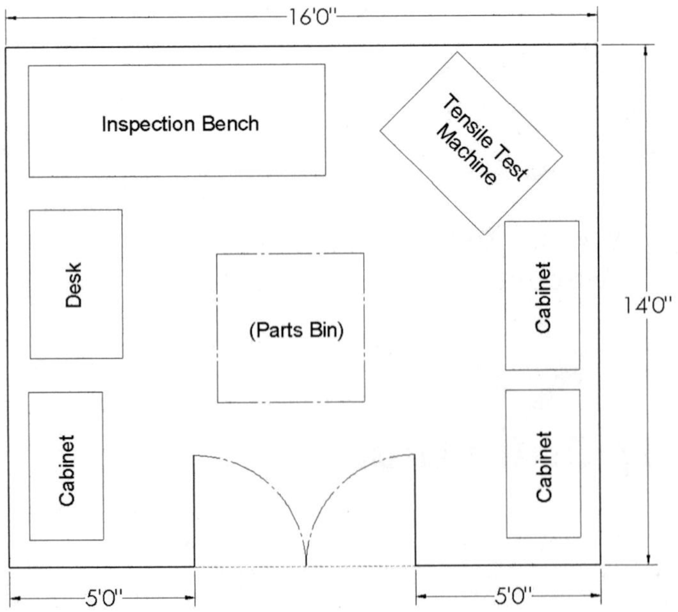

Hide the angular dimension, drag the box and text into position, and fix one of the corner points of the box.

The drawing is now complete, as shown in **Figure 9.42**. Dimensions can be added to precisely place the objects in the room, if required. (It will be necessary to delete the fixed relations of the points if these dimensions are to be added.)

In order to more easily move objects around the room, you can create a block entity that includes both the object shape and the text.

Delete the Fixed relation setting the location of the Inspection Bench, so that both it and its label are free to move. With the Ctrl key depressed, select the four lines and the text that comprise the Inspection Bench. From the main menu,

FUTURE STUDY

Industrial Engineering

In this example, we created a floor plan to see if the proposed lab space would accommodate the equipment required for the lab. Consider the challenge of planning the layout of a manufacturing plant with hundreds of thousands of square feet of floor space and hundreds of machines. If not well planned, the operations of the plant will be crippled by inefficiencies in the ways that raw materials and parts move through the plant. Efficient plant layout is one of the functions of industrial engineers.

Industrial engineers perform many other functions toward the goal of improving operations. These functions might include monitoring and improving the quality of finished products, streamlining material handling and product flow, or redesigning work cells for better efficiency.

While the word "industrial" refers to the manufacturing environment where industrial engineers have traditionally worked, the skills of industrial engineers are now being widely used in service sector businesses as well. For example, many hospitals use industrial engineers to help improve quality and efficiency. Package-delivery companies, facing monumental logistics challenges associated with delivering packages worldwide under extreme time pressure, also use the services of industrial engineers.

Industrial engineers also are involved with product design, usually from the standpoint of *ergonomics,* the consideration of human characteristics and limitations in the design of products. In this area, there is overlap with the functions of the industrial designer.

Most engineering students in other disciplines take some coursework in industrial engineering topics. These topics could include engineering economy, quality control, project management, and ergonomic design.

select Tools: Block: Make, as shown in Figure 9.43. Click the check mark to create the block and close the Make Block PropertyManager.

FIGURE 9.43

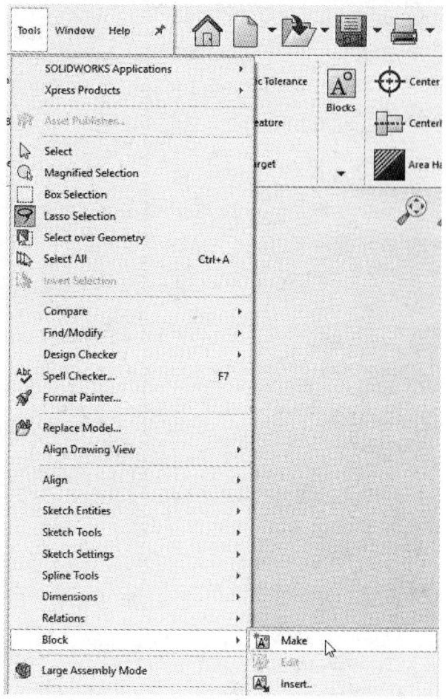

A block is a grouping of drawing entities that can now be treated as a single entity. We can now move or rotate the group of entities together.

FIGURE 9.44

Click and hold a corner of the Inspection Bench, and drag it to a new location, as shown in Figure 9.44.

Note that the entire block (lines and text) move together. If you later decide to separate the entities, the block must be exploded. This can be done by clicking on any part of the block to select it and selecting Tools: Block: Explode from the main menu.

Locate the Inspection Bench as desired, and reapply a fixed relation to locate the block. Save the drawing if desired, and close it.

9.2 Finding the Properties of 2-D Shapes

The SOLIDWORKS program can be used to determine the properties of areas. For example, the areas of complex shapes can be determined, and the locations of *centroids* and the values of *moments of inertia* of areas can be computed.

FIGURE 9.45

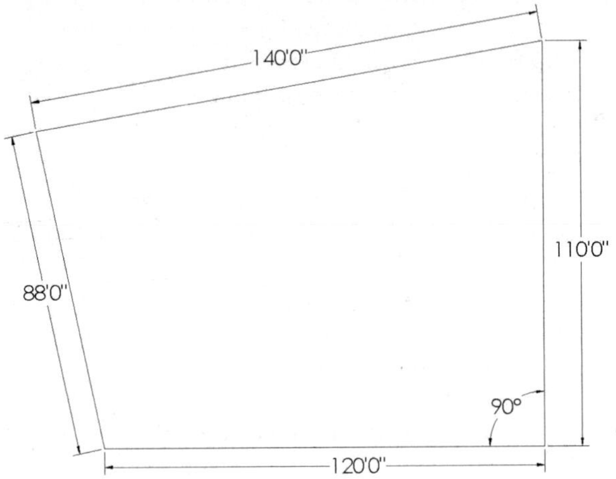

9.2.1 Calculating the Area of a Shape

Consider the building lot shown in **Figure 9.45**. Suppose that we want to find the acreage of this lot. The first step is to draw the lot. In order to fit the drawing on an 8-1/2 × 11 inch sheet, the minimum scale factor possible will be 1320 inches (110 feet) divided by 8.5 inches = 155. We will use a scale factor of 1:200.

Open a new drawing. Choose an A-Landscape paper size without a sheet format displayed. Set the drawing scale to 1:200 and the units to feet & inches. Draw and dimension the lot shown in Figure 9.45. Fix one corner of the lot, and the drawing should be fully defined.

FIGURE 9.46

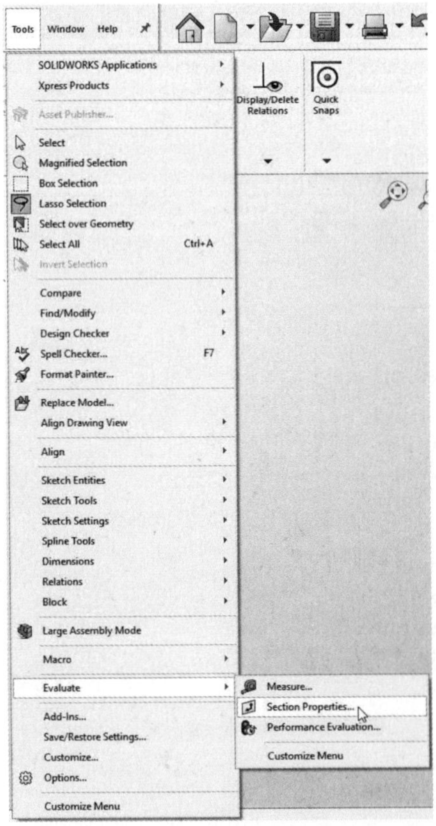

Press the Esc key to deselect any points or lines. Select Tools: Evaluate: Section Properties from the main menu, as shown in Figure 9.46. Click the Options button, make sure the Scientific Notation box is unchecked, and check the Use custom settings option. Set the units to feet and the number of decimal places to 6, as shown in Figure 9.47. Click OK.

FIGURE 9.47

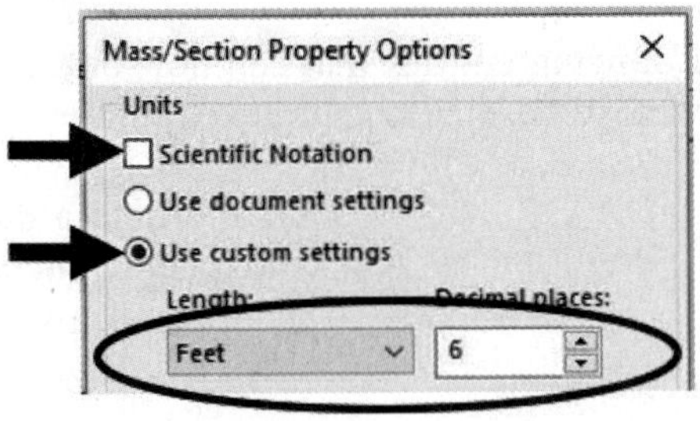

The area will be displayed in the units selected, in this case square feet. However, the area displayed will be that of the *drawing* area, not the actual area. Since the area of the drawing in square feet will be small, a large number of decimal places will be needed for accuracy.

The area displayed is 0.318910 square feet, as shown in Figure 9.48. To determine the area of the lot, the calculated area must be adjusted to account for the scale. Since area has the units of length squared, the calculated area must be multiplied by the scale factor squared:

FIGURE 9.48

Area = $(0.318910 \text{ ft}^2)(200^2) = 12{,}756 \text{ ft}^2$

To convert to acres, the conversion factor of 43,560 ft^2/acre must be applied:

Area = $(12{,}756 \text{ ft}^2)/(43{,}560 \text{ ft}^2/\text{acre}) = 0.29$ acres

9.2.2 Calculating the Section Properties of a Shape

In mechanics of materials, it is often necessary to compute the moment of inertia of the cross section of a structural member. For simple shapes, such as a rectangle or circle, easy formulas can be used for this computation. For compound shapes, such as a T-beam constructed from two rectangular shapes, the calculations are lengthier.

Open a new drawing. Choose A-Landscape paper without a sheet format displayed. Draw and dimension a rectangle, as shown in Figure 9.49.

Add and dimension the second rectangle shown in Figure 9.50, making sure to snap the first point of the new rectangle to the edge of the first rectangle.

FIGURE 9.49

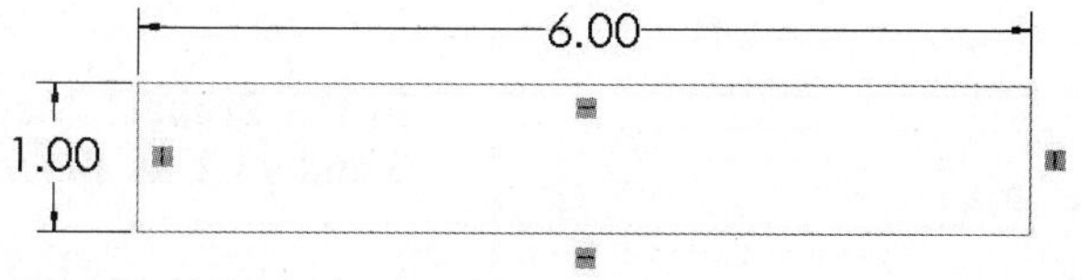

FIGURE 9.50

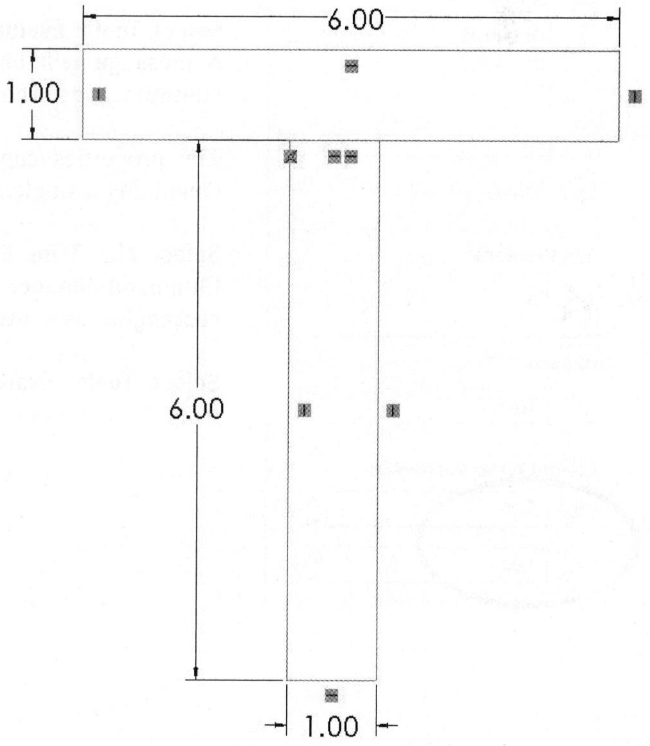

FIGURE 9.51

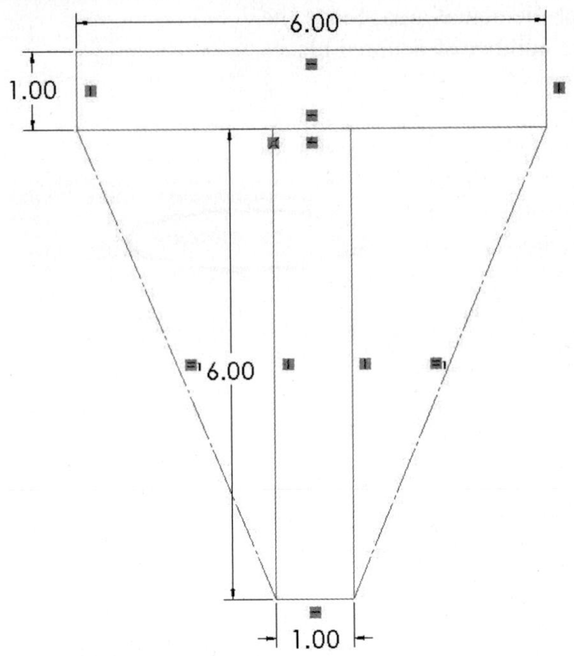

To center the second rectangle relative to the first, we could add either additional dimensions or relations. Add the centerlines shown in Figure 9.51, and add an Equal relation between them.

To compute the area or moments of inertia of the shape, it is not necessary to locate the drawing on the page. However, if we want to know the location of the centroidal axes, then we need to set a point at a known location.

Select the lower-left point of the second rectangle, as shown in Figure 9.52.

FIGURE 9.52

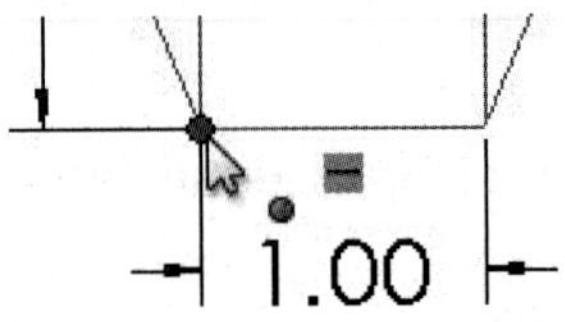

FIGURE 9.53

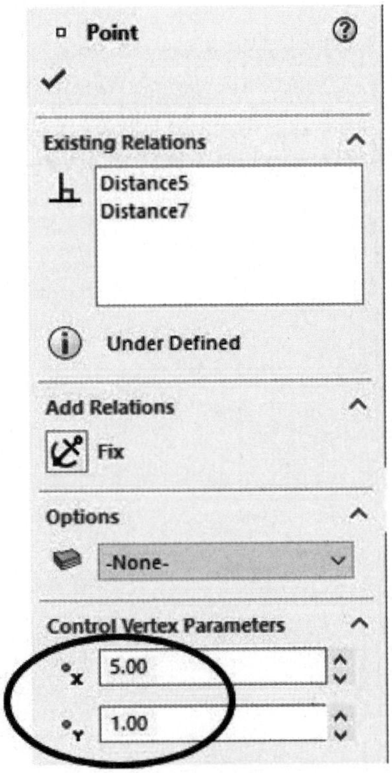

In the PropertyManager, set the coordinates of the point to x = 5 and y = 1, as shown in Figure 9.53. Fix the point with a relation.

In drawings, the lower-left corner of the page is the origin of the x-y coordinate system.

Select Tools: Evaluate: Section Properties from the main menu. A message will be displayed that the sketch has intersecting contours. Close the Section Properties dialog box.

The properties can be computed only for a sketch or drawing containing a single, closed contour.

Select the Trim Entities Tool from the Sketch group of the CommandManager. Trim away the overlapping portions of both rectangles, as shown in Figure 9.54.

Select Tools: Evaluate: Section Properties from the main menu.

FIGURE 9.54

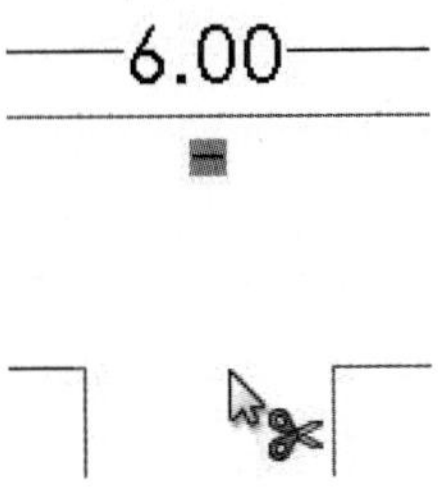

FUTURE STUDY

Mechanics of Materials

The study of mechanics of materials involves determining *stresses*, *strains*, and *deformations* in bodies subjected to various loadings. Stress is the force per unit area acting on a point in a body, and the calculation of stress is necessary to predict failure of a structure. Strain is a measure of geometrical changes within a body. Stress and strain are related by properties of the material used. Deformations are related to strains in the body. While strain applies at a given point, deformation is the dimensional change of the entire body. For example, consider a 1×6-inch wooden plank resting on sawhorses near its ends. If the span of this beam is 6 feet, and a 175-pound person stands in the middle, how much will the plank move downward?

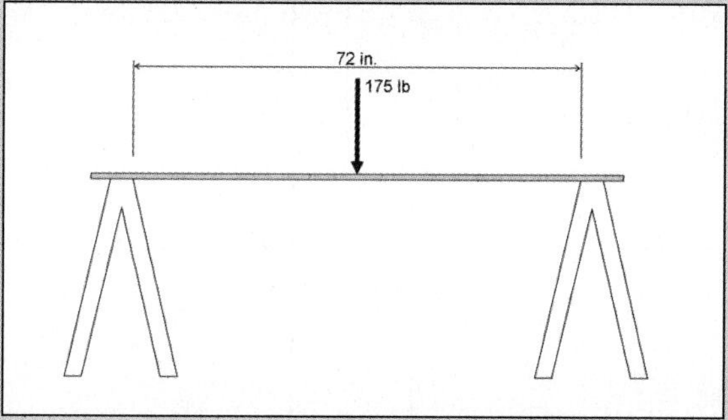

Using mechanics of materials concepts, it can be determined that the deflection of such a beam is:

$$\delta = \frac{PL^3}{48EI}$$

where:

- P = force = 175 lb
- L = span = 72 in
- E = *modulus* of the material. A typical value for wood = 1,500,000 lb/in^2
- I = moment of inertia of the cross section

For a rectangular shape, the moment of inertia is:

$$I = \frac{1}{12} bh^3$$

So for our plank, with base b = 6 inches and height h = 1 inch, I = 0.50 in^4, and the deflection = 1.81 inches. If we could carefully place the plank so that it is resting on the 1-inch edge, then the values of b and h are switched. The moment of inertia increases to 18 in^4 and the deflection reduces to 0.05 inches. Therefore, the *bending stiffness* of the plank has increased by a factor of 36 by changing its orientation. The moment of inertia is increased by moving material to the greatest possible distance away from the *centroidal axis* (the axis passing through the centroid, or center of area, of the section). With wood construction, this is done by aligning the members appropriately, as with floor beams that are placed with their long dimensions perpendicular to the floor. In steel construction, wide-flange beams maximize bending stiffness by placing most of the material in the flanges, as far from the centroidal axis as possible.

When complex shapes are used, calculation of the moment of inertia can be cumbersome. First, the centroid must be located, and then the moments of inertia of simple regions of the shape must be calculated and adjusted for their distances away from the centroidal axis. Finally, the moments of inertia of the individual regions are summed.

The Section Properties Tool can be a useful tool for calculating and/or checking the value of moment of inertia.

FIGURE 9.55

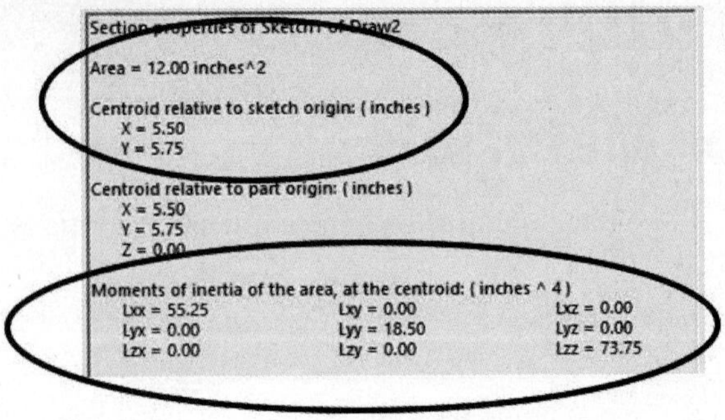

The section properties are displayed, as shown in **Figure 9.55**.

In addition to the area and centroid, the moments of inertia of the area are calculated. The moment labeled "Lxx" is the moment of inertia about an axis through the centroid of the cross section, parallel to the x axis. The location of this axis is given by the y coordinate of the centroid, 5.75 inches. Since a point at the bottom of the section was set at y = 1 inch, the axis is 4.75 inches above the bottom of the section, as shown in **Figure 9.56**. The moment of inertia about this axis is 55.25 in^4. The moment of inertia labeled "Lyy" is about the axis shown in **Figure 9.57**. The value of the moment of inertia about this axis is 18.50 in^4.

FIGURE 9.56

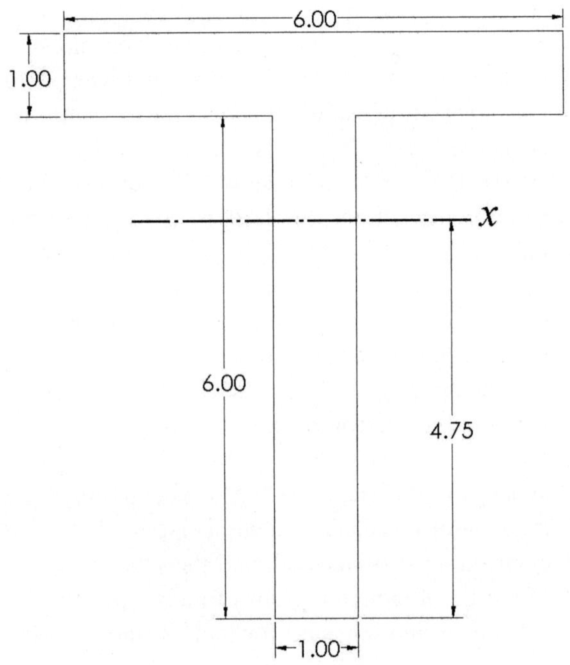

FIGURE 9.57

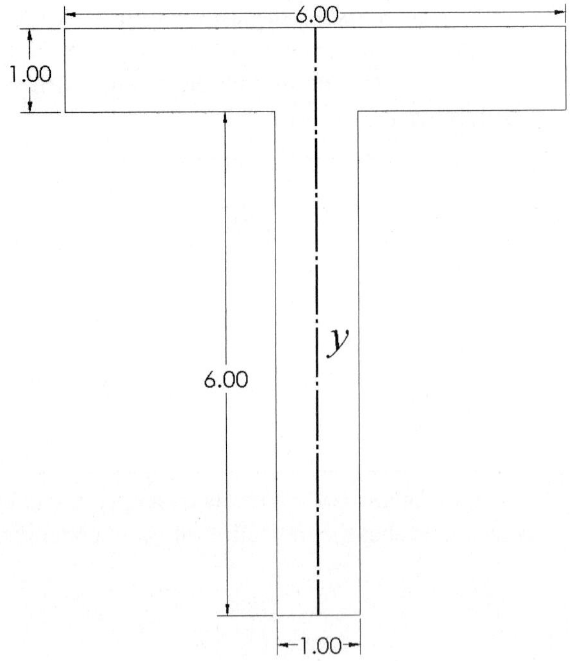

PROBLEMS

P9.1 Place the items from section 9.1 into the space shown here, which includes two 18-inch-square columns.

FIGURE P9.1

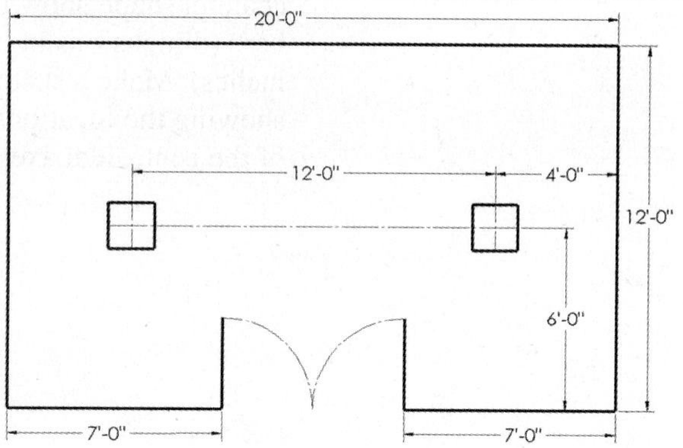

P9.2 Sketch a floor plan of your room or apartment, showing the locations of furniture.

P9.3 A builder desires to construct a house on the irregular lot shown here. The desired house will be rectangular in shape, 55 feet by 40 feet, with the front of the house 55 feet wide. The front of the house is to be parallel to the street. If local regulations require that there be 10 feet between property lines and any point on the house, can the proposed house be built on the lot?

FIGURE P9.3

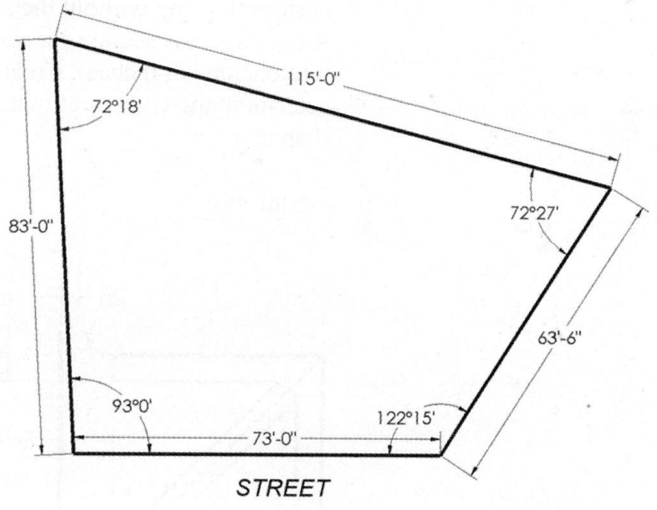

Note: Three of the angular dimensions will be driven dimensions, since the length of all sides and one angle defines the lot shape. However, when property is surveyed, all lengths and angles are measured. Trigonometry is then used to ensure that the dimensions are consistent to within a certain tolerance. This method allows for incorrectly measured or recorded lengths and angles to be detected. The angles are measured in degrees and minutes (1 minute = 1/60 degree).

Hint: Use the Offset Entities Tool to offset the property lines 10 feet inward. You can then check to see if the house plan fits within the offset lines.

P9.4 Find the number of square feet in the lot described in **P9.3**.

(Answer: 6506 ft^2)

P9.5 Find the location of the centroid and the principal moments of inertia of the channel shape shown here (dimensions in inches). Make a sketch showing the locations of the centroidal axes.

FIGURE P9.5

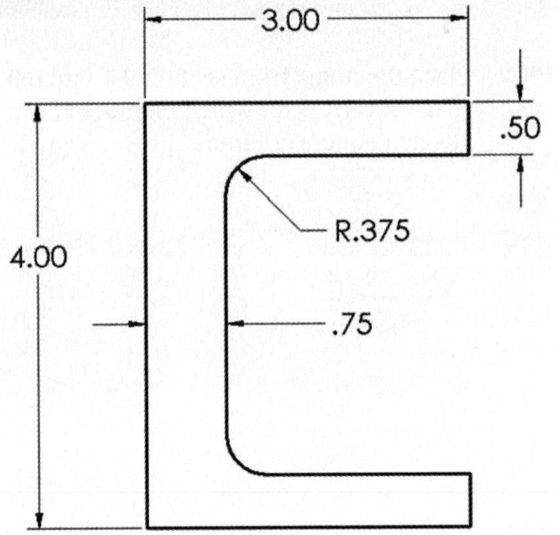

P9.6 Frames for industrial equipment are sometimes built out of prefabricated aluminum extrusions. These extrusions are cut to length and assembled with custom fasteners, providing engineers with the ability to construct custom framing without the need for complicated fixturing and welding. A typical cross section for such an extrusion is shown in **Figure P9.6** (dimensions in inches). Create a drawing of this cross section, and use it to determine the cross-sectional area and principal moments of inertia of the shape.

FIGURE P9.6

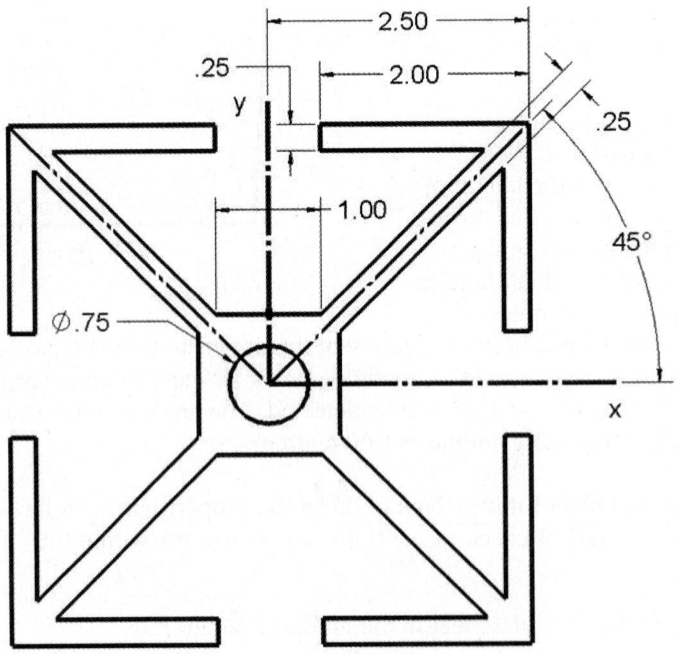

P9.7 a. Fiber-reinforced composite plates are often made by stacking up layers of material with the fibers oriented in specified directions. Consider a 6-inch by 12-inch rectangular plate that is to be made up of two layers with fibers oriented at +30 degrees relative to the long axis of the plate, and two layers with fibers oriented at –30 degrees, as illustrated in **Figure P9.7A**. If the fiber-reinforced material comes on a roll that is 24 inches wide, with the fibers oriented along the length of the roll, determine the length L that is required for a rectangular portion of material from which the four required layers can be cut (see the example in **Figure P9.7B**). Since the material is expensive, try to place the layers in a manner that reduces the amount of scrap. Calculate the percentage of material that will be scrap.

FIGURE P9.7A

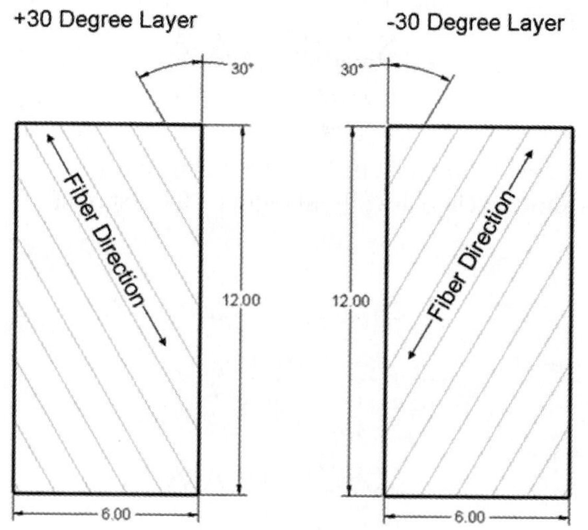

FIGURE P9.7B

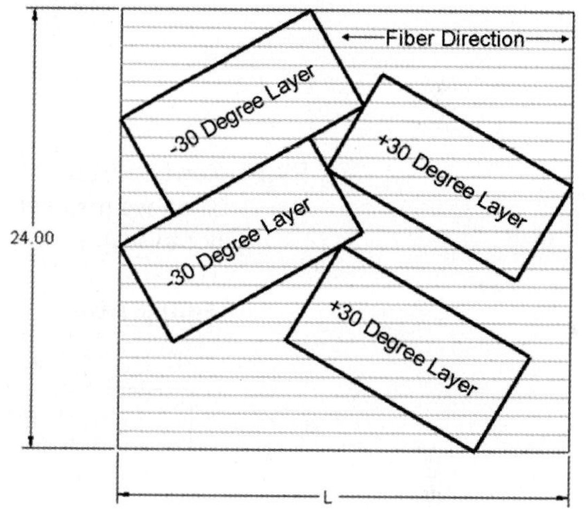

b. Suppose you are asked to evaluate a suggestion that the material be cut into rectangular sections from which two complete plates can be made (that is, four +30 layers and four –30 layers can be cut from the rectangular section). Can you develop a cutting pattern that will reduce the scrap percentage calculated in part a?

P9.8 A foam pad for an industrial product is shown in **Figure P9.8** (dimensions in inches). The pad will be mass-produced by die-cutting it from an 11-inch by 11-inch sheet of material. The die-cutting process requires that the parts must be separated by at least 0.1 inch from each other and the edge of the sheet. Design a layout for a die to cut as many parts as possible from the sheet, and use the information from the Section Properties Tool to determine the percentage of the raw material that will be scrap.

FIGURE P9.8

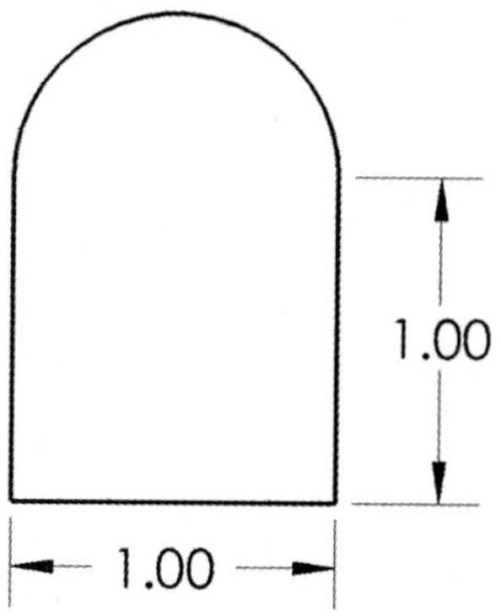

P9.9 A standard 2×10 is used as a floor joist in a house. A plumber drills a 2.5″ diameter hole along the centerline of the joist, as shown schematically in **Figure P9.9** (shown not to scale). Quantify the weakening of the joist due to the hole by computing the area moment of inertia about the centroidal x-axis at Section A-A both with and without the hole. Express your answer as a percent reduction in the area moment of inertia. Hint: Be sure to use the *actual* dimensions of a standard 2×10 in your calculations. (See the Future Study box on Mechanics of Materials for a description of how area moment of inertia relates to the structural integrity of the joist.)

FIGURE P9.9

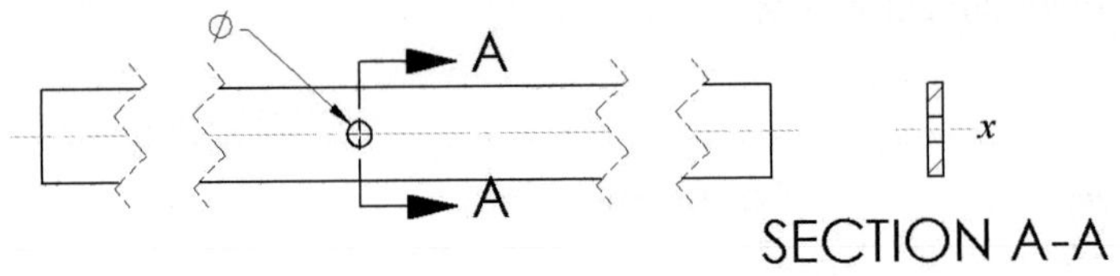

P9.10 Rework **Problem P9.9** if the hole in the joist is drilled a distance of 3″ above the centerline, as shown schematically in **Figure P9.10** (shown not to scale).

FIGURE P9.10

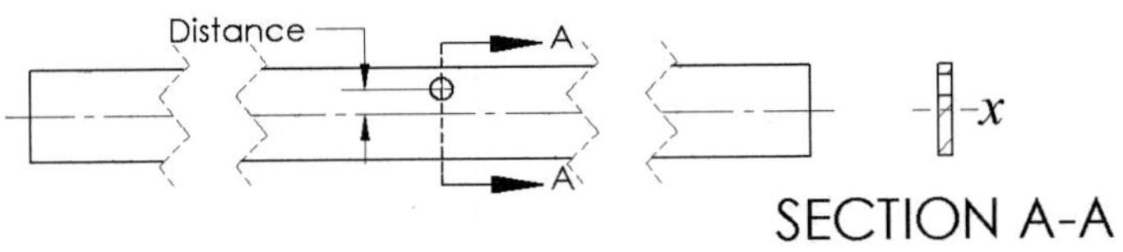

CHAPTER 10

Solution of Vector Problems

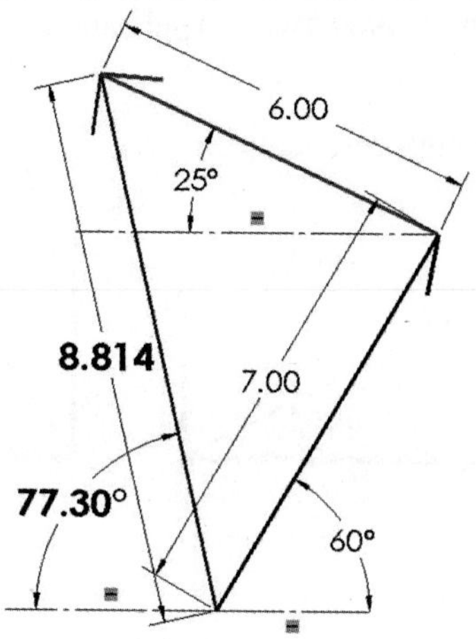

Introduction

Many engineering problems involve the manipulation of vectors. A vector is a representation of a quantity that is defined by both a magnitude and a direction, as opposed to a scalar quantity that can be defined by a magnitude only. For example, the speed of an object is a scalar quantity, while velocity is defined by both the speed and the direction of the object's motion, and therefore is a vector quantity. The SOLIDWORKS® 2-D drawing environment allows for easy graphical solution of vector problems.

Chapter Objectives

In this chapter, you will:

- learn how to add vector quantities graphically,
- work with driving and driven dimensions,
- learn to solve for any two unknowns (magnitudes and/or directions) in a vector equation, and
- perform position analysis for some common mechanisms.

10.1 Vector Addition

Consider the two forces acting on the hook as shown in **Figure 10.1**. We would like to find the *resultant force*, or the single force that affects the hook in an equivalent manner to the two forces. The resultant force is the vector sum of the two vectors **A** and **B**. (Vector quantities are usually denoted by bold type or by a bar or arrow over the symbol: **A** or $\bar{A}$ or $\vec{A}$.)

FIGURE 10.1

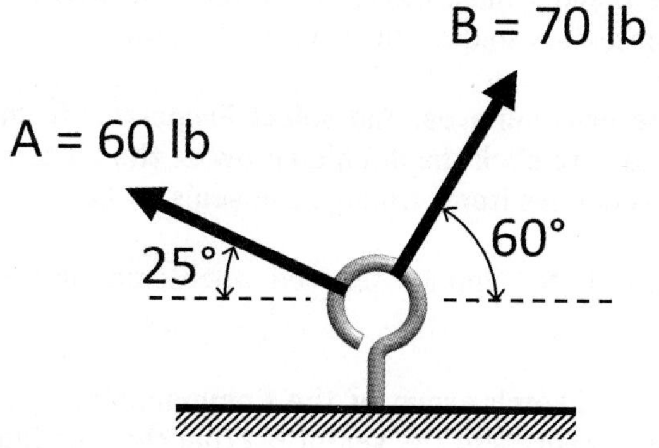

SOLIDWORKS is a registered trademark of Dassault Systémes SolidWorks Corporation.

FIGURE 10.2

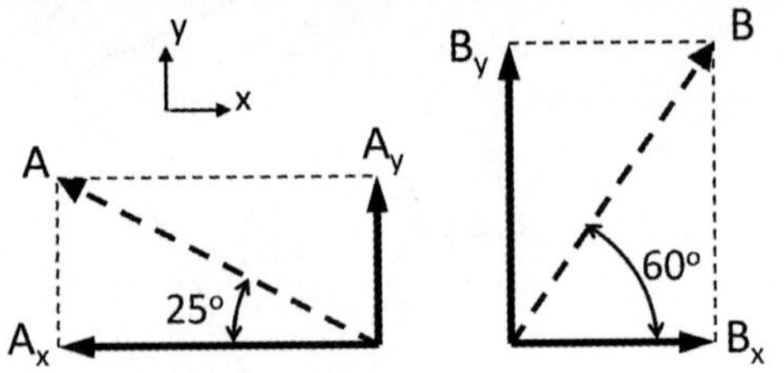

Analytically, the vectors can be added by adding the *components* of the two vectors. If we define an x-y coordinate system, then each vector can be broken into its x and y components, as shown in **Figure 10.2**.

The components of the resultant vector **R** are found by adding the components of **A** and **B**:

	x component	y component
A	–60 cos (25°) = –54.38 lb.	60 sin (25°) = 25.36 lb.
B	70 cos (60°) = 35.00 lb.	70 sin (60°) = 60.62 lb.
R	–19.38 lb.	85.98 lb.

From the components, the magnitude and direction of **R** can be determined (**see Figure 10.3**):

$$R = \sqrt{(19.38)^2 + (85.98)^2} = 88.14 \text{ lb.}$$

$$\theta = \tan^{-1}\left(\frac{85.98}{19.38}\right) = 77.3 \text{ deg}$$

FIGURE 10.3

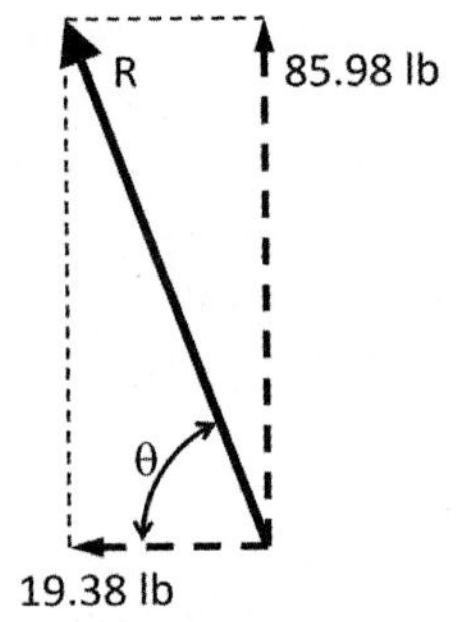

10.2 Vector Addition with SOLIDWORKS

Open a new drawing. Choose A-Landscape as the paper size, with the "Display sheet format" box unchecked. Close the Model View Manager if it opens automatically.

We must now choose scales: one for the magnitude of the vectors and another for the drawing. The vectors must be drawn to scale so that the length of a vector is proportional to its magnitude. If we choose to let 1 inch represent 10 pounds of force, then our two vectors will be 6 and 7 inches long. To fit onto our 8-1/2 × 11-inch drawing, it will probably be necessary to draw the vectors at one-half scale.

FIGURE 10.4

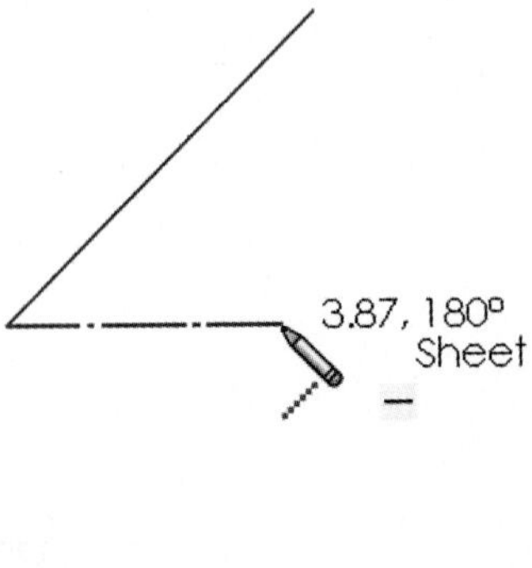

Right-click anywhere in the drawing area, and select Properties from the menu (it may be necessary to click the double arrow at the bottom of the menu to show the Properties item). Change the scale to 1:2.

(The default scale is defined in the template selected when creating the drawing.)

Select the Line Tool from the Sketch group of the CommandManager, and drag out a diagonal line to represent the Vector B. From the starting point of this vector, drag out a horizontal centerline, as shown in Figure 10.4.

If sketch relations are not visible, show them by selecting View: Hide/Show: Sketch Relations from the main menu. Also, to make the lines representing the vectors stand out from the dimension lines, select the Options Tool and under the Document Properties tab, select Line Font: Sketch Curves. Choose a heavier line thickness, as shown in Figure 10.5, and click OK.

FIGURE 10.5

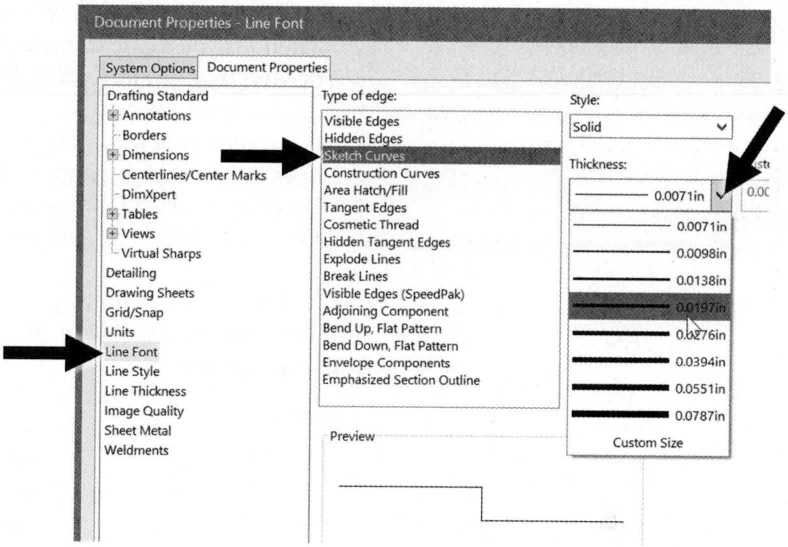

Dimension the length of the vector line as 7 inches, and the angle between the vector line and the horizontal centerline as 60 degrees, as shown in Figure 10.6. If desired, increase the font size of the dimensions by selecting the Options Tool and under the Document Properties tab, choose Dimensions: Font and set the font size to a larger point size.

FIGURE 10.6

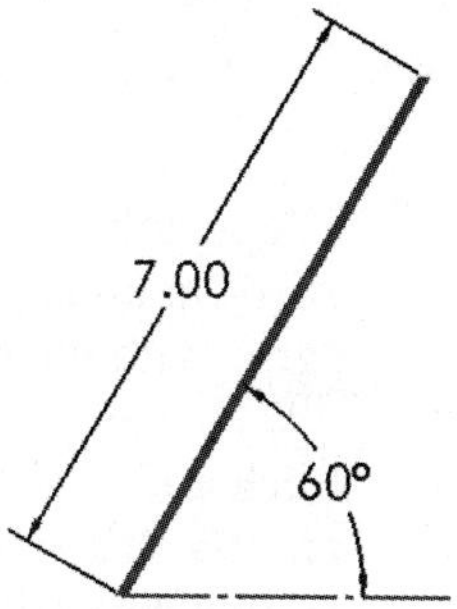

To add vectors graphically, we set them "tip-to-tail." That is, the second vector starts at the end of the first vector.

Draw the second vector line diagonally from the end of the first vector. (Be sure not to align this vector line with any of the dashed lines that appear. Doing so will add a relation to the line that will have to be deleted before the vector's magnitude and direction can be defined.) Add another horizontal construction line, and add the 6-inch and 25-degree dimensions as shown in Figure 10.7.

FIGURE 10.7

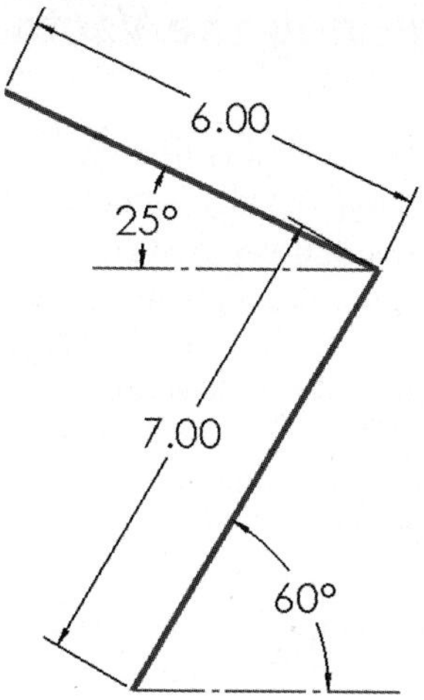

FIGURE 10.8

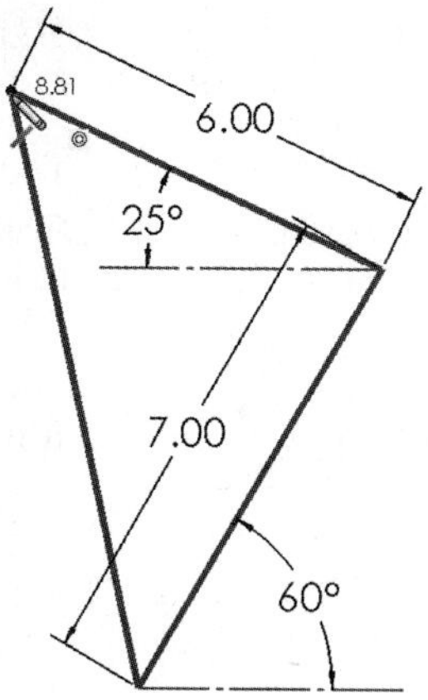

The resultant vector extends from the starting point of the first vector to the ending point of the second vector.

Add the line representing the resultant vector, as shown in Figure 10.8, snapping to the endpoints. Add a length dimension to this line.

FIGURE 10.9

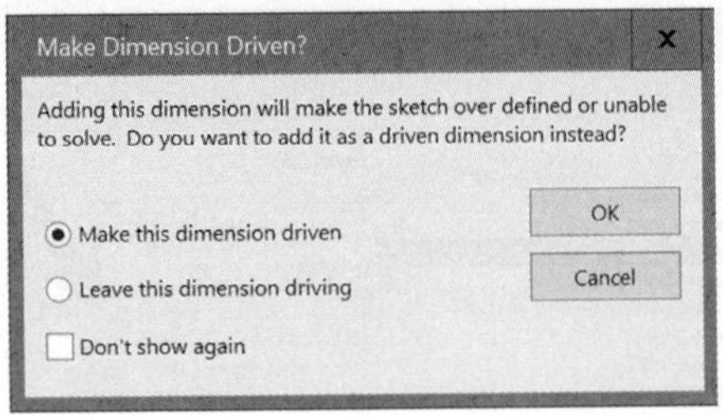

When you add this dimension, you will get a message that adding this dimension will make the drawing over defined, and asking if you want to make the dimension driven or leave it as driving (Figure 10.9). Click OK to make the dimension driven.

A driving dimension helps control the size and position of the drawing or sketch entities, while a driven dimension does not. By default, driven dimensions are shown gray. We will see how to change this option in the next section.

FIGURE 10.10

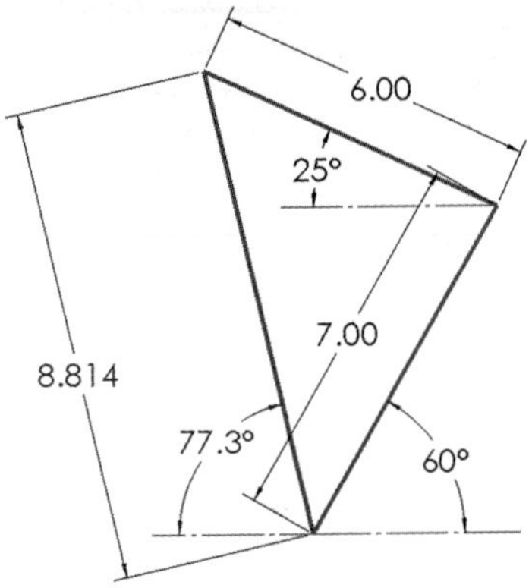

Add a horizontal centerline and the angular dimension of the resultant vector, as shown in Figure 10.10. Click OK to make the dimension driven.

FIGURE 10.11

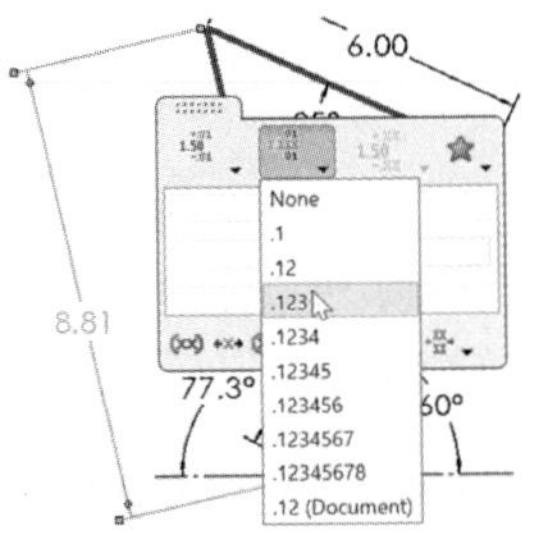

To change the display of the resultant vector length to 3 decimal places, select that dimension and move the cursor over the Dimension Palette that appears next to the dimension. Select 3 places (.123) as the unit precision, as shown in Figure 10.11. Also, set the angular dimension of the resultant vector to 1 or 2 decimal places.

The length of the resultant vector is 8.814 inches, corresponding to a magnitude of 88.14 pounds. This value and the angle of 77.3 degrees, relative to the horizontal, agree with the results found analytically.

Click and drag the starting point of the first vector so that the sketch is centered on the drawing sheet. (You will probably need to move the endpoints of the centerlines and the angular dimensions, since the lengths of the centerlines are not defined.) Add a fixed relation to this point. All vector lines should now be black, as the drawing is fully defined.

10.3 Modifying the Vector Addition Drawing

The drawing just created is fine for calculating the magnitude and direction of the resultant vector, but visually is not clear. Vectors are usually shown with arrowheads to indicate their directions. Also, the dimensions of the resultant vector can be shown differently to make them stand out from the input dimensions, and the resultant vector itself can be shown in a different color. We will modify the drawing to make it more understandable, and to allow easy modifications of the input quantities.

We will first add arrowheads to the vectors. We will use a simple arrowhead: a single line segment.

FIGURE 10.12

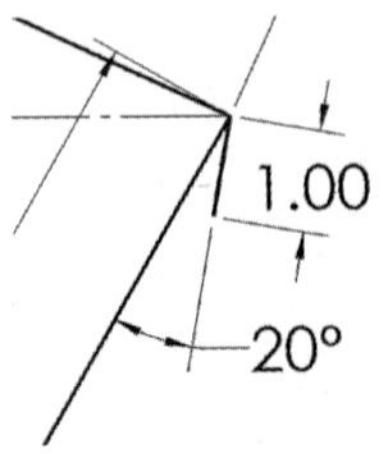

Add a line to the end of the 7-inch vector line. (Make sure that the new line is neither horizontal nor vertical.) Add a length dimension of 1 inch and angular dimension of 20 degrees, as shown in Figure 10.12.

Hide the 1-inch and 20-degree dimensions.

Add arrowhead lines to the other vector lines. Add 20-degree angular dimensions between each arrowhead line and its corresponding vector line. Instead of adding 1-inch dimensions to each arrowhead line, select all three lines and add an equal relation. This will allow the lengths to be easily changed. Hide the dimensions associated with the arrowheads, as shown in Figure 10.13 (shown with the sketch relations hidden).

Right-click anywhere on the CommandManager. In the list of available toolbars, select Formatting and Line Format, as shown in Figure 10.14. Click OK. Modify the text of the dimensions associated with the resultant vector by selecting them and changing the point sizes from the Formatting toolbar. Make the font of these dimensions larger than those of the other vectors. You can also make these text items bold by clicking the "B" Tool on the Formatting toolbar, as shown in Figure 10.15.

To change the default color of the driven dimensions from gray to black, select the Options Tool. Under the System Properties tab, select Colors, and change the color of Dimensions, Non Imported (Driven) to black (or to some other color, if desired). To make the resultant vector easily identifiable, select both the vector and arrowhead line associated with the resultant and choose Line Color from the Line Format toolbar, as shown in Figure 10.16. Choose a new color, and click OK.

The completed vector drawing is shown in **Figure 10.17**.

Often, we prefer to define the direction of each vector by an angle measured from a common reference. Typically, the +x axis is used as the reference.

FIGURE 10.13

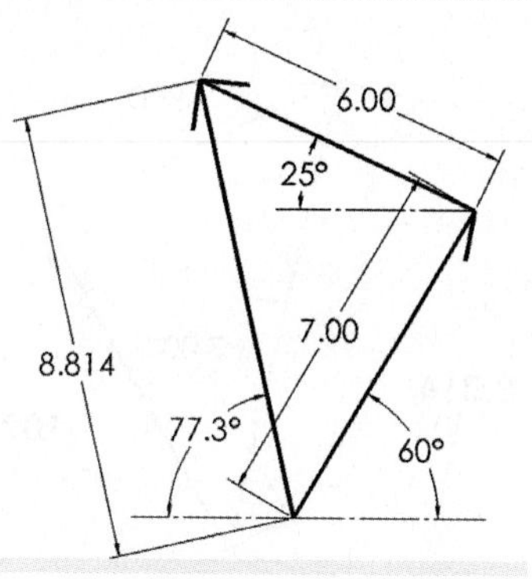

FIGURE 10.14

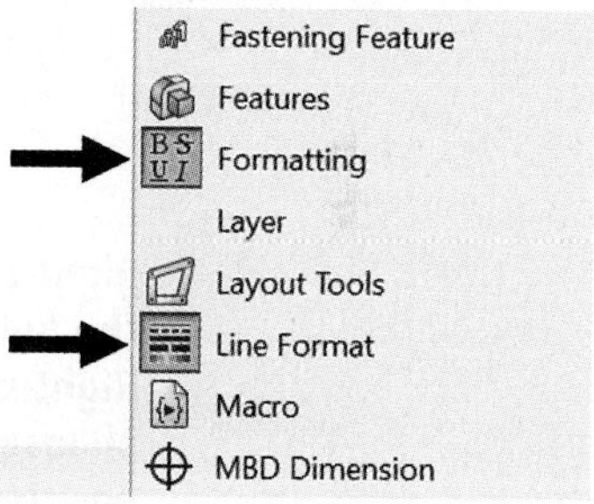

FIGURE 10.15

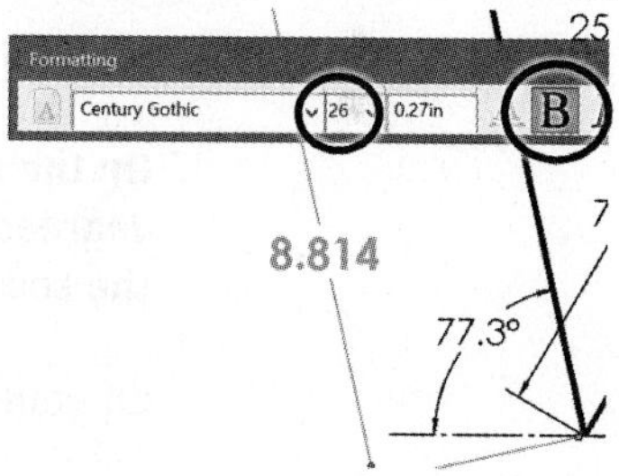

FIGURE 10.16

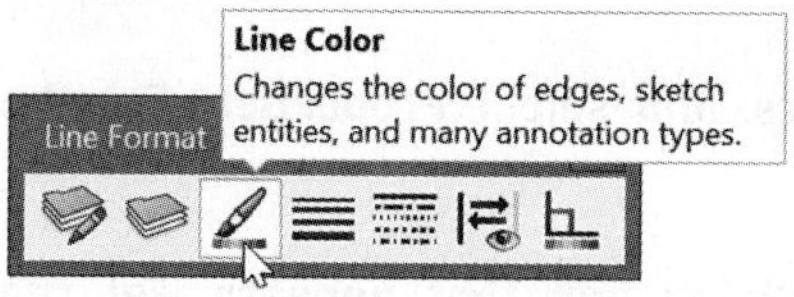

FIGURE 10.17

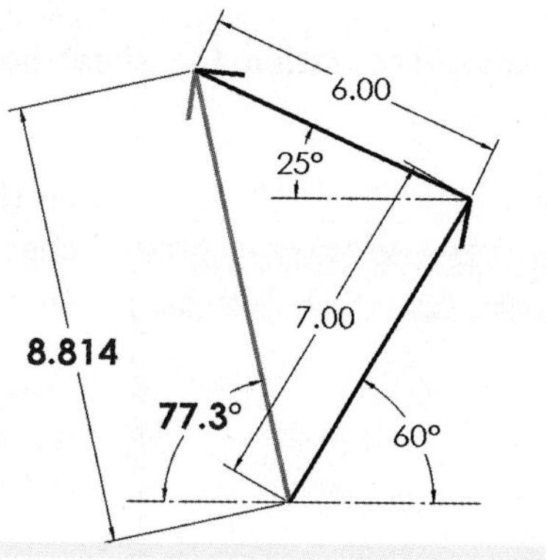

FIGURE 10.18

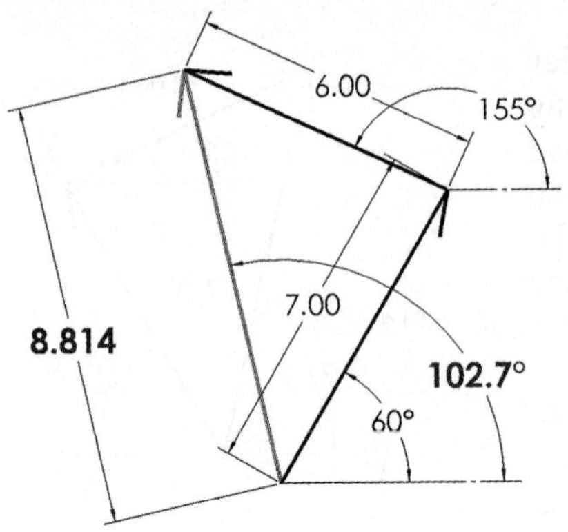

Delete the 25- and 77.3-degree dimensions and associated reference lines. Add new centerlines and dimensions as shown in Figure 10.18. Add the 155-degree dimension first so that the 102.7-degree dimension is the driven one. Save the drawing.

Dimensions can be changed from driven to driving or from driving to driven by right-clicking on the dimension and checking or un-checking Driven. Of course, if too many dimensions are driving, the drawing will be over defined.

One of the main benefits of using the SOLIDWORKS program to add vectors is that the vector drawing can be easily modified to solve different problems. We will use our drawing to add a different pair of vectors.

Let's use this drawing to add the two vectors shown in Figure 10.19. In order to keep the results from the previous problem, we will copy our vector drawing to a new sheet.

Right-click the sheet name (Sheet1) near the bottom of the screen and select Copy. Right-click the sheet name again and choose Paste. Click OK to accept the default position of the new sheet, which is labeled Sheet1(2). Note that you can rename a sheet by right-clicking the sheet's tab and selecting Rename.

FIGURE 10.19

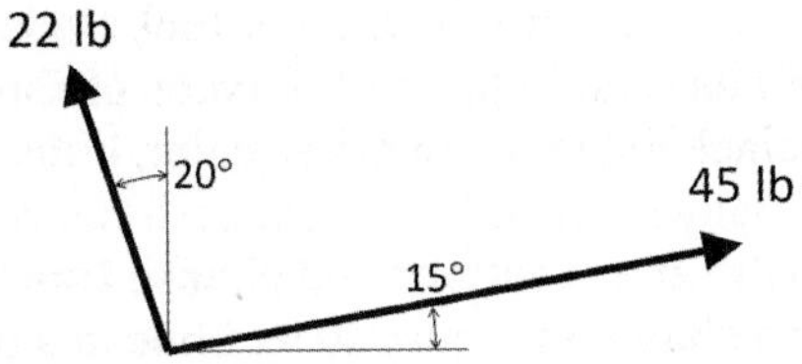

We will let 1 inch equal 1 pound for the vector's scale.

On the new sheet, change the dimensions of the input vectors to 22 inches and 110 degrees (from the horizontal) for the first vector, and 45 inches and 15 degrees for the second vector.

Of course, the drawing now extends beyond the edges of the sheet.

Right-click anywhere within the sheet borders, and select Properties. Set the scale to 1:5.

FIGURE 10.20

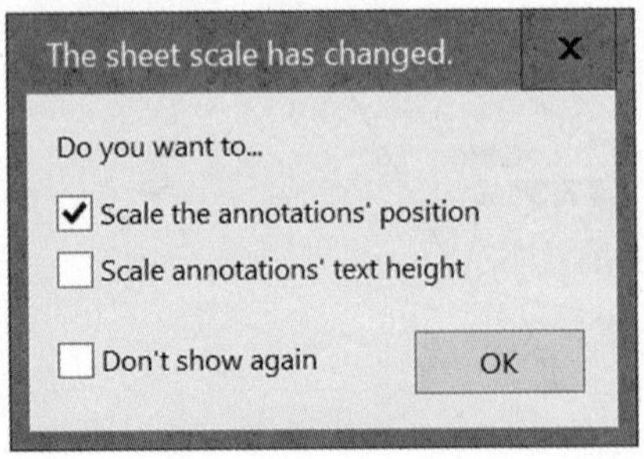

In the dialog box that appears, set the check boxes as shown in Figure 10.20 and click OK.

This will allow the relative position of the dimensions to stay the same, without changing the size of the text.

If necessary, delete the Fixed relation on the starting point, drag the drawing into the sheet boundaries, and then re-apply the Fixed relation. Drag the dimensions into desired positions, as shown in Figure 10.21. If desired, show the 1-inch dimension defining the lengths of the arrowheads, change it to 4 inches, and re-hide the dimension. If the 41.96-degree angle changes to 221.96 degrees, click and drag the endpoint of the horizontal centerline to the right of the starting point, so that the angle is measured relative to the +x axis. Save the drawing.

FIGURE 10.21

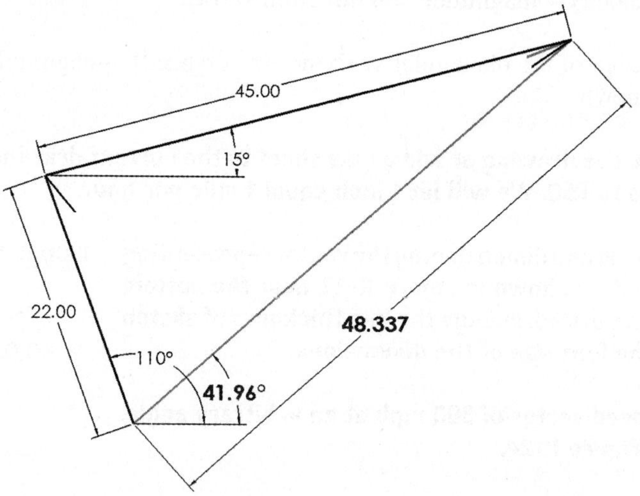

The resultant vector has a magnitude of 48.3 pounds, and is oriented at 42 degrees counterclockwise (CCW) from the +x axis.

10.4 Further Solution of Vector Equations

In the previous example, two vectors were added to find the resultant vector's magnitude and direction. In a vector equation, any two unknowns (magnitudes and/or directions) can be determined. The following examples illustrate this concept.

A small plane can travel at an airspeed of 300 miles per hour. The flight path is to be at a heading of 15 degrees. (Heading is the angular direction measured CW from due North.) The wind is blowing from the WNW, as shown in **Figure 10.22**, at 60 mph. Find the plane's ground speed and the direction of the plane's travel relative to the air.

FIGURE 10.22

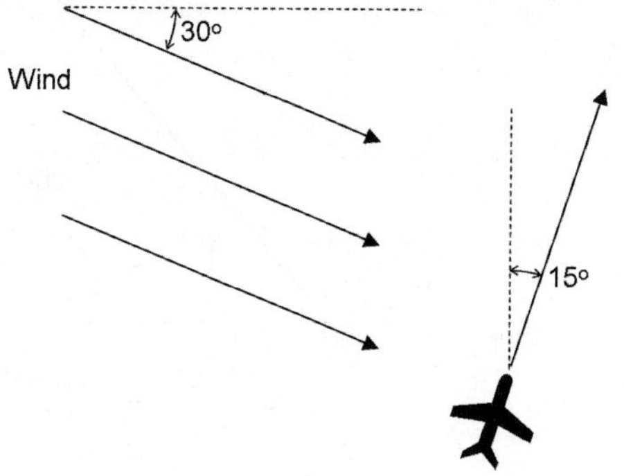

The vector equation for this problem is:

$$\mathbf{V}_{plane} = \mathbf{V}_{air} + \mathbf{V}_{plane/air}$$

where:

$\mathbf{V}_{plane}$ = absolute velocity of the plane (ground speed)—magnitude unknown and direction known

$\mathbf{V}_{air}$ = wind velocity—magnitude and direction known

$\mathbf{V}_{plane/air}$ = velocity of the plane relative to the air (airspeed)—magnitude known and direction unknown

Open a new A-size drawing or add a new sheet to the current drawing and set the drawing scale to 1:50. We will let 1 inch equal 1 mile per hour.

Begin by drawing and dimensioning the vector representing the wind speed, as shown in Figure 10.23, near the bottom of the sheet. If desired, change the line thickness of sketch curves and the font size of the dimensions.

FIGURE 10.23

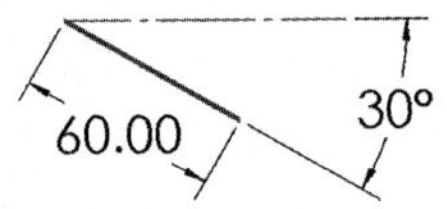

Add the airspeed vector of 300 mph at an arbitrary angle, as shown in Figure 10.24.

Add the resultant (ground speed) vector at 15 degrees from vertical as shown in Figure 10.25. If desired, change the color of the resultant vector line to make it easy to distinguish from the airspeed and wind vectors.

FIGURE 10.24

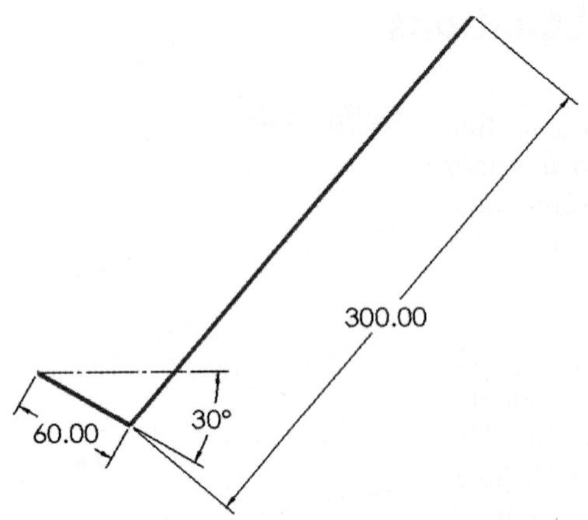

FIGURE 10.25

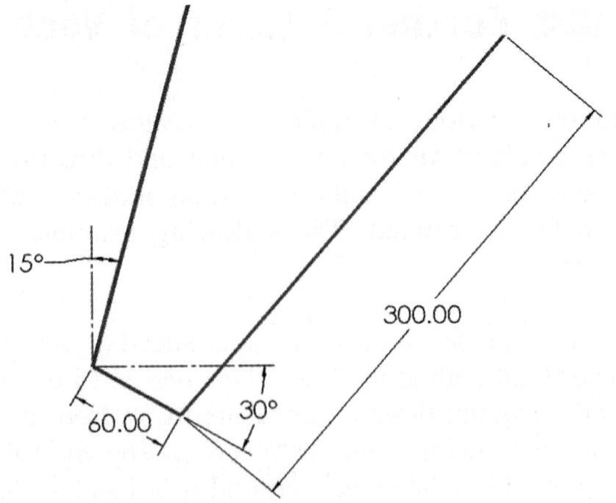

Select the endpoints of the ground speed and airspeed vectors, as shown in Figure 10.26. In the PropertyManager, click Merge to bring these points together, as shown in Figure 10.27.

FIGURE 10.26

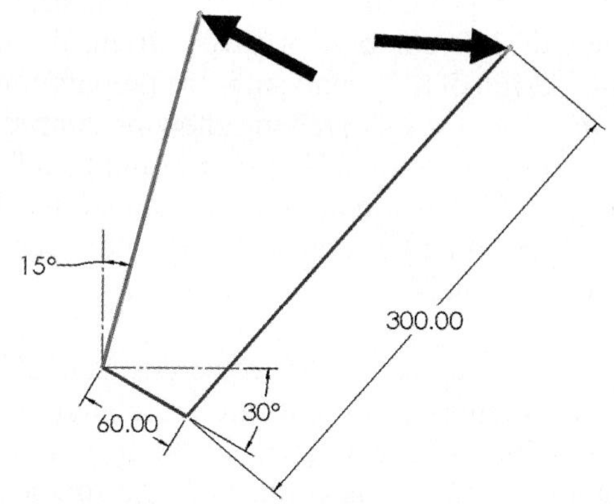

FIGURE 10.27

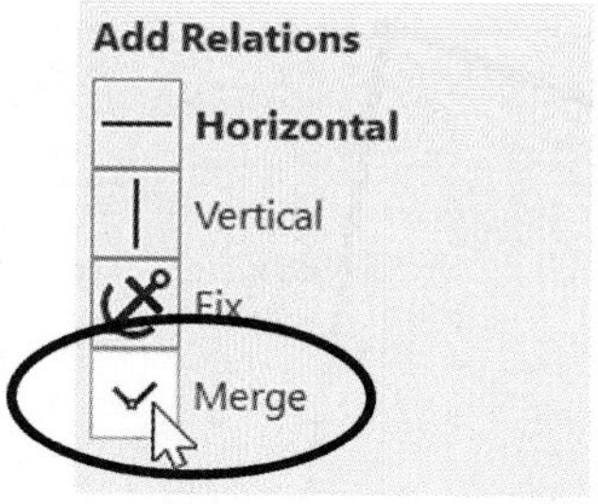

Fix the starting point of the first vector, so that the drawing is fully defined. Add driven dimensions for the ground speed magnitude and the direction of the airspeed, as shown in Figure 10.28. If desired, add arrowheads to make the directions of the vectors easier to visualize.

FIGURE 10.28

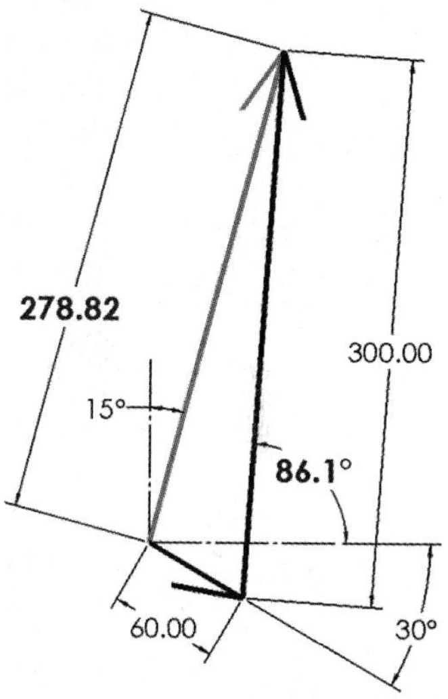

The result is that the ground speed is 279 miles per hour. Relative to the wind, the plane must fly about 4 degrees east of due north to achieve the desired course.

Variations of this problem can be easily solved by changing the input quantities. For example, consider the case where the wind is blowing from due west at 70 mph.

Change the dimensions defining the wind speed vector, as shown in Figure 10.29. Make the wind speed vector directly from west to east by changing the value of the angle to zero or by deleting the 30-degree dimension and adding a horizontal relation to the line.

FIGURE 10.29

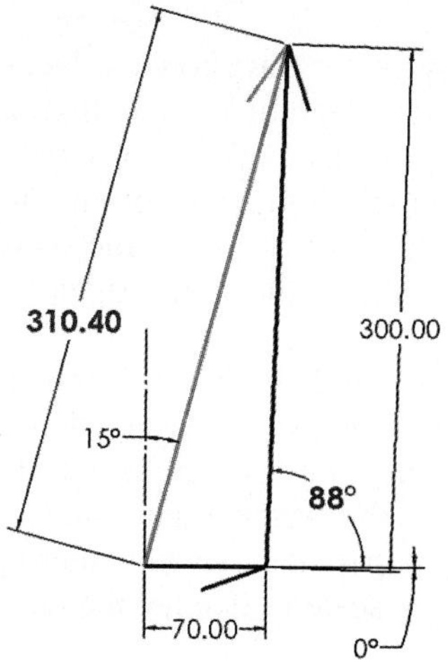

Figure 10.29 shows the resulting vector equation. The ground speed is now 310 miles per hour, as the wind contributes to the east-to-west component of the plane's travel. Note that even if the decimal places for angles is set as one, the 88-degree dimension is shown only as an integer. By default, trailing zeros are not shown for angles. To change this setting, choose Options: Document Properties: Dimensions: Angle Under Zeros, set the option for Trailing zeros: Dimension to "Show."

10.5 Kinematic Sketch of a Simple Mechanism

FIGURE 10.30

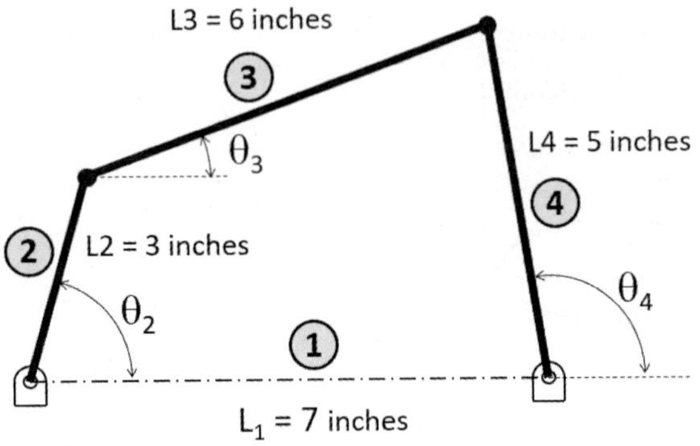

In Chapter 11, we will look at the application of the SOLIDWORKS program to model mechanisms, with assemblies of 3-D component parts. Often, the first step in the design of a mechanism is the preparation of a *kinematic sketch*, a 2-D drawing showing simplified representations of the members. For example, a four-bar linkage, which is a common mechanism used in many machines, can be represented by four lines, as shown in **Figure 10.30**.

In a *kinematic analysis*, the velocities and accelerations of the links and points are calculated, based on the velocity and acceleration of the driving link. For example, if Link 2 is connected to a motor and rotated about its pivot point at a constant velocity, the angular velocities and accelerations of the other two moving links and the translational velocities and accelerations of points on those links can be calculated. The accelerations are important because force is proportional to acceleration. If the accelerations are known, the forces acting on the members and joints can be calculated. (Note: Although there are only three moving members, this mechanism is referred to as a four-bar linkage because it is connected to a fixed or ground link, which is usually called Link 1.)

The first step in any kinematic analysis is a *position analysis*. For a given position of the driving link (the angular position of Link 2), the positions of the other links must be calculated. Obviously, these positions can be calculated using trigonometry. For many mechanisms, a position analysis using trigonometry is surprisingly complex. A graphical solution is often utilized. The SOLIDWORKS program is an excellent tool for graphical position analyses, in that dimensions can be easily changed and the effects on the rest of the linkage can be determined.

FIGURE 10.31

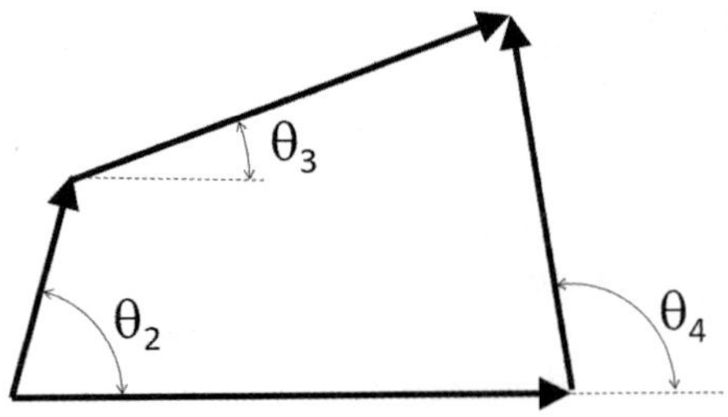

The links are often represented by vectors, as shown in **Figure 10.31**. In this case, all of the vector magnitudes (lengths) are known, as well as two of the vector angles. Link 1 is horizontal, and we will be performing the analysis for a given orientation of Link 2, the driving link. Therefore, we can solve the vector equation for the two unknown quantities, the angles θ_3 and θ_4.

In the previous sections, we used a SOLIDWORKS drawing to create our sketches. For this example, we will sketch within a SOLIDWORKS part file. For creating a quick sketch, the part file is often preferable because the sketches automatically scale. That is, since the sketch is not required to fit a paper size, we can enter the dimensions in any scale and simply change the viewing scale to display the entire sketch. (Sketches can also be created in assembly files. Often, a *layout sketch* within an assembly is used to size and/or position the components of a mechanism.) To print a copy of a sketch created in a part file, the screen capture utility can be used to copy and paste the sketch to a word processing file. The advantages of using a drawing file to create a sketch are that the drawing can be printed to the precise scale and with a title block, if desired, and the drawing can be formatted and annotated more easily.

Open a new part file. Select the Front Plane.

Beginning at the origin, draw three lines at arbitrary orientations, as shown in Figure 10.32, without adding vertical, horizontal, or perpendicular relations.

Add a centerline representing Link 1, and add a Horizontal relation to this line, as shown in Figure 10.33.

Add length dimensions, as shown in Figure 10.34. Be sure to add dimensions oriented with the links, not horizontal or vertical dimensions. If desired, activate the Line Format toolbar and set the thickness of the lines representing the links to a greater thickness.

Add the angular dimension between Links 1 and 2, setting it to 90 degrees, as shown in Figure 10.35. The sketch is now fully defined.

FIGURE 10.32

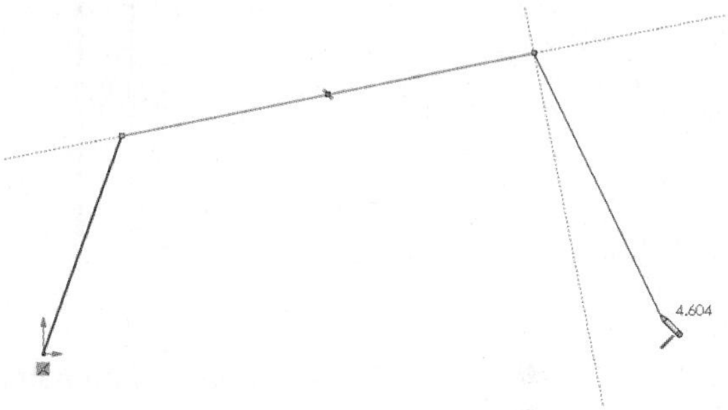

FIGURE 10.33

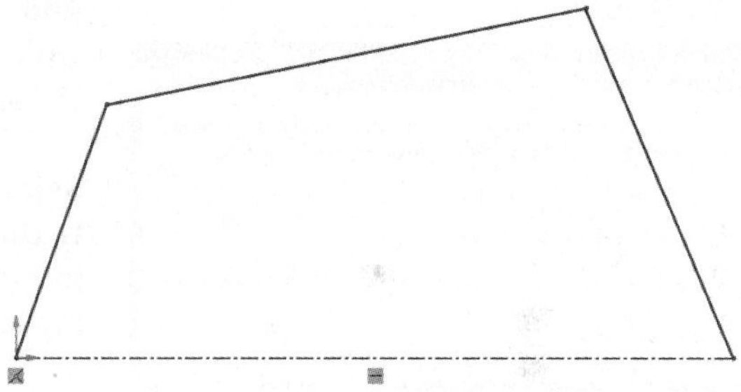

FIGURE 10.34

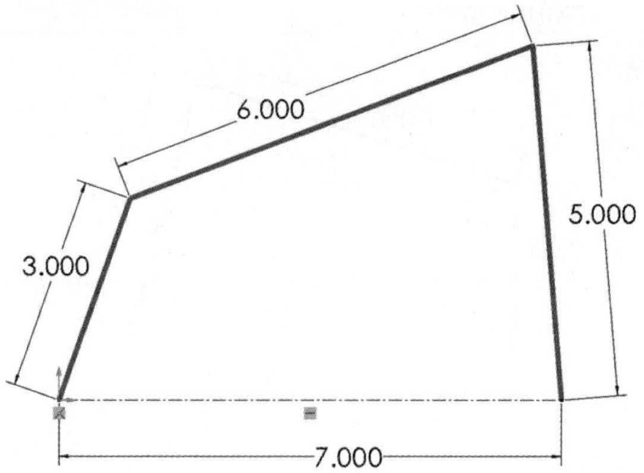

FIGURE 10.35

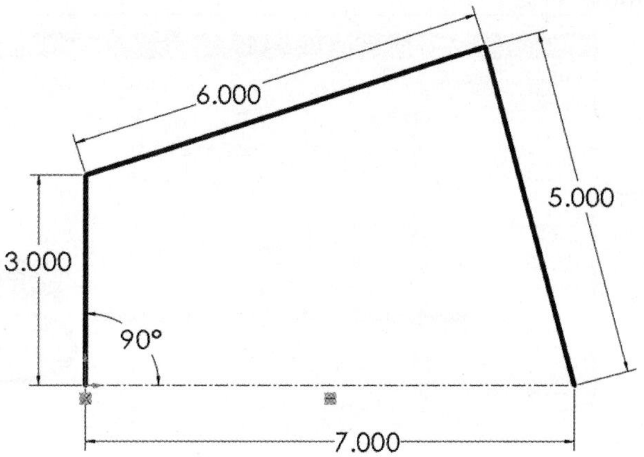

Add horizontal centerlines from which the angular positions of Links 3 and 4 will be measured, as shown in Figure 10.36.

FIGURE 10.36

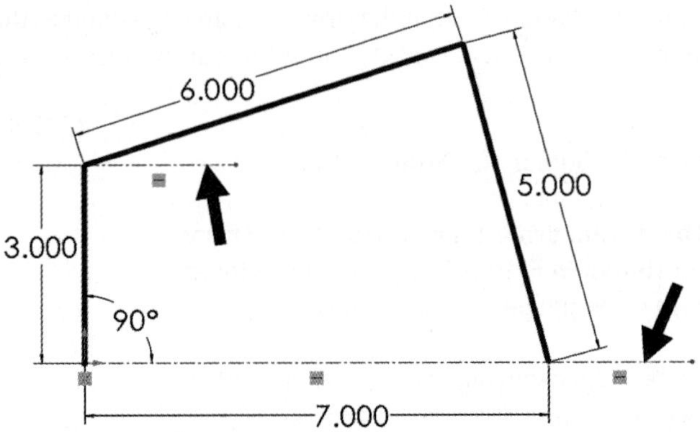

Add an angular dimension between Link 3 and the adjacent horizontal centerline. A box will appear with the message that this dimension will over define the drawing. Select OK to make the dimension driven, as shown in Figure 10.37.

FIGURE 10.37

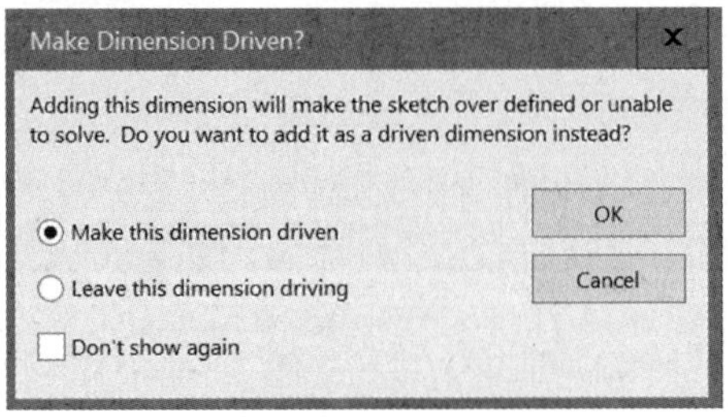

Add an angular driven dimension defining the position of Link 4. In the PropertyManager, set the decimal places of these dimensions to 2.

Select the Options Tool. Under System Options: Colors, scroll to Dimensions, Non Imported (Driven) and select Edit, as shown in Figure 10.38. Change the color of the driven dimensions to make them easy to distinguish from the driving dimensions.

The completed sketch is shown in **Figure 10.39**.

FIGURE 10.38

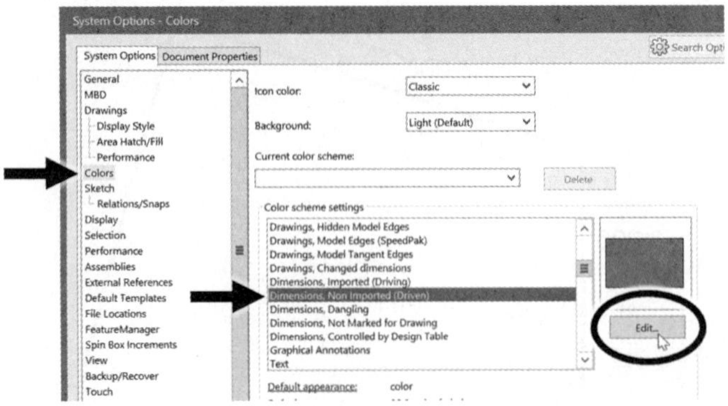

FIGURE 10.39

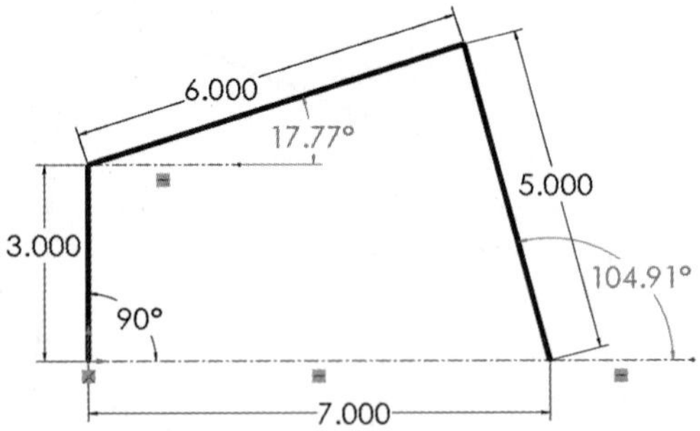

When driven dimensions are shown in a different color, it makes clear that these dimensions do not control the position of the mechanism. Double-clicking on these driven dimensions will not allow their values to be edited.

The advantages of modeling mechanisms in the SOLIDWORKS environment are seen when multiple variations of the mechanisms are to be found. For example, if we need to find θ_3 and θ_4 for a value of $\theta_2 = 45$ degrees, we need to change only that dimension.

Double-click on the 90-degree dimension and change its value to 45 degrees.

Notice that the driven dimensions θ_3 and θ_4 are updated accordingly, as shown in **Figure 10.40**. As noted earlier, if the 28.50-degree dimension shows only one decimal place, select the Options Tool, and under Document Properties: Dimensions: Angle, set the option for Zeros: Trailing zeros: Dimension to "Show."

FIGURE 10.40

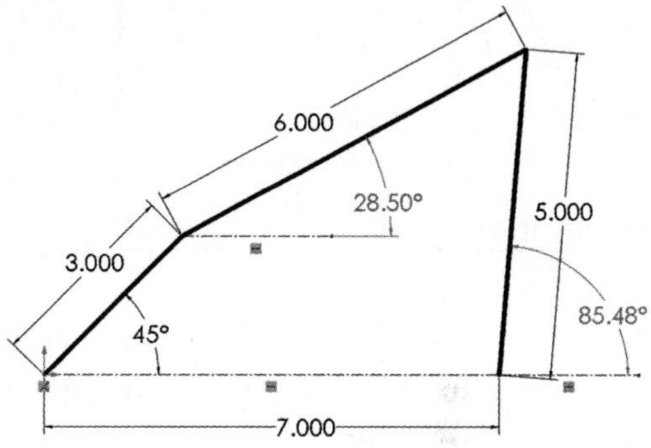

A driving dimension can be changed to a driven dimension by right-clicking and selecting Driven, as shown in **Figure 10.41**. (The right-click menu is quite long when selected in this context; you may need to scroll down to find the Driven item toward the bottom of the menu.) You can also make a dimension driven or driving from the Other tab in the PropertyManager.

FIGURE 10.41

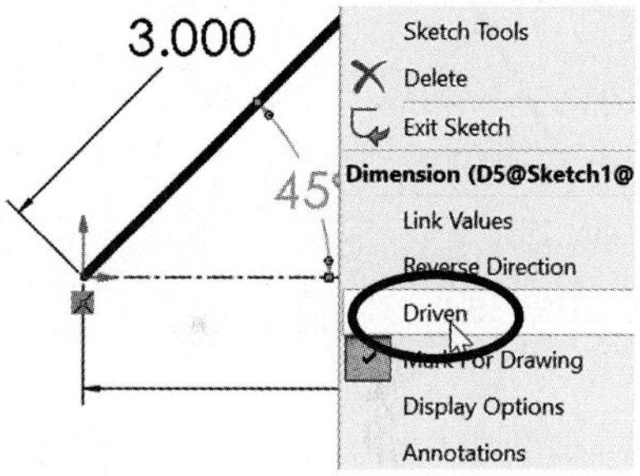

Change the 45-degree dimension to Driven.

We can now click and drag the endpoint of Link 2, as shown in **Figure 10.42**, to investigate the range of possible motions. This mechanism is classified as a crank-rocker because one of the members that pivots about a fixed point can rotate 360 degrees (a crank), while the other member that pivots about a fixed point oscillates back and forth (a rocker). We notice that as the crank, Link 2, revolves, there are two positions for which Link 4, the rocker, is vertical. If we want to determine the positions of the crank for which this condition applies, then we need to make θ_4 a driving dimension.

FIGURE 10.42

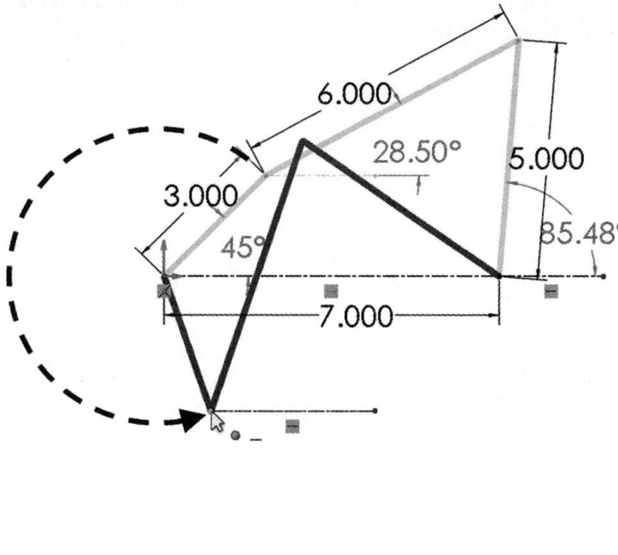

FIGURE 10.43

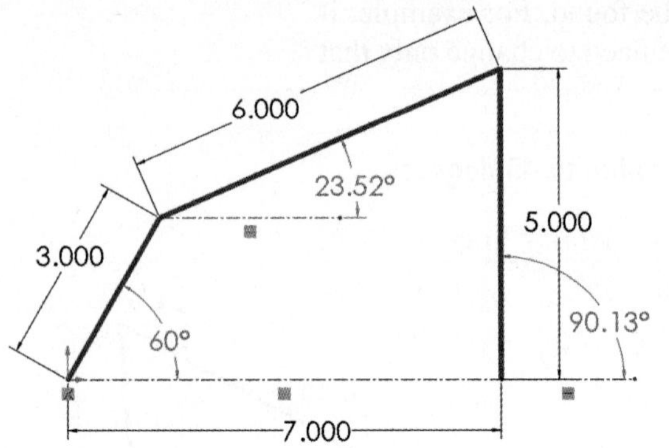

Drag one of the links until Link 4 is almost vertical, as shown in Figure 10.43.

Right-click on the dimension defining the angular position of Link 4 and clear the Driven check mark. Change the dimension to 90 degrees. Set the display of the angular position of Link 2 to two decimal places.

FIGURE 10.44

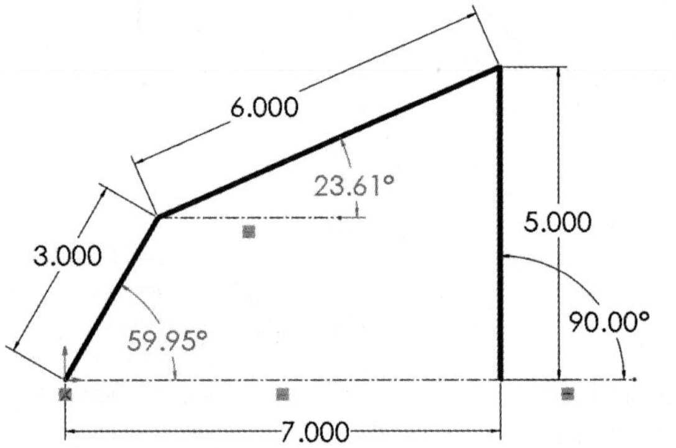

We see in **Figure 10.44** that θ_2 equal to 59.95 degrees results in Link 4 being vertical.

When dragging Link 2 through its full range of motion, we found that there are two configurations for which Link 4 is vertical.

FIGURE 10.45

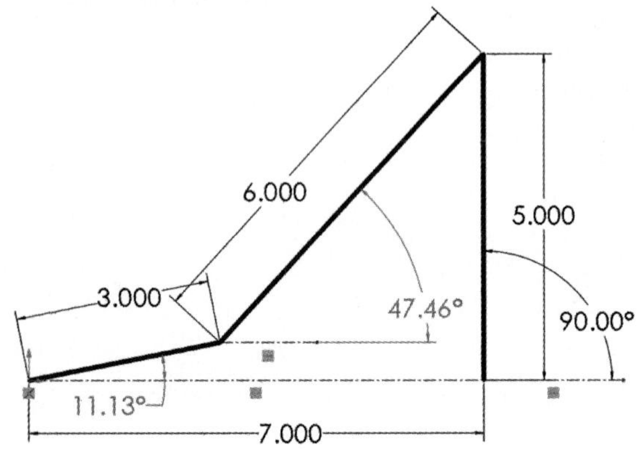

Set the 90° angle defining Link 4 to Driven. Drag the mechanism to approximately the position shown in Figure 10.45, and set the dimension back to Driving. Set the dimension to 90 degrees.

We see that $\theta_2 = 11.13$ degrees is another solution for which the rocker is vertical.

We can now examine the effect of changing the length of one of the links.

Set the mechanism back to its original position, with θ_2 as the driving dimension with a value of 90 degrees and θ_4 as a driven dimension, as shown in Figure 10.46.

Double-click the 6-inch dimension and change its value to 4 inches, as shown in Figure 10.47.

FIGURE 10.46

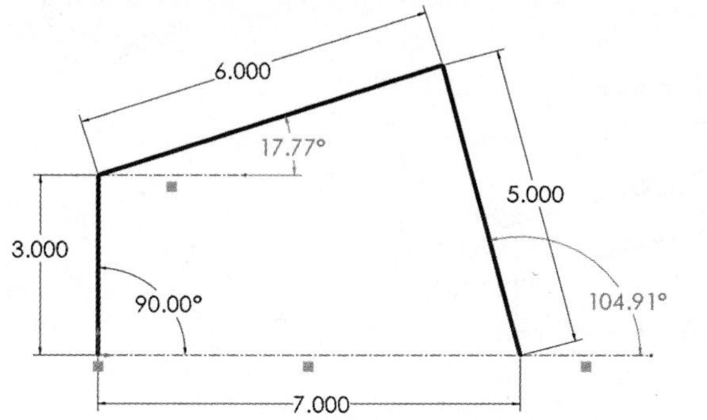

FIGURE 10.47

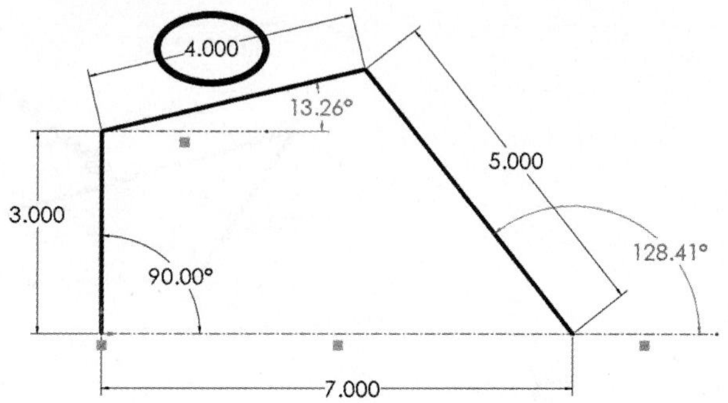

Change the 90-degree dimension to 150 degrees.

Note that the position of the links remains unchanged, and the message shown in **Figure 10.48** is displayed. To understand why the position is invalid, we can drag the links to determine their range of motion.

FIGURE 10.48

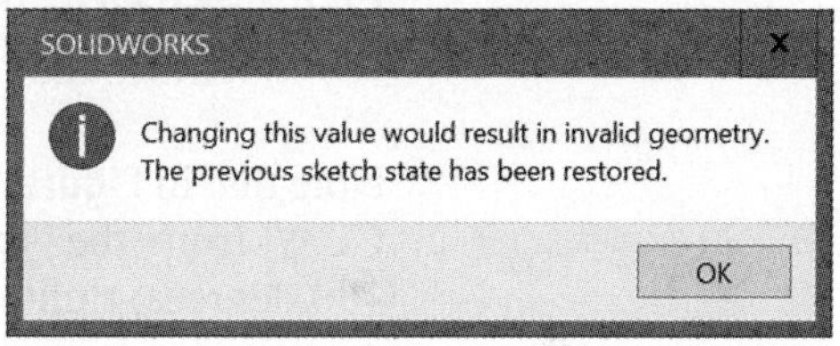

Click OK to clear the message box. Use the Undo Tool to restore the previous configuration, and set the 90-degree dimension to Driven. Drag the endpoint of Link 2 as far as possible in the counterclockwise direction, as shown in Figure 10.49.

FIGURE 10.49

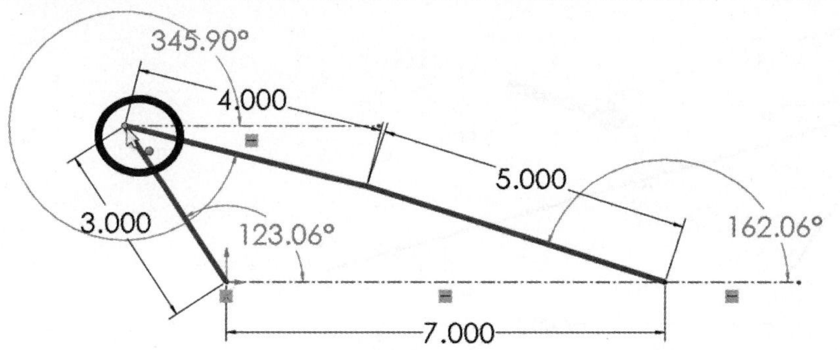

We find that when θ_2 reaches about 123 degrees, Links 3 and 4 are aligned, preventing Link 2 from rotating further. This condition defines a *toggle position* of the mechanism. The mechanism is called a *double-rocker*, since neither link that pivots around a fixed point can rotate 360 degrees.

To find the precise location of the toggle position, a relation between Links 3 and 4 can be added.

Select Links 3 and 4, and add a collinear relation. The value of θ_2 shown in Figure 10.50, 123.20 degrees, defines the toggle position.

FIGURE 10.50

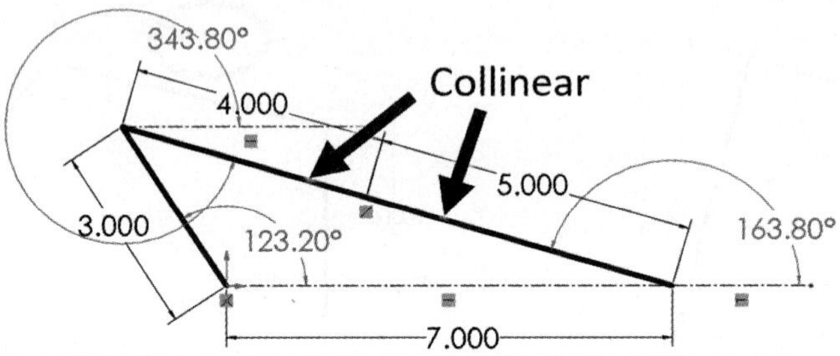

A second toggle position exists at $\theta_2 = 236.80$ degrees (–123.20 degrees).

If you want to move the mechanism, it is necessary to either click on the collinear relation icon on the drawing and press the Delete key, select either Link 3 or Link 4 and delete the collinear relation from the PropertyManager, or use the Undo key to remove the relation.

Note that in **Figure 10.50**, the angle defining θ_3 is shown as 343.80 degrees measured CCW from the +x axis. If you would prefer to show this angle as less than 180 degrees, right-click the dimension and choose Display Options: Explementary Angle. This will cause the angle to be displayed as 16.20 degrees measured CW from the +x axis, as shown in **Figure 10.51**.

FIGURE 10.51

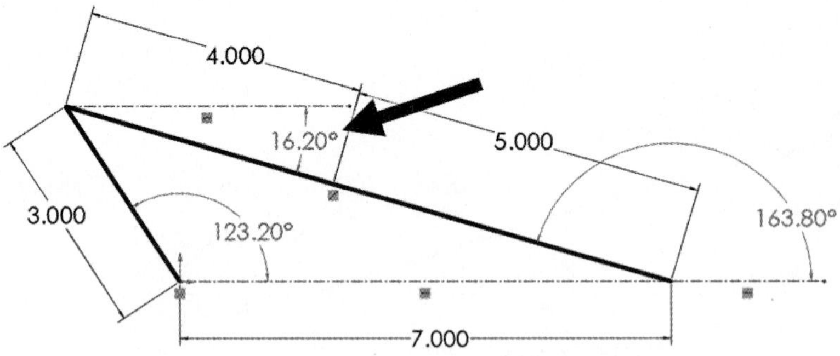

PROBLEMS

P10.1–10.6 For each problem, create a drawing to find the resultant vector from the addition of the vectors shown in **Figures P10.1** through **P10.6**. Report the resultant's magnitude and its direction angle, measured counterclockwise from the +x axis.

FIGURE P10.1

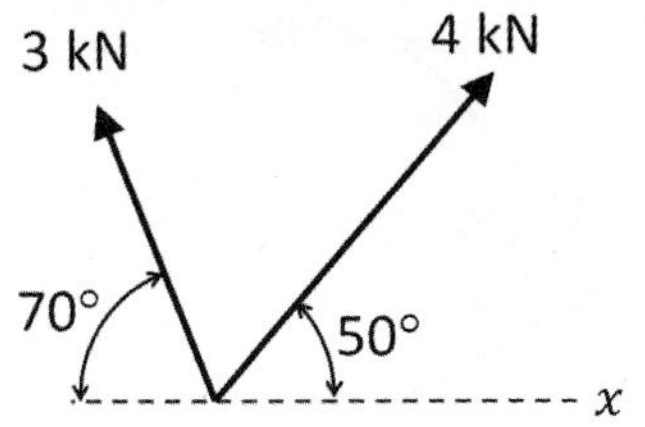

FIGURE P10.2

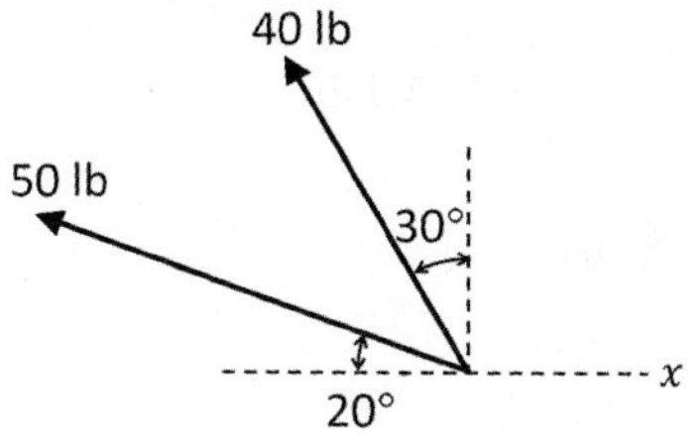

FIGURE P10.3

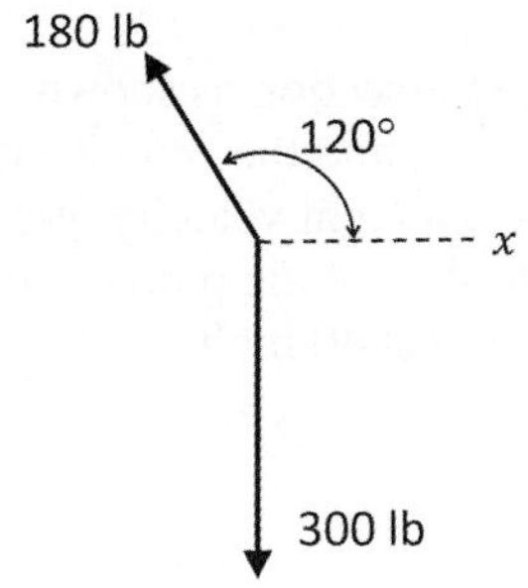

FIGURE P10.4

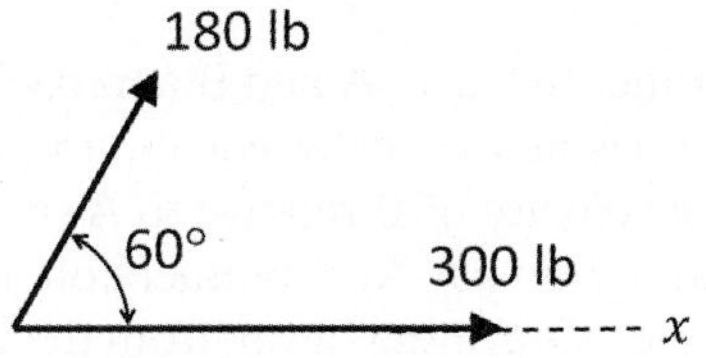

FIGURE P10.5

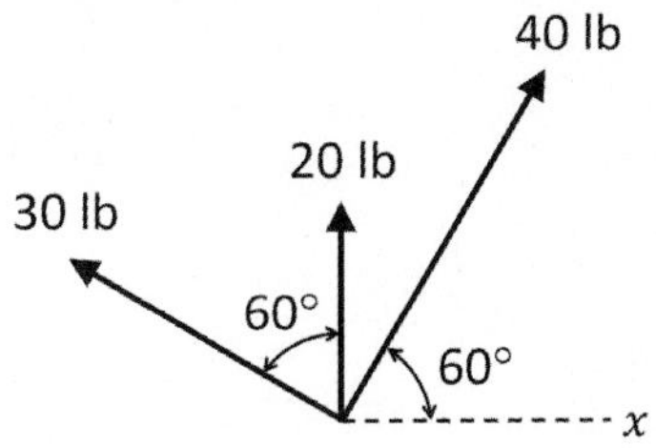

FIGURE P10.6

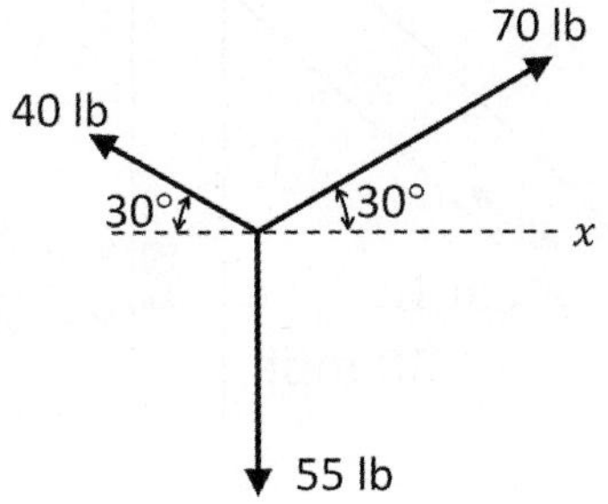

P10.7 The resultant of these two vectors is a vector with direction along the +x axis. Make a drawing to find the magnitudes of vector A and the resultant vector.

P10.8 The resultant of these two vectors is a vector with direction along the +x axis, with a magnitude of 80 pounds. Make a drawing to find the angles θ_A and θ_B.

FIGURE P10.7

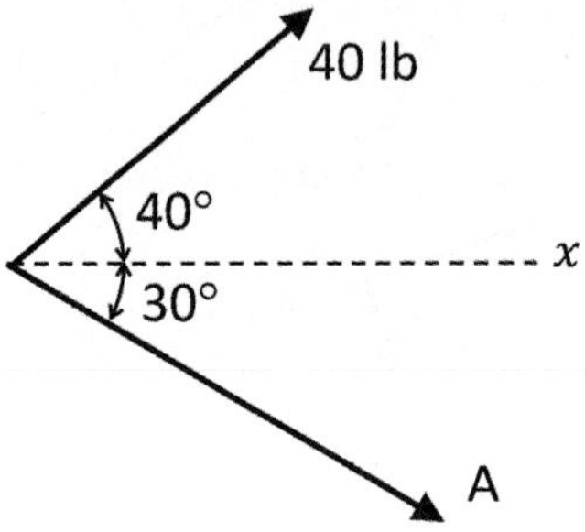

FIGURE P10.8

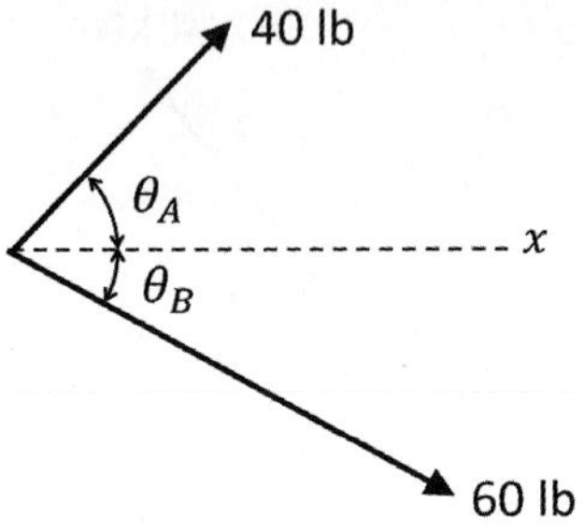

P10.9 Consider two cars A and B approaching an intersection as shown in **Figure P10.9**. Using the vector equation $\mathbf{v}_B = \mathbf{v}_A + \mathbf{v}_{B/A}$, find the velocity $\mathbf{v}_{B/A}$, which is the velocity of B relative to A (that is, the apparent velocity of car B when viewed from car A). Comment on the result. Repeat the problem to find $\mathbf{v}_{B/A}$ if car B is travelling away from the intersection at 40 mph.

FIGURE P10.9

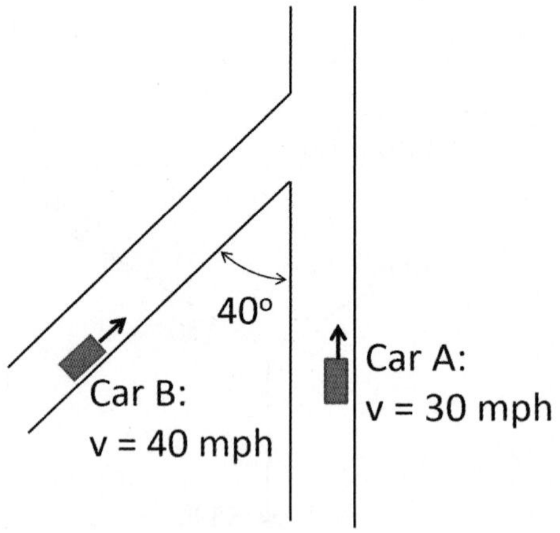

CHAPTER 11

Analysis of Mechanisms

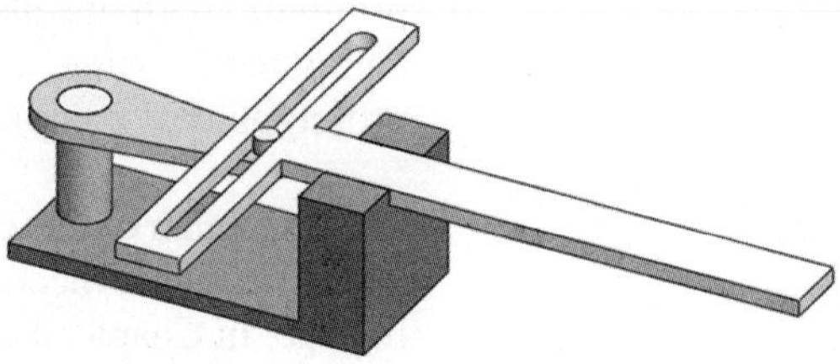

Introduction

A *mechanism* is an assemblage of *links* and *joints* that are connected together to achieve a desired motion task. The links provide the mechanical structure of the mechanism, while the joints provide the ability for the mechanism to move. Some typical kinds of mechanisms used by mechanical engineers are shown in **Figure 11.1**.

FIGURE 11.1

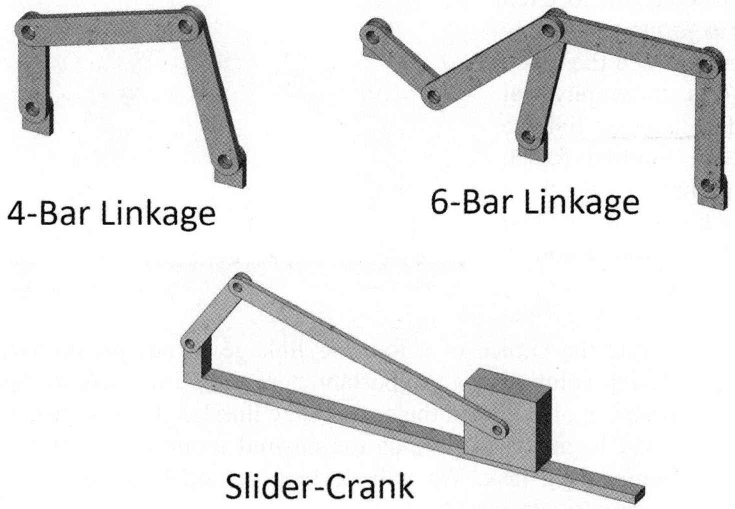

The ways in which we can connect links with joints and provide a mechanism with the ability to move are seemingly endless. Also, the links can take on any shape and size we desire, and the motion may change in complicated and seemingly unpredictable ways as we modify the links. The design of a mechanism may seem like a daunting task.

Chapter Objectives

In this chapter, you will:

- learn some basic terminology associated with mechanism design,
- use the assembly modeling techniques learned in Chapter 6 to develop a working model of a mechanism,
- perform a motion simulation of a mechanism, and
- export a motion simulation as a video file.

SOLIDWORKS is a registered trademark of Dassault Systémes SolidWorks Corporation.

Computer-aided design (CAD) packages have become valuable tools in the design and analysis of mechanisms. The ability for the design engineer to adjust the size, shape, and interconnectivity of links and joints and quickly assess the impact on the mechanism has accelerated the design cycle of mechanisms. The SOLIDWORKS® program, with its ability to represent geometric constraints between structural components using assembly mates, is ideally suited to the design and virtual prototyping of complex mechanisms.

The remainder of this chapter will be devoted to the use of the SOLIDWORKS program in the design of mechanisms through a case study involving the design of a four-bar linkage. In Chapter 10, we evaluated the motion of a four-bar linkage in a 2-D sketch, with lines representing the links. In this chapter, we will model the links as 3-D parts, and use a motion study to visualize the motion of the linkage.

11.1 Approaching Mechanism Design with SOLIDWORKS Assemblies

Consider the four-bar linkage, a classic mechanism used in engineering, shown in **Figure 11.2**.

It consists of three structural *links*, connected to each other and to fixed pivot points by *pin joints* that allow for rotating motion between the links. Though it only has three physical links, it is called a four-bar linkage because there is an implied fourth structural link that represents the fixed frame to which the linkage is attached and connects the fixed ground points, as shown in **Figure 11.3**.

FIGURE 11.2

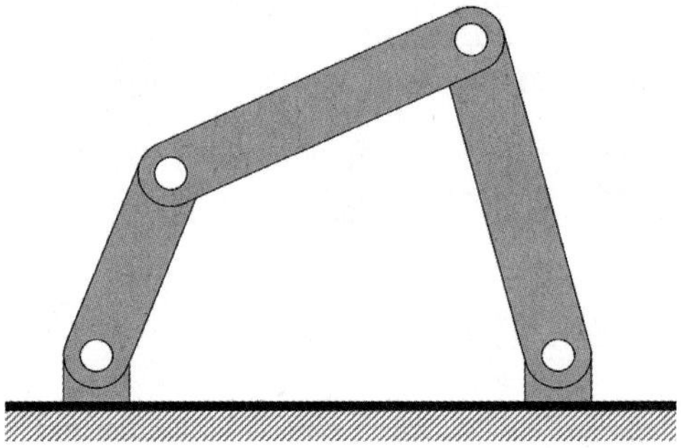

FIGURE 11.3

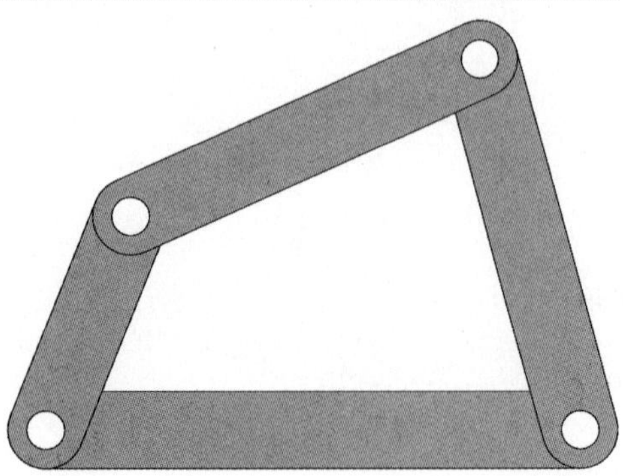

While the choice of a four-bar linkage as our preliminary design solution is an important step, the *parametric design problem* of selecting the appropriate link lengths and ground pivot locations to give us the desired motion is a difficult engineering task. We will develop a model to aid in this parametric design phase.

The features that we will employ are the assembly capabilities. The development of assembly models and the definition of assembly mates were covered in Chapter 6. Mated assemblies are an ideal tool for use in mechanism design, since the joints that provide the physical relationships between the links are analogous to the mates that define the geometric relationships between parts.

Think about two links connected by a pin joint, as shown in **Figure 11.4**.

FIGURE 11.4

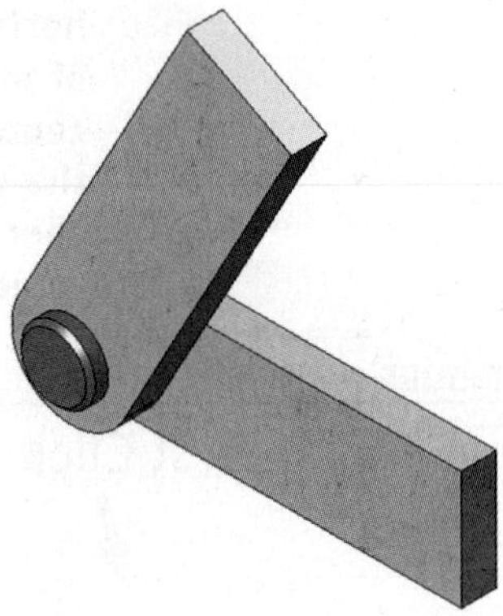

The pin through the holes in the links allows for rotation between the links, giving a "scissors" action. The mated assembly representation of this type of link/joint assembly would involve two parts, with two mates serving the same geometric purpose as the pin joint in the physical linkage:

- The front face of one link is aligned with the back face of the other link using a coincident mate; this allows one link to "slide by" the other without any interferences between the surfaces. (While a physical linkage would have a low-friction spacer to separate the links slightly, it is not necessary to model the spacer to simulate the allowed motion of the joint.)

- The hole in one link is aligned with the hole in the other link using a concentric mate; this keeps the holes aligned at their central axes.

By capturing the essential geometric constraints that underlie a pin joint, this sequence of two mates imparts the same motion capabilities to the solid model that are seen in the physical mechanism.

Using this simple two-link mechanism as a building block, complex mechanisms can be assembled and virtually tested. Mechanism motion can be tested and debugged without the need for physical prototypes to be constructed. The following section will step through the development of a model of a four-bar linkage.

11.2 Development of Part Models of Links

In this section, a step-by-step tutorial will lead you through the development of the part models required to make a "working" assembly model of a four-bar linkage. Four links, similar in shape but of different lengths, will be constructed.

Open a new part. Select the Front Plane. Choose the Slot Tool from the Sketch Group of the CommandManager, as shown in Figure 11.5. In the PropertyManager, set the type of slot to Centerpoint Straight Slot (this option can also be selected directly from the pull-down menu beside the Slot Tool), check the Add Dimensions box, and make sure that the type of dimensioning is Center-to-Center, as shown in Figure 11.6.

FIGURE 11.5

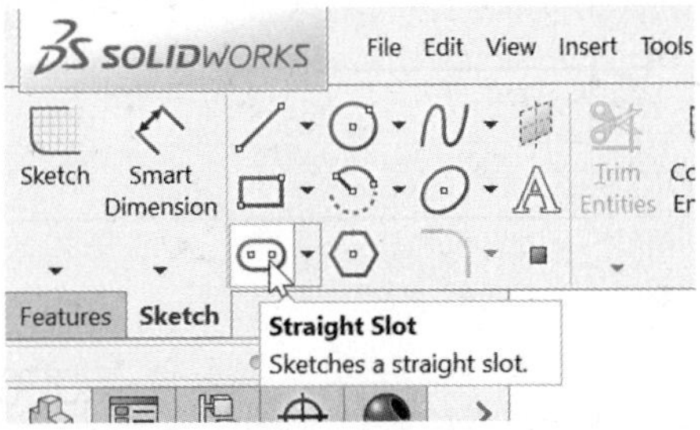

FIGURE 11.6

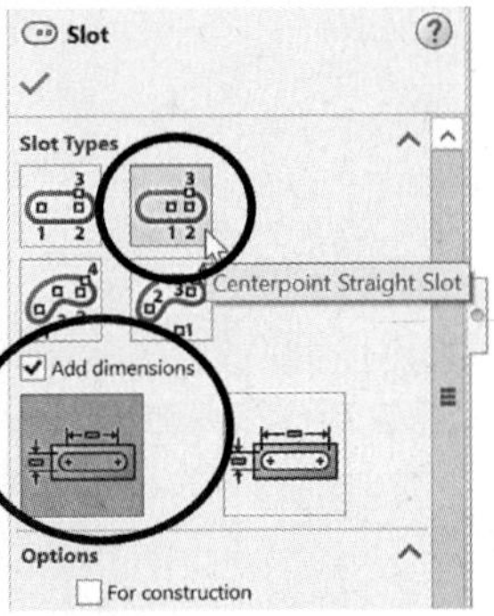

Click on the origin to place the center of the shape, and drag the cursor horizontally, as shown in Figure 11.7. Note that a centerline connecting the centers of the arcs is created as the cursor is being dragged. Click to place the end of the centerline, and drag the cursor upward, as shown in Figure 11.8. Click to complete the shape, and the dimensions will be added, as shown in Figure 11.9. Click the check mark to turn off the Slot Tool. Double-click each dimension and set the values to 7 inches and 1 inch, as shown in Figure 11.10.

FIGURE 11.7

FIGURE 11.8

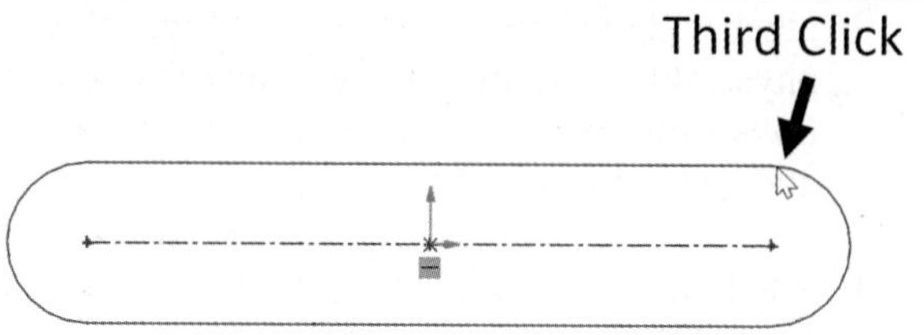

FIGURE 11.9

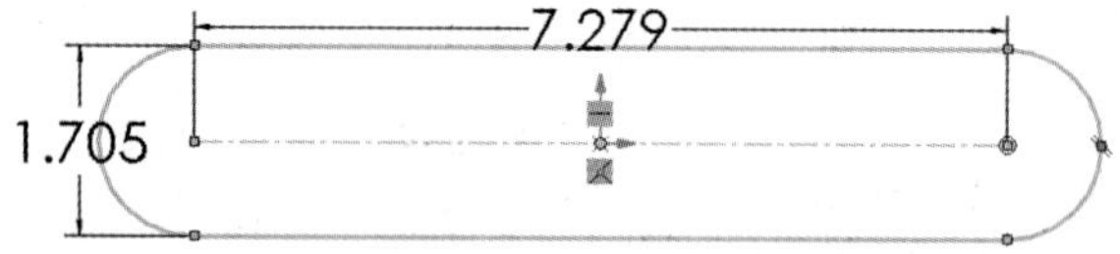

FIGURE 11.10

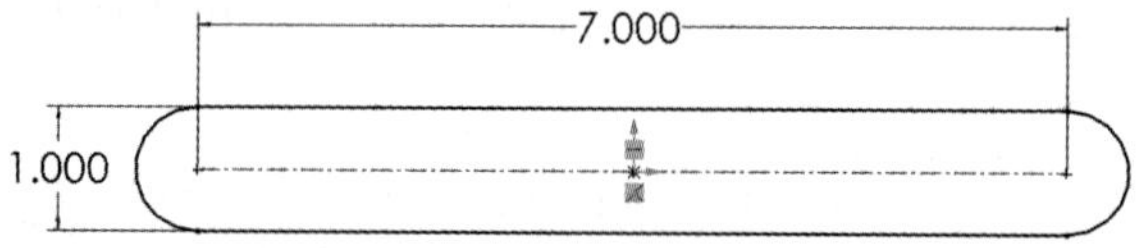

We could now extrude the link and add the holes in a separate operation, but instead we will add the circles representing the holes to our sketch and complete the link as a single extruded feature.

Select the Circle Tool. Add circles centered at both ends of the centerline, as shown in Figure 11.11. Add an equal relation between the two circles, and add a 0.5-inch diameter dimension to one of the circles, as shown in Figure 11.12.

FIGURE 11.11

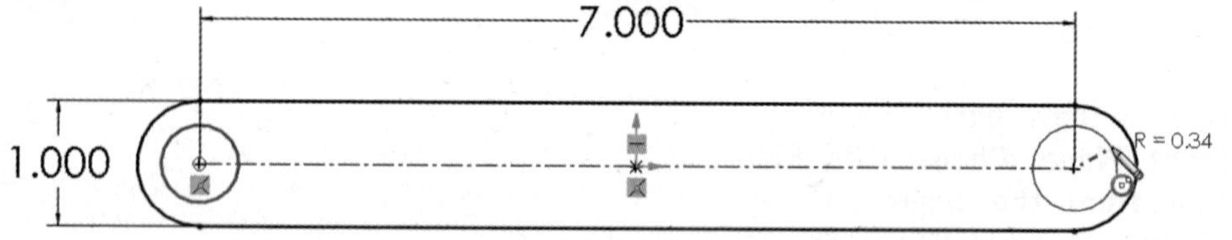

FIGURE 11.12

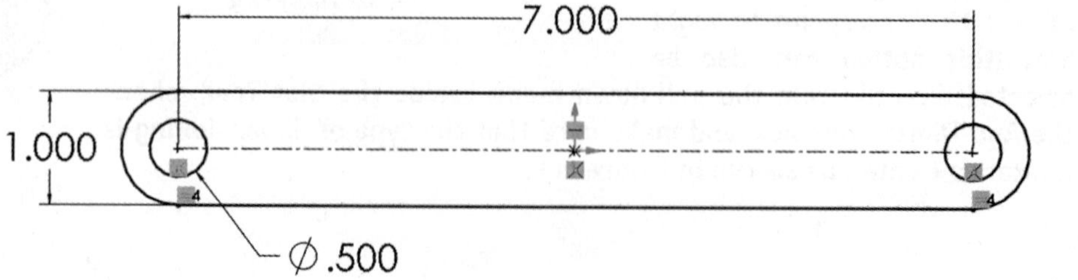

Select the Extruded Boss/Base Tool from the Features group of the CommandManager. Set the extrusion depth to 0.25 inches, and click the check mark to complete the extrusion.

The completed part is shown in **Figure 11.13**. Note that our sketch included more than one closed contour. In previous chapters, when we have had a sketch with multiple closed contours, we have had to select the contours that we wanted to extrude. However, in this case, one of the contours (the overall shape) completely surrounded the other closed contours (the circles), and the interior contours did not touch or overlap each other. When this is the case, by default the outer contour is extruded and the interior contours are not.

FIGURE 11.13

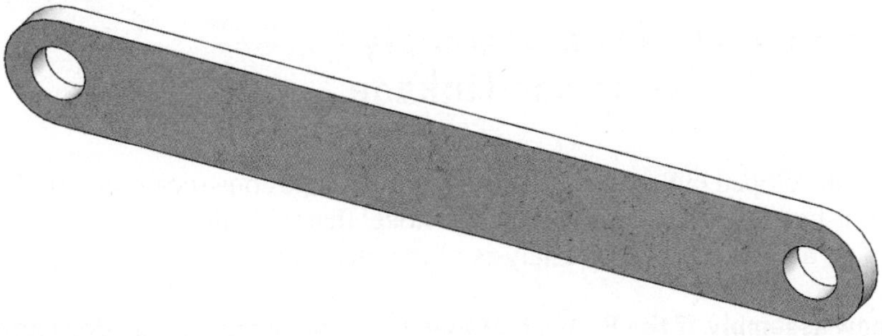

Save the part as "Link1."

Each of the other three links will be identical except for its length. Rather than creating each of the other links individually, we can either make different configurations of this part, as explained in Chapter 3, or modify the length of the part and save each modification with a different name. We will use the latter method.

Double-click Boss-Extrude1 in the FeatureManager. The dimensions defining this feature will be displayed, as shown in Figure 11.14.

FIGURE 11.14

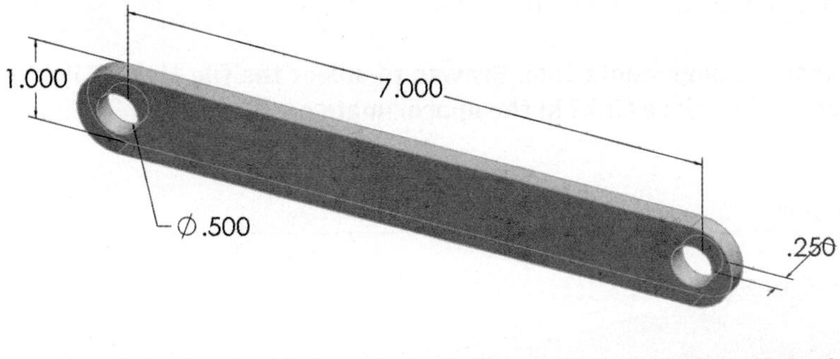

Double-click the 7-inch length and change it to 3 inches. Rebuild the model. The modified link is shown in Figure 11.15.

Choose File: Save As from the main menu. Save this part as "Link2."

FIGURE 11.15

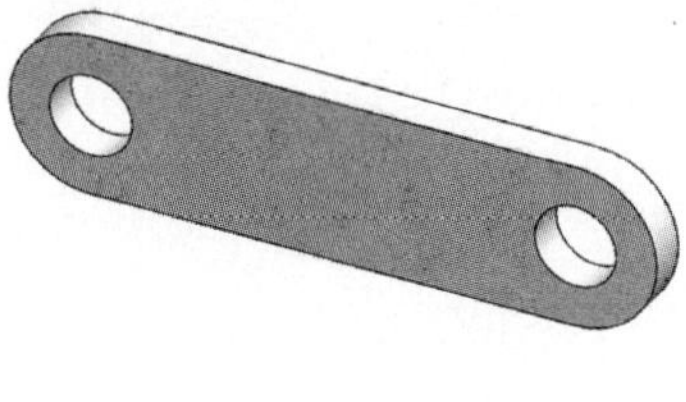

Create the other two links, using the lengths as shown in Table 11.1, and save them using the names "Link3" and "Link4."

TABLE 11.1 Lengths of the Four Link Parts

	Length
Link1	7 in.
Link2	3 in.
Link3	6 in.
Link4	5 in.

11.3 Development of the Assembly Model of the Four-Bar Linkage

The part models developed in Section 11.2 will now be used to construct an assembly model of the four-bar linkage. It is this assembly model that will allow us to perform parametric design and simple motion analysis of the mechanism.

FIGURE 11.16

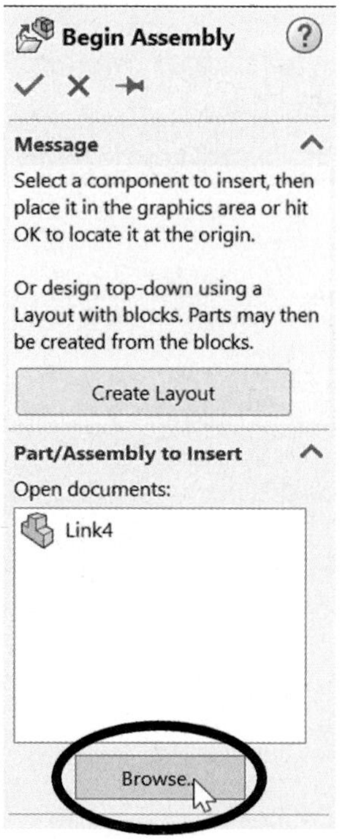

Open a new assembly. If the PropertyManager shown in Figure 11.16 does not appear, open it by selecting the Insert Components Tool from the Assembly group of the CommandManager. (The Begin Assembly PropertyManager will open whenever a new assembly is started as long as the box labeled "Start command when creating new assembly" is checked.) Any part or assembly files that are currently open will be shown in the PropertyManager; any other file can be selected by selecting Browse and navigating to the desired file's location. Select Browse, select the file Link1, and select Open. Click the check mark to place this part.

This first part will be fixed in space—note the "(f)" beside its name in the FeatureManager. We refer to it as the "ground link" since all motion of the other links will be relative to this fixed link.

Select the Insert Components Tool. Browse to select the file Link2. Click in the graphics area to place Link2 in the approximate position shown in Figure 11.17.

FIGURE 11.17

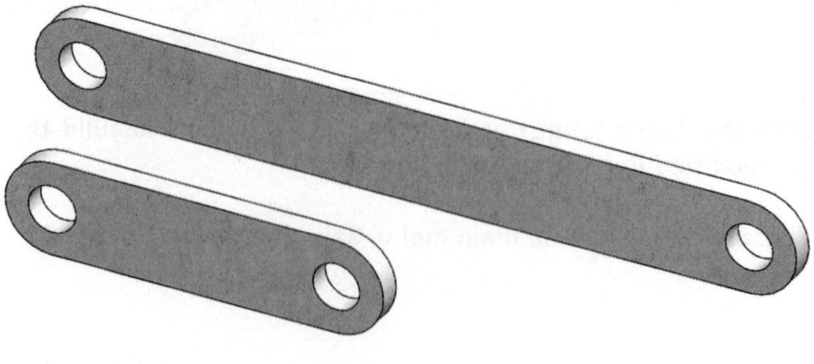

Note that there is a (-) symbol beside Link2 in the FeatureManager. This indicates that the link is "floating"; that is, it is free to move within the assembly. You can confirm this by clicking and dragging the part within the graphics area.

A coincident mate aligning the faces of these links will be defined.

Press Esc to cancel any selections made when moving Link2.

Select the Mate Tool, as shown in Figure 11.18, to initiate a new mate.

FIGURE 11.18

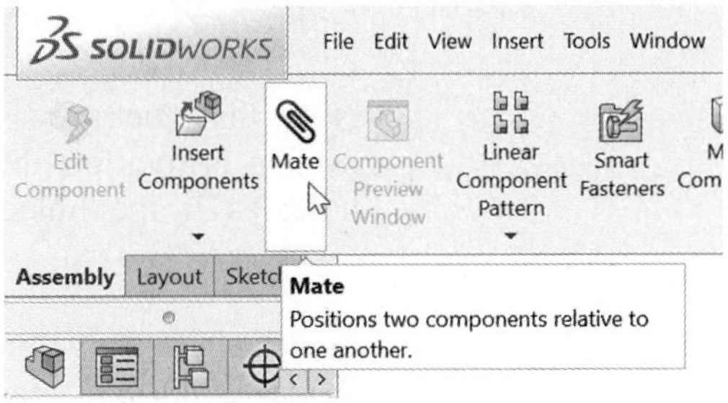

The Mate PropertyManager will open.

Using the Rotate View Tool, rotate the parts so that the back faces of the links can be seen. Select the back face of Link2, as shown in Figure 11.19.

Switch to the Trimetric View, and select the front face of Link1, as shown in **Figure 11.20**.

FIGURE 11.19

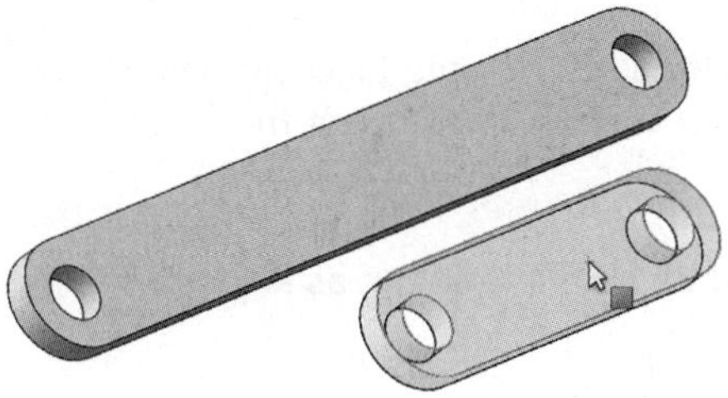

FIGURE 11.20

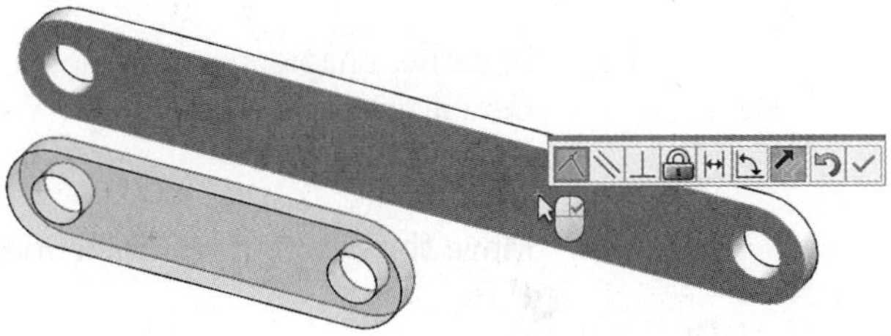

Link2 will automatically move toward Link1 so that the selected faces show a coincident mate.

FIGURE 11.21

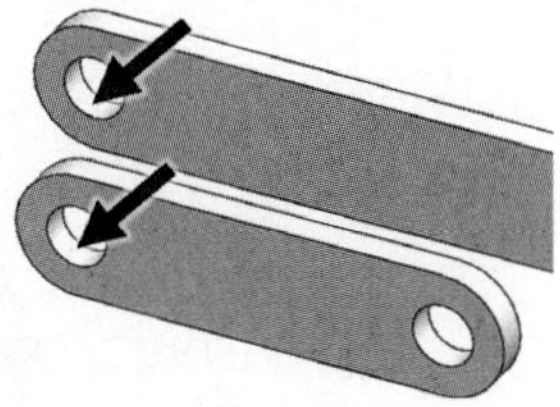

Click the check mark in the PropertyManager or the pop-up box to apply the mate.

A concentric mate between the holes must also be added to simulate the kinematic constraint of the pin joint.

Select the inside surfaces of the holes at the left end of the links, as shown in Figure 11.21.

FIGURE 11.22

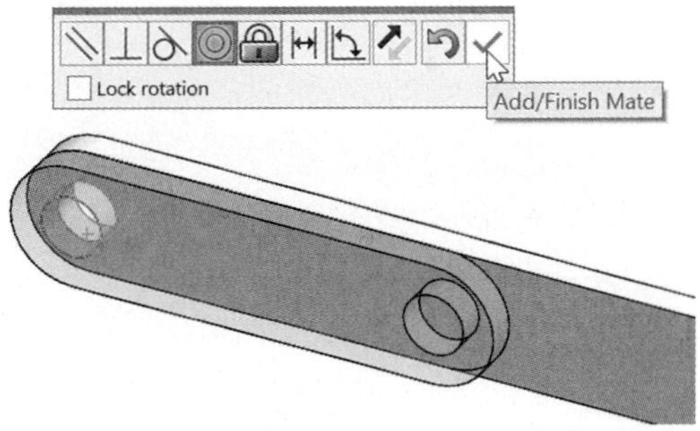

A concentric mate will be previewed, as shown in Figure 11.22. Click the check mark in the PropertyManager or the pop-up menu to apply the mate. Click the check mark again to close the Mate PropertyManager.

The kinematic constraints of the pin joint are now fully defined. Link2 is still free to move, as long as the movements do not violate the mates that we have placed on it. We can experiment to see the type of motion still allowed under the constraints we have imposed.

Click and drag Link2. Confirm that you can rotate the link through a full 360 degrees. Leave Link2 in the approximate position shown in Figure 11.23.

FIGURE 11.23

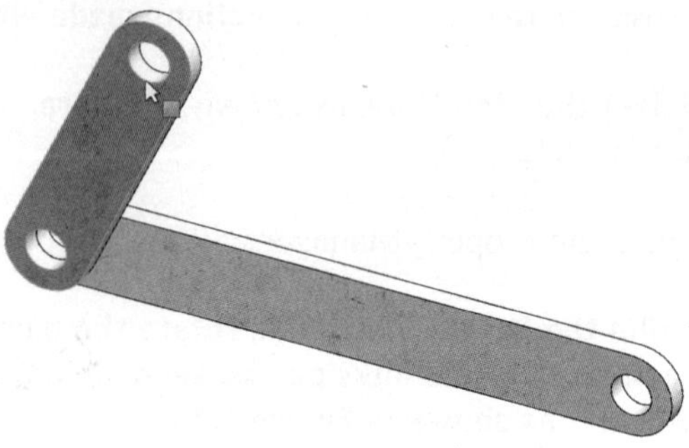

Note that the only unconstrained motion is rotation about the mated hole, as if the links were pinned together.

Select the Insert Component Tool. Select Link4 and place it into the assembly.

Select the Mate Tool.

In earlier chapters, we learned a shortcut method of selecting faces not visible from the current view orientation. We will use that technique to select the back face of Link4.

Move the cursor over Link4. Right-click and pick Select Other, as shown in Figure 11.24.

FIGURE 11.24

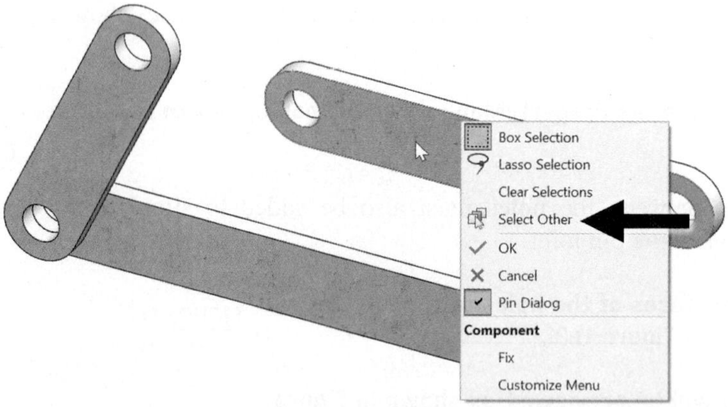

FIGURE 11.25

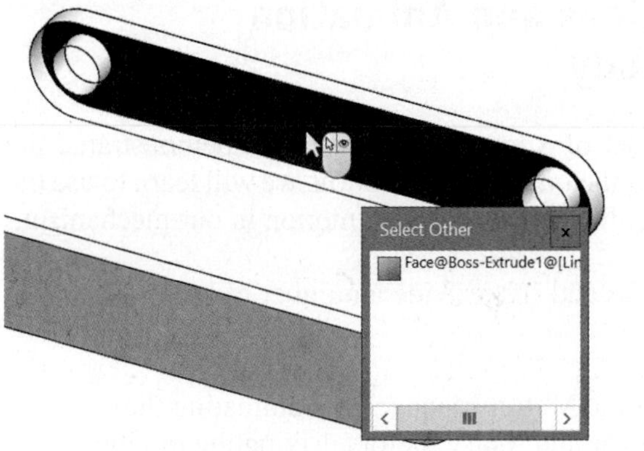

The back face of the link is highlighted, as shown in Figure 11.25. Click the left mouse button to accept the selection. Select the front face of Link1. A coincident mate will be previewed; click the check mark to accept the mate.

Add a concentric mate to the inner faces of corresponding holes in Link1 and Link4. Close the Mate PropertyManager.

Click and drag Link4 to the approximate position shown in Figure 11.26.

FIGURE 11.26

Insert Link3 into the assembly. Add three mates:

1. **A coincident mate between the back face of Link3 and the front face of either Link2 or Link4,**
2. **A concentric mate between the inner faces of the corresponding holes of Link2 and Link3, and**
3. **A concentric mate between the inner faces of the corresponding holes of Link3 and Link4.**

Note that only one coincident mate is required. The coincident mate between the back face of Link3 and the front face of one of the other links completely defines the z-direction location of Link3.

The completed assembly is shown in **Figure 11.27**.

FIGURE 11.27

Click on Link2, and drag to rotate it. The link will rotate through a full 360°, and the motion of the other links can be observed. Save the assembly with the file name "Linkage." If prompted to do so, select Rebuild and save the document.

11.4 Creating Simulations and Animation with a Motion Study

In Section 11.3, we created a model of a four-bar linkage, and demonstrated the motion by manually dragging one of the links. In this section, we will learn to use the simulation tools built into the Assembly mode to create motion in our mechanism.

The SOLIDWORKS program and its add-ins provide a number of levels of motion simulation. These include:

- Animation, available through the MotionManager, for animating the motion of assemblies with linear and rotary motors driving the motion,
- Basic Motion, available through the MotionManager, for studying the motion of assemblies with the effects of contact between members, gravity, and springs in addition to motors, and
- SOLIDWORKS Motion, an add-in package, used for performing physical simulations with quantitative analysis of velocities, accelerations, and forces. A tutorial covering the use of SOLIDWORKS Motion is available on the book's website, *www.mhhe.com/howard2019.*

This section covers simple animation with the Animation Tool. Section 11.5 will demonstrate more complex simulations using the Basic Motion Tool.

At the bottom left-hand corner of the assembly window, note that there are two tabs: a Model tab and a Motion Study tab (labeled Motion Study 1). Note that if the Motion Study tab is not available, it can be added by selecting View: User Interface: MotionManager from the main menu.

Click on the Motion Study 1 tab, as shown in Figure 11.28.

FIGURE 11.28

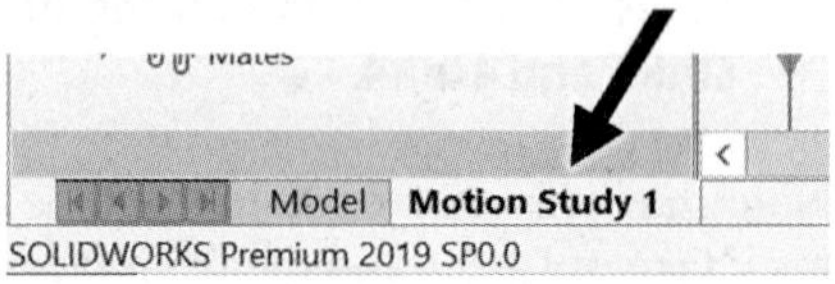

FIGURE 11.29

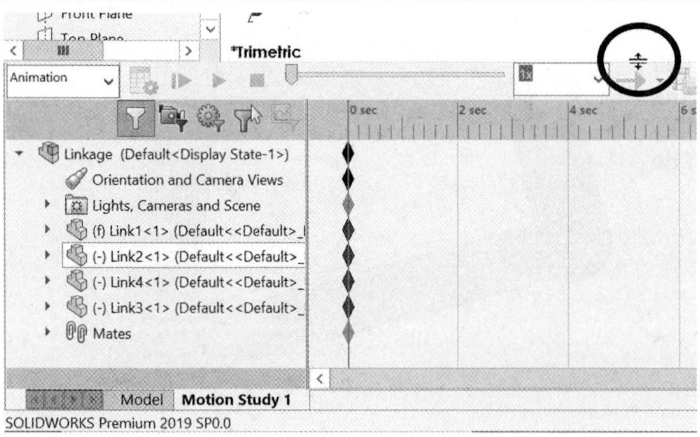

The MotionManager will appear, as shown in **Figure 11.29**. The MotionManager allows us to add simulation components such as motors, springs, and gravity to our model, lets us define the duration and resolution of our simulation, and gives us the ability to output animation files of our simulation. Also in **Figure 11.29**, note that you can click and drag on the border between the graphics area and the MotionManager to show more or less of the MotionManager.

If desired, you can provide more space on the screen for the graphics area and the MotionManager by selecting View: Workspace: Widescreen. This moves the CommandManager to the left side of the screen, as shown in **Figure 11.30**, and allows the PropertyManager to be moved around the screen. To return to the regular screen configuration, select View: Workspace: Default. (Note that you can change the workspace only if the box labeled Lock CommandManager and toolbars under the Customize Menu is unchecked.)

FIGURE 11.30

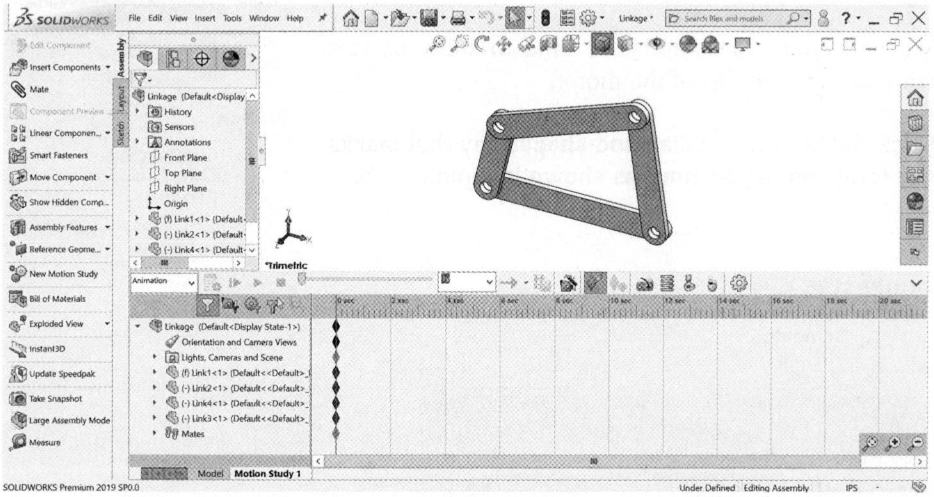

The first step in our simulation will be to add a rotary motor to drive Link2.

Press the Esc key to deselect any selected part. Click on the Motor Tool, as shown in Figure 11.31.

FIGURE 11.31

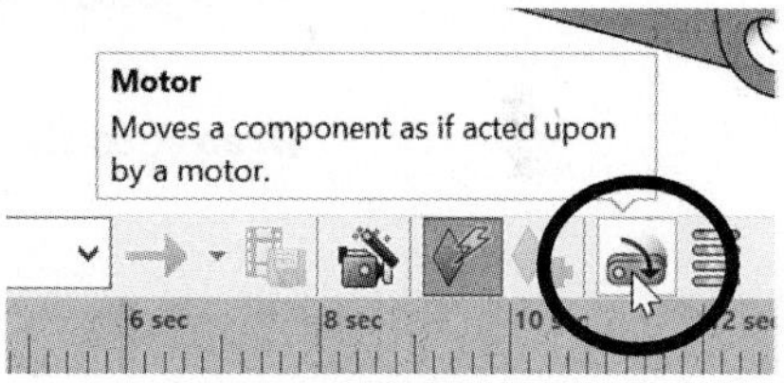

The Motor PropertyManager will open, as shown in **Figure 11.32**, with the Component/Direction box highlighted.

Select the front face of Link2 and a arrow showing the direction of the motor's rotation will appear, as shown in Figure 11.33.

FIGURE 11.32

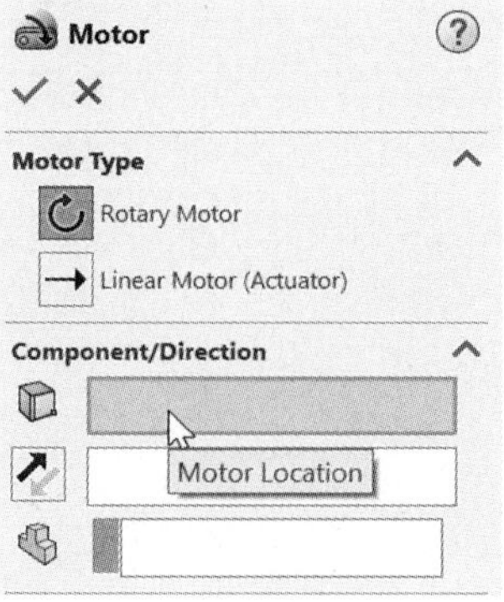

FIGURE 11.33

This adds a motor to the link, with the "drive shaft" of the motor oriented perpendicular to the selected face. The type and speed of the motion can also be set; we will use the default constant speed motion.

Change the speed to 10 RPM, as shown in Figure 11.34. The direction of the rotation can be reversed by clicking the arrows beside the direction box; in this case the default counterclockwise direction is fine. Click the check mark to close the PropertyManager.

FIGURE 11.34

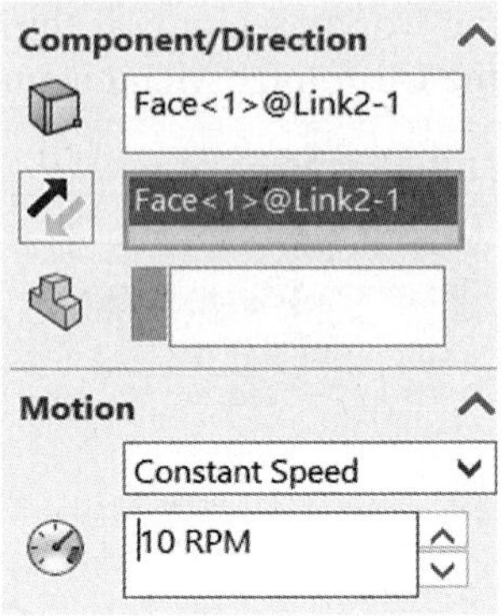

We will adjust the simulation time to show us two complete revolutions of the motor.

Click and hold on the diamond-shaped key that marks the total simulation time, as shown in Figure 11.35.

FIGURE 11.35

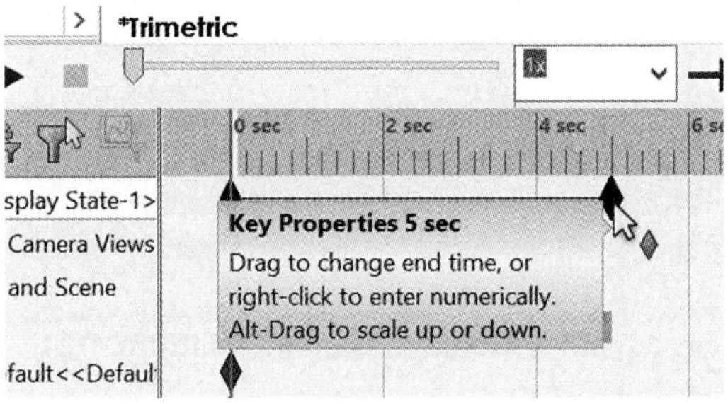

Drag it to the right, stopping at the 12-second mark, as shown in Figure 11.36.

FIGURE 11.36

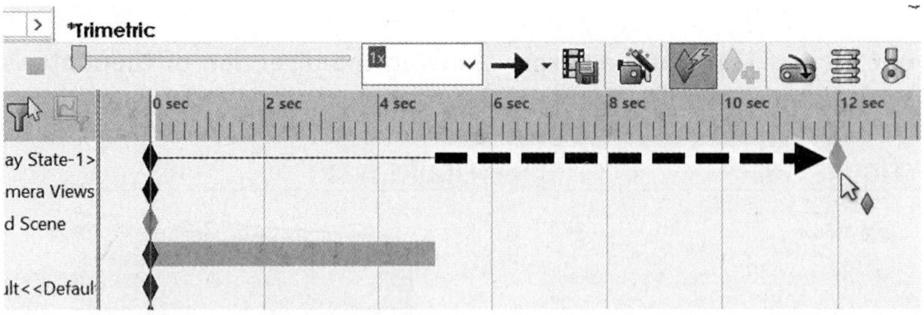

Click on the Motion Study Properties Tool, as shown in Figure 11.37.

FIGURE 11.37

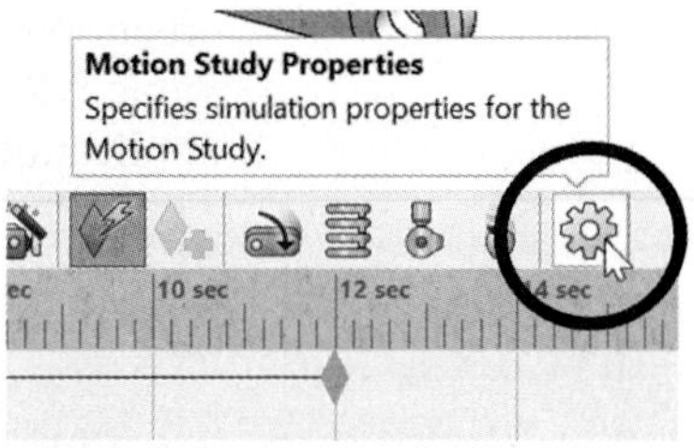

By default, the resolution of the simulation is set to 8 frames per second; this means that the simulation will produce a "snapshot" of the motion every 0.125 seconds (or every 7.5° of motor rotation, since our speed is 10 RPM). The higher we set the frames per second, the smoother the motion will appear in our simulation, but both more memory and more computation time will be required.

Set the Frames per second to 60 (by either moving the scroll wheel, clicking the up arrow, or typing in the value directly), as shown in Figure 11.38, and click the check mark to close the PropertyManager. This will create a "snapshot" of motion at every 1° of motor rotation.

FIGURE 11.38

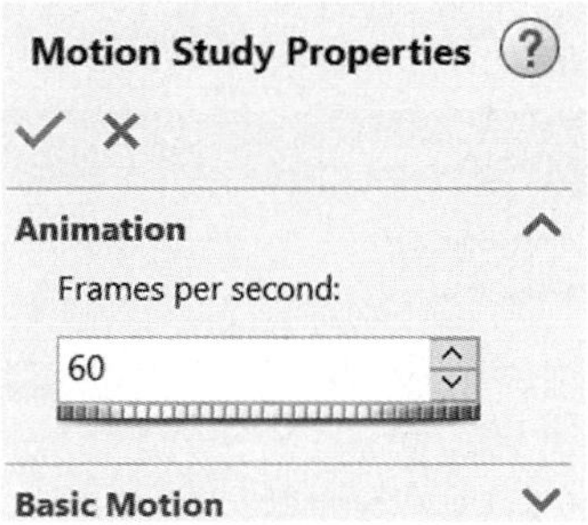

Save the assembly.

Click the Calculate Tool to compute the simulation, as shown in Figure 11.39. The simulation will be computed and displayed to the screen.

As long as no modifications are made to the model, the simulation need only be computed once; after this, it may simply be replayed, using the Play from Start, Play, and Stop controls, shown in **Figure 11.40**. The speed of the playback can be changed to a multiple of the actual speed (for example, .5X = one-half speed) or to the total playback duration in seconds. Note that the position of the linkage can be displayed for any time value of the simulation by dragging the Timeline slider bar to the desired time. The Playback Mode pull-down menu allows for options such as looping the animation continuously or playing the animation forward and backward continuously. We will use the Save Animation Tool to create a video file of the simulation.

FIGURE 11.39

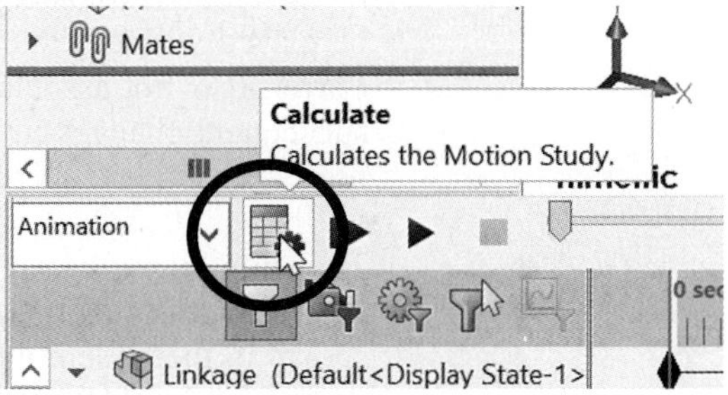

FIGURE 11.40

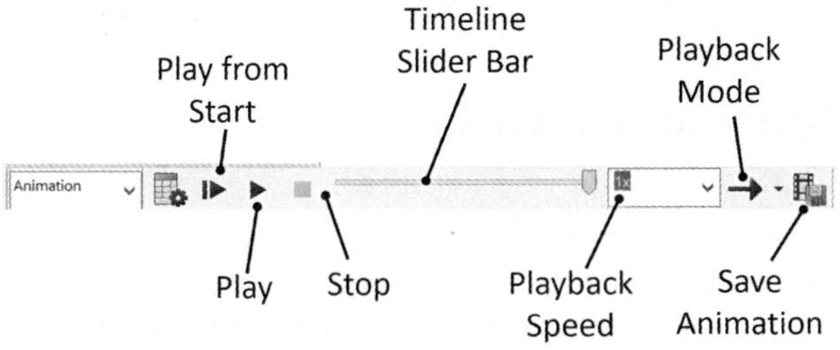

Click the Save Animation Tool.

FIGURE 11.41

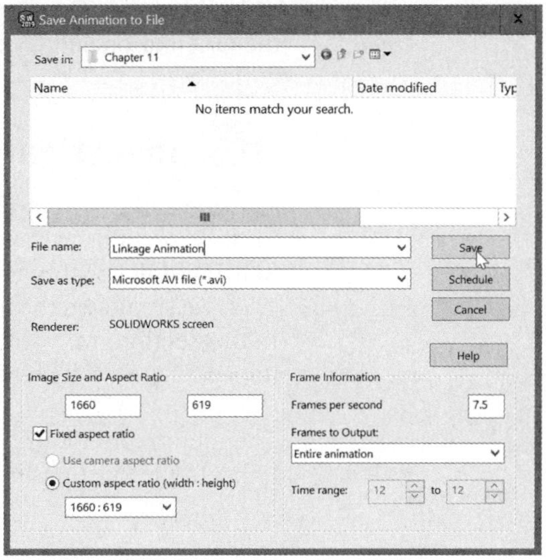

The Save Animation to File dialog box will appear, as shown in **Figure 11.41**. This allows us to select the appropriate directory and filename for our video file. It also allows us to set some parameters that control file size and image quality, such as number of frames per second to be saved as we create our video file. Of course, while setting the frames per second to a higher value will result in a smoother motion, it will also increase the file size. For this application, we will accept the default values.

Browse to the appropriate file location, enter the name as Linkage Animation, and click Save.

The Video Compression dialog box will appear, as shown in **Figure 11.42**. This dialog box allows us to select the image resolution of the video file. Higher-resolution images will result in larger file sizes. Similarly, reducing the number of frames between Key Frames will result in a smoother playback but will increase file size. For this application, we will accept the default values.

FIGURE 11.42

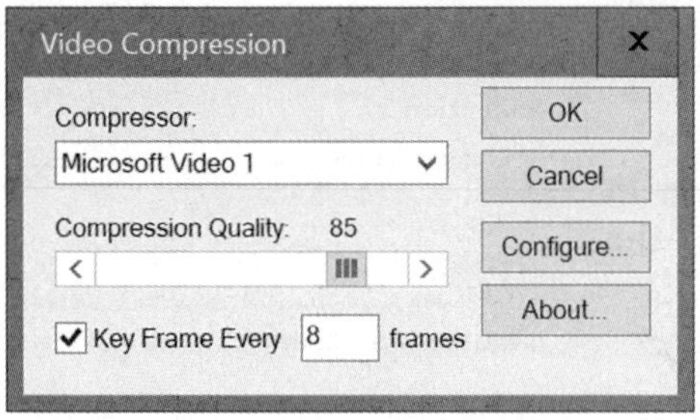

Click OK to close the Video Compression dialog box.

The simulation will be replayed again in the graphics window. While this is occurring, the video file will be written at the frames per second and compression level we specified.

As with the video created in Chapter 3 and the animation of the assembly explosion created in Chapter 6, the resulting file is an AVI file. It can be played with any standard video playback software such as the Windows Media Player. It is not associated in any way with the SOLIDWORKS software or files; subsequent changes in your part or assembly files will not be reflected in the AVI file.

The animation file is essentially a "screen capture" of the Assembly window during the motion simulation; the motion will appear using the model orientation and scale shown in the Assembly window. In order to optimize the quality of the animation, the model should be oriented, centered, and scaled appropriately prior to creating the animation. If desired, the MotionManager and the FeatureManager can be collapsed prior to creating the animation file, in order to give the largest window area for the screen capture. The view can be changed during the animation by placing keys on the Orientation and Camera Views timeline, as shown in Chapter 3.

11.5 Investigating Mechanism Design

In this section, we will exploit the parametric modeling capabilities of the SOLIDWORKS progam to further investigate the design of mechanisms. In Section 11.4, we simulated the motion of the four-bar linkage using some given link lengths. In this section, we will modify the lengths and use motion simulation to predict the impact of the design changes on the motion of the mechanism.

The motion profile of a four-bar linkage is controlled by the relative lengths of the links. In the existing linkage, Link2 is capable of rotating through a full 360°. Such

FUTURE STUDY

Machine Dynamics and Machine Design

In this chapter, we have examined a four-bar linkage and performed some simple motion analysis. If we were designing this mechanism for an engineering application, many more questions would remain:

- How fast will the output link oscillate, if we know the speed of the input link?
- What size motor would be required to drive the input link?
- How will we transmit the rotational power of the motor to the input link?

Providing the answers to these questions would generally require the expertise of a mechanical engineer. The analysis of link speeds and accelerations is classified as a kinematics problem; extending this analysis to the sizing of motors is classified as a kinetics problem. Most engineering curricula include basic courses in physics and dynamics, which address the essentials of kinematic and kinetic analysis. However, because of the great emphasis in mechanical engineering on applying these principles to mechanisms and machinery, many mechanical engineering curricula include upper-level courses in the advanced application of kinematic and kinetic analysis. Such a course is often called a *machine dynamics* course, and may include the study not only of linkages, but also of the dynamics of gears, cams, and other mechanical devices.

The question of power transmission requires further insight beyond kinematic and kinetic analysis. The choice of transmission also requires the investigation of various alternatives, such as a geared transmission, a system of belts and pulleys, or a chain and sprocket drive. The selection and sizing of the appropriate transmission system requires knowledge not only of kinematic and kinetic analysis, but also of the application of stress analysis to the transmission components. This type of analysis is often covered in mechanical engineering curricula in *machine components* or *machine design* courses.

a link is called a “crank”; in a four-bar linkage a crank link will exist when the following equation is satisfied:

$$L + S < P + Q$$

where:

L = the length of the longest link
S = the length of the shortest link
P, Q = the length of the other two links.

In our linkage, L = 7 inches, S = 3 inches, P = 5 inches, and Q = 6 inches. Since the condition is satisfied, at least one link (Link2) is a crank.

We will now modify our mechanism, and use a motion study to determine the impact of the redesign on the motion of the linkage.

Click on the Model tab, as shown in Figure 11.43. Double-click on Link3 in the modeling window, change the length from 6 to 4 inches, as shown in Figure 11.44, and rebuild the model.

The modified model is shown in **Figure 11.45**.

FIGURE 11.43

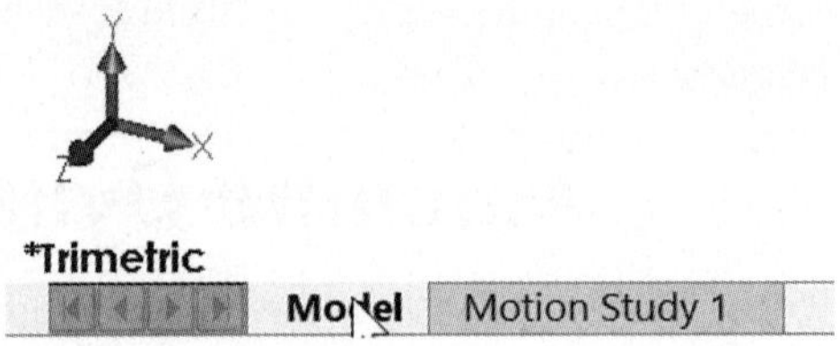

FIGURE 11.44

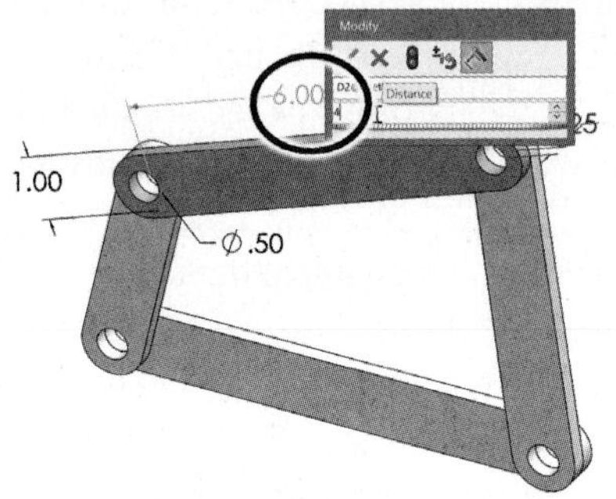

FIGURE 11.45

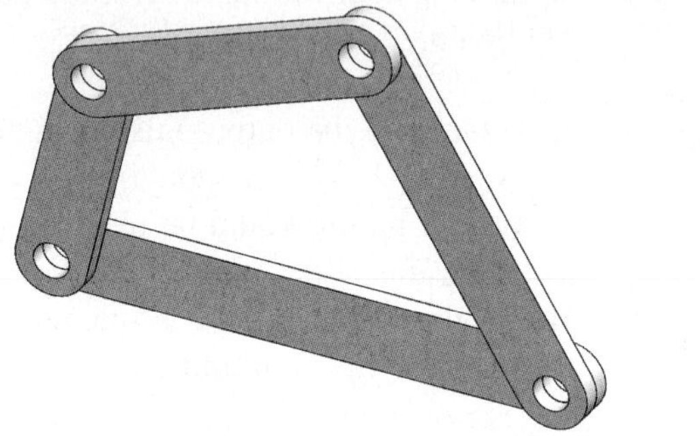

Click on the Motion Study 1 tab to bring up the MotionManager. The Update Initial Animation State dialog box will appear (Figure 11.46). Click Yes to close the box.

FIGURE 11.46

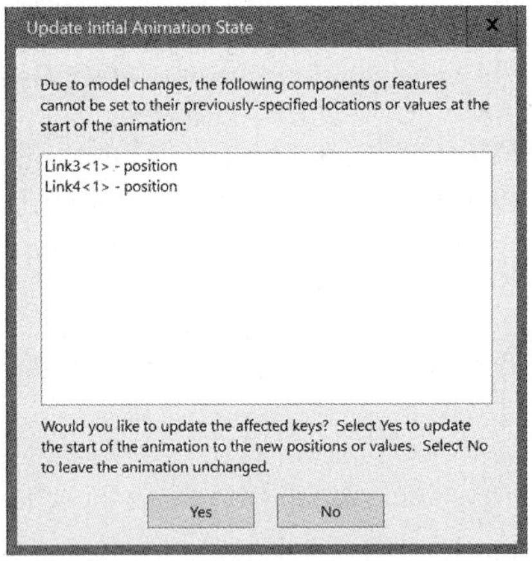

The model may temporarily appear to be disassembled, as in **Figure 11.47**. Disregard this; the links will be properly positioned when the simulation is calculated.

FIGURE 11.47

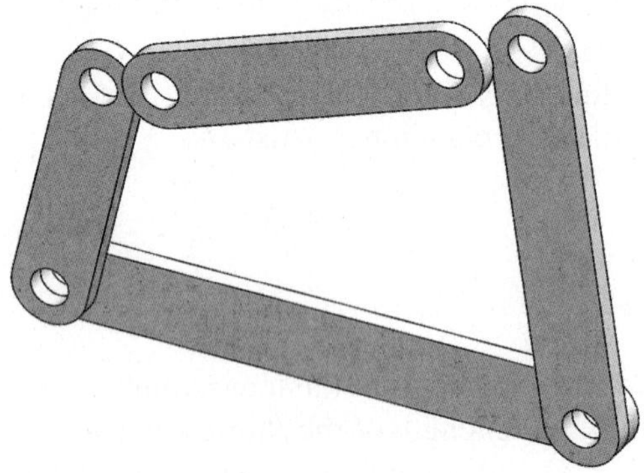

In order to better simulate the physics behind the motion of the mechanism, we will use a higher-level motion simulation tool called Basic Motion.

Use the Type of Study pull-down menu to change the study type from Animation to Basic Motion (Figure 11.48). Select Motion Study Properties and set the frame rate to 60 frames per second. Click the Calculate Tool to generate the motion simulation.

FIGURE 11.48

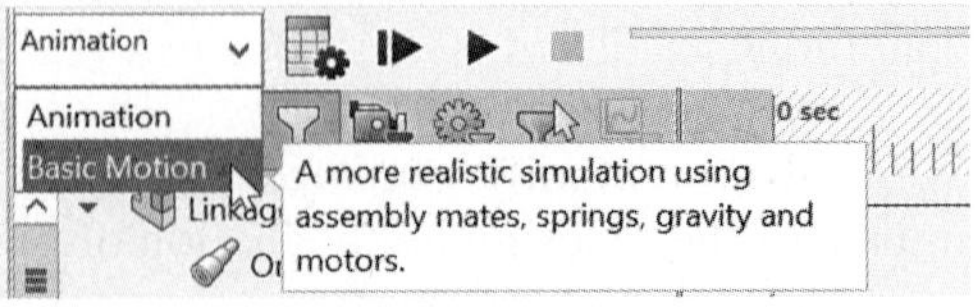

The linkage begins to move as before, until the Basic Motion simulation detects a position in which the linkage "locks up," as shown in **Figure 11.49**. Notice that in this position, Link3 and Link4 are aligned, and no further rotation of Link2 is possible. The position in which the mechanism "locks up" is called a *toggle position* (this was also demonstrated in Chapter 10). The presence of a toggle position is sometimes undesirable, since it prevents full rotation of the input link. Sometimes, however, toggle positions are designed into a linkage, so that it can be used as a clamp or fixturing device. Checking our equation with these new dimensions, we now have L = 7 inches, S = 3 inches, P = 5 inches, and Q = 4 inches. The equation is no longer satisfied, which is further evidence of the presence of a toggle position.

FIGURE 11.49

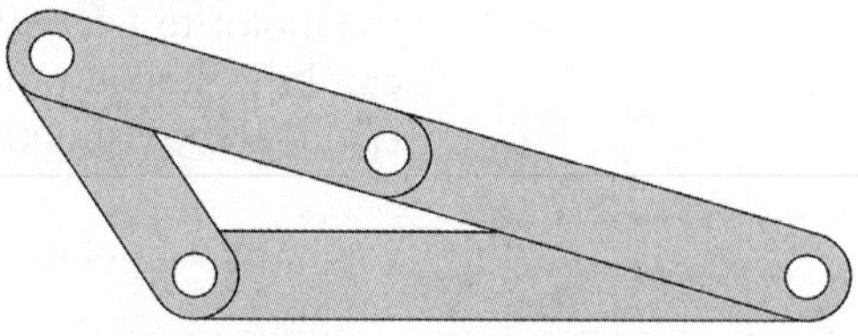

A second toggle position also exists in our mechanism. We will find it by further modifying our motion simulation.

Click on the Timeline slider bar at the top of the MotionManager, which shows the current time of the simulation, as shown in Figure 11.50, and drag it to zero, as shown in Figure 11.51. Right-click on the motor (RotaryMotor1) in the MotionManager tree and select Edit Feature, as shown in Figure 11.52. Click the Reverse Direction button in the PropertyManager, as shown in Figure 11.53. Click the check mark to apply the change.

FIGURE 11.50

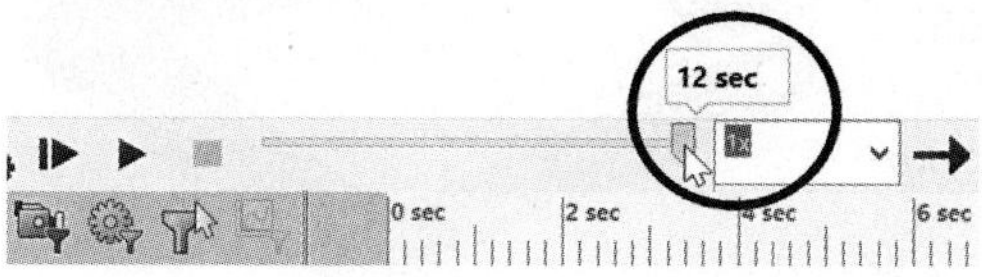

FIGURE 11.51

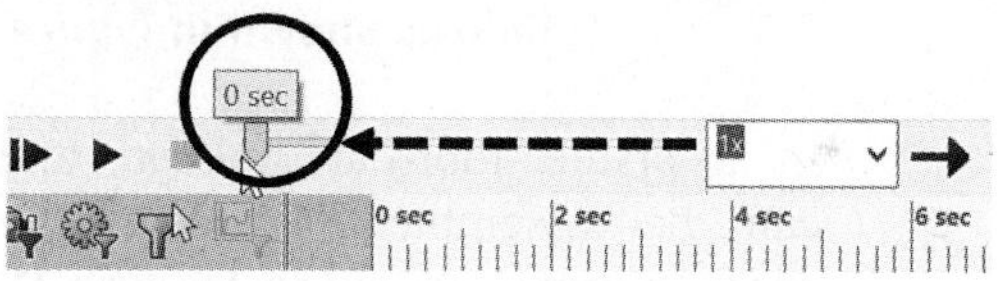

FIGURE 11.52

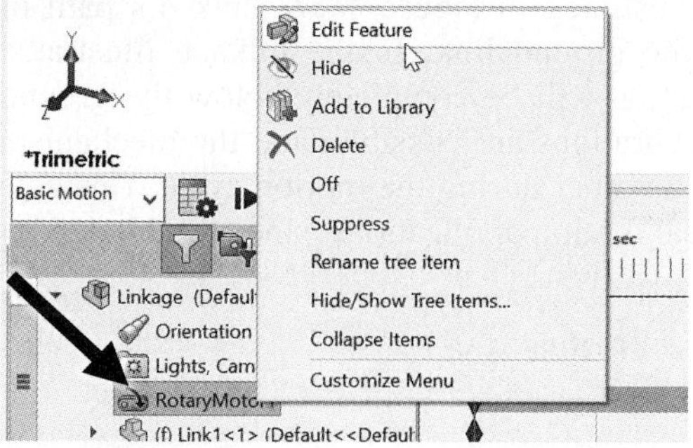

FIGURE 11.53

Moving the time back to zero before making this change causes the modification to the motor to take effect at the beginning of the simulation. The motor parameters can be edited at other time points in the simulation, causing the motor to change speed and/or direction during the simulation.

Click the Calculate Tool to compute the new simulation.

FIGURE 11.54

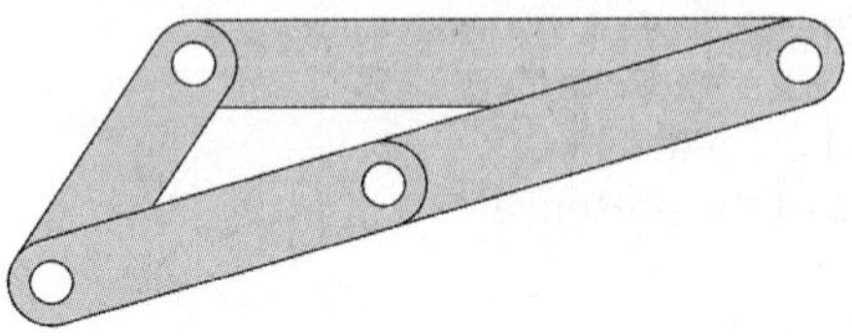

The mechanism will again stop at a toggle position, as shown in **Figure 11.54**.

We have examined the motion case of a four-bar linkage where L + S < P + Q (where Link2 could rotate through 360°), and also the case where L + S > P + Q (where we found toggle positions). We will now explore the special case where L + S = P + Q.

FIGURE 11.55

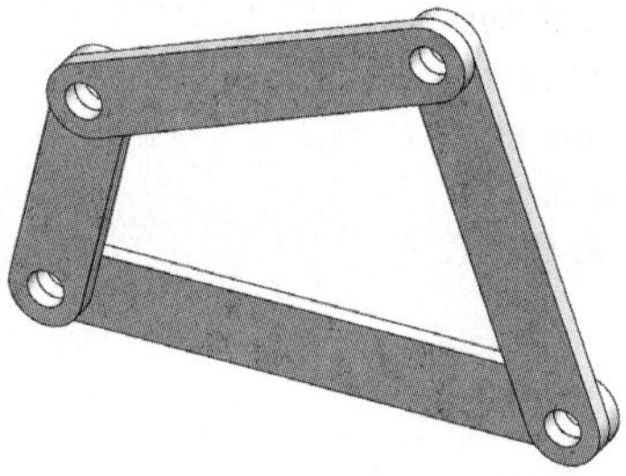

Click on the Model tab to return to the assembly. Change the length of Link3 to 5 inches and rebuild.

The rebuilt model is shown in **Figure 11.55**.

Click and drag on the links, and you will see that the moving links can be moved so that they are all below the ground link, as shown in Figure 11.56.

FIGURE 11.56

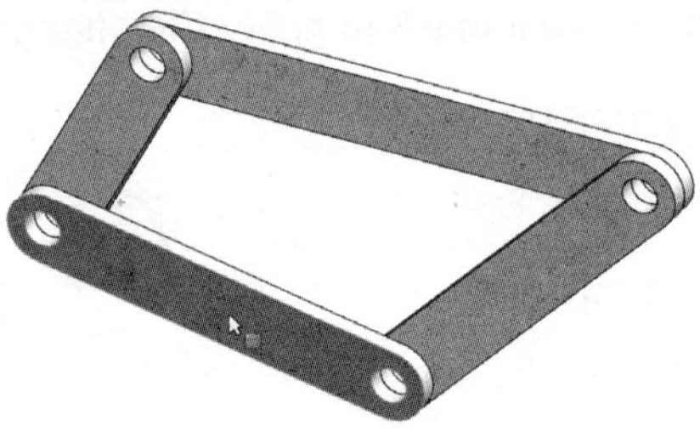

In a stable four-bar linkage, the links can be assembled in either of two configurations, shown in **Figures 11.57** and **11.58**. Once assembled, the configuration is set. To change from one configuration to the other, it is necessary to disassemble one joint, rearrange the links, and re-assemble the joint. In the linkage illustrated in **Figure 11.57**, Link 4's path of oscillation will be completely above the ground link. In the linkage illustrated in **Figure 11.58**, Link 4's path of oscillation will be completely below the ground link. In an unstable linkage, both configurations are possible, and the mechanism may switch from one configuration to the other during the motion cycle. Unstable mechanisms are generally avoided in real motion applications, since their behavior can be unpredictable.

FIGURE 11.57

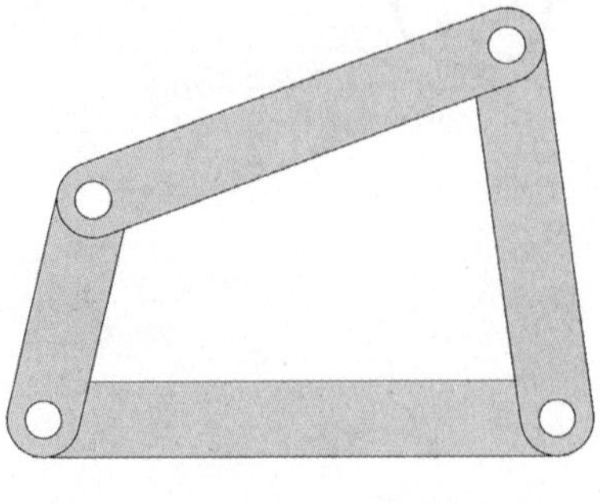

FIGURE 11.58

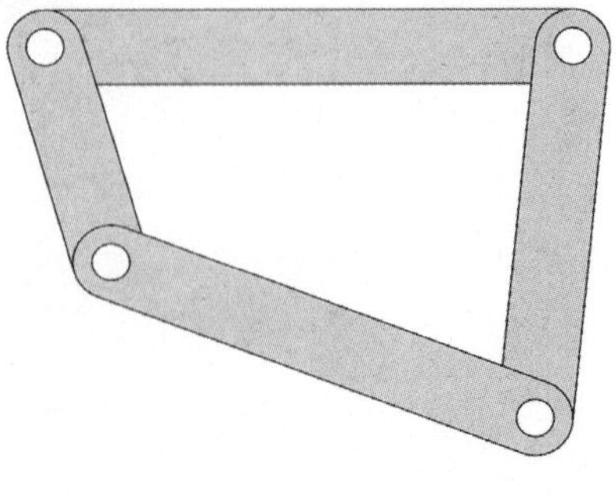

PROBLEMS

P11.1 A type of mechanism used in engineering systems is the slider-crank (see **Figure P11.1A**). In the slider-crank, the input link (crank) rotates continuously through a full 360 degrees, while the output slider slides along a fixed surface. Among other things, the slider-crank is the working schematic for a single cylinder of an internal combustion engine. Create a working assembly model of this mechanism, using the dimensions shown in **Figure P11.1B**. The dimensions are mm. (Hint: The bottom of the slider is always coincident with the ground plane.)

FIGURE P11.1A

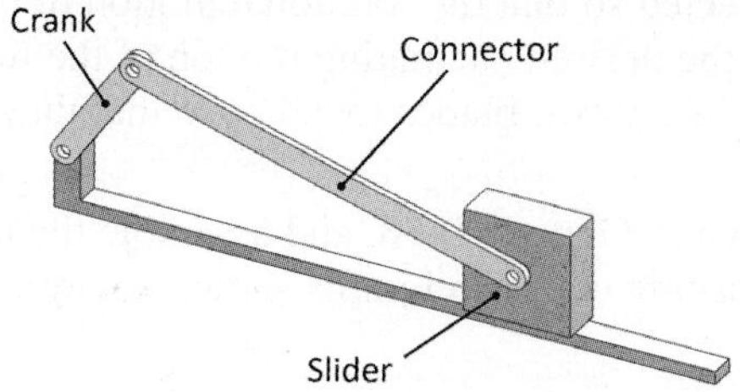

FIGURE P11.1B

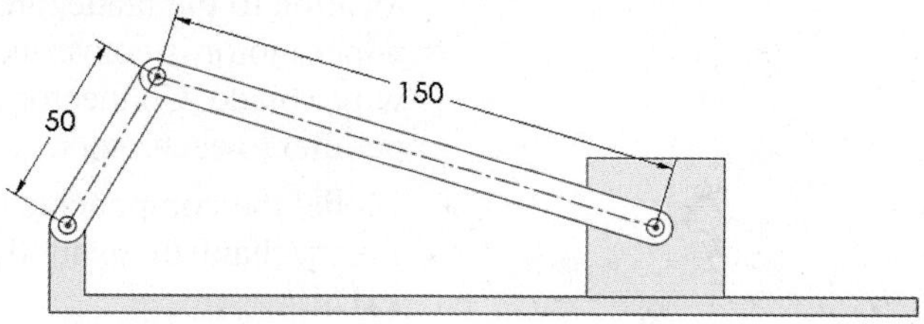

P11.2 The mechanism described in **Problem 11.1** can be modified by offsetting the pivot location of the crank from the path of the slider, creating an *offset slider-crank,* as shown in **Figure P11.2A**. This type of mechanism is sometimes called a *quick-return* mechanism, since the back-and-forth motion of the slider is different in one direction from the other. Create a working model of this mechanism, using the dimensions shown in **Figure P11.2B**. The dimensions are mm.

FIGURE P11.2A

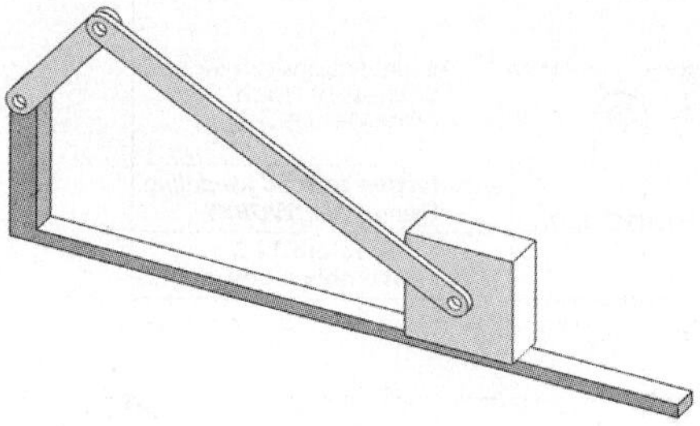

FIGURE P11.2B

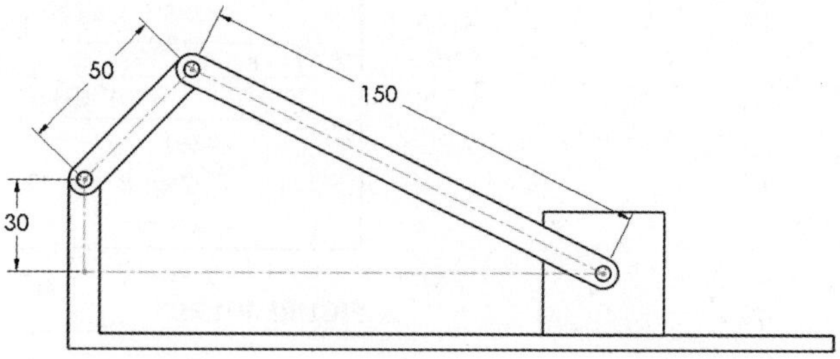

P11.3 Additional links can be added to a four-bar linkage to modify the motion of the links. A common engineering example is the *six-bar linkage.* **Figure P11.3A** shows a model of a six-bar linkage that simulates the operation of a car's windshield wipers. The Crank is attached to the wiper motor, which rotates at a constant speed. The lengths of the Crank and Connector 1, along with the attachment location to the blade, are selected so that the rotational motion of the wiper motor is converted to the desired oscillating motion of the first wiper blade. Connector 2 links the two blades together so that they move parallel to each other.

FIGURE P11.3A

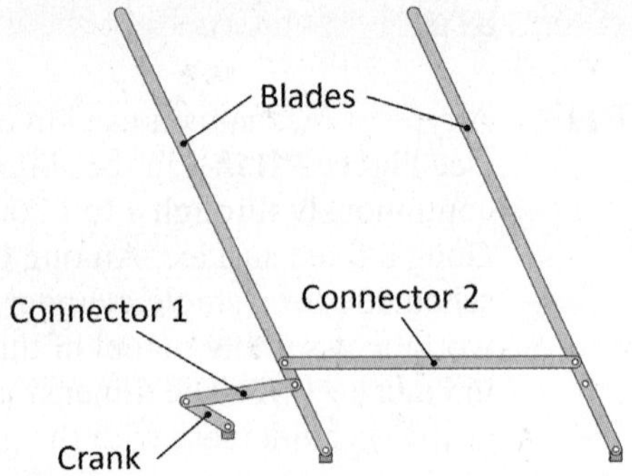

Model the components shown in **Figure P11.3B**, and assemble them into the mechanism, using the spacing of ground joints shown in **Figure P11.3C**.

FIGURE P11.3B

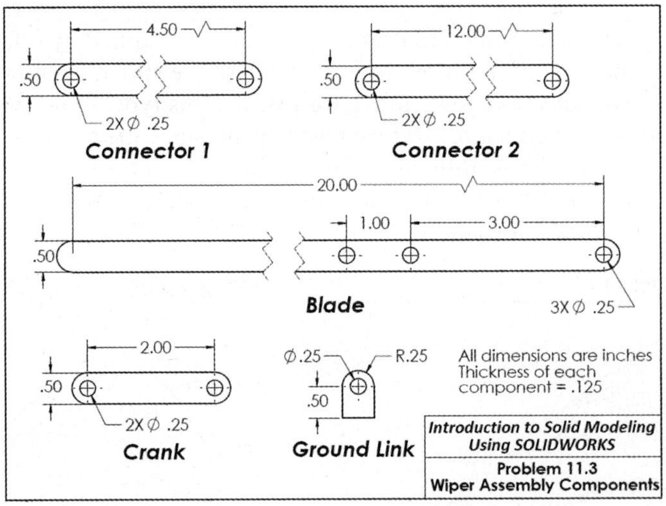

FIGURE P11.3C

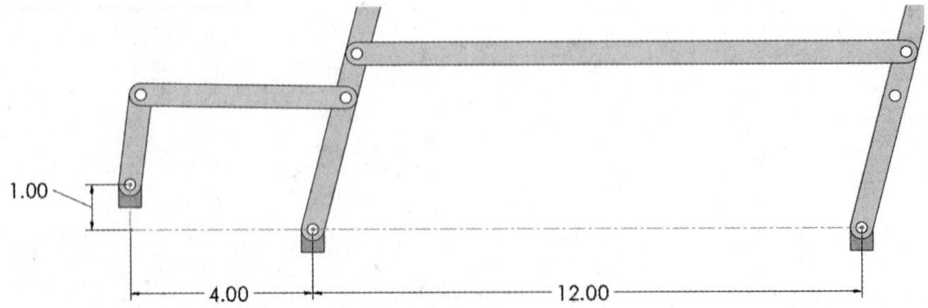

P11.4 In the 1700s, noted engineer James Watt devised and patented a mechanism for generating straight-line motion with a rotational input. In this mechanism, shown in **Figure P11.4A**, the center point of Link 3 will trace out a straight line in space as the input link (Link 2) rotates, as illustrated in **Figure P11.4B**. Originally designed for guiding the stroke in a steam engine piston, this mechanism is currently used to guide axle motion in automotive suspension applications.[1] Develop a working model of the *Watt Straight Line Mechanism.* Use a length of 8 inches for Links 2 and 4, and a length of 4 inches for Link 3. The distance between the fixed pivot points is 16 inches. (Note: This is considered a *double rocker mechanism*; the input link is *not* able to rotate through a full 360 degrees.)

FIGURE P11.4A

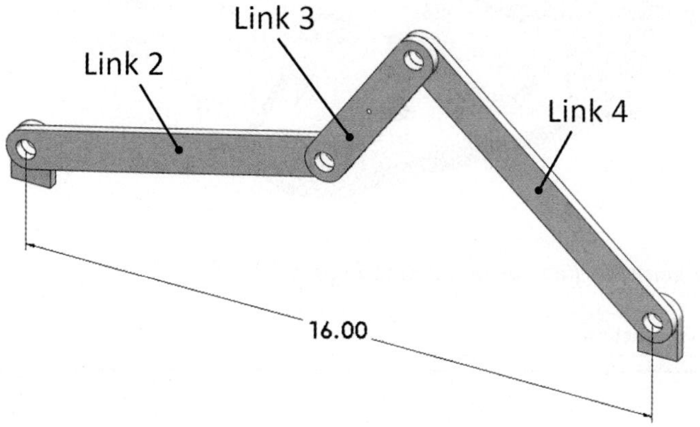

FIGURE P11.4B

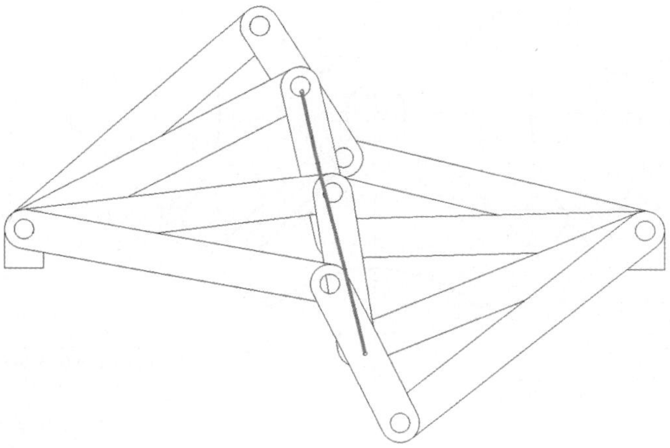

[1] Robert I. Norton, *Design of Machinery,* 2nd ed. (New York: McGraw-Hill, 1999).

P11.5 A *geneva mechanism,* which is illustrated in **Figure P11.5A**, is used to transform constant rotational motion into intermittent motion. Among other uses, this type of mechanism is used to control the motion of an *indexing table* in an assembly line. An indexing table will remain stationary for a period of time, and then rotate a fixed amount.

Model the components necessary to create a model of the geneva mechanism:

FIGURE P11.5A

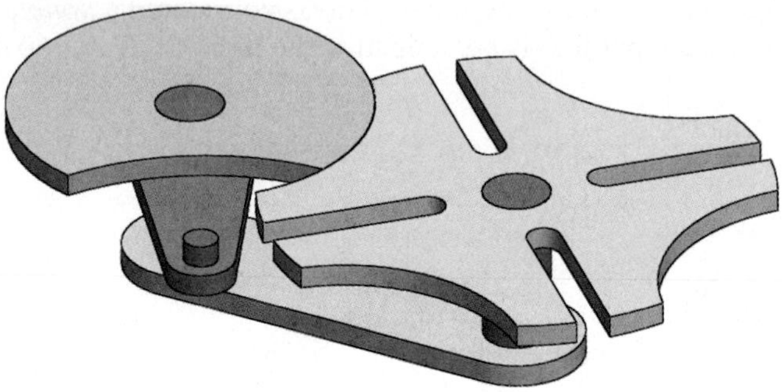

1. The *geneva wheel,* as detailed in **Figure P11.5B**.

FIGURE P11.5B

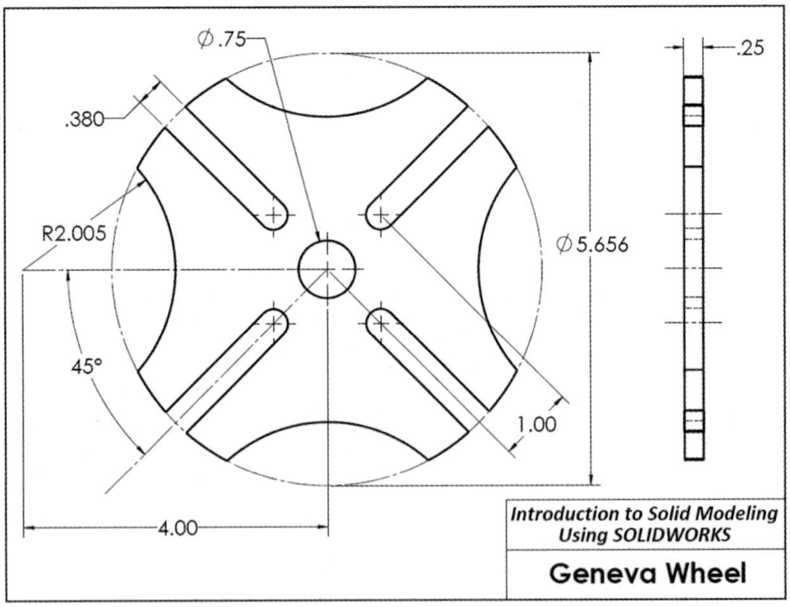

2. The *locking disk,* as detailed in **Figure P11.5C**.

FIGURE P11.5C

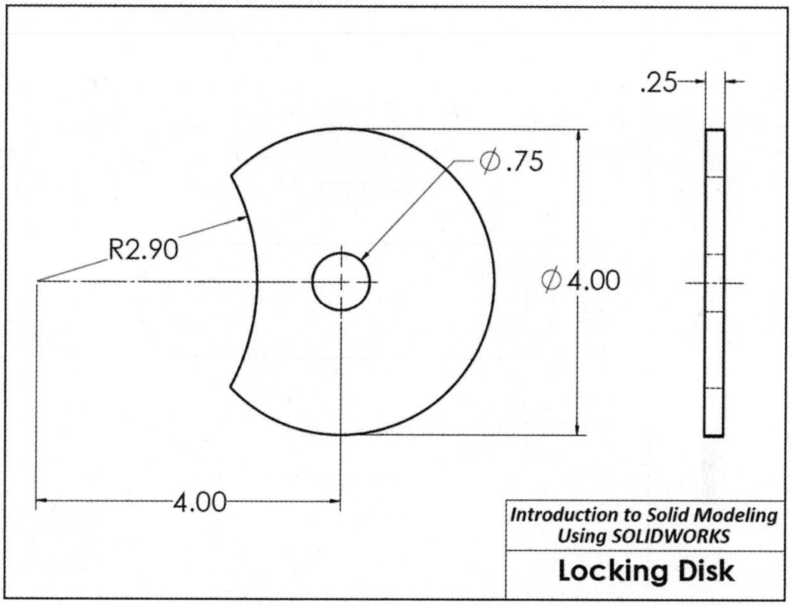

3. The *crank,* as detailed in **Figure P11.5D**.

FIGURE P11.5D

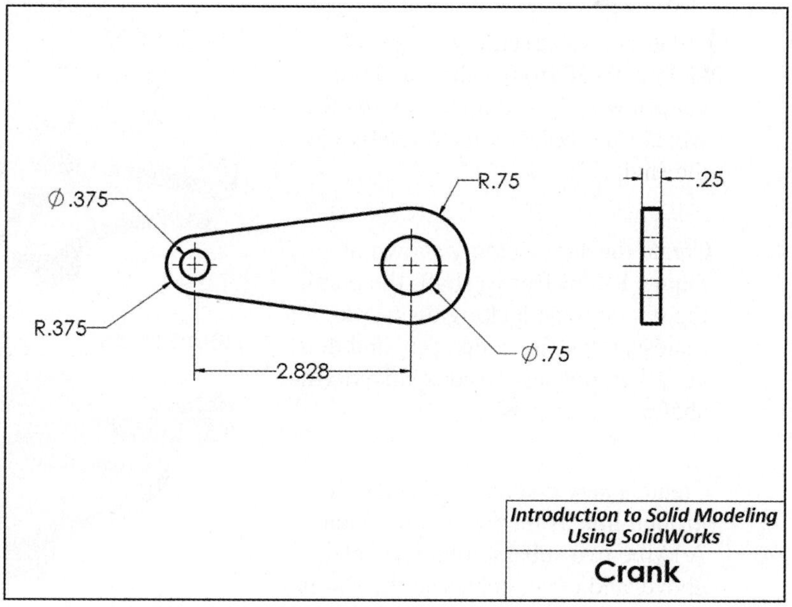

4. The *ground link,* as shown in **Figure P11.5E**.

FIGURE P11.5E

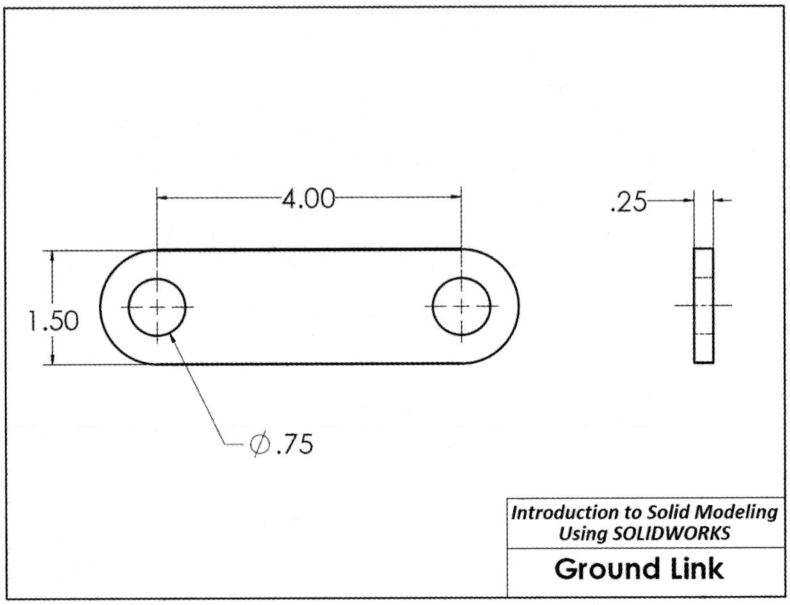

5. A cylindrical shaft, 0.75 inches in diameter by 2 inches long.
6. A cylindrical pin, 0.375 inches in diameter by 0.50 inches long.

Create the subassembly shown in **Figure P11.5F** from a shaft and the geneva wheel. Add mates so that the wheel is completely fixed relative to the shaft.

FIGURE P11.5F

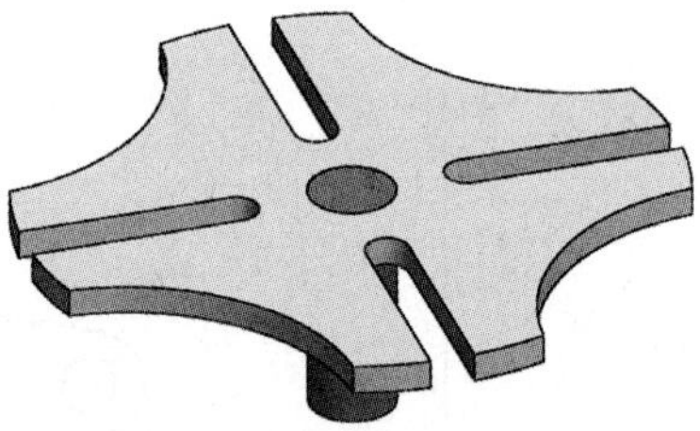

Create the subassembly shown in **Figure P11.5G** from a shaft, the crank, the pin, and the locking disk. Add mates so that the crank, pin, and disk are all completely fixed relative to the shaft.

FIGURE P11.5G

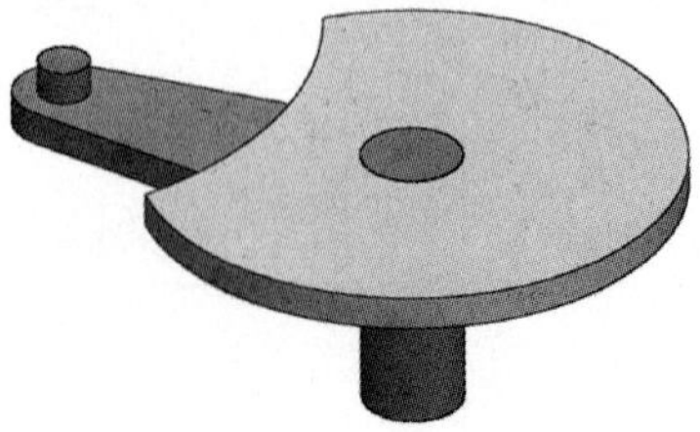

Create a new assembly with the ground link as the fixed component. Add the two subassemblies created above, and create mates so the shafts can rotate within the ground link. Since the clearance between the locking disk and the geneva wheel is relatively small, it is important to align the components so that they are not overlapping. One way to do this is to add a mate between the Front Planes of the subassemblies, as shown in **Figure P11.5H**. After the components are aligned, suppress or delete the mate so that the subassemblies can move.

FIGURE P11.5H

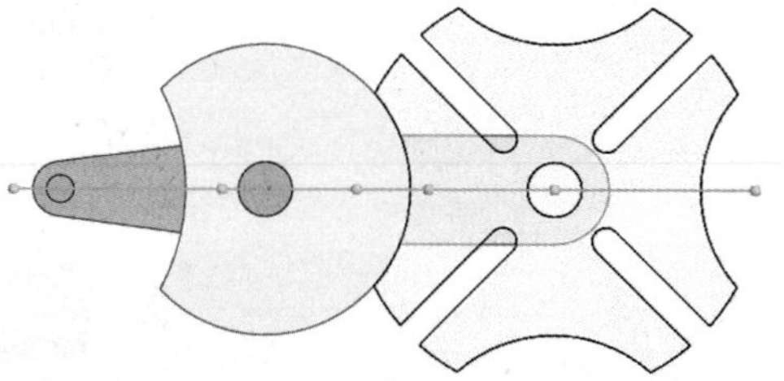

To view the motion, click on the Move Component Tool (**Figure P11.5I**). In the PropertyManager, click the Physical Dynamics option (**Figure P11.5J**). Drag the crank in a circle to view the motion, as shown in **Figure P11.5K**.

FIGURE P11.5I

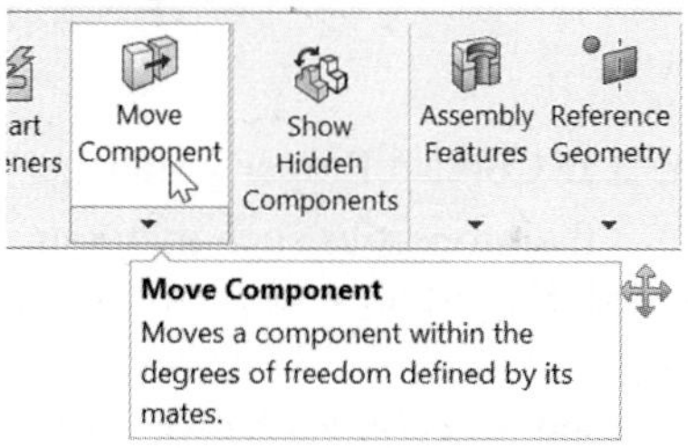

FIGURE P11.5J

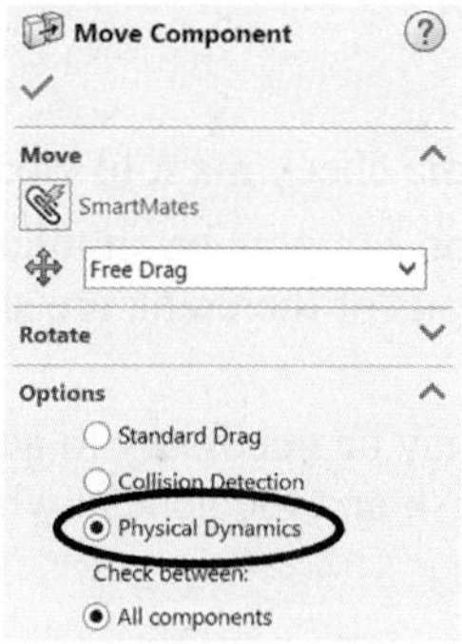

FIGURE P11.5K

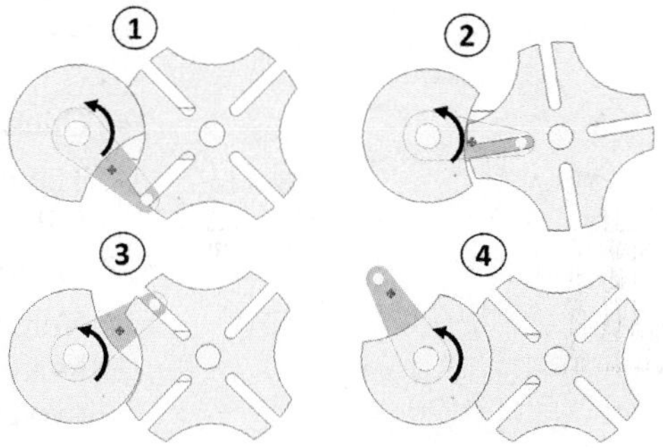

P11.6 Develop a motion simulation of the geneva mechanism model created in **Problem 11.5**. In order to do this, perform the following steps:

a. Create a motion study, and add a rotary motor to the crank subassembly.

b. Use the pull-down menu to change the motion type to Basic Motion.

c. Add a 3-D Contact between the pin and the geneva wheel by:

- Selecting the Contact Tool from the MotionManager (**Figure P11.6A**)

FIGURE P11.6A

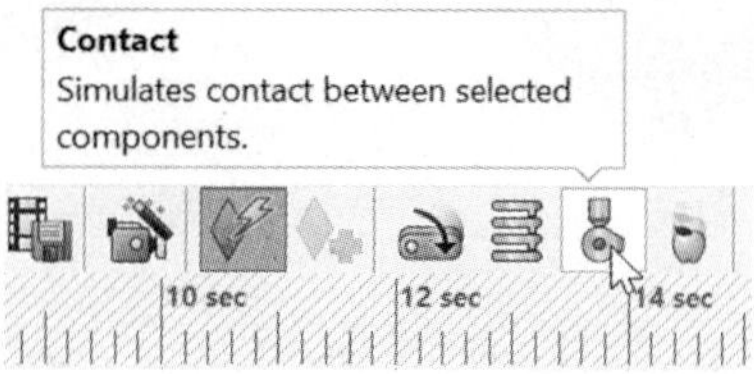

- Clicking on the pin (from the crank subassembly) and the geneva wheel part to define a contact between these components in the PropertyManager (**Figure P11.6B**)

FIGURE P11.6B

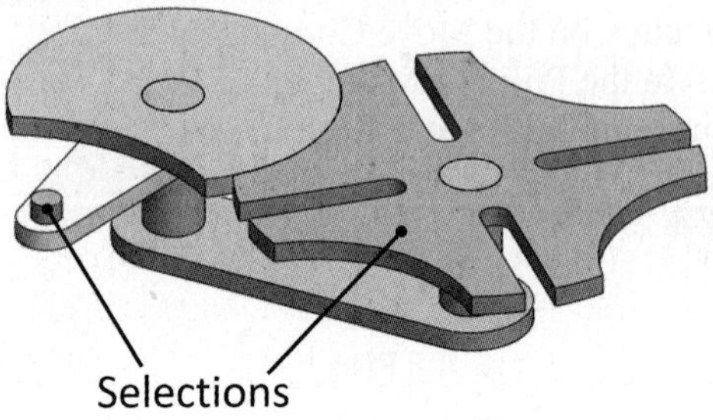

- Clicking the check mark to close the 3-D Contact PropertyManager

  The motion can now be simulated. Set the properties such that four full rotations of the crank are simulated. Export an AVI file of the motion.

  Note: It may be necessary to add a second 3-D contact between the locking disk and the geneva wheel to prevent overlapping of those two parts.

P11.7 A Scotch yoke mechanism is used to convert continuous rotary motion into reciprocating linear motion. The mechanism is shown in **Figure P11.7A**. Create a simulation of the Scotch yoke mechanism by performing the following steps.

FIGURE P11.7A

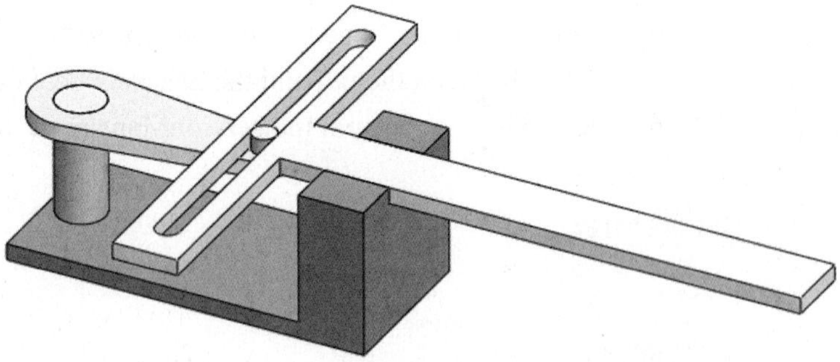

a. Using the crank, shaft, and pin parts from **Problem 11.5**, create the subassembly shown in **Figure P11.7B**. All parts should be fully constrained in the assembly. Save the assembly file as "Scotch crank."

FIGURE P11.7B

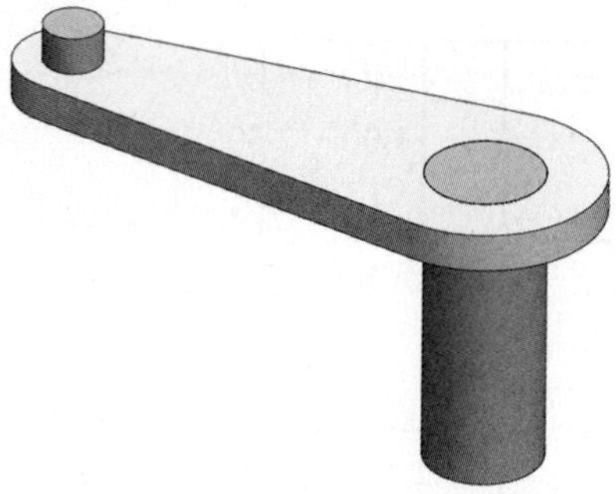

b. Sketch the slotted T-shaped part, as shown in **Figure P11.7C**. Extrude it to a height of 0.25 inches, and save the part file as "yoke."

FIGURE P11.7C

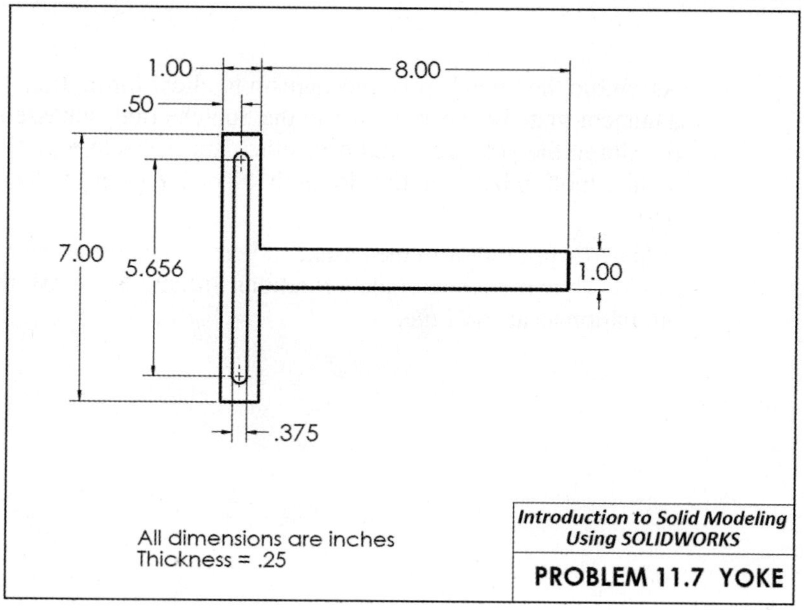

c. Create the base part shown in **Figure P.11.7D**. Save it with the name "Scotch base."

FIGURE P11.7D

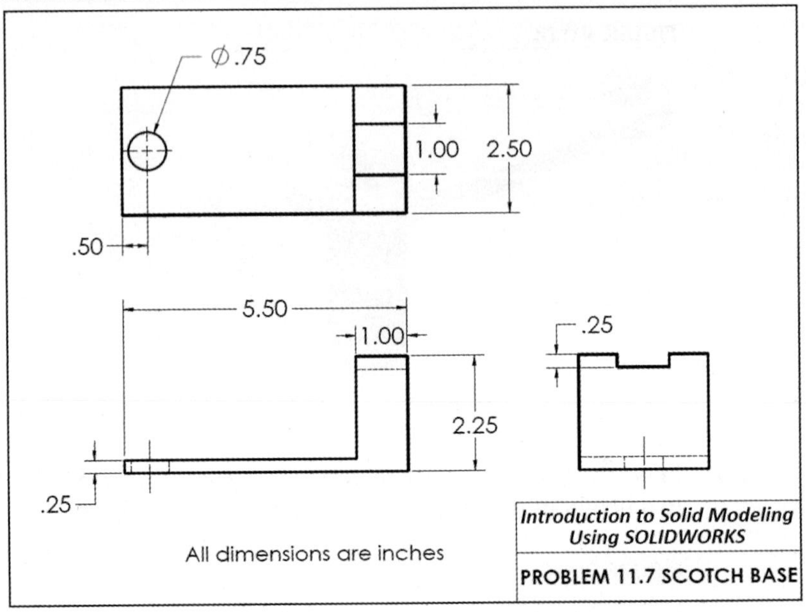

d. Assemble the Scotch yoke mechanism to allow for motion. (Hint: Use a tangent mate between the pin in the Scotch crank subassembly and the slot in the yoke part, and use coincident mates to establish the sliding motion between the slot in the Scotch base part and the yoke part.)

e. Add a 10-rpm motor to the crank, and create an animation of the motion showing two complete rotations of the crank. Save the animation as an AVI file.

CHAPTER 12

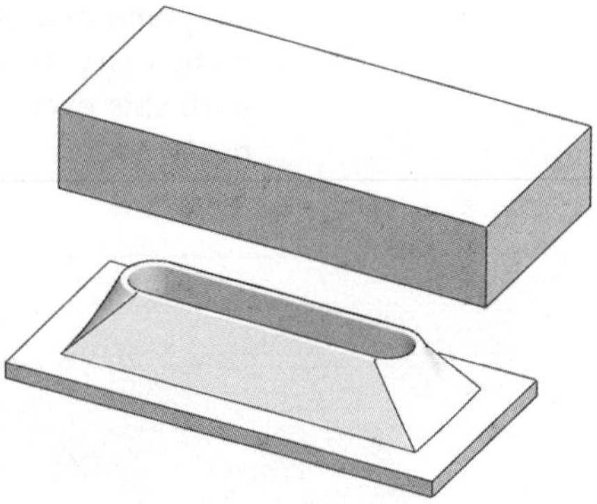

Design of Molds and Sheet Metal Parts

Introduction

A design engineer must always consider the method of manufacture when designing any part. Failure to do so may result in part designs that are more expensive to make than necessary, have high scrap rates, or cannot be made at all.

Some manufacturing processes create unique challenges from a solid modeling standpoint. For example, when designing a mold, the shape of the part to be molded must be removed from the interior of the mold, usually with the dimensions adjusted to allow for shrinkage of the part during the cool-down portion of the molding cycle. Sheet metal parts are cut from flat material, and then bent into the final shape. Therefore, the part definition must include both the flat shape and the finished geometry. The SOLIDWORKS® program has specialized tools for working with molds and sheet metal parts.

Chapter Objectives

In this chapter, you will:

- create a cavity within a mold base,
- create and modify two mold halves that are linked to the mold base with a cavity,
- make a simple sheet metal part,
- learn how to show a sheet metal part in either the flat or bent state, and
- make a drawing of a sheet metal flat pattern.

12.1 A Simple Two-Part Mold

In this exercise, we will make a mold to produce a simple cylindrical part. We will begin by making the part itself.

Open a new part. Draw and dimension a 3-inch diameter circle in the Right Plane, centered at the origin. Extrude the circle using a

SOLIDWORKS is a registered trademark of Dassault Systémes SolidWorks Corporation.

midplane extrusion, as shown in Figure 12.1. Set the total depth of extrusion to 5 inches. Create a new folder called "Cylinder Mold" to contain the files associated with this exercise. Save the part with the file name "Cylinder." Do not close the part window.

FIGURE 12.1

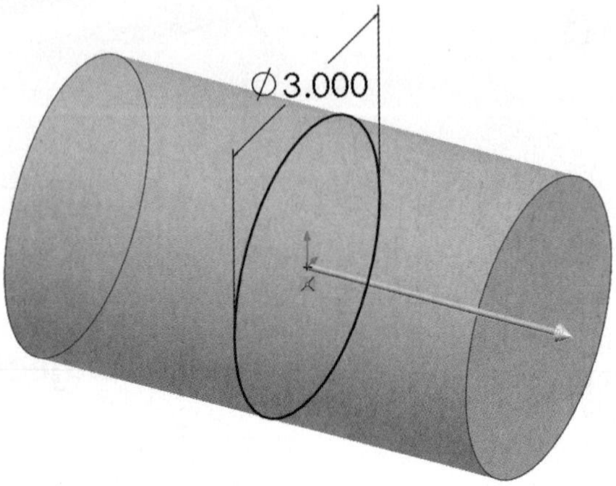

FIGURE 12.2

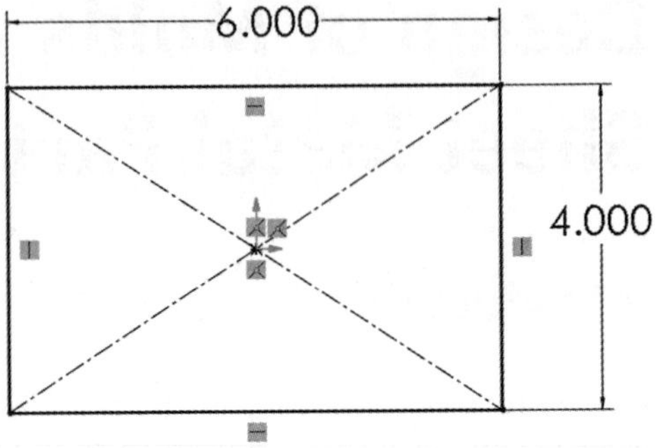

The next step is to create a *mold base*, from which a cavity in the shape of the part will be removed. The mold base must be large enough to completely enclose the part.

Open a new part. Draw and dimension a 4-inch by 6-inch rectangle in the Top Plane centered about the origin (Figure 12.2).

Extrude the rectangle with a midplane extrusion, as shown in Figure 12.3. Set the total depth of extrusion to be 4 inches. Save this part with the file name "Mold Base." Do not close the part window.

We will now place the part within the mold base.

From the main menu, select File: Make Assembly from Part, as shown in Figure 12.4.

FIGURE 12.3

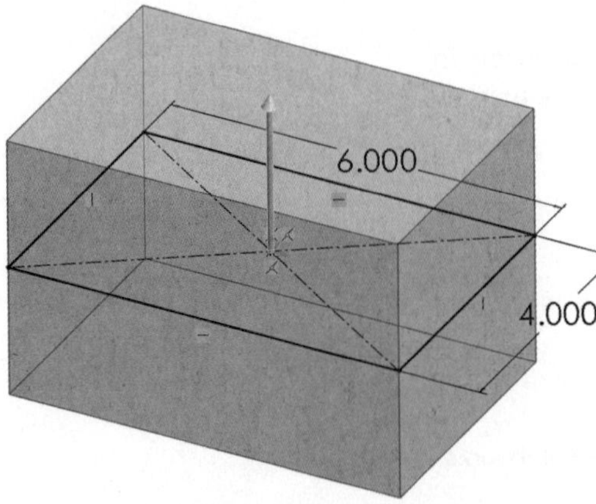

FIGURE 12.4

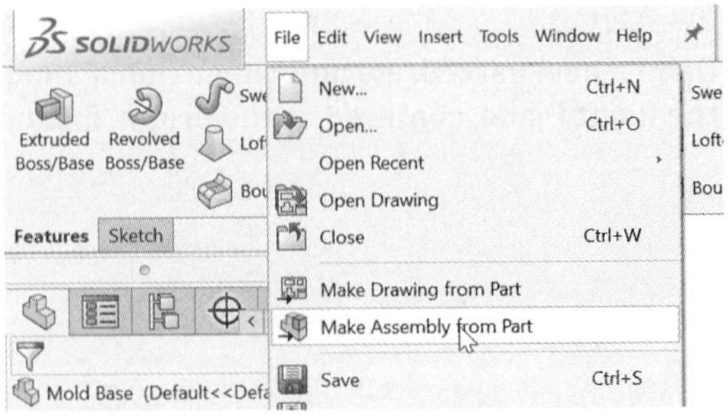

A new assembly window will open, with the Insert Component Tool active.

Select the Mold Base from the list of open documents, as shown in Figure 12.5. Click the check mark to place the part into the assembly, with the origin of the part placed at the origin of the assembly.

FIGURE 12.5

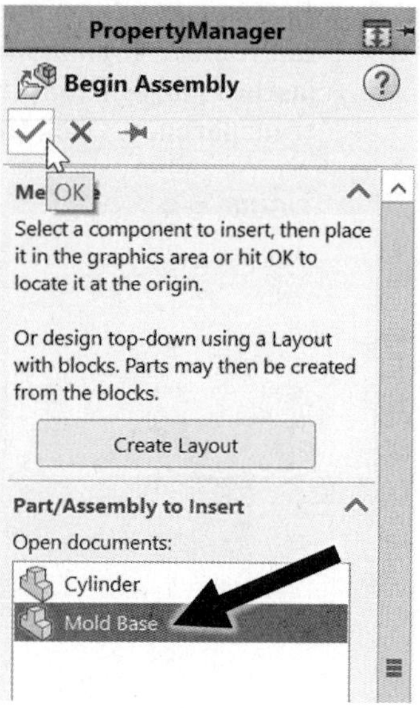

Note that placing the origin of the first component of an assembly at the origin of the assembly itself is not required, since the first component will be fixed and subsequent parts will be located relative to the first part. However, as noted in earlier chapters, placing the first part in this manner is good practice, since the Front, Top, and Right Planes of the assembly, as well as the origin, can be used as mating entities.

Select the Insert Components Tool. Select the Cylinder part from the list of open files.

Click the check mark to place the cylinder into the assembly, with the origin of the cylinder placed at the origin of the assembly. Save this assembly with a file name of "Mold Assembly."

The cylinder is now centered within the base, as shown in the wireframe view of Figure 12.6. Rather than work in a wireframe mode, it is helpful to display the base as transparent.

FIGURE 12.6

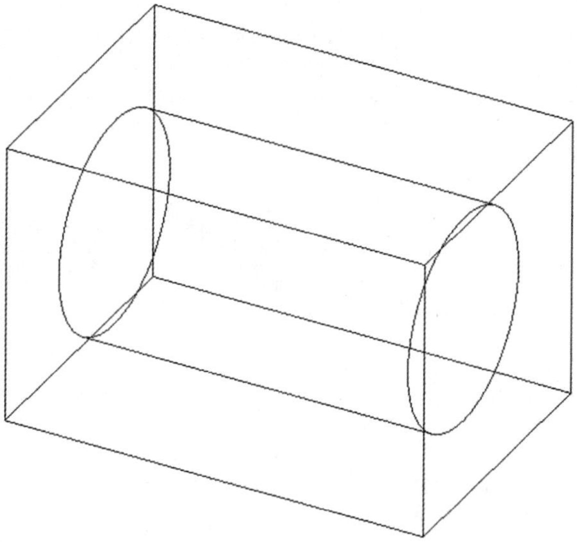

FIGURE 12.7

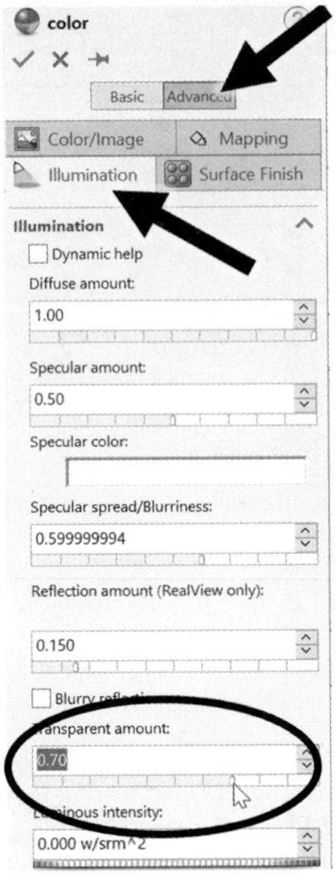

Right-click on the part name "Mold Base" in the FeatureManager. Select Appearance: Appearance. In the PropertyManager, select the Advanced tab. Select Illumination, and move the Transparent amount slider bar (as in Figure 12.7) to the right to display the base in varying degrees of transparency. Click the check mark to apply the desired transparency.

FIGURE 12.8

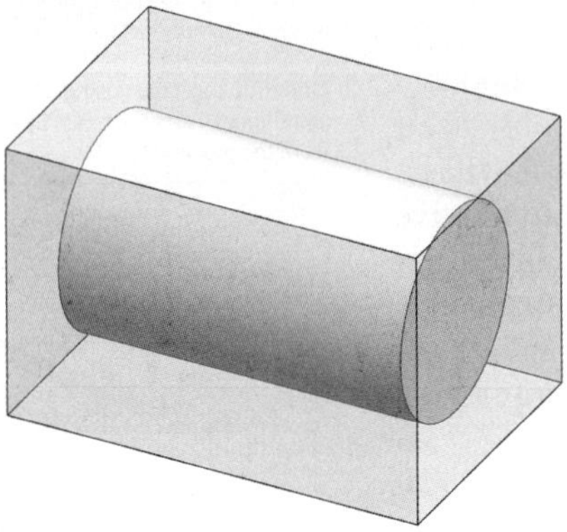

The cylinder can now be seen within the base, as shown in **Figure 12.8**.

We will now create the cavity. Since the cavity will be created in the base, the base must be selected for editing.

Select the Mold Base from the FeatureManager. Choose the Edit Component Tool from the Assembly group of the CommandManager, as shown in Figure 12.9.

From the main menu, select Insert: Molds: Cavity, as shown in Figure 12.10.

FIGURE 12.9

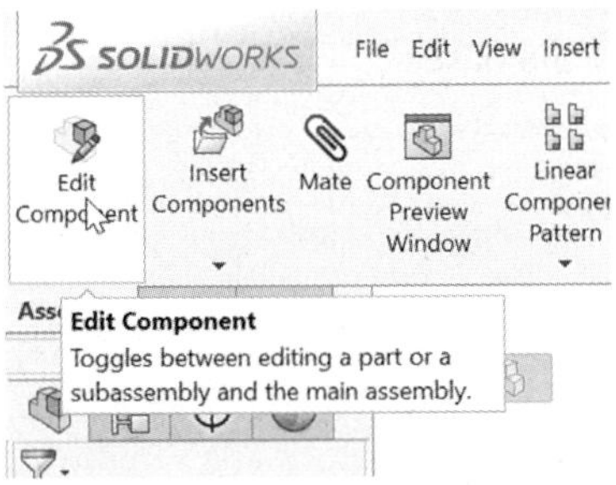

FIGURE 12.10

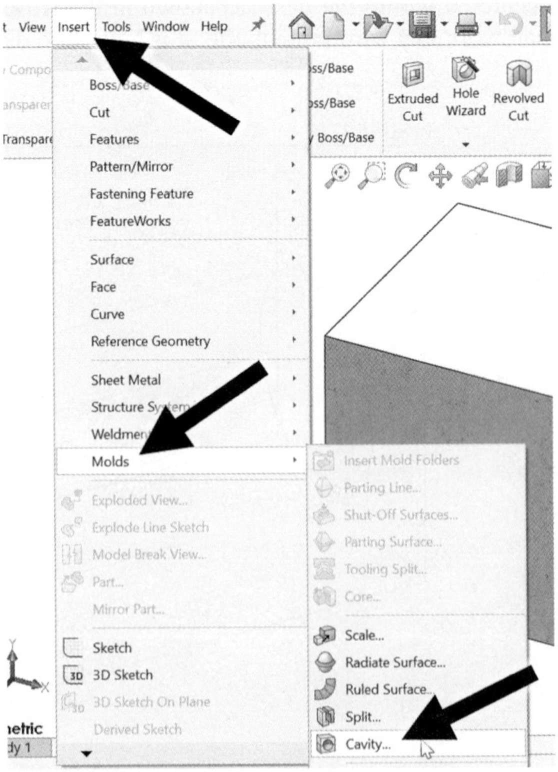

In the PropertyManager, select the Cylinder from the FeatureManager as the component defining the cavity. Set the "Scale about" option to "Component Origins" and the scale to 3%, as shown in Figure 12.11. Click the check mark to create the cavity. Click on the Edit Component Tool to end the editing of the base. Save the assembly.

FIGURE 12.11

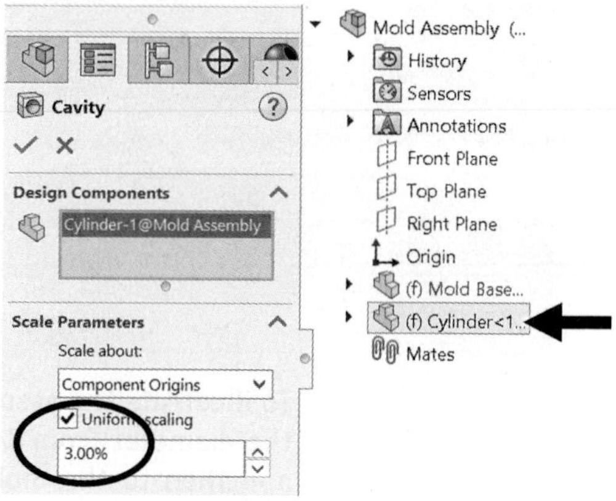

The scale factor causes the cavity to be larger than the finished part. Most molding materials shrink during cure or cooling, so the scale factor allows for that shrinkage.

The cavity can now be seen within the base, as shown in Figure 12.12. If you look closely, you can see that there are gaps between the edges of the part and the corresponding edges of the cavity, because of the shrink factor.

FIGURE 12.12

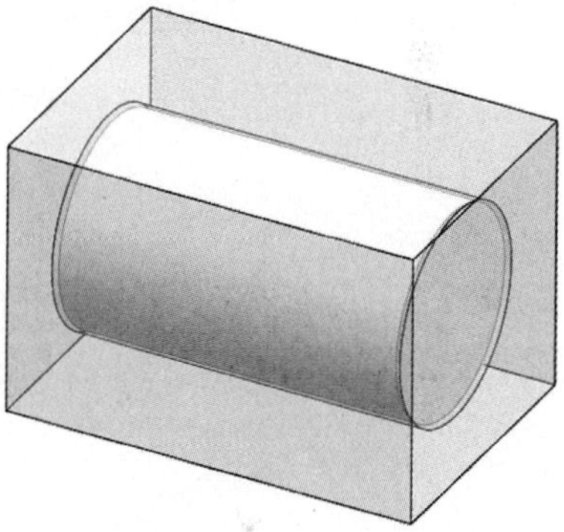

The base, which now includes the cavity, needs to be split into two halves. For this simple mold, the two halves will be identical, and so we could cut away half of the original base part. However, if the two mold halves will be different (as in the next exercise), then copies of the base must be made. This procedure of creating *derived parts* is illustrated here.

Select the Mold Base from the FeatureManager. From the main menu, select File: Derive Component Part, as shown in Figure 12.13. Click the check mark to place the new part at the origin.

FIGURE 12.13

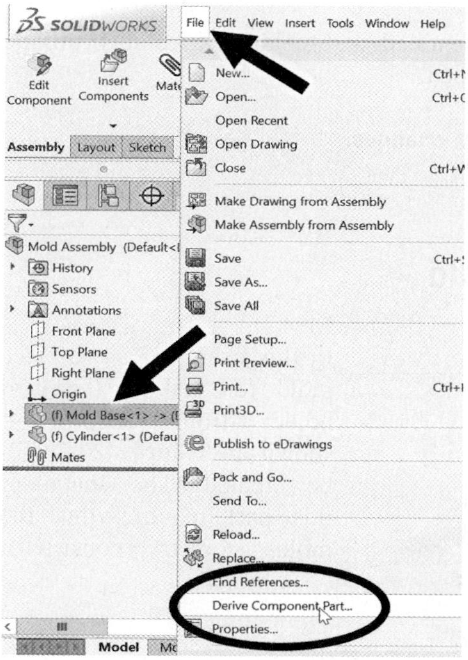

A new part window is opened, and a copy of the base (including the cavity), is created. The advantage of creating a derived part rather than simply saving a copy of the base part is that associativity is maintained. That is, if a change is made to the cylinder part, then the cavity in the assembly, the mold base, and the derived mold half part are all updated.

FIGURE 12.14

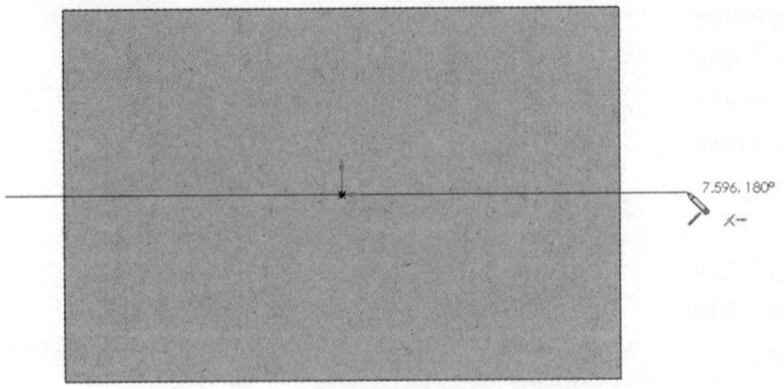

Open a sketch on the front face of the new part. Select the Line Tool, and draw a line completely through the part, passing through the origin, as shown in Figure 12.14. Extrude a cut with a type of Through All, with the direction to cut arrow pointing up.

The resulting part is shown in **Figure 12.15**.

Save the part with a file name of "Mold Half."

To illustrate the associativity of the parts, open the Cylinder part file and change the diameter from 3 to 2 inches. Rebuild the part. Switch to the mold assembly, and then to the mold half. At each step, the new cylinder diameter will be reflected, as shown in Figure 12.16.

FIGURE 12.15

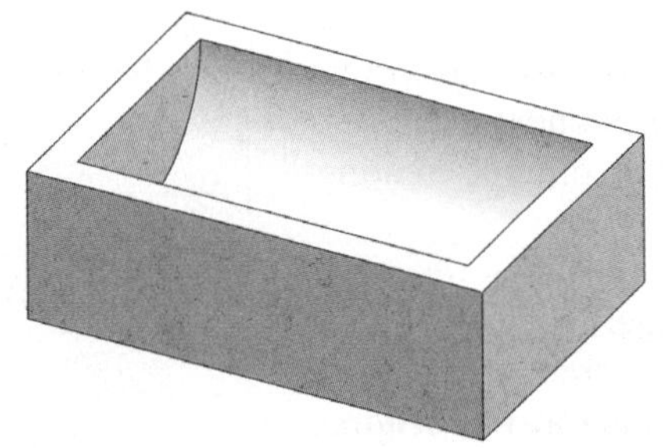

FIGURE 12.16

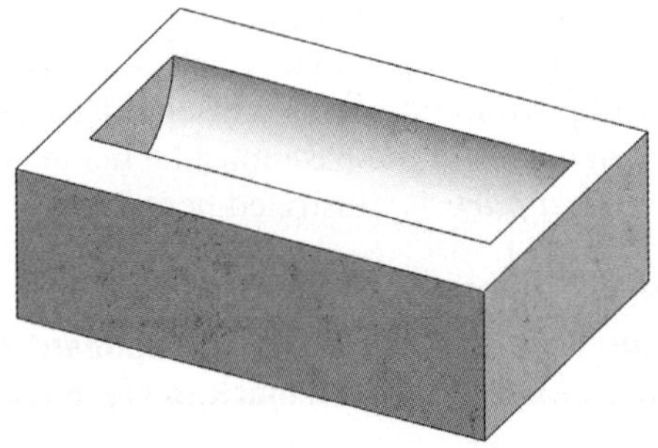

Two of these mold halves can now be assembled, and mold-level features (fill and vent ports, alignment pins, etc.) can be added.

Close all windows without saving the last changes.

12.2 A Core-and-Cavity Mold

FIGURE 12.17

In this exercise, a two-piece mold for making the card holder from Chapter 4, shown in **Figure 12.17**, will be created. The shape of this part requires that the mold geometry consist of

FIGURE 12.18

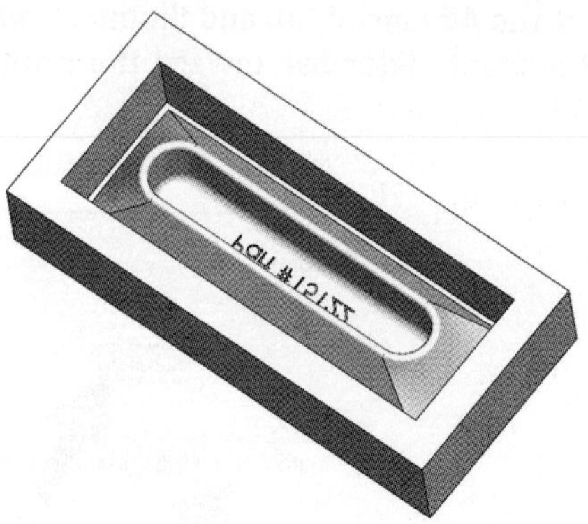

FIGURE 12.19

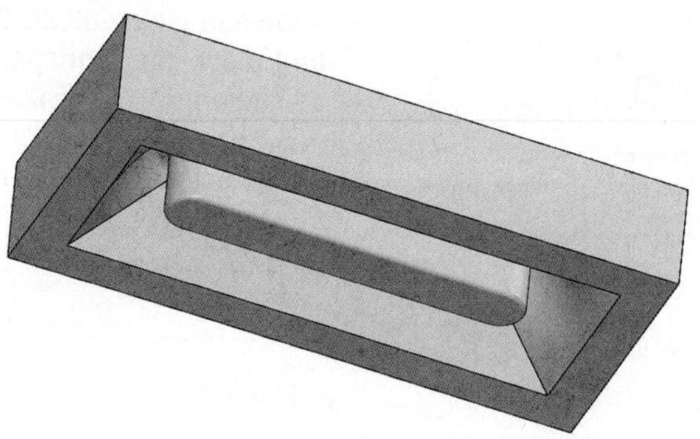

a *core* half (**Figure 12.18**), with features protruding outward from the parting line, and a *cavity* half (**Figure 12.19**), with features cut inward from the parting line.

FIGURE 12.20

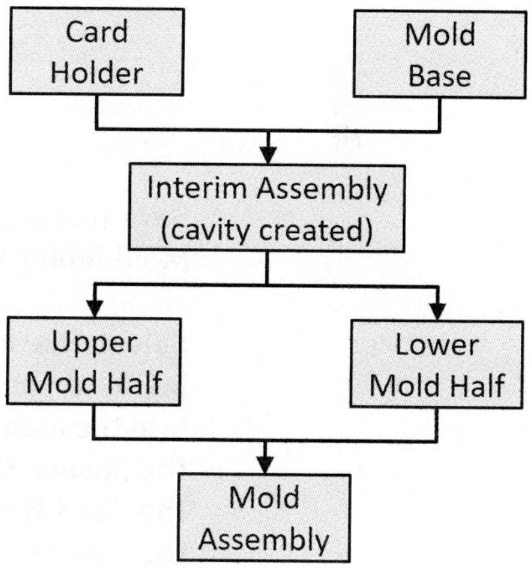

Our procedure will be similar to the one followed in the previous section. Since there are several files involved in this procedure, it is helpful to consider the process steps, which are illustrated in **Figure 12.20**. An interim assembly will be made from the mold base and part, so that the cavity can be placed in the mold base. From that assembly, copies of the mold base (with the cavity) will be derived. These copies will be modified to become the two mold halves. Finally, the two mold halves will be brought together into an assembly.

The first step of the process will be to create the mold base, which must be sized so that the card holder part fits completely within its boundaries.

Create a new folder called "Core and Cavity Mold," and copy the card holder part into the folder. Open a new part. In the Top Plane, draw and dimension a rectangle centered about the origin, as shown in Figure 12.21.

FIGURE 12.21

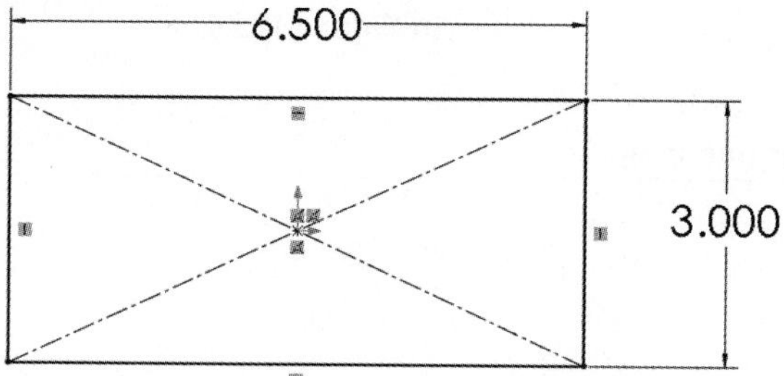

The dimensions shown are each 1 inch greater than the corresponding dimensions of the card holder, allowing for 1/2-inch clearance between the part and the mold edges.

The height of the mold base will be selected to allow for 1/4 inch above and below the part.

Extrude the rectangle upward a distance of 1.5 inches.

FIGURE 12.22

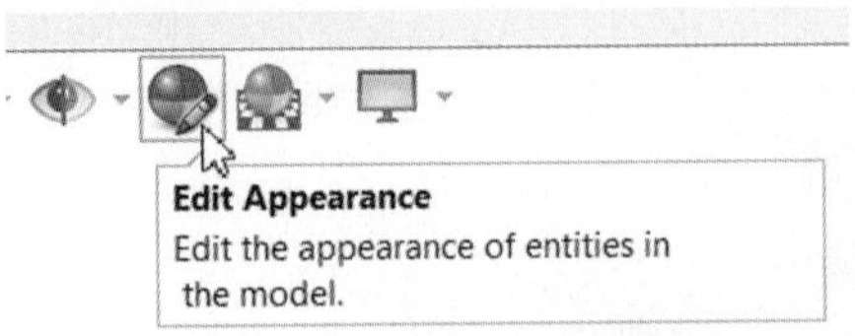

Choose the Edit Appearance Tool from the Heads-Up Toolbar, as shown in Figure 12.22. Select the Advanced tab and Illumination, and move the Transparent amount slider bar toward the right, as shown in Figure 12.23.

The transparent part is shown in Figure 12.24.

FIGURE 12.23

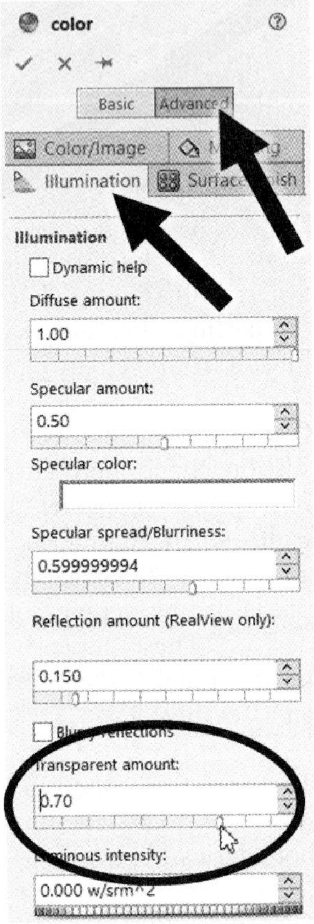

FIGURE 12.24

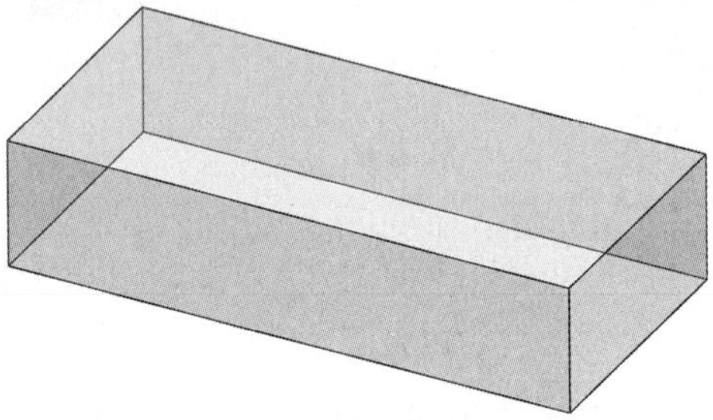

Save this part as "Mold Base" in the Core and Cavity Mold folder that you created earlier.

Select File: Make Assembly from Part from the main menu to open a new assembly. Insert the Mold Base into the assembly by clicking the check mark. Choose the Insert Components Tool and browse to locate the Card Holder. Click to place it in the assembly in the approximate position shown in Figure 12.25.

We will now align the card holder with the mold base.

Rotate the view so that the bottom surfaces of the card holder and the mold base are visible, as shown in Figure 12.26.

FIGURE 12.25

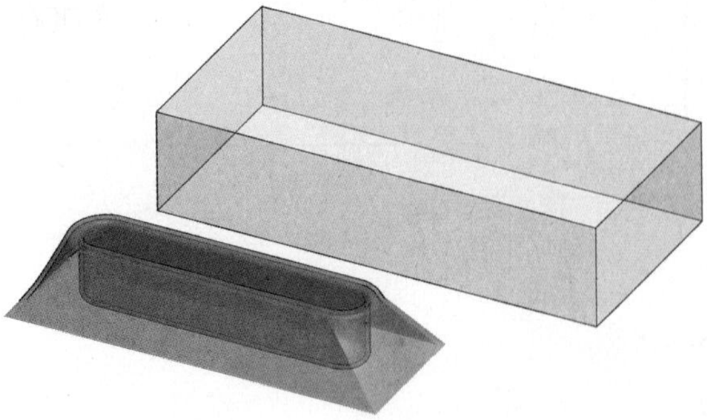

FIGURE 12.26

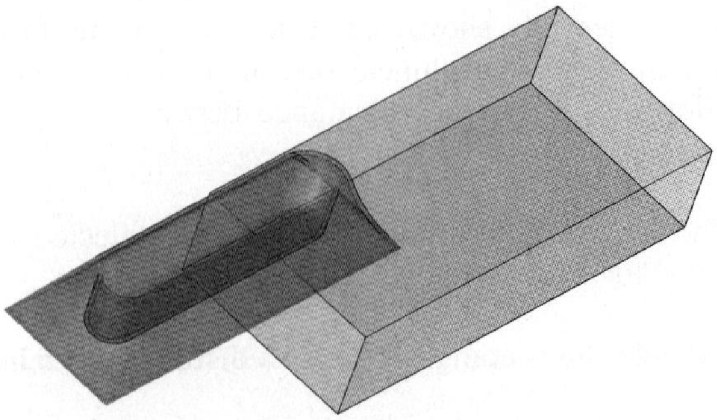

Choose the Mate Tool. Select the bottom surface of the card holder (Figure 12.27), and then the bottom surface of the mold base. A coincident relation will be previewed. Select the Distance Mate icon from the PropertyManager (Figure 12.28). Set the offset distance between the two surfaces to be 0.25 inches. If necessary, check the Flip Dimension box so that the bottom surface of the card holder is above the bottom surface of the mold base, as previewed in Figure 12.29. Click the check mark to apply the mate.

FIGURE 12.27

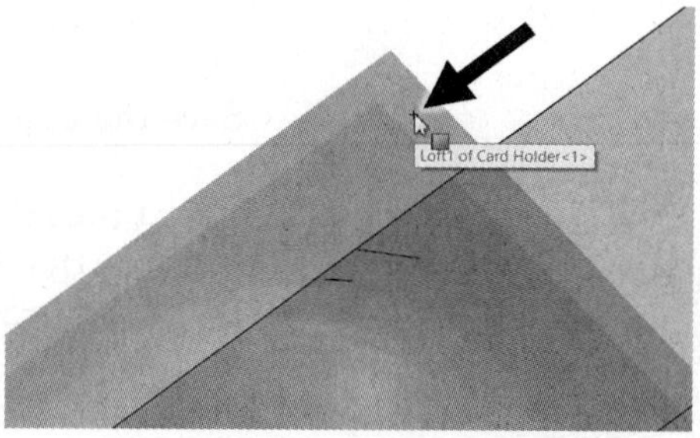

FIGURE 12.28

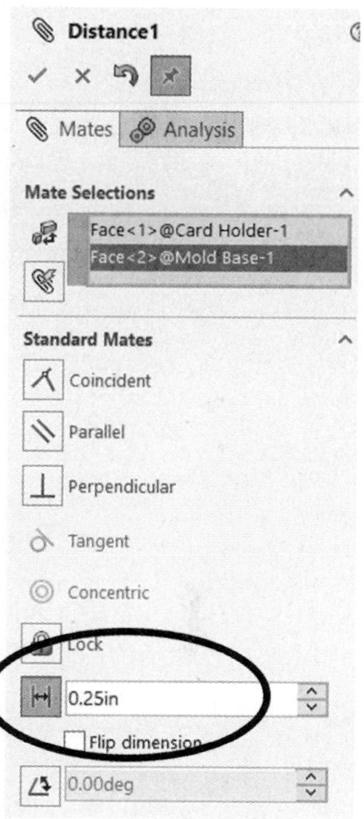

Expand the fly-out FeatureManager by clicking on the arrow beside the assembly name, and then expand the entries for both the base and the card holder by clicking the arrow beside each part's name. Select the Front Planes of both parts as shown in Figure 12.30. Add a coincident mate, which is previewed in Figure 12.31.

Add a coincident mate between the Right Planes of the card holder and mold base. Close the Mate PropertyManager.

FIGURE 12.29

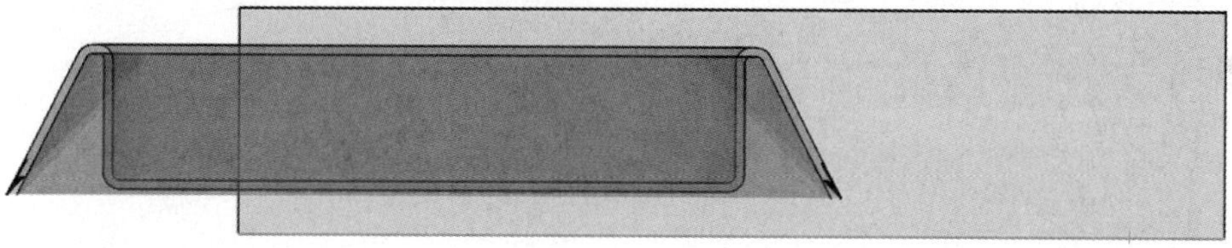

FIGURE 12.30

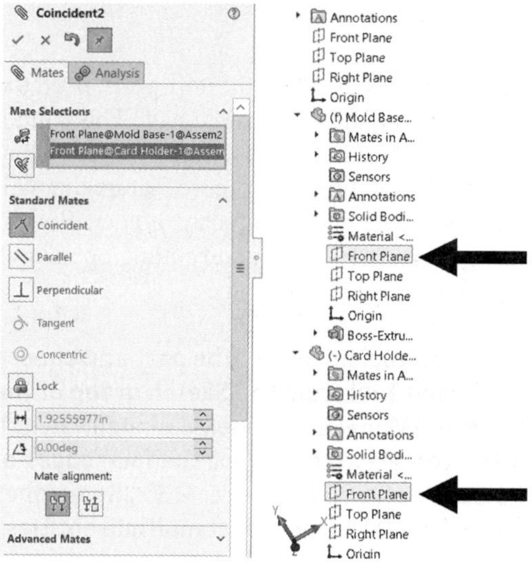

FIGURE 12.31

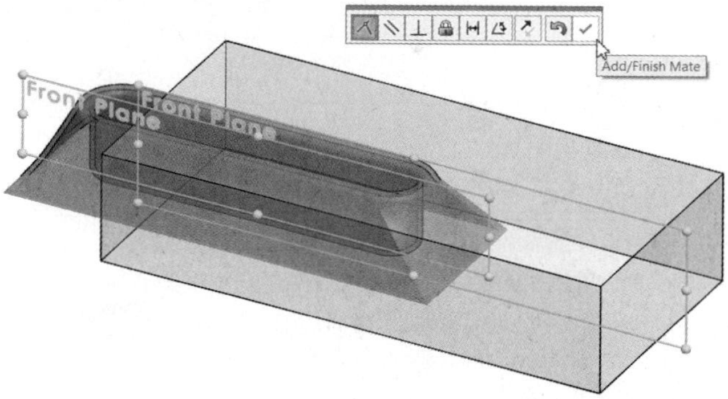

FIGURE 12.32

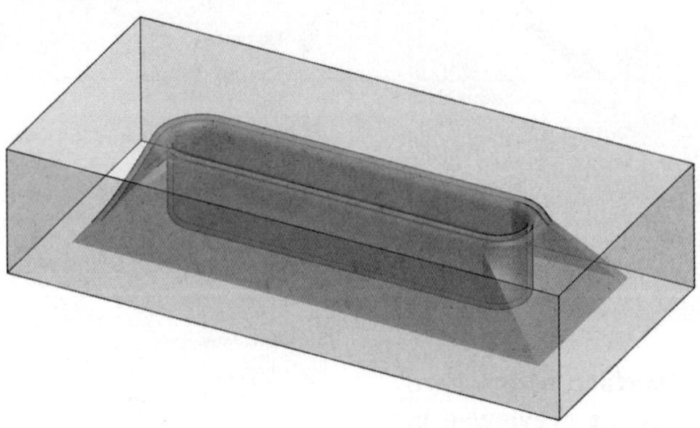

The card holder is now placed within the mold base, as shown in Figure 12.32.

Save the assembly as "Interim Mold Assembly."

Select the Mold Base from the FeatureManager. Choose the Edit Component Tool from the Assembly group of the CommandManager.

From the main menu, select Insert: Molds: Cavity. In the PropertyManager, select the Card Holder as the component defining the cavity. Set the "Scale about" option to Component Centroids and the scale to 1%, as shown in Figure 12.33. Click the check mark to create the cavity. Click the Edit Component Tool to end the editing of the mold base. Save the assembly and mold base.

Select the mold base from the FeatureManager. From the main menu, select File: Derive Component Part. Click the check mark to place the new part.

Display the part in the Wireframe mode, as shown in Figure 12.34.

FIGURE 12.33

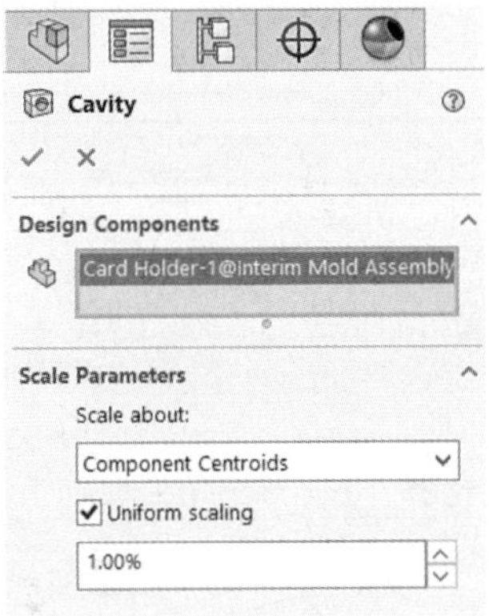

FIGURE 12.34

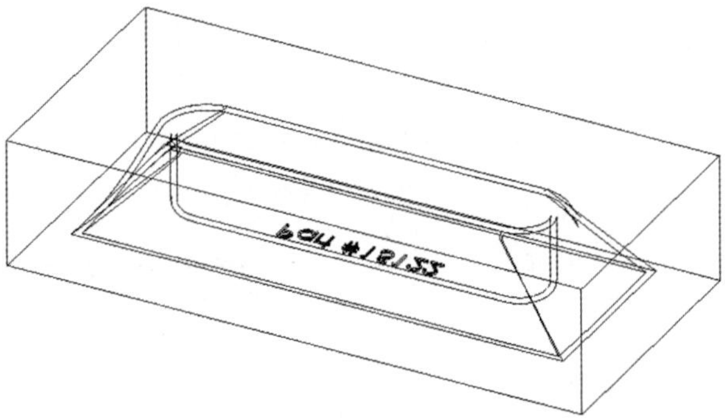

FIGURE 12.35

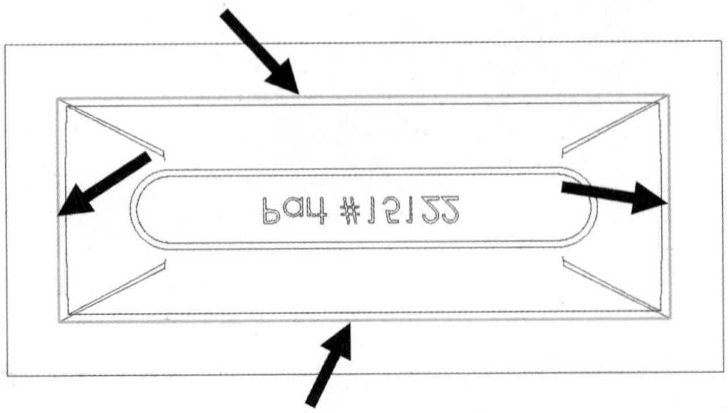

We will now cut away the top portion of the mold to create the lower mold half. This cannot be done with a simple extruded cut, since that would also cut through the core feature that will form the underside of the card holder. Rather, we will perform two separate cutting operations to achieve the desired geometry.

Select the top surface of the part and choose the Sketch Tool from the Sketch group of the CommandManager to open a sketch. Switch to the Top View and select the four edges of the cavity shown in Figure 12.35. (Remember to use the Ctrl key to select multiple entities.)

Select the Convert Entities Tool, which will create lines from the selected edges.

Extrude a cut with a type of "Up to Next." This will cut away the top of the mold, but only to the cavity, as shown in Figure 12.36.

FIGURE 12.36

In order to cut away the material around the edges of the mold half, it is necessary to use a different type of cut: one that cuts down to the parting surface.

Open a new sketch on the top surface. Convert the eight edges shown in Figure 12.37 into lines, using the Convert Entities Tool. (Note: if you select the top surface and then select the Convert Entities Tool, then the outer edges of the surface will be converted. It is still necessary to individually select and convert the inner edges of the surface.)

FIGURE 12.37

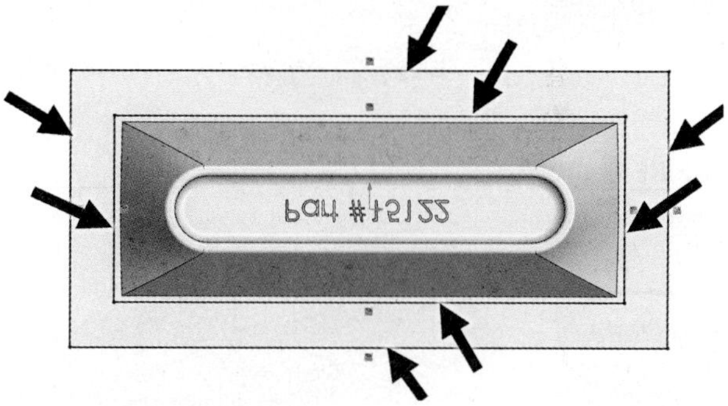

Extrude a cut, with a type of "Up to Surface." For the surface, choose the bottom surface of the cavity, which corresponds to the parting line, as shown in Figure 12.38.

The completed mold half is shown in Figure 12.39.

Save this part as "Mold Half Lower."

FIGURE 12.38

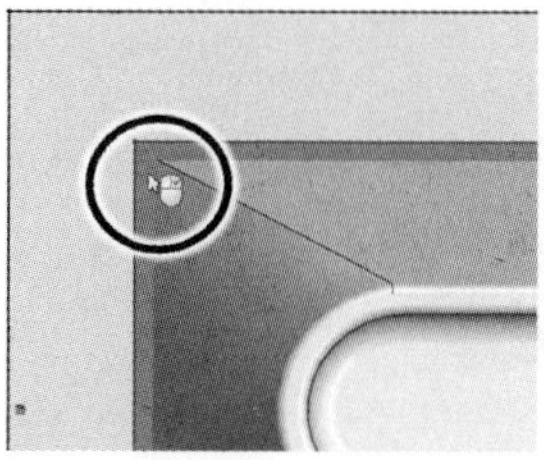

FIGURE 12.39

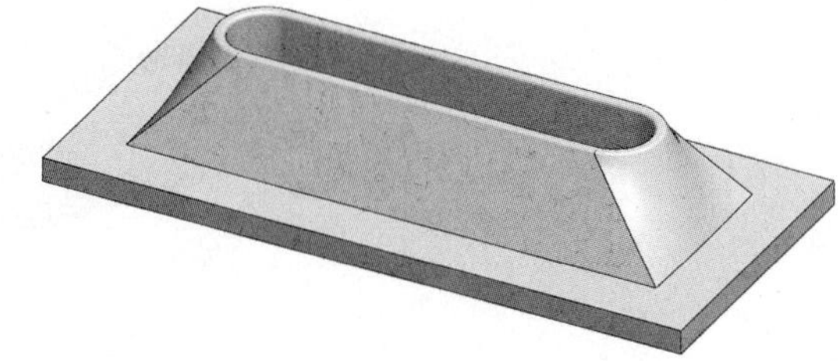

We will now create the upper mold half.

Switch back to the assembly containing the mold base and the card holder (Interim Mold Assembly). With the mold base selected, select File: Derive Component Part. Click the check mark.

In the new part, select the Section View Tool from the Heads-Up View Toolbar. Click the check mark to accept the Front Plane as the section plane. Zoom in to the bottom of the cavity, and click on the bottom surface to select it, as shown in Figure 12.40. Open a sketch on this surface by selecting the Sketch Tool from the Sketch group of the CommandManager.

FIGURE 12.40

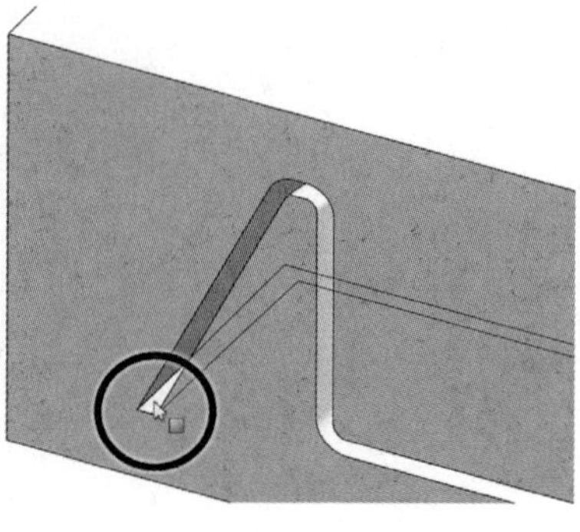

FIGURE 12.41

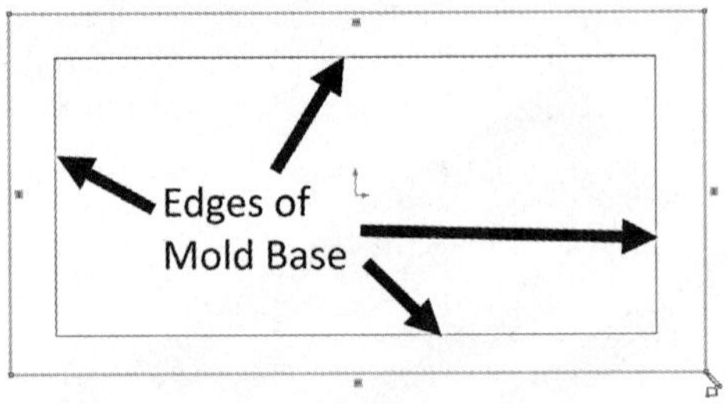

Turn off the Section View and switch to the Top View. Using the Corner Rectangle Tool, drag out a rectangle such that the entire part is enclosed within the rectangle, as shown in Figure 12.41.

Select the Extruded Cut Tool, and set the type of cut to Through All. Make sure that the direction of the cut is downward. Rotate the view so that the bottom of the part is visible, as shown in Figure 12.42, and click the check mark.

FIGURE 12.42

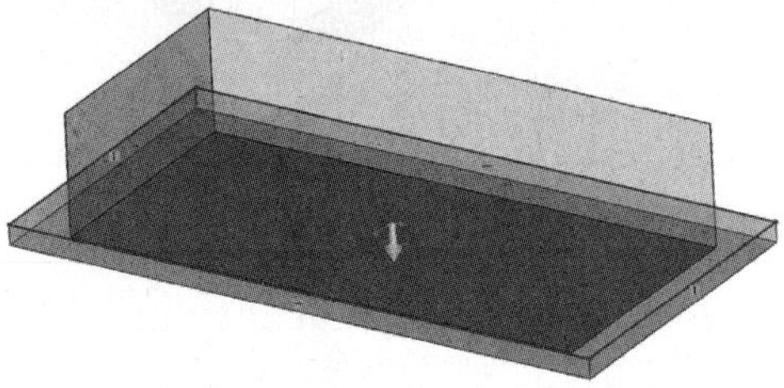

Since this cut will produce two separate solid bodies (the mold half and part of the core), you are prompted to identify which of the bodies you want to keep. In our case, we do not want to keep the core with the upper mold half. In the dialog box, checking the Selected bodies option allows you to choose which of the solid bodies you want to keep, and which will be deleted.

Choose the Selected bodies option. Check the box for Body 1 (the mold half) and clear the box for Body 2 (the core), as shown in Figure 12.43. The selected body is highlighted. Click OK to complete the cut.

FIGURE 12.43

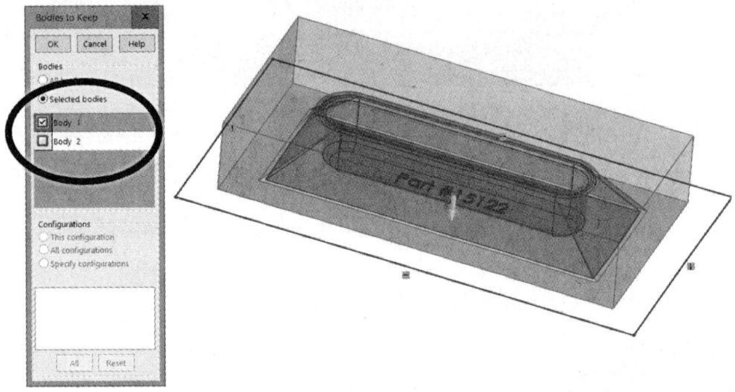

The resulting part is shown in Figure 12.44.

FIGURE 12.44

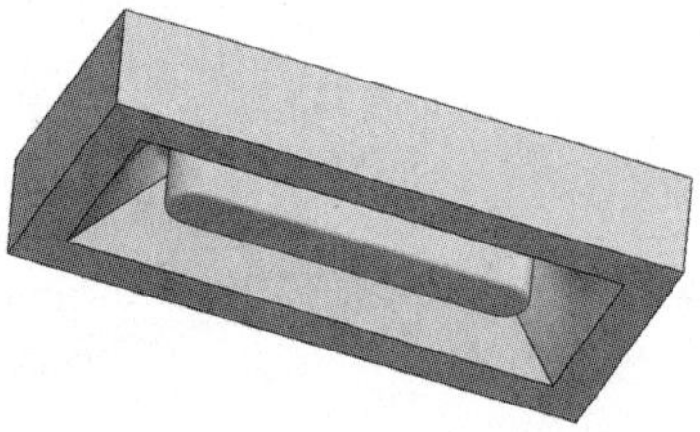

Save this part as "Mold Half Upper."

Open a new assembly. Insert the two mold halves, as shown in Figure 12.45. Make sure to place the lower mold half at the origin of the assembly.

FIGURE 12.45

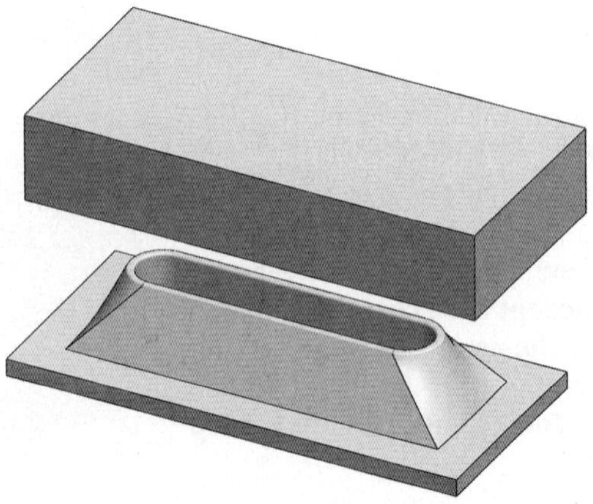

FUTURE STUDY

Materials and Processes

Our simple two-piece mold includes a cavity the shape of the finished part. If we created this mold for making a few prototype parts from a room-temperature curing material, such as polyurethane, then we could add holes for bolting the halves together, a hole for pouring in the material, and a vent hole, and our mold would be functional.

Most molds require many other features, however, and tooling design is an important function at any manufacturing company. If a plastic part is to be injection-molded, then the injection points and vent locations must be carefully designed so that the molten plastic fills the cavity completely. The plastic's melting and cooling temperatures and its resistance to flow must be considered when designing both the part and the mold. The tolerances required for the finished part might require that a filler be added to the material for dimensional stability. Ejector pins might need to be added to help remove the part from the mold. (Note that in our example, even though the part walls are tapered, the shrinkage of the part onto the core will result in forces that will need to be overcome in order to remove the part from the lower mold half.) Since the plastic must be injected hot but allowed to cool before removal from the mold, cooling lines for circulating water are usually added to the mold halves.

Other materials have different processing requirements. Composite materials used for automotive materials are mostly compression molded, in which the raw material is placed between two mold halves and formed by applying pressure with a hydraulic press. Lower-quantity parts can be produced by resin transfer molding, in which dry fabric is placed in a mold and liquid resin is pumped in under low pressure.

Often the choice of a process depends on the quantity of parts to be made. A high-quality tool for injection molding can cost tens of thousands of dollars, but if this cost can be spread over 100,000 or 1,000,000 parts, then the fast cycle times resulting from a good mold design can result in significant cost savings.

Add three mates to align the mold halves together, as shown in Figure 12.46.

Choose the Section View Tool from the Heads-Up View toolbar.

Click the check mark to accept the Front Plane as the section plane.

FIGURE 12.46

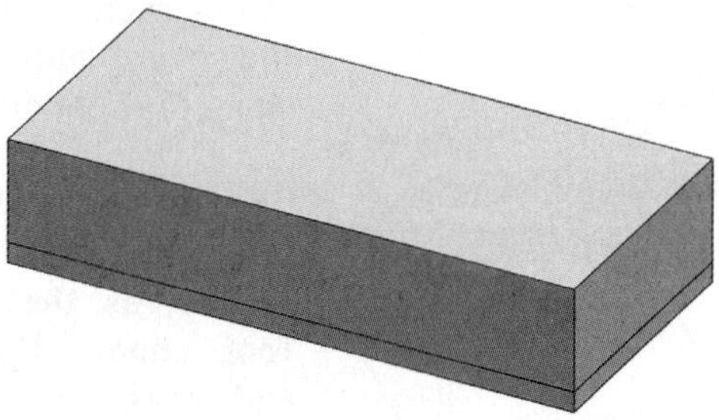

FIGURE 12.47

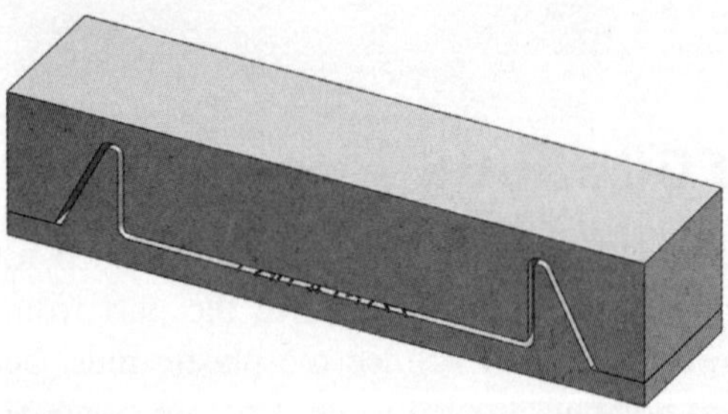

The mold cavity can now be clearly seen (**Figure 12.47**).

Save the file with the name "Card Holder Mold Assembly."

Suppose that we now want to change the thickness of the molded part. Often, if a material change is specified, the thickness will need to be changed, since the flow of the material in the mold is a limiting factor on the thickness.

Open the Card Holder part file. Right-click on the shell feature in the FeatureManager. Select Edit Feature, and increase the thickness from 0.06 to 0.125 inches, as shown in Figure 12.48. Rebuild the model.

FIGURE 12.48

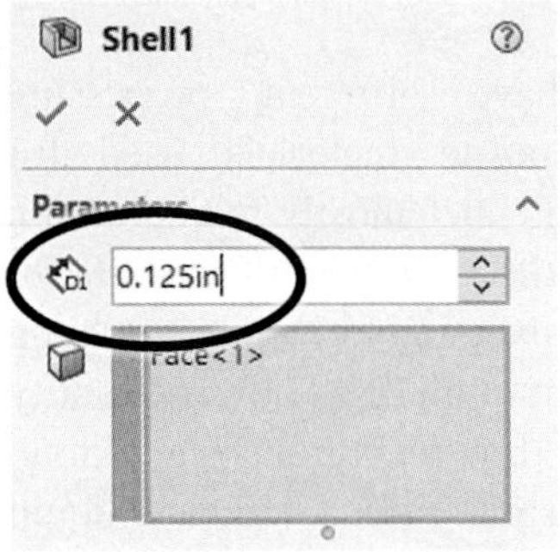

FIGURE 12.49

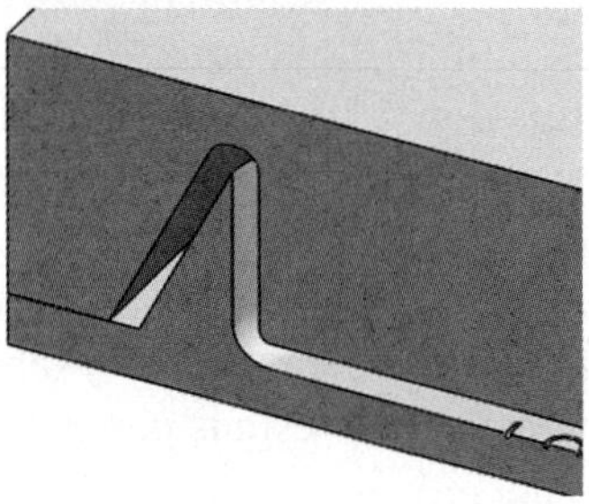

Open, or switch to, the interim mold assembly and then to the mold assembly. Click Yes to rebuild the models at each step.

The updated mold assembly is shown in **Figure 12.49**. Of course, the mold we created is not usable without some way to get material into the mold. The manner in which this is done depends on the molding process, as discussed in the Future Study box. Many other features may also be required for the mold to be usable. Mold design is a very specialized field, combining mechanical design with material science. However, an essential part of any mold design is the creation of the mold cavity and the separation of the core and cavity mold parts, such as we have done in these exercises.

Close all windows without saving the last changes made.

12.3 A Sheet Metal Part

FIGURE 12.50

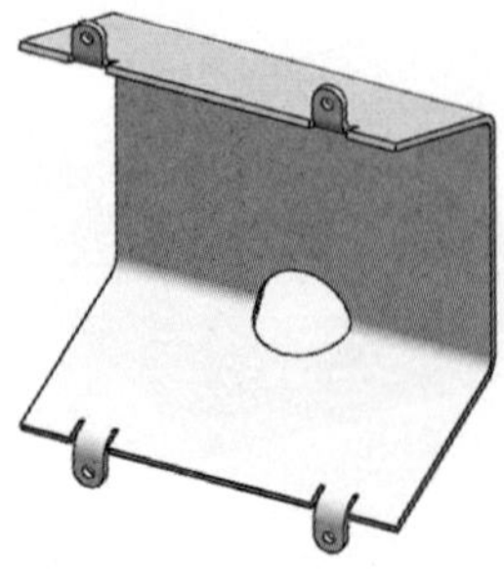

In this exercise we will create the sheet metal part shown in **Figure 12.50**. The part can be shown in either the bent or flat state.

Open a new part. Click on the units (IPS) on the Status Bar and select MMGS (millimeter, gram, second) as the unit system. From the Options Tool, choose Document Properties: Dimensions and set the option for trailing zeros of dimensions to Remove. (Recall that integer values of mm are typically shown without a decimal point.)

FIGURE 12.51

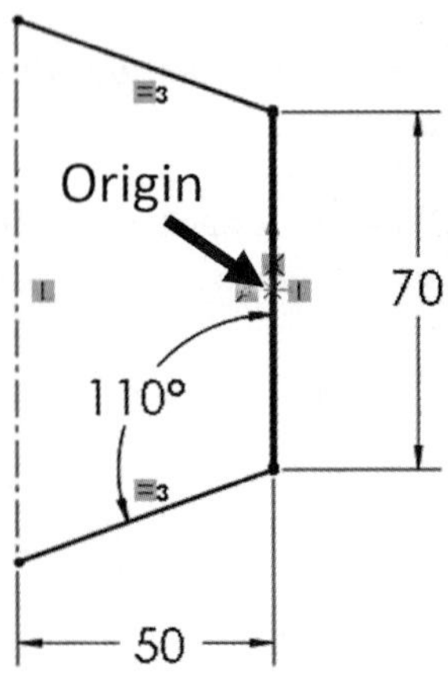

Select the Right Plane. Sketch and dimension the three lines shown in Figure 12.51. Draw the

vertical line first, choosing the Midpoint Line from the available line tools. This tool allows you to begin the line at the origin and then drag out the line equally in both directions. Press Esc to turn off the Midpoint Line Tool and then select the regular Line Tool and add the two diagonal lines.

Use the vertical centerline to align the endpoints of the diagonal lines. Add the three dimensions shown in Figure 12.51 and add an Equal relation between the two diagonal lines. The sketch should now be fully defined.

Note that we are not adding radii to the sharp corners. In sheet metal parts, bends are added as separate features to a part.

FIGURE 12.52

Select the Extruded Boss/Base Tool from the Features group of the CommandManager. Set the type to Mid Plane, and the extrusion depth to 100 mm. Since the sketch is open, a thin-feature extrusion will be created. Set the thickness to 2 mm, as shown in Figure 12.52.

Change the direction of the thickness if necessary so that the dimensions apply to the outside of the part, as shown from the Right View in Figure 12.53. Click the check mark to complete the extrusion.

FIGURE 12.53

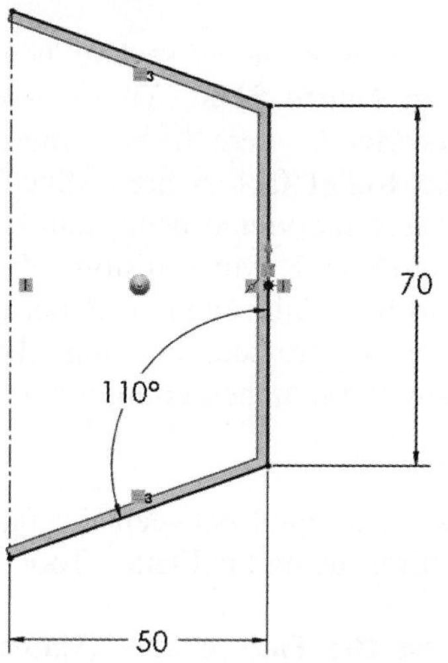

There are a number of tools that are specific to sheet metal parts. We can access these tools through the Sheet Metal toolbar or by adding the Sheet Metal tools to the CommandManager.

Right-click on one of the CommandManager tabs. In the list of tool groups that is displayed, click on Sheet Metal, as shown in Figure 12.54.

The Sheet Metal tools are now included in the CommandManager, as shown in **Figure 12.55**.

Select the Sheet Metal tab. Select the Insert Bends Tool, as shown in Figure 12.56.

FIGURE 12.54

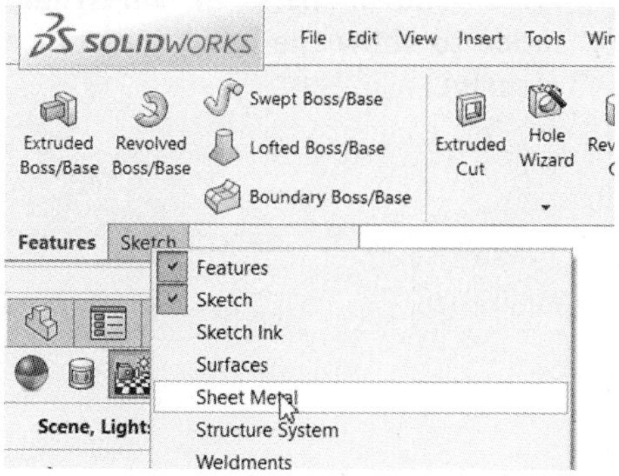

FIGURE 12.55

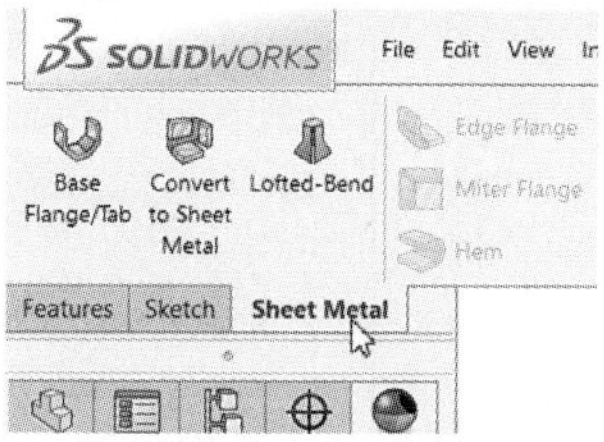

FIGURE 12.56

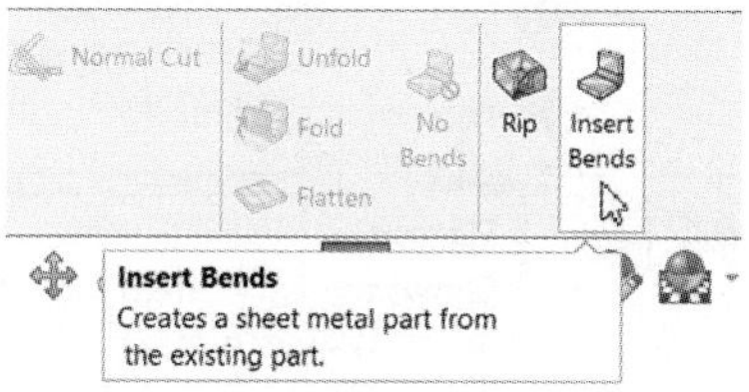

FIGURE 12.57

Select the middle face as the one that will remain flat during the bending, as shown in Figure 12.57. Set the radius to 4 mm, expand the Bend Allowance parameters, and set the K-factor to 0.5. Leave the auto relief checked with the type as Rectangular and the factor set to 0.5, as shown in Figure 12.58. Click the check mark to complete the operation.

FIGURE 12.58

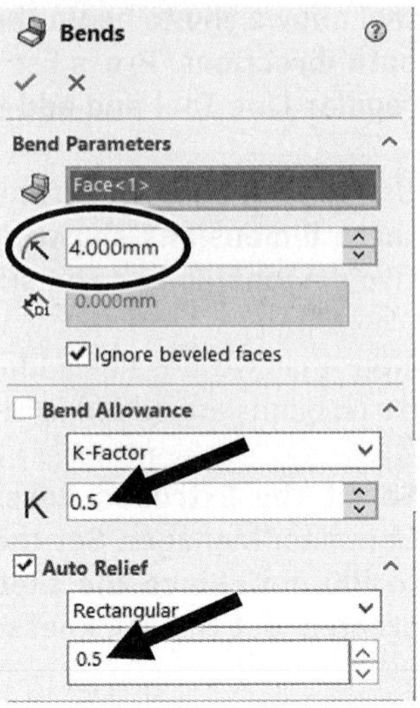

The resulting bent geometry is shown in **Figure 12.59**.

FIGURE 12.59

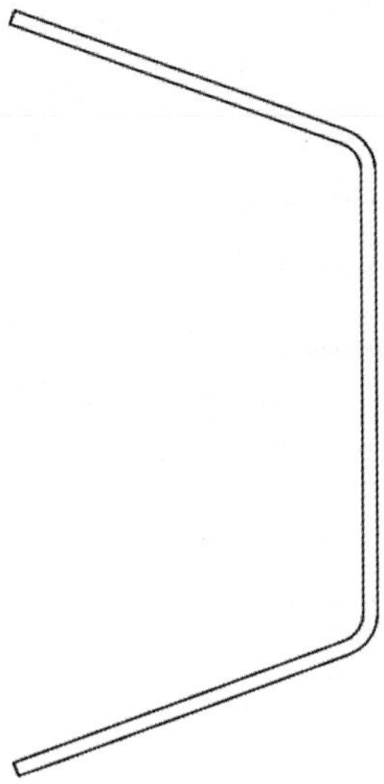

The K-factor is used in converting the bent geometry into the flat geometry. A K-factor of 0.5 means that the length of the flattened metal will be calculated based on the arc length at the mid-thickness of the metal. This is typical for relatively large bend radii. If extremely tight bends are desired, then the K-factor may need to be adjusted to get a better correlation between the flat geometry and bent geometry dimensions. The effect of the auto relief setting will be seen later when we add the tabs to the part.

FIGURE 12.60

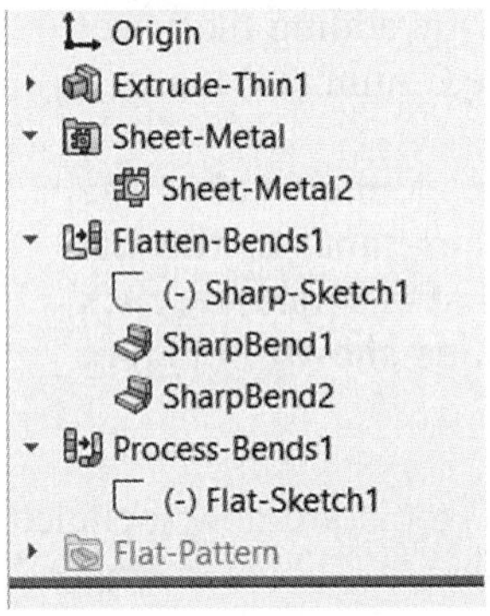

The FeatureManager now shows several new items, as shown in **Figure 12.60**. These new items include Sheet-Metal, where the bend radii, K-factor, and Auto Relief factors are defined, Flatten-Bends1, where individual bends can be edited, and Process-Bends1, which restores the bends when the part is rebuilt. The Flat-Pattern, which is shown as suppressed, contains the information to show the flattened configuration of the part.

The part can now be toggled between the flat and bent configurations using the Flatten Tool.

FIGURE 12.61

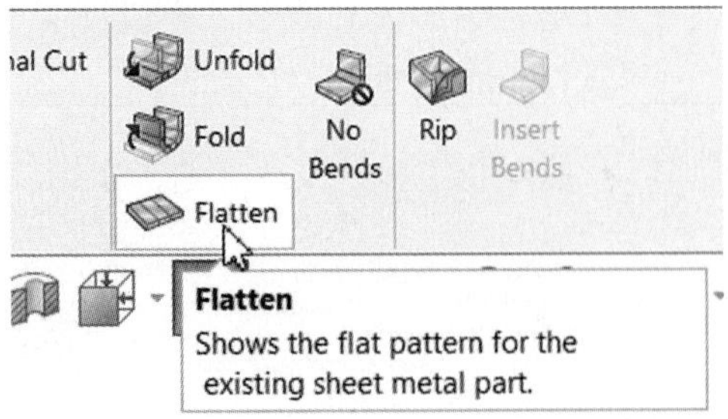

FIGURE 12.62

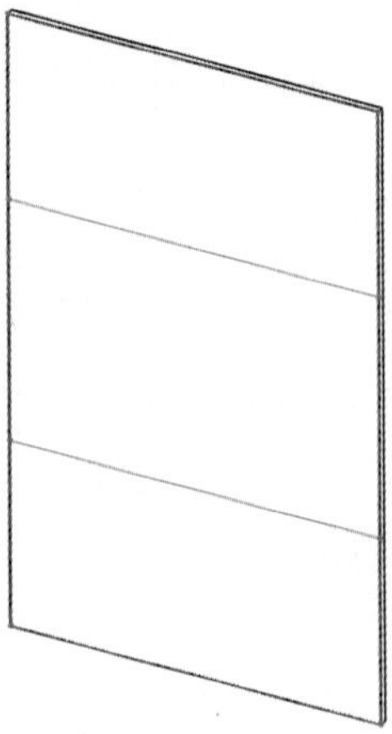

Click on the Flatten Tool (Figure 12.61) to show the part in the flattened state, as shown in Figure 12.62 (shown here in wireframe mode to show the bend lines more clearly).

Notice that the Flat-Pattern is now active in the FeatureManager. Expanding the Flat-Pattern shows the bend lines and individual bend properties, as shown in **Figure 12.63**.

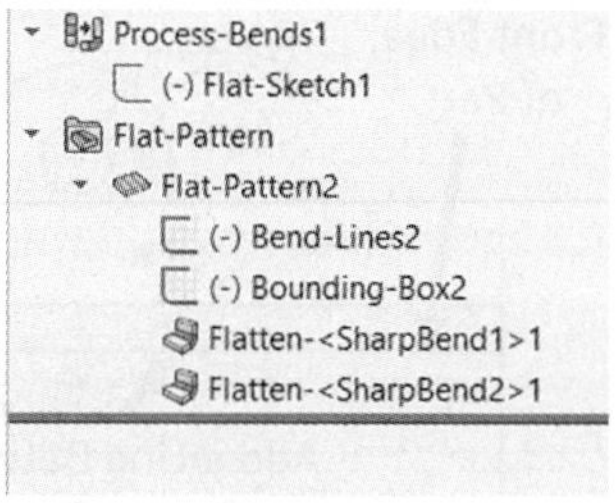

FIGURE 12.63

Click the Flatten Tool again to display the part in the bent state.

With the part in the bent configuration, the bends themselves can be suppressed with the No Bends Tool, as shown in **Figure 12.64**. This tool allows you to toggle between the geometry with sharp corners and the geometry with the bends shown.

Select the No Bends Tool, as shown in Figure 12.64. Note that the bend properties are grayed out in the FeatureManager, and that the rollback bar is above the bend properties, as shown in Figure 12.65.

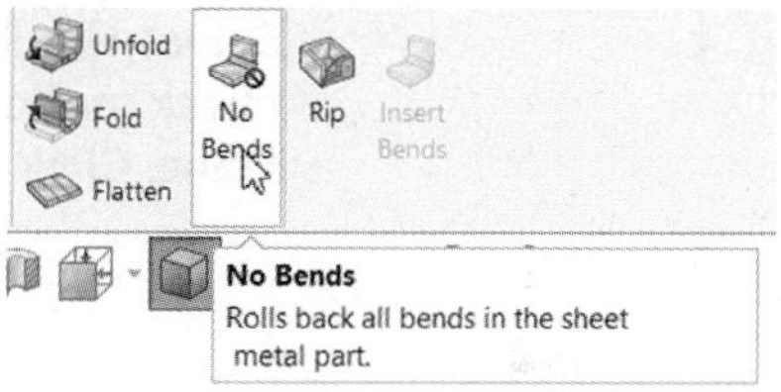

FIGURE 12.64

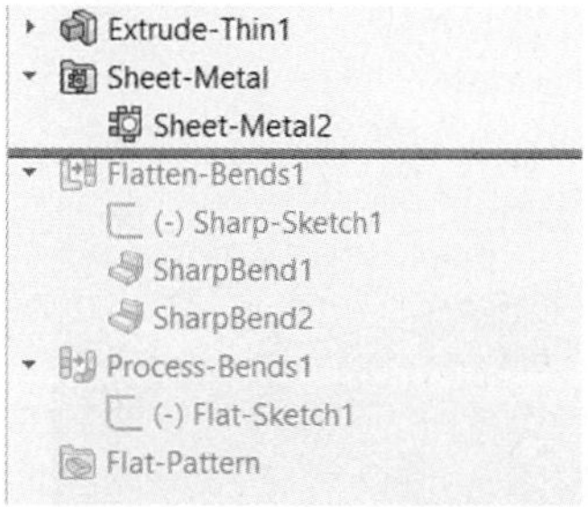

FIGURE 12.65

We will now add the tabs in the no-bend configuration and then bends will be added with parameters that were set earlier. To add the tabs, we need to create a new plane. Since the tabs are to allow the part to mount flush to another surface, the new plane needs to correspond to the front edge of the part.

Select: Reference Geometry: Plane from the Features group of the CommandManager. If there are any items already selected, clear them by right-clicking in any white space and selecting Clear Selections.

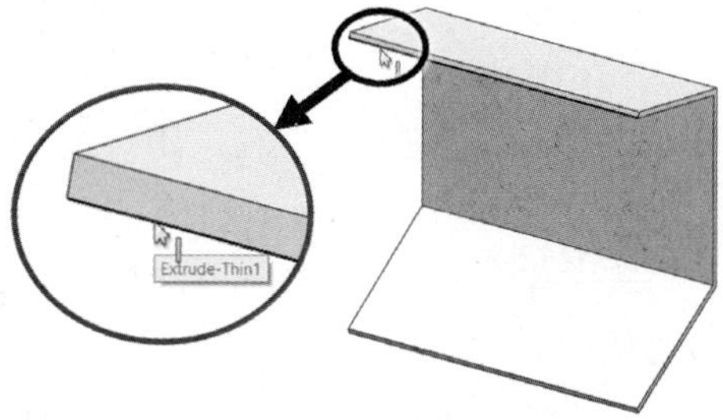

FIGURE 12.66

Select the front edge near the top of the part, as shown in Figure 12.66, and then the front edge near the bottom of the part, as shown in Figure 12.67. Click the check mark to create the plane.

View the plane from the Right View to make sure that it is in the correct location, as shown in **Figure 12.68**.

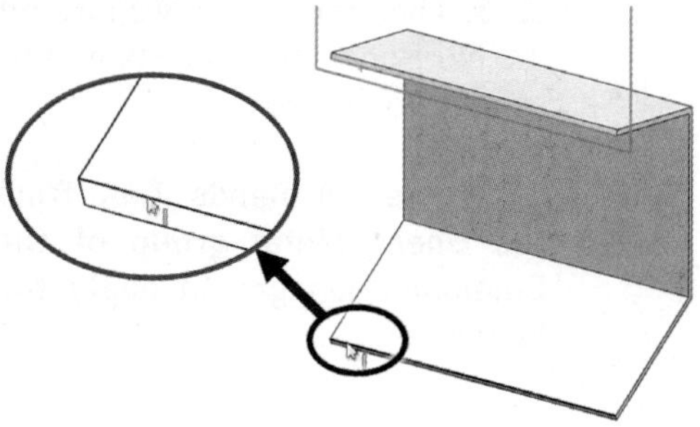

FIGURE 12.67

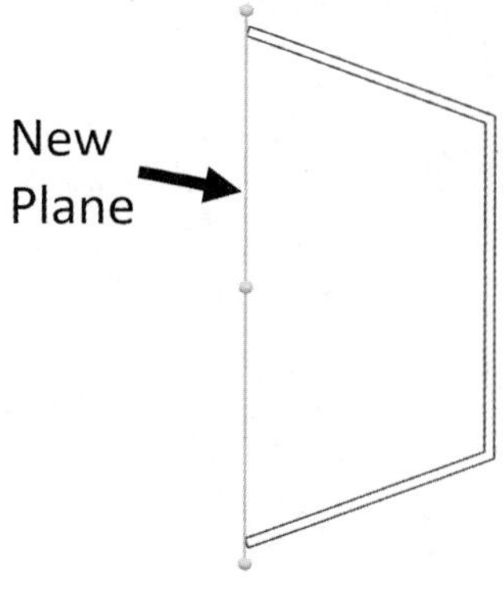

FIGURE 12.68

FIGURE 12.69

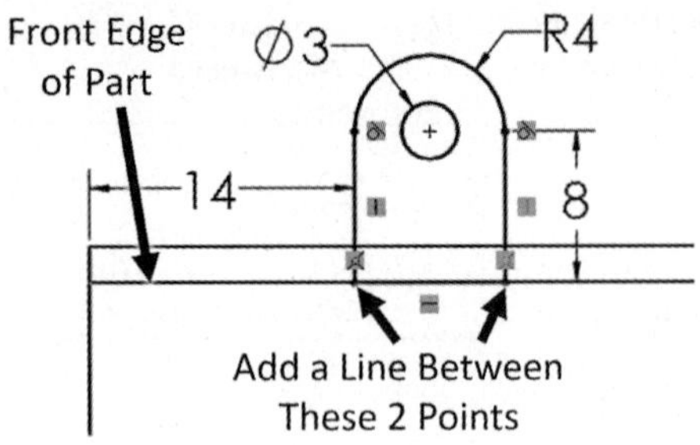

Select the new plane. Sketch and dimension the first tab, near the upper-left corner of the part, as shown in Figure 12.69 (shown in wireframe mode and with the construction plane hidden for clarity). Make sure to add a line along the bottom of the tab so that a closed contour is created.

Extrude the tab toward the rear of the part by clicking the Reverse Direction arrows, as shown in Figure 12.70. Check the "Link to thickness" box to set the thickness as the same value that was chosen when the part was defined as a sheet metal part. Click the check mark to complete the extrusion. With the tab (Boss-Extrude1) selected, choose the Mirror Tool. Select the Right Plane as the Mirror Plane, as shown in Figure 12.71. Click the check mark. Select the Mirror Tool again, and select the Top Plane as the Mirror Plane. Click the check mark.

FIGURE 12.70

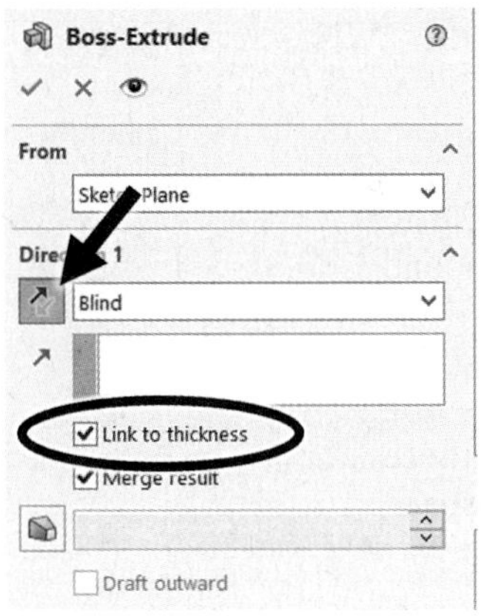

FIGURE 12.71

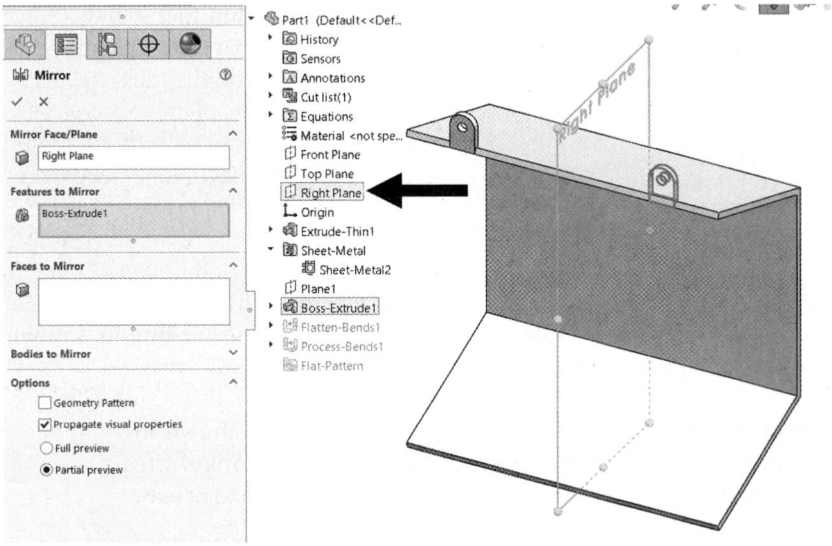

FIGURE 12.72

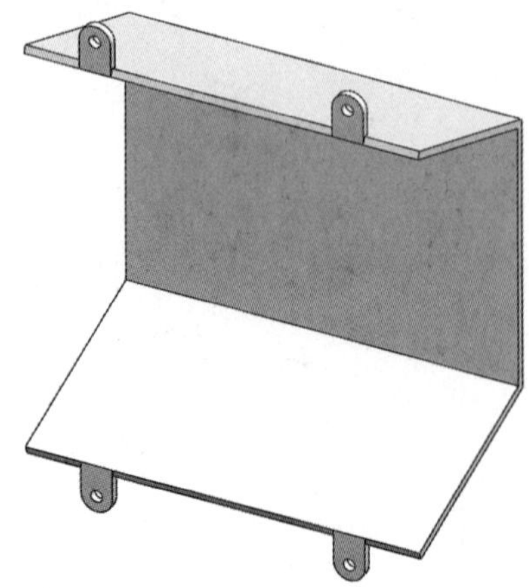

Note that the tabs have sharp corners rather than bends, as shown in **Figure 12.72**. This is because the tabs were added with the No Bends Tool selected. In the FeatureManager, note that the tabs (Boss-Extrude1 and the mirror features) appear before the bend properties, as shown in **Figure 12.73**. Therefore, when we turn off the No Bends Tool, bends will be applied to the tabs.

FIGURE 12.73

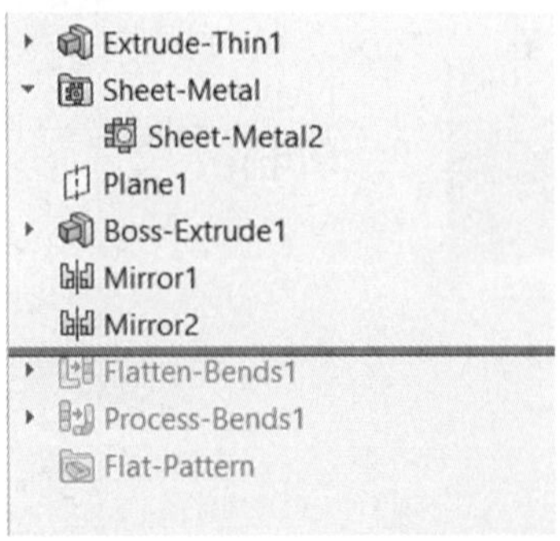

Click the No Bends Tool from the Sheet Metal group of the CommandManager to apply the bends.

The tabs now appear with bends, as shown in **Figure 12.74**. The metal around the bends was cut per the Auto-Relief parameters set earlier. (The width of the cuts is 0.5 times the thickness, and the cuts extend 0.5 times the thickness beyond the end of the bends.)

FIGURE 12.74

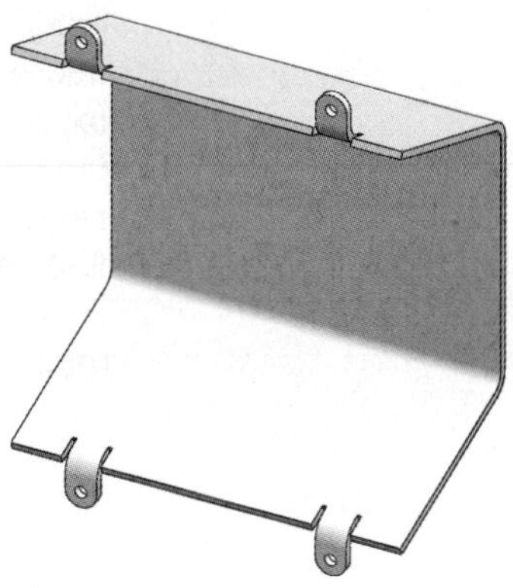

The cutout in the part must be added in the flattened state, so that it can be cut as a circle.

Drag the Rollback Bar to just below Flatten-Bends in the FeatureManager, as shown in Figure 12.75. To display the bend edges on the flat pattern, select View: Display: Tangent Edges Visible from the Main Menu. On the front surface sketch and dimension the 25-mm circle as shown in Figure 12.76, using a centerline and midpoint relations to place the circle.

Cut the hole with a through-all extruded cut. Show the part in the bent shape by dragging the Rollback Bar to the end of the FeatureManager, as shown in Figure 12.77. The finished part appears in Figure 12.78, with the tangent edges hidden.

FIGURE 12.75

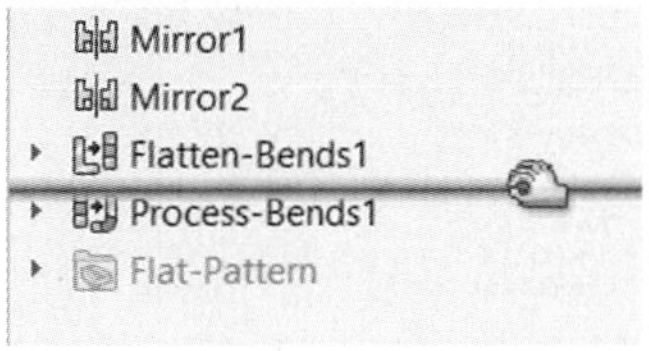

FIGURE 12.76

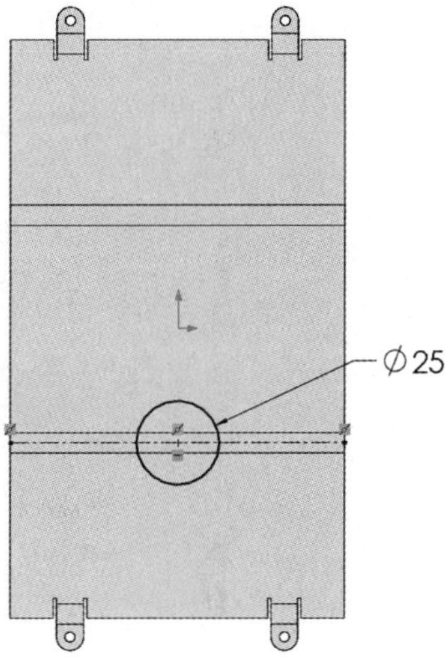

Save the part as "Sheet Metal Part."

FIGURE 12.77

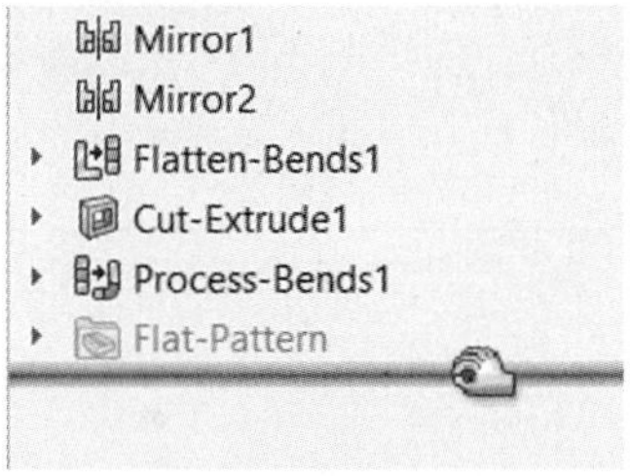

FIGURE 12.78

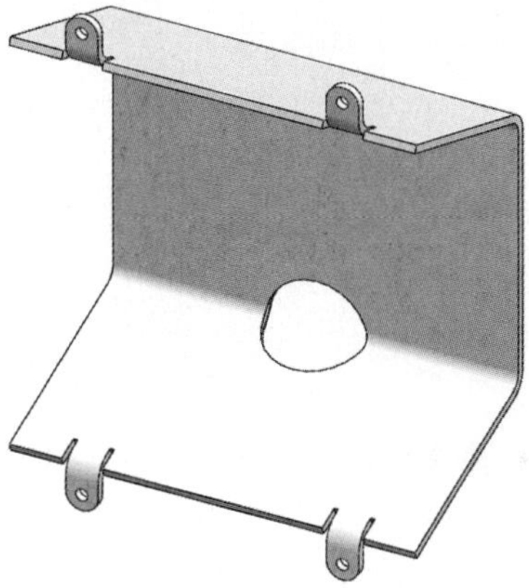

While drawings of sheet metal parts showing their final dimensions are often made, a manufacturing drawing will show the dimensions of the flat pattern and *bend notes* detailing how the pattern is bent into the final configuration. We will now make a manufacturing drawing of the part that we just modeled.

Open a new A-size drawing, with a title block if desired. The Sheet Metal Part should be highlighted in the Model View Manager. Click the Next arrow, and choose the Flat pattern as the view, as shown in Figure 12.79. Check the Preview box, and move the cursor into the drawing area to see a preview of the view placement. If necessary, change the rotation of the view to 90 degrees so that the orientation is as shown in Figure 12.80. Click to place the view. Click the check mark or press Esc to end the insertion of views.

FIGURE 12.79

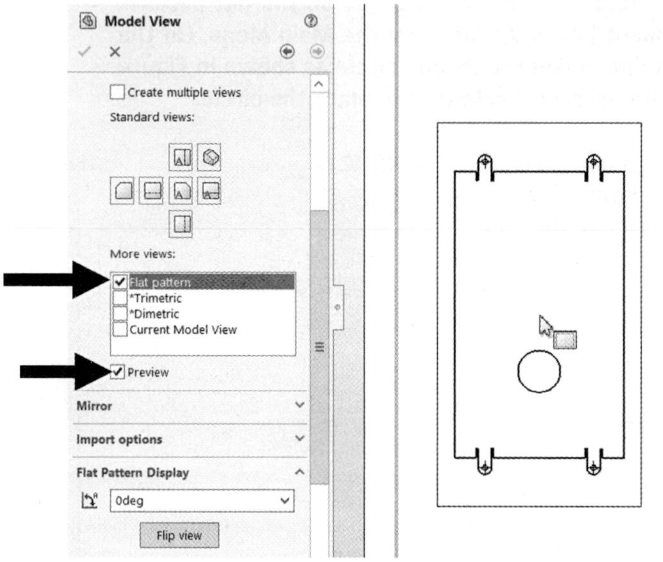

FIGURE 12.80

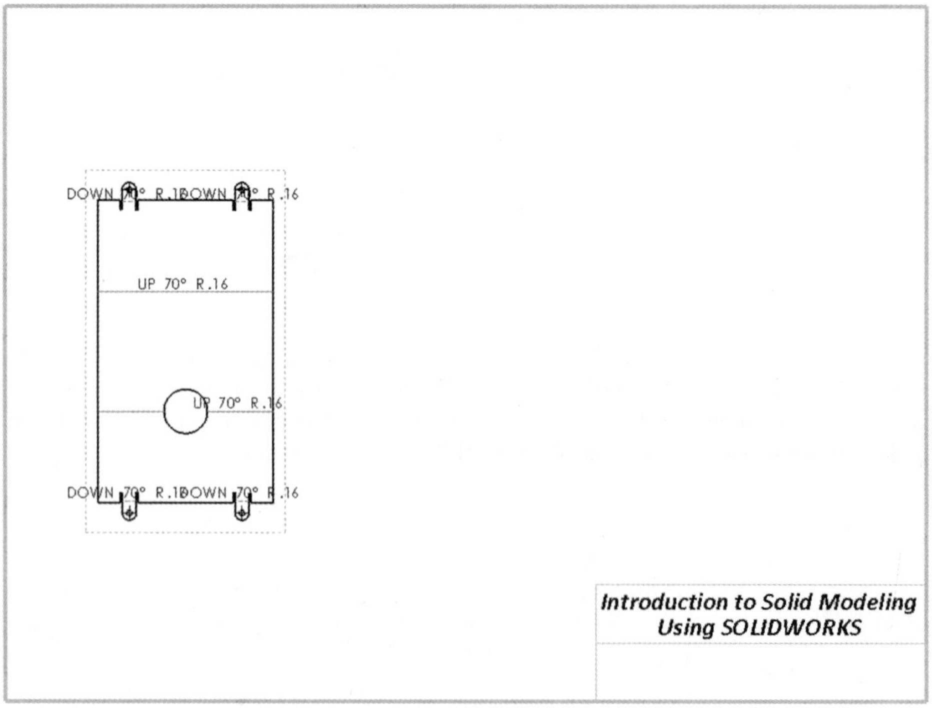

Select the Options Tool. Under Document Properties, select Units and set the unit system to MMGS, with one decimal place for length dimensions. Select Dimensions and set the option for the display of trailing zeros to Remove. Select Sheet Metal, and set the bend note style to With Leader, as shown in Figure 12.81. If desired, change the font size for Annotations to make the bend notes larger. Click OK. The notes are now shown with leaders pointing to the bend lines, as shown in Figure 12.82.

FIGURE 12.81

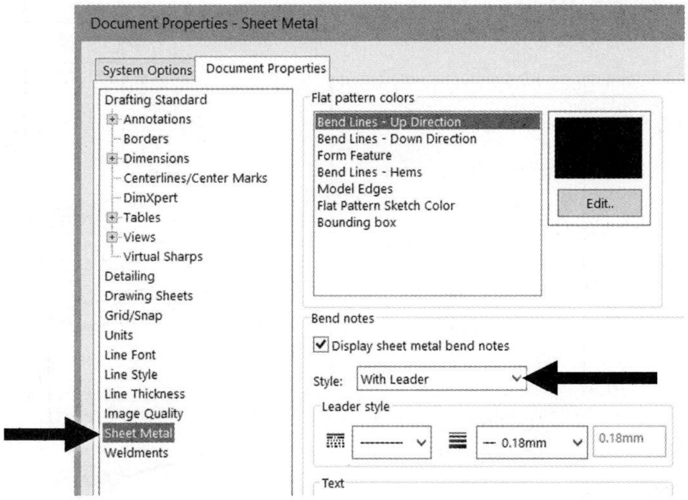

FIGURE 12.82

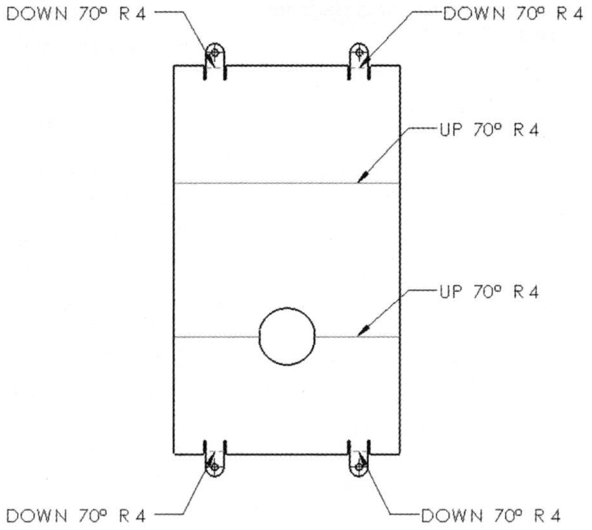

Each bend note shows the bend radius and angle of the bend, as well as the direction of the bend (up or down) relative to the plane of the flat pattern. Importing dimensions into the drawing will not work correctly for most dimensions since the part dimensions are relative to the bent configuration. Instead, dimensions are added manually with the Smart Dimension Tool.

Add dimensions and notes as shown in Figure 12.83. Change the display of the dimensions that are dependent on the bend parameters to show one decimal place. Add a detail view to show the dimensions of one of the tabs and a trimetric view to show the bent part. Change the dimension and note font sizes as desired. Save the drawing file.

FIGURE 12.83

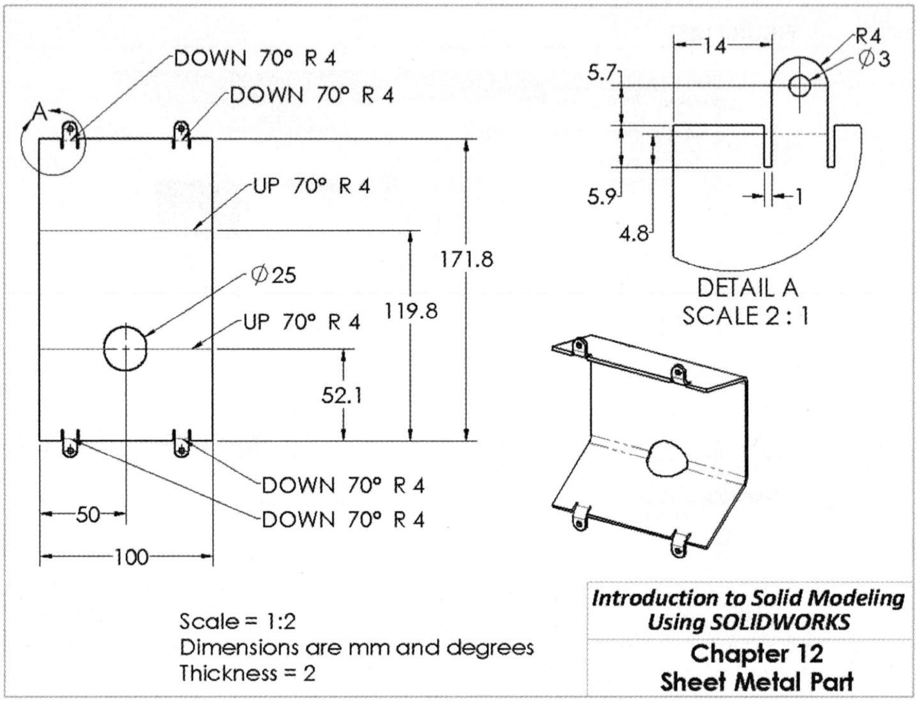

PROBLEMS

P12.1 Open the interim assembly of the mold base and card holder that was created in section 12.2 (**Figure P12.1**). Perform an interference detection on the assembly. How are the interferences found related to the shrinkage factor used to create the cavity?

FIGURE P12.1

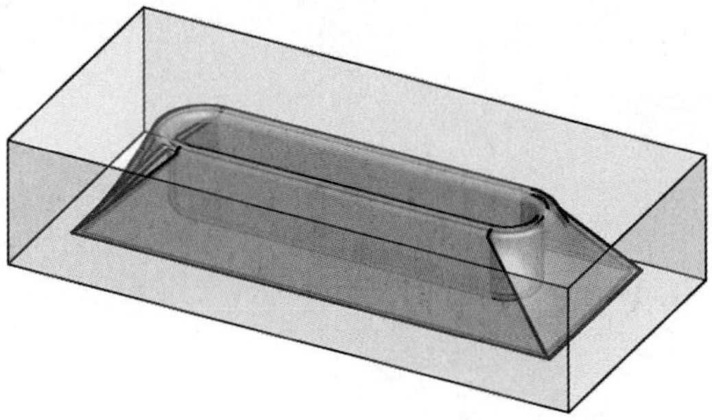

P12.2 Create a two-piece mold for the flange part of Chapter 1, allowing a 2% shrink factor. **Figure P12.2A** shows the lower mold half. Note that the material forming the holes is contained in the lower mold half. **Figure P12.2B** shows the upper mold half. Note that the chamfer feature is included in the upper mold half.

FIGURE P12.2A

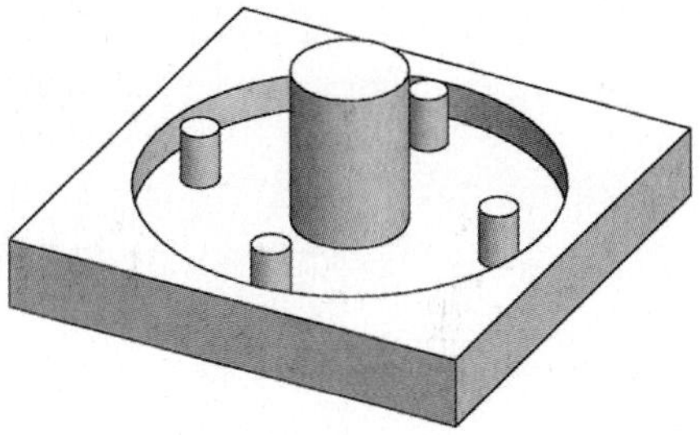

Hint: The easiest way to cut away material from the mold halves is to use *revolved cuts*. **Figure P12.2C** shows the sketch defining the "cutting tool" to be revolved around the centerline to create the lower mold half.

FIGURE P12.2B

FIGURE P12.2C

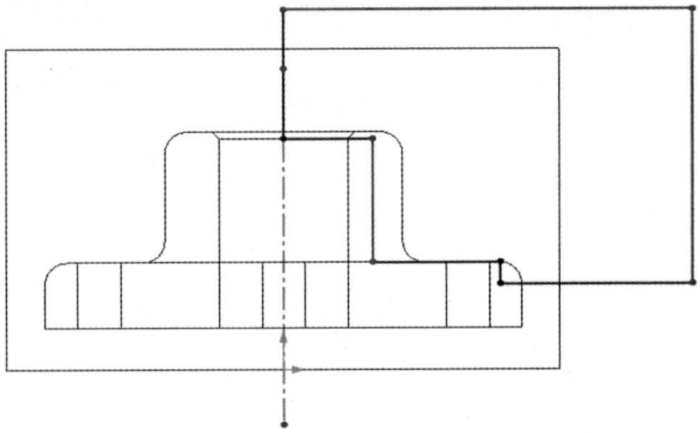

P12.3 Create an assembly of the two mold halves of **Problem 12.2** and the flange. Show a section view of the assembly, with the mold halves transparent, as shown in **Figure P12.3**. Comment on the interference that can be seen between the flange and the mold halves.

FIGURE P12.3

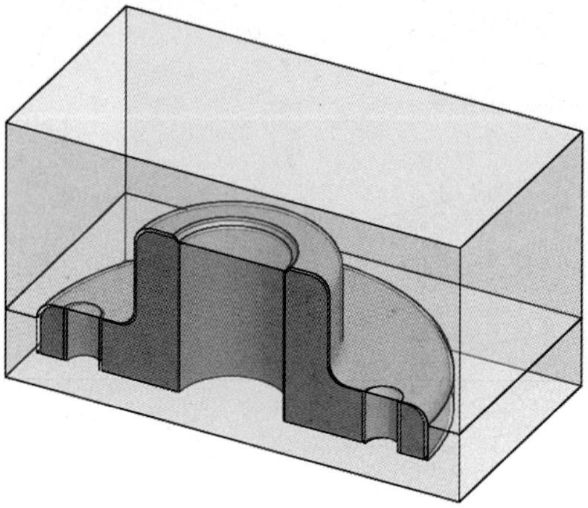

P12.4 Create a model of the sheet metal part shown in **Figure P12.4**. The metal thickness is 0.060 inches, and the bend radii are 0.125 inches. All dimensions shown are in inches. The overall dimensions are to the outer surfaces of the part.

FIGURE P12.4A

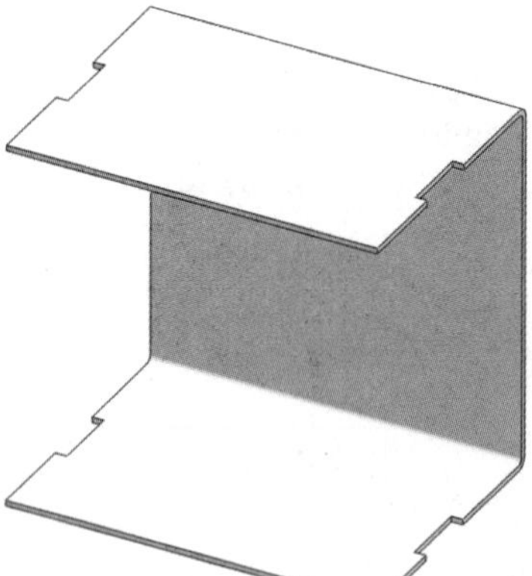

FIGURE P12.4B

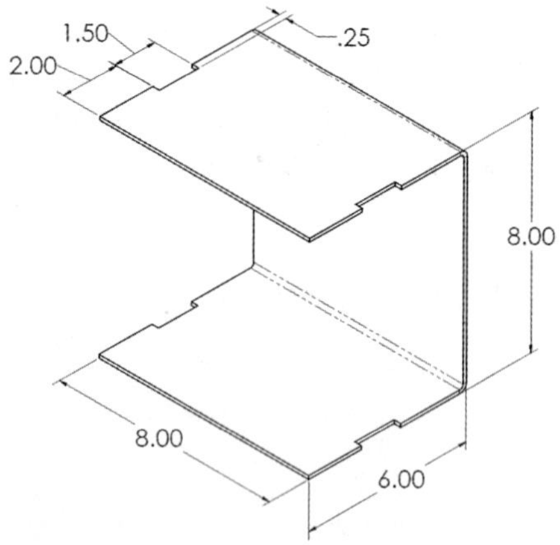

P12.5 Create a model of the sheet metal part shown in **Figure P12.5**. The metal thickness is 0.060 inches, and the bend radii are 0.125 inches. All dimensions shown are in inches.

FIGURE P12.5A

FIGURE P12.5B

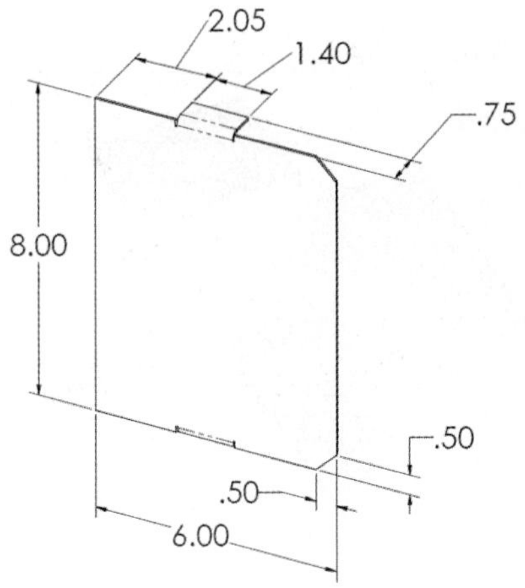

P12.6 Create the box shown from the parts created in exercises **Problem 12.4** and **Problem 12.5**. Since the sides of the box are mirror images of each other, it will be necessary to have two different part files. Open the part created in **Problem 12.5**, and select one of the vertical faces. Then select Insert: Mirror Part from the main menu. Click the check mark, and a new part file is created, linked to the original file. Save this file with a new name, and then assemble the three parts to form the box.

FIGURE P12.6

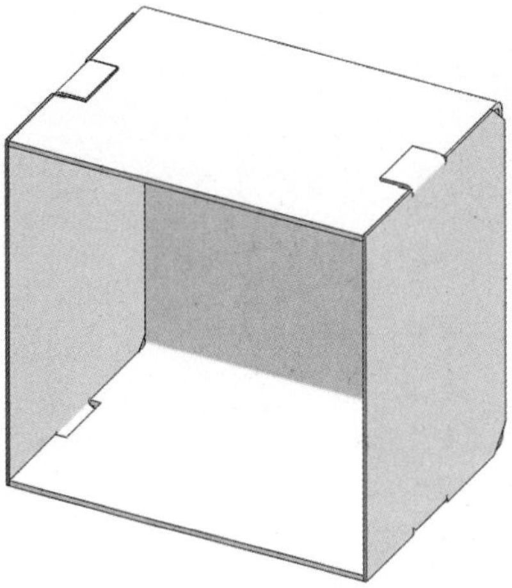

P12.7 Design a sheet metal bracket similar to the one shown in **Figure P12.7A**. The bracket is intended to be used to join a wood 2 × 4 (actual dimensions 1.5 by 3.5 inches) to a 4 × 4 (3.5 by 3.5 inches) post, as shown in **Figure P12.7B**, with nails or screws. Make a dimensioned drawing of the flat pattern of the part, including the bend notes.

FIGURE P12.7A

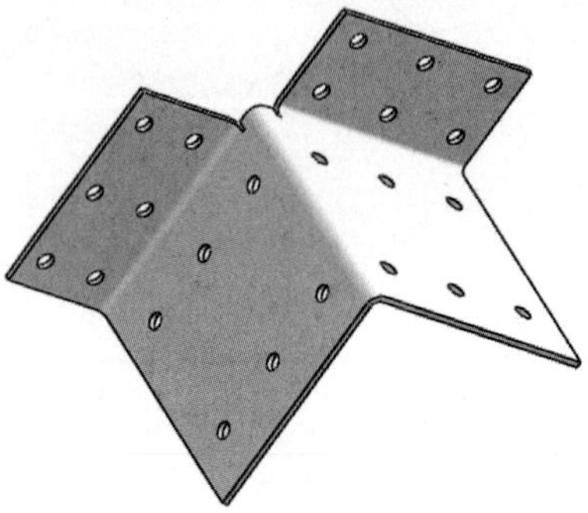

FIGURE P12.7B

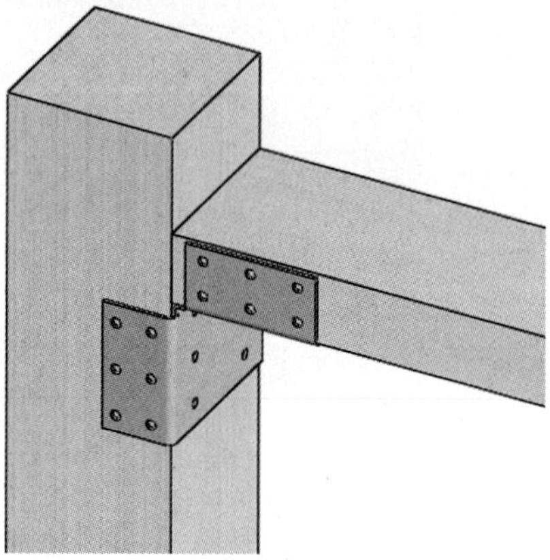

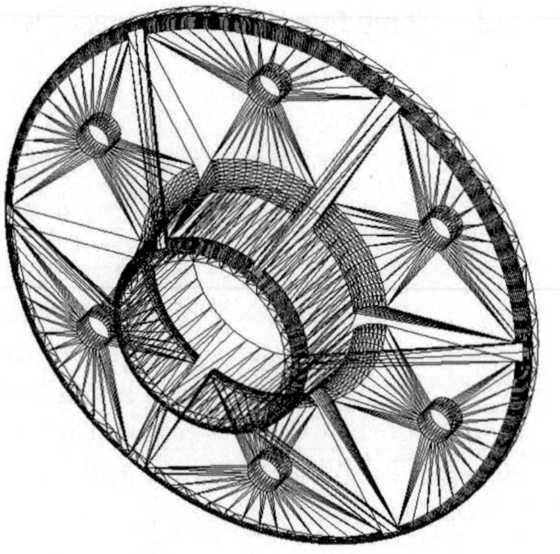

CHAPTER 13

The Use of SOLIDWORKS® to Accelerate the Product Development Cycle

Chapter Objectives

In this chapter, you will:

- be introduced to the most popular 3-D printing processes,
- learn how to create an .stl file to prepare a solid model for 3-D printing,
- be introduced to the capabilities and limitations of finite element analysis, and
- learn about Product Data Management software.

Introduction

Any company that designs and/or manufactures products has seen an increase in global competition over the past decades. To remain competitive, there is great pressure to develop new products. As a result, product models have shorter lives, and development costs are spread over a fewer number of units.

With these increased pressures, there has been a greater emphasis on improving and accelerating the product development process. While different industries and companies have their own unique procedures, there are some activities that are typically a part of a modern product development cycle:

- Physical prototypes are used early and often.
- Computer analysis is used extensively to complement physical testing.
- Engineering functions are performed simultaneously as much as possible, necessitating better teamwork and data sharing.

Solid modeling is an important tool in the product development process. The solid model becomes the common database used for a variety of engineering functions. In this chapter, we will explore two of the most common uses of solid modeling in product design: 3-D printing and finite element analysis. We will also consider the challenge of managing and controlling the data produced during the product development cycle.

SOLIDWORKS is a registered trademark of Dassault Systémes SolidWorks Corporation.

13.1 3-D Printing

3-D printing, which is also widely known as Additive Manufacturing (AM), is a technology that is closely allied with solid modeling. In additive manufacturing, parts are built up layer-by-layer using equipment driven by computer models of the part being built. The layer-by-layer building process is called an additive process to contrast it with traditional subtractive machining processes, which start with solid material stock and create the part shape by cutting away material.

Additive manufacturing technologies can be grouped into the following seven broad categories, as defined by ASTM:

- vat polymerization,
- powder bed fusion,
- material extrusion,
- binder jetting,
- material jetting,
- sheet lamination, and
- directed energy deposition.

Vat polymerization was the first AM technology to be commercially introduced. In these types of processes, parts are built using a liquid resin that cures by exposure to a certain wavelength of light. A light source is used to selectively cure the resin, as driven by a computer model of the part. *Stereolithography* was the first commercially available form of this process, and is still the most widely used vat polymerization technology. In this process, a photopolymer, a liquid resin that cures under the application of certain wavelengths of light, is cured layer-by-layer by a precisely guided laser beam. The SLA process is illustrated in **Figure 13.1**. The part

FIGURE 13.1

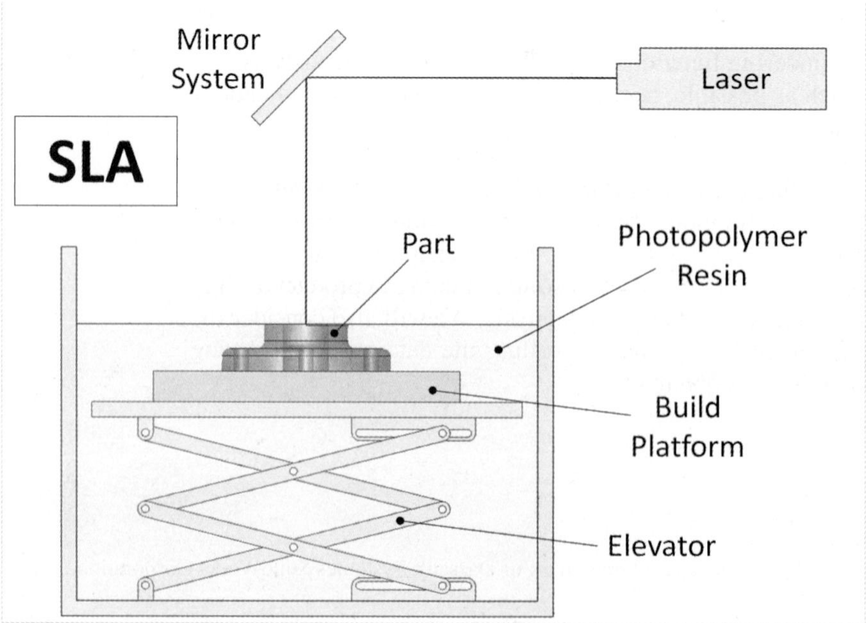

is submerged in a vat of photopolymer, supported on an elevator platform. The fabrication of each layer begins with the elevator lowering by the thickness of a layer, typically about 0.005 inches. The part is then covered by uncured resin. A mirror system directs the laser beam across the area to be cured for that particular cross section. Laser power and the speed of the beam across the resin surface are controlled to ensure a complete cure of the resin while maximizing the speed of the process. When all of the layers are completed, which is typically several hours after the beginning of the build, the elevator lifts the essentially finished part out of the resin. Cleaning the part with a solvent and curing in a UV oven to complete the cure of the photopolymer are usually the only steps necessary to finish the part. Other forms of vat polymerization, which use light projectors to cure the layers of material, are currently in production. These offer build speeds much faster than traditional beam-steered laser systems, as entire layers can be cured at one instant.

Powder bed fusion processes are those in which a fine powder of material is exposed to an energy source, which melts the material into a solid part. The most common example of this process is called Selective Laser Sintering (SLS). In this process, powder is *sintered*, or fused together, by the heat from a precisely guided laser beam. The SLS process is illustrated in **Figure 13.2**. As each layer is completed, the build platform drops by the thickness of one layer, and a fresh layer of powder is spread over the part and leveled. The laser is then directed onto the surface of the new powder layer, sintering the powder in the desired areas. A major advantage of this and other powder bed fusion processes is that many engineering materials, including nylon, polycarbonate, and various metals can be used. Another advantage of these processes is that for complex shapes, the loose powder supports the part as it is being built. In most other AM processes, supporting structures must be built for complex parts, and then removed in a post-processing operation.

FIGURE 13.2

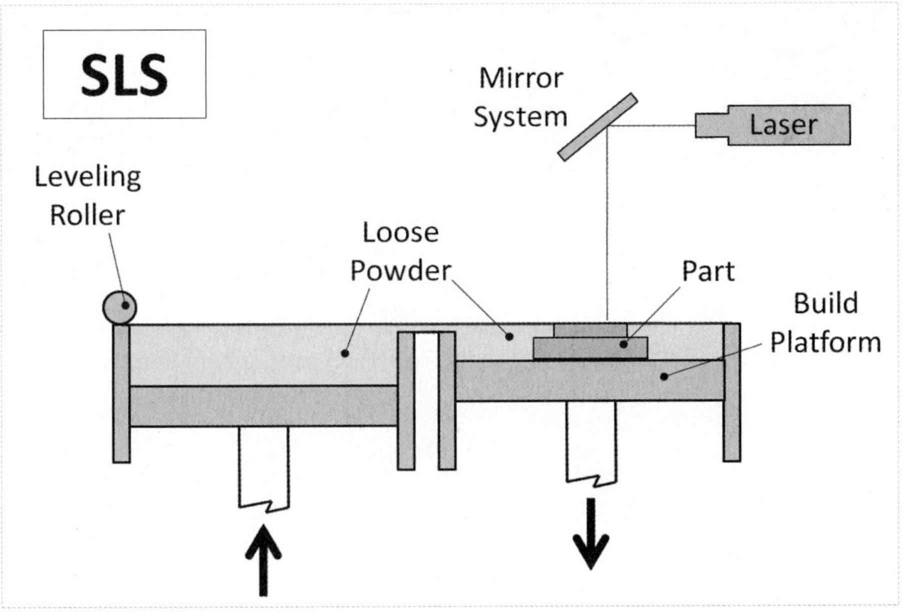

Material extrusion processes are perhaps the most commonly seen throughout industry, academic institutions, and even some homes today. The most common form of this technology is called Fused Deposition Modeling (FDM). The FDM process is illustrated in **Figure 13.3**. The FDM process can be thought of as precisely directing a fine hot-glue gun to build up a part. The thermoplastic material (such as ABS, polycarbonate, or PLA) is fed from a spool. In large industrial systems, there are actually two feed spools and extruder heads, as the material for building support structures is different from the part material. The support material is formulated to break away from the part easily, or to dissolve in water or another solvent. The introduction of FDM was especially significant in that no laser or hazardous materials were required, allowing the machine to be placed in an office environment (although the machines should be used in well-ventilated areas to minimize fumes from the molten plastic).

FIGURE 13.3

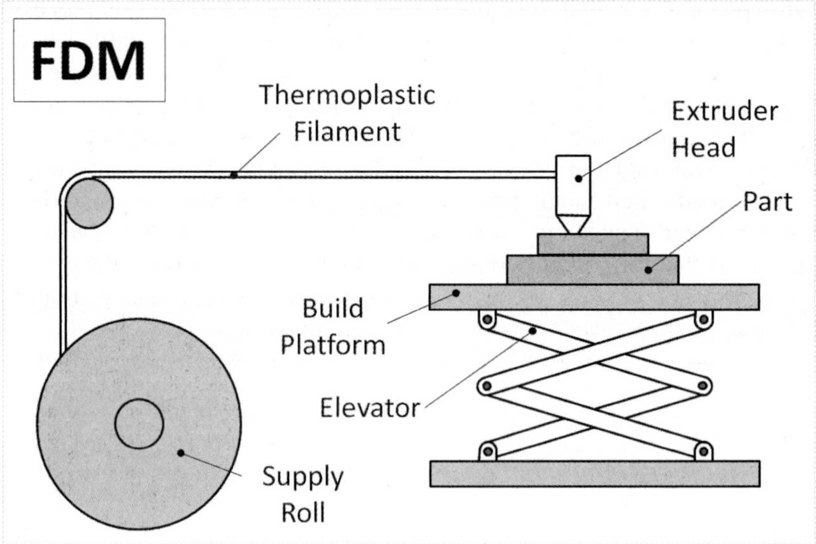

Binder jetting processes can be thought of as being similar to powder bed fusion processes, except that instead of using a laser to sinter the powder together, an inkjet printer head is used to spread binder fluid onto the surface of the part. A powder-based concept modeler is shown in **Figure 13.4**. Notice the cross section of the ribbed flange part from Chapter 5 being printed.

FIGURE 13.4

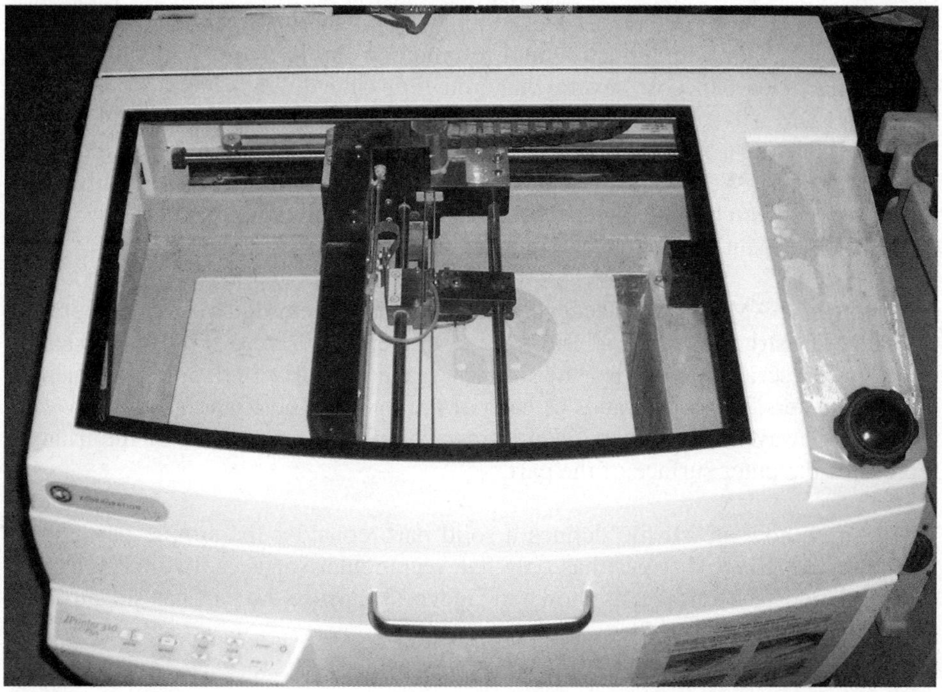

© William E. Howard

Material jetting processes are similar in concept to traditional ink jet printing, except drops of material are deposited in layers on an indexing platform rather than ink on paper. The layers are then cured, generally with exposure to a UV light source. This process has the advantage of being well-suited to production of parts from multiple different materials, as multiple print heads filled with different materials can be used to print on a single layer before curing.

In *sheet lamination processes*, the individual layers are cut to shape using some form of subtractive manufacturing, generally a laser or other computer-guided cutting system. These layers are then fused together into a solid part, using either adhesives or ultrasonic welding. The materials available range from paper to plastics to sheet metal.

Directed energy deposition is an emerging technology for additive manufacturing of metal parts. It uses a very high-powered energy source, such as an electron beam, to deposit and fuse together metal material provided in the form of a powder or a wire. In some sense, it can be thought of as producing a solid metal part through computer-controlled welding. These systems are very expensive to purchase and operate. They do produce very high quality functional metal parts.

Additive manufacturing techniques have long been used for prototyping of mechanical components, but as build speed has increased and material properties have improved these techniques are finding acceptance in the production of end-use components for both industrial and commercial applications. While production quality machines can cost several hundred thousand dollars, low-cost machines based on FDM and even SLA technology can be found for a few thousand dollars. Extremely low-cost machines based on FDM technology can be found for a few hundred dollars. Web-based businesses where hobbyists can upload solid model files and have them built using AM processes are becoming popular, and have made the term "3-D printing" a household term.

Most domestic AM machines accept as input a type of file called a stereolithography (.stl) file. The structure of an .stl file is quite simple: the surfaces of a solid model are broken into a series of triangles, the simplest planar area. Each triangle is defined by four parameters: the coordinates of each of the three corners, and a *normal vector* that points away from the part, identifying which of the two faces of the triangle represents the outer surface of the part.

To illustrate how an .stl file defines a solid part, consider the simple part shown in **Figure 13.5**. Since this part has only flat, rectangular surfaces, two triangles can exactly define each surface, as shown in **Figure 13.6**. Since the part has six surfaces, the part can be described by 12 triangles.

FIGURE 13.5

FIGURE 13.6

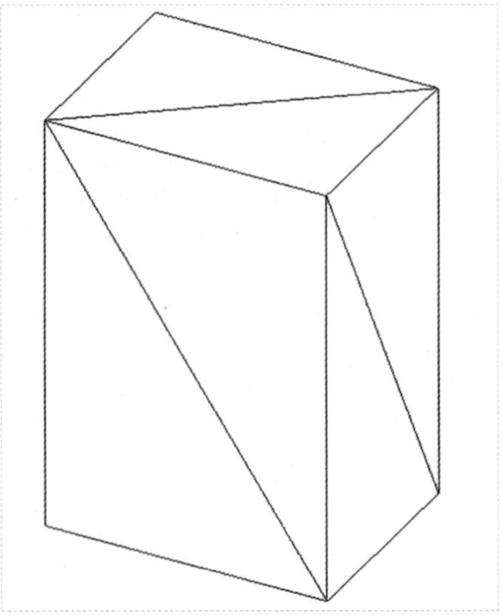

To illustrate how an .stl file is created for a part with curved surfaces, consider the ribbed flange part modeled in Chapter 5, which is shown in **Figure 13.7**.

To create an .stl file from a SOLIDWORKS part, select File: Save As from the main menu and select STL as the type of file from the pull-down menu, as shown in **Figure 13.8**.

FIGURE 13.7

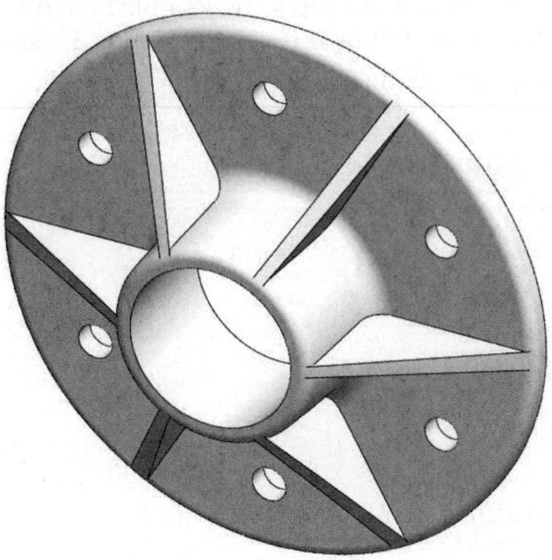

FIGURE 13.8

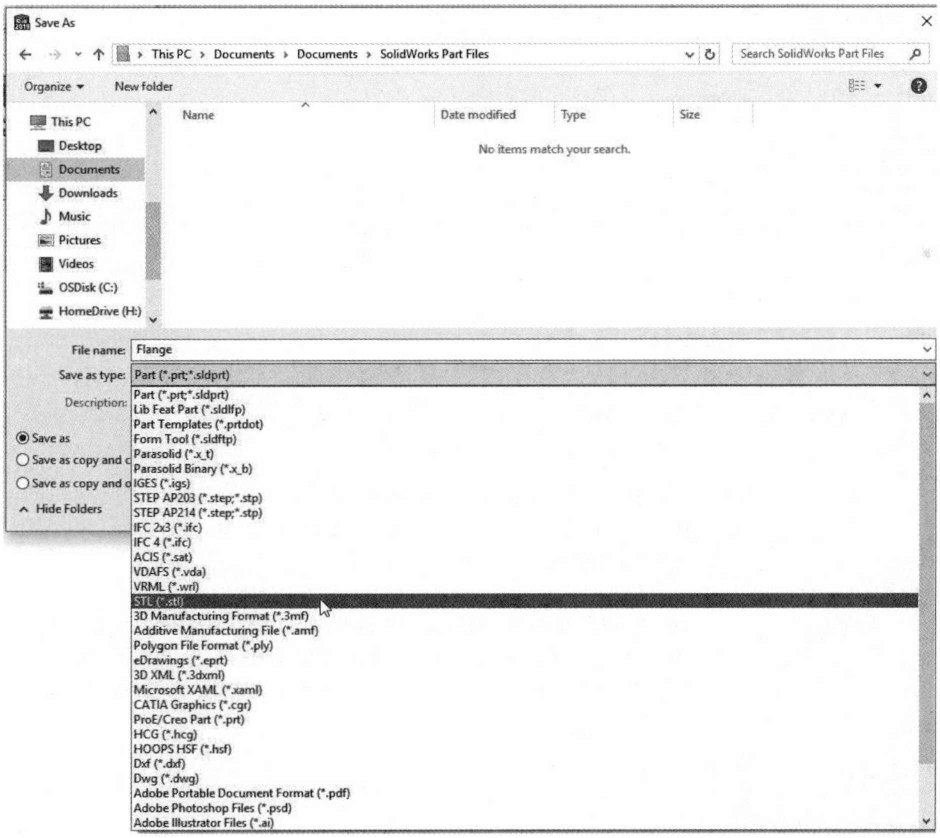

Before saving the file, select the Options button. This opens the Export Options dialog box, and allows you to adjust the resolution of the .stl file, as shown in **Figure 13.9**. If the Preview option is selected, then the triangles of the .stl file are shown displayed in the part window.

FIGURE 13.9

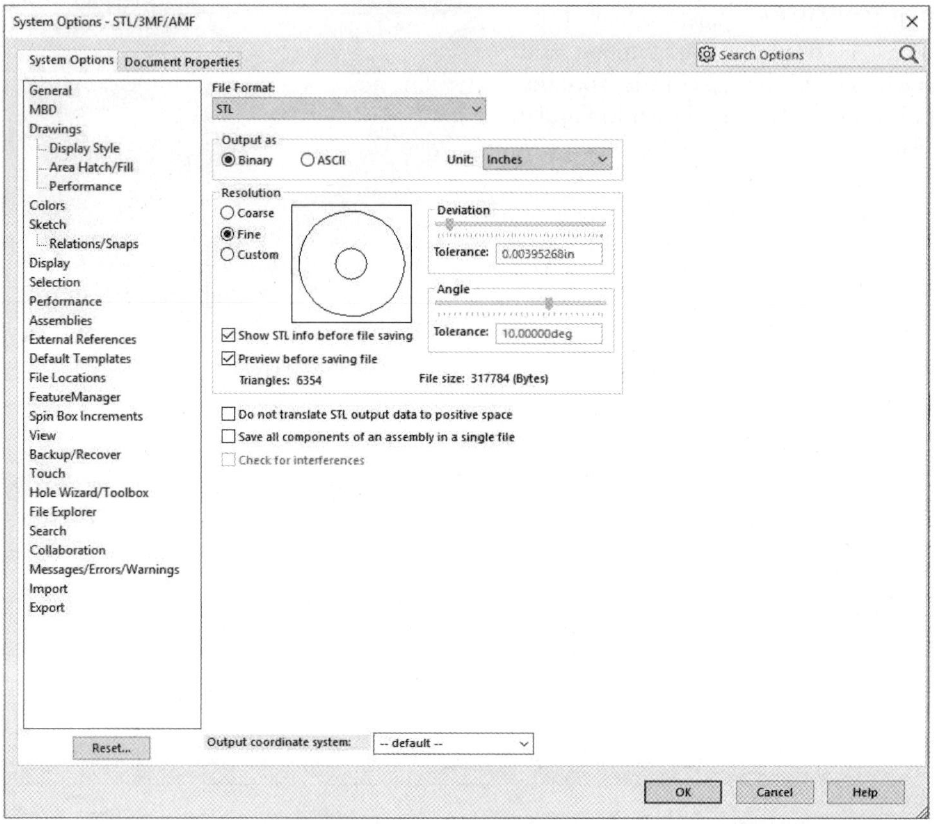

FIGURE 13.10

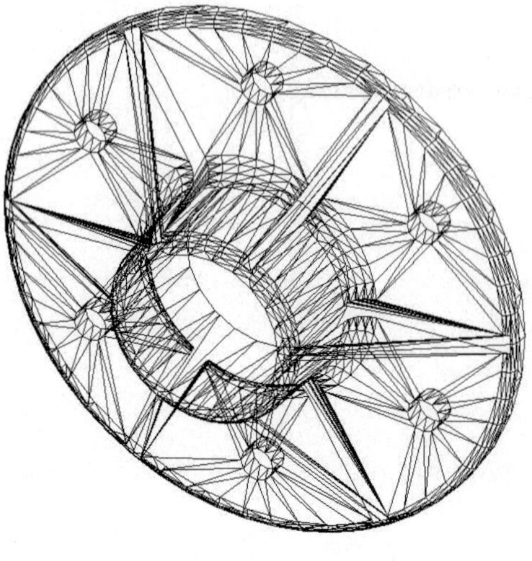

The triangles created for a resolution set to Coarse are shown in **Figure 13.10**. Note the rough approximation of the fillets by the triangles.

The number of triangles created and the file size are displayed in the Export dialog box. For the Coarse resolution, 2326 triangles are used, and the size of the .stl file is 116 kb.

If the resolution is changed to Fine, then many more triangles are added to the curved surfaces, resulting in a smoother approximation of the actual shape, as shown in **Figure 13.11**. In this example, 6354 triangles are created, resulting in a file size of 318 kb. This file size is small enough to be easily transferred by e-mail or temporary storage media.

FIGURE 13.11

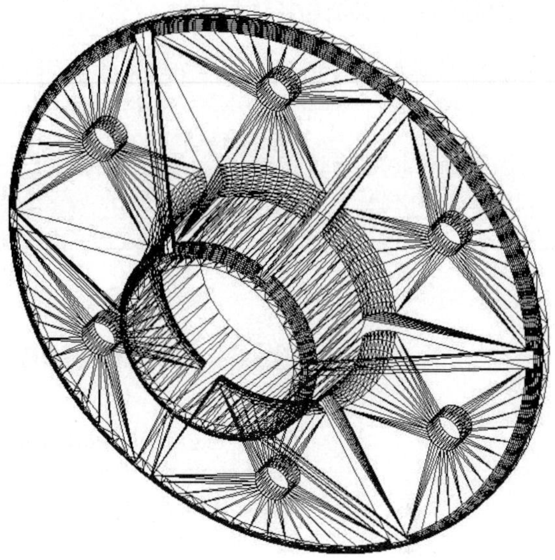

Usually, the Fine quality is sufficient. For parts where greater resolution is desired, the Custom quality can be selected, and the slider bars moved toward the right. The only trade-off for this higher quality is larger file sizes. However, since the transfer of large files via e-mail or Internet has become easier in recent years, file size is usually not a major problem.

When the .stl file is received by the computer controlling the AM machine, it is "sliced" into cross sections by machine-specific software. Some machines allow for the thickness of the sections to be adjusted. Thinner layers result in greater accuracy, but of course longer build times. Most machines allow for multiple parts to be made simultaneously.

A model of the ribbed flange being produced using two popular low-cost 3-D printing processes is shown in **Figures 13.12** and **13.13**. **Figure 13.12** shows the model being produced using a stereolithography process. **Figure 13.13** shows the model being printed using an FDM process. Both machines use the .stl file type described in this chapter.

FIGURE 13.12

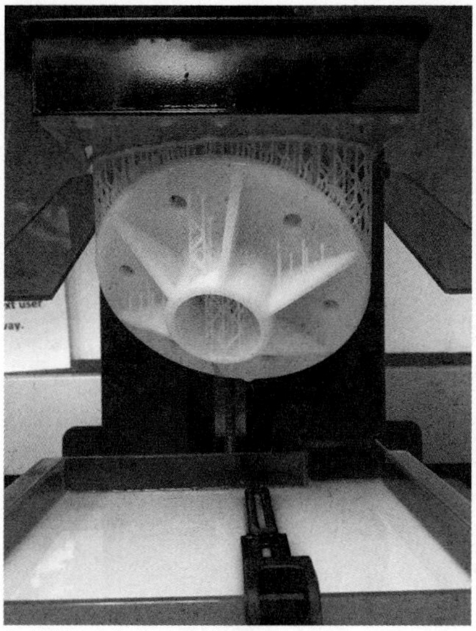

© William E. Howard

FIGURE 13.13

Courtesy of Angel Chukwu

VIDEO EXAMPLE 6

In this section, we introduced methods of 3-D printing, and described the process of preparing a file for 3-D printing.

An example of this process, used to generate the 3-D printed part shown in **Figures 13.12** and **13.13**, is shown in a video at the book website, *www.mhhe.com/howard2019*.

13.2 Finite Element Analysis

Finite Element Analysis (FEA) is a simulation tool that can help engineers predict the behavior of a part or assembly when subjected to its working environments. Typical FEA applications include structural analysis, thermal analysis (as illustrated by the temperature distribution plot on the cover of this book), and fluid flow analysis. FEA works by breaking up a part or assembly into a number of small pieces (*elements*) that are joined together at *nodes*. Equations governing the behavior of each element are assembled into a series of simultaneous equations. For structural analysis, the equations relate the applied forces to the displacements of the nodes. From these displacements, the *stresses* (forces per unit area) are calculated for each element. From the stresses, the factor of safety against failure of the structure can be predicted. The finite element method can also be applied to fluid flow analysis, heat transfer, and many other applications.

Solid modeling has enabled the increased usage of FEA, because it automates the task that in the past required the most time: creating the geometric model and the finite element mesh. A finite element mesh can be created from a solid model with only a few mouse clicks. **Figure 13.14** shows a mesh created from the bracket model of Chapter 3.

FIGURE 13.14

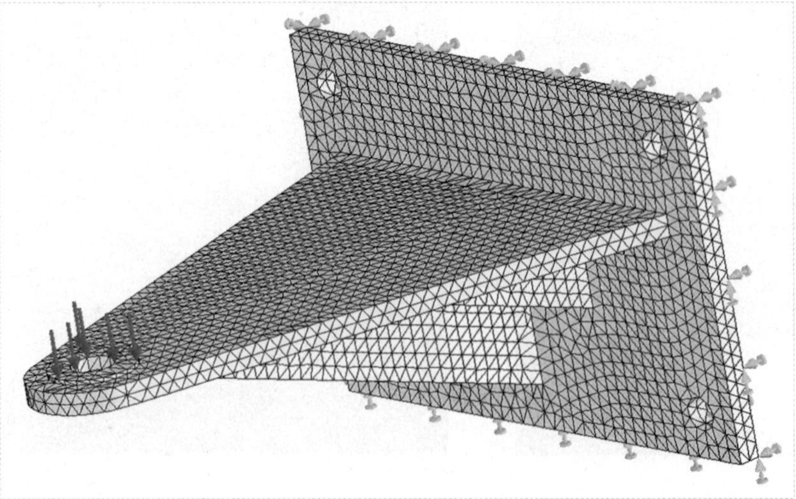

A 50-pound load will be applied to the end of the flange. After the mesh is created, the loads and *boundary conditions* are applied. Boundary conditions are the displacements controlled by external influences. For example, if we assume that the bolts attaching the bracket to the wall cause the back face of the bracket to be perfectly fixed, then we would apply corresponding constraints to the movements of the nodes on that face. The stresses resulting from the 50-pound load are shown as stress contours in **Figure 13.15**.

FIGURE 13.15

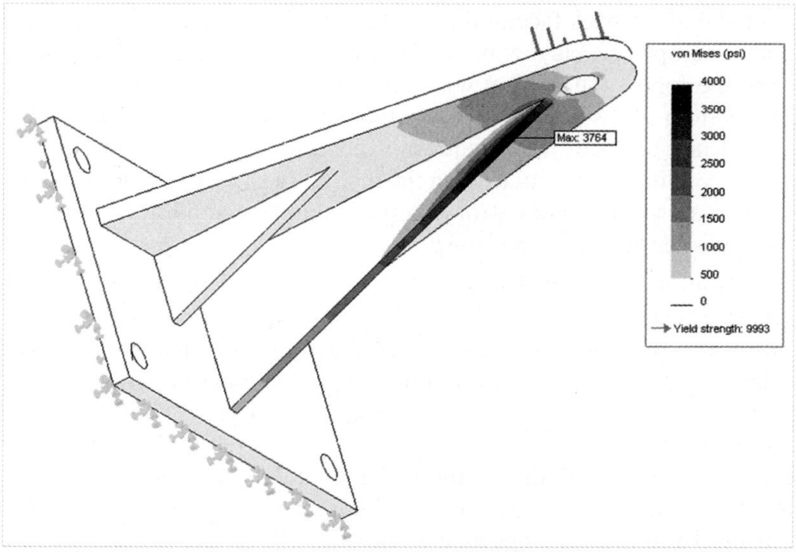

As FEA has become easier to use, it has also become easier to *misuse*. While the software takes care of the number-crunching, a knowledgeable engineer is required to set up the analysis and interpret the results.

Among the common errors made by inexperienced users of FEA are:

- *Inappropriate choice of elements.* Solid modeling software with a built-in FEA mesh generator allows a mesh of solid elements to be easily created from a solid model. However, in many cases solid elements are not the best choice. Relatively thin structures subjected to bending loads are usually better approximated with shell elements. A frame made up of welded structural members may require tens of thousands of solid elements to produce an acceptable grid, while a few beam elements will produce better results with substantially less calculation time.
- *Use of only linear behavior.* Linear analysis is based on the assumption that structural response varies linearly with loading. For example, under the assumption of linear behavior, the deflections and stresses of the bracket when loaded to 100 pounds are exactly double those produced by a 50-pound load. There are two types of nonlinearities possible, however, that would violate this assumption. Geometric nonlinearity is present when the structure's stiffness is significantly different in the displaced configuration than it is in the original configuration. Material nonlinearity is present when the material is stressed

beyond its yield point. When the material has yielded, it may not break, but will deflect further with little or no additional loading. Both of these nonlinearities require iterative solutions. That is, rather than applying all of the load at once, an increment of the load is applied and the equations are solved. The equations are then reformulated based on the results of the first solution, and solved again. This process is repeated until the solution is found. While this procedure requires much more processing time, it is necessary to obtain an accurate solution for many problems.

- *Inappropriate boundary conditions.* In the example of the analysis of the bracket, it was assumed that the back face of the bracket was fixed to a rigid wall. Is this a reasonable assumption? Maybe, but if the screws holding the bracket stretch slightly, then the top of the bracket will separate from the wall, placing more force on the bottom of the bracket. Most actual structural restraints are neither perfectly fixed nor perfectly free, requiring the engineer to use judgment in specifying the conditions to be placed on the model. In some cases, the analysis may be performed several times with different boundary conditions to obtain the limits of possible structural response.
- *Misinterpretation of results.* In **Figure 13.15**, the stresses displayed are the von Mises stresses. These stresses are calculated based on a specific failure criterion (the von Mises criterion) that is an excellent predictor of the yielding of ductile materials, but may be inappropriate for predicting the failure of many materials.

These common errors are not presented to discourage the use of FEA, but rather to encourage the proper use of the method. An engineer should have a good understanding of mechanics of materials and at least an introduction to FEA theory before using FEA for any important application. A tutorial for conducting a simple analysis of the bracket from Chapter 3 is available on the book's website, *www.mhhe.com/howard2019.*

13.3 Product Data Management

We have mentioned earlier that solid modeling supports concurrent engineering, since many engineering functions can work from the same database (the solid model). Allowing multiple users to work on the same model or drawing, however, creates challenges, as well. How do companies manage their part files and drawings to control who can make changes?

Before answering this question, it is helpful to consider how companies managed paper drawings and other data before solid modeling was introduced. When a draftsman completed a drawing, it was checked and then went through a *release* process in which it was approved and signed off by various department representatives (design engineering, manufacturing engineering, safety, etc.). Copies were made and distributed, and then the released drawing was stored in a vault. Most medium-sized and large companies had a *configuration management* department that controlled the release process and subsequent changes. When a change was requested, a formal engineering change order process was followed. When the change was approved, it was amended to the drawing, or the drawing was modified and given a revision number, if the change was significant. It should be noted that many drawings would eventually have dozens of associated change orders, so managing this process was an important job. When a user of the drawing, say a stress analyst about to begin an FEA model, needed a copy, the configuration management department provided the latest version, including all change orders. Similarly, the Bill of Materials for each part was maintained and provided to the purchasing and manufacturing departments. In addition to controlling the drawing changes, the configuration management department would maintain *drawing trees* that showed how drawings related to each other.

A change in a part could affect an assembly at the next level. Other data might have been controlled but not necessarily linked to the drawings. For example, a stress analysis report might have been released and stored in the vault, but a person looking at the drawing would probably not know that the analysis had been done, or where to find it. Similarly, in industries where weight is important, mass property reports were produced, but usually not linked to drawings.

The introduction of solid modeling presented many new challenges to configuration management. Some of the features that make solid modeling so exciting to design engineers—associativity between parts, assemblies, and drawings, ability to use the solid model for many functions—could cause huge problems from a data management perspective. That is, one person could make a seemingly simple change to a component and that change would propagate throughout an entire assembly without the person being aware of its effect. Now that some companies model major systems with solid modeling, controlling who can make changes is vitally important.

Product Data Management (PDM) software is used to perform some of the tasks of the configuration management department, while streamlining communications between various departments. PDM entails two broad categories of functions:

- Data management, the control of documents (part files, drawings, stress reports, etc.), and
- Process management, the control of the way in which people create and modify the documents.

The data management function is similar to that of the paper-based configuration management department, except that the released drawings are now electronic files and are stored in a *virtual* vault instead of a physical vault. (Actually, backup tapes and disks of the virtual vault are often stored in a fireproof physical vault.) Since the part and drawing information is stored in a relational database, immediate location and retrieval of files is possible. This helps to reduce redundancy, especially among standard components such as rings and fasteners. A design engineer who needs to specify a fastener can easily determine if there is already a similar fastener with a part number assigned in the system. If there is, then the purchasing department will not have an additional item to buy.

Process management is the control of active procedures: who generates the data and how the data are transferred from one group to another. One important feature of PDM systems is work history management. In a paper-based system, having old, outdated drawings around is an invitation for trouble, since they can be used by mistake. But by destroying old drawings, the history of the modifications made to that drawing can be easily lost as well. PDM software can track the change history of a part, an important function in Total Quality Management (TQM), while protecting against accidental usage.

What does the future hold? Most manufacturing companies implemented Manufacturing Resource Planning (MRP) systems long before PDM became popular. MRP systems allow the tracking of raw materials, work in progress, and finished inventory in the plant. The move to reduce inventories and adopt just-in-time raw materials deliveries necessitated the adoption of MRP systems. In many ways, implementing PDM to engineering functions is analogous to implementing MRP for manufacturing.

The logical next step is to integrate all product development data into a single system referred to as Product Lifecycle Management (PLM). As with any new large-scale system, implementation is expensive and time-consuming, so these systems are not yet widely used in industry.

An example of the complexity of engineering data and the challenges of managing that data is the delay announced in 2006 of the production of the Airbus 380 jumbo jet. Both Airbus and its competitor in the commercial aircraft market, Boeing, have suppliers around the world. Coordinating the work of these suppliers is a monumental challenge. On October 3, 2006, Airbus CEO Christian Streiff announced that delivery of the 550-seat A380 would be significantly delayed because of data translation problems between engineers in Germany, who were designing and building the wiring harness for the plane, and engineers in France, where the final assembly of the plane was taking place. As a result of the errors, the wiring harnesses would not fit correctly—a major problem for a plane with hundreds of miles of wiring. The problems were expected to cost Airbus over $6 billion in profits.[1]

13.4 Some Final Thoughts

The widespread adoption of solid modeling has been part of a revolution in the way that products are designed and developed. As many companies have thrived with new technology, others that have not kept up have not been able to compete and have been forced to close. Although no one can predict the future, one thing seems certain: the pressure on companies to develop new products faster and better will not lessen.

On a more personal scale, the same concept has held true for many engineers. An engineer who is adverse to change and reluctant to learn new tools is at a great competitive disadvantage to his or her peers.

The good news is that most people who enter the engineering profession do so because they have the curiosity to want to learn new things. Keeping that curiosity alive will allow an engineer to have a rewarding career that is always fresh and interesting.

[1] "PLM: Boeing's Dream, Airbus' Nightmare," Mel Duvall and Doug Bartholomew, *Baseline*, February 1, 2007, Vol. I number 69, Ziff Davis Media.

APPENDIX A

Recommended Settings

Recommended changes to the default settings of the SOLIDWORKS® program are summarized in this appendix. These changes are introduced at different points in the text, but some users might prefer to change all of the settings at one time or to apply settings for a new installation of SOLIDWORKS without going back through the early chapters of the book. For those users, step-by-step instructions for changing and saving the recommended settings are presented here, and a list summarizing the recommended settings is shown at the end of this appendix.

A.1 System Settings

FIGURE A.1

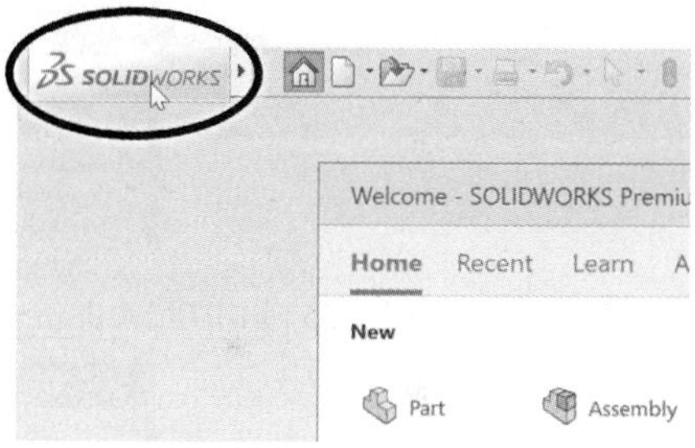

Open a new SOLIDWORKS session. By default, the Welcome window will appear, from which we will usually specify a new document or select an existing document. The Welcome window can be dismissed by pressing the Esc key or by selecting an icon or menu item. Click on the SOLIDWORKS logo in the upper-left corner of the screen, as shown in Figure A.1. This will cause the Main Menu to be displayed. Click the push pin icon, as shown in Figure A.2, so that the menu is always displayed.

FIGURE A.2

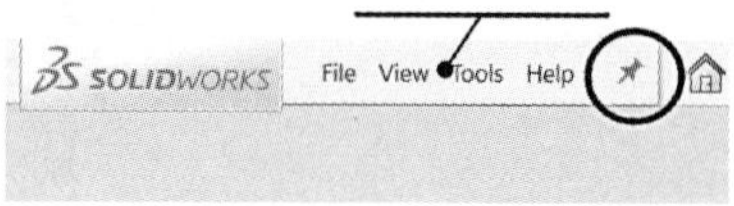

Choose the Options Tool, as shown in Figure A.3. Under System Options: Drawings: Display Style, select Hidden lines visible as the display style for new views and Removed as the option for tangent edges in new views, as shown in Figure A.4. Under Colors, choose

FIGURE A.3

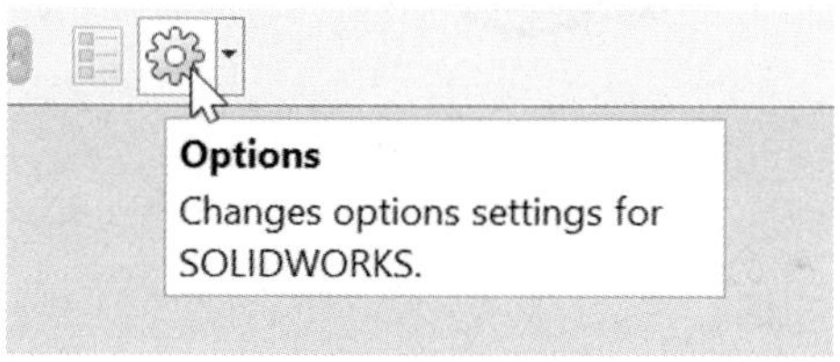

FIGURE A.4

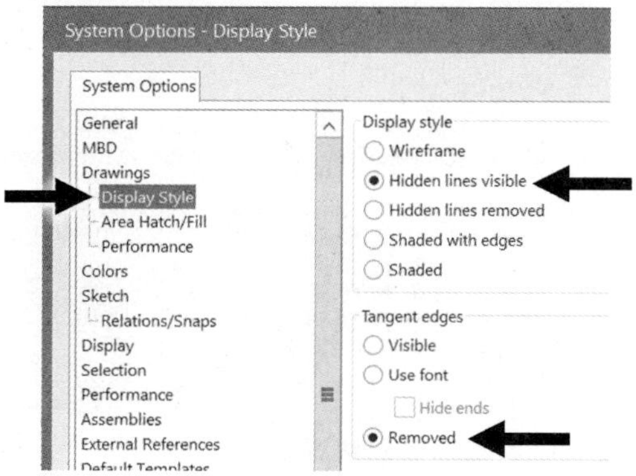

SOLIDWORKS is a registered trademark of Dassault Systémes SolidWorks Corporation.

VIDEO EXAMPLE 7

The settings covered in this appendix are demonstrated in a video on the book's website: *www.mhhe.com/howard2019.*

Classic as the Icon color and Green Highlight as the color scheme, as shown in Figure A.5. While still in the Colors options, browse to the color setting for Drawings, Paper Color, and select Edit, as shown in Figure A.6. Choose white as the paper color, as shown in Figure A.7, and click OK. (Note that the Current color scheme box will be blank after making this change.) Under Display, select

FIGURE A.5

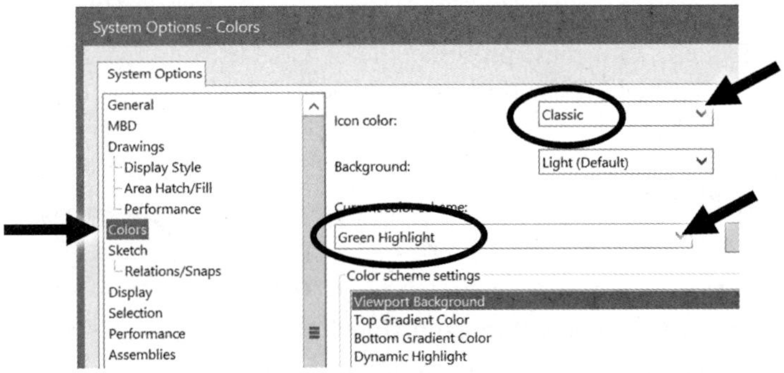

FIGURE A.6

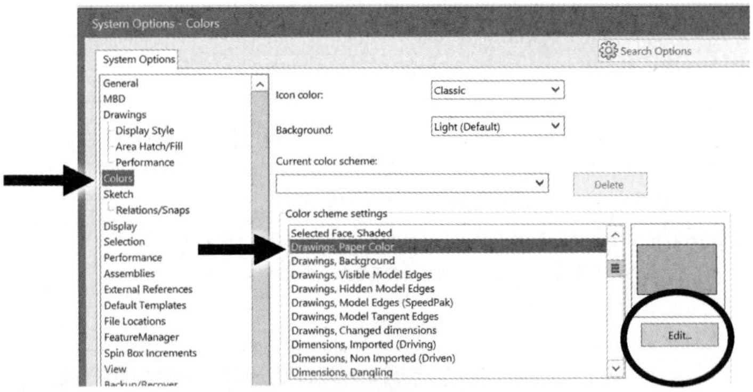

FIGURE A.7

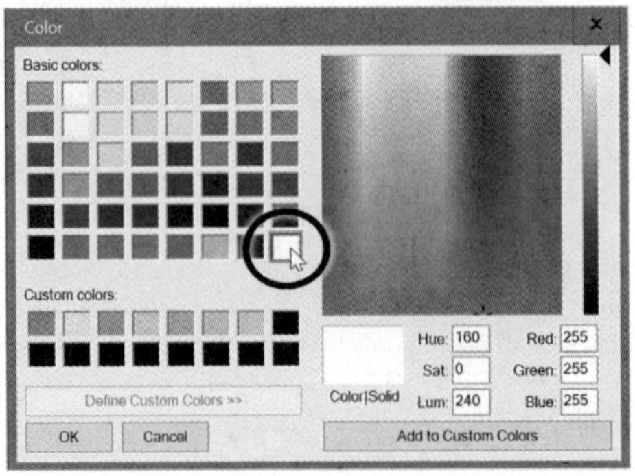

Removed as the option for part/assembly tangent edge display, as shown in Figure A.8. Also, near the bottom of the dialog box, change the Projection Type for the four view viewport to Third Angle, as shown in Figure A.9. Click OK to close the Options window.

FIGURE A.8

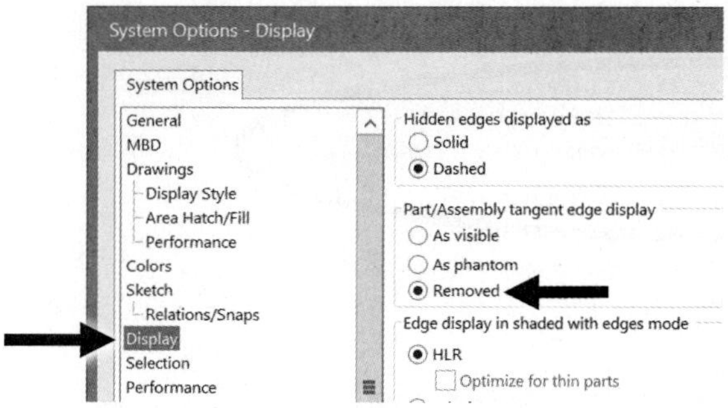

FIGURE A.9

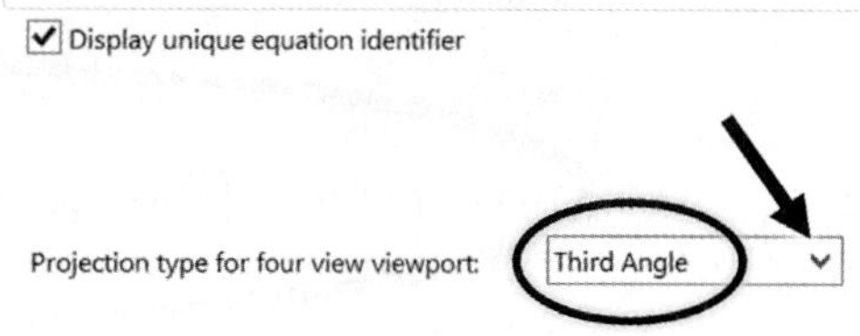

All of the changes made so far have been under System Options. These changes will be effective for all new documents. In the following sections, we will make changes under Document Properties, which apply only to the current document that is open, and learn how to save those changes so that they apply whenever we open a new part, drawing, or assembly file.

A.2 Part Settings

Choose the New Document Tool, as shown in Figure A.10. If this is the first time that you have opened SOLIDWORKS, then you will be prompted to choose a unit system and drafting standard. Select IPS and ANSI, as shown in Figure A.11, and click OK. Choose Part, as shown in Figure A.12, and click OK.

FIGURE A.10

FIGURE A.11

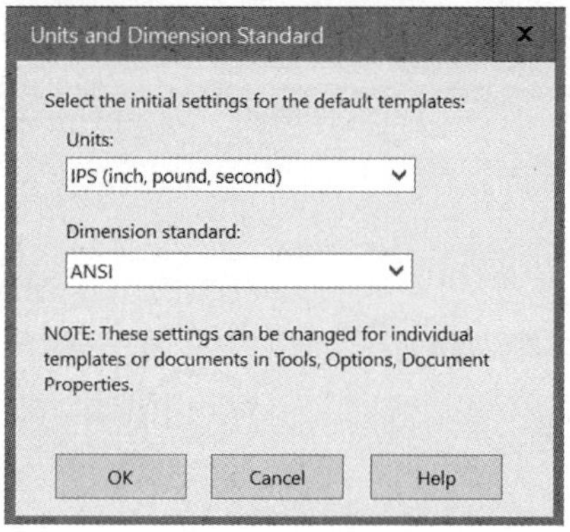

FIGURE A.12

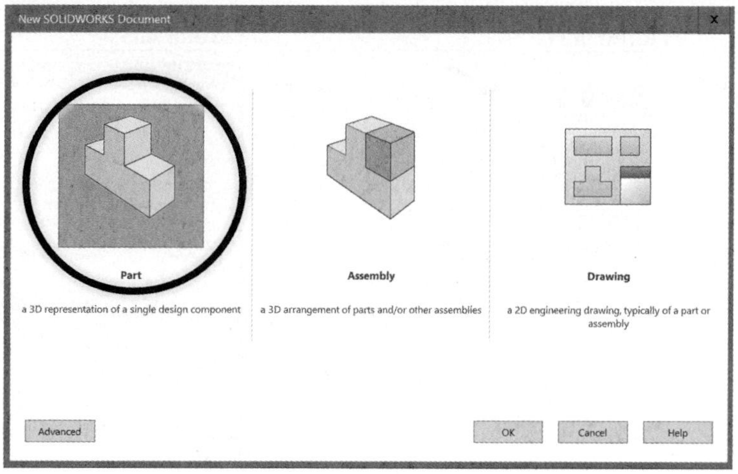

Right-click any tab of the CommandManager. In the list of CommandManager tabs, click to clear all of the tabs except for Features and Sketch, as shown in Figure A.13.

FIGURE A.13

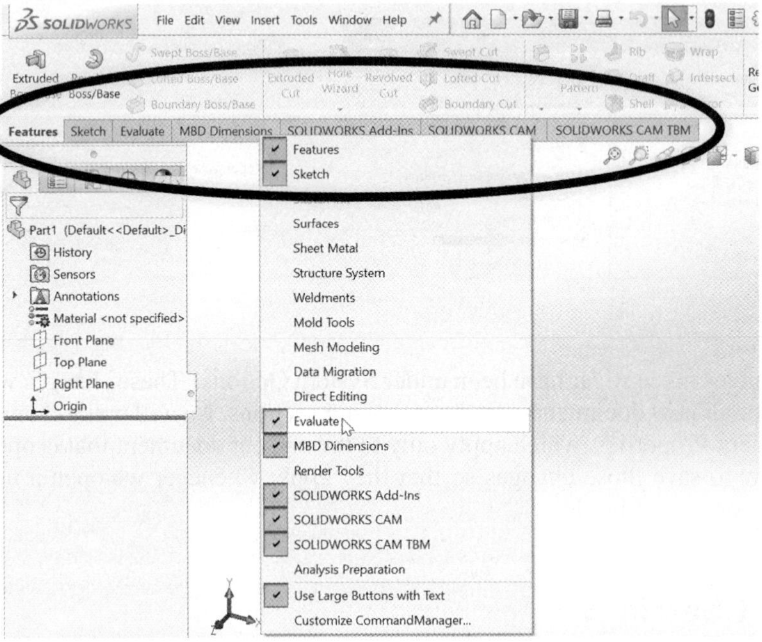

Select Customize from the pull-down menu beside the Options Tool, as shown in Figure A.14. Clear the box labeled "Show in shortcut menu," as shown in Figure A.15. If you check the box labeled "Lock CommandManager and toolbars," then accidental rearrangement of the user interface is prevented. You can also change the size of icons. Larger icons may be preferred, depending on the size of your screen. Select the Commands tab. Locate the Rotate View Tool from the View

FIGURE A.14

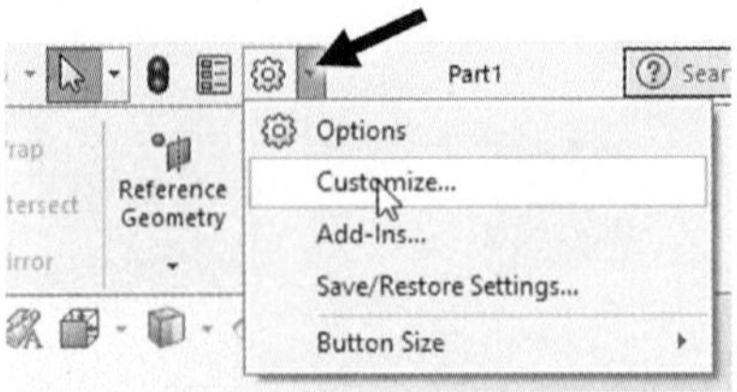

FIGURE A.15

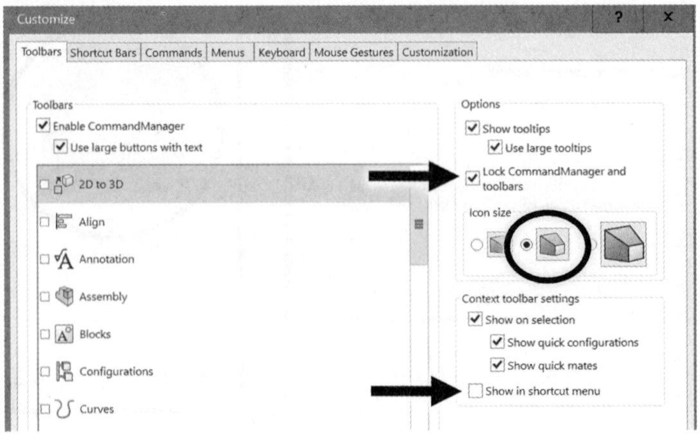

group, as shown in Figure A.16, and click and drag it onto the Heads-Up View Toolbar, as shown in Figure A.17. Repeat for the Pan Tool, also found in the View group, as shown in Figure A.18, and the Trimetric View Tool, found in the Standard Views group, as shown in Figure A.19. Click OK to close the Customize window.

FIGURE A.16

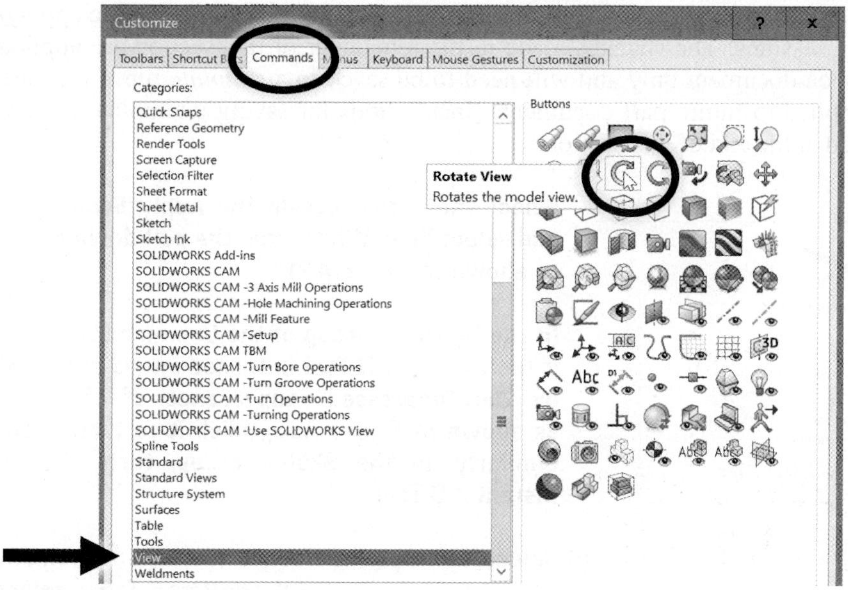

FIGURE A.17

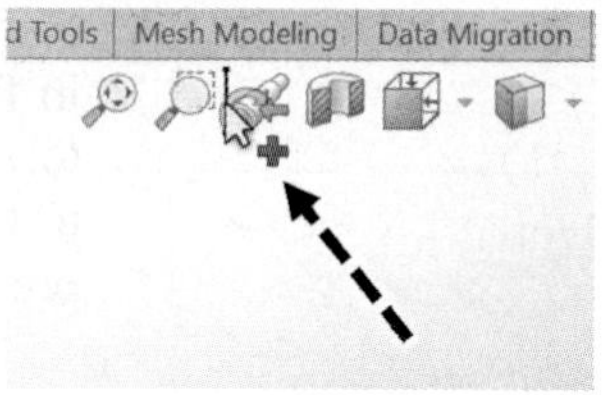

FIGURE A.18

FIGURE A.19

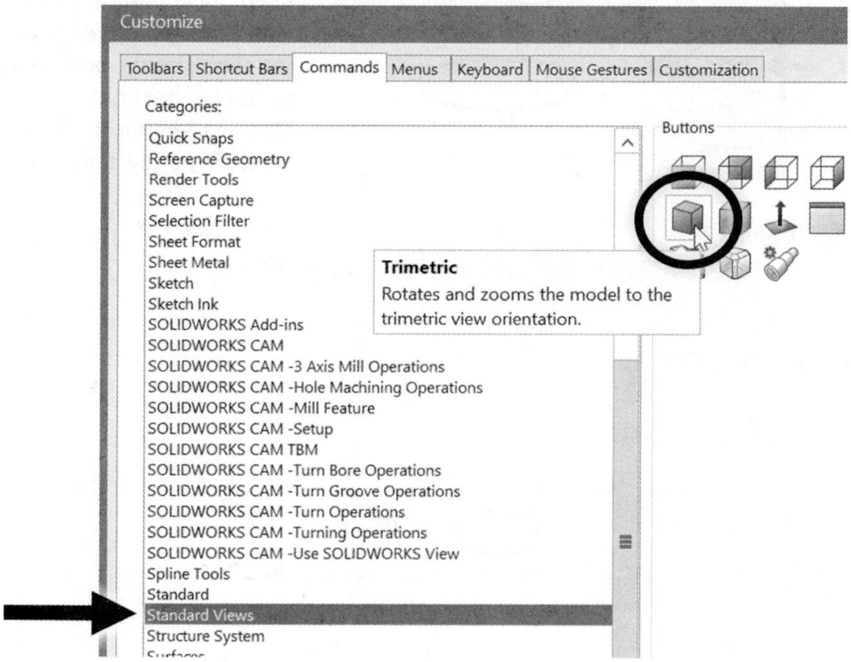

The Heads-Up View Toolbar should now have the tools shown in **Figure A.20**. (The locations of the added tools are not critical.)

FIGURE A.20

Note that all of the settings made so far are automatically saved and will be applied in future sessions. The changes made in the remainder of this section are applied to the open document only and will need to be saved in a *template* file if they are to be applied to future part documents. Instructions for saving a template file are presented at the end of this section.

FIGURE A.21

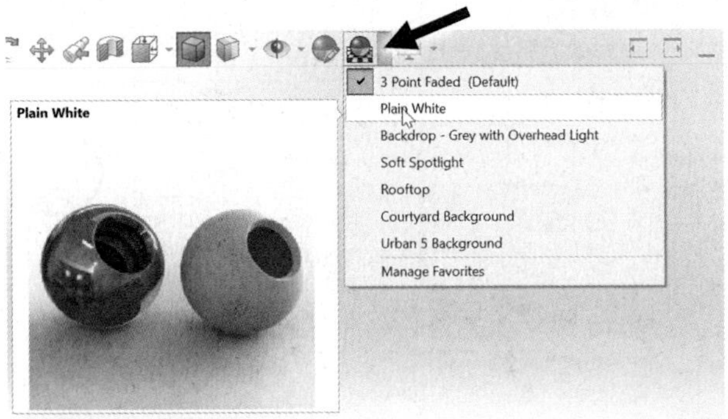

Select the arrow beside the Apply Scene Tool and select Plain White from the pull-down menu, as shown in Figure A.21.

In the Features Group of the CommandManager, if the Instant 3-D Tool is turned on (as indicated by the "depressed" appearance of the tool, as shown in Figure A.22), click it to turn it off. Similarly, in the Sketch group, turn off the Instant 2-D Tool.

Select the Options Tool. Under the Document Properties Tab, select Drafting Standard and set the overall standard to ANSI, as shown in Figure A.23. Under Dimensions, set the Primary precision to three decimal places (.123), as shown in Figure A.24. (If you see a message

FIGURE A.22

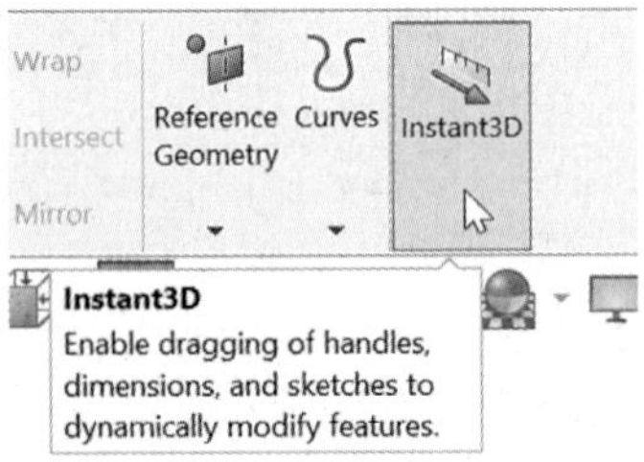

FIGURE A.23

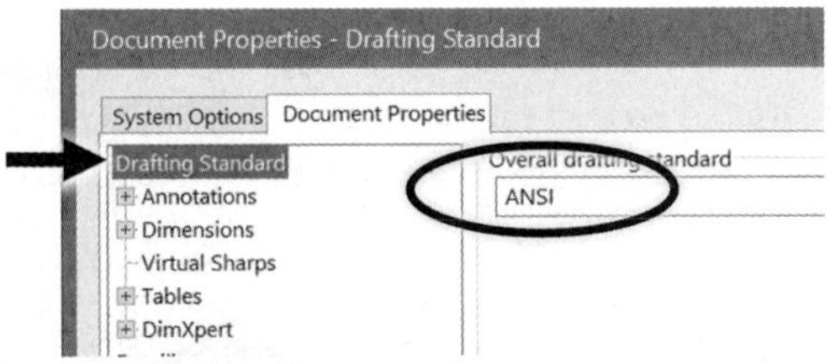

FIGURE A.24

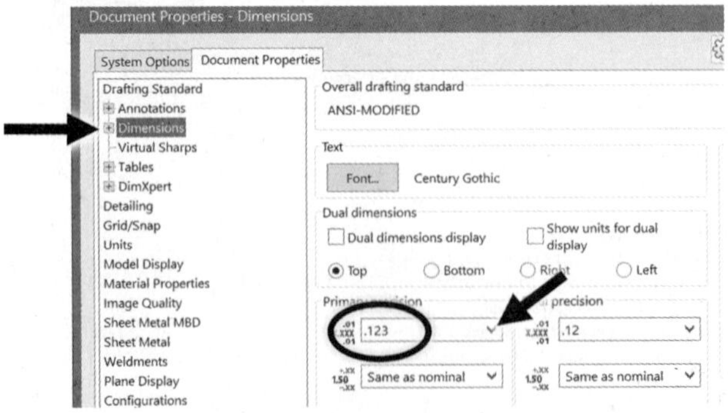

that the drafting standard has been changed to ANSI-Modified, ignore it.) Under Grid/Snap, click to check the box labeled "Display grid," as shown in Figure A.25. Under Units, set the Unit system to IPS, the decimals for length units to .123, and the decimals for angles to None, as shown in Figure A.26. Click OK to close the Options window.

As noted earlier, these last changes apply only to the open part document. To save these settings for future use, they must be stored in a template.

From the Main Menu, select File: Save As. Change the file type to Part Templates, as shown in Figure A.27. The file directory will automatically change to the one where the templates are stored. Click on the file named Part to select it, and click Save. You will be prompted to confirm that you are overwriting an existing file; click OK.

FIGURE A.25

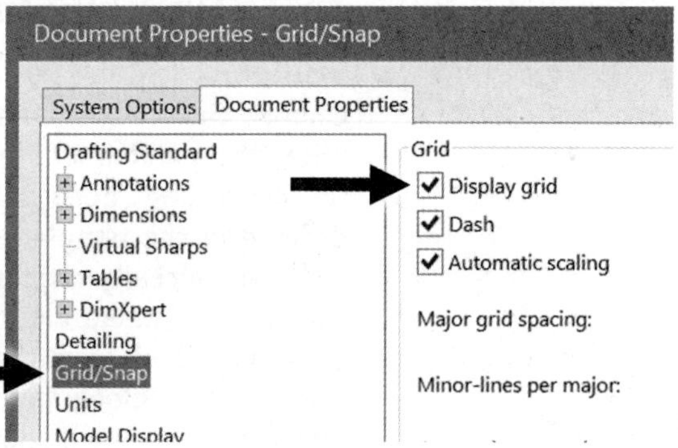

FIGURE A.26

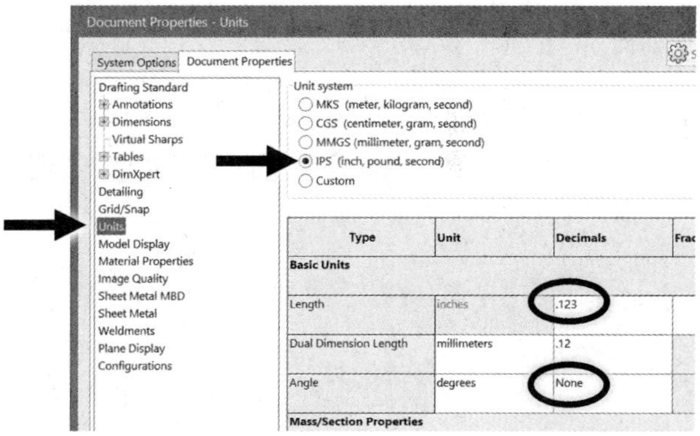

FIGURE A.27

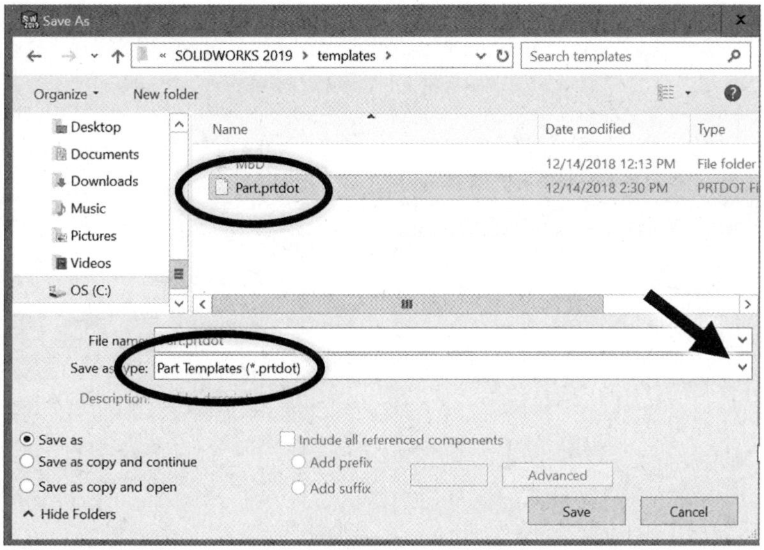

A.3 Drawing Settings

Choose the New Document Tool, choose Drawing, as shown in Figure A.28, and click OK. A dialog will prompt you to select a sheet size, as shown in Figure A.29. (If this prompt does not appear, then there is already a sheet size and format defined in the template, and you will have an opportunity to change it later.) If you have already created a title block that you want to use for most drawings that you will make, clear the box labeled "Only show standard formats" and select your title block from the list, checking the box labeled "Display Sheet format" so that the title block appears on the drawing. Otherwise, choose the A-Landscape sheet (8-1/2 × 11-inch sheet oriented with the long side horizontal), clear the check box labeled "Display sheet format," and click OK.

If the Model View Command opens, click the X to close it, as shown in **Figure A.30**.

FIGURE A.28

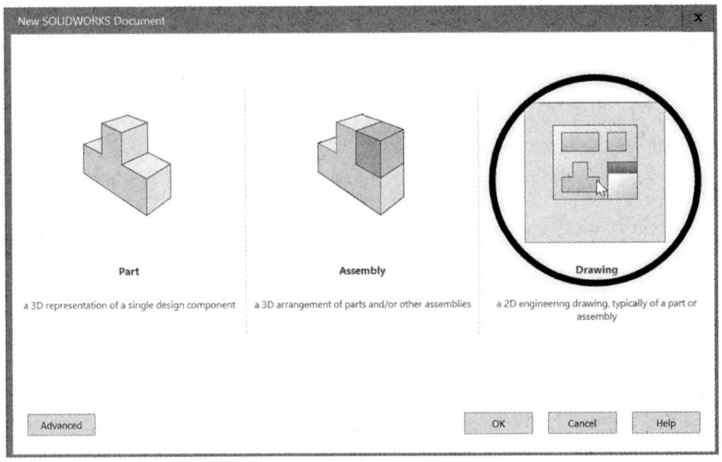

FIGURE A.29

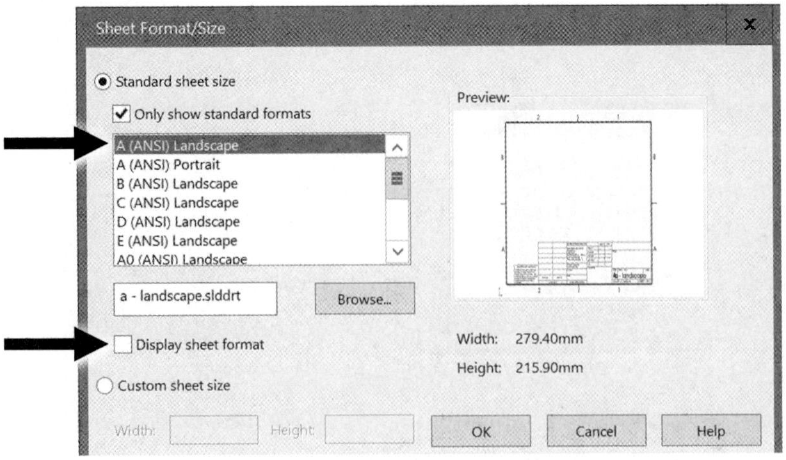

FIGURE A.30

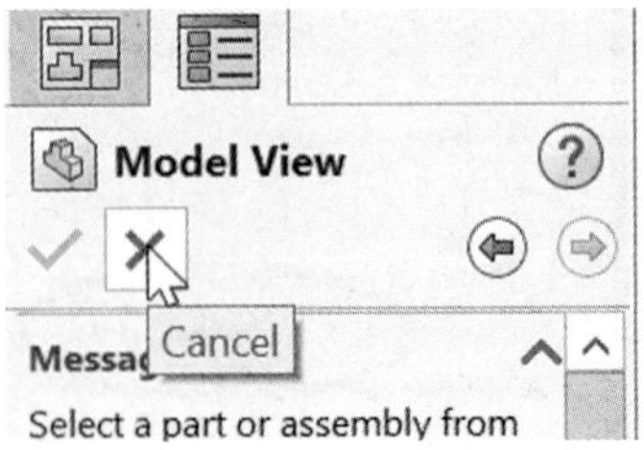

Right-click any tab of the CommandManager. In the list of CommandManager tabs, click to clear all of the tabs except for View Layout, Annotation, and Sketch.

Select the Options Tool. Under the Document Properties Tab, select Drafting Standard and set the overall standard to ANSI. Under Dimensions, set the Primary precision to two decimal places (.12). Under Detailing, check the box for auto-insertion of Centerlines, as shown in Figure A.31. Under Units, set the Unit system to IPS, the decimals for length units to .12, and the decimals for angles to None. Click OK to close the Options window.

FIGURE A.31

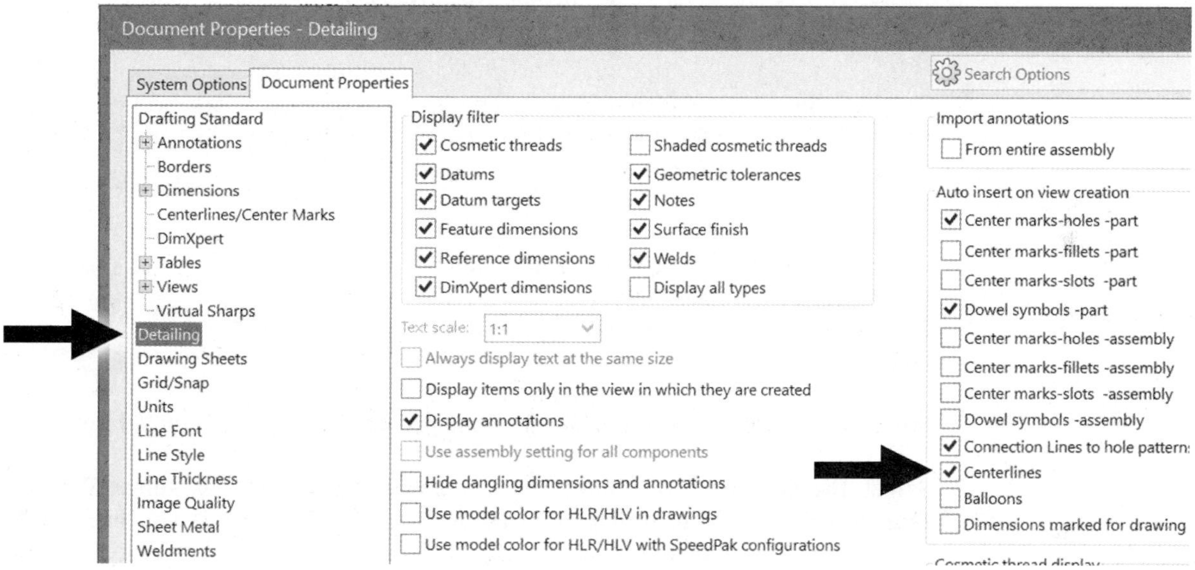

Right-click in the drawing area and choose Properties. Note that you may need to click on the double-arrow at the bottom of the menu to display the Properties option. To show the Properties option whenever you choose the menu, choose Customize Menu and check Properties. Set the projection type to Third Angle, as shown in Figure A.32. If you were not prompted for a sheet size earlier and wish to specify a custom title block, you may now choose it from the list. Click Apply Changes (or Cancel if you made no changes) to close the Properties window.

FIGURE A.32

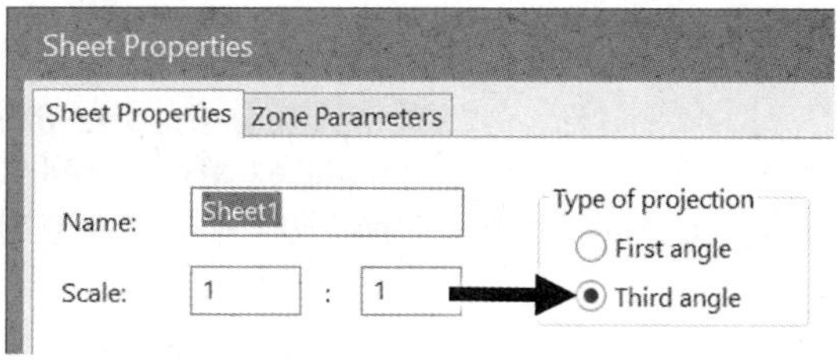

Before saving the settings in a drawing template file, note that if you chose a sheet format other than a plain sheet, or if a sheet format had been stored earlier in the default template, then the sheet format is stored in the FeatureManager as Sheet Format1 under the Sheet1 entry, as shown in Figure A.33. If you save the template with this entry, then the specified sheet format will be loaded for each new drawing. If you delete the Sheet Format1 entry, then you will be prompted to specify a sheet size every time a new drawing is created.

FIGURE A.33

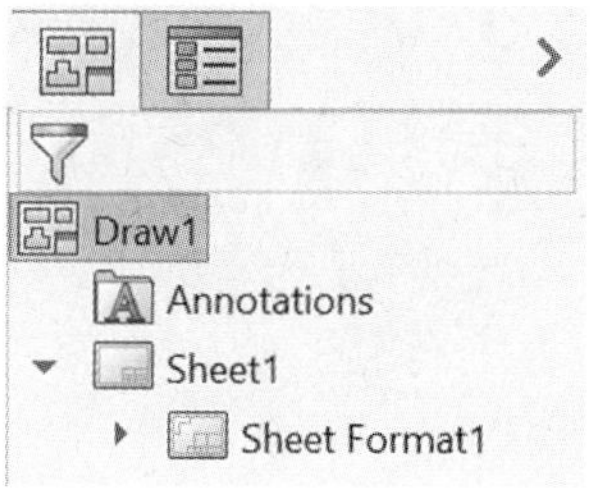

If you want to be prompted to select a sheet size for every new drawing, then delete the Sheet Format1 from the FeatureManager.

From the Main Menu, select File: Save As. Change the file type to Drawing Templates. The file directory will automatically change to the one where the templates are stored. Click on the file named Drawing to select it, and click Save. You will be prompted to confirm that you are overwriting an existing file; click OK.

A.4 Assembly Settings

Choose the New Document Tool, choose Assembly, and click OK. If the Begin Assembly Command opens, click the X to close it, as shown in Figure A.34.

FIGURE A.34

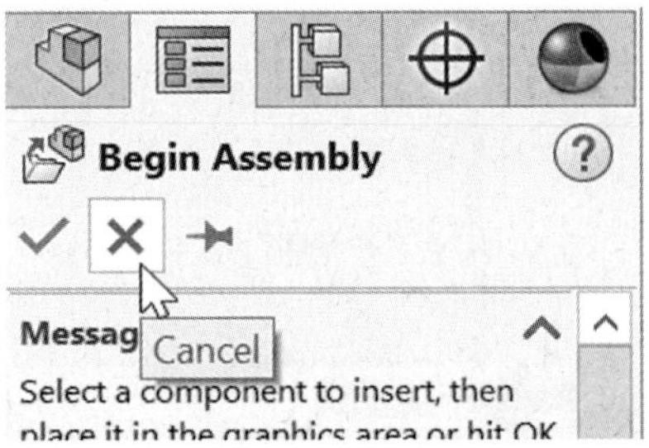

Right-click any tab of the CommandManager. In the list of CommandManager tabs, click to clear all of the tabs except for Assembly and Sketch.

From the pull-down menu beside the Apply Scene Tool, select Plain White.

Select the Options Tool. Under the Document Properties Tab, select Drafting Standard and set the overall standard to ANSI. Under Dimensions, set the Primary precision to three decimal places (.123). Under Units, set the Unit system to IPS, the decimals for length units to .123, and the decimals for angles to None. Click OK to close the Options window.

From the Main Menu, select File: Save As. Change the file type to Assembly Templates. The file directory will automatically change to the one where the templates are stored. Click on the file named Assembly to select it, and click Save. You will be prompted to confirm that you are overwriting an existing file; click OK.

The next time that you save a document, the default directory will be the one where the templates are saved. Make sure to change the directory to the one where you want to save the file.

Your computer is now set to match the configuration used in the book. If you would like to back up these settings or copy them to another computer, go on to the next section.

A.5 Backing Up and Transferring Settings

The settings that you have specified and stored can be easily copied for backup purposes or to transfer the settings to another computer. Except for the document-specific settings that were stored in the templates, settings are stored in the Windows registry. While editing the registry directly is not recommended, there is a tool for copying settings. To access this tool, select the Start Menu in Windows, browse to the SOLIDWORKS Tools 2019 folder, and select Copy Settings Wizard 2019, as shown in **Figure A.35**. The wizard will prompt you to save your settings, as shown in **Figure A.36**. After specifying the location to store the settings file and choosing the settings to save, as shown in **Figure A.37**, an executable file

FIGURE A.35

FIGURE A.36

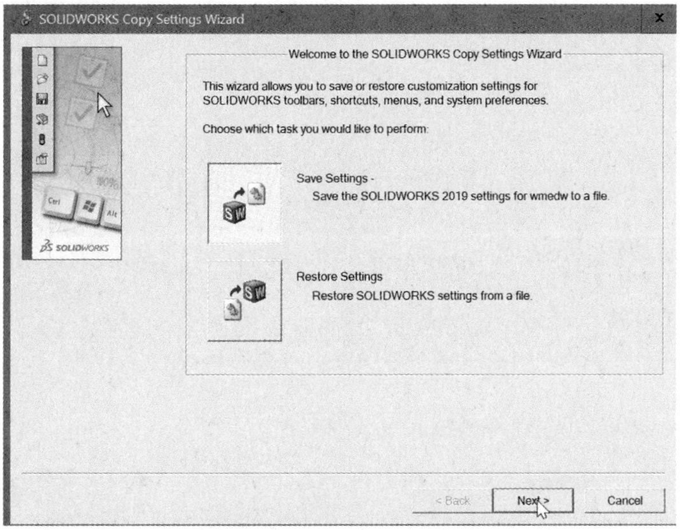

FIGURE A.37

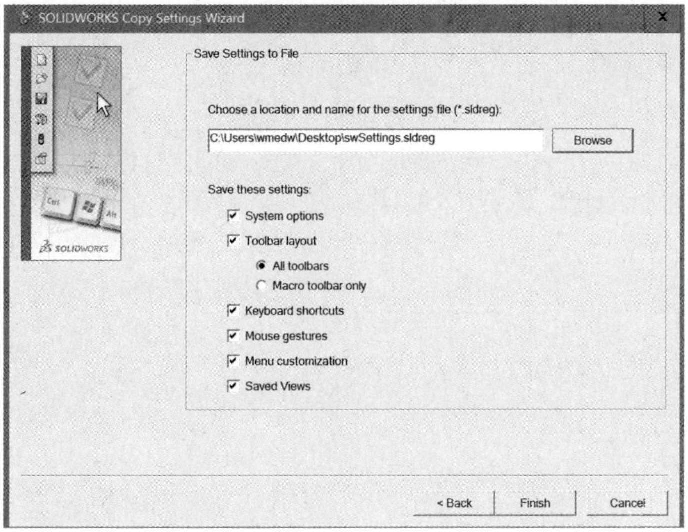

FIGURE A.38

is made, the icon for which is shown in **Figure A.38**. To apply these settings to a different computer or after a re-installation, double-click the file and choose Restore Settings. After browsing to the location of the settings file (**Figure A.39**), you will be prompted to apply the settings to the current user (the usual choice for a personal computer) or to all users of the computer, as shown in **Figure A.40**.

FIGURE A.39

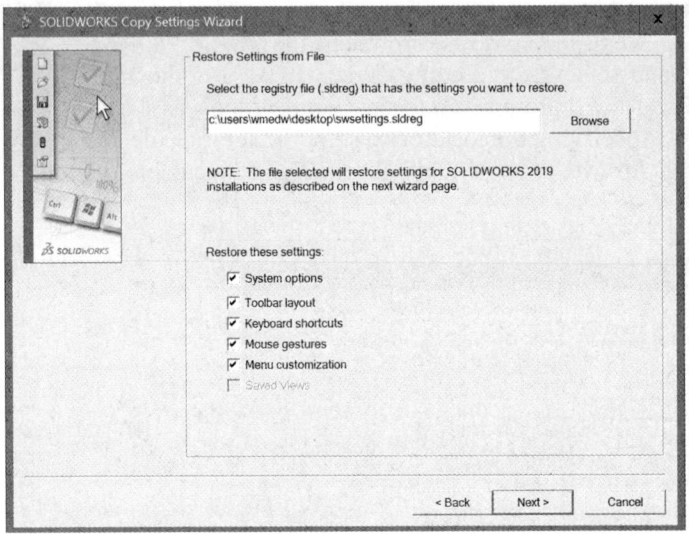

FIGURE A.40

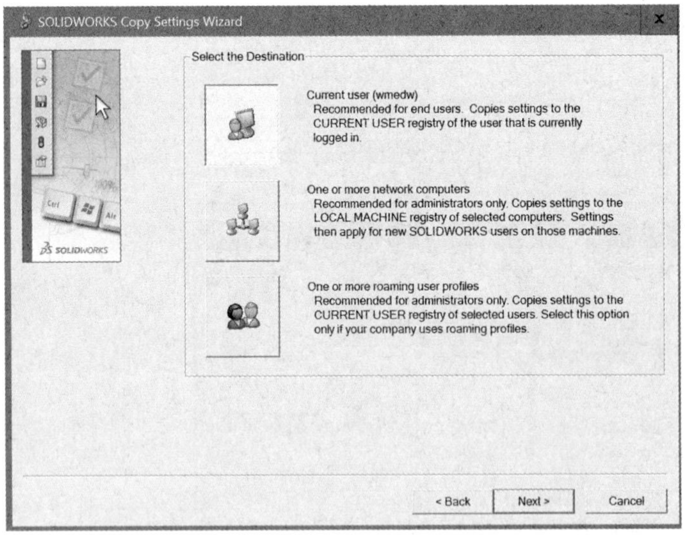

In order to back up your templates, you must first determine where they are stored. To do so, open SOLIDWORKS and choose Options: System Options: File Locations. Choose Document Templates from the list, as shown in **Figure A.41**. The location of the templates will be displayed. You can then browse to this folder and copy the three template files, which are shown in **Figure A.42**. (The files may be in a hidden directory, in which case you will need to change the folder options to display hidden files. In Windows 7 and 10, this is done from the Control Panel. Choose Appearance and Personalization: Folder Options: Show hidden files and folders.) To install these templates to a new computer, find the location of the template files from the System Options and copy the desired template files to that location.

FIGURE A.41

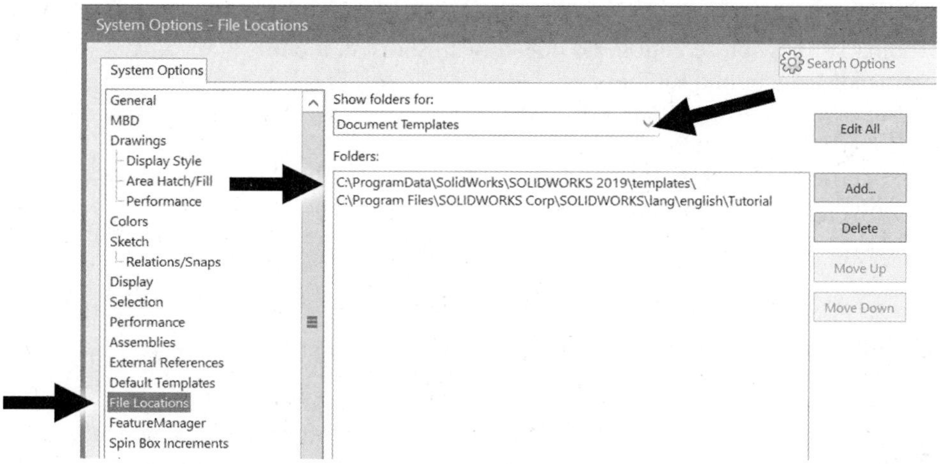

FIGURE A.42

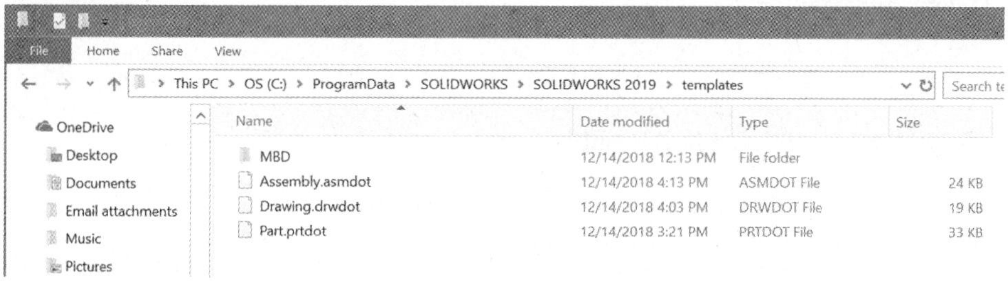

A.6 Summary of Recommended Settings

WITHIN PART ENVIRONMENT (*Stored in Part template):

Main menu: Always displayed with pushpin
CommandManager: Sketch and Features groups displayed; others hidden
CommandManager: Features group: Instant 3D off
CommandManager: Sketch group: Instant 2D off
Options: System Options: Colors: Green highlight scheme, Classic Icon colors
Options: System Options: Display/Selection: Tangent edge display removed
Options: System Options: Display/Selection: Third angle projection
Heads-Up Toolbar: Add Rotate View, Pan, and Trimetric tools
Customize: Context toolbars: Show on selection, do not show in shortcut menu
Customize: Lock CommandManager and toolbars
*Options: Document Properties: Drafting Standard: ANSI
*Options: Document Properties: Grid/Snap: Display grid
*Options: Document Properties: Units: IPS system, Length decimals = .123, Angle decimals = None
*Heads-Up Toolbar: Apply Scene: Plain white

WITHIN DRAWING ENVIRONMENT (Stored in Drawing template):**

CommandManager: View Layout, Sketch, and Annotation groups displayed; others hidden
Options: System Options: Drawings: Display Style: Hidden lines visible, tangent edges hidden
Options: System Options: Colors: Drawings, Paper Color: white
**Options: Document Properties: Drafting Standard: ANSI
**Options: Document Properties: Detailing: Auto-insert of centermarks and centerlines
**Options: Document Properties: Units: IPS system, Length decimals = .12, Angle decimals = None
**Sheet Properties: Third-angle projection

WITHIN ASSEMBLY ENVIRONMENT (*Stored in Assembly template):**

CommandManager: Assembly and Sketch groups displayed; others hidden
***Options: Document Properties: Drafting Standard: ANSI
***Options: Document Properties: Units: IPS system, Length decimals = .123, Angle decimals = None
***Heads-Up Toolbar: Apply Scene: Plain white.

APPENDIX B

The SOLIDWORKS Interface: Use and Customization

The SOLIDWORKS® user interface allows for commands to be accessed in a number of different ways, and the interface can be customized to reflect the preferences of the user.

In this book, we access most commands from the CommandManager. When any part, drawing, or assembly is open, the CommandManager can be toggled on and off by selecting Customize from the pull-down menu beside the Options Tool, as shown in **Figure B.1**, and checking/unchecking the box labeled "Enable CommandManager" under the Toolbars tab, as shown in **Figure B.2**. In order to move the toolbars and/or the CommandManager as will be discussed below, the check box labeled "Lock CommandManager and toolbars" must be cleared.

FIGURE B.1

FIGURE B.2

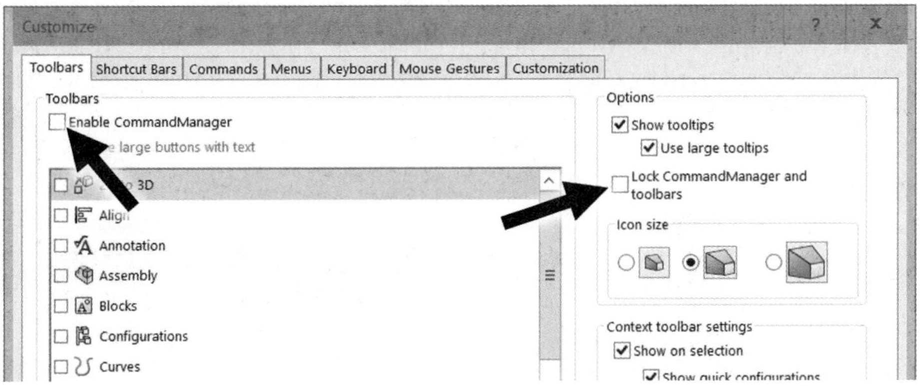

SOLIDWORKS is a registered trademark of Dassault Systémes SolidWorks Corporation.

When the CommandManger is turned off, commands can be accessed though the main menu or through toolbars. When modeling parts, we had the Features and Sketch groups of the CommandManager active. We can activate those toolbars by clicking the check box next to the name of each, as shown in **Figure B.3**.

If these toolbars have not been activated before, then they appear "docked" at the left and right edges of the screen, as shown in **Figure B.4**. Otherwise, they will appear at the position where they were last displayed.

FIGURE B.3

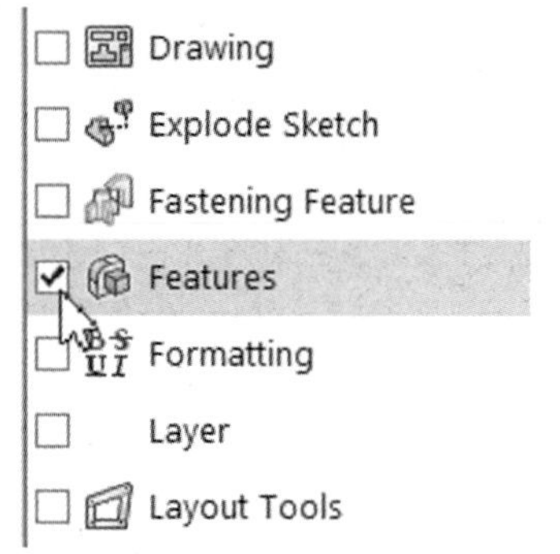

FIGURE B.4

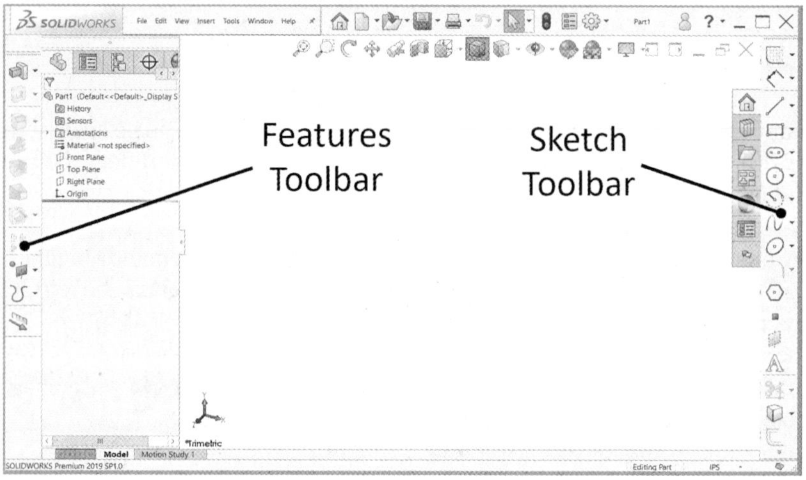

FIGURE B.5

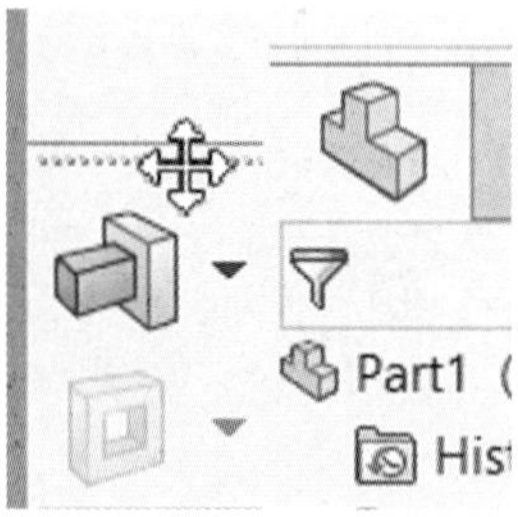

If a toolbar is docked, it can be moved from its default position by clicking and dragging the "handle" at the top or left of a docked toolbar, as shown in **Figure B.5**. (If the handle is not present, then the "Lock CommandManager and toolbars" checkbox shown in **Figure B.2** is probably checked.) The toolbar can be docked to another position along any edge of the screen, as shown in **Figure B.6**, or "floated" over the graphics area, as shown in **Figure B.7** (an exception is the Head-Up View Toolbar, which can be edited or turned off, but which is always positioned at the top of the graphics area). When floated, the toolbar's handle disappears and the toolbar can be moved by clicking and dragging its title bar. The toolbars can also

FIGURE B.6

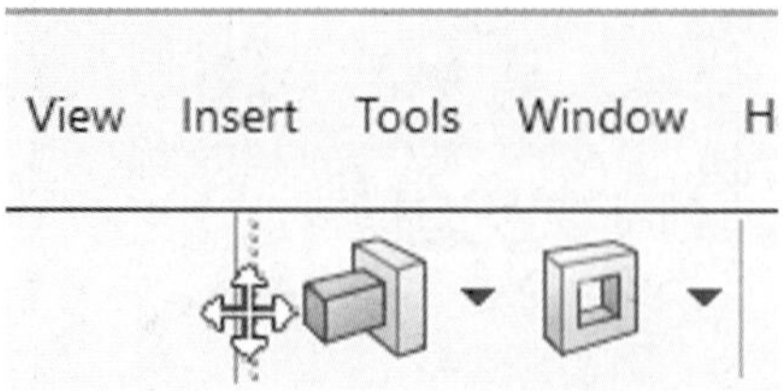

FIGURE B.7

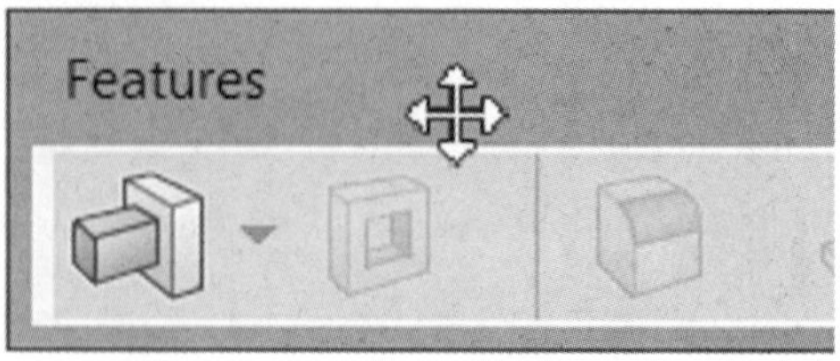

be re-sized by clicking and dragging an edge of the toolbar, as shown in **Figure B.8** (although the re-sizing is somewhat limited by the way in which tools are grouped within a toolbar). Clicking the X in the upper-right corner of the toolbar closes that toolbar, and it can be reopened as described in the next paragraph.

FIGURE B.8

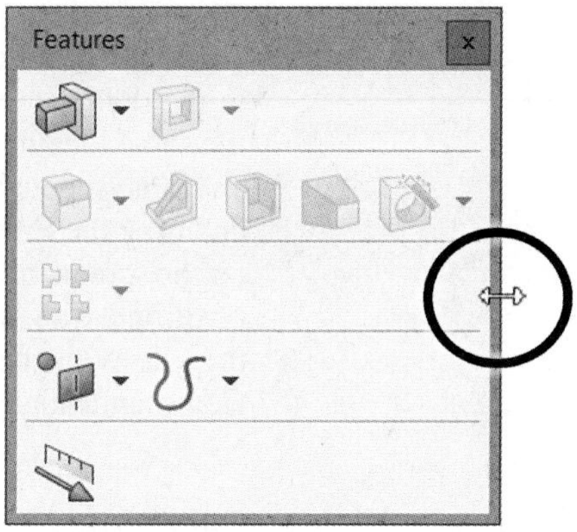

There are many other toolbars available. To see the complete list, select Customize from the pull-down menu beside the Options Tool (or right-click any toolbar or the CommandManager). Any active toolbar will have a check mark beside its name (or its icon will show as "depressed" if the list is accessed by right-clicking). Note that if the toolbar is docked in a position where there is not enough space for all tools to be displayed, as in **Figure B.9** for the Standard Views toolbar, then a double-arrow at the bottom or right end of the toolbar indicates that not all tools are shown. Clicking on the double-arrow displays the hidden tools, as shown in **Figure B.10**.

The advantage of using toolbars instead of the CommandManager is that all tools are available with a single mouse click, while with the CommandManager it is often necessary to click once to change the group of tools and then click again to select the desired tool. For new users, the CommandManager is preferred because many tools have text labels.

As with toolbars, the CommandManager can be moved to other locations on the screen. If the CommandManager is moved or turned off accidentally, then the default position can be restored by selecting View: Workspace: Default from the main menu. Users with wide screen monitors may want to try selecting View: Workspace: Widescreen. The widescreen mode places the CommandManager at the left side of the screen,

FIGURE B.9

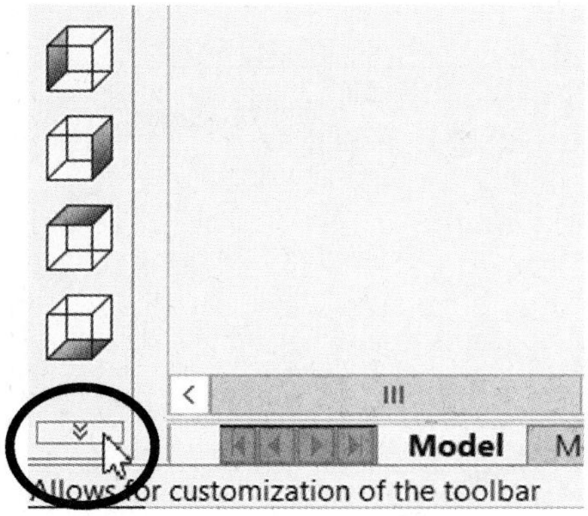

FIGURE B.10

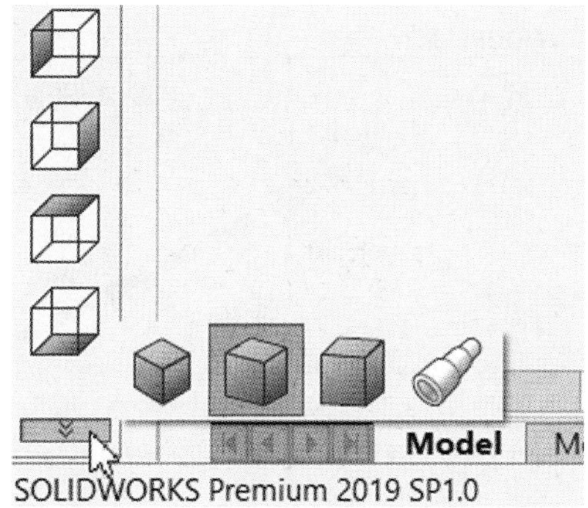

as shown in **Figure B.11** (with the toolbars closed), creating a better aspect ratio for the graphics area. When the widescreen mode is activated, the PropertyManager is allowed to float on the screen. The ability to move the PropertyManager is available in the default workspace configuration as well, but its default position is to occupy the same location as the FeatureManager. The FeatureManager is displayed unless an entity is selected, in which case the PropertyManager is displayed. When you click and drag the PropertyManager to a new position, several "docking" positions appear on the screen, as shown in **Figure B.12**. Dragging the PropertyManager to the docking position at the top of the FeatureManager causes the two to share the same space, as they do with the default workspace configuration. To return to the layout with the CommandManager at the top of the screen, select View: Workspace: Default.

FIGURE B.11

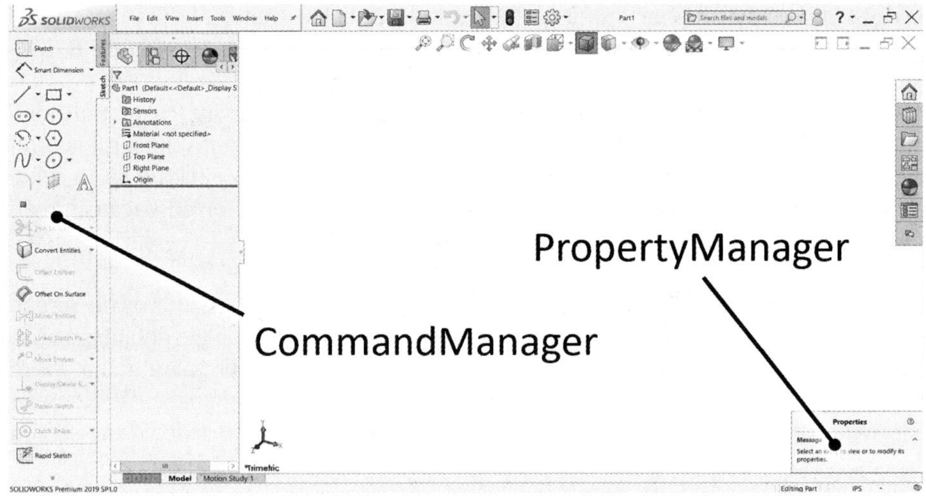

FIGURE B.12

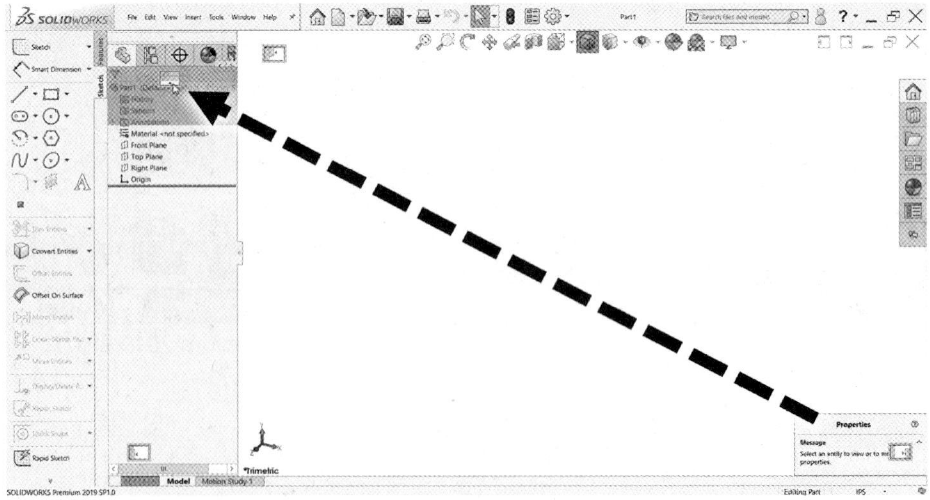

Tools can be added and removed from toolbars and the CommandManager. When Customize is selected from the pull-down menu beside the Options Tool, active toolbars and all CommandManager groups can be edited. To remove a tool, simply click and drag it into the graphics area. For example, in **Figure B.13**, the Plane Tool, which is used to create planes in a 3-D sketch, is removed by dragging it from its position in the Sketch group of the CommandManager into the graphics area. As shown in **Figure B.14**, the other tools in the group are re-ordered after the Plane Tool has been removed. (Note that removing this tool from the CommandManager does not mean that the tool itself has been deleted; it can be accessed from the Main Menu under Tools: Sketch Entities.) To add a tool, you must first locate it in the Customize box under the Commands tab. Commands are listed in groups. For example, you might want to add a Centerline Tool so that you do not have to select it from the pull-down menu of the Line Tool each time you want to use it. The Centerline Tool is located in the Sketch group, as shown in **Figure B.15**. To move it onto the CommandManager, click and drag it to the desired position. A plus sign will appear when the cursor has been moved to a position where

FIGURE B.13

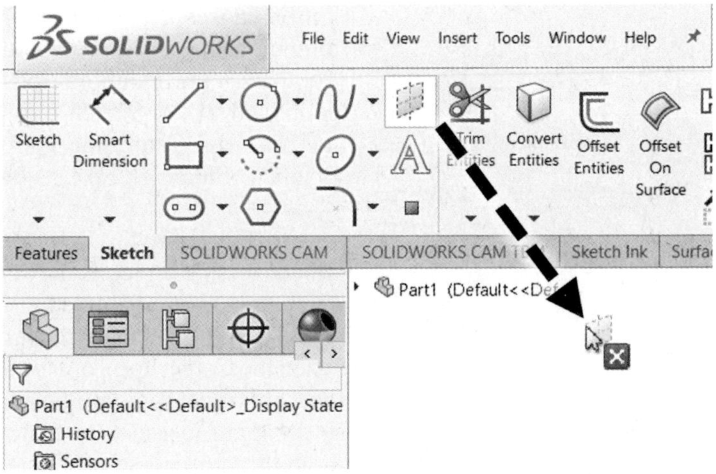

FIGURE B.14

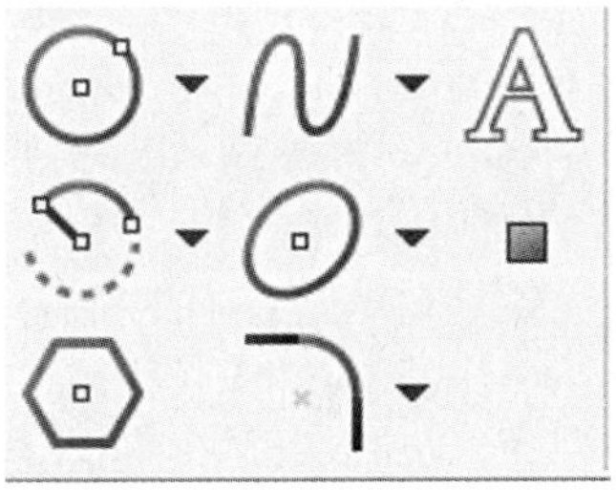

FIGURE B.15

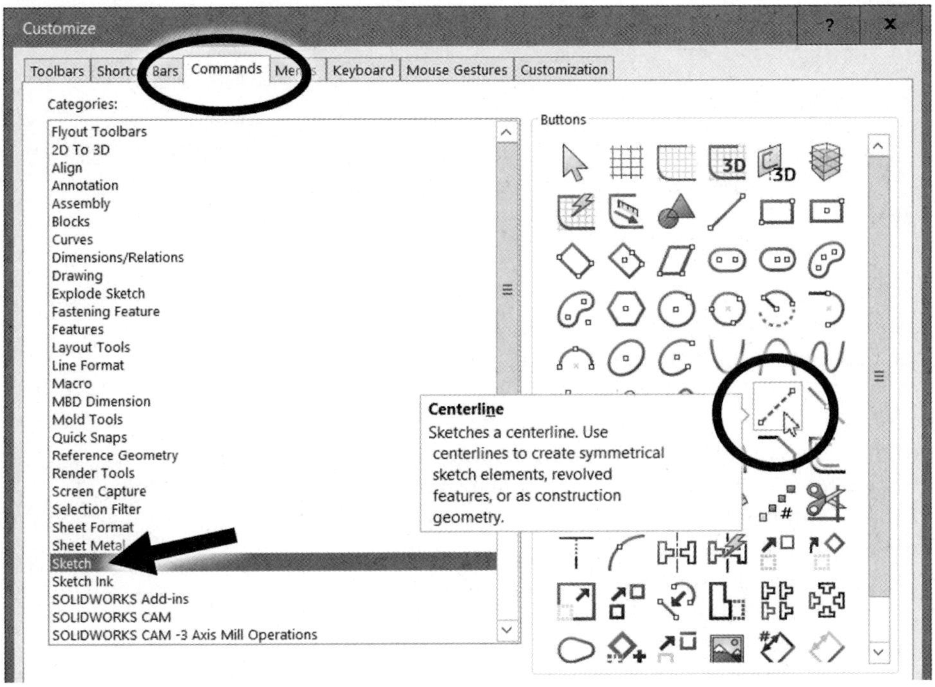

FIGURE B.16

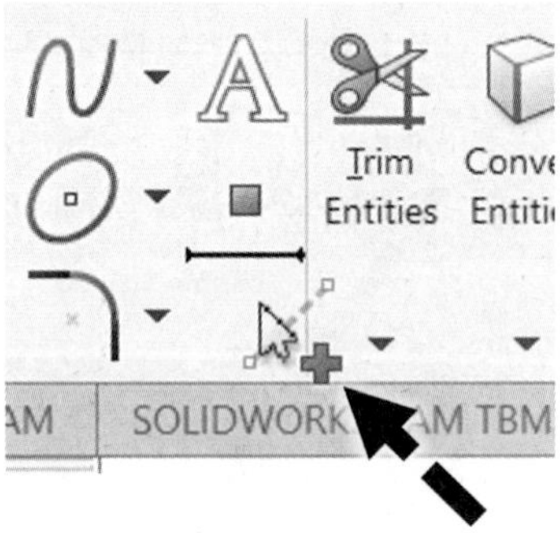

FIGURE B.17

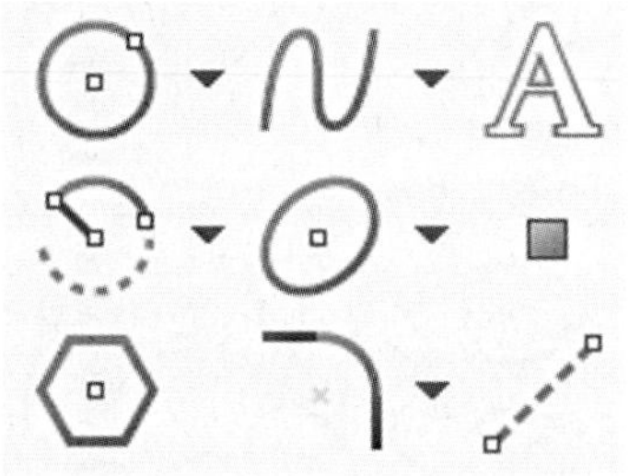

the tool can be placed, as shown in **Figure B.16**. Releasing the mouse button causes the tool to be placed, as shown in **Figure B.17**. Note that all tools with pull-down menus are included in the group named "Flyout Toolbars." For example, the individual Line and Centerline Tools are contained in the Sketch group, while the Line Tool that contains both the Line and Centerline Tools in a pull-down menu is contained in the Flyout Toolbars group.

In a toolbar, only the icon appears. In the CommandManager, there is a mix of tools with icons only and those with text labels. When you drag a tool into a group of tools in the CommandManager, then the appearance of the new tool matches those of the other tools in the group. For example, when the Centerline Tool is moved into a group of tools without text, then the Centerline Tool will be added without text. However, the appearance of the text label can be customized for each tool. By right-clicking on the tool, you can select Show Text, as shown in **Figure B.18**. This will cause the text to appear beside the tool. Tools with text to the side can be "stacked" in a column of tools. Right-clicking again allows you to choose Text Below, as shown in **Figure B.19**. The result, as shown in **Figure B.20**, is that the tool icon occupies a width of the CommandManager by itself, and no other tools can be placed above or below it. Of course, if all tools were displayed this way, there would not be enough room on the screen for all of the tools to be shown. However, with a wide-screen monitor, there is usually plenty of room to add and customize tools as desired.

FIGURE B.18

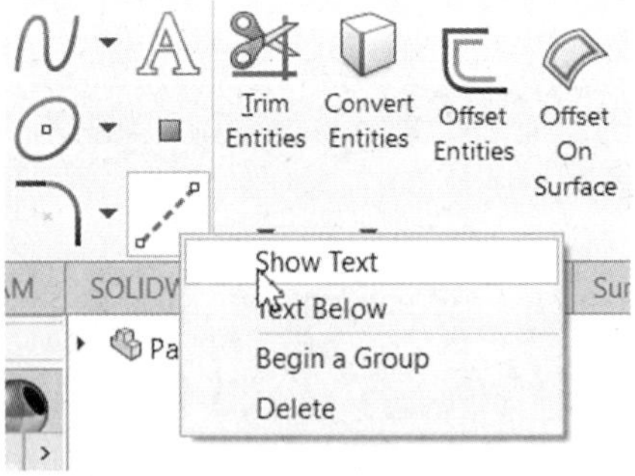

FIGURE B.19

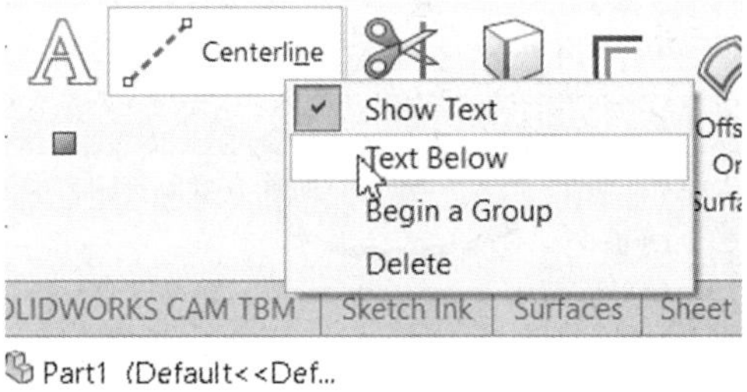

It should be noted that tools can be mixed and matched onto toolbars and CommandManager groups. That is, tools other than sketch tools can be placed on the Sketch toolbar or the Sketch group of the CommandManager, and tools can be duplicated. For example, the Evaluate group of the CommandManager contains the Mass Properties Tool. You may find that you use this tool often but do not use the other Evaluate tools regularly. Rather than display the Evaluate

FIGURE B.20

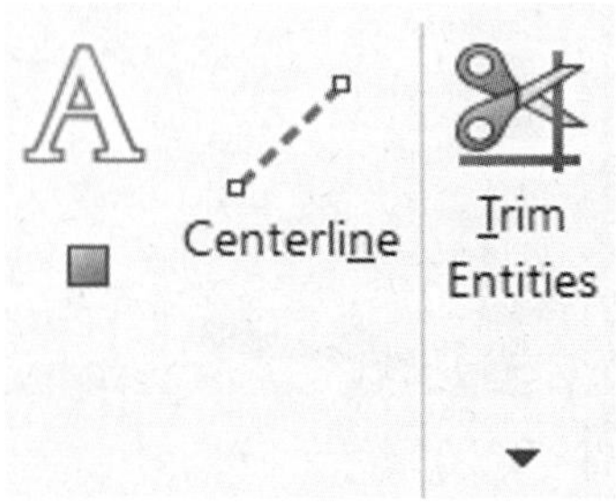

group or access the Mass Properties Tool from the main menu when needed, the tool can be added to another group of the CommandManager, most logically the Features group. To do this, locate the tool from the Tools group of the Command list and drag it onto the Features group of the CommandManager, as shown in **Figure B.21**. The Mass Properties Tool is now available from the Features group, as shown in **Figure B.22**, as well as from the Evaluate group of the CommandManager.

In addition to the default groups of the CommandManager, additional groups can be added by clicking on the New Tab icon at the far right of the tabs, as shown in **Figure B.23**. A list of available groups is shown, along with the option of creating a custom group of commands. For example, if Standard Views is chosen from the list, then a Standard Views group is added to the CommandManager. The order of the tabs can be changed by clicking and dragging a tab to a new location. For example, the Standard Views tab just added can be moved to the left, as shown in **Figure B.24**. A group added in this manner can be removed by right-clicking its tab and selecting Delete, as

FIGURE B.21

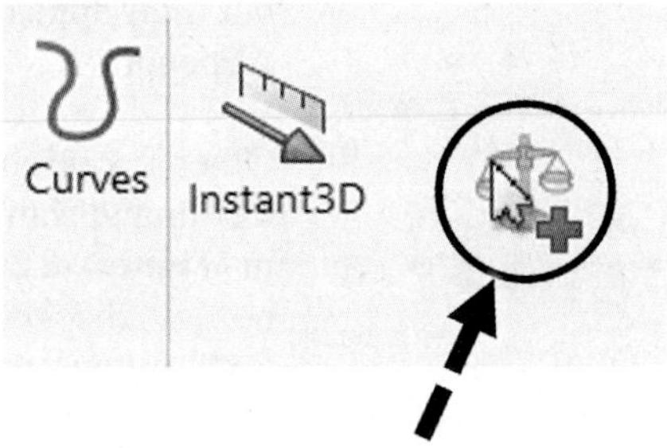

FIGURE B.22

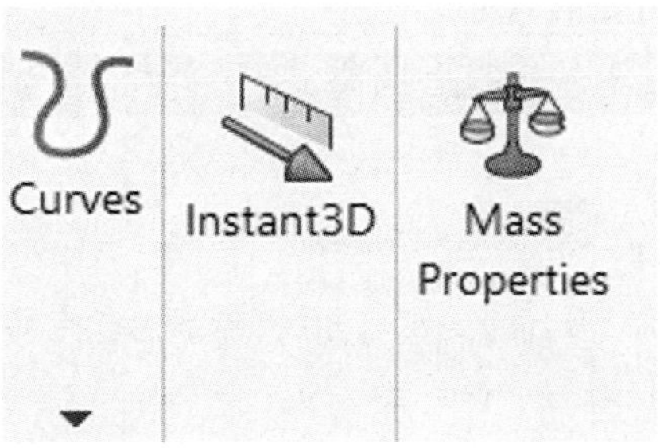

FIGURE B.23

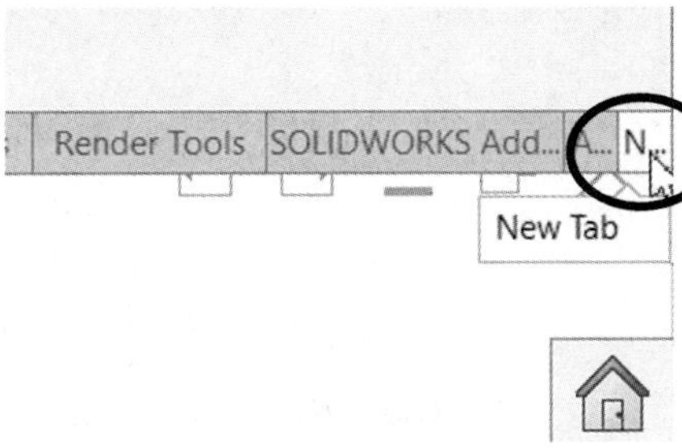

FIGURE B.24

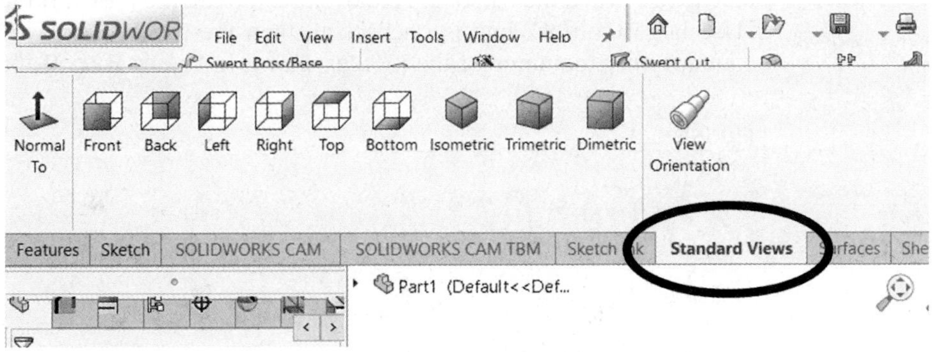

FIGURE B.25

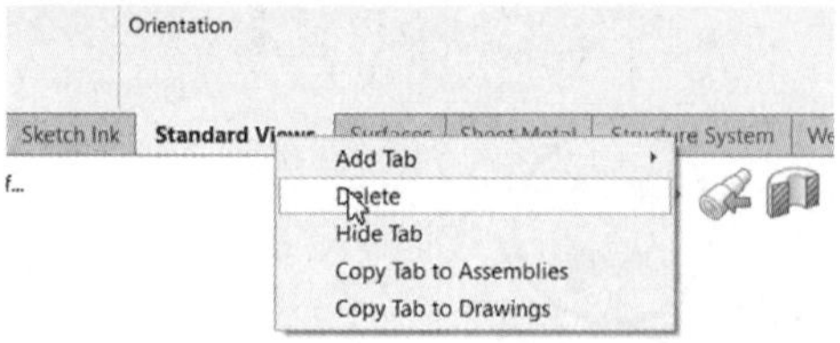

shown in **Figure B.25**. The default groups cannot be deleted, but their appearance can be toggled on or off as discussed in Appendix A.

Other options for displaying toolbars and the CommandManager are controlled by the checkboxes shown in **Figure B.26**. When the "Enable CommandManager" box is checked, the box labeled "Use large buttons with text" allows the display of text labels with some tools as described previously. When this box is unchecked, the CommandManager contains unlabeled icons similar to those of a toolbar, as shown in **Figure B.27**.

FIGURE B.26

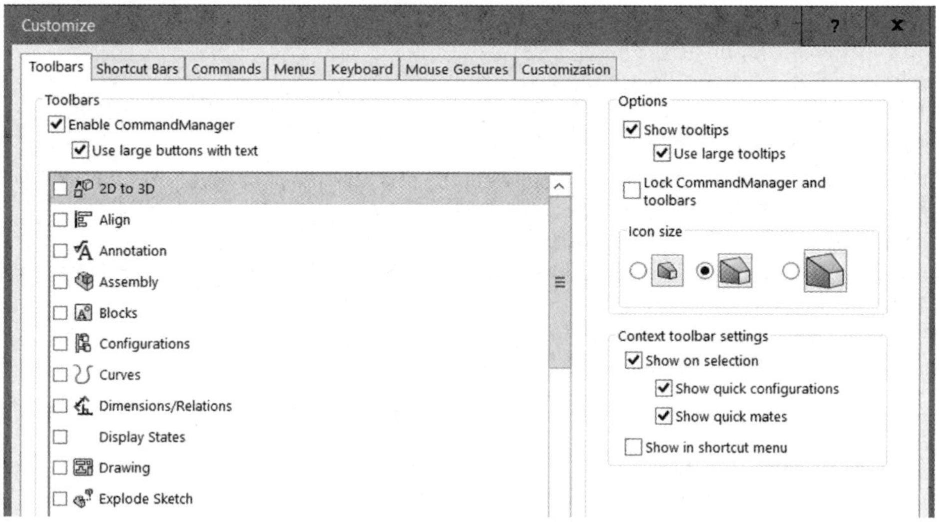

FIGURE B.27

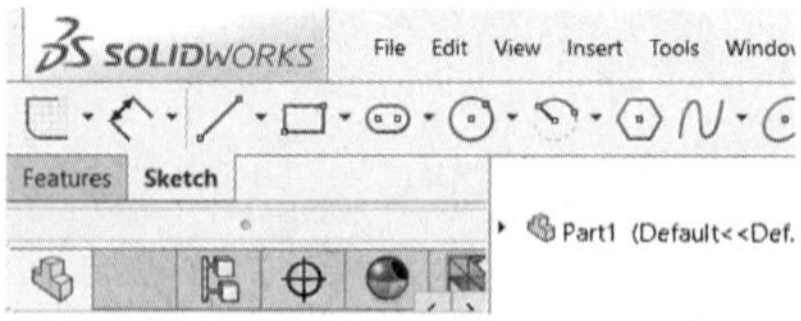

The Icon size option is self-explanatory; large buttons in the toolbars and CommandManager are easier to interpret but take up more room on the screen. Tooltips are the descriptions of each tool that appear when the cursor is held over the tool momentarily. If the "Use large tooltips" box is unchecked, then the tooltip displayed contains only the name of the tool, as shown in **Figure B.28**. If the

FIGURE B.28

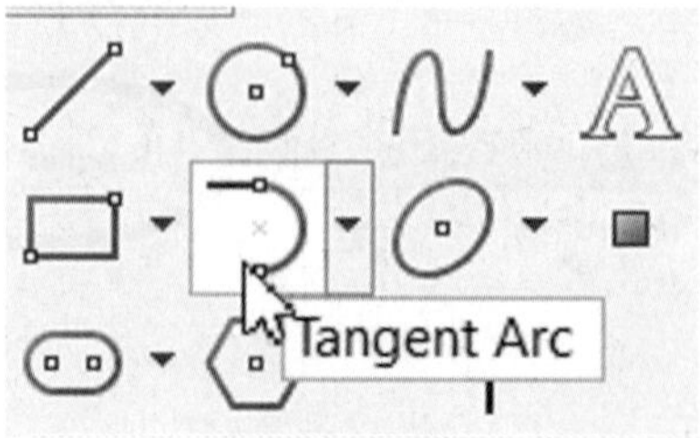

box is checked, then a more complete description is displayed, as shown in **Figure B.29**.

Context toolbars appear when certain entities are selected when the box labeled "Show on selection" is checked. For example, if a flat surface is selected, then the menu shown in **Figure B.30** is displayed. Although the icons are small, holding the cursor over an icon displays its function, as shown in **Figure B.31**. (This is true even if the tooltips are turned off.) We use these context toolbars sparingly in the text but have found the tools to open a sketch on the selected surface or to change the view orientation to be normal to the selected surface to be handy shortcuts. Below the "Show on selection" box is one labeled "Show in shortcut menus." A shortcut menu is the menu that appears when you right-click a feature, either in the graphics area or in the FeatureManager. For example, right-clicking the feature name shown in **Figure B.32** causes the menu shown to be displayed. If the "Show in shortcut menus" option is enabled, then the context toolbar is shown at the top of the menu, as shown in **Figure B.32**. If that option is cleared, then many of the commands from the context toolbar are displayed instead in the shortcut menu, as shown in **Figure B.33**.

FIGURE B.29

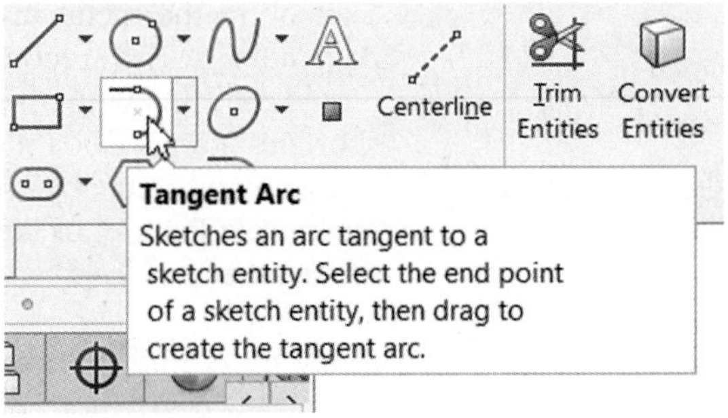

FIGURE B.30

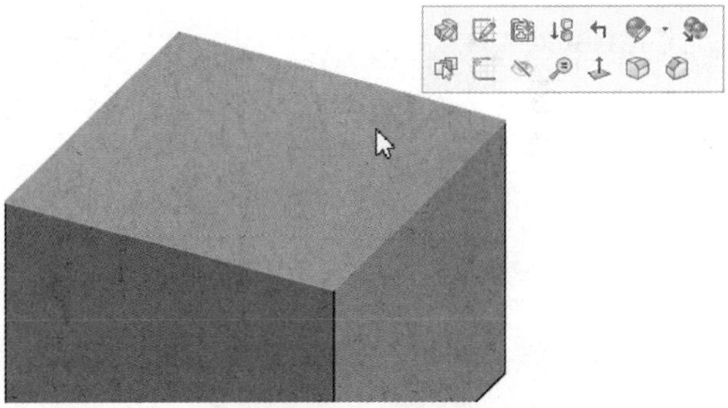

FIGURE B.31

FIGURE B.32

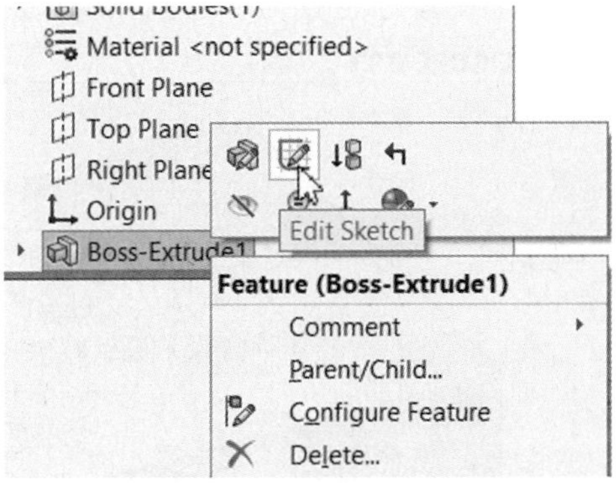

FIGURE B.33

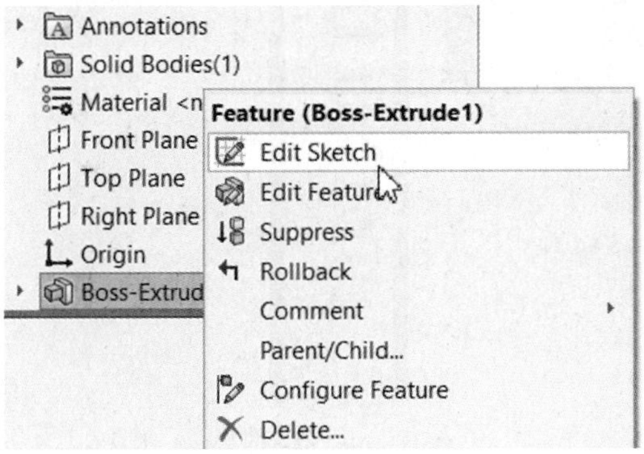

Most new users will probably find it easier to select a command such as "Insert Sketch" in the menu instead of from a small icon. Therefore, in the text and in Appendix A we recommend leaving the "Show in shortcut menus" option turned off. Leaving the "Show on selection" box checked and the "Show in shortcut menus" box unchecked allows either selection option to be utilized: a command can be selected from the context toolbar by selecting (left-clicking) a feature, or the command can be selected from a menu by right-clicking the feature.

Many users find keyboard shortcuts to be useful. For example, when you want to zoom out to see more of your model, pressing the Z key allows you to zoom out without having to access any menu. Similarly, holding the Shift key while pressing the Z key allows you to zoom in. The F key scales the view so that the entire model can be seen (zoom to fit).

FIGURE B.34

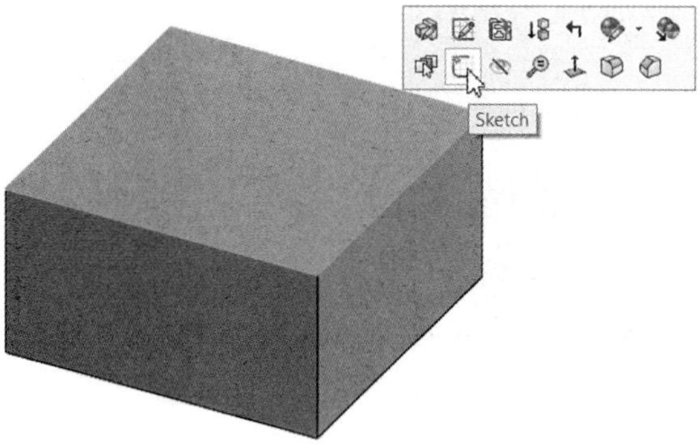

One keyboard shortcut that is worth mentioning is the Shortcut Bar, which is accessed by pressing the S key. The Shortcut Bar is context-sensitive; that is, it displays sketching tools if a sketch is open and feature tools otherwise. As an example, suppose that we want to add a hole from the top surface of the block shown in **Figure B.34**. We can open a sketch by selecting the Sketch Tool from the context toolbar. With the sketch open, pressing the S key causes the menu in **Figure B.35** to be displayed. The Circle Tool can be selected and a circle added. Pressing the S key again allows the

FIGURE B.35

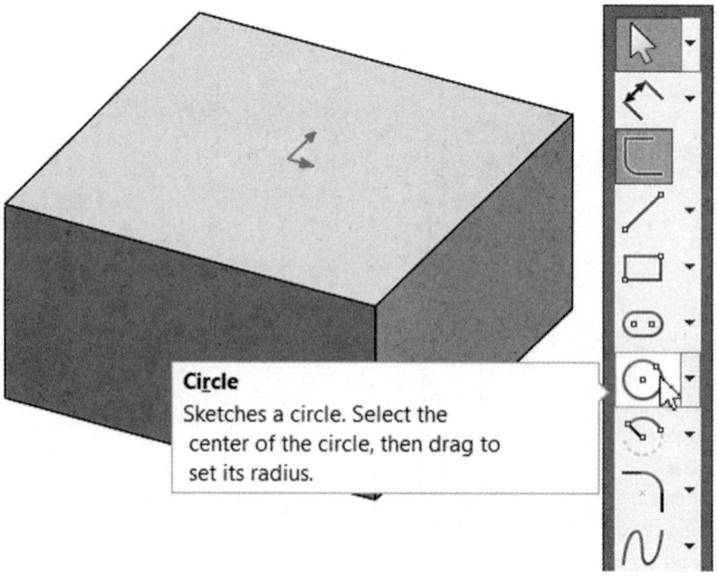

selection of the Smart Dimension Tool, as shown in **Figure B.36**, and the sketch can be dimensioned. The sketch can then be closed, also from the shortcut menu, as shown in **Figure B.37**. With the sketch closed, pressing the S key results in a menu of feature tools to be displayed, from which the Extruded Cut Tool can be selected, as shown in **Figure B.38**. Note that when the S key is pressed, the Shortcut Bar appears at the location of the cursor. This makes it a very efficient way to select commands, minimizing the mouse movements between selections. The Shortcut Bar can be customized like any other toolbar by right-clicking on the toolbar and choosing Customize. However, the default settings contain the most commonly used sketch and features tools, and most users will find these to be sufficient.

FIGURE B.36

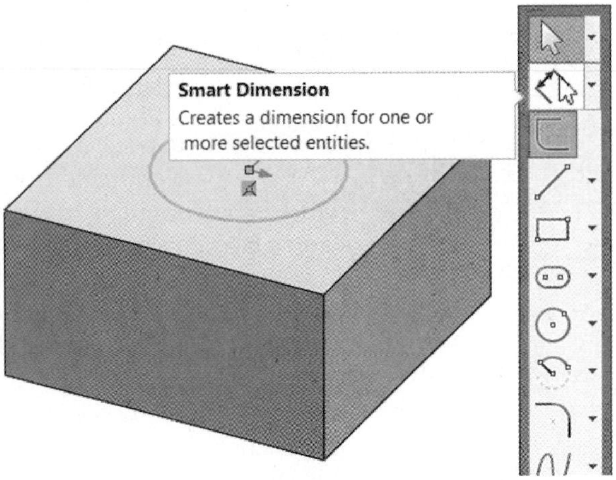

FIGURE B.37

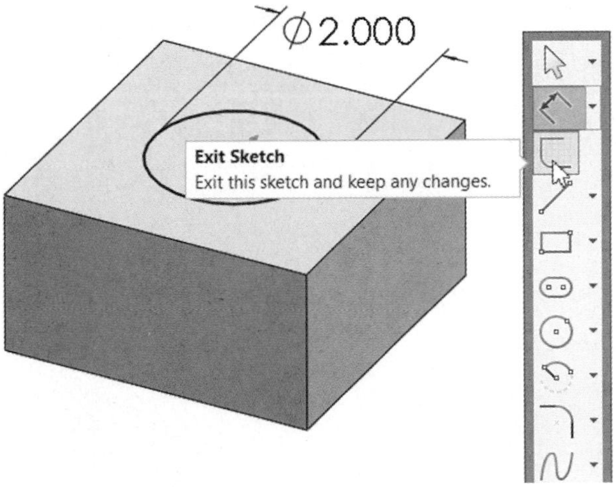

FIGURE B.38

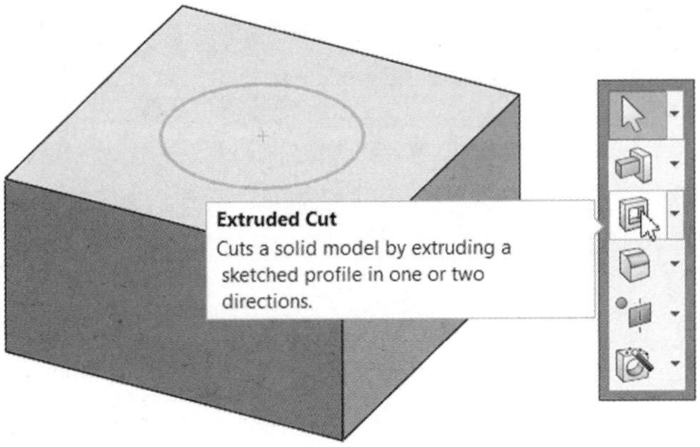

FIGURE B.39

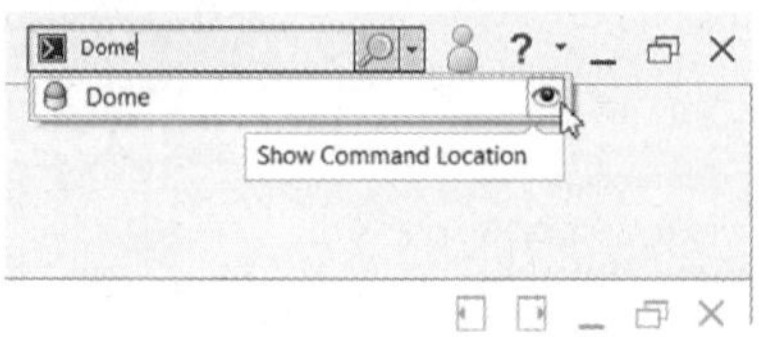

Just to the right of the Main Menu is the Command Search. (It may be necessary to collapse the Main Menu by clicking on the push pin at the right end of the menu in order for the Command Search to be displayed.) The Command Search allows you to access commands by typing in a few letters, as shown in **Figure B.39**. This is a handy way to access commands that you do not use often. By clicking the Show Command Location icon beside a command, the command's location will be shown, either in the CommandManager, an open toolbar, or, as shown in **Figure B.40**, in a menu.

To view all of the keyboard shortcuts, select Customize from the pull-down menu beside the Options Tool and select Keyboard. Selecting the option "Commands with Keyboard Shortcuts" from the menu makes browsing the keyboard shortcuts easier, and clicking the "Shortcut(s)" column heading organizes the shortcut keys alphabetically, as shown in **Figure B.41**. Shortcuts can be removed or reassigned, but this should be done with caution as many of the shortcuts, such as copy and paste,

FIGURE B.40

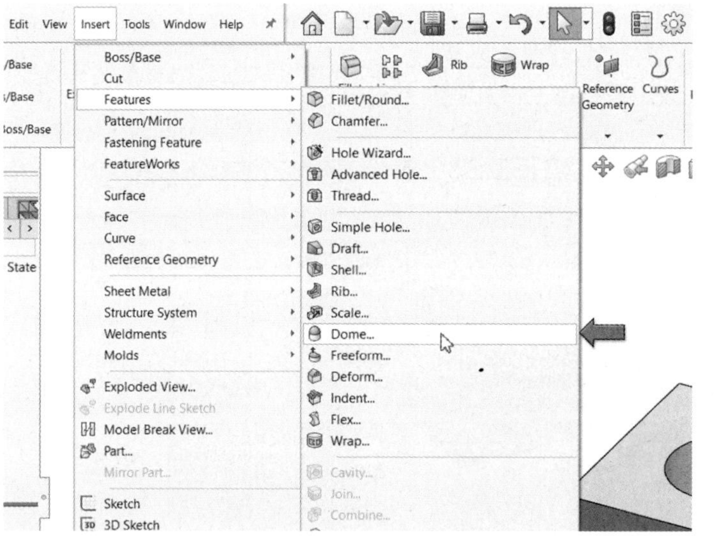

FIGURE B.41

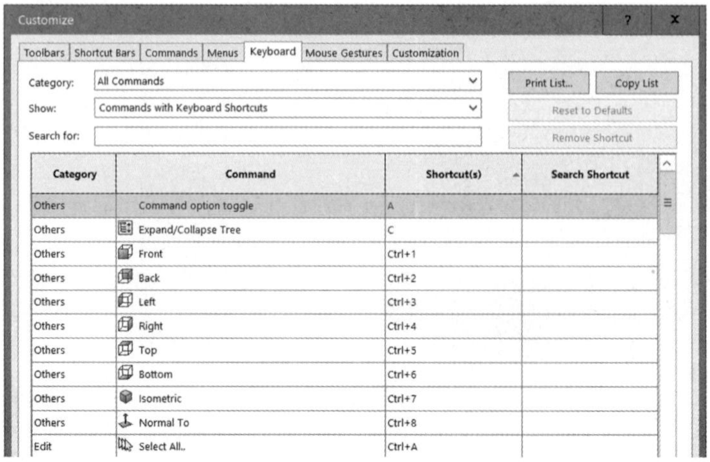

match standard Windows shortcuts. A list of some of the most handy keyboard shortcuts is shown here:

KEYS(S)	COMMAND
Viewing Tools	
F	Zoom to fit: allows entire model to be seen
Z	Zoom out
Shift + Z	Zoom in
Space Bar	Orientation: displays a pop-up menu of standard view orientations
P10	Toolbars: toggles display of the toolbars and CommandManager to provide a larger graphics area
Selection Filter Tools	
E	Filter Edges: allows only edges to be selected
X	Filter Faces: allows faces to be selected
F6	Toggle Filters: toggles the selected filter off and on
File Tools	
Ctrl + N	New File
Ctrl +O	Open File
Ctrl + P	Print
Ctrl + S	Save File
Other Commands	
S	Shortcut Bar

A note about the selection filter tools, which are often used in assembly mates to assist in selecting the proper entity: when a selection filter is active, a "filter" icon appears next to the cursor, as shown in **Figure B.42**. When a selection filter is active, then only the type of entity specified by the filter can be selected. Occasionally, a user will press a key by accident (most likely the X key), which causes a filter to become active, preventing any other selections. If this happens, pressing the F6 key clears the filter.

FIGURE B.42

Another way of selecting some commands is with mouse gestures. When mouse gestures are enabled, then holding down the right mouse button while moving the mouse slightly in any direction displays a circular menu like the one shown in **Figure B.43**. If the mouse is then moved to any of the segments of the circle, then the corresponding command is executed. Mouse gestures are controlled by selecting Customize

FIGURE B.43

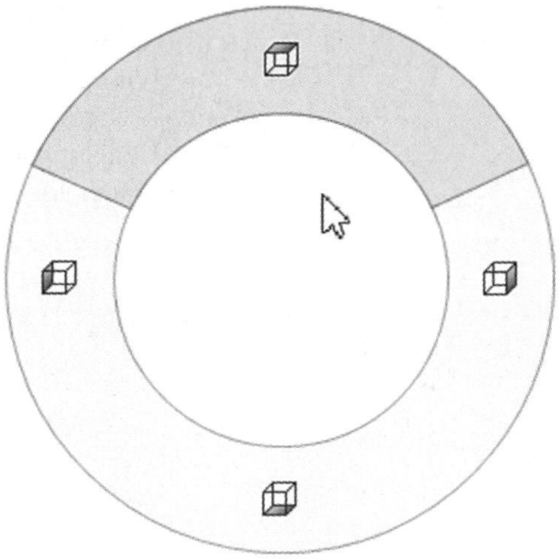

from the pull-down menu beside the Options Tool and selecting the Mouse Gestures tab, as shown in **Figure B.44**. When the Mouse Gestures tab is selected, a window showing the gestures is opened. Notice that the commands are context-dependent; that is, different commands are active if a part, assembly, drawing, or sketch is active. If mouse gestures are enabled, then the menu can be set to include either four commands, as shown in **Figure B.43**, or from two to twelve commands. **Figure B.45** shows the commands that are active when eight commands are selected and a sketch is active. Mouse gestures may seem awkward at first, but with a little practice some users will find them useful.

FIGURE B.44

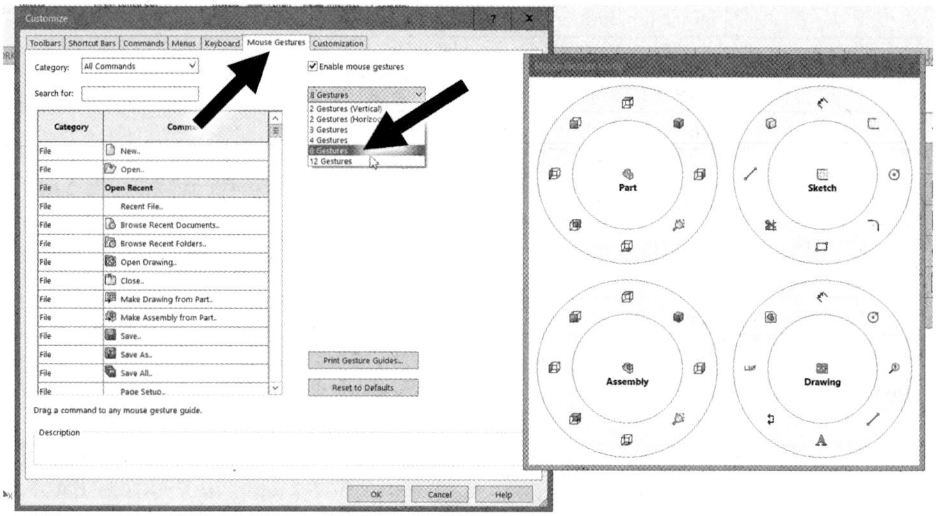

FIGURE B.45

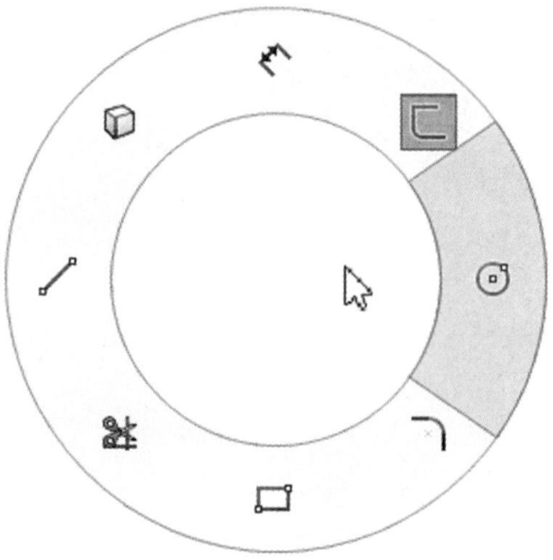

In this Appendix, we have seen that the SOLIDWORKS User Interface allows for multiple ways to select commands and for a great deal of customization to fit a user's preferences. In this book, we focus primarily on the CommandManager and the Heads-Up Toolbar to select commands, with a minimal amount of customization (such as adding the Rotate View, Pan, and Trimetric View Tools to the Heads-Up Toolbar), and we recommend that new users limit customization. Experienced users will no doubt find many individual preferences for setting up and using the interface.

Index

G

H

I

J

K

L

M